POULTRY SCIENCE

(Animal Agriculture Series)

Cover montage. *Upper* (left to right): Leghorn hen (Courtesy, H&N, Redmond, WA); conveying eggs (Courtesy, FACCO, Padova, Italy); broilers (Courtesy, National Broiler Council, Washington, DC). *Center* (left to right): Turkeys (Courtesy, USDA); ducks (Courtesy, California Polytechnic State University, San Luis Obispo). *Lower:* Geese (Courtesy, Schiltz Foods, Inc., Sisseton, SD).

Picture page x. (Courtesy, Alabama Poultry & Egg Assn., Cullman, AL)

Other books by M. E. Ensminger
available from Interstate Publishers, Inc.:

The Stockman's Handbook
Stockman's Handbook Digest
Animal Science
Animal Science Digest
Beef Cattle Science
Sheep and Goat Science
Swine Science
Dairy Cattle Science
Horses and Horsemanship

The Stockman's Handbook and *Stockman's Handbook Digest* present the "why" as well as the "how." They contain, under one cover, the pertinent things that a livestock producer needs to know in the daily operation of a farm or ranch. They cover the broad field of animal agriculture, concisely and completely, and wherever possible in tabular and outline form.

Animal Science and *Animal Science Digest* present a perspective or panorama of the far-flung livestock industry, whereas each of the other books presents specialized material pertaining to the specific class of farm animals indicated by its respective title.

POULTRY SCIENCE

(Animal Agriculture Series)

78   by
M. E. Ensminger, B.S., M.A., Ph.D.

Formerly: Assistant Professor in Animal Science
University of Massachusetts

Chairman, Department of Animal Science
Washington State University

Consultant, General Electric Company
Nucleonics Department (Atomic Energy Commission)

Currently: President
Consultants-Agriservices
Clovis, California

President
Agriservices Foundation
Clovis, California

Collaborator
U.S. Department of Agriculture

Adjunct Professor
California State University-Fresno

Adjunct Professor
The University of Arizona-Tucson

Distinguished Professor
University of Wisconsin-River Falls

Third Edition

INTERSTATE PUBLISHERS, INC.
Danville, Illinois

POULTRY SCIENCE, Third Edition.
Copyright © 1992 by Interstate Publishers, Inc.
All rights reserved.

Printed in the United States of America.

Editions:
First   1971
Second 1980
Third   1992

Order from

Interstate Publishers, Inc.
510 North Vermilion Street, P.O. Box 50
Danville, IL 61834–0050
Phone: (800) 843-4774
FAX: (217) 446-9706

Library of Congress Catalog No. 91–74154 ✓

ISBN 0–8134–2929–3

1 2 3
4 5 6
7 8 9

To my son, JOHN
Each of us was disadvantaged
I because I had too little; he because he had too much
Of the two, his was the greater disadvantage

ABOUT THE AUTHOR

Dr. M. E. Ensminger is President of Agriservices Foundation, a nonprofit foundation serving world agriculture. Also, he is Adjunct Professor, California State University-Fresno; Adjunct Professor, The University of Arizona-Tucson; Distinguished Professor, University of Wisconsin-River Falls; Collaborator, U.S. Department of Agriculture; and Honorary Professor, Huazhong Agriculture College-Wuhan, People's Republic of China. Dr. Ensminger (1) grew up on a Missouri farm; (2) completed B.S. and M.S. degrees at the University of Missouri, and the Ph.D. at the University of Minnesota; (3) served on the staffs of the University of Massachusetts, the University of Minnesota, and Washington State University; and (4) served as Consultant, General Electric Company, Nucleonics Department (Atomic Energy Commission).

Dr. Ensminger is the author of 21 widely used books that are translated into several languages and used throughout the world.

Among Dr. Ensminger's honors and awards are: Distinguished Teacher Award, American Society of Animal Science; Washington State University named and dedicated the *Ensminger Beef Cattle Research Center*, in recognition of his contributions to the University; and an oil portrait of him was placed in the 300-year-old gallery of the famed Saddle and Sirloin Club, which is recognized as the highest honor that can be bestowed on anyone in the livestock industry.

PREFACE TO THE THIRD EDITION

Since 1950, changes in poultry meat and egg production and processing have paced the whole field of agriculture. Practices with all species, and in all phases of poultry production—breeding, feeding, management, housing, marketing, and processing—have become very highly specialized. The net result is that more poultry products have been made available to consumers at favorable prices and per capita consumption has increased.

Science and technology will continue to be the great multipliers in the poultry industry. The following developments will usher the U.S. poultry industry into the 21st century—and beyond.

1. **Increased biotechnology.** The application of biotechnology, along with improved breeding, feeding, and management, will make for increased production and efficiency of layers, broilers, turkeys, ducks, and geese. In the future, scientists will be able to engineer birds genetically that are resistant to some of the most costly diseases. The goal is to improve the overall health of poultry without compromising desirable production traits like growth, feed efficiency, egg production, and meat quality.

2. **Increased mass production, economical products, and year-round availability of products.** Because poultry products can be hatched on schedule, handled as a flock, and adapted to confinement, mass production methods can be applied economically to produce poultry meat and eggs on a continuous basis, very efficiently and at a highly competitive price basis. Today, the industry is so efficient that more food in the form of poultry products can be produced with fewer raw materials and less energy than is required by any animal source other than some phases of aquaculture.

3. **More contract and integrated production, and fewer and bigger integrators.** Poultry production has become so highly specialized that farmers contract with large integrated companies to produce one kind or phase of production only; for example, the production of started pullets, eggs, broilers, or turkeys. Generally, the integrators supply chicks, poults, feed, litter, fuel, and medication, and the contract growers provide housing and the care of the birds. Most contract growers are on an incentive basis; they receive a graduated fee (bonus) that compensates them for superior management and bird performance.

Because capital requirements for modern poultry operations are so great and the margin of profit per bird so small, individual growers cannot afford the financial risk of selling poultry or eggs on the open market. As a result, the vast majority of poultry products are produced under contract. Also, in the future, there will be fewer and bigger integrators.

4. **Increased laborsaving devices and mechanization.** Higher priced labor, along with more sophisticated equipment, is making, and will continue to make, for increased mechanization all along the line, from production through processing and marketing.

5. **Improved housing and environmental control.** Physics, engineering, and physiology are being combined in such a way as to bring about improved houses, incubators, and brooders.

6. **Sustainable agriculture will be in vogue.** In the future, we shall no longer operate like there is no tomorrow. Instead, we shall be good stewards of nature, which we shall preserve for future generations. The accolades of the future will be accorded for such things as pollution control, clean air, soil and water conservation, energy conservation, and preservation of rain forests.

7. **Increased attention to poultry behavior and welfare.** We now know that controlled poultry environment must embrace far more than an air-conditioned chamber, along with ample feed and water. Producers need to concern themselves more with the natural habitat of poultry, and to breed birds for adaptation. In the future, greater knowledge of poultry behavior and breeding for adaptation will make for improved poultry welfare and profits, for all three—behavior, welfare, and profits—are on the same side of the ledger.

8. **Increased food safety.** American consumers care little about what they put on their backs, but they are greatly concerned about what goes into their stomachs. In the future, poultry producers, processors, and marketers will extol the safety and nutritive values of foods, rather than quantity.

9. **Improved quality of products.** No other country in the world produces as high-quality poultry meat and eggs as the United States. Yet, further improvements can and will be made.

10. **Increased consumption of poultry products.** Per capita consumption of poultry meat will increase because it (a) is available fresh or frozen, (b) is easy to prepare, (c) is easily combined with other foods, (d) is lean, nutritious, tender, and easy to chew, (e) is mild in flavor, (f) is easy to digest, and (g) is a good buy when compared with other animal products.

Eggs are also easy to prepare, nutritious, digestible, and economical. Following a decline in consumption, it appears that consumption has leveled off at about 240 eggs per capita, annually.

In the future, it is predicted that the consumption of poultry meat will increase due to both increased human population and increased per capita consumption; and that there will be a modest increase in total egg consumption due to population increase, rather than increased per capita consumption. Increased per capita consumption of poultry meats will be achieved with the development of (a) more cut-up products, and (b) more further processed products such as poultry hot dogs and weiners; more deboned, filleted, smoked, or formed-into-patties meats; more convenience foods; and increased shelf life.

11. **Increased business acumen.** Bigger operations increased competition will require more business acumen.

I am grateful to all those who contributed so richly to this revision of *Poultry Science*. They were a great team! Audrey Ensminger provided invaluable professional help, encouragement, and book design and layout. Dr. Lawrence A. Duewer, Agricultural Economist USDA, ERS, provided many of the statistical facts and figures; Janetta Shumway, assisted by Jo Schepers, processed the manuscript and provided editorial assistance; Randall and Susan Rapp set the type and proofread the copy; and Margo Williams did the art and pasteup work. Additionally, a host of individuals, associations, and companies provided pictures and made other notable contributions, which are gratefully acknowledged throughout the book.

M. E. Ensminger

Clovis, California
1992

CONTENTS

THE POULTRY INDUSTRY

Fig. 1–1. Nature's way—a setting hen served as both the incubator and the brooder. (Courtesy, Euribrid B. V., Boxmeer, Holland)

CHAPTER

1

The term *poultry* applies to a rather wide variety of birds of several species, and it refers to them whether they are alive or dressed (slaughtered and prepared for market). The term applies to chickens, turkeys, ducks, geese, swans, guinea fowl, pigeons, peafowl, ostriches, pheasants, quail and other game birds. *The study of birds which are not classed as poultry is known as ornithology.*

There are about 600,000 species of animals in the world, of which 10,000 species are birds. The most highly developed animals are mammals, which include humans and the four-footed farm animals, and which are distinguished by the presence of hair and mammary glands. Birds are distinguished by the covering of feathers.

DOMESTICATION AND EARLY USE OF POULTRY

Primitive people persuaded chickens to live and produce near their abodes. It is not known exactly when this happened, but it's obvious that chickens were domesticated at a remote period. The keeping of poultry was probably contemporary with the keeping of sheep by Abel and the tilling of the soil by Cain. Chickens were known in ancient Egypt, and they had already achieved considerable status at the time of the Pharaohs, because artificial incubation was then practiced in crude ovens resembling some that are still in use in that country.

Fig. 1–2. Wild turkeys. Turkeys were native to the United States and Mexico. (Courtesy, University of Massachusetts, Amherst)

The use of poultry and eggs as food goes back to very early times in the history of primitive people. Methods of slaughter and preparation for consumption have varied with succeeding civilizations and cultures. Not until fairly recent times did these operations become a matter of great commercial importance, or of serious concern to consumers, public health officials, and governments alike.

TRANSFORMATION OF THE AMERICAN POULTRY INDUSTRY

The American poultry industry had its humble beginning when chickens were first brought to this continent by the early settlers. Small home flocks were started at the time of the establishment of the first permanent homes at Jamestown in 1607. For many years thereafter, chickens were tenderly cared for by the farmer's wife, who fed them on table scraps and the unaccounted-for grain from the crib.

As villages and towns were established, and increased in size, the nearby farm flocks were also increased. Surplus eggs and meat were sold or bartered for groceries and other supplies in the nearby towns. Eventually, grain production to the West, the development of transportation facilities, the use of refrigeration, and artificial incubation further stimulated poultry production in the latter part of the 1800s.

Since World War II, changes in poultry and egg production and processing have paced the whole field of agriculture. Practices in all phases of poultry production—breeding, feeding, management, housing, marketing and processing—have become very highly specialized. The net result is that more products have been made available to consumers at favorable prices, comparatively speaking, and per capita consumption and value of production have soared. This is shown in Fig. 1–3. Note that the combined value of production from broilers, eggs, turkeys, and chickens in 1990 was $14.9 billion. Of the combined total, 56% was from broilers, 27% from eggs, 16% from turkeys, and 1% from other chickens.

A brief chronology of the development of the poultry industry in the United States is given in Table 1–1.

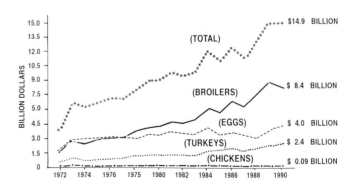

Fig. 1–3. Value of production: broilers, eggs, turkeys, chickens and total, United States 1972–1990. (Source: *Poultry Production and Value 1990 Summary*, USDA, National Agricultural Statistics Service, April 1991, p. 3)

TABLE 1–1
CHRONOLOGY OF THE DEVELOPMENT OF THE POULTRY INDUSTRY IN THE UNITED STATES[1]

Year	Event
1607	Small flocks were established in the Jamestown Colony.
1828	First Single Comb White Leghorns imported into the U.S.
1830	Breeding programs involving red chickens began around Narragansett Bay, in Rhode Island.
1840	First census of poultry taken in the U.S. Birds from China began to enter the U.S.
1842	First recorded American incubator.
1850	Large-scale raising of ducks began on Long Island, New York.
1862	Morrill Agricultural Land Grant Act passed, enabling states to establish Colleges of Agriculture.
	Congress authorized creation of USDA.
1869	First trapnest patent granted.
1870	Toe punch introduced for identification in breeding programs.
	First commercial hatchery established in the U.S.
1872	A commodity exhange for the trading of cash eggs was chartered.
1873	Pekin ducks first imported into the U.S.
1874	First American Standard of Excellence issued, known as Standard of Perfection.
	Chicken wire first introduced.
1875	Connecticut established the first state experiment station.
1878	First commercial egg drying operation established.
1887	Hatch Act passed, establishing agricultural experiment stations.
1889	Artificial light first used to stimulate egg production.
1892	First long-distance express shipment of baby chicks (Stockton, New Jersey to Chicago, Illinois), marking the beginning of the commercial hatchery industry.
	Henry Ford built first automobile.
1895	Commercial feed industry began in Chicago.
1898	Battery fattening introduced by Swift and Company.
1899	Frozen eggs first marketed.
1900	First egg-breaking operation established.
1901	Dry mash feeding first recommended to the public.
1903	Cornell gasoline brooder developed by Professor James Rice of Cornell University.
1904	First official egg laying contest started in America.
1905	First chickens raised in batteries.
	First battery brooder developed by Charles Cyphers.
1906	First U.S. county agent appointed.
1909	Electric candler developed.
1911	First U.S. egg laying contests established at Storrs, Connecticut and Mountain Grove, Missouri. Researchers "coined" the word *vitamin* to describe certain feed components.
1912	Casimar Funk proposed the "vitamine theory," stating that, "special substances which are of the nature of organic basis, called vitamines," are needed for animal life.
1913	First oil brooder marketed.
1914	Vitamin A discovered. Smith-Lever Agricultural Extension Act passed by Congress, establishing Cooperative Extension Service.
1918	First USDA federal-state grading programs for poultry established.
	U.S. Post Office shipped chicks by mail.
1922	First soybean meal processed.
1923	Electrically heated incubator developed by Ira Petersime. USDA issued first tentative classes, standards, and grades of eggs.
	English investigator reported sex linkage and sex determination of chicks at hatching.
	Commercial broiler industry started on the Delmarva Peninsula, near Ocean View, in Sussex County, Delaware.
1926	USDA inaugurated Federal Poultry Inspection Service.
1928	Sex-linked crosses advertised.
	USDA began inspection of poultry products for wholesomeness.
1929	Layers kept in individual cages at Ohio Agricultural Experiment Station.
	Pelleting of poultry feed initiated.
1931	Laying cages first given general publicity. All-night lights for layers used at Ohio Agricultural Experiment Station.
1932	Forced molting for commercial egg production started in Washington.
1933	Riboflavin isolated, with milk established as a valuable source of it.
1934	USDA issued standards for eggs, with legal status.

(Continued)

TABLE 1-1 *(Continued)*

Year	Event
1935	Artificial insemination of poultry introduced.
	First successful incrossbred-hybrid chickens produced for egg production.
	Pyridoxine discovered.
	First sulfa drug announced.
1936	Vitamin E isolated.
1937	Vitamin A, which was first identified in 1914, synthesized.
1940	Mechanical poultry dressing initiated.
	Hy-Line Poultry Farms, Des Moines, Iowa, marketed its first Hy-Line layers, thereby applying inbred-hybrid corn breeding principle to egg production.
	USDA announced Beltsville Small White turkey.
1941	Newcastle disease reported in U.S.
1942	Debeaker developed.
1946	Chicken-of-Tomorrow contest initiated by the Great Atlantic and Pacific Tea Company.
1947	Antibiotics first used in treatment of poultry diseases.
1948	Crystalline B-12 isolated.
1949	New nutritional developments discovered, including animal protein factor, vitamin B-12, antibiotics, hormones, and surfactants.
1954	USDA discovered parthenogenesis (reproduction without male fertilization) in the fowl.
1956	Col. Harland Sanders began franchise operations of Kentucky Fried Chicken.
	Integration and contracting began in the egg business.
1957	U.S. Congress authorized the Poultry Products Inspection Act.
1962	Corporate egg farms became a viable part of the poultry industry.
1969	University of Delaware began growing broilers in wire cages.
	Col. Sanders' Kentucky Fried Chicken began integrating, acquiring a broiler producer and a processor.
1970	Economists estimated that vertical integration, a form of contract production, involved 95% of the nation's broilers, 85% of the turkeys, and 30% of the table eggs.
1971	Exotic Newcastle disease discovered in southern California, ultimately requiring the condemnation and destruction of more than 3,000,000 birds, with indemnities of more than $5 million paid.
1972	Marek's vaccine approved.
1974	Selenium was approved by the Food and Drug Administration for use at the rate of 0.1 part per million (ppm) in complete feed for growing chickens to 16 weeks of age, for breeder hens producing hatching eggs, and at the rate of 0.2 ppm in complete rations for turkeys.
1981	For the first time in the U.S., the cash receipts from marketing poultry exceeded the cash receipts from marketing hogs.
1987	FDA increased the maximum allowance of selenium in complete feeds for chickens, turkeys, and ducks from 0.1 ppm to 0.3 ppm.
1990	Americans ate more broilers and turkeys than ever before because of (1) lower prices in relation to other meats, and (2) new and further processed products.
	Poultry production became specialized to the point that most producers contracted with large integrated companies to produce eggs, broilers, and turkeys. The companies supplied the chicks/poults, feed, litter, fuel, and medication, and the contract growers provided housing and labor for the care of the birds.

[1]Adapted by the author from *American Poultry History, 1823–1973*, American Poultry Historical Society, Inc., Chapter 20, Chronology.

A discussion of some of the most important changes in the poultry industry follows:

1. **Changes in breeding methods.** Standard-bred chickens decreased as modern breeding methods were applied.

The poultry geneticist discovered that family, as well as individual bird records, are needed to develop high egg production. From this base, breeders created certain strains for high egg production and feed efficiency.

When breeding for broiler production, hybrid vigor is obtained by systematic matings that may involve crossing different breeds, different strains of the same breed, or the crossing of inbred lines. Many of the strains used as sires trace their ancestry to the broad-breasted Cornish breed. The main objective is the improvement of broiler growth rate to 7 to 8 weeks of age, although improvement in other economic factors is sought.

Breeding for egg production differs from breeding for broiler production in that the individual methods most useful for improving growth rate have little value in selecting to improve egg production, because egg production is of low heritability. In breeding for egg production, high-producing families are selected. Then, either of two types of crosses are made: (a) crossing of inbred lines, or (b) using strains which are not inbred.

2. **Changes in hatcheries.** In the beginning, hatching was done according to nature's way—by a setting hen hovering over eggs. Then came the first American incubator, patented in 1844, followed by the U.S. Post Office acceptance of chicks for shipment by mail in 1918. Hatcheries became larger in size and fewer in number. In 1934, ½ billion

chicks were hatched in 11,000 hatcheries. In 1989, 6.3 billion chicks, including both broiler and egg type, were hatched in 376 hatcheries each with an average capacity of 1.4 million eggs; and 289 million turkey poults were hatched in 82 hatcheries each with an average capacity of 527,000 eggs.

3. **Changes in egg production.** A hundred years ago, a hen produced about 100 eggs per year. In 1990, the U.S. average was 252 eggs per hen (Fig. 1–4). Formerly, eggs were sold largely on an ungraded basis. Today, most of them are candled for interior quality, weighed, cartoned, and sold according to size and quality. In many modern egg-grading plants, efficient, power-operated weighing machines speed grading and move the eggs through the marketing system with dispatch.

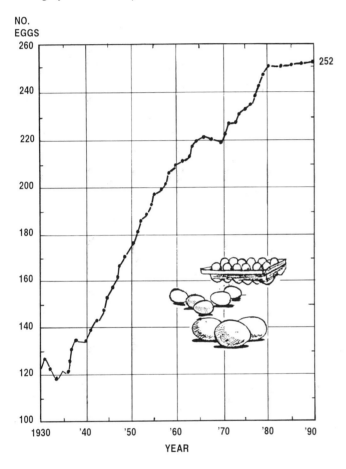

Fig. 1–4. Changes in U.S. annual egg production per hen. (Source: *Agricultural Statistics 1990*, p. 353, Table 528)

Beginning in the early 1930s, three important changes took place in relation to egg production: (a) with the greater emphasis on commercial size flocks, the light breeds and strains of chickens gradually replaced general or dual-purpose breeds for egg production; (b) as the technique for "sexing" chicks became perfected, only the female chicks of egg-type breeds were sold by the hatcheries to layer operations; and (c) feeding, breeding, management, and disease control practices were improved so that more eggs were produced per layer, thereby requiring fewer layers to provide the eggs necessary to supply the market demands.

Fig. 1–5. Superlayer perched on the 448 eggs that she laid nonstop. (Courtesy, University of Missouri, Columbia)

In recent years, the total number of layers and the total egg production in the United States declined rather sharply, perhaps due to the cholesterol scare (see Fig. 1–6).

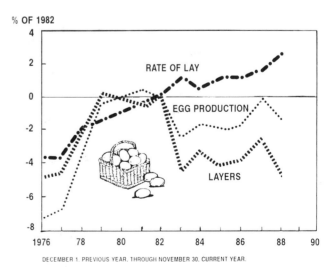

Fig. 1–6. Eggs: Rate of lay, production, and number of layers. (Source: *1989 Agricultural Chartbook*, USDA, Agricultural Handbook No. 684, p. 92, Chart No. 206)

4. **Changes in chicken meat production.** Prior to 1930, chicken meat was mainly the by-product of egg production. Birds which were no longer producing eggs at a satisfactory rate were sold for meat purposes, mainly in the fall of the year. Cockerels raised with the pullets were disposed of as fryers, or roasters, at weights of 3 to 8 lb.

In 1934, 34 million broilers were produced in the United States. By 1960, this figure had increased to 1.8 billion; and in 1989, 5.5 billion broilers were produced in the United States.

Modern broiler production is so concentrated, and so highly commercialized, that the industry might properly be classed as a poultry meat factory, rather than a farming operation. Very little land is required beyond the space necessary for a broiler house and a driveway. Chicks, feed, and other items used in production are purchased, or are obtained, from another division of an integrated operation of which the broiler production is a part. The operation is highly specialized, mechanized, and carried on within the limits of the broiler house. Also, the operation is characterized by large numbers.

Similar progress has been made in the processing of poultry. In most areas of the country, poultry processing has become a highly industrialized, large-scale operation using modern mechanical equipment and sanitary methods.

Fig. 1–7. Poultry meat has changed! This shows packaged Cornish hen (left) and young turkey. (Courtesy, USDA)

5. **Changes in turkey production.** Like the situation in chicken production, the production of turkeys was mostly a small sideline enterprise until 1910. At that time, 870,000 farmers raised 3⅔ million turkeys, or an average of four turkeys per farm. In 1987, farmers raised 240 million turkeys, and flocks of more than 50,000 birds were common.

6. **Changes in number and size of poultry farms.** In 1910, more than 5½ million farms in the United States (88% of the 6.4 million farms in the nation) kept chickens. The average size flock in the United States numbered 50 laying hens.

In recent years, U.S. poultry farms have become fewer and larger. In 1989, the average egg farm had 154,000 hens. The average number of broilers raised per farm rose from about 34,000 birds in 1959 to about 158,000 in 1987. During this same period, turkey numbers rose sharply from 80 million birds in 1959 to about 243 million in 1987; and average farm output increased from 900 birds in 1959 to 33,000 in 1987. This transition to bigness in turkey operations is portrayed in Fig. 1–8. This trend to bigness will continue.

OPERATIONS

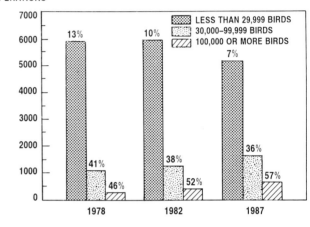

Fig. 1–8. Number of turkey growout operations by size category and share of total raised. (Source: *Livestock and Poultry Situation and Outlook Report*, USDA, ERS, LPS–46, February 1991, p. 39, Fig. A–2)

7. **Changes in ownership and organization.** As poultry operations grew in size and efficiency, they vertically integrated (chiefly with feed companies and processors) to secure more credit; and contractual production became commonplace. Today, 95% of the broilers are produced under contract, and the remaining 5% are raised on integrator-owned farms. Also, it is estimated that 90% of the turkeys, and 89% of the eggs are produced under some kind of integrated or contract arrangement.

8. **Changes in labor requirements.** Poultry producers have achieved remarkable efficiency, primarily through increased confinement production and mechanization. In 1935 to 1939, it required 8½ work-hours to produce 100 lb of broiler; in 1985 to 1989, it required a minimum of 0.1 work-hour to produce 100 lb of broiler; and in this same period of time, the labor requirements to produce 100 lb of turkey were lowered from 23.7 work-hours to 0.2 work hour (Table 1–2). But no such progress has been made in lowering the labor requirements for the production of red meats, with the result that poultry producers have achieved a very real advantage.

TABLE 1-2
POULTRY LABOR REQUIREMENTS[1]

Type of Poultry	Work Hours/Hundredweight Production[1]		No. Birds Cared for in 1990 by One Worker in the Most Efficient Operations[2]
	1935–39	1985–89	
	(hours)	(hours)	
Layers	1.7/100 eggs	0.2/100 eggs	Laying hens (cage) .... 40,000[3]
			Laying hens (floor) .... 15,000[3]
Broilers	8.5	0.1	 75,000
Turkeys	23.7	0.2	 40,000

[1]Source: Data on "work hours/hundredweight production," USDA.
[2]Estimates by M. E. Ensminger.
[3]Does not include time devoted to egg processing.

9. **Changes in feed efficiency.** There has been a marked lowering of feed required to produce a unit of eggs, turkeys, and broilers since 1940. In 1940, it required 7.4 lb of feed to produce a dozen eggs; in 1990, it took only 3.75 lb. In 1940, it required 4.7 lb of feed to produce 1 lb of weight gain of broilers; in 1990, it took only 1.9 lb. In 1940, it required 4.5 lb of feed to produce 1 lb of weight gain of turkeys; in 1990, it took only 2.7 lb.

10. **Changes in geography of production and processing.** Layer, broiler, and turkey production in the United States remains concentrated in what have long been the leading states for each type of enterprise. But there is a growing interest in moving broiler production closer to grain areas and major consumer markets.

11. **Changes in marketing.** In the movement of poultry meat and eggs from the producer to the consumer, fewer agencies are being used and there is more direct marketing.

12. **Changes in proportion of federally inspected slaughter.** Prior to passage of the Poultry Inspection Act of 1957, relatively little poultry was slaughtered under federal inspection. Today, all commercial poultry slaughter is under federal inspection.

13. **Changes in processing.** In 1947, New York dressed chicken (only the blood and feather removed) accounted for 80% of the chickens and 75% of the turkeys marketed. By 1963, only about 11% of the chickens and 5% of the turkeys were marketed that way. By the mid-1960s, 88% of the broilers were ice-packed and ready-to-cook, whereas 86% of the turkeys were marketed frozen. Today, practically all chickens and turkeys are eviscerated. Not only that, much of the meat is cut up or further processed — changes designed to lessen labor in the kitchen and to add variety to menus. Also, there is a big trend toward ready-to-eat items and TV dinners.

POULTRY PRODUCTS—BASIC FOR GOOD NUTRITION

Most nutrition programs can be improved by the inclusion of eggs and/or poultry meat. Poultry products are of the highest nutritional quality and are marketed at prices that even the poor can afford.

EGGS

Of all the foods available to people, the egg most nearly approaches a perfect balance of all the nutrients. This is evidenced by the fact that the egg is the total source of nutrition for the developing embryo. Unlike mammals, the chick embryo cannot secure needed nutrients from the reserves of the mother; rather, it lives in a closed system which must contain all the food needed for development. If a single nutrient were to be lacking, the developing embryo would die.

Eggs contain an abundance of proteins, vitamins, and minerals. The protein fraction of eggs is highly digestible and of high quality, having a biological value of 94 on a scale of 100, the highest rating of any food. The reason for this high quality of protein is that egg proteins are complete proteins; that is, they contain all the essential amino acids required to maintain life and promote growth and health. Additionally, eggs are a rich source of iron; phosphorus; trace minerals; vitamins A, E, and K; and all the B vitamins, including vitamin B–12. Eggs are second only to fish-liver oils as a natural source of vitamin D. Eggs are moderate from the standpoint of calorie content, a medium-size egg containing about 77 calories. Table 1–3 illustrates the nutritive value of eggs.

TABLE 1–3
SOME OF THE ESSENTIAL NUTRIENTS IN TWO MEDIUM EGGS
(0.24 lb, or *108 g*) WITHOUT SHELL[1]

Nutrient	Function	Percentage of Daily Dietary Recommendations[2]
		(%)
Protein	To build muscles and body tissues	16.3
Iron	To build red blood cells and help promote good health	17.3
Vitamin A	To help give normal vision and clear, healthy skin	22.0
Thiamin (Vitamin B–1)	To promote growth, good appetite, and a healthy nervous system	8.3
Riboflavin (Vitamin B–2)	To promote growth and good health	14.8
Vitamin B–12	To help prevent and cure pernicious anemia	1.0 mcg[3]
Vitamin D	To help calcium in building bones and teeth	25.0

[1]From the Poultry and Egg National Board, Chicago, IL.

[2]Recommendations for girls, ages 16–19.

[3]No daily recommendations established.

Notwithstanding the high nutritive value of eggs, millions of Americans routinely eat eggs for breakfast each morning simply because they like them. Additionally, eggs are used in numerous baked and processed foods to enhance flavor and texture.

But per capita egg consumption by Americans has been declining steadily for many years! In 1945, the final year of World War II, the U.S. per capita average exceeded 400 eggs, nearly double the current rate of 234.

POULTRY MEAT

Poultry meat is supplied chiefly by chickens and turkeys, although ducks, geese, guinea fowl, quail, pheasants, squabs (pigeons), and other fowl contribute thereto. Poultry meat is economical, and quick and easy to prepare and serve. Also, it has a number of desirable nutritional properties.

Nutritionally, people eat poultry meat for its high content of high-quality protein and its low-fat content. Turkey and chicken meat is slightly higher in protein and slightly lower in fat than beef and other red meats (see Table 1–4). Additionally, the protein is a rich source of all the essential amino acids, as shown in Table 1–5. The close resemblance of the amino acid content of poultry meat to the amino acid profiles of milk and eggs serves to emphasize the latter point.

TABLE 1–4
COMPARISON OF NUTRIENT COMPOSITION OF COOKED TURKEY, CHICKEN, AND BEEF[1]

Kind of Meat	Protein	Fat	Moisture	Food Energy Calories
	(%)	(%)	(%)	(per 3½ oz or *100 g*)[2]
Turkey (mature, roasted and boned):				
Breast (white meat)	32.9	3.9	62.1	176
Leg (dark meat)	30.0	8.4	60.5	204
Chicken (16 weeks old, roasted and boned):				
Breast (white meat)	32.4	5.0	61.3	182
Leg (dark meat)	29.2	6.5	62.7	185
Beef (cooked and boned):				
Round steak	31.3	6.4	61.2	183
Rump roast	29.6	7.1	62.0	190
Hamburger	27.4	11.3	60.0	219

[1]Source: *Foods & Nutrition Encyclopedia*, Ensminger, *et al*, pp. 912, 914.

[2]Standard portion size.

TABLE 1–5
COMPARISON OF AMINO ACID COMPOSITION OF VARIOUS ANIMAL FOODS[1]

Amino Acid	Percentage of the Food						
	Turkey	Chicken[2]	Eggs	Beef	Pork	Lamb	Milk
Arginine . . .	1.31	1.47	0.86	1.12	1.09	1.17	0.10
Cystine	0.20	0.40	0.28	0.24	0.22	0.22	0.03
Histidine . . .	0.60	0.44	0.34	0.49	0.54	0.46	0.09
Isoleucine . .	1.00	0.90	0.90	0.87	0.83	0.82	0.18
Leucine	1.53	1.45	1.09	1.43	1.28	1.25	0.33
Lysine	1.81	1.65	0.87	1.43	1.32	1.31	0.27
Methionine . .	0.52	0.40	0.42	0.39	0.42	0.39	0.08
Phenylalanine	0.74	0.88	0.69	0.68	0.70	0.66	0.16
Threonine . .	0.80	0.88	0.70	0.68	0.87	0.83	0.15
Tryptophan . .	0.18	0.18	0.24	0.19	0.24	0.22	0.05
Valine	1.02	1.47	1.05	0.97	0.85	0.85	0.23

[1]Adapted by the author from *Protein Resources and Technology: Status and Research Needs*, edited by M. Miner, N. S. Scrimshaw, and D. I. C. Wang, The Avi Publishing Company, Inc., Westport, CT, 1978, pp. 393, 396, and 525, Tables 21.4, 21.10, and 25.2.

[2]Average of values for whole chicken (white meat and dark meat).

HEALTH PROBLEMS ATTRIBUTED TO POULTRY PRODUCTS

Americans are becoming more and more health and diet conscious. Unfortunately, much of the nutritional information that they receive is not authoritative, and the facts are not presented objectively. Worse yet, passions and prejudices sometime trigger such charges as "poultry products cause allergies and heart disease." Such accusations must be answered by more than simple denial. To this end, sections on "Allergies" and "Cholesterol, Animal Fat, and Heart Disease" follow.

ALLERGIES

Occasionally, a child exhibits an allergic sensitivity to eggs. In most cases, the white of the egg is the portion that creates the reaction, and the yolk is generally readily tolerated. In highly sensitive individuals, the diet must be totally devoid of egg. However, these cases are rare, and most infants and children allergic to eggs can follow a diet that provides for some heat-treated or cooked eggs.

CHOLESTEROL, ANIMAL FAT, AND HEART DISEASE

In recent years, the attention of the American public has been focused on heart disease — the leading killer in the United States. The nation's newspapers and airwaves devote much space and time to its causes, treatment, and prevention.

Numerous types of heart disease contribute to the yearly death toll of about one million persons in the United States; among them, hypertension, cerebrovascular disease (stroke), congestive heart failure, and atherosclerosis. Diet has been implicated in a number of these diseases, and much attention has been given to the role of animal fats in atherosclerosis. *Atherosclerosis is a form of heart disease wherein a buildup of soft, amorphorus lipids and connective tissue develops on the walls of the arteries of the heart.* If these deposits become sufficiently large, clots may form and subsequently decrease the diameter of the arterial lumen — in some cases, totally blocking blood flow. When blood flow to the heart is greatly impaired, a heart attack will ensue.

The cholesterol-heart disease debate first began when, in 1953, Dr. Ancel Keys, at the University of Minnesota, reported a positive correlation between the consumption of animal fat and the occurrence of atherosclerosis.[1] Subsequent research indicated that individuals with high serum cholesterol levels had a higher rate of atherosclerosis than people with normal levels. Increased serum cholesterol levels can be induced in susceptible individuals when animal fats (saturated fats) and foods high in cholesterol (notably eggs) are consumed. Thus, most research implicates cholesterol as a primary cause of heart disease. Yet, it is noteworthy that certain African tribes whose diets consist almost entirely of animal products do not have elevated serum cholesterol levels. Additionally, it has been shown that the consumption of dietary cholesterol and saturated fats cause a temporary elevation of serum cholesterol in healthy individuals, but, after a short time, the levels return to normal.

That the form of dietary fat is implicated in atherosclerosis is fairly well documented by unbiased, controlled experiments; but it must be emphasized that research indicates that it is not the sole cause. Rather, a number of factors enter into the cause of heart disease, many of which play far greater roles than cholesterol; among them, stress, heredity, hypertension, diabetes mellitus, smoking, lack of exercise, and obesity.

When evaluating the risk of heart disease posed by the consumption of animal products, one must weigh the benefits against the hazards. As shown in Table 1–6 (see next page), most of the countries with the highest life expectancies are noted for their egg production. The nutrients supplied by eggs and meat provide well-balanced nutrition; hence, poultry products must not be eliminated from the diet lest nutrient deficiencies arise. Rather, a well-planned diet, along with exercise and a minimum of stress, constitutes the best prevention against heart disease.

[1]Keys, A., *Journal of Chronic Diseases*, Vol. 4, 1956, p. 364.

TABLE 1–6
PER CAPITA EGG PRODUCTION OF COUNTRIES
HAVING THE HIGHEST LIFE EXPECTANCY AT BIRTH FOR MALES

Longevity Rank	Country	Life Expectancy[1]	Per Capita Egg Production[2]
		(yrs)	(no.)
1	Japan	76.0	337.3
2	Greece	75.0	207.3
3	Iceland	75.0	261.0
4	Israel	75.0	408.6
5	Jamaica	75.0	82.3
6	Brunei Darus	74.0	139.0
7	Cyprus	74.0	186.4
8	Netherlands	74.0	742.6
9	Spain	74.0	336.2
10	Sweden	74.0	263.5
11	Switzerland	74.0	114.5
12	Australia	73.0	193.7
13	Barbados	73.0	106.1
14	Canada	73.0	201.9
15	Italy	73.0	200.0
16	Norway	72.7	219.3
17	United States	72.7	276.2
18	Malta	72.5	314.3

[1]*The World Almanac and Book of Facts 1991*, World Almanac, Pharos Books, New York, NY.

[2]*FAO Production Yearbook 1989*, Food and Agriculture Organization of the United Nations, Vol. 43, p. 279, Table 103.

POULTRY PRODUCTS TO FEED THE HUNGRY

It is not easy to present a precise picture of the extent of undernutrition and malnutrition in the world. Intake of energy and protein over a period of time appears to be the best indicator of dietary sufficiency. This information is presented in Table 1–7 and Fig. 1–9, which show the per capita calories and proteins per day of the geographic areas of the world.

Through the ages, eggs and poultry meat have been a basic food. In the future, they will become increasingly important in meeting the challenge for feeding the hungry people of the world. Nevertheless, the food vs feed argument will wax hot. Can we afford to feed grains to animals while denying them to the starving? There are many arguments for and against the practice of feeding animals to produce human food. However, poultry products have proven to be extremely valuable as a source of food for the following reasons:

1. **Poultry convert feed to food efficiently.** Table 1–8 (see p. 12) lists the feed to food efficiency rating by species of animal. Broilers have the most favorable protein efficiency ratio of any animal; and when compared to the mammalian livestock species grown for meat, they have the most favorable feed conversion ratio. Protein and feed efficiencies of turkeys and layers are also favorable when compared to the other types of livestock.

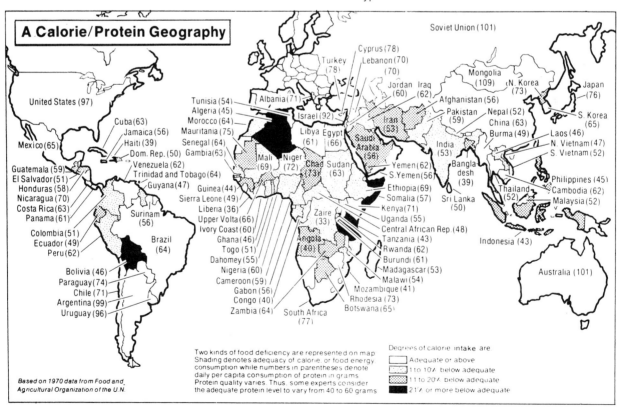

Fig. 1–9. World geography of calories and proteins. (Courtesy, the *New York Times*)

TABLE 1-7
PER CAPITA CALORIES AND PROTEIN PER DAY
IN GEOGRAPHICAL AREAS OF THE WORLD[1]

Area	Per Capita Calories Per Day	Per Capita Grams of Protein Per Day
	(no.)	(g)
Europe	3,397	101.5
U.S.S.R.	3,394	105.6
North and Central America	3,370	94.8
Oceania	3,126	91.4
South America	2,622	66.2
Asia	2,485	60.7
Africa	2,299	57.4
World Total	2,694	70.3

[1]Source: *FAO Production Yearbook 1988*, United Nations, Rome, Italy, Vol. 42, pp. 291–294. Data for 1984–86.

2. **The poultry industry is dynamic.** Because of the short periods required for growing and marketing, the poultry industry can adjust rapidly to a variety of economic factors, e.g., feed availability, numbers of birds on feed, costs, etc. Other livestock enterprises, notably the cattle industry, necessitate relatively long periods to adjust to market changes because of length of time required for the animals to mature and reproduce.

3. **Poultry feeds not commonly used for human consumption.** A large number of by-product feeds are fed to poultry; among them, blood meal, fish meal, and meat and bone meal. Additionally, poultry do not compete to any appreciable extent with the hungry people of the world for food grains, such as rice or wheat. Instead, they eat feed grains—like field corn, grain sorghum, barley, and oats—for which there is little or no demand for human consumption.

4. **Layers provide a continuous source of food.** Unlike meat animals that must be fed for a period of time before a usable product can be attained, layers produce eggs throughout the year. Thus, the layer can produce several times her weight in eggs throughout her life, while the products derived from meat animals are restricted in their final market weight.

5. **Vegetarians consume eggs.** In countries such as India and Pakistan, the eating of meat is taboo by religious precept. But the consumption of eggs is acceptable and is a major source of animal protein consumed by the people of these countries.

6. **Poultry products are inexpensive.** Poultry meat and eggs are among the best buys in the supermarket.

7. **Poultry manure can be used as a fertilizer or a feed.** Poultry manure is an extremely rich source of nitrogen and organic material. Hence, it is highly regarded as a fertilizer. Also, in recent years, poultry wastes have proven to be a valuable feed for ruminants. Thus, feed may be used twice; once through the birds, thence by the ruminant.

Fig. 1–9a. Eggs provide a continuous source of food. (Courtesy, American Egg Board, Park Ridge, IL)

TABLE 1-8
FEED TO FOOD EFFICIENCY RATING BY SPECIES OF ANIMALS, RANKED BY PROTEIN CONVERSION EFFICIENCY
(Based on Energy as TDN or DE and Crude Protein in Feed Eaten by Various Kinds of Animals Converted into Calories and Protein Content of Ready-to-Eat Human Food)

Species	Unit of Production (on foot)	Feed Required to Produce One Production Unit				Dressing Yield		Ready-to-Eat; Yield of Edible Product (meat and fish deboned and after cooking)				Feed Efficiency[1] (lb feed to produce one lb product)		Efficiency Rating			
														Calorie Efficiency[5]		Protein Efficiency[6]	
		Pounds	TDN[2]	DE[3]	Protein	Percent	Net Left	As % of Raw Product (carcass)	Amount Remaining from One Unit of Production	Calorie[4]	Protein[4]	(%)	(ratio)	(%)	(ratio)	(%)	(ratio)
		(lb)	(lb)	(kcal)	(lb)	(%)	(lb)	(%)	(lb)	(kcal)	(lb)	(%)	(ratio)	(%)	(ratio)	(%)	(ratio)
Broiler	1 lb chicken	2.4[7]	1.94[8]	3,880	0.21[8]	72[9]	0.72	54[10]	0.39	274	0.11	41.7	2.4:1	7.1	14.2:1	52.4	1.9:1
Dairy cow	1 lb milk	1.11[7]	0.90[8]	1,800	0.10[8]	100	1.00	100	1.00	309	0.037	90.0	1.11:1	17.2	5.8:1	37.0	2.7:1
Turkey	1 lb turkey	5.2[7]	4.21[8]	8,420	0.46[8]	79.7[9]	0.797	57[11]	0.45	446	0.146	19.2	5.2:1	5.3	18.9:1	31.7	3.2:1
Layer	1 lb eggs (8 eggs)	4.6[7]	3.73[8]	7,460	0.41[8]	100	1.00	100[12]	1.00[12]	616	0.106	21.8	4.6:1	8.3	12.1:1	25.9	3.9:1
Rabbit	1 lb fryer	3.0[13]	2.20	4,400	0.48	55[13]	0.55	79[13]	0.43	301	0.08	35.7	2.8:1	6.8	14.6:1	16.7	6.0:1
Fish	1 lb fish	1.6[14]	0.98	1,960	0.57	65[15]	0.65	57[16]	0.37	285	0.093	62.5	1.6:1	14.5	6.9:1	16.3	6.1:1
Hog (birth to 200 lb) ...	1 lb pork	4.9[17]	3.67	7,340	0.69	70[18]	0.70	44[19]	0.31	341	0.088	20.4	4.9:1	4.6	21.5:1	12.7	7.8:1
Beef steer (yearling finishing period in feedlot) ...	1 lb beef	9.0[17]	5.85	11,700	0.90	58[18]	0.58	49[19]	0.28	342	0.085	11.1	9.0:1	2.9	34.2:1	9.4	10.6:1
Lamb (finishing period in feedlot)	1 lb lamb	8.0[17]	4.96	9,920	0.86	47[18]	0.47	40[19]	0.19	225	0.052	12.5	8.0:1	2.3	44.1:1	6.0	16.5:1

[1]Feed efficiency as used herein is based on pounds of feed required to produce 1 lb of product. Given in both percent and ratio.

[2]TDN pounds computed by multiplying pounds feed (column to left) times percent TDN in normal rations. Normal ration percent TDN taken from M. E. Ensminger's books and rations, except for following: dairy cow, layer, broiler, and turkey from Agricultural Statistics 1974, p. 358, Table 518. Fish based on averages recommended by Michigan and Minnesota Stations and U.S. Fish and Wildlife.

[3]Digestible Energy (DE) in this column given in kcal, which is 1 Calorie (written with a capital C), or 1,000 calories (written with a small c). Kilocalories computed from TDN values in column to immediate left as follows: 1 lb TDN = 2,000 kcal.

[4]Lessons on Meat, National Live Stock and Meat Board, 1965.

[5]Kilocalories in ready-to-eat food = kilocalories in feed consumed, converted to percentage. Loss = kcal in feed ÷ kcal in product.

[6]Protein in ready-to-eat food = protein in feed consumed, converted to percentage. Loss = pounds protein in feed ÷ pounds protein in product.

[7]Agricultural Statistics 1974, p. 358, Table 518. Pounds feed per unit of production is expressed in equivalent feeding value of corn.

[8]Since pounds feed (column No. 3) per unit of production (column No. 2) is expressed in equivalent feeding value of corn, the values for corn were used in arriving at these computations. No. 2 corn values are TDN, 81%; protein, 8.9%. Hence, for the dairy cow 81% × 1.11 = 0.9 lb TDN; and 8.9% × 1.11 = 0.1 lb protein.

[9]Marketing Poultry Products, 5th ed., by E. W. Benjamin, et al., John Wiley & Sons, 1960, p. 147.

[10]Factors Affecting Poultry Meat Yields, University of Minnesota Sta. Bull. 476, 1964, p. 29, Table 11 (fricassee).

[11]Ibid., p. 28, Table 10.

[12]Calories and protein computed basis per egg; hence, the values herein are 100% and 1.0 lb respectively.

[13]Based on information in Commercial Rabbit Raising, Ag. Hdbk., No. 309, USDA, 1966, and A Handbook on Rabbit Raising, by H. M. Butterfield, Washington State College Ext. Bull. No. 411, 1950.

[14]Data from report by Dr. P. J. Schaible, Michigan State University, Feedstuffs, April 15, 1967.

[15]Industrial Fishery Technology, edited by M. E. Stansby, Reinhold Pub. Corp., 1963, Ch. 26, Table 26-1.

[16]Ibid. Reports that, "Dressed fish averages about 73% flesh, 21% bone, and 6% skin." In limited experiments conducted by A. Ensminger, it was found that there was a 22% cooking loss on filet of sole. Hence, these values—give 57% yield of edible fish after cooking, as a percent of the raw, dressed product.

[17]Estimates by the author.

[18]Ensminger, M. E., The Stockman's Handbook, 5th ed., Sec. XIV.

[19]Allowance made for both cutting and cooking losses following dressing. Thus, values are on a cooked, ready-to-eat basis of lean and marbled meat, exclusive of bone, gristle, and fat. Values provided by National Live Stock and Meat Board (personal communication of June 5, 1967, from Dr. W. C. Sherman, Director, Nutrition Research, to the author), and based on data from The Nutritive Value of Cooked Meat, by R. M. Leverton and G. V. Odell, Misc. Pub. MP–49, Appendix C, March 1958.

WORLD POULTRY DISTRIBUTION

It is important that poultry producers have a global perspective concerning poultry in order to know which countries are potential competitors and what's ahead.

The production of poultry is worldwide. Figs. 1–10, 1–11, and 1–12, along with Tables 1–9, 1–10, and 1–11, give pertinent details pertaining to the leading poultry-producing countries of the world.[2]

■ **Leading chicken-producing countries** – China holds a commanding lead in chicken numbers, followed by the United States and the U.S.S.R., respectively. In chickens, per capita, however, the United States ranks first in this group, with the U.S.S.R. ranking second. Japan produces more eggs per capita than any of the ten leading chicken-producing countries, with the U.S.S.R. ranking second.

■ **Leading turkey-producing countries** – The United States ranks first in turkey numbers by a considerable margin. The U.S.S.R. ranks a solid second in turkey production, followed by third ranking Italy and fourth ranking France.

■ **Leading duck-producing countries** – China ranks first in duck production by a wide margin; China produces more ducks than all the other countries combined. Bangladesh, Indonesia, and Vietnam rank second, third, and fourth, respectively, in duck production.

LEADING CHICKEN-PRODUCING COUNTRIES—1989

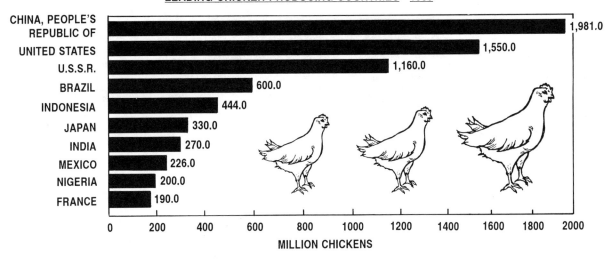

Fig. 1–10. Leading chicken producers of the world, 1989.

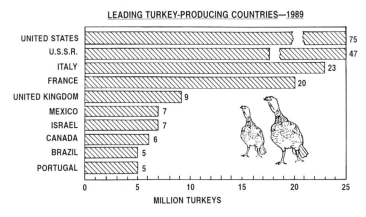

Fig. 1–11. Leading turkey producers of the world, 1989.

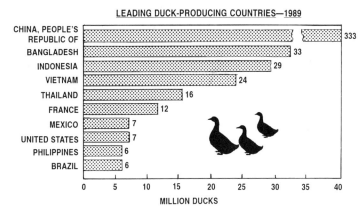

Fig. 1–12. Leading duck producers of the world, 1989.

---

[2]**Note well:** The poultry data for these figures and tables came from the *FAO Production Yearbooks*. The bird and per capita numbers of each species are much too low for the United States. But, since FAO is the only source of data for all countries and all species of poultry, the author reasoned that a similar method of collecting data was likely used by FAO in all countries, and that the data would show the relative ranking of countries by species of poultry.

TABLE 1–9
LEADING CHICKEN-PRODUCING COUNTRIES OF THE WORLD, 1989

Country	Chickens[1]	Egg Production[2]	Human Population[3]	Size of Country[4]		Chickens per Capita[5]	Eggs per Capita	Chickens per	
	(thousands)	(millions)	(thousands)	(sq mi)	*(sq km)*			(sq mi)	*(sq km)*
China, People's Republic of	1,981,000	117,241	1,117,173	3,705,390	*9,596,960*	1.8	104.9	534.6	*206.4*
United States	1,550,000	68,510	248,082	3,618,770	*9,372,614*	6.3	276.2	428.3	*165.4*
U.S.S.R.	1,160,000	80,690	288,750	8,649,496	*22,402,195*	4.0	279.5	134.1	*51.8*
Brazil	600,000	18,966	147,404	3,286,470	*8,511,957*	4.1	128.7	182.6	*70.5*
Indonesia	444,000	7,500	177,839	735,268	*1,904,344*	2.5	42.2	603.9	*233.2*
Japan	330,000	41,522	123,116	145,856	*377,767*	2.7	337.3	2,262.5	*873.6*
India	270,000	18,483	836,339	1,266,595	*3,280,481*	0.3	22.1	213.2	*82.3*
Mexico	226,000	14,607	86,737	761,604	*1,972,554*	2.6	168.4	296.7	*114.6*
Nigeria	200,000	5,293	109,153	356,667	*923,768*	1.8	48.5	560.8	*216.5*
France	190,000	15,362	56,115	220,668	*571,530*	3.4	273.8	861.0	*332.4*
World	10,574,000	598,519	5,205,242	50,460,631[6]	*130,693,034*	2.0	115.0	209.6	*80.9*

[1]*FAO Production Yearbook*, 1989, Vol. 43, pp. 250–251. Table 91. Primarily layers. FAO reports that data for chickens do not seem to represent the total number of these birds.

[2]*FAO Production Yearbook*, 1989, Vol. 43, pp. 280–281. Metric tons of eggs converted to number of eggs by estimating the average weight of one egg at 58 grams. FAO reports that some countries have no statistics on egg production; so, some statistics represent estimates made by FAO.

[3]*FAO Production Yearbook*, 1989, Vol. 43, pp. 63-79. Table 3.

[4]*The World Almanac and Book of Facts 1991*, World Almanac, Pharos Books, New York, NY.

[5]Chickens per capita computed by dividing column 2 by column 4.

[6]*FAO Production Yearbook*, 1989, Vol. 43, p. 47. Table 1.

TABLE 1–10
LEADING TURKEY-PRODUCING COUNTRIES OF THE WORLD, 1989

Country	Turkeys[1]	Human Population[2]	Size of Country[3]		Turkeys per Capita[4]	Turkeys per	
	(thousands)	(thousands)	(sq mi)	*(sq km)*		(sq mi)	*(sq km)*
United States[5]	75,000	248,082	3,618,770	*9,372,614*	0.30	20.7	*8.0*
U.S.S.R.	47,000	288,750	8,649,496	*22,402,195*	0.16	5.4	*2.1*
Italy	23,000	57,525	116,303	*301,225*	0.40	197.8	*76.4*
France	20,000	56,115	220,668	*571,530*	0.36	90.6	*35.0*
United Kingdom	9,000	57,417	94,226	*244,045*	0.16	95.5	*36.9*
Mexico	7,000	86,737	761,604	*1,972,554*	0.08	9.2	*3.6*
Israel	7,000	4,509	7,847	*20,324*	1.55	892.1	*344.4*
Canada	6,000	26,248	3,558,096	*9,215,469*	0.23	1.7	*0.7*
Brazil	5,000	147,404	3,286,470	*8,511,957*	0.03	1.5	*0.6*
Portugal	5,000	10,264	36,390	*94,250*	0.49	137.4	*53.1*
World	234,000	5,205,242	50,460,631[6]	*130,693,034*	0.04	4.6	*1.8*

[1]*FAO Production Yearbook*, 1989, Vol. 43, pp. 250–251, Table 91. FAO reports that data for turkeys do not seem to represent the total number of these birds.

[2]*FAO Production Yearbook*, 1989, Vol. 43, pp. 63–79. Table 3.

[3]*The World Almanac and Book of Facts 1991*, World Almanac, Pharos Books, New York, NY.

[4]Turkeys per capita computed by dividing column 2 by column 3.

[5]All data for 50 states.

[6]*FAO Production Yearbook*, 1989, Vol. 43, p. 47. Table 1.

TABLE 1–11
LEADING DUCK-PRODUCING COUNTRIES OF THE WORLD, 1989

Country	Ducks[1]	Human Population[2]	Size of Country[3]		Ducks per Capita[4]	Ducks per	
	(thousands)	(thousands)	(sq mi)	(sq km)		(sq mi)	(sq km)
China, People's Republic of	333,000	1,117,173	3,705,390	9,596,960	0.30	89.9	34.7
Bangladesh	33,000	112,585	55,598	143,999	0.29	593.6	229.2
Indonesia	29,000	177,839	735,268	1,904,344	0.16	39.4	15.2
Vietnam	24,000	65,662	128,401	332,559	0.37	186.9	72.2
Thailand	16,000	54,945	198,456	514,001	0.29	80.6	31.1
France	12,000	56,115	220,668	571,530	0.21	54.4	21.0
Mexico	7,000	86,737	761,604	1,972,554	0.08	9.2	3.6
United States	7,000	248,082	3,618,770	9,372,614	0.03	1.9	0.8
Philippines	6,000	60,970	115,831	300,002	0.10	51.8	20.0
Brazil	6,000	147,404	3,286,470	8,511,957	0.04	1.8	0.7
World	527,000	5,205,242	50,460,631[5]	130,693,034	0.10	10.4	4.0

[1]*FAO Production Yearbook*, 1989, Vol. 43, pp. 250–251. Table 91. FAO reports that data for ducks do not seem to represent the total number of these birds.

[2]*FAO Production Yearbook*, 1989, Vol. 43, pp. 63–79. Table 3.

[3]*The World Almanac and Book of Facts 1991*, World Almanac, Pharos Books, New York, NY.

[4]Ducks per capita computed by dividing column 2 by column 3.

[5]*FAO Production Yearbook*, 1989, Vol. 43, p. 47. Table 1.

U.S. POULTRY INDUSTRY

Poultry producers need to know the leading states, from the standpoints of both their competition and markets; they need to know the relative importance of species of poultry; they need to know the components of the poultry industry, how the process of getting poultry meat and eggs to the consumer involves the coordinated efforts of producers, processors, and marketing specialists; they need to know the importance of poultry as a source of farm income; they need to know the per capita consumption of poultry products; they need to know the role of poultry in scientific research; they need to know of the many industrial uses of poultry and eggs, and their by-products; and they need to ponder the future of the poultry industry.

LEADING STATES

Tables 1–12, 1–13, and 1–14 (next page) show the 10 leading states of the United States in number of eggs, broilers, and turkeys produced. From these tables, the following deductions can be made.

1. From an overall standpoint, the Southeast ranks high as a poultry area.

2. California is the leading egg-producing state by a considerable margin. Indiana, Pennsylvania, Ohio, Georgia, and Arkansas follow in the order listed.

3. The Southeast—Arkansas, Georgia, Alabama, North Carolina, and Mississippi—completely dominates broiler production. The top 10 broiler states listed in Table 1–13 produced 83% of the nation's broilers in 1989.

4. North Carolina, Minnesota, and California hold a sizable lead in number of turkeys produced (Fig. 1–13 and Table 1–14). It is difficult to understand why these three states should lead the nation in turkey production, for few states could be more dissimilar in climate, crops, and population density.

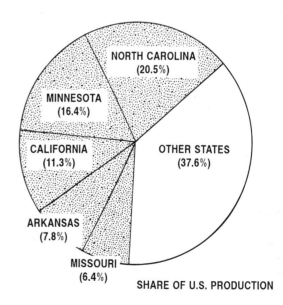

Fig. 1–13. Leading turkey producing states in 1990. (Source: *Livestock and Poultry Situation and Report*, USDA, ERS, LPS–46, February 1991, p. 38, Fig. A–1)

TABLE 1–12
TEN LEADING STATES IN NUMBER OF
EGGS PRODUCED, 1990 AND U.S. TOTAL[1]

State	Egg Production
California	7,472,000,000
Indiana	5,445,000,000
Pennsylvania	4,976,000,000
Ohio	4,667,000,000
Georgia	4,302,000,000
Arkansas	3,620,000,000
Texas	3,317,000,000
North Carolina	2,986,000,000
Florida	2,586,000,000
Minnesota	2,499,000,000
U.S. Total	67,832,000,000

[1]*Poultry Production and Value, 1990 Summary*, 1991, USDA, p. 7.

TABLE 1–13
TEN LEADING STATES IN NUMBER OF OF COMMERCIAL
BROILERS PRODUCED, 1990 AND U.S. TOTAL[1]

State	Broiler Production
Arkansas	951,200,000
Georgia	854,500,000
Alabama	846,900,000
North Carolina	540,300,000
Mississippi	413,000,000
Texas	338,100,000
Maryland	265,400,000
Delaware	231,700,000
California	231,100,000
Virginia	195,900,000
U.S. Total	5,864,650,000

[1]*Poultry Production and Value, 1990 Summary*, 1991, USDA, p. 6.

TABLE 1–14
TEN LEADING STATES IN NUMBER OF
TURKEYS PRODUCED, 1990 AND U.S. TOTAL[1]

State	Turkey Production
North Carolina	58,000,000
Minnesota	46,300,000
California	32,000,000
Arkansas	22,000,000
Missouri	18,000,000
Virginia	17,000,000
Indiana	13,700,000
Iowa	8,800,000
Pennsylvania	8,430,000
South Carolina	5,500,000
U.S. Total	283,000,000

[1]*Poultry Production and Value, 1990 Summary*, 1991, USDA, p. 9.

RELATIVE IMPORTANCE OF DIFFERENT SPECIES OF POULTRY

Table 1–15 indicates the relative importance of different species of poultry. This shows that both chicken and turkey production have increased dramatically in recent years, while duck numbers have nearly doubled and geese numbers have declined.

TABLE 1–15
U.S. POULTRY PRODUCTION[1]

Year	Species of Poultry			
	Chickens (broilers)	Turkeys	Ducks[2]	Geese[2]
	(millions)	(millions)	(millions)	(millions)
1930	34[3]	17	11.3	3.9
1940	142	33	11.9	1.3
1950	631	44	11.5	1.1
1960	1,795	85	11.1	0.9
1970	4,151	116	10.7	0.7
1980	3,963	165	15.1	0.6
1990	5,517	283	21.6	0.5

[1]USDA sources.
[2]Calculated from available data.
[3]Data for 1934.

COMPONENTS OF THE POULTRY INDUSTRY

The process of getting eggs and poultry meat from the birds to the consumer involves the coordinated efforts of many people, from production through marketing. The poultry business is unique among the livestock industries in the United States in that it is highly integrated; many firms are organized in such manner as to control every level of production—all the way from the laying of the egg or the production of the meat through the ultimate promotion and marketing of the finished product, be it an egg or a loaf of turkey pastrami. However, the chain of events in producing the poultry product and getting it to the consumer can be broken down into three primary divisions or stages: (1) producers, (2) processors, and (3) market specialists.

PRODUCERS

Because poultry operations are highly integrated, the average poultry production division is very specialized. In general, poultry production operations can be broken down as follows:

1. **Breeding flocks.** Breeding flocks provide the eggs necessary to supply industry demand for layers and young meat birds. Fertility and rate of lay are extremely important factors in evaluating breeding stock. The nutrient requirements for these birds are high and exacting.

2. **Hatchery.** Eggs produced by breeders are transported to hatcheries where they are incubated and hatched. The newly hatched birds are then shipped to grow-out operations.

3. **Grow-out operations.** When the newly hatched birds are to be used as replacement breeders, the management protocol becomes more involved, necessitating carefully planned feeding, vaccinating, and lighting programs. In cases where the newly hatched birds are from meat-type strains, grow-out operations feed a high-quality ration to promote growth as quickly and as inexpensively as possible.

4. **Market egg producing operations.** The types of market egg producing operations are extremely varied, ranging from cage laying to floor pen arrangements. The eggs collected in these operations are generally cleaned, graded, and packaged on the farm, thereby eliminating the processor and enabling the marketed product to go directly from the farm to the consumer.

Fig. 1–14. Modern egg handling is highly automated. This shows automatic vacuum lift placing 30 eggs into a filler-flat for 30 dozen cases. (Courtesy, DEKALB AgResearch, Inc., DeKalb, IL)

5. **Game bird, sport, and hobby operations.** In general, these types of operations are small and represent sidelines of the farm operator. The market for game birds raised for game reserves (e.g., quail, pheasant, chukar, etc.) has been increasing in recent years as more and more people have the money and time for recreation. There has always been a large number of poultry hobbyists who raise birds for a number of purposes—show, fighting, or sport flying (e.g., pigeons).

PROCESSORS

The second division of the poultry industry involves the processing of the various products from poultry. The branches of this division are:

1. **Egg-breaking industry.** Most of the market quality eggs are processed at the farm level. However, there is a demand for eggs that are too small, that have broken shells, or that are of low quality. These eggs are taken to egg-breaking plants where the broken eggs are marketed in either frozen, liquid, or dry form to such industries as the baking industry.

2. **Packinghouse industry.** The packinghouse industry utilizes almost every part of a slaughtered bird. Inedible or condemned parts are processed and sold as fertilizer, livestock feed, or as raw materials for other industries.

3. **Processed meat industry.** Poultry meat is now being used in such processed meats as hot dogs and cold cuts.

MARKETING SPECIALISTS

Once poultry products have been processed for marketing, they must be packaged, distributed, and promoted in such a way as to achieve maximum consumer demand. Marketing of poultry products is a diverse field with each aspect requiring the services of a specialist in that particular area. The following market specialists are involved in the marketing of poultry products:

1. **Market forecasters.** These specialists accumulate a large volume of data from which to predict future trends in supply and demand. For example, if there is too great a volume of grade A eggs on the market, resulting in depressed prices, these specialists recommend possible alternatives to decrease the surplus.

2. **Research and development specialists.** These specialists are responsible for improving the quality and acceptability of poultry products and for developing new products with consumer appeal.

3. **Advertisers.** These specialists study the wants and desires of consumers and tailor the marketing of poultry products to meet these needs. Packaging and promotion can turn a relatively unknown product with low demand into one with great appeal.

4. **Public relations and education experts.** These specialists are playing an ever-increasing role in marketing. Their primary function is to promote the poultry industry as a whole rather than the individual products. Their main responsibility is to educate the public relative to the nutritional and economic merits of poultry products.

POULTRY AS A SOURCE OF FARM INCOME

Poultry producers in the United States sold eggs, broilers, and turkeys valued at $15.3 billion in 1989, representing 18.3% of the total livestock receipts and 9.6% of the total cash farm income that year (see Table 1–16, next page). In addition to accounting for an increasing percentage of the cash farm income, the cash income from poultry has been received by fewer and fewer poultry producers because the big got bigger. Further, the production of different kinds of poultry has gradually been concentrated in certain geographic areas where farmers have the greatest comparative advantage in the production of one or more kinds of poultry or poultry products.

TABLE 1–16
U.S. FARM INCOME BY COMMODITY 1989[1]

Commodity	Rank by Commodity Group	Income	Percent of Commodity Group
		(mil $)	(%)
Livestock and poultry:			
Cattle and calves	1	36,676	43.8
Dairy products	2	19,401	23.2
Poultry and eggs	3	15,346	18.3
Hogs	4	9,426	11.3
Sheep and lambs	5	489	0.06
Other	6	2,386	2.8
Total		83,724	100.0
Crops:			
Feed crops	1	16,656	22.1
Oil-bearing crops	2	12,172	16.1
Vegetables and melons .	3	11,340	15.0
Fruits and tree nuts . . .	4	9,020	12.0
Food grains	5	6,139	10.7
Cotton (lint and seed) . .	6	4,740	6.3
Tobacco	7	2,381	3.2
Other	8	11,068	14.7
Total		75,449	100.0

[1]Source: *Agricultural Statistics 1990*, USDA, p. 391, Table 575.

PER CAPITA CONSUMPTION OF POULTRY PRODUCTS

Poultry meat and eggs are used chiefly for human food. In 1976, the U.S. per capita consumption of chicken and turkey on a ready-to-cook basis totaled 42.4 and 9.1 lb, respectively. Additionally, in that same year 271 eggs were consumed per person.

Fig. 1–15 and Tables 1–17 and 1–18 illustrate the trends of per capita consumption of poultry meat and eggs. As noted, the consumption of broilers and turkeys, on a per capita basis, has increased sharply while the consumption of certain other types of meat, notably beef and pork, has declined. One alarming trend for the poultry producer is the steady decline in the use of eggs. This trend can be largely

POUNDS OF POULTRY

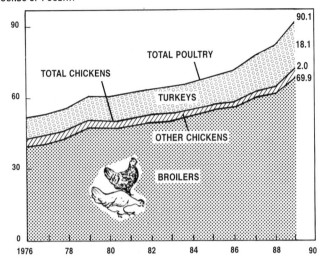

POULTRY IS READY-TO-COOK WEIGHT. EGGS ARE SHELL EGGS PLUS SHELL-EGG EQUIVALENT OF EGG PRODUCTS.

NUMBER OF EGGS

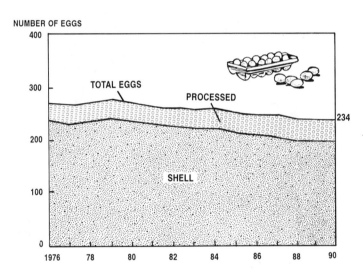

Fig. 1–15. Per capita consumption of poultry and eggs. (Source: *1989 Agricultural Chartbook*, USDA, Agriculture Handbook No. 684, p. 91, Chart 204. 1990 figures from USDA sources.)

TABLE 1–17
U.S. PER CAPITA CONSUMPTION
OF MEAT AND FISH, 1960–1990[1]

Commodity	1960		1970		1980		1990	
	(lb)	*(kg)*	(lb)	*(kg)*	(lb)	*(kg)*	(lb)	*(kg)*
Meats:								
Chicken (ready-to-cook)	27.8	*12.6*	40.5	*18.4*	49.7	*22.6*	71.9	*32.7*
Turkey (ready-to-cook)	6.2	*2.8*	8.0	*3.6*	10.5	*4.8*	18.1	*8.2*
Beef	64.3	*29.2*	84.1	*38.2*	76.4	*34.7*	67.5	*30.7*
Pork	60.4	*27.4*	62.0	*28.2*	57.3	*26.0*	49.5	*22.5*
Fish	10.3	*4.7*	11.8	*5.4*	12.8	*5.8*	15.9[2]	*7.2*

[1]Source: USDA.

[2]Data for 1989.

attributed to the cholesterol controversy and the development of egg substitutes that are low in fat and have no cholesterol.

TABLE 1-18
PER CAPITA CONSUMPTION OF EGGS, 1965-1989[1]

Year	Per Capita Number of Eggs Consumed
1965	314
1970	311
1975	277
1980	272
1985	256
1990	234

[1]Source: USDA.

POULTRY IN SCIENTIFIC RESEARCH

Poultry, especially chickens, have been used extensively as biological tools through which humans have been able to learn more about themselves and other animals. Birds can be effectively used in almost every area of biomedical research; in nutrition, genetics, embryology, behavior, and diseases and vaccines.

NUTRITION STUDIES

The chick is much more sensitive to deficiencies of each of several different nutrients in the diet than the laboratory rat. In the early days of nutrition research, this great sensitivity was a handicap and made it very difficult to keep chicks alive on the kinds of diets which would support a rat reasonably well. Eventually, this higher sensitivity proved of great value in discovering new information relative to vitamins, minerals, and amino acids, and in their more accurate estimation. Also, chicks have the advantages of being cheap and readily available, and large numbers of them can be hatched at the same time so as to provide the accuracy which goes with large numbers. Additionally, many scientists feel that the nutritional needs of the human are more like that of the chicken than the rat.

Much of the early work on vitamins involved chicks and/or eggs. For example, the need for vitamin K in blood clotting was first demonstrated in chicks. Many of the early names given to specific vitamins reflected the involvement of chicks or eggs in research; for example: (1) antineuritic factor (thiamin); (2) ovoflavin (riboflavin); (3) egg white injury factor (biotin); and (4) chick antidermatitis factor (pantothenic acid).

Chicks have been routinely used to demonstrate the needs for and functions of many of the essential elements and fatty acids. Additionally, egg protein is commonly used as the standard by which the nutritional qualities of other proteins are evaluated.

Since the young chick is in a very active metabolic state, requiring a well-balanced, nutritious diet for optimum growth and development, it is commonly used in bioassays to determine the dietary levels of certain micronutrients. A deficiency in any particular nutrient will result in morphological changes and/or reduced growth.

GENETIC STUDIES

Birds make ideal experimental animals for genetic studies for the following reasons:

1. They are readily adapted to artificial insemination. Thus, one male can "fertilize" many females from one easily obtainable ejaculate.

2. Females can produce a large number of embryos in a short period.

3. The incubation of poultry is short in comparison to mammals.

4. Birds achieve sexual maturity at a young age. Therefore, generations are relatively short.

EMBRYOLOGICAL STUDIES

Probably more is known about the development of the chick embryo than any other species of animal. Unlike mammals, the chick develops separately from its mother, and the fertilized egg is a readily accessible tool in teaching and research.

BEHAVIORAL STUDIES

Without doubt, more behavioral studies have been conducted with poultry than with any other class of animal. The pioneering work on imprinting (socialization) was done with goslings, by the Austrian zoologist, Lorenz. Social order, known as "peck order," was first observed in chickens. The mating behavior in chickens and turkeys, involving a chain reaction between the male and female is well known. Of all abnormal animal behaviors, cannibalism in chickens is most common. The effect of prolactin (a hormone) on the broodiness behavior of poultry is of scientific and practical interest.

(Also see Chapter 9, Poultry Behavior and Environment.)

POULTRY IN MEDICINE

The use of eggs for research in human and veterinary medicine, and in the preparation of serums, vaccines, and pharmaceutical products, is invaluable.

In addition to their nutritive value in the prevention and treatment of diseases, eggs have other medical uses. Oil of egg (obtained from the yolks) is used by druggists in the preparation of certain ointments and emulsions; egg albumen is sometimes injected with cocaine to prolong anesthesia; and albumen is a remedy for burns and an antidote for certain poisons, especially arsenic.

Bacteriological laboratories use large quantities of eggs in the preparation of culture media.

The developing chick embryo is used for the study of various viruses, bacteria, and protozoa, for which purpose it is satisfactory and far less expensive than use of animals which they commonly infect. The incubated egg is also of value in the diagnosis of such diseases as spinal meningitis.

Fertile eggs have become very important in the production of immunizing sera and vaccines against a long list of human and animal diseases, including encephalomyelitis (sleeping sickness), measles, small pox, etc.

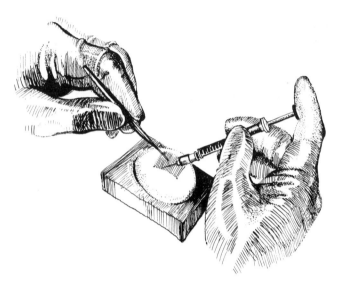

Fig. 1–16. Fertilized eggs are required for the production of many of our vaccines today. The germs which cause many diseases, and particularly forms of viruses, can live and multiply only in living tissue, such as chick embryos. Thus, injection of disease material into chick embryos has made possible large-scale production of many vaccines. The developing chick embryo can be observed through the "window" in the egg.

POULTRY PRODUCTS IN INDUSTRY

Science and technology have teamed up to make many uses of poultry and eggs, and their by-products. Among such industrial uses are: inedible eggs used in the preparation of animal feed and fertilizers; egg whites used in the making of pharmaceuticals, paints, varnishes, adhesives, printer's ink, photography, bookbinding, wine clarification, leather tanning, and textile dyeing; egg yolks used in the making of cake mixes, soap, paints, shampoos, leather finishing, and bookbinding; eggshells used in making mineral mixes and in fertilizer; feathers used in animal feed, fertilizer, millinery goods, pillows, cushions, mattresses, dusters, and insulation material; poultry offal used in animal and mink feeds; and endocrine glands used in making biological products. Also, it is noteworthy that dilutents containing egg yolk are widely used in artificial insemination.

FUTURE OF THE POULTRY INDUSTRY

Science and technology have made for great expansion of the poultry industry in recent years. Still further advances will be made, but it is reasonable to expect that the vast majority of these will come among the producers who are less efficient at this time—that a slower advancement will be made among the top 10%.

Some added poultry production will be needed to take care of our expanding human population.

Per capita consumption increases of poultry meat and eggs will depend largely on how economically they are produced and sold in comparison with other similar foods with which they must compete. It also depends on the new uses developed for poultry products and their convenience as food items.

It appears reasonable to expect that the following transitions, most of which are well under way, will continue in the future:

1. **Production units will be larger and more commercial.** In the future, it appears likely that layers, broilers, and turkeys will be produced in larger units.

2. **More integration-ownership will come.** It appears that large and well-financed commercial feed companies, processors, co-ops, and others will be doing more integrating through ownership in the future.

3. **More specialization will evolve.** An increasing number of poultry producers will specialize in just one kind or phase of production only; for example, in the production of started pullets. Fewer and fewer of them will produce any crop, or diversity in any way whatsoever. Additionally, there will be increased specialization in the feed industry; that is, one local elevator and even some feed manufacturers will produce primarily one kind of feed.

4. **Laborsaving devices and mechanization will increase.** Higher priced labor, along with more sophisticated equipment, will make for increased mechanization all the way along the line, from production through processing and marketing.

5. **Improved housing and environmental control will come.** Physics, engineering, and physiology will be combined in such a way as to bring about improved houses, brooders, incubators, and laborsaving equipment.

6. **Bird density will increase.** With improved housing, along with better environmental control and mechanization, bird density will be increased.

7. **Growth rate and feed efficiency of broilers will increase.** A reasonable goal in modern broiler production is a straight-run bird weighing near 4.4 lb in about 45 days. Through the application of improved breeding, feeding, and management this can be achieved. Hand in hand with more rapid growth rate there will be greater feed efficiency. Already, young chickens have been grown on less than a pound of feed per pound of gain by the use of high-energy rations. In the future, chemistry, physiology, and nutrition will be combined in the formulation and manufacture of poultry

rations that will bring about greater efficiency of feed utilization.

8. **Egg production per layer will increase.** Some hens among the better strains now lay more than 300 eggs per year. There is increasing evidence that this trait (egg production) may have plateaued in high-producing strains. If this is true, perhaps the time has come when breeders of egg producers should pay more attention to body size and egg size. Without doubt most of the advances in egg production in the future will come in raising the level of the lower producers.

9. **Livability will increase.** Drugs and vaccines have helped increase livability; and they will continue to be used in the future. However, it is expected that breeding stock will be selected for greater livability.

10. **Quality of products will be improved.** No other country in the world produces as high-quality poultry meat and eggs as the United States. Yet, further improvements can and will be achieved.

11. **More specialty meat and egg products will evolve.** In particular, attention will be focused on improved shelf life, and on convenience foods—the kinds that require a minimum amount of preparation on the part of the cook.

12. **Improved processing will come.** Despite the strides already made in poultry processing, more improvements lie ahead. Deboning will be perfected, better methods of extending shelf life will be developed, and improved by-product utilization will come.

13. **Marketing costs and efficiency will increase.** The better operations are now very efficient from the standpoint of marketing. However, it is apparent that the less efficient poultry producers can still bring about improvements in this regard.

14. **Production geared to consumption will improve.** Without doubt, the greatest problem facing the poultry producer today is the ability to overproduce. As a result, from time to time, the market is flooded and prices plummet, thereby making for severe losses. Without doubt, some solution—through marketing orders or quotas, or as a result of the controls exerted by fewer and larger owners—will evolve.

QUESTIONS FOR STUDY AND DISCUSSION

1. Define the following terms: (a) poultry, and (b) ornithology.

2. Trace the domestication of poultry and the transformation of the poultry industry in the United States.

3. List chronologically what you consider to be the five most important events in the development of the U.S. poultry industry (see Table 1–1).

4. List and discuss five of the most important changes in the U.S. poultry industry.

5. Why is the egg considered to be one of the best balanced foods available?

6. How does poultry meat compare nutritionally with beef?

7. What health problems may be associated with the consumption of poultry products? How would you counter the argument that poultry products are hazardous to health?

8. How can poultry production be justified when so many people are going hungry?

9. What are the three leading chicken-producing countries?

10. Of the ten leading turkey-producing countries, which country has the largest number of turkeys per capita?

11. Which country produces the most ducks?

12. List the top five states of the United States for (a) egg production, (b) broiler production, and (c) turkey production. What are the regional trends associated with each of these types of production?

13. Discuss the relative importance of the different species of poultry in the United States.

14. List and discuss the various types of poultry producers.

15. What are the various branches of the poultry processing industry?

16. What are the roles of market specialists in the poultry industry?

17. What percent of U.S. farm income is derived from poultry and eggs?

18. Discuss the recent trends of per capita consumption of poultry meat and eggs in the United States.

19. Outline the roles poultry play in scientific research.

20. Discuss the future of the poultry industry. What trends do you foresee? Does the industry have a bright future?

SELECTED REFERENCES

Title of Publication	Author(s)	Publisher
American Poultry History, 1823–1973	Edited by O. A. Hanke, J. K. Skinner, J. H. Florea	American Poultry Historical Society, Inc., Mount Morris, IL, 1974
Commercial Chicken Production Manual, Fourth Edition	M. O. North, D. D. Bell	Van Nostrand Reinhold Co., New York, NY, 1990
Diseases of Poultry, Ninth Edition	Edited by B. W. Calnek, *et al.*	Iowa State University Press, Ames, 1991
Nutrition of the Turkey	M. L. Scott	M. L. Scott of Ithaca, Ithaca, NY, 1987
Poultry Breeding and Genetics	Edited by R. D. Crawford	Elsevier Science Publishers, Amsterdam, The Netherlands, 1990
Poultry: Feeds & Nutrition, Second Edition	H. Patrick, P. J. Schaible	The Avi Publishing Company, Inc., New York, NY, 1980
Poultry Husbandry, Third Edition	M. A. Jull	McGraw-Hill Book Company, Inc., New York, NY, 1951
Poultry Husbandry I	C. J. Price	Food and Agriculture Organization of the United Nations, Rome, Italy, 1969
Poultry Husbandry II	C. J. Price, J. E. Reed	Food and Agriculture Organization of the United Nations, Rome, Italy, 1971
Poultry Keeping in Tropical Areas	W. Thomann	Food and Agriculture Organization of the United Nations, Rome, Italy, 1968
Poultry Meat and Egg Production	C. R. Parkhurst, G. J. Mountney	Van Nostrand Reinhold Co., New York, NY, 1987
Poultry Production, Twelfth Edition	M. C. Nesheim, R. E. Austic, L. E. Card	Lea & Febiger, Philadelphia, PA, 1979
Poultry Products Technology, Second Edition	G. J. Mountney	The Avi Publishing Company, Inc., Westport, CT, 1976
Poultry Science and Practice, Fifth Edition	A. R. Winter, E. M. Funk	J. B. Lippincott Company, Chicago, IL, 1960
Practical Poultry Management, Sixth Edition	J. E. Rice, H. E. Botsford	John Wiley & Sons, Inc., New York, NY, 1956
Processing of Poultry	Edited by G. C. Mead	Elsevier Science Publishers, Ltd., Barking, Essex, England, 1989

Historical

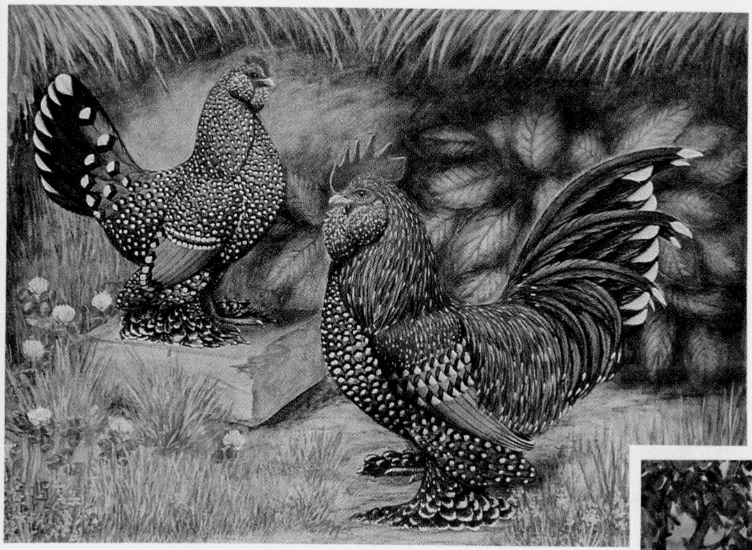

Plate 1. Mille Fleur Booted Bantams. Exhibition chickens. (Courtesy, *Poultry Tribune*, Mount Morris, IL)

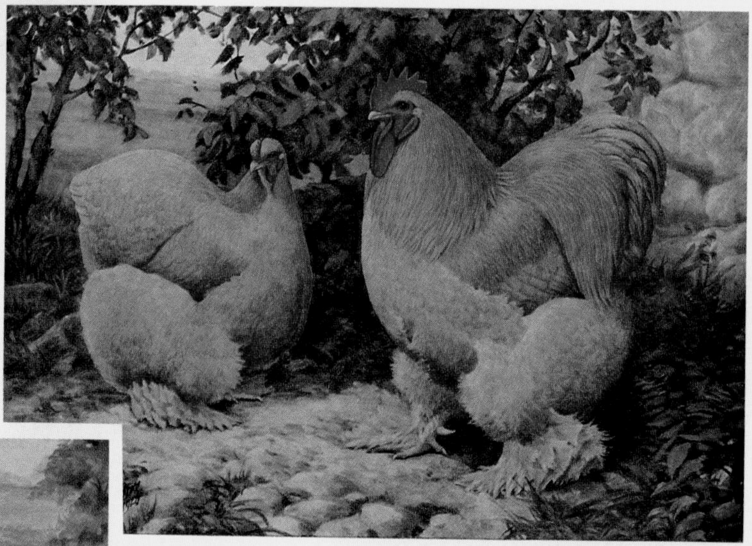

Plate 2. Buff Cochins. Exhibition chickens. (Courtesy, *Poultry Tribune*, Mount Morris, IL)

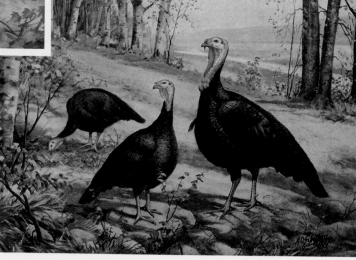

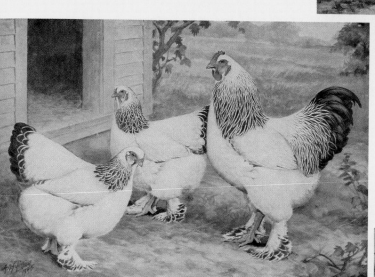

Plate 3. Light Brahmas. Exhibition chickens. (Courtesy, *Poultry Tribune*, Mount Morris, IL)

Plate 4. Black Turkeys. (Courtesy, *Turkey World*, Mount Morris, IL)

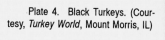

Breeders

Plate 5.   Red hens which lay brown eggs. (Courtesy, H&N, Redmond, WA)

Plate 8.   Broiler breeders. (Courtesy, North Carolina State University, Raleigh)

Plate 6.   Nest box in a broiler breeder house. (Courtesy, North Carolina State University, Raleigh)

Plate 9.   Broiler breeders on slotted floor. (Courtesy, California Polytechnic State University, San Luis Obispo)

Plate 7.   Hens nesting. (Courtesy, USDA)

Plate 10.   Gathering eggs. Hatching eggs are gathered more frequently than eggs intended for table use. (Courtesy, USDA)

Layers

Plate 11.   Layers with continuous flow watering. (Courtesy, Big Dutchman, Zeeland, MI)

Plate 14.   Replacement layer pullets raised in cages. (Courtesy, California Polytechnic State University, San Luis Obispo)

Plate 12.   Three-deck layer house. (Courtesy, Big Dutchman, Zeeland, MI)

Plate 15.   Automated egg packing. (Courtesy, North Carolina State University, Raleigh)

Plate 13.   Pampered layers in cages, with automatic feeding, watering, and gathering of eggs. (Courtesy, California Polytechnic State University, San Luis Obispo)

Plate 16.   Niagara with baffles. (Courtesy, FACCO, Padova, Italy)

Broilers

Plate 17. Broiler breeder pullet dark-out house showing intake light traps and black curtain covering plexiglass. (Courtesy, North Carolina State University, Raleigh)

Plate 20. Caged chicks feeding. (Courtesy, FACCO, Padova, Italy)

Plate 18. Newly hatched broiler chicks in broiler house on wood shaving litter. (Courtesy, Leo S. Jensen, Department of Poultry Science, University of Georgia, Athens)

Plate 21. Interior of broiler house. (Courtesy, University of Georgia, Athens)

Plate 19. Chicks being brooded in a typical broiler house. (Courtesy, North Carolina State University, Raleigh)

Plate 22. Broilers feeding on a computer-formulated ration conveyed to them by long chain feeders. (Courtesy, USDA)

Plate 23. Turkey poult. (Courtesy, National Turkey Federation, Reston, VA)

Plate 26. Turkey grower house in California. (Courtesy, University of California, Davis)

Plate 24. Three-week old poults in typical brooder house. (Courtesy, North Carolina State University, Raleigh)

Plate 27. Curtain-sided turkey house with sun porch. (Courtesy, North Carolina State University, Raleigh)

Plate 25. Bulk feed truck delivering feed to turkeys on the range. (Courtesy, University of Georgia, Athens)

Plate 28. Broad-Breasted White turkey gobbler. (Courtesy, The National Turkey Federation, Reston, VA)

Ducks

Plate 29. Duck farm at Cherry Valley where more than one million ducks are produced each year. (Courtesy, Cherry Valley Farms, Ltd., Rothwell, England)

Plate 32. Intensive commercial growing of ducks. (Courtesy, Cherry Valley Farms, Ltd., Rothwell, England)

Plate 30. Breeder ducks in lay. (Courtesy, Cherry Valley Farms, Ltd., Rothwell, England)

Plate 33. Pekin ducks at Ward Duck Co., La Puente, California. (Courtesy, California Polytechnic State University, San Luis Obispo)

Plate 31. Brooding young ducklings. (Courtesy, Cherry Valley Farms, Ltd., Rothwell, England)

Plate 34. Growing ducklings on the range. (Courtesy, Cherry Valley Farms, Ltd., Rothwell, England)

Geese

Plate 35. Young goslings in starter house. (Courtesy, Schiltz Foods, Inc., Sisseton, SD)

Plate 36. Computers are used extensively in poultry production. (Courtesy, Monsanto Agriculture Company, St. Louis, MO)

Poultry Business

Plate 37. Feed truck filling a feed tank adjacent to a poultry house. (Courtesy, North Carolina State University, Raleigh)

Plate 38. Broiler chicken futures take flight at Chicago Mercantile Exchange. "The Famous Chicken" autographs a picture for a trader on the floor of the Chicago Mercantile Exchange, where broiler chicken futures were introduced on February 7, 1991. (Courtesy, Chicago Mercantile Exchange)

Processing

Plate 39.   USDA inspector examining broiler carcasses on the processing line. (Courtesy, North Carolina State University, Raleigh)

Plate 42.   Roast turkey. (Courtesy, Nicholas Turkey Breeding Farms, Sonoma, CA)

Plate 40.   Broiler. (Courtesy, USDA)

Plate 41.   Cut-up parts of a broiler. (Courtesy, University of Maryland, College Park)

Plate 43.   Roast duckling. (Courtesy, Cherry Valley Farms, Ltd., Rothwell, England)

AVIAN
ANATOMY AND
PHYSIOLOGY;
THE EGG

CHAPTER

Fig. 2–1. Eggs—a marvel of nature and one of the most complete foods known. (Courtesy, Alabama Poultry &

Contents   *Page*

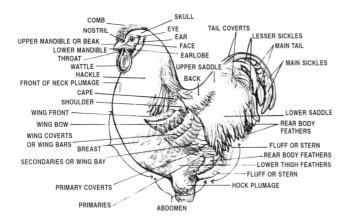

Fig. 2–2. Parts of the male chicken.

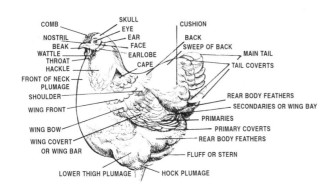

Fig. 2–3. Parts of the female chicken.

If poultry producers are to achieve maximum bird performance, they must have a basic understanding of the bird's structural makeup and of how the various systems of the body function (Figs. 2–2 and 2–3). Through this knowledge, they can manipulate the body functions of the birds to attain maximum production. One example of how the knowledge of physiology has been adapted to poultry production is the manipulation of light to promote egg production. Additionally, the structure of the digestive tract of the fowl is very different from that of the ruminant (e.g., sheep, cattle) or of the mammalian nonruminant (e.g., horse, pig); hence, feeds and procedures must be designed accordingly.

Anatomy—literally meaning "to cut apart"—is the branch of biological science dealing with the structure and form of living organisms. Physiology is the branch of biological science dealing with how the various structures of the organisms function. Thus, as a team, the anatomist and the physiologist correlate the intricate interrelationships that make organisms "tick."

EVOLUTION AND CLASSIFICATION OF BIRDS

From an evolutionary viewpoint, birds are merely warm-blooded, feathered, flying reptiles. More than 250 million years ago, the primeval ancestors of reptiles came out of their watery habitat and adapted to the hot, dry climate of their new terrestrial environment.

In the Triassic period of the Mesozoic era, early forms of the subclass *Archosauria* appeared. From this subclass, the group *Dinosauria* evolved. Fossil evidence indicates the existence of a flying dinosaur called the pterodactyl. Its 30-ft wingspan and large body probably made this reptile the heaviest flying animal ever to inhabit the earth.

Approximately 180 million years ago, another group—*Archaeopteryx*—evolved from the *Archosauria*. This group eventually became birds as we know them today. The developmental link between the *Archaeopteryx* and the *Archosauria* has not yet been found; but it has been hypothesized that the animal must have been a scaled, two-legged avireptile because, by the time *Archaeopteryx* had evolved, feathers had replaced the scales characteristic of reptiles.

Today, true birds are grouped in the *Aves*, subclass *Neornithes*. The subclass is further divided into two superorders: (1) *Ratitae* (referring to a flat, raftlike breastbone), a division encompassing the more primitive birds, such as the ostrich, emu, and kiwi; and (2) *Carinatae* (referring to a keellike breastbone), a division encompassing the rest of the birds. The balance of this chapter is devoted to the second of these suborders. The zoological classification of birds follows:

Phylum . . . . *Chordata*
Subphylum . . . . *Vertebrata*
Class . . . . *Aves*
Subclass . . . . *Neornithes*
Superorder . . . . *Ratitae*
Superorder . . . . *Carinatae*

Within the superorder *Carinatae*, there are 2,810 genera, 8, 616 species, and over 28,000 subspecies. The genus-species classification of domesticated birds follows:

Chickens . . . . *Gallus domesticus*
Duck . . . . *Anas domestica*
Goose . . . . *Anser domesticus*
Guinea fowl . . . . *Numida meleagris*
Pigeon . . . . *Columba domestica*
Turkey . . . . *Meleagris gallopavo*

INTEGUMENT

In poultry, the skin and feathers collectively form the integument; that is, the outer protection of the body. They (1) protect the body from injury, (2) help to maintain a relatively constant body temperature, (3) aid in flight, and (4) act as receptors for sensory stimuli.

SKIN

The skin is rather thin and relatively free of secretory glands with the exception of the uropygial or oil gland (preen gland) which is located on the upper part of the tail. Oil produced by this gland is collected in the beak and distributed on the feathers to act as a kind of water repellant. This oil is of particular importance for aquatic species.

In yellow-skinned chickens, the yellow color of the skin and shanks is due to several xanthophylls deposited in the fat layers below the skin, which are derived from the feed.

Several specialized structures consist of exposed areas of skin; among them, the comb, wattles, snood (in turkeys), earlobes, beak, claws, and spurs (see Figs. 2–2 and 2–3). The comb, wattles, and snood (in turkeys) are sensitive to the effects of sex hormones and consequently serve as indicators of secondary sex characteristics. Male hormones cause these appendages to become enlarged. The comb is the fleshy protuberance on top of the head. It is usually red and occurs in a number of shapes, being classified as (1) single comb, (2) rose comb, (3) pea comb, (4) cushion comb, (5) buttercup comb, (6) strawberry comb, or (7) V-shaped comb (see Fig. 2–4). The wattles, which are usually red in color, are pendulous growths of flesh at either side of the base of the beak and upper throat. The snood in turkeys is a fleshy protuberance at the base of the upper beak. The earlobe is a fleshy patch of bare skin below each ear which varies in color, depending on the breed. The beak, claws, and spurs are horny, keratinized structures. Additionally, the exposed parts of the legs and feet are covered with hard scales.

FEATHERS

Feathers are epidermal outgrowths which form the external covering or plumage of birds. At hatching, birds are covered by down feathers — soft, fine, fluffy, plumulelike feathers. These feathers are rapidly replaced by a coarser type of feather. Adult feathers can be classified into three types:

1. **Contour feathers.** These outermost feathers can be divided into four distinct parts: (a) quill, (b) shaft or rachis, (c) fluff or undercolor, and (d) web. The quill and shaft are continuous and hollow, tapering to a fine point at the distal end of the feather. The web is formed by barbs which contain small barbules that interlock with other barbs, thereupon forming a continuous, uniform series (see Fig. 2–5). The fluff or undercolor is a series of barbs having no barbules. The absence of barbules causes this area of the feather to take on a scattered, downy appearance.

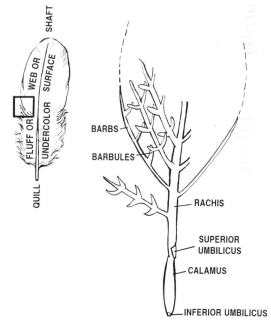

Fig. 2–5. Structure of contour feathers.

2. **Plumules.** These feathers form a soft downy undercoat. The rachis is short, and the barbs and barbules radiate freely.

3. **Filoplume.** These feathers have a short, flexible, hair-like rachis with barbs confined to the apex.

Feathers are distributed on the skin in well-defined tracts (see Figs. 2–2 and 2–3). Through this ordered arrangement of tracts, flight is facilitated and body heat conserved. In periods of cold weather, muscles attached to the feathers cause them to become erect in relation to the skin, thereupon creating a thicker and more efficient insulation.

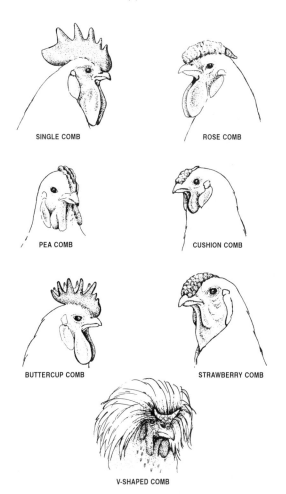

Fig. 2–4. Male heads showing different comb types.

STRUCTURAL SYSTEMS

The two physiological systems most involved with the structural integrity of the fowl are the skeletal system and the muscular system. Because flying involves a considerable expenditure of energy, it is essential that the structure of the fowl be designed so as to maximize the efficiency of energy utilization.

SKELETAL SYSTEM

Poultry are bipeds; that is, they stand on two legs. The skeletal system of the chicken is shown in Fig. 2–6.

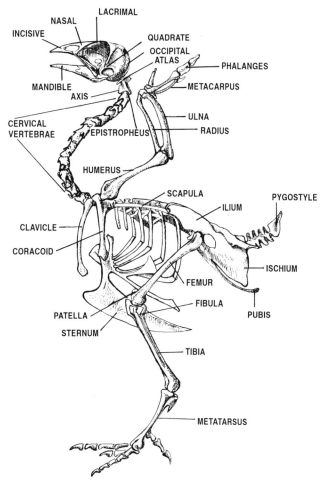

Fig. 2–6. Skeletal system of the chicken.

The basic skeletal arrangement is generally analogous to that of mammals. Yet, there are several differences. Birds possess a pair of extra bones in the shoulder area, called the coracoids. This pair of bones facilitates wing movement and offers additional support for the wing. Also, several morphological differences from mammals are evident in the spine. The cervical vertebrae (neck bones) form an S-shaped column connecting the body to the head. When a flying bird lands, considerable pressure is exerted throughout the body,

and this S-shaped conformation acts as a spring to minimize the impact on the head. Another difference from the mammalian spine is that the vertebrae along the trunk and body of the bird are fused together. This rigid conformation of the back provides considerable support for the wings.

The skeletal system is intimately connected to the respiratory system; many bones are pneumatic. Pneumatic bones are hollow and are connected to the respiratory system, thereby serving as a reservoir for air and reducing the weight of the bird for flight. The skull, humerus, keel, clavicle, and lumbar and sacral vertebrae are all part of this system. In fact, if one were to cut off the inflow of air through the trachea but open one of these bones, e.g., the humerus, the bird would continue to breathe.

Egg laying places a great demand on the hen for calcium since eggshells consist primarily of calcium carbonate. To facilitate mobilization of calcium in the body for this type of production, birds have what are termed medullary bones. The marrow cavity of these bones is filled with interlacing spicules of bone. The spaces between the spicules are filled by red marrow and blood sinuses. In pullets, medullary bone – involving the tibia, femur, pubic bone, sternum, ribs, toes, ulna, and scapula – develops about 10 to 14 days prior to the laying of the first egg. At the onset of lay, the medullary bone enables the hen to mobilize calcium rapidly – so fast that if the hen is fed a diet very low in calcium she will lose 40% of the calcium in her skeleton after laying six eggs. Medullary bone is generally absent in males or nonlaying females. However, administration of estrogen, the female hormone, can initiate the formation of medullary bone.

MUSCULAR SYSTEM

Birds, like mammals, possess three distinct types of muscles: (1) smooth, (2) cardiac, and (3) skeletal. Smooth muscle is the type of muscle that lines many of the organs over which the bird has no voluntary control, e.g., gastrointestinal tract. Cardiac muscle is the type of muscle found in the heart. To a large degree, smooth muscle and cardiac muscle are self-regulating. That is, they need no outside (extrinsic) stimulus for the initiation of contraction. The third type of muscle, skeletal muscle, constitutes most of the muscle mass in the body. It is responsible for executing most voluntary movements.

The skeletal muscles of poultry contain three types of muscle fibers – red fibers, white fibers, and intermediate fibers. The red fibers predominate in what is commonly called *dark meat*. These fibers contain large quantities of myoglobin, an iron-containing, oxygen-carrying compound very similar to hemoglobin. The white fibers, collectively forming *white* or *pale meat,* contain relatively little myoglobin. Intermediate fibers represent a type of fiber possessing some characteristics of both red and white fibers.

Red fibers abound in muscles which are used continuously. They receive more blood and contain more fat and myoglobin than white fibers, thereby favoring the aerobic (with oxygen) production and utilization of energy which is conducive to prolonged activity. White fibers, on the other hand, are rich in glycogen, a sugar-rich compound that is readily broken down in anaerobic (without oxygen) conditions needed to sustain brief spurts of activity. Thus, one would expect the muscles used in flight by a good flying bird to contain more red fibers than those of a poor flier. This is, in fact, the case when the pigeon (a flying bird) is compared to the chicken (a poor flier). The pectoralis muscle (a breast muscle) of the pigeon contains about 40 times as much myoglobin as in the same muscle of the chicken.

CIRCULATORY SYSTEM

The circulatory system of birds functions much like that of mammals. The avian heart consists of four chambers — right atrium, right ventricle, left atrium and left ventricle. Incoming deoxygenated blood is received in the right atrium and is subsequently passed to the right ventricle. Contraction of the heart pushes the blood from the right ventricle to the lungs where oxygen is picked up by the blood and carbon dioxide removed. Freshly oxygenated blood travels from the lungs to the left atrium and left ventricle. Upon contraction of the left ventricle, blood is pushed through the arterial system where it eventually reaches the target cells, gives off its oxygen, and picks up waste products which will be excreted ultimately. The deoxygenated blood then returns to the heart through the venous system, thence the process is repeated.

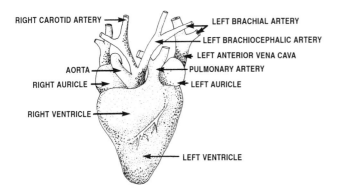

Fig. 2–7. The heart of the chicken; antero-ventral view.

Also, when the demand for oxygen and nutrients is great, such as in flying, the heart rate increases dramatically from the rate sustained when the bird is at rest — on the order of a 5- to 14-fold increase.

There is considerable variation in heart rates between species (see Table 2–1).

One of the notable differences between the blood of birds and mammals is the structure of the erythrocyte (red blood cell). Erythrocytes in birds are nucleated and contain mitochondria and endoplasmic reticulum — features which are absent in mammalian erythrocytes. The life span of erythrocytes of chickens averages from 28 to 35 days, about one-half that of humans. Erythrocytes constitute 20 to 40% of the volume of blood in chickens, with males generally having a higher percentage of erythrocytes than females.

TABLE 2–1
HEART RATE IN BIRDS

Type of Bird	Heart Rate Beats/Minute
Canary	1,000
Quail	500–600
Chicken	350–475
Turkey	200–275
Pigeon	220
Goose	200

In the 1-week-old chick, blood represents about 8.7% of the body weight. This percentage steadily decreases as the bird becomes older, declining to about 4.6% of body weight at maturity. Blood volume remains fairly constant from day to day. A 5.5 lb chicken will generally have about 240 ml of blood (roughly the equivalent of 0.5 pt).

RESPIRATORY SYSTEM

Because of the extremely heavy demands for energy in flight, birds have developed a respiratory system that permits the greatest exchange of oxygen per unit time of any animal; yet the lungs of birds are smaller than those of mammals in relation to body size.

To accomplish this rapid uptake of oxygen, the anatomy and physiology of the avian respiratory system differ markedly from mammalian systems. The first difference is the role of the lungs. In mammals, the diaphragm muscle controls the expansion and contraction of the lungs. Birds have no diaphragm, and the lungs do not expand and contract upon inspiration and expiration, respectively. Rather, they act solely as organs in which gas exchange in the blood takes place.

Birds possess an extensive air sac system wherein air is stored. Most birds have eight air sacs: the median cervical sac, median clavicular sac, and paired cranial thoracic, caudal thoracic, and abdominal sacs (see Fig. 2–8). In a few birds, there are two cervical sacs, thereby increasing the number of air sacs to nine.

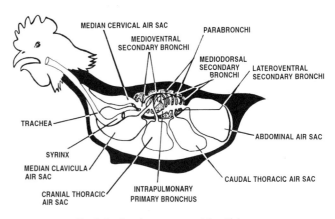

Fig. 2–8. Respiratory system of the chicken.

When a bird inhales, the inspiratory muscles increase the volume of the body cavity creating a subatmospheric pressure in the air sacs which subsequently draws fresh air through the lungs and into the air sacs. During exhalation, the expiratory muscles decrease the volume of the body cavity forcing air out of the air sacs, back through the lungs, and out of the body.

DIGESTIVE SYSTEM

The digestive system of poultry differs considerably from that of other nonruminant animals. Birds have no teeth; hence, there is no chewing. The esophagus empties directly into the crop, where the feed is stored and soaked. From the crop, the feed passes to the proventriculus (or grandular stomach), the thick-walled organ immediately in front of the gizzard. Here it is stored temporarily while digestive juices are copiously secreted and mixed with it. Thence, it passes to the gizzard, a very muscular organ, which normally contains stones or grit, where it is crushed and ground. Then the feed moves through the small intestine, the cecae, and large intestine to the cloaca.

Digestion in the fowl is rapid, requiring only about 2½ hours in the laying hen and 8 to 12 hours in the nonlaying hen for feed to pass from the mouth to the cloaca.

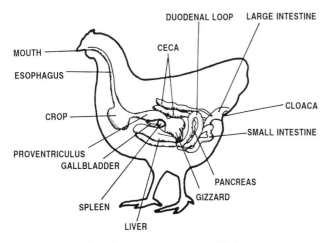

Fig. 2–9. The digestive system of the hen.

PROCESS OF DIGESTION

Digestion, taken in a narrow sense of the word, can be defined as the process whereby proteins, fats, and complex carbohydrates are broken down into units small enough to be absorbed. This process is accomplished primarily through the action of digestive enzymes.

Enzymes are organic catalysts produced by certain cells within the body which speed biochemical reactions at ordinary body temperatures without being used up in the process. Enzymatic activity is responsible for most of the chemical changes occurring in feeds as they move through the digestive tract. A summary of the enzymes involved in the digestive process of poultry is presented in Table 2–2.

Many of the digestive enzymes are stored in an inactive form. When they are in the inactive form, they are called *zymogens* or *proenzymes*. Once secreted into a favorable environment for digestion, generally governed by pH, these inactive enzymes "turn on" and perform their specific digestive functions.

PHYSIOLOGY OF DIGESTION

The discussion of the physiology of digestion centers on various regions of the digestive tract rather than on the specific organs of the bird. These regions are the oral region, pharyngeal and esophageal region, gastric region, pancreatic region, hepatic region, intestinal region, and ceca-large intestine-cloaca region.

ORAL REGION

Two physical processes occur in the oral region of birds—prehension and the initiation of deglutition.

Prehension can be defined as the act of bringing food into the mouth. Deglutition is the act of swallowing. The manner in which birds pick up food and swallow it depends largely on whether they have a soft palate. Birds lacking a soft palate (e.g., chicken, goose, duck, turkey) pick up feed with their beaks, mix the feed with saliva, and then raise their heads and extend their necks, thereupon allowing the feed to progress downwards by gravity and negative pressure in the esophagus. Birds having a soft palate (e.g., pigeons) utilize this structure to help move the feed to the back of the mouth and force it down the esophagus. Thus, these birds can drink with their heads in a downward position.

The tongue of birds is a rigid, heavily cornified structure in contrast to the labile, soft tongue found in mammals. Taste buds are located at the base of the tongue, but are few in number, a fact which explains the relative absence of taste sensitivity in birds.

The development of the salivary glands in poultry depends on the eating habits of the particular species being studied. Birds that consume aquatic feed have poorly developed salivary glands. Conversely, birds that eat dry feed have, by necessity, well-developed salivary glands. In some species, such as the sparrow, saliva contains amylase—an enzyme that breaks down starch. However, most domesticated birds do not secrete amylase.

The uses of saliva in digestion are manyfold, including the following:

1. **Lubricant.** These secretions act as aids in mastication, the formation of the bolus, and swallowing. Without this moisture, swallowing would be extremely difficult.

2. **Enzymatic activity.** The enzyme alpha-amylase (ptyalin) is found in the saliva of some birds. It serves to break $\alpha 1, 4$ glucosidic linkages in starch and glycogen.

3. **Buffering capacity.** A large quantity of bicarbonate is secreted in saliva, thus serving as a buffer in the ingesta.

4. **Taste.** Saliva solubilizes a number of the chemicals in the feed which, once in solution, can be detected by the taste buds.

TABLE 2–2
DIGESTIVE PROCESSES IN POULTRY

Region	Secretion (secreted by)	Enzyme	Enzyme Acts on, or Function	End Product of Digestion	Comments
Mouth	Saliva (salivary glands)	Amylase (ptyalin)	Starch, dextrins Lubricates the food	Dextrins Glucose	Amylase is secreted in some birds, but most domestic birds lack this enzyme.
Crop	Mucus		Lubricates and softens food		
Proventriculus	Gastric juice and acids (chiefly (HCl) (walls of stomach) Mucus	Pepsin Lipase (in carnivores) Amylase	Protein Fat	Proteoses, poly-peptides, peptides Higher fatty acids and glycerol Coating of stomach lining and lubrication of food	
Gizzard			Grinding	Ground foods Reduced particle size	Grit in the gizzard increases the motility and grinding action of the gizzard and the digestibility of coarse feed.
Duodenum (small intestine)	Pancreatic juice (pancreas) Bile (liver)	Trypsin Chymotrypsin Amylopsin (amylase) Steapsin (lipase) Carbozypeptidase Collagenase Cholesterol esterase	Proteins, proteoses, peptones, and peptides Starch, dextrins Fats Peptides Collagen Cholesterol Fats	Peptones, peptides Amino acids Maltose, dextrins Higher fatty acids and glycerol Amino acids and peptides. Peptides Cholesterol esterified with fatty acids Emulsion of fats (soap, glycerol)	
Small intestine	Intestinal juice (secreted by intestinal wall)	Peptidase (erepsin) Sucrase (invertase) Maltase Lactase Polynucleotidase	Peptides Sucrose Maltose Lactose Nucleic acid	Amino acids and dipeptides Glucose and fructose Glucose Glucose and galactose Mononucleotides	
Ceca		A limited amount of microbial activity	Cellulose, poly-saccharides, starches, sugars	Volatile fatty acids Microbial protein B vitamins Vitamin K	

5. **Protection.** The membranes within the mouth must be kept moist in order to remain viable. Saliva provides one means by which this is accomplished.

PHARYNGEAL AND ESOPHAGEAL REGION

The pharynx is the structure which controls the passage of air and feed. Unlike mammals, birds have no sharp demarcation where the mouth ends and the pharynx begins. However, when the neck is extended in the process of eating, there is a change in the position of the trachea which prevents the passage of food down it.

The esophagus is a muscular tube extending from the pharynx to the cardia of the stomach. The musculature and

innervation of the esophagus are such that peristaltic waves move the bolus. *Peristalsis is the coordinated contraction and relaxation of smooth muscles creating a unidirectional movement which pushes the bolus through the digestive tract.*

At the junction of the cervical segment and the thoracic segment of the esophagus in birds, there is a differentiated outpouching of the esophagus called the crop. If the bird has been starved, feed will bypass this structure and go directly to the proventriculus and gizzard. As feeding progresses, the crop begins to fill and acts as a storage organ. In the crop, there is limited digestion due to the presence of salivary amylase mixed in the bolus and a small amount of fermentation. Limited absorption of glucose and volatile fatty acids in the crop has been demonstrated in some birds. The size and shape of the crop is dependent upon the eating habits of the bird. Birds consuming large amounts of grain tend to have large bilobed crops while birds that primarily consume insects have rudimentary crops or, in some cases, no crop at all. Some birds, notably the pigeon and the dove, have the ability to produce a milklike secretion in the crop which can be regurgitated and fed to their young.

GASTRIC REGION

Gastric digestion in birds is carried out in two separate and distinct organs—the proventriculus and the gizzard.

The proventriculus is a small organ, through which ingested feed passes rapidly. Its main function is that of gastric fluid secretion. The fluids secreted by the proventriculus are very similar to those in the stomach of the nonruminant, containing both pepsin and hydrochloric acid. Very little churning and mixing of feed occurs in this organ.

The function of the gizzard is the mechanical action of mixing and grinding the feed. Since the bird has no teeth and swallows its feed whole, this muscular organ—sometimes called the "hen's teeth"—acts primarily as an organ for mastication (the process of grinding feed). Here fluids secreted by the proventriculus are mixed in the ingesta during grinding. Grit, such as small pieces of granite, is often added to poultry rations to increase the digestibility of whole grains or grains with a minimal amount of processing. Grit stimulates motility in the gizzard as well as provides additional surface for grinding. When feed is provided in mash form, the benefits of grit are minimal.

PANCREATIC REGION

The pancreatic region involves the pancreas and the pancreatic duct—a duct leading from the pancreas to the small intestine.

The pancreas, an accessory organ of digestion, is a glandular structure that plays an essential role in the digestive physiology of poultry. The pancreas—being both an endocrine and exocrine gland—serves two physiologically distinct functions. The endocrine function is that of the secretion of the hormones, insulin and glucagon. The exocrine function deals with the production and secretion of fluids that are necessary for digestion within the small intestine.

Many of the pancreatic enzymes are stored and secreted in an inactive form which become activated at the site of digestion. Trypsinogen is a proteolytic enzyme that is activated in the small intestine by enterokinase, an enzyme secreted from the intestinal mucosa. When activated, trypsinogen becomes trypsin. Trypsin, in turn, can then activate chymotrypsinogen to chymotrypsin.

The nucleases, lipases, and pancreatic amylase are secreted in their active form. Many of the enzymes require a specific environment before they will function. For example, amylase requires a pH of about 6.9 and the presence of inorganic ions before it will digest complex carbohydrates (see Table 2–3).

TABLE 2–3
COMPOSITION OF PANCREATIC SECRETIONS

Item	Functions
Proteolytic enzymes:	Splits proteins into peptides and amino acids
Trypsinogen	
Chymotrypsinogen A	
Chymotrypsinogen B	
Procarboxypeptidase A	
Procarboxypeptidase B	
Collagenase	Breakdown of collagen
Lipolytic enzymes:	Breakdown of lipids
Prophosphoralipase A	
Pancreatic lipase	
Cholesterol esterase	Esterification of cholesterol
Nucleolytic enzymes:	Breakdown of nucleic acids
Ribonuclease	
Deoxyribonuclease	
Amylolytic enzymes:	Breakdown of starches
Pancreatic amylase	
Cations:	Buffers; cofactors; osmotic regulators
Sodium	
Potassium	
Calcium	
Magnesium	
Anions:	Buffers; osmotic regulators
HCO_3	
Cl^-	
$SO_4^=$	
$HPO_4^=$	
Proteins:	Buffers
Albumin	
Globulin	

HEPATIC REGION

The hepatic region embraces the liver, gallbladder, and bile duct.

In addition to the pancreas and salivary glands, the liver is an indispensable accessory organ of the gastrointestinal tract. From the stomach and small intestine, most of the absorbed nutrients travel through the portal vein to the liver—the largest gland in the body. The liver not only plays an important part in nutrient metabolism and storage, but also forms bile, a fluid essential for lipid absorption in the small intestine. The numerous physiological functions of the liver follow:

1. Secretion of bile.
2. Detoxification of harmful compounds.
3. Metabolism of proteins, carbohydrates, and lipids.
4. Storage of vitamins.
5. Storage of carbohydrates.
6. Destruction of red blood cells.
7. Formation of plasma proteins.
8. Inactivation of polypeptide hormones.

The primary role of the liver in digestion and absorption is the production of bile. Bile facilitates the solubilization and absorption of dietary fats and also aids in the excretion of certain waste products such as cholesterol and by-products of hemoglobin degradation. The greenish color of bile is due to the end products of red blood cell destruction — biliverdin and bilirubin. Bile contains a number of salts resulting from the combination of sodium and potassium with bile acids. These salts combine with lipids in the small intestine to form micelles. *Micelles are colloidal complexes of monoglycerides and insoluble fatty acids that have been emulsified and solubilized for absorption.* When the micelle has been formed, the lipid can be digested and the resulting products (fatty acids and glycerol) can cross the mucosal barrier of the small intestine and enter the lymphatic system. Bile salts, however, do not travel with the lipid; rather, they are recycled into the enterohepatic circulation.

The volume of bile production is highly variable. A bird that has been starved produces little bile. Conversely, a bird fed a high-fat ration will produce substantial quantities in order to keep up with absorptive requirements. Generally, the volume of bile is dependent on (1) blood flow, (2) nutritive state of the bird, (3) type of ration being fed, and (4) the enterohepatic bile salt circulation.

In many animals, the gallbladder is the storage site for bile. Several species of mammals and birds, however, do not have gallbladders; among them, horses, rats, gophers, deer, elk, moose, giraffes, camels, elephants, pigeons, and doves.

INTESTINAL REGION

The small intestine is divided anatomically into three sections — duodenum, jejunum, and ileum. The first segment, the duodenum, originates at the distal end of the gizzard. It is difficult to differentiate the jejunum from the ileum in birds, and many researchers collectively call these two segments the lower intestine. The length of the small intestine varies according to the eating habits of birds. Carnivorous birds (meat eaters) have substantially shorter intestines than herbivorous birds (plant eaters). This can be explained by the fact that meat products are more readily digested and absorbed than plant products.

Throughout the luminal surface of the small intestine lies an extensive network of fingerlike projections called villi. Each villus contains a lymph vessel called a lacteal and a series of capillary vessels. On the surface of the villi are a great number of microvilli which provide further surface area for absorption.

Three types of motility can be observed in the small intestine. The first type is called *pendular motion*. These waves do not advance down the intestine. Rather, they are merely a localized shortening and lengthening of the intestine which produces a mixing action. *Segmentation contractions* are the second type of intestinal motility. These intestinal movements are ringlike contractions at regular intervals which periodically relax, whereupon the area that had been previously relaxed contracts. This type of motility provides a means of mixing in addition to the pendular contractions. *Peristalsis*, a form of motility that has been previously discussed, is the third type of intestinal motility, providing a means for movement of chyme (intestinal contents) down the tract.

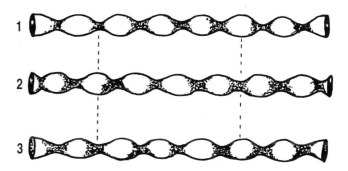

Fig. 2–10. Segmentation waves provide a mixing and churning action by alternately relaxing and contracting in localized areas.

DIGESTION AND ABSORPTION IN THE SMALL INTESTINE

The small intestine is the primary organ of digestion and absorption. Specialized enzymes present in the various segments of this long organ provide fast, effective means of breaking down carbohydrates, lipids, and proteins for subsequent absorption.

Carbohydrates

The digestion and absorption of most carbohydrates occur in the small intestine. Here, such enzymes as sucrase and maltase split carbohydrates into monosaccharides, whereupon absorption takes place. The region of the greatest absorption of sugars is in the jejunum. Glucose and galactose are absorbed through an active transport mechanism. Sodium ion concentration within the intestinal contents has been shown to be critical in this mechanism. A high Na^+ concentration facilitates rapid absorption of these sugars while a low Na^+ concentration reduces the rate of absorption. Some pentoses and hexoses are absorbed through diffusion — a process considerably slower than that of active transport.

Lipids

Lipids are digested and absorbed primarily in the upper part of the small intestine, but considerable absorption can take place as far down as the ileum. When lipids, emulsified by bile salts, come into contact with the various lipases that are found in the duodenum, they are broken down into monoglycerides and fatty acids. Short-chain fatty acids are then absorbed directly into the mucosa of the small intestine and are transported to the portal circulation. Monoglycerides and insoluble fatty acids are emulsified by bile salts, forming micelles. By attaching to the surface of epithelial cells, the micelles enable these components to be absorbed into the mucosal cells. Once inside these cells, the long-chain fatty acids are reesterfied to form triglycerides. Tryglycerides then combine with cholesterol, lipoproteins, and phospholipids to form chylomicrons — minute fat droplets. The chylomicrons are then passed into the lymphatic circulatory system.

Proteins

Although protein digestion is initiated in the proventriculus and gizzard, most digestion and absorption occur in the small intestine. Numerous pancreatic and intestinal enzymes split proteins into their constituent amino acids, which are subsequently absorbed.

Amino acid absorption is not clearly understood; but an active transport mechanism involving Na^+, similar to that of glucose absorption, is implicated. Amino acids are rapidly absorbed in the duodenal and jejunal segments, but are poorly absorbed in the ileum.

Minerals and Vitamins

Mineral absorption occurs throughout the small and large intestines, with the rate of absorption depending on a number of factors — pH, carriers, etc. Numerous mechanisms of mineral absorption have been elucidated. Many minerals, for example iron and sodium, require active transport systems. Others, such as calcium, utilize both carrier proteins and diffusion mechanisms.

Most of the vitamins are absorbed in the upper portion of the intestine, with the exception of vitamin B-12 which is absorbed in the lower intestine. Water-soluble vitamins are rapidly absorbed, but the absorption of fat-soluble vitamins relies heavily on the fat absorption mechanisms which are generally slow.

CECA, LARGE INTESTINE, AND CLOACA

The cecum (plural ceca) is a blind-ended tube found at the junction of the small and large intestines. In grain-eating birds, there are two large ceca, while in some other types of birds there may be only one rudimentary pouch, or none at all.

The large intestine, or colon, is extremely short in birds and is very similar in structure to the small intestine. It is generally believed that the large intestine in birds does not play any significant role in digestion and absorption.

In birds, all waste products — both urinary and fecal — empty into a structure called the cloaca, which leads to the vent. Thus, urinary and fecal waste products are mixed together.

EXCRETORY SYSTEM

The kidneys in birds are rather large and elongated and are situated along the fused backbone. Each kidney consists of three lobes (cranial lobe, middle lobe, and caudal lobe) which empty into a ureter leading to the urodeum of the cloaca from which urine is excreted. The primary functions of the kidney are twofold: (1) to filter the blood so as to remove water and waste products therefrom, and (2) to reabsorb any nutrients (e.g., glucose or electrolytes) which might be recycled for additional use. Because the kidneys control the absorption and excretion of water and electrolytes, they are the primary control center for maintaining the proper osmotic balance of body fluids.

The urine of birds is cream colored. Much of it consists of a thick, pasty mucoid material which contains uric acid.

Unlike mammals whose urine contains urea primarily, birds excrete uric acid as the primary nitrogen metabolite. Uric acid is synthesized in the liver and is excreted via the urine. It comprises 60 to 80% of the total urinary nitrogen. Birds can, however, produce urea to a limited extent as an end product of purine metabolism and the catabolism of arginine; but the level of urea found in the urine is insignificant as compared to the uric acid content.

Because urine is transported to the cloaca, feces and urine are excreted from the body together. In fact, some urine may pass into the large intestine where additional water can be reabsorbed.

REPRODUCTIVE SYSTEM

The reproductive physiology of poultry is markedly different from that of mammals; the most obvious difference being that the egg is fertilized in the infundibulum, supplied with nutrients, surrounded by a shell, and expelled from the body in birds, while the fertilized egg remains *in utero* until birth in mammals.

MALE REPRODUCTIVE SYSTEM

The male reproductive system of birds is extremely simple, consisting of two testes, each having an epididymis and vas deferens that lead to the copulatory organ.

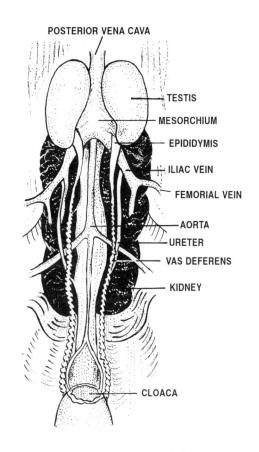

POSTERIOR VENA CAVA

TESTIS

MESORCHIUM

EPIDIDYMIS

ILIAC VEIN

FEMORIAL VEIN

AORTA

URETER

VAS DEFERENS

KIDNEY

CLOACA

Fig. 2–11. The male reproductive and urinary systems.

The male bird is unique among domestic animals in that the testes are located along the backbone within the abdominal cavity. In mammals, the testes are located in a sac called the scrotum which hangs from the body. This sac keeps the testes slightly cooler than the interior of the abdominal cavity because spermatogenesis cannot take place in mammals at normal body temperature. However, birds have evolved a mechanism, not yet completely understood, that permits their intra-abdominal testes to be fully functional.

The process of removing the testes from young males to promote meat quality for market birds is called caponizing, and castrated birds are called capons. Because incisions must be made in the abdominal wall to remove the testes, the operation is more dangerous than the castration of mammals where the testes hang away from the body and can be easily removed. Because broilers are now marketed at about 45 days of age, well before they reach sexual maturity, caponizing is rarely performed. However, in some metropolitan areas where certain ethnic groups are located, a very limited market for capons may exist.

The copulatory apparatus for the turkey and chicken consists of two papillae and a rudimentary copulatory organ that is located at the vent. In ducks and geese, this organ is fairly well developed and is erectile in nature.

The gross appearance of avian sperm is somewhat different from mammalian sperm. It has a long cylindrical head with a pointed acrosome, a short midpiece, and a long tail. Because the fowl has no seminal vesicles or prostate gland, the volume of seminal fluid is very low. The seminal fluid for avian sperm contains no fructose, citrate inositol, phosphoryl choline, or glyceryl phosphoryl choline — compounds commonly found in mammalian semen. Chickens normally produce about 1 ml of whitish semen per ejaculate, and turkeys produce about 0.5 ml of yellowish or brownish semen per ejaculate. Although turkeys produce only about one-half the volume of semen as chickens, the sperm concentration is about twice that of chickens. Thus, chickens yield roughly the same number of sperm per ejaculate as turkeys — about 1.75 to 3 billion.

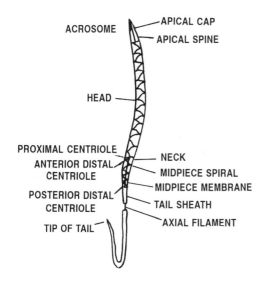

Fig. 2-12.   Structure of avian sperm.

FEMALE REPRODUCTIVE SYSTEM

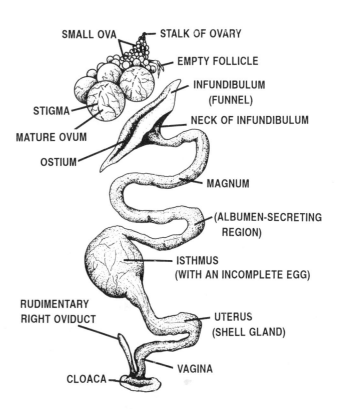

Fig. 2-13.   The reproductive tract of the female chicken.

Females of most animals have two functional ovaries — a right one and a left one. But the hen has only one functional ovary, the left one, which is situated in the body cavity near the backbone. At the time of hatching, the female chick's left ovary contains up to approximately 3,600 to 4,000 tiny ova from which full-sized yolks may develop when the hen matures.

Each yolk (ovum) is enclosed in a thin-walled sac (follicle) which is attached to the ovary by a stalk. This sac contains the vast network of blood vessels which supply the yolk materials.

When a pullet reaches sexual maturity, or comes into egg production, some of the ova develop to mature yolks. When mature, the yolk is released from the follicle by rupture of the follicle wall along a line called the stigma. Soon after its release, the yolk is picked up, or engulfed, by the funnel of the oviduct.

The oviduct is a coiled, folded tube about 20 to 30 in. long occupying a large part of the left side of the abdominal cavity. It is divided into five rather clearly defined regions, each of which plays a specific role in the completion of the whole egg. A normal hen requires slightly over 24 hours to complete an egg. Within 30 minutes after the egg is laid, another yolk is released from the ovary for laying the following day. The functions of each of the five parts of the oviduct are set forth in Table 2–4.

TABLE 2–4
FUNCTIONS OF THE OVIDUCT

Part	Approximate Time Egg Spends in Section	Functions
Infundibilum (funnel)	15 minutes	Picks up yolk from the body cavity after it is released from the follicle. If live sperm are present, fertilization occurs in this section.
Magnum (albumen-secreting region)	3 hours	Thick white (albumen) is deposited around the yolk. This layer later forms the chalaziferous layer, the chalaza, and inner thin and thick white.
Isthmus	1¼ hours	Inner and outer shell membranes are added and some water and mineral salts. These membranes give some protection to the egg contents from outside contamination.
Uterus (shell gland)	21 hours	During the first part of the egg's stay in the shell gland, water and minerals pass through the shell membranes into the white, inflating the egg and giving rise to the outer layer of thin white. Soon after the egg is inflated, the shell gland starts to add calcium over the shell membranes, continuing this process until just prior to laying. If the shell is going to be colored, pigment is added in this section.
Vagina	Entire time from ovulation to laying is slightly more than 24 hours.	The egg passes into this section just prior to laying. Its function is not known.

THE EGG

The bird egg is a marvel of nature. It is one of the most complete foods known, as evidenced by the excellent balance of proteins, fats, carbohydrates, minerals, and vitamins which it provides during that 20-day in-the-shell period when it serves as the developing chick's only source of food. Also, the egg is one of the few foods that is produced in prepackaged form. Not only that, it is the reproductive cell (ovum) of the hen. Upon fertilization by the male's reproductive cell (sperm), the egg will develop into a chick when incubated properly.

Pertinent facts about eggs of different species are presented in Table 2–5.

TABLE 2–5
EGG PRODUCTION OF POULTRY

Species	Age of Sexual Maturity	Eggs/Year	Egg Size	
	(mo.)	(no.)	(lb)	*(g)*
Chicken:				
Light-type	5–6	240	0.13	*58*
Broiler-type		170		
Turkey	7	105	0.19	*85*
Goose	24	15–60	0.47	*215*
Duck (Pekin)	7–8	110–175	0.18	*80*
Pheasant	8–10	40–60	0.07	*32*
Quail (Bobwhite)	8–10	150–200	0.02	*9*
Pigeon	6	12–15	0.04	*17*
Guinea fowl	10–12	40–60	0.09	*40*

FECUNDITY

The term *fecundity* is used to describe the inherent capacity of an organism to reproduce rapidly. In higher animals, reproduction is possible only after the ovum (female gamete) is fertilized or united with the spermatozoon (male gamete). In chickens, fertilization is not a necessary preliminary to egg laying. Thus, the hen can lay eggs continuously without being mated or without being stimulated by the presence of a male. This biological phenomenon has been advantageously utilized in producing infertile eggs for food. Infertile eggs are of more economic value for food then fertile eggs, because there is no danger of loss through development of the embryo.

Details of mating, fertilization, and inheritance of some characters are presented in Chapter 4, Poultry Breeding.

EGG SIZE

There is an enormous range in size among the eggs of different species of birds. Ostrich eggs average about 3.1 lb, whereas hummingbird eggs weigh only 0.001 lb. The average size of eggs laid by a number of domestic species of birds is shown in Table 2–5.

There is also considerable variation in egg size within a species. Thus, in chickens, the eggs of Dark Brahmas are more than twice as heavy as those of Japanese Bantams. The eggs of most of the heavy laying breeds weigh from 0.1 to 0.14 lb, with the eggs of the large meat-producing breeds at the upper end of size and those of the smaller game birds at the lower end.

Also, the size of the eggs laid by one individual may differ widely from those laid by another of the same species and breed. An understanding of the various factors that influence egg size is important because, economically, it is not unusual for the price of medium size eggs to be 5 to 10¢ lower per dozen than the price of large eggs.

The most important factors affecting egg size are:

1. **Breeding.** Egg size is an inherited trait. Although environmental factors may result in smaller eggs, the upper limit in egg size is determined when the pullet is hatched.

For more than 70% of all eggs to Grade Large or better, a strain with an average egg weight of 25 oz per dozen is needed.

2. **Age of bird.** When pullets start to lay, 80% or more of their eggs will be under 21 oz per dozen. Egg size should increase gradually until the birds are around 12 to 14 months of age, following which some decrease in egg size may be expected.

3. **Clutch order.** The order of an egg in the clutch affects its weight. The first egg of a clutch is usually the heaviest, and there is usually a progressive decrease in the weight of the rest of the eggs of the clutch.

4. **Total eggs laid in a year.** There is a tendency toward a decline in egg size with the total number of eggs laid in a year.

5. **Age at maturity and season of hatch.** Delay in maturity will usually result in larger eggs at the start of production. Also, season of hatch when birds are grown in natural daylight affects both maturity and size of eggs at the start. Late fall- and winter-hatched birds will be subjected to an increasing day length during the critical age period of 8 to 20 weeks if no supplementary lights are used. This will stimulate early maturity and result in a prolonged period of small eggs.

6. **Temperature.** The size of eggs declines during the hot summer months. Fortunately, temperatures favoring large eggs also favor high egg production. Thus, anything that can be done in the summer to keep layers cool will benefit egg size. This includes insulation, proper ventilation, reflective paint on the roof, foggers, and plenty of cool water.

7. **Type of housing.** Caged birds lay eggs about 1/2 oz per dozen larger than birds on the floor, and the birds on slats lay about 10% more large eggs than birds on the floor with part litter.

8. **Feed and water.** Normally, feed is a minor factor in egg size so long as a well-balanced ration is fed. However, underfeeding (as a result of feeders being empty for lengthy periods, or of timid birds not getting enough to eat when feeder space is limited), a deficiency of needed minerals (especially calcium), or a very low level of protein will result in smaller eggs. The dietary fat in the ration may increase egg size slightly.

If water intake is too low because the water is too cold, too hot, too unpalatable, or because water is not available, egg size will be hurt along with egg production.

9. **Disease.** Some diseases affect egg size dramatically, and in some cases the effect persists for months after the layers appear to be healthy. Both Newcastle disease and infectious bronchitis will cause a drop in egg production, with many of the eggs being small and often odd shaped. With bronchitis, especially, small egg size may continue for months.

10. **Grain fumigants.** Ethylene dibromide, an ingredient of a fumigant commonly used on seed oats, will decrease egg size. A drop in egg size to as small as 16 oz per dozen from an initial 23.8 oz average has been reported.

EGG SHAPE

Eggs differ considerably in shape. Although many are truly ovate, some are nearly spherical, whereas others are elongated. Some eggs are almost equally pointed or rounded at both ends; others taper sharply from the large end to the small end. The eggs laid by birds of the same species resemble each other in shape, but they are not identical; nor are all eggs of a particular bird alike.

Aristotle (384–322 B.C.) believed that the cock hatched from the more pointed chicken egg and the hen from the rounder type. Early 19th century naturalists argued that egg contours indicated the general body form of the bird that would develop within. Somewhat later, natural selection advocates theorized that, through adaptation, the eggs of different birds had assumed the shape most likely to ensure the survival of each species in the particular environment.

It is now generally agreed that physiological factors largely determine the diversities in the form of the egg, but that the shape may be modified by certain conditions. Among the causations of egg shape variations within species, or of a particular bird, are the following:

1. **Pressure exerted by the oviduct muscles.** The walls of the oviduct contain two layers of muscles; (a) the inner circular layer, which moves the egg forward; and (b) the outer longitudinal layer, which expands the oviduct. Muscle tones and difference in the degree of coordination between these muscles likely account for many of the minor variations in the shape of the egg.

2. **Volume of albumen and size of isthmus.** It has been postulated that an elongated egg might be formed if a large volume of albumen is secreted and then forced through a narrow isthmus.

3. **Breed and flock variation.** The eggs laid by birds of the same breed resemble each other in shape, but there may be enormous variation within a single flock.

4. **Heredity.** Chickens may be bred and selected for a specific egg shape with comparative ease; for example, they may be bred and selected for round eggs or long eggs.

5. **Commencement of laying.** The first eggs laid by a pullet are likely to be atypical in shape.

6. **Cycle of laying.** The first egg of a cycle is usually longer and narrower than the second egg.

7. **Pause in laying.** The first egg laid after a pause of 7 days is longer and narrower than the last egg preceding the pause.

EGG COLOR

The eggs of the domestic hen may be white, many shades of brown, or yellow. One breed lays blue green eggs. Sometimes very small, dark flecks are present on the shell, especially if it is brown.

Among domestic fowl, the color of the eggs is peculiar to the breed, although tinted eggs occasionally appear in breeds that ordinarily lay white eggs. Of the four races recognized in the United States, only the Mediterranean (comprised of Leghorns, Minorcas, Anconas, Black Spanish, and Blue Andalusian) lays white eggs. The other three races—Asiatic, English, and American—lay tinted eggs, with the exception of two or three breeds. Cochin China hens lay eggs that range from bright yellow to dark yellow—speckled with small red dots. Langshans lay dark yellow eggs. Brahmas lay reddish yellow eggs. Most of the continental European breeds lay white eggs. The Araucana of South America lays light bluish green eggs. Hens normally lay eggs of the same color, but there may be considerable variation in color among hens of the same breed.

Colored eggs occur because pigment is deposited in the shell as it is formed in the uterus. The hereditary nature of such pigmentation is evidenced by the manner in which egg color is identified with species and breed. However, the mode of inheritance of shell color is difficult to determine, because most varieties of fowl are the product of many years of crossbreeding.

Egg color often assumes economic importance, as there are numerous local prejudices in favor of certain shell tints.

STRUCTURE OF THE EGG

A schematic side view of an egg is shown in Fig. 2–14, with the various parts labeled in their normal position.

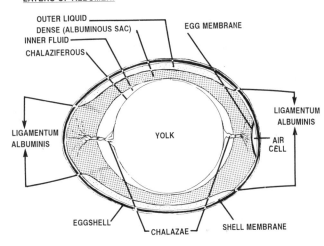

Fig. 2–14. Structure of the egg.

The protective covering, known as the shell, is composed primarily of calcium carbonate, with 6,000 to 8,000 microscopic pores permitting transfer of volatile components. The air cell, located in the large end of the egg, is formed when the cooling egg contracts and pulls the inner and outer shell membranes apart. The cordlike chalazae hold the yolk in position in the center of the egg. As shown, the yolk is surrounded by membrane, known as the vitelline membrane. The germinal disc, a normal part of every egg, is located on the surface of the yolk. Embryo formation begins here only in fertilized eggs.

COMPOSITION OF THE EGG

The chemical composition of the egg is given in Table 2–6.

TABLE 2–6
CHEMICAL COMPOSITION OF THE EGG

	Whole Egg	White	Yolk	Shell
	(percent of whole egg)			
	100	58	31	11
	(%)	(%)	(%)	(%)
Water	65.5	88.0	48.0	—
Protein	11.8	11.0	17.5	—
Fat	11.0	0.2	32.5	—
Ash	11.7	0.8	2.0	96.0
Total	100.0	100.0	100.0	96.0

Cholesterol ($C_{27}H_{45}OH$) is present in all eggs, as it is in the bodies of all birds, all animals, and all humans. Cholesterol is essential to cellular metabolism. However, high levels of cholesterol have been associated with many types of arterial diseases in people.

Extensive tests conducted by the U.S. Department of Agriculture in 1988 and 1989 showed that eggs produced in the United States have an average cholesterol content as follows: extra large eggs, 230 mg; large eggs, 213 mg; medium sized eggs, 180 mg.

EGG ABNORMALITIES

Periodically, a deviation in the mechanics of egg laying will create abnormal eggs. Some of these abnormalities are as follows:

1. **Double-yolked eggs.** This is the result of two ova ripening at the same time.

2. **Blood spots.** When a small blood vessel breaks in the ovary or oviduct, blood spots are formed in the egg.

3. **Meat spots.** Meat spots are degenerated blood clots from the ovary or oviduct that are deposited in the egg.

4. **Yolkless eggs.** Occasionally, foreign material may get into the oviduct and stimulate the secretion of albumen in much the same manner as the yolk.

5. **Dented eggshells.** When one egg is kept too long in the uterus, a second egg may pass down the tract and actually touch the first egg, thereupon creating an indentation in the shell of the second egg.

6. **Soft-shelled eggs.** These eggs result when no shell is secreted.

EGG LAYING

After the yolk is released from the follicle, it is picked up by the infundibulum and subsequently passed into the magnum. In the magnum, the egg begins to take shape as the chalazae (a ropylike substance) and three of the four layers of albumen become attached to the yolk. The albumen deposited in the magnum first appears to be homogenous, but as the egg passes through this organ, it is continually turned which gives rise to the various layers of egg white. The egg spends a total of about 3 hours in the magnum.

From the magnum, the egg is passed into the isthmus where the inner and outer shell membranes are added along with some mineral salts. After spending about 1¼ hours in the isthmus, the egg travels to the uterus or shell gland. Here, additional water and minerals pass through the shell membranes into the white, inflating the egg and giving rise to the fourth layer of albumen (outer layer of thin white). It is also in this organ where the shell—composed primarily of calcium carbonate—is formed. If the eggshell is to be colored, the pigments are deposited in the uterus. Additionally, a waxy substance responsible for the bloom of eggs is secreted in the uterus. When this substance dries, it seals the openings of the porous eggshell. The egg spends about 21 hours in the uterus.

From the uterus the egg passes through the vagina and out the cloaca. One of the interesting physiological phenomena in the hen is the fact that as the egg passes down the oviduct, it travels small end first; however, immediately prior to expulsion, the egg is turned 180° so that the large end is exposed first. This turning is believed to aid the muscles of the tract in expelling the egg.

INTEGRATION OF BODY PROCESSES

The bird is an extremely complex biological organism, requiring considerable information to be passed to and from its different physiological systems. This integration of messages is accomplished through two modes of transport—neural (via nerves) and humoral (via blood). The nervous systems provide electrical impulses which stimulate or inhibit various body functions. Humoral controls are provided via the endocrine glands and their secretions called hormones. *A hormone can be defined as a chemical released by a specific area of the body that is transported to another region within the animal where it elicits a physiological response.*

NERVOUS SYSTEM

The nervous system can be divided into two anatomical systems—the somatic nervous system, and the autonomic nervous system. The somatic nervous system enables the body to adapt to stimuli from the external environment. Various stimuli, such as touch, are perceived by specialized receptors within this system, and the body responds accordingly. The autonomic system involves the maintenance of homeostasis—the internal environment of the body. This is the system that controls the gastrointestinal tract.

The autonomic system can be further divided into the *sympathetic autonomic nervous system* and the *parasympathetic autonomic nervous system*. The sympathetic system is generally associated with the traditional "fight or flight" response, and the parasympathetic system is usually associated with routine integration of normal activity.

When the sympathetic system is stimulated, there is a need for large amounts of blood in peripheral tissues, such as skeletal muscle. In order to accommodate this need, blood is shunted from such areas as the gastrointestinal tract. Heart rate and respiratory rate increase to accommodate the increased demand for oxygen.

SPECIAL SENSES

One means by which a bird can react to the environment is through the use of its special senses.

The eyes of birds are located on the sides of the head. This lateral placement, plus the great mobility of the head, gives birds an extremely wide field of sight. Additionally, the eyes of birds make up a larger proportion of the head than those of mammals, and the eyeballs of birds are flatter than the generally spherical eyeballs of mammals. Visual acuity in the bird is not much greater than that of humans, but the bird has the ability to process visual cues much faster than humans. Thus, while birds do not see images much clearer than humans, they can see the images more rapidly. All diurnal birds possess color (chromatic) vision.

The structure of the ears in poultry resembles that of reptiles more than that of mammals. Poultry have an acute sense of hearing and use auditory sounds to signal behavioral stimuli—warnings of danger, communication between the mother hen and her chicks, etc.

Neither the sense of smell nor the sense of taste is well developed in poultry. Certain odors and tastes can be detected, but for the most part, these senses are rather primitive. For example, pigeons are insensitive to the odor of a strong ammonia solution, and chicks show no real preference or dislike for sweet feed or mildly alkaline solutions. However, chicks will not consume water that contains in excess of 0.9% salt.

ENDOCRINE SYSTEM

The endocrine system is composed of a number of glands that produce, store, and secrete hormones. Fig. 2–15 shows the location of these glands, and Table 2–7 (next page) lists the hormones produced by the various glands and the physiological function of the hormones. Hormones can be classified into two broad categories, according to structural properties: (1) protein hormones, and (2) steroid hormones.

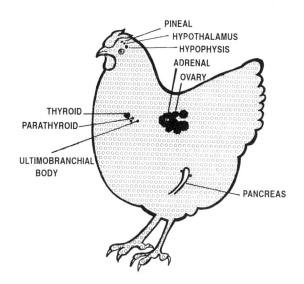

Fig. 2–15. The endocrine glands of the chicken and their location in the body.

TABLE 2–7
HORMONES OF THE FOWL

Gland	Hormone	Type of Hormone	Functions of Hormone
Hypophysis (pituitary gland):	Gonadotropic hormones		
1. Adenohypophysis (anterior lobe)	1. Follicle stimulating hormone (FSH)	Protein	Stimulates growth of ovarian follicles in females. Maturation of sperm in males.
	2. Lutenizing hormone (LH)	Protein	Triggers ovulation in females. Acts on Leydig cells of testes to produce androgen.
	3. Prolactin	Protein	Causes broodiness in chickens. Initiates crop-sac secretion in pigeons.
	Growth hormone (GH) or somatotropin (STH).	Protein	Growth promotion. Protein synthesis.
	Adrenocorticotropin (ACTH)	Protein	Stimulation of the adrenal cortex and release of the adrenal corticoids.
	Thyrotropin (TSH)	Protein	Stimulation of the thyroid glands to (1) release thyroxin, and (2) absorb iodine.
	Melanotropin (MSH)	Protein	Function is not known in birds.
	Oxytocin (storage)	Protein	Stimulates uterine tissue.
2. Neurohypophysis (posterior lobe)	Vasotocin (storage)	Protein	Antidiuretic hormone. May initiate the contraction of the uterus that begins oviposition.
Hypothalamus	Oxytocin	Protein	See Neurohypophysis.
	Vasotocin	Protein	See Neurohypophysis.
	Releasing factors for: 1. LH 2. FSH 3. TSH 4. ACTH		Stimulates the adenohypophysis to release its hormones.
Thyroid glands	Thyroxin and triiodotyrosine	Protein	Affects metabolic rate. Affects feather growth and color.
Ultimobranchial glands	Calcitonin	Protein	Calcium metabolism. May play a role in the regulation of serum phosphorus.
Parathyroid glands	Parathyroid hormone (PTH)	Protein	Calcium mobilization and also phosphorus metabolism.
Adrenals:	Aldosterone	Steroid	Electrolyte and water metabolism.
1. Cortex	Corticosteroids	Steroid	Carbohydrate, fat, and protein metabolism.
	Catecholamines		
2. Medulla	1. Adrenaline (epinephrine)	Protein derivative	Initiates sympathetic neural responses.
	2. Noradrenaline (norepinephrine)	Protein derivative	Neural transmitter.
Pancreas	Glucagon	Protein	Carbohydrate, fat, and protein metabolism.
	Insulin	Protein	Carbohydrate, fat, and protein metabolism.
Testes (male)	Testosterone	Steroid	Secondary sex characteristics. Sexual behavior, Spermatogenesis.
Ovary (female)	Estrogens: estradiol, estriol, estrone	Steroid	Secondary sex characteristics. Affects growth and fat deposition. May be involved with the growth and development of the follicle. Involved in albumen synthesis.
	Progesterone	Steroid	Involved in albumen synthesis. Antagonistic to ovulation.
Pineal gland	Melatonin	Protein	Functions unclear in poultry.

Oxytocin and vasotocin are synthesized in the hypothalamus and are transported along nerve tracts to the posterior lobe of the pituitary where they are subsequently stored. They, like all of the hormones, are secreted into the blood when called upon to promote their physiological roles.

Hormones carry out their regulatory roles through several types of feedback systems as seen in Fig. 2–16. Two types of feedback systems—negative feedback and positive feedback—are utilized.

NEGATIVE FEEDBACK

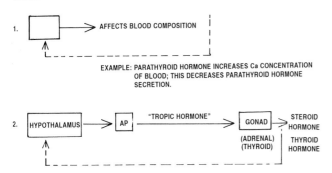

EXAMPLE: PARATHYROID HORMONE INCREASES Ca CONCENTRATION
OF BLOOD; THIS DECREASES PARATHYROID HORMONE
SECRETION.

EXAMPLE: ANDROGEN INHIBITION OF HYPOTHALAMUS WHICH
INHIBITS GONADOTROPIN SECRETION.

POSITIVE FEEDBACK

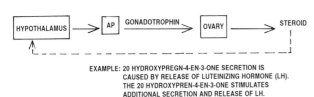

EXAMPLE: 20 HYDROXYPREGN-4-EN-3-ONE SECRETION IS
CAUSED BY RELEASE OF LUTEINIZING HORMONE (LH).
THE 20 HYDROXYPREN-4-EN-3-ONE STIMULATES
ADDITIONAL SECRETION AND RELEASE OF LH.

Fig. 2–16. Feedback systems.

NEGATIVE FEEDBACK

Control of the various feedback systems depends on the circulating levels of various metabolites, hormones, and nutrients. For example, if the level of calcium in the blood drops, the parathyroid gland secretes a hormone that mobilizes body calcium and subsequently increases the concentration of calcium in the blood. This increase in the blood calcium then decreases the secretion of parathyroid hormone. Thus, the endocrine gland initially reacts to the absence of a particular compound—a negative feedback.

Another example of negative feedback involves a three-step process. If the metabolic state of the bird calls for the secretion of a hormone, such as thyroxin, the hypothalamus is stimulated to secrete the releasing factor for thyrotropin. This releasing factor stimulates the anterior pituitary to secrete thyrotropin. Thyrotropin travels via the bloodstream to the thyroid which is then stimulated to produce thyroxin. The thyroxin travels to the target cells, and metabolites then circulates and inhibits the production of the releasing factor in the hypothalamus.

POSITIVE FEEDBACK

In positive feedback mechanisms, a hormone is secreted which promotes the production of a second compound; thence the second compound travels back to the original endocrine gland and stimulates the production of even more of the initial hormone. For example, the hypothalamus is stimulated to produce a releasing factor that causes the anterior pituitary to secrete LH. LH travels to the ovary, stimulating it to produce a certain steroid which stimulates the hypothalamus to secrete more releasing factors.

HUNGER AND APPETITE

Hunger is the physiological desire for feed following a period of fasting. Appetite, on the other hand, is a learned or habitual response to the presence of feed. A bird that is extremely hungry may not have an appetite for a type of feed that it deems undesirable. Conversely, if the feed is of a desirable nature, the bird may have an appetite for it in spite of the fact that it is not hungry.

HYPOTHALAMIC CONTROL OF APPETITE

The hypothalamus (derived from the terms *hypo* meaning below, and *thalamus*—a region of the brain)—a structure in the ventral region of the diencephalon—has been implicated as one of the major control centers of appetite regulation. Within the hypothalamus, certain areas can be differentiated. Two of these are of particular importance in the regulation of appetite. The first area is that of the lateral hypothalamus. It is commonly called the *feeding center*, because, upon stimulation of this region, the bird commences to eat whether or not it is hungry. If this area is damaged, the bird loses all desire to eat and eventually starves. The ventro-medial area of the hypothalamus functions as the *satiety center*. Stimulation of this region will depress appetite. If the ventro-medial nuclei are destroyed, there is no inhibition of feed intake, and the bird has an uncontrollable appetite. It is believed that there is a chronic activity in the lateral hypothalamus which is kept in check by the inhibitory influence of the ventro-medial area.

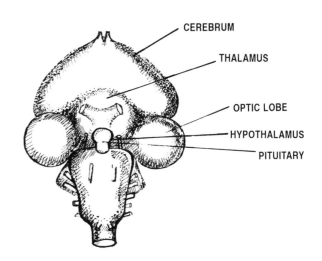

Fig. 2–17. Brain of the fowl showing location of important parts, including hypothalamus.

Several theories have been advanced as to the exact physiological mechanism which triggers the hypothalamus to tell the bird when to eat. While each theory has its merits, there is not conclusive proof in favor of any one of them. In the long run, it seems probable that a combination of a number of factors will provide the answer.

The two theories concerning the hypothalamic control of appetite that have received the most attention are (1) the chemostatic hypothesis, and (2) the thermostatic hypothesis. The chemostatic hypothesis reasons that the hypothalamus is sensitive to circulating blood nutrient levels, such as sugar or lipid. When these levels become too low, the hypothalamus sends signals to begin feeding. Once the blood nutrient level is elevated, stimuli from the feeding center are inhibited and the bird feels full. The second theory of appetite control, the thermostatic hypothesis, theorizes that the hypothalamus plays an important role in heat regulation within the body, and that a decrease in hypothalamic temperature will induce feeding.

INFLUENCE OF VOLUME ON APPETITE REGULATION

Appetite in birds is governed to a certain degree by the volume of food eaten. There are volume receptors in the crop and esophagus which can alter feed intake when stimulated.

THERMOREGULATION

Birds are warm-blooded animals — homeotherms. Hence, they must maintain a constant body temperature for normal physiological functions. If the bird is in a hot environment, the body must give off heat and utilize cooling mechanisms. If the bird is in a cold environment, the body must produce heat through metabolic processes and use insulatory mechanisms to keep this heat from escaping. Table 2–8 shows the deep body temperature of several species of poultry.

TABLE 2–8
DEEP BODY TEMPERATURE OF POULTRY AT REST
AND IN THERMONEUTRAL CONDITIONS

Species	Temperature	
	(°F)	(°C)
Chicken	106.7	41.5
Turkey	106.2	41.2
Duck	107.8	42.1
Goose	105.8	41.0
Pigeon	108.0	42.2

ADAPTATION TO HEAT STRESS

Birds have no sweat glands and must, therefore, use other means of dissipating heat. The comb and wattle areas have highly vascularized, exposed skin. Additionally, respiratory mechanisms allow for some cooling, and the feathers act as a means of protecting the body from exposure to heat.

The primary cooling mechanism is respiratory in nature. When air is inhaled, heat is emitted from the nasal membranes and other areas of the respiratory tract. This heat is then lost when the air is exhaled.

Heat can be transferred from the inner body to the surface through the following mechanisms:

1. **Conduction.** In this mechanism, heat is transferred from molecule to molecule in the body and is eventually lost when the heat reaches the outermost portions. Conduction is increased by the birds crouching to the ground and squashing the breast feathers to facilitate heat loss to the soil.

2. **Convection.** In this mechanism, heat is transferred from heat-producing tissues to the blood. The blood then travels to the skin, resulting in an increased skin temperature and a loss of heat to the atmosphere.

3. **Countercurrent heat exchange.** Arterial blood is warmer than venous blood. Thus, when blood reaches the limbs, some heat is lost to the environment and additional heat is given up to the venous blood returning from the limbs.

(Also, see Chapter 9, Poultry Behavior and Environment, section headed "Thermoregulation.")

ADAPTATION TO COLD STRESS

Birds are very well adapted to cold, due primarily to their highly efficient insulation provided by feathers. Adult chickens have been known to survive temperatures as low as –58°F for 1 hour, and pigeons have survived temperatures as low as –148°F for 1 hour.

When birds are exposed to cold stress, several compensatory mechanisms are used. The first line of defense is the plumage. Feathers are erected to provide more efficient protection from the environment and to conserve body heat.

Birds shiver in response to cold. This activity increases the metabolic rate of the body to produce additional heat.

EFFECT OF LIGHT ON POULTRY

Poultry are extremely sensitive to photoperiodism (length of daylight). Research has shown turkeys to be more sensitive to the length of daylight than chickens.

The wavelength of the light can affect the amount of stimulation. Orange to red (664 to 740 nm) wavelengths yield the most satisfactory results. Since incandescent light bulbs emit these wavelengths, they are entirely satisfactory for light stimulation programs.

EFFECT OF LIGHT ON SEXUAL MATURITY
AND REPRODUCTION

When growing pullets are exposed to increasing duration of light, sexual maturity is stimulated. The increasing duration of light stimulates the hypothalamus to secrete gonadotropic releasing factors which, in turn, stimulate the anterior pituitary to release LH and FSH. If egg production is started at too early an age, the eggs will be too small to be profitable. Thus, most producers utilize a growing program designed to suppress sexual maturity in pullets until they reach an age and size when marketable-sized eggs can be produced.

Birds in lay require only about 0.5 to 1.0 footcandle of light. Light intensities greater than 1 footcandle are wasteful if artificial light is used. Generally, a 14 to 16 hour light program is used. The length of light exposure should never be decreased for birds in production.

CLUTCHES

Hens lay eggs in a certain time pattern called a clutch. For example, a hen may lay an egg on each of four consecutive days and on the fifth day fail to produce an egg. Generally, each successive egg of a clutch becomes smaller and smaller. The weight of the yolks remains relatively constant, but the amount of albumen decreases as the clutch progresses. Clutches are determined by the hormonal cycles in the hen and are highly variable, ranging from 1 day to as many as 200 days. A hen having a three- to four-egg clutch has an ovulation cycle of about 26 hours. The 2-hour difference over the 24 hours in 1 day is called lag. Since most lighting programs are on a 24-hour cycle and since LH and FSH secretion in poultry is extremely sensitive to light, a bird with a 26-hour cycle will be thrown off synchrony periodically. This is one factor which may be responsible for the nonlaying days that terminate a clutch. The "day off" may allow the hen to get back into synchrony. Most hens on a 14 hours of light-10 hours of darkness program lay their eggs in the morning between 14 to 16 hours after the beginning of the dark period.

Because FSH and LH affect gonadal development in males, it is important to monitor light exposure for them, also. Twelve to fourteen hours of light are generally required to achieve maximum stimulation of growth and development in the testes of young cockerels.

MOLTING

Birds normally shed their old feathers and grow new feathers once a year through a process called molting. In some cases, birds may molt twice a year, and, in rare cases, once every 2 years.

Molting is controlled by the gonads and the thyroid gland and is associated with a drop in estrogen levels and a decreasing rate of egg production. Egg production is not greatly affected by the process of molting, but molting is prolonged when the birds are kept in production. High producers tend to molt late; but once production ceases, molting is rapid.

The loss of feathers is an orderly process with the head feathers being shed first, followed by the feathers on the neck, body, wing, and finally, the tail.

Several factors affect the onset and length of molting; among them, (1) weight and physical condition of the bird, (2) length of light exposure, (3) nutrition of the bird, and (4) environmental influences, such as temperature and humidity. Thus, if one drastically reduces the amount of light or starves the birds in such a way as to knock them out of production, molting can be induced or speeded up.

QUESTIONS FOR STUDY AND DISCUSSION

1. Define anatomy and physiology. What is the distinction between the two terms?

2. What is meant when birds are referred to as "flying, feathered reptiles"?

3. List the genus and species for (a) chickens, (b) turkeys, (c) geese, and (d) ducks.

4. What is the integument of birds?

5. What is the function of the preen gland? What causes the yellow color of the skin and shanks of yellow-skinned chickens?

6. List the different types of combs.

7. List and describe the three types of feathers of adult birds.

8. How does the skeletal systems of birds differ physiologically and anatomically from that of mammals?

9. What are the three types of muscles found in birds?

10. What are the three types of muscle fibers found in skeletal muscle? What kind of muscle abounds in good flying birds?

11. What percentage of the total body weight of the adult bird consists of blood?

12. How do the red blood cells of chickens differ from those of mammals?

13. Diagram the respiratory system of birds. What are the functions of the lungs? What are the functions of the air sacs?

14. Describe how the digestive system of poultry differs from that of other nonruminant animals.

15. Define (a) enzymes, and (b) zymogens.

16. Trace the path of grain as it goes through the digestive tract of a bird. To what mechanical and chemical processes is the grain exposed?

17. List the functions of the liver.

18. Why did nature ordain that carnivorous birds have shorter intestinal tracts than herbivorous birds?

19. In what part(s) of the digestive system is each of the following absorbed: amino acids, minerals, and vitamins?

20. What are the primary functions of the kidneys?

21. What is the primary nitrogenous metabolite in the urine of birds?

22. Diagram the male and female reproductive tracts of birds.

23. How does avian sperm differ from mammalian sperm?

24. Which ovary is functional in the chicken?

25. Outline the sequences of ovulation and the laying of an egg.

26. Discuss the differences among species of birds in (a) age of sexual maturity, and (b) eggs laid per year.

27. From the standpoint of utilizing eggs for food, in what way is it advantageous that chickens can lay eggs continuously without being mated?

28. List and discuss the most important factors affecting egg size.

29. List and discuss the causations of egg shape variations within species.

30. Does egg color ever assume economic importance? What is the most desired egg color on the Boston market?

31. Diagram the structure of the egg.

32. Discuss the cholesterol content of eggs.

33. Define hormone.

34. The nervous system can be divided into two anatomical systems. What are they, and how do they serve to integrate the body?

35. Does the fowl have a well developed sense of taste? Explain.

36. Differentiate between negative and positive feedback.

37. Define hunger and appetite. How is appetite regulated?

38. List the three mechanisms of heat transfer in the body. What is the primary mechanism for dealing with heat stress?

39. What wavelengths of light provide the most stimuli to birds?

40. What is a clutch?
41. Describe the process of molting.

SELECTED REFERENCES

Title of Publication	Author(s)	Publisher
Anatomy of the Domestic Birds	Edited by R. Nickel, *et al.*	Springer-Verlag, New York, NY, 1977
Avian Egg, The	A. L. Romanoff A. J. Romanoff	John Wiley & Sons, Inc., New York, NY, 1963
Avian Embryo, The	A. L. Romanoff	The Macmillan Company, New York, NY, 1960
Avian Physiology, Third Edition	Edited by P. D. Sturkie	Springer-Verlag, New York, NY, 1976
Physiology and Biochemistry of the Domestic Fowl, Vols. 1, 2, and 3	Edited by D. J. Bell B. M. Freeman	Academic Press, New York, NY, 1971
Reproduction in Farm Animals, Fourth Edition	Edited by E. S. E. Hafez	Lea & Febiger, Philadelphia, PA, 1980

Fig. 3–1. Newly hatched chicks off to a good start in life. (Courtesy, Maple Leaf Mills, Ltd., Ontario, Canada)

INCUBATION AND BROODING

CHAPTER

3

Eggs have been incubated by artificial means for thousands of years. Aristotle, writing in the year 400 B.C., told of Egyptians incubating eggs by the hot-bed method in which decomposing manure furnished the heat. The Egyptians also constructed large brick incubators which they heated with fires in the rooms in which the eggs were incubated. The Chinese developed artificial incubation as early as 246 B.C. They developed a method in which they burned charcoal to supply the heat. Also, they used dung heaps for heating purposes. These ancient incubation methods were often practiced on a large scale. There are reports of single locations having as many as 3,600 eggs.

For many years, the application of artificial incubation principles was a closely-guarded secret, passed from one generation to the next. The hatchery operator determined the proper temperature by placing an incubating egg in the socket of his/her eye. Temperature changes were effected in the incubator by moving the eggs, by adding additional eggs to use the heat of embryological development of older eggs, and by regulating the flow of fresh air through the hatching area. Apparently humidity was no problem, for primitive incubators were located in highly humid areas, and the heat source, such as decomposing manure, often furnished moisture. Turning was done as frequently as five times daily, after the fourth day of incubation.

The construction, use, and patent of artificial incubators in the United States dates from 1844. The Smith incubator, which was virtually a large room with fans for forcing heated air to all parts of the chamber, was patented in 1918. It was the forerunner of today's efficient, large-scale incubators, used for the hatching of chickens, turkeys, ducks, and other eggs.

The artificial incubation of poultry was followed by artificial brooding, but the latter was somewhat later in developing. Natural brooding was commonplace until about 1930. However, today it is most unusual to see a mother hen brooding chickens, turkeys, ducks, or other young poultry.

The brooding period is usually considered to extend from the time poultry hatch, or are received from the hatchery, until they are about 8 to 10 weeks old.

Today, incubators with capacities of more than 500,000 eggs, equipped with sophisticated controls to maintain optimum conditions for hatchability, are rather common; and by using banks of incubators, as many as one million chicks a week can be hatched. Also, commercial brooding of poultry has become larger and more scientific. It is not uncommon to find houses and equipment so arranged that one caretaker is responsible for as many as 50,000 chicks.

POULTRY REPRODUCTION

Poultry reproduction is markedly different from the reproduction of mammals. The most obvious differences are that the egg is fertilized in the infundibulum, supplied with nutrients, surrounded by a shell, and expelled from the body, while the fertilized egg in mammals remains *in utero* until birth. Also, in higher animals, reproduction is possible only after the ovum (female gamete) is fertilized or united with the spermatozoon (male gamete). In chickens, fertilization is not a necessary preliminary to egg laying; a hen can lay continuously without being mated. However, except for parthenogenesis, fertilization of poultry eggs is a requisite to reproduction.

Table 3–1 is a summary of pertinent information relative to the reproduction of several species of poultry.

PARTHENOGENESIS

Parthenogenesis, the development of unfertilized eggs, occurs in both chickens and turkeys — especially the latter. By genetic selection, the incidence of parthenogenesis in turkeys has been increased to over 40% in eggs of experimental flocks. Most of the parthenogenetic embryos die, but about 1% of them complete development and hatch. All of those that hatch have been diploid males, but testes weights are low and only about 20% of them produce semen.

INCUBATION

Incubation is the act of bringing an egg to hatching. It may be either natural or artificial. In the beginning, incubation was by nature's way — by a setting hen hovering over eggs. Incubation independent of the hen is known as *artificial incubation.*

Fig. 3–2. White aluminum, fully automatic incubator being set with hatching eggs; suitable for incubation of chicken, turkey, and duck eggs. (Courtesy, Robbins Incubator Co., Denver, CO)

TABLE 3–1
REPRODUCTION OF POULTRY[1]

Species	Incubation Period	Eggs/Year	Fertility	Hatchability of Fertile Eggs	Age of Sexual Maturity
	(days)	(no.)	(%)	(%)	(mo)
Chicken:					
Light-type	21	240	90	90	5–6
Broiler-type	21	170	85	81	6
Turkey	28	105	90	90	7
Goose	28–32	15–60	85	65–70	24
Duck (Pekin)	28	110–175	85	65–70	7–8
Pheasant	24	40–60	90	85	8–10
Quail (Bobwhite)	23–24	150–200	85	80	8–10
Pigeon	18	12–15	90	85	6
Guinea fowl	27	40–60	90	80	10–12

[1]Average figures presented.

With some species, the natural incubation of eggs remains the most efficient. With chickens, turkeys, and ducks, however, artificial methods have been perfected which give results superior to those achieved by natural means.

One broiler breeder hen will produce about 150 offspring in a year, weighing about 600 lb at market time. Without artificial incubation, it would not be possible to capitalize on this reproductive capacity. Artificial incubation has freed the breeding hen from incubating eggs and enabled her to work full-time and year-long producing hatching eggs.

HATCHABILITY DETERMINING FACTORS

Hatchability refers to the percentage of eggs hatched. It may be reported as either (1) the percentage of fertile eggs hatched, or (2) the percentage of chicks hatched from all eggs placed in incubation. The percentage based on fertile eggs is more precise, provided adequate techniques are used to distinguish between early mortality of the embryo and infertility of the egg. However, the percentage of salable chicks hatched from all eggs set is often a useful measure, because it prevents errors in estimates of hatchability that arise from inaccurate determinations of fertility.

A number of factors influence the hatchability of eggs; among them, fertility, genetics, nutrition, diseases, egg selection, and handling of hatching eggs.

FERTILITY

Fertility refers to the capacity to reproduce. It is the factor which determines the number of offspring that may be obtained from a given number of eggs. If the male mates with a hen, the male sperm unites with the ova which is found on the yolk. Unless one or both birds are sterile, such an egg will be fertile and should hatch. Factors influencing fertility include the number of females mated to one male, the age of breeders, the length of time between mating and saving eggs for hatchability, and management practices.

It is estimated that 10%, or more, of the eggs set in the United States each year are infertile (see Table 3–1). Such eggs are not only a loss to the industry, but they occupy valuable incubator space and require time-consuming labor and handling.

Some eggs laid within 24 hours after mating may be fertile, but generally it requires 2 weeks after the flock is mated before satisfactory fertility may be obtained. The removal of males from a flock is followed by decline in fertility within 1 week for chickens and 2 weeks for turkeys. Few, if any, fertile eggs will be produced in chickens after 3 weeks and in turkeys after 6 weeks. If the males in the mating are changed, the eggs laid within a few days are fertilized by the new males and there is little overlapping in progeny.

GENETICS

Both research workers and hatchery operators are aware that hatchability is an inherited trait. Thus, the hatchery should select strains that possess high fertility and hatchability. Among the genetic factors affecting hatchability are the following:

■ **Inbreeding**—Close breeding, without rigid selection for hatchability, has been shown to lower hatchability in both chickens and turkeys.

■ **Crossbreeding and incrossbreeding**—Although the results of crossing pure breeds or incross breeds will depend upon the characters or genes carried by the parent stock, such crossing usually results in increased hatchability.

■ **Lethal and semilethal genes**—More than 30 lethal or semilethal genes are known in poultry. These genes cause the death of the developing embryo before the end of incubation or soon after hatching.

■ **Egg production**—Eggs laid by hens producing at a high rate are more fertile and possess higher hatchability than eggs laid by low producers.

■ **Age**—Hatchability tends to be highest during the first laying year for both chickens and turkeys.

NUTRITION

The egg must contain all the nutrients needed by the embryo at the time it is laid by the hen, since there is no further contact with the mother once the egg is completed. Therefore, breeder hens must be fed rations which supply adequate quantities of the nutrients needed for embryo development. The nutrients most likely to affect hatchability are the vitamins and trace minerals. For high hatchability and good development of young, breeder hens (chickens, turkeys, ducks, geese, and so forth) require greater amounts of vitamins A, D, E, B-12, riboflavin, pantothenic acid, niacin, and of the mineral manganese, than birds kept for commercial laying. Birds intended as breeders should be started on special breeder rations at least a month before hatching eggs are to be saved.

DISEASES

Hatching eggs from healthy flocks produce the most chicks.

Pullorum disease, caused by *S. pullorum*, once common, has been eradicated from most commercial flocks. Transmission of this disease is chiefly by the egg. Control is based on a regular testing program of breeding stock to assure freedom from infection.

Hatching eggs should be produced under MG (*Mycoplasma gallisepticum*)-negative and MS (*Mycoplasma synovial*)-negative programs.

Mycoplasma gallisepticum is the organism responsible for PPLO, a respiratory disease affecting the air sac of chickens. The organism is egg-transmitted; and the accepted method for control is by elimination of breeder flocks containing carrier birds, accomplished by the serum agglutination test.

MS, which is caused by *Mycoplasma synovial*, is also a respiratory disease, but the respiratory tract is seldom involved in sickness or death. The organisms locate in the synovial fluids of the hock and joints of the footpads, and in some cases the wing joints. These areas become swollen and inflamed. The disease is egg-transmitted. Two methods of control are available: (1) eradication in the breeders by the serum agglutination test, and (2) heat treatment (at 115°F) of the hatching eggs before they are placed in the incubator.

Certain other diseases affect hatchability, even though the disease organisms may not pass into the egg and affect the embryo directly. Newcastle disease and infectious bronchitis, for example, may affect egg shape and shell porosity. Eggs from hens affected by these diseases usually do not hatch as well.

EGG SELECTION

Certain physical characteristics of eggs are related to hatchability; among them, size, shape, shell quality, and interior quality.

■ **Size of egg**—The size of egg is related to hatchability in all species of poultry. Extremely large or very small eggs do not hatch well.

■ **Shape of egg**—Eggs that deviate greatly from normal shape do not hatch well. Since the shape of egg is inherited, it follows that hatchability can be increased by selection.

■ **Shell quality**—The quality of the shell is related to hatchability. Eggs possessing strong shells hatch better than eggs with thin shells. The kind of shell depends upon breeding, nutrition, and weather. Some strains or families produce eggs with thick, strong shells, whereas others lay eggs with thin, weak shells. The amount of calcium and vitamin D in the ration affects the shell. Also, eggs produced in hot weather have thinner shells than those produced when the weather is more moderate.

■ **Interior quality**—There is evidence that eggs which show quality when candled before incubation, i.e., show movement of yolk and well-centered position of the yolks, hatch more chicks than eggs showing a weak or low-quality condition.

HANDLING HATCHING EGGS

The following practices for handling eggs are recommended:

1. **Gather frequently.** Generally, hatching eggs are gathered more frequently than eggs intended for table use. When temperatures are normal, three to four gatherings a day will suffice. However, when temperatures are extremely hot or cold, hatching eggs should be gathered every hour. Frequent gathering reduces the likelihood of contaminating eggs from contact with nesting material and feces, and prevents chilling in winter and overheating in summer.

2. **Clean soiled eggs.** It is best to use only clean eggs for hatching purposes. Realistically, however, hatching eggs are usually valuable; hence, soiled eggs should be cleaned if possible. The following washing procedure is recommended:

    a. Keep the washing water temperature lower than the egg temperature.

    b. Use a sanitizer-detergent at the recommended level.

    c. Use a nonrecycle washer.

    d. Cool the eggs quickly after washing.

3. **Sanitize.** For effective sanitizing, eggs should be treated within 1 to 2 hours after they are gathered. Several decontaminants may be used; among them, the following:

    a. **Formaldehyde gas.** Triple strength (made by mixing 0.6 g of potassium permanganate with 1.2 cc of Formalin [37.5%] for each cubic foot of space in the room) is usually recommended, with the fumigating time in an airtight chamber not exceeding 20 minutes.

Fig. 3–3. Fumigating chamber for hatching eggs. (Courtesy, Indian River International, Nacogdoches, TX)

b. **Quaternary ammonia.** It is sprayed on eggs in a lukewarm water solution containing 200 ppm. Its chief advantage is that it may be sprayed on the eggs as they are picked up from the nest.

c. **Chlorine dioxide.** It may be used as a spray, or as a dip, on the eggs soon after gathering.

d. **Ozone (O_3).** When generated at 100 ppm, ozone (O_3) is an effective sanitizer.

It is noteworthy that chlorine dioxide, used as a spray or dip, gives results superior to fumigation with formaldehyde gas. Moreover, it is easier to use.

Any sanitizer should be administered as soon as possible after the eggs are laid. Following laying, the number of bacteria on the shell increases tremendously; and some penetrate the shell.

4. **Hold minimum time.** Hatchery eggs should be held for as short a period of time as possible, for hatchability decreases as the time of holding is increased. A rule of thumb is: for every day eggs are held or stored after 4 days, hatching time is delayed 30 minutes and hatchability is reduced 4%. At the most, they should not be held longer than 10 days. Commercial hatcheries usually set twice a week.

5. **Holding temperature.** When the egg is laid and its temperature drops, embryological development ceases. As soon as possible, hatching eggs should be cooled to a temperature of 65°F.

When it is necessary to hold hatching eggs longer than 7 days, it is recommended that they be warmed to 100°F for 1 to 5 hours early in the holding period.

6. **Humidity.** High humidity in a hatchery tends to (a) prevent evaporation and an enlargement of the egg's air cell, and (b) improve hatchability. A relative humidity of 75 to 80% is recommended.

7. **Position.** Hatching eggs are best held with the small ends up from the standpoint of increased hatch. However, most hatchery operations do not consider this holding position practical, because eggs are incubated with the large ends up; hence, eggs that are stored with the small ends up must be inverted.

8. **Turning.** If it is necessary to hold hatching eggs for more than 7 days, they should be turned. This prevents the yolk from sticking to the shell. Turning can be done by tipping the egg cases sharply. It is recommended that the cases be turned in this manner twice daily.

INCUBATOR OPERATION

There are seven factors of major importance in incubating eggs artificially: temperature, humidity, ventilation, position of eggs and turning, testing (candling), setters and hatchers, and incubation time. Of these, temperature is the most critical. In natural conditions, heat is furnished by the body of the setting hen. This temperature is usually slightly lower than that of the nonbroody hen's average temperature of 106°F.

Fig. 3–4. Trays of hatching eggs in a large "room-type" incubator. (Courtesy, Babcock Industries, Ithaca, NY)

1. **Temperature.** The fertile egg will resume development when it is placed in the incubator. But maintenance of the proper temperature is of prime importance for good hatchability. Depending on the type of incubation, optimum temperatures range from 99° to 103°F. In the usual forced-air machine, the temperature should be maintained at about 99.5°F.

Overheating is much more critical than underheating; it will speed up rate of development, cause abnormal embryos in the early stages, and lower the percentage of hatchability. Fig. 3–5 shows the effect of temperature on the percentage of fertile eggs.

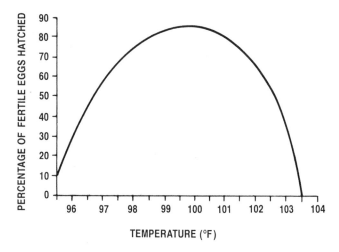

Fig. 3–5. The effect of incubation temperature on percentages of fertile eggs hatched; with relative humidity 60%, oxygen 21%, carbon dioxide below 0.5%. To convert to Celsius, see Appendix. (From The University of Connecticut College of Agriculture Extension Service publication entitled "Incubation and Embryology of the Chick," p. 3)

2. **Humidity.** Humidity is of great importance for normal development of the chicken embryo. During embryonic development (incubation), moisture is lost from the egg contents through the shell. This increases the size of the air cell, which after 19 days of incubation occupies about one-third of the egg.

Although a variation of 5 to 10% is acceptable, the relative humidity of the air within an incubator for the first 18 days should be about 60%. During the last 3 days, or the hatching period, it should be nearer 70%. Best results are obtained with turkey eggs when the humidity is 2 to 3% higher than these figures. Lower humidity than recommended causes excess evaporation of water, while high humidity prevents the evaporation of sufficient amounts of water from the egg. In both cases, hatchability is reduced.

3. **Ventilation.** As the embryo develops, it uses oxygen and gives off carbon dioxide. Thus, sufficient ventilation within the incubator is required to assure an adequate supply of oxygen and the proper removal of carbon dioxide.

The best hatching results are obtained with 21% oxygen in the air—the normal oxygen level in the atmosphere. The embryo will tolerate a carbon dioxide level of 0.5%, but it will die if this level reaches 5%.

Since the normal oxygen and CO_2 concentrations present in air seem to represent an optimum gaseous environment for incubating eggs, no special provision to control these gases is necessary other than to maintain adequate circulation of fresh air at the proper temperature and humidity.

4. **Position of eggs and turning.** The embryo head must occupy a position in the large end of the egg for proper hatching. Thus, the egg must be incubated large end up as gravity orients the embryo with its head uppermost. Somewhere between the fifteenth and sixteenth day, the head of the embryo is near the air cell.

Eggs should be turned from 3 to 5 times a day between the second and the eighteenth day. The purpose of this turning is to prevent the germ spot from migrating through the albumen and becoming fastened to the shell membrane. That is, turning the eggs prevents an adhesion between the chorion and the shell membrane.

Proper turning consists of rotating the eggs back and forth, not in one direction (a 30° to 45° angle is best).

A hen sitting on a nest of eggs turns them frequently, using (a) her body as she settles on the eggs and (b) her beak as she reaches under her body.

Modern incubators are equipped with turning devices that are able to rotate egg trays through a 90° angle. Also, they are equipped with timing mechanisms; and usually the eggs are turned every hour, which is probably more often than necessary.

5. **Testing (candling).** Under some circumstances, it may be advisable to check incubating eggs for fertility or embryo mortality. This is done by candling the eggs, using a 75-watt blue bulb. With suitable equipment, infertile eggs may be detected after 15 to 18 hours of incubation. The second test may be made 14 to 16 days after incubation, at which time the dead embryos may be removed.

Fig. 3–6. Candling duck eggs for fertility or embryo mortality. (Courtesy, Cherry Valley Farms, Ltd., Rothwell, England)

6. **Setters and hatchers.** In commercial hatcheries two separate incubators are used during the incubation process. The bulk of the incubation (usually through the nineteenth day) is done in *setters* while the end of the process is in *hatchers*. The main reason for separate setters and hatchers are:

a. The machines are kept in two separate rooms, thereby (1) isolating the down, egg debris, and microorganisms that accompany hatching from the eggs in the hatchery, and (2) permitting the hatchers to be cleaned, disinfected, and fumigated without disturbing the eggs in the setters.

b. The temperature of the hatchers is lowered to 98°F because there is some evidence that the hatch may be slightly improved by the lower temperature.

c. The hatchers are equipped with special chick-holding trays, not needed in setters.

7. **Incubation time.** The incubation period varies for different species of birds. Generally speaking, the larger the egg, the longer the incubation period. Also, the incubation period may vary with the temperature and humidity of the incubator. The normal incubation periods for several species of birds follow:

Common Name	Incubation Period (in days)
Ostrich	42
Swan	35–40
Muscovy duck	33–35
Goose	28–32
Turkey	28
Duck	28
Peafowl	28
Guinea	27
Pheasant	24
Quail	23–24
Chicken	21
Parakeet	19
Pigeon	18
Canary	13

EMBRYONIC DEVELOPMENT

Structural development begins soon after fertilization (before laying) and continues during incubation. Very early, the embryo becomes differentiated into three distinct cell layers out of which the various organs and systems of the body develop.

■ **Development of the egg before laying and soon thereafter** (see Figs. 3–7 and 3–8)—Soon after fertilization, cell division proceeds. The blastoderm spreads out over the yolk and becomes differentiated into the following two layers of cells by a process called gastrulation:

1. The *ectoderm*, the first layer, forms the skin, feathers, beak, claws, nervous system, lens and retina of the eye, and the lining of the mouth and vent.

2. The *entoderm*, the second layer, produces the linings of the digestive tract and the respiratory and secretory organs.

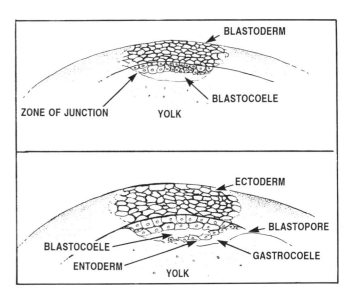

Fig. 3–7. Gastrulation in form with telolecithal egg containing a large amount of yolk—a bird egg. These schematic diagrams show the effect of yolk on gastrulation. In the chick, the great amount of yolk effectively prevents the formation of an open blastopore.

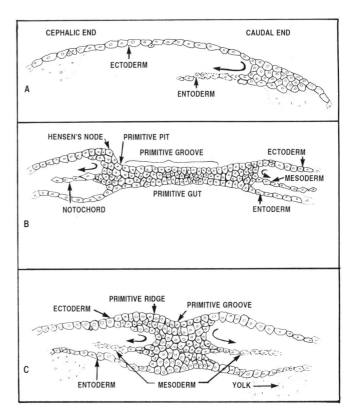

Fig. 3–8. Transverse sections of early development of chick embryo, showing differentiation into three distinct cell layers out of which the various organs and systems of the body develop.

A—Longitudinal section of the blastoderm of a preprimitive streak chick showing ectoderm formation.

B—Longitudinal section of the embryo following approximately 17 hours of incubation, showing all three cell layers—ectoderm, entoderm, and mesoderm.

C—Cross section of the embryo in the primitive streak stage, following approximately 36 hours of incubation, showing turning in of the cells at the primitive groove to enter the mesodermal layers.

■ **Development during incubation**—Soon after incubation, a third germ layer, the mesoderm, originates. It gives rise to the bones, muscles, blood, and reproductive and excretory organs.

The embryologist is little interested in the development of the chick beyond the fourth day (see Fig. 3–9), because from there on only growth takes place. But the producer continues to be very much interested because the objective is the hatching of a vigorous live chick; hence, it is important to know more about the development which occurs throughout the entire period of incubation.

Figs. 3–7 and 3–8 show the development of the three cell layers, which give rise to all the organs, systems of the body. Fig. 3–9 shows successive changes in the position of the chick embryo and its embryonic membranes beginning with the fifth day of incubation and extending to just before

hatching. Table 3–2 summarizes the important embryonic developments from before laying until hatching.

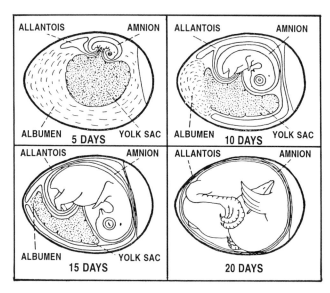

Fig. 3–9. Successive changes in the position of the chick embryo and its embryonic membranes. (From A. L. Romanoff, Cornell Rural School Leaflet, September 1939)

Fig. 3–10. Newly hatched chick. (Courtesy, Euribrid B. V., Boxmeer, Holland)

TABLE 3–2
IMPORTANT EVENTS IN EMBRYONIC DEVELOPMENT

Stage or Period	What Takes Place
Before egg laying	Fertilization, division, and growth of living cells, segregation of cells into groups of special function.
Between laying and incubation	No growth; stage of inactive embryonic life.
During incubation:	
First day:	
16 hours	First sign of resemblance to a chick embryo.
18 hours	Appearance of alimentary tract.
20 hours	Appearance of vertebral column.
21 hours	Beginning of formation of nervous system.
22 hours	Beginning of formation of head.
23 hours	Appearance of blood islands-vitelline circulation.
24 hours	Beginning of formation of eye.
Second day:	
25 hours	Beginning of formation of heart.
35 hours	Beginning of formation of ear.
42 hours	Heart begins to beat.
Third day:	
50 hours	Beginning of formation of amnion.
60 hours	Beginning of formation of nasal structure.
62 hours	Beginning of formation of legs.
64 hours	Beginning of formation of wings.
70 hours	Beginning of formation of allantois.
Fourth day	Beginning of formation of tongue.
Fifth day	Beginning of formation of reproductive organs and differentiation of sex.
Sixth day	Beginning of formation of beak and egg-tooth.
Eighth day	Beginning of formation of feathers.
Tenth day	Beginning of hardening of beak.
Thirteenth day	Appearance of scales and claws.
Fourteenth day	Embryo turns its head toward the blunt end of egg.
Sixteenth day	Scales, claws, and beak becoming firm and horny.
Seventeenth day	Beak turns toward air cell.
Nineteenth day	Yolk sac begins to enter body cavity.
Twentieth day	Yolk sac completely drawn into body cavity; embryo occupies practically all the space within the egg except the air cell.
Twenty-first day	Hatching of chick.

EMBRYO COMMUNICATION

In some species of birds, eggs in a clutch tend to hatch at about the same time despite the fact that some eggs have been laid earlier and incubated longer than others. Vince, of England, using Bobwhite quail and the Coturnix quail, showed that this phenomenon is due to signals transmitted from egg to egg.[1] Apparently these signals consist of vibrations or clicks made by movement of the embryos. This communication synchronizes hatching time, a feature that might be of considerable survival value in the wild. Slow clicking accelerates development, while fast clicking retards it. In quail, with an incubation period of about 16 to 18 days, embryo development can be accelerated by a full day and the young will still hatch fully mature.

Acceleration or retardation of hatching can be caused by use of artificial clicks. In chickens, clicking has been shown to advance the time of hatch, but not to retard it.

Thus, it has been demonstrated that there is communication between embryos—that they "talk" to each other, and that nervous control of development does take place.

YOLK MAKES FOR A GOOD START IN LIFE

Newly-hatched chicks can be shipped long distances without food or water. The yolk is largely unused by the embryo and is drawn into the body of the chick on the nineteenth day, just before it hatches. It is highly nourishing and provides proteins, fats, vitamins, minerals, and water for the first several hours of the chick's life. The yolk is gradually used up during the first 10 days of life of the chick.

HATCHERIES

Few phases of the poultry industry have undergone more changes than has the hatchery business. In the early 1900s, the "old cluck" was forced to yield her pleasant job of spending 3 weeks on the nest "expecting." The incubator in the small hatchery gradually took over the job. As hatcheries developed, they grew in size and improved efficiency, and the old setting hen was practically eliminated.

Perhaps it is fair to say that hatcheries have exerted more influence on improving the general level of the poultry industry than any other segment of the business. This has been achieved through their own breeding programs or through the breeding programs of the poultry breeders with whom they have a franchise agreement.

CHANGES IN HATCHERIES

The major changes in the hatching industry in recent years follow:

1. **Fewer and larger hatcheries.** Back in 1934, there were 11,405 chicken hatcheries in the United States; in 1989, there were 376—only 3.3% remained. But, in this same period of time, the average size of hatchery increased nearly ninefold, so that there was a substantial excess of capacity in the hatchery industry. The following figures from the U.S. Department of Agriculture point up the trend toward fewer and larger chicken hatcheries:

Year	Number of Hatcheries	Total Egg Capacity	Average Egg Capacity per Hatchery
1934	11,405	276,287,000	24,000
1953	8,233	616,976,000	80,000
1965	2,365	471,318,000	199,000
1975	797	416,000,000	521,000
1981	538	466,096,000	466,000
1989	376	524,615,000	1,395,000

The increased size of hatcheries becomes obvious when it is realized that on January 1, 1989, U.S. chick hatcheries with incubators capable of holding 500,000 or more eggs accounted for 60.4% of the nation's capacity. A similar trend has occurred in turkey poult hatchery capacity. On January 1, 1989, 38 of the 82 turkey hatcheries had a capacity of 500,000 or more eggs, and these 38 hatcheries accounted for 46.3% of the poult hatching capacity in the United States.

2. **Year-round business.** Formerly, most hatcheries operated in the spring of the year only. In 1937–38, slightly under 2.0 hatches of chicks were taken off per year. In 1989, more than 11 chicken hatches and more than 6 turkey hatches were obtained during the year. (This is known as turnover; it is obtained by dividing hatch by capacity.) The primary reasons for extending the hatching season to a more nearly year-round business are (a) less seasonality in the production of eggs and turkeys, and (b) the development of a year-round commercial broiler industry.

3. **Geographical shift.** Hatcheries have followed production, geographically speaking. Thus, with the shift of the broiler industry to the South and South Atlantic states, the total output of hatcheries has also increased in this area.

4. **Franchise agreements have come in.** In 1937–38, franchising was practically unknown in the hatchery business. Today, many hatcheries have franchise agreements with breeders who use advanced methods to produce superior strains of birds. Under these agreements, the breeders furnish the eggs and the hatcheries agree to sell breeders' strain of birds in specified areas.

5. **Less custom-hatching.** Today, very little custom-hatching is done; only a limited amount of custom hatching of turkey poults remains.

6. **More integration.** Many hatcheries have become a part of broader businesses which involve poultry growing and processing, or sale of supplies. This may involve the selling of poultry feeds, medicines, equipment, farm supplies, and farm products.

7. **Fewer started chicks sold.** With the growth of larger and more specialized poultry farms and their improved brooding technology, fewer and fewer started chicks (chicks 2 to 4 weeks of age) are being sold.

8. **Business methods.** Today, hatchery business is big business, and it must be treated as such. A complete set of records must be kept in every successful hatchery. Such records should show all financial transactions. Additionally, there should be a record of inquiries received, the name and address of customers, the name and grade of chicks sold to each customer, the date of each sale, and a record of all hatching eggs purchased.

---

[1]Vince, M. A., "Artificial Acceleration of Hatching in Quail Embryos, *Animal Behavior*, Vol. 14, 1966, pp. 389–394.

Further information on the business aspects is presented in Chapter 17, Business Aspects of Poultry Production; hence, the reader is referred thereto.

9. **Marketing chicks.** The hatchery has three common channels through which chicks are sold: (a) local sales at the hatchery; (b) mail orders direct to customers; and (c) wholesale. Local chick sales are the most satisfactory and the least expensive. In every community, there is a demand for high-quality chicks that can be supplied by the local hatchery. The wholesale method is the least satisfactory and most apt to lower chick quality. It alleviates contact between the producer and the customer, an arrangement which is important from the standpoint of improvement of quality. The mail order business is dependent upon advertising.

10. **Advertising.** Among the advertising techniques used by successful hatcheries are the following: (a) reputation chicks, which make for satisfied customers and bring new customers; (b) a neatly kept plant, including well-kept and well-painted buildings, neat premises, and cleanliness and orderliness within the hatchery; (c) advertisements on signs, in newspapers and magazines, on the radio and television; (d) literature and catalogs; and (e) news releases.

Fig. 3–11. Eggs from breeder farms being placed in racks at the hatchery, with a code identifying their farm of origin. This rack holds 6,480 eggs. (Courtesy, Delmarva Poultry Industry, Inc., Georgetown, DE)

HATCHERY MANAGER

The manager can make or break any hatchery. Good managers are scarce and hard to come by, and the chain hatchery operator's greatest concern. A good manager can build a profitable hatchery enterprise even where the conditions are unfavorable, but a poor manager can ruin the best hatchery. The traits of a good manager are: knowledge of the poultry business as well as the hatchery business, business ability, a strong personality, leadership, promptness and dependability, a good organizer, and willingness to work. Additional traits of a good manager are presented in Chapter 8, Poultry Management; hence, the reader is referred thereto.

HATCHERY BUILDING

Certain general requirements of hatchery buildings should always be considered. It is with these that the ensuing discussion will deal. Once buildings are constructed, there is a practical limit to the changes that can be made in remodeling. Consequently, it is most important that very careful consideration be given to the following requisites:

1. **Egg-chick flow through hatchery.** Hatcheries should be designed so that the hatching eggs may be taken in at one end of the building and the chicks removed from the other. A knowledgeable architect should be employed to draw the details and write the specifications.

2. **Separate rooms.** Every commercial hatchery should have separate rooms for the office, fumigation, egg grading and traying, egg holding (see Fig. 3–11), incubators, display chicks, display equipment and feed, chick sorting and boxing room, and a room for chick boxes and other supplies.

3. **Construction.** The building which houses the hatchery should be constructed of adequately durable material which will withstand outside temperature fluctuations reasonably well. A well-drained concrete floor is conducive to sanitation in the hatchery. Also, the building should be kept in good repair, be well painted, and present an attractive appearance.

4. **Lighting.** The hatchery should be well lighted. Good lighting makes for visibility and convenience of the caretakers, and makes a good impression on customers.

5. **Ventilation.** Ventilation refers to the changing of air — the replacement of foul air with fresh air. A well-ventilated hatchery makes possible proper incubator ventilation. Forced air circulation is necessary in larger hatcheries to remove foul air and harmful humidity, without excess heat loss or creation of drafts.

6. **Heating and cooling.** It is common practice to install heating and cooling units in the same ventilating system. During cold weather, the heating unit operates; during hot weather, the cooling functions.

7. **Humidity.** Generally additional humidity should be supplied. The recommended humidity for the various rooms follows: egg-holding room, 75 to 80%; incubator and hatching rooms, 50%; and chick-holding room, 60%.

8. **Battery brooders.** Most hatcheries use battery brooders for holding surplus chicks which accumulate as a result of weather conditions, order cancellations, or oversettings.

HATCHERY EQUIPMENT

Fig. 3–12. Racks containing eggs are placed in incubators for about 19 days. Fans help control temperature and humidity to provide maximum hatchability. The eggs are turned regularly to insure proper incubation. (Courtesy, Delmarva Poultry Industry, Inc., Georgetown, DE)

Fig. 3–13. At the end of about 19 days, the eggs are transferred from the incubators to hatching trays. There the chicks crack their shells and emerge. (Courtesy, Delmarva Poultry Industry, Inc., Georgetown, DE)

Many different kinds of equipment are necessary in a hatchery, with the size and number of these pieces of equipment determined by many factors. Generally, most modern hatcheries are equipped with the following:

1. Water softeners and heaters.

2. Egg handling equipment, including carts and conveyors.

3. Egg grading and washing equipment, including vacuum egg lifts, automatic egg graders, and egg washers.

4. A standby electric plant for use in case of power failure.

5. Modern incubators featuring forced-draft air circulation, small amount of floor space in relation to egg capacity, durable cabinet material, automatic temperature control, automatic humidifiers, carbon dioxide recorders, elimination of egg traying, mechanical egg turners, separate hatchers, efficient cooling, shortened egg transfer from setting trays to hatching trays, and various emergency devices.

6. Other hatchery equipment, including such items as egg candlers, test thermometers, chick box racks, chick graders, sexing equipment, tray washers, dubbing shears, debeaking equipment, detoers, automatic syringes, Marek's vaccination equipment, *Mycoplasma gallisepticum* test equipment, and a waste disposal system.

HATCHERY SANITATION

(Also see Chapter 10, Poultry Health, Disease Prevention, and Parasite Control, section headed "Hatchery Management and Sanitation.")

In order to reduce the possibility of disease, as well as make a good impression upon customers and potential customers, it is important that certain sanitation practices be rigidly adhered to in a hatchery; among them, isolation and disease-free chicks, clean eggs, sanitizing, separate rooms, and proper waste disposal.

1. **Isolation and disease-free chicks.** The hatchery should take every precaution to avoid bringing in infection. Generally speaking, this calls for banning or controlling visitors, showers and clothing changes for employees, and obtaining eggs from disease-free stock. In particular, every effort should be made to avoid MG (*Mycoplasma gallisepticum*) and MS (*Mycoplasma synovial*).

2. **Clean eggs.** Only clean eggs should be used for hatching, and they should be collected and stored in clean equipment and containers. Although hatching eggs may be washed, there is always the hazard of introducing microorganisms into them.

3. **Sanitizing eggs.** All hatching eggs should be sanitized as soon as possible after collection. (See earlier section on "Handling Hatching Eggs," Point 3, Sanitize.)

Fumigation of eggs and incubators is an essential part of a hatchery sanitation program. When improperly done, fumigation can be a hazard; hence, it should always be done by, or under the supervision of, an experienced person.

Routine preincubation fumigation of hatching eggs on the farm is highly recommended for eliminating Salmonella infection from poultry flocks. Each egg entering the hatchery should have been subjected to preincubation fumigation as soon as possible after its collection from the nest.

High levels of formaldehyde gas will destroy Salmonella organisms on shell surfaces if used as soon as possible after the eggs are laid. An inexpensive cabinet for enclosing the eggs is required. Fans circulate the gas during the fumigation process and then exhaust it from the cabinet. Eggs for fumigation should be placed on racks so that the gas can reach the entire surface. Plastic trays used for washing market eggs are ideal for this purpose.

A high level of formaldehyde gas is provided by mixing 1.2 cubic centimeters (cc) of formalin (37% formaldehyde) with 0.6 g of potassium permanganate ($KMnO_4$) for each cu ft of space in the cabinet. An earthenware, galvanized, or enamelware container having a capacity at least 10 times the volume of the total ingredients should be used for mixing the chemicals. The gas should be circulated within the enclosure for 20 minutes, then expelled to the outside.

Humidity for this method of preincubation fumigation is not critical; but temperature should be maintained at approximately 70°F. Extra humidity may be provided in dry weather. Eggs should be set as soon as possible after fumigation and extra care taken to ensure that they are not exposed to new sources of contamination.

Eggs should be routinely refumigated after transfer to the hatchery to destroy organisms that may have been introduced as a result of handling. Recommendations for loaded-incubator fumigation vary widely, depending upon the make of the machine. Therefore the method, concentration, and duration of fumigation should be in accordance with the manufacturer's instructions.

Empty hatchers should be thoroughly disinfected and fumigated prior to each transfer of eggs from setters.

4. **Separate rooms.** Hatcheries should be so designed that there are separate rooms for egg receiving, incubation, hatching, chick-holding, and waste disposal.

5. **Proper waste disposal.** Waste disposal may be a problem and a major source of infection in the hatchery unless it is properly handled. Incineration is an effective means of disposal of hatchery waste from the standpoint of sanitation and disease control, but it is an unprofitable method. The larger hatcheries now process poultry waste into livestock feeds, known as hatchery by-product. It consists of infertile eggs, eggs with dead embryos, and unsalable sexed male chicks. Hatchery by-product is high in protein and calcium.

SEXING CHICKS

Today, chick sexing is considered a normal hatchery service and is routinely offered. Leghorn cockerels are not worth growing for broilers because of their small size and feed requirements per pound of gain. Hence, they are usually destroyed and incorporated in the hatchery by-product. Chicks are sexed by the following methods:

1. **Sex-linked genes.** Sex-linked genes for (a) rate of feathering or (b) color pattern, are being used by breeders to facilitate sexing at day-old. Each calls for an examination based on rate of feathering or down coloring as a result of special parental matings. These two methods of sexing visually are:

a. **Feather sexing.** Day-old chicks may be sex separated accurately by examining the relative length of the primary and covert feathers of the wing, with the females carrying genes for fast feathering and the males carrying genes for slow feathering. It is noteworthy, however, that slow-feathering males usually do not feather well in the brooder house, particularly during hot weather, often resulting in slower growth and increased cannibalism.

b. **Color sexing.** Usually, gold and silver genes are used for color sexing at day-old; gold, buff, or red chicks are females, and white or light yellow chicks are males. Chicks can be sexed rapidly and accurately in this manner. Where this method of sexing is used for broiler production, it is noteworthy that the color differences are also evident in the mature broiler; females are gold or brown, males are white. But the pinfeathers are usually white; hence, there is no processing problem.

2. **Cloacal identification.** Examination of the cloacal wall is widely used by hatcheries to sex newly hatched chicks.

3. **Examination of the rudimentary male organs.** Although it takes considerable skill to be accurate, the rudimentary copulatory organ, or male process, can be identified at hatching and used to identify sexes. Because so many chick sexors of Japanese origin use this method, it is commonly referred to as the *Japanese method*.

By using a special instrument, it is possible to see the testes of day-old chicks through the intestinal wall.

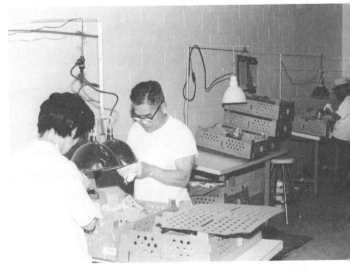

Fig. 3–14. Sexing by skilled sexors, using the *Japanese method*. (Courtesy, Hubbard Farms, Walpole, NH)

DEBEAKING AND DUBBING

Most large hatcheries now offer debeaking and dubbing services.

Debeaking of day-old broiler chicks or replacement pullets by hatcheries is now a rather common practice for preventing cannibalism. When beaks are properly cut at hatching, they will not usually regrow to the point where picking is a serious problem during the broiler-growing period. However, replacement pullets, if debeaked at hatching, will usually need to be debeaked again.

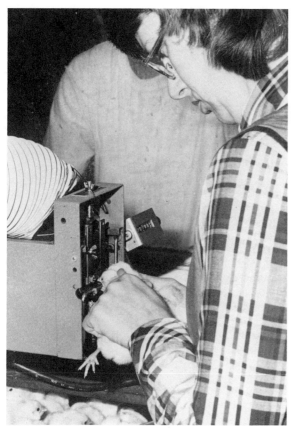

Fig. 3–15. Debeaking chicks. (Courtesy, Foster Farms, Livingston, CA)

Dubbing is the term commonly used to designate the removal of a part of the comb of day-old chicks. The large combs of Single-Comb White Leghorn hens are sometimes injured, particularly when the birds are housed in cages. Such injuries can be prevented by dubbing. This procedure is relatively harmless when performed at an early age; and the comb that develops will be smaller and less likely to be injured when the hen is an adult.

VACCINATION

The immunity passed to chicks from the egg protects them for the first few days of life and often makes vaccination ineffective. For this reason, most vaccines are administered after chicks have left the hatchery. There is one exception: Today, most hatcheries vaccinate day-old chicks for Marek's disease, especially if the chicks are intended as replacement pullets.

DELIVERING CHICKS

Provided they are not overheated or chilled during shipment, chicks may be transported for up to 3 days time without feed or water (except what comes from the yolk), and still be expected to arrive in good condition. With broiler chicks, however, most hatcheries attempt to deliver the chicks as expeditiously as possible in order to get them on feed and water immediately after they hatch, as this practice has been shown to reduce early mortality and make for a better start of the broiler chick.

BROODING

Brooding refers to the care of young poultry from the time of hatching or from the time received from the hatchery, until they are about 8 to 10 weeks old.

Wherever chicks are raised—whether for broilers or replacement layers—artificial brooding of some kind is necessary. No phase of the poultry business is so important as brooding—it's the part that makes for a proper start in life.

Good brooding and rearing practices bring out the good qualities inherited by chicks, whereas, poor brooding practices can ruin chicks of the best breeding. Success in breeding pullets for layers is indicated by a uniform flock of fast-growing, well-feathered, healthy pullets. Success in brooding broilers results in rapid and uniform growth, good efficiency of feed utilization, fast feathering, and a low rate of mortality.

Brooding is still largely an art, although information is gradually being accumulated which will eventually make artificial brooding a science.

Requisite to successful brooding is that the producer start with healthy chicks secured from a reliable hatchery, purchase chicks bred for the specific job intended, and start chicks at the right time.

BROODING REQUIREMENTS

Certain environmental conditions are requisite to successful brooding, especially temperature control, ventilation, moisture control, and adequate space.

TEMPERATURE CONTROL

There are optimum temperatures for chicks of different ages. When chicks are a day old, the temperature should be from 90° to 95°F at the level of the chicks on the floor. The temperature is usually lowered about 5°F each week until a temperature of 70° to 75°F is reached, or until the chicks are 6 to 8 weeks old and fully feathered. A reliable thermometer will enable the caretaker to provide comfortable temperatures. If in doubt, however, the chicks will tell you (see Fig. 3–16). If the chicks huddle or crowd together and cheep, they are too cold. If they move away from the source of heat or from under the brooder, or if they pant and hold their wings away from their bodies, they are too warm. Chilling chicks may result in their piling and smothering. Too cool temperatures also cause diarrhea and increase susceptibility to infectious disease. Too high temperatures result in reduced appetite and retarded growth.

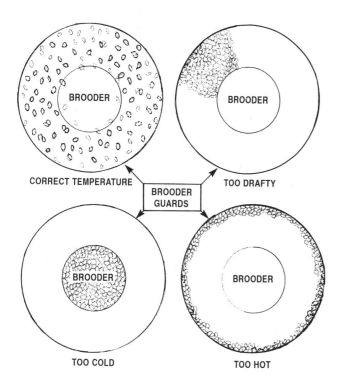

Fig. 3–16. The chicks will tell you when they are comfortable.

VENTILATION

Ventilation is required to provide fresh air for the chicks — to remove carbon dioxide and carbon monoxide, to remove ammonia and other fumes, and to aid in keeping the litter dry. Gas, oil, and other flame-type brooders that burn the oxygen out of the air require more ventilation than other types of brooders. A good ventilating system provides plenty of fresh air without drafts. Chicks huddling in certain areas or spots may indicate floor drafts. A strong ammonial odor means that there is not enough air movement. If the floor litter is relatively dry and the air in the house has little or no ammonia or other odors, the ventilation is adequate.

MOISTURE CONTROL

A fairly humid environment, around 50 to 60% relative humidity, is desirable and conducive to good feathering. A very dry atmosphere will cause poor feathering.

Too much moisture in the brooder house may also cause trouble. If the litter or walls and ceiling become wet, there is too much moisture in the brooder house and efforts should be made to reduce it. The litter must be kept relatively dry. By increasing ventilation, the condition of the litter can be improved and kept dry. If moisture condenses on the ceiling and walls until it drips, better insulation and ventilation are needed. Wet litter may lead to an outbreak of diseases.

SPACE REQUIREMENTS

Each broiler chick should have ¾ to 1 sq ft of floor space. Each Leghorn pullet chick to be reared to laying age should have from 1½ to 2 sq ft of space, and heavy-breed layer replacements should have 2½ to 3 sq ft.

The amount of space required under the brooder (or hover) will vary from 7 to 10 sq in. per chick. More space is required under electric brooders than with those having other types of heat, and more brooder space may be required during cold than during mild weather. As a general rule, no more than 500 chicks should be placed under each individual brooder or hover, regardless of size.

BROODER HOUSES AND EQUIPMENT

In modern commercial production, chicks are reared in specially designed and constructed brooder houses, with 30,000 to 50,000 or more birds brooded as a unit and cared for by one person. Only farm flock owners, part-time producers, and poultry fanciers (those who breed poultry for exhibition purposes) brood chicks (1) in buildings that are used for other purposes part of the year (such as part of the laying house, or the garage, or one end of the machine shop), or (2) in movable colony houses, built on skids.

Presently, there is little data to define the one best system of housing. Many systems are being tried, with consideration given to floor, battery, and cage rearing; degree of environmental control; degree of isolation; choice of construction material; density of birds; and choice of equipment. Also, some variations are being made between the brooding of layer replacement pullets and broilers.

Concrete floors in brooder houses generally give best results because they are easily cleaned and can be sanitized. Adequate insulation for the area should be used.

Presently, two basic systems are being used for brooding replacement pullets: (1) a two-step system involving conventional floor brooding until the chicks are 7 to 8 weeks of age, at which time they are shifted to cages; and (2) the use of various systems of cage rearing from day-old to ready-to-lay. The second system is meeting with varied success. Total management is more critical and demanding when pullets are reared in cages.

To date, fewer broilers than replacement pullets, percentagewise, are being raised in cages, primarily because of downgrading at slaughter as a result of breast blisters, wing fractures, and leg damage. Also, in comparison with litter housing, batteries make for more feather and vent picking. However, with cages, there are fewer parasites, no litter cost, stocking density can be increased — the number of birds in a given area can be doubled, and feed cost may be lowered.

Scientists and manufacturers are making great effort to develop a satisfactory broiler production cage.

The type of equipment will vary between floor brooding and cage rearing. Further, the requirements for brooders, waterers, and feeders are determined by the number of birds, the physical shape of the building, and the type and amount of insulation. Forty foot wide houses require four lines of mechanical pan feeders, but additional hanging feeders must be supplemented if trough feeders are used. Thirty-two-foot houses with maximum bird density can be adequately fed with four lines of mechanical trough feeders or three lines of pan feeders.

Modern brooding systems are heated by oil, natural gas, bottled gas, or electricity. There is more risk of fire with open-flame brooders and they must be kept in good operating condition. Catalytic and electric brooders present less fire hazard. The risk of fire with hot water systems is reduced since the furnace unit is located outside the brooding area. The brooder operator should choose the system which (1) utilizes the least costly fuel, and (2) minimizes the fire hazard.

The brooding operations of replacement pullets, broilers, and turkeys are usually distinct and separate; that is, they are rarely conducted by the same operator. Thus, each of them is presented separately in the discussion that follows, even though there are similarities.

BROODING REPLACEMENT PULLETS

Almost all pullets are now raised in confinement. Hence, the discussion that follows will pertain thereto.

■ **Floor**—Replacement pullets may be grown on solid floors or on wire floors. In addition to being convenient, wire floors alleviate later problems of adjustment to wire floors of laying cages.

Successful pullet rearing requires adequate floor space. Overcrowding may result in a breakdown of the social order, feather picking, cannibalism, high mortality, and excessive culls. Optimum insulation and forced air circulation can reduce floor space requirements by as much as 50% over uninsulated conventional housing. Suggested floor space requirements are shown in Table 3–3.

■ **Brooder**—First week brooding temperatures should be at least 90° to 95°F, with a 5°F weekly reduction until 70°F is reached, or until heat is no longer required. Chilling or overheating young birds should be avoided. Cage rearing requires that the entire house be uniformly comfortable. Any one of several types of commercial brooders will work satisfactorily if properly used. Gas, oil, or electric (hover-type) brooders, or infrared, are most common. Each of the brooding systems offers certain advantages that make it desirable. The availability and cost of fuel must be considered also in choosing the type of brooder to use. Chicks should not be crowded during the brooding period. Where hovers are used, 8 to 10 sq in. of hover space per chick should be provided.

■ **Litter**—Wood shavings, straw, rice hulls, or other commercial litter may be used. The litter selection should be on the basis of what is most convenient and economical.

■ **Feeders**—It is important that adequate feeder space be provided. The amount of space will vary according to age, cage vs floor feeding, and trough vs pan feeders. Table 3–4 (next page) gives recommended feeder space per 1,000 birds.

Fig. 3–17. Chicks brooding in a typical broiler house. (Courtesy, P. R. Ferket, North Carolina State University, Raleigh)

TABLE 3–3
MINIMUM FLOOR SPACE REQUIREMENTS FOR PULLETS[1]

Age of Pullets	Space/Bird							
	Controlled Environment				Uncontrolled Environment			
	In Cages		On Floor		In Cages		On Floor	
	(sq ft)	(sq m)	(sq ft)	(sq m)	(sq ft)	(sq m)	(sq ft)	(sq m)
1 day to 7 weeks	0.20	0.018	0.50	0.045	0.30	0.027	0.75	0.067
7 to 11 weeks	0.25	0.022	0.70	0.063	0.40	0.036	1.00	0.090
11 to 22 weeks	0.40	0.036	1.20	0.108	0.55	0.049	1.80	0.162

[1]Adapted by the author from *Production of Commercial Layer Replacement Pullets*, University of Georgia College of Agriculture-Athens, p. 3, Table I.

TABLE 3–4
RECOMMENDED FEEDER SPACE FOR PULLETS[1]

Age	Feeder Space/1,000 Birds[2]		
	Cages	Floor	
	Trough Feeders	Trough Feeders	Pan Feeders
1 day to 10 weeks	92 ft (1.1"/bird)	10 feeder lids	10 feeder lids
10 days to 10 weeks	92 ft (1.1"/bird)	78 ft (.94"/bird)	22 (46 bird/pan)
10 to 22 weeks	167 ft (2.0"/bird)	117 ft (1.4"/bird)	31 (32 bird/pan)

[1]Adapted by the author from *Production of Commercial Layer Replacement Pullets*, University of Georgia College of Agriculture-Athens, p. 5, Table III.

[2]To convert feet to meters, multiply by 0.305.

■ **Waterers** — The value of adequate water space is evidenced by the fact that water is consumed at about twice the rate of feed. Even sick "off feed" birds will continue to drink, provided that water is easily accessible to them. Inadequate water space can cause or accelerate health problems, particularly for growing birds. Recommended water space is given in Table 3–5.

■ **Management schedule** — The successful brooding of replacement pullets necessitates a well-programmed and well-executed management schedule. The following management schedule is divided into two parts: (1) before the chicks arrive, and (2) after the chicks arrive.

1. **Before chicks arrive:**

a. **Clean thoroughly.** Thoroughly clean the brooder house — floor, walls, and overhead.

b. **Disinfect.** Disinfect the brooder house, using one of the commercial disinfectant materials according to manufacturer's directions. For the disinfectant to be most effective, the brooder house must first be thoroughly cleaned.

c. **Set up and make a trial run with brooders.** Set up and start the brooders 2 or 3 days before the chicks arrive to make sure they are properly adjusted and working satisfactorily. This will reveal any missing or malfunctioning parts, permit temperature adjustment, and help remove moisture from the house and litter.

d. **Clean all equipment.** All equipment, such as feeders and waterers, should be thoroughly cleaned and disinfected before the chicks arrive.

e. **Provide clean, fresh litter.** Put clean, fresh litter 3 to 6 in. deep on the floor before the brooders are set up.

f. **Install brooder guards.** If brooder guards are used, one which is 14 to 18 in. high will be sufficient. Allow about 36 in. between the outer edge of the brooder and the brooder guard. This guard will keep the chicks confined to the brooder area and result in their eating and drinking faster. A brooder guard made of poultry netting is satisfactory for brooding during the summer months. The brooder guard should be removed by the end of the first week.

2. **After chicks arrive:**

a. **Brooder temperature.** Follow the manufacturer's recommendations.

b. **Vaccination.** A vaccination program which is appropriate for the area should be followed. Variations and conditions in management systems often dictate the

TABLE 3–5
RECOMMENDED WATERER SPACE FOR PULLETS[1]

Age	Waterer Space/1,000 Birds			
	Cages			Floor
	Trough Waterers		Cup	Trough Waterers
	(ft)	(m)		
1 to 10 days	—	—	45 cups	15 1-gal fountains
10 days to 10 weeks	40	12.2	55 cups	20 ft (6.1 m) trough
10 to 22 weeks	80	24.4	84 cups	40 ft (12.2 m) trough

[1]Adapted by the author from *Production of Commercial Layer Replacement Pullets*, University of Georgia College of Agriculture-Athens, p. 5, Table III.

necessity of variation. For guide purposes, a pullet vaccination program is presented in Table 3–6.

TABLE 3–6
OUTLINE OF SUGGESTED VACCINATION PROGRAM[1]

Vaccination	Age	Method of Administration
Marek's disease	1 day	Subcutaneous
Newcastle[2]	14 days	Spray, eye drop, drinking water
	4 weeks	Spray, eye drop, drinking water
	14 weeks	Drinking water, internasal, spray
Bronchitis		(given along with Newcastle)
Fowl pox[3]	1 day	Wing web
	8–10 weeks	Wing web
Epidemic tremor	12–18 weeks	Drinking water, inoculate 10% of birds
Laryngotracheitis	12–14 weeks	Modified eye drop
Fowl cholera	12–18 weeks (optional)	Subcutaneous

[1]*Production of Commercial Layer Replacement Pullets*, University of Georgia College of Agriculture-Athens, p. 7, Table IV.

[2]Potential velogenic viscerotrophic Newcastle challenge would require modification of the above program.

[3]In endemic areas, pigeon or modified fowl pox may be used.

c. **Feeding.** Since growth rate is not critical in White Leghorn pullets, the dietary energy level of the ration is usually lower than that of broiler diets. However, the young bird eats mainly to satisfy its energy requirements; hence, the intake of other nutrients is largely dependent upon the energy level of the feed. The importance of nutrient balance is more apparent when it is noted that the young bird consumes about 10 g of feed a day during the first week, and no more than 70 g a day by the twenty-second week (based on a ration containing 1,360 calories of metabolizable energy per pound).

The feeding objective in commercial pullet production is to manage nutrient intake so that the birds develop their full genetic potential. This is primarily accomplished by avoiding excessive body fat and early maturity. Obesity may reduce the total number of eggs, while early maturity tends to cause more small eggs and thus fewer large market eggs.

Feeding during the growing and development stage (8 to 22 weeks) usually follows either one or two basic programs:

(1) Full feed with a controlled lighting program.

(2) Controlled feed intake, using a high-energy feed and appropriate lighting control.

Further information on the nutritive requirements and feeding programs for replacement pullets is contained in the following two chapters of this book:

Chapter 6, Fundamentals of Poultry Nutrition.

Chapter 7, Poultry Feeding Standards, Ration Formulation, and Feeding Programs.

d. **Water.** Plenty of clean, fresh water should be provided at all times. Water should be distributed so that the chicks can drink conveniently.

e. **Light.** Some light should be left burning all night, except for 1 hour of darkness, during the first few days as this may be helpful in keeping birds from crowding or piling.

■ **Roosts**—If replacement pullets are to be housed on floors as adults, low roosts should be installed when the chicks are about 4 weeks of age. About 4 to 5 in. of roosting space per chick is recommended. If the replacement pullets are to be placed in cages, roosts during the growing period are optional.

BROODING BROILERS

Commercial broiler production is big business; and almost all of it is under some type of contract between the poultry company and the grower. In a typical grower contract, the contractor (the company) supervises the growing of the broilers, and furnishes chicks, feed, and medication. The grower provides housing, equipment, and labor. The contractor decides when to market the birds and to whom they shall be sold. The contractor also outlines the management program to be followed and specifies the type of housing and equipment to be used.

(Also see Chapter 13, Broilers [Fryers], Roasters, and Capons.)

■ **Housing**—The first requirement for growing broilers is adequate housing. Since broiler production is essentially a chick-brooding operation, the house must control such factors as temperature, moisture, air movement, and light. It should also provide for efficient installation and operation of brooding, feeding, watering, and other equipment.

Currently, most broiler houses that are being constructed are 32 to 40 ft in width and of sufficient length to give the desired housing capacity. Most modern houses have a minimum capacity of 10,000 to 50,000 broilers. Housing should be economical yet substantial enough to last many years without costly maintenance. Also, it should be sufficiently insulated so that it will heat easily in the winter and have sufficient ventilation (natural and mechanical) for cooling birds in the summer. Whatever the location, houses should be capable of maintaining a temperature of nearly 70°F throughout the year.

A great deal of research is underway on growing broilers in cages. If a workable system of cage rearing is developed, it could drastically change housing and space requirements for broilers.

■ **Brooders**—The type of brooder to use varies with the source of fuel available. Natural gas, butane or propane gas, and electrical brooders are the most common types.

Manufacturer's recommendations for number of chicks per brooder are usually based on day-old chicks. As chicks grow larger, they need additional space. For this reason, only about two-thirds of the rated chick capacity should be started under one brooder. The manufacturer's directions should be followed in the operation of the brooder. Brooder guards to keep chicks confined to the brooding area are recommended by some manufacturers. Corrugated cardboard 14 to 18 in. high may be used in winter, not only to confine the chicks but to aid in preventing floor drafts. Poultry netting may be used to replace cardboard for summer brooding. The guards should extend out about 3 ft from the edge of the brooder to allow the chicks to move out from under the brooder if they become too hot. The guard may be removed after about the first week.

■ **Feeding equipment** — Most broilers are started on chick box lids or similar temporary feeders. This allows them to find the feed more easily. Feeds should be placed in the feed troughs at the same time as in the box lids. Temporary feeders may be removed as soon as chicks are eating from the troughs.

Mechanical feeders are generally used on the large broiler establishments. They reduce labor and allow the grower to care for more birds. However, the mechanical feeding system does not reduce the necessity for frequent and regular observation of the birds. The manufacturer's instructions should be followed relative to the installation and operation of mechanical feeders. Usually about 2 in. of feeder space is recommended per bird. If mechanical feeders are used, it is best to install them on a winch with pulleys and cable so that the entire feeder can be pulled up to the ceiling during clean-out.

When trough-type feeders are used, provide 2 in. of trough space per bird through 6 weeks of age, and 3 in. thereafter to market time at 6 to 7 weeks of age. When circular-type feeders are used, 20% less feeder space per bird will be sufficient.

Bulk feed storage bins are also a necessary part of feeding equipment. The bins are usually located outside the house. An auger is used to move the feed from the bin into the house.

■ **Waterers** — Adequate watering space is essential. A 1-gal water fountain should be provided for each 100 broilers the first week to 10 days. (But do not remove all the gallon waterers until all birds are drinking from the automatic waterers.) Following this, each broiler should have 1 in. of water space when troughs are used, with 20% less water space when circular waterers are used. This will provide sufficient watering space even during hot weather. Arrange the waterers so that the birds will not have to go more than 10 to 12 ft for water.

■ **Vaccination** — Protecting broilers by vaccinating against diseases common to the area is good insurance against costly disease losses. Marek's vaccine is usually administered at the hatchery by subcutaneous injection of each chick. Whether or not broilers are vaccinated for other diseases is determined by the threat of exposure to specific diseases. In Arkansas, which is the leading broiler producing state of the nation, most producers vaccinate against infectious bronchitis and Newcastle. Also, some producers vaccinate against fowl pox during the fall.

Where birds are immunized against both Newcastle and infectious bronchitis, it should be done sometime between the first and fourteenth day of age. Both vaccines may be given at the same time. Intraocular or intranasal dust, spray, and water-type vaccines are available; hence, the method that has given the most suitable results should be used. Booster vaccinations of Newcastle may be given when the birds are 4 to 5 weeks old.

■ **Litter** — It is recommended that broilers be provided with dry, absorbent material 2 to 4 in. deep. Wood shavings, sawdust, straw, rice hulls, and other products may be used. The choice of the litter should be made on the basis of what is most convenient and economical.

■ **Light** — All-night lights (0.35 to 0.50 footcandle at bird level) will give a big assist in keeping the birds from crowding in corners or piling up in case of disturbance during the night and will tend to increase feed consumption, especially during hot weather. Some producers set time clocks so that lights are off 1 hour each night (after the first 5 days) to condition birds in case of power interruption.

■ **Sanitation** — Starting the chicks out in a clean house with clean equipment and clean litter reduces the possibility of exposing them to many infective organisms that might have accumulated from previous broods of chicks. Washing waterers each day will reduce contamination and stimulate water consumption. Dead birds should be removed from the house promptly and either burned or placed in a disposal pit. Ventilation openings should be covered by small mesh wire in order to keep birds out of the house. Wild birds may carry diseases and parasites.

■ **Management practices** — Much of the success or failure of producing broilers depends upon the preparation made before the chicks arrive, including the following:

1. **Clean thoroughly.** Clean the brooder house thoroughly several days before chicks are to be delivered. Remove all the old litter and manure and sweep out dirt and dust. Cobwebs and dust should be brushed or washed off walls and ceilings. Clean and disinfect all equipment such as brooders, feeders, and waterers. Rinse equipment with clean water after disinfection. Then, spread 2 to 4 in. of clean, dry litter on the floor. Do not use wet, moldy, or partially decayed material as litter. The most commonly used litter is wood shavings, but other types of litter may be used.

If previous broods were relatively free from infectious disease, and if the contractor so permits, old litter may be reused. If litter is reused, about 1 in. of new litter should be placed on top of the old, or at least the areas beneath the brooders should have new litter added.

2. **Have brooders ready.** Set up the brooders and start them operating a day or more before chicks are to be delivered, and adjust them for proper operation. Use a thermometer to check the temperature. The brooder temperature should be 95°F at the edge of the hover, 2 in. above the litter. In controlled environment houses, however, satisfactory results can be obtained with starting temperatures as low as 88°F.

3. **Place brooder guards, feed troughs and waterers.** Some producers use brooder guards, others do not. If guards are used, they should be in place before the chicks

arrive. Box lids or small feeders are often used before chicks are large enough to eat from automatic feeders. The arrangement illustrated in Fig. 3–18 is suggested.

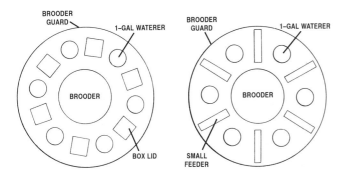

Fig. 3–18. Arrangements around the hover; using box lids, small feeders, and 1-gal waterers.

When the chicks are around 5 to 7 days old, the brooder guard may be removed. Also, at that time, it is usually possible to switch from small founts and feeders or box lids to automatic equipment.

4. **Keep records.** Accurate records should be kept of feed consumption, mortality, vaccination dates or medication given. This information will be helpful in preventing and controlling trouble.

5. **In case of trouble.** If the birds go off feed or if there is a sudden increase in mortality, immediate steps should be taken to determine the trouble. It is best to have a positive disease diagnosis before any treatment is started. To this end, birds should be sent to a diagnostic laboratory.

BROODING TURKEY POULTS

Large-scale turkey production calls for large-scale brooding; and, to be successful, it must be properly done. The successful brooding of turkey poults necessitates that attention be given to the same factors and principles as are

Fig. 3–19. Three-day-old turkey poults in brooder. Note (1) ring around the poults, and (2) gas hoods overhead. (Courtesy, California Polytechnic University, San Luis Obispo)

involved in the successful brooding of chicks, with consideration given to difference in size and certain species peculiarities—for example, the difficulty in getting poults to eat and drink.

■ **Brooder house**—Turkey houses are often used both for brooding and rearing to market, in a manner similar to many broiler houses. Width usually varies from 24 to 40 ft, and length from 100 to 600 ft. Clear-span, rigid-frame, gable-roof houses 40 ft wide and 300 ft long are popular. For cold-weather brooding, insulation of side walls and ceiling is needed.

In long houses, partitions with no more than 1,000 to 1,500 turkeys per pen are recommended. This arrangement is useful in separating turkeys by age and sex, and in controlling stampedes and disease outbreaks. Additional turkey brooder house requisites are: electric wiring for lighting and debeaking, water piped in, and exclusion of rats and wild birds. (The latter necessitates concrete foundations, tight-fitting doors and windows, and covering all openings with ¾-in. mesh, or smaller, wire.)

■ **Brooders**—There are many styles and sizes of brooders; and they may be heated by gas, oil, or electricity. Commercial turkey growers often use hot water or hot air types of equipment with a central heating system.

■ **Temperature**—Young poults must be kept warm and dry. For the first 2 weeks, the temperature 3 in. above the floor at the edge of the hover should be 95°F (35°C) for dark poults and about 100°F (37.8°C) for white poults; and temperature may be reduced about 5°F each week until heat is no longer needed. Temperature near the floor outside the brooding area should be about 70°F (21°C). A guard should be placed around the brooder for the first 2 weeks to prevent crowding and smothering in the corners of the house.

■ **Ventilation**—Mechanical ventilation is highly desirable in all brooder houses, especially in those 40 ft or more in width. Trouble from respiratory diseases is invited when the air in the brooder house is low in oxygen and contaminated with exhaust gases from the brooder stoves.

■ **Floor space**—For brooding poults to 8 weeks of age in naturally ventilated brooder houses, allow 1¼ sq ft of floor space for large-type poults and 1 sq ft for small-type poults. In insulated force-ventilated houses, 0.8 to 1 sq ft will be adequate for large-type poults to 8 weeks of age.

For growing turkeys to fryer-roaster age (12 to 13 weeks for large-type turkeys, or 15 to 16 weeks for small- or medium-type turkeys) allow 2½ sq ft for each bird.

■ **Floor**—Poults may be started on covered or uncovered litter, on asphalt roofing (not tar paper), on wire or slat floors, or in batteries.

To start the poults on covered litter, proceed as follows: distribute about 2 in. of suitable litter material on the brooder house floor; on top of this litter, under the hover and extending 3 ft beyond, place a layer of strong, rough-surface paper and enclose it by a brooder ring or poult guard; after 5 or 6 days, remove the paper and increase the litter to a depth of 4 in., then allow the poults on the litter.

To start poults on asphalt roofing, proceed as follows: place the asphalt roofing under the hover and extending 3 ft beyond; sprinkle a thin layer of sand on top of the roofing; after 3 weeks, either remove the roofing or cover it with litter.

Wire and slat floors make the use of litter unnecessary, and they tend to prevent such filth-borne diseases as coccidiosis, hexamitiasis, and blackhead. Plenty of heat and good ventilation are essential. Also, it is necessary to prevent floor drafts, which can be done with solid partitions underneath, about every 20 ft.

■ **Litter** — For starting poults on uncovered litter for the first 5 or 6 days, either use softwood shavings or bright straw that is free from chaff. Following this, litter may be chosen largely according to availability and price, with consideration given to shavings, wheat straw, beardless barley straw, peat moss, shredded cane, rice hulls, and processed flax straw. If free of harmful molds, peanut hulls, crushed corncobs, and shredded corn stover make good litter. The following materials are unsatisfactory for use as poult litter: splinty shavings, sawdust, oat hulls, cottonseed hulls, dried beet pulp, and the straw of rye, oats, and bearded barley.

Good litter should be put down to begin with. Then wet or caked litter should be removed at intervals and replaced with clean litter.

To promote sanitation, feeders and waterers should either be moved frequently or placed on wire- or slat-covered platforms after the poults are about 3 weeks old.

■ **Feeders and waterers** — The watering and feeding facilities for turkey poults are very similar to those used for baby chicks, but generally they are somewhat larger, and, of course, more space is needed per bird.

Start with small trough-type metal or wooden feeders placed like spokes in a wheel, part under the hover and part outside, but within the brooder ring. Provide 2 linear in. of feeding space per bird. In order to encourage eating, heap the trough with a little mash in a few paper plates or egg flats for the first day or two, and place some glass marbles on the feed to attract the poults. After 7 to 10 days, use larger feeders, but retain about 2 linear in. of feeder space per bird during the remainder of the brooding period.

For watering, start with one circular 1- to 2-gal glass or metal baby-poults-size waterer, with a narrow drinking space not over 1¼ in. wide and 1¼ in. deep, for each 50 poults; or as an alternative, start with one baby-poult-size automatic water trough 4 ft long for each 80 poults. Place the waterers around the edge of the hover and put a few colored glass marbles in each one. As an extra precaution, some growers add one quart-size waterer per 50 poults for the first 2 days.

After about 2 weeks, replace the baby-poult-size waterers with larger waterers that provide a water depth of 1½ in. and about ½ linear in. of watering space per bird. Gradually move the small waterers near the larger ones to assist the poults in making the change.

Wash and rinse waterers daily.

■ **Lighting** — For the first 2 weeks of brooding in all types of houses, the room should be well lighted day and night at 10 to 15 footcandles at poult level, preferably with a 7½- to 15-watt light bulb installed under the hover, also.

After 2 weeks, only dim light at 0.5 footcandle is needed at night, none in the daytime. In windowless houses, the light intensity can be reduced gradually to about 1 footcandle during the 16-hour day and 0.5 footcandle during the 8-hour night. The dim light at night may help to prevent piling and stampeding.

QUESTIONS FOR STUDY AND DISCUSSION

1. Describe the artificial incubation of eggs by the Egyptians thousands of years ago.

2. How does poultry reproduction differ from the reproduction of mammals?

3. What is parthenogenesis? How do you explain this phenomenon?

4. List and discuss the major factors that influence the hatchability of eggs.

5. List and discuss the factors of major importance in incubating eggs artificially.

6. Describe the embryonic development of the egg.

7. How do embryos of some species communicate or "talk" to each other? Of what practical importance is this phenomenon?

8. How does the yolk make for a good start in life for a chick?

9. It is generally agreed that hatcheries have exerted more influence on improving the general level of the poultry industry than any other segment of the business. How have they accomplished this?

10. List and discuss the major changes that have taken place in the hatching industry in recent years.

11. Describe a typical modern hatchery.

12. List and describe some of the sophisticated features of a modern incubator.

13. List and summarize the sanitation practices which should be rigidly adhered to in a hatchery.

14. Describe the methods of sexing chicks.

15. What is meant by each of the following: debeaking, dubbing?

16. Why is vaccination of chicks during the first days of life often ineffective?

17. Define *brooding*.

18. List and discuss the chief requisites for successful brooding.

19. Discuss the major differences between the brooding of (a) replacement pullets, (b) broilers, and (c) turkeys

.SELECTED REFERENCES

Title of Publication	Author(s)	Publisher
Anatomy of the Domestic Birds	Edited by R. Nickel, *et al.*	Springer-Verlag, New York, NY, 1977
Avian Egg, The	A. L. Romanoff A. J. Romanoff	John Wiley & Sons, Inc., New York, NY, 1963
Avian Physiology, Third Edition	Edited by P. D. Sturkie	Springer-Verlag, New York, NY, 1976
Commercial Chicken Production Manual, Fourth Edition	M. O. North D. D. Bell	Van Nostrand Reinhold Co., New York, NY, 1990
Dukes' Physiology of Domestic Animals, Ninth Edition	Edited by M. J. Swenson	Cornell University Press, Ithaca, NY, 1977
Poultry Meat and Egg Production	C. R. Parkhurst G. J. Mountney	Van Nostrand Reinhold Co., New York, NY, 1988
Poultry Production, Twelfth Edition	M. C. Nesheim R. E. Austic L. E. Card	Lea & Febiger, Philadelphia, PA, 1979
Reproduction in Domestic Animals, Third Edition	Edited by H. H. Cole P. T. Cupps	Academic Press, New York, NY, 1977
Reproduction in Farm Animals, Fourth Edition	Edited by E. S. E. Hafez	Lea & Febiger, Philadelphia, PA, 1979

Fig. 3–20. Duck egg packing at the hatchery, where more than 250,000 eggs a week are handled. (Courtesy, Cherry Valley Farms, Ltd., Rothwell, England)

Fig. 3–21. Grading day-old ducklings. (Courtesy, Cherry Valley Farms, Ltd., Rothwell, England)

Fig. 3–22. Pedigreed hatching eggs. (Courtesy, Indian River International, Division of Hy-Line Indian River Company, Nacogdoches, TX)

Fig. 3–24. Hatching tray full of newly hatched chicks. (Courtesy, North Carolina State University, Raleigh)

Fig. 3–23. Incubator in a commercial broiler hatchery. (Courtesy, North Carolina State University, Raleigh)

Fig. 3–25. Baby chicks. (Courtesy, California Polytechnic State University, San Luis Obispo)

Fig. 4–1. Broiler breeders. Note (1) the nest boxes on the left, (2) the chain feeders, and (3) ⅓ solid floor and ⅔ slotted floor. (Courtesy, California Polytechnic University, San Luis Obispo)

POULTRY BREEDING

CHAPTER
4

Breeding of poultry differs from the breeding of four-footed farm animals primarily in that (1) it is more flexible due to greater numbers and more rapid reproduction; (2) it has passed through the total presently known cycle of breeding methods more than any other animal species; and (3) it is concentrated in fewer hands. Also, more has been accomplished from the scientific application of genetics to poultry than from its application to other classes of animals. In developing high-performing genetic stocks, breeders have applied Mendelism, quantitative inheritance, inbreeding, hybridization, and even blood-typing techniques.

The typical commercial breeding establishment is directed by a person with a knowledge of methods of record keeping, high-speed computers, and the application of statistics in the evaluation of genetic stocks and progress from selection.

The commercial breeding of economically important poultry is directed toward three broad objectives: (1) increased product output per bird; (2) increased efficiency of production; and (3) improved quality of the product. Improvement in fertility, hatchability, growth rate, body conformation, egg yield, meat yield, feed conversion, egg quality, meat quality, and viability are all facets of these three objectives.

MENDELISM

The essence of Mendelism is that inheritance is by particles or units (called genes), that these genes are present in pairs—one member of each pair having come from each parent—and that each gene maintains its identity generation after generation. Thus, Mendel's work with peas laid the basis for two of the general laws of inheritance: (1) the law of segregation, and (2) the independent assortment of genes. Later genetic principles were added; yet all the phenomena of inheritance, based upon the reaction of genes, are generally known under the collective term, *Mendelism.*

SOME FUNDAMENTALS OF HEREDITY

It is not intended that all of the diverse field of genetics and animal breeding be covered in this chapter. Rather, the author presents a condensation of the known facts in regard to the field and summarizes their applications to practical poultry operations.

THE GENE AS THE UNIT OF HEREDITY

Genes determine all the hereditary characteristics of poultry, from the body type to the color of the feathers. They are truly the fundamental unit of genetics.

The bodies of all animals are made up of millions or even billions of tiny cells, microscopic in size. Each cell contains a nucleus in which there are a number of pairs of bundles called chromosomes. In turn, the chromosomes carry thousands of pairs of minute particles, called genes, the basic hereditary material which determines comb type, skin color, and all other characteristics.

The exact number of chromosomes in the fowl is not known. However, recent counts indicate that there are 39 pairs in the chicken and 41 in turkeys. Also, in all avian species there are two distinct sizes of chromosomes. Thus, chickens and turkeys have five pairs of large chromosomes in which consistent features are recognized. Additionally, each species has *microchromosomes*, which are very small and without distinguishable features. It is not known whether the latter carry genetic information.

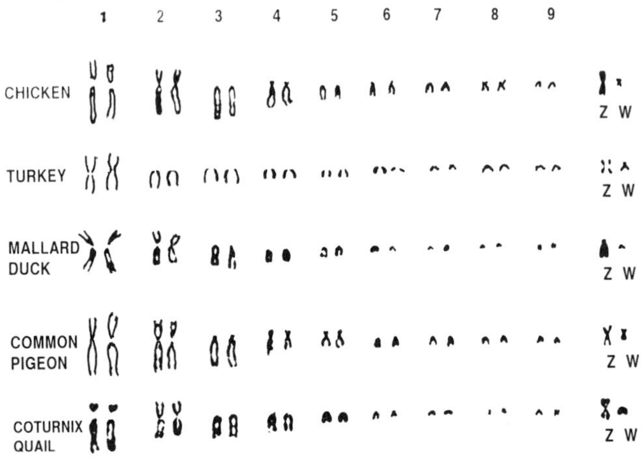

Fig. 4–2. Partial karyotypes of the nine largest pairs and the ZW sex chromosomes of the chicken, turkey, duck, pigeon, and coturnix quail. The chicken has 78 chromosomes; the number for the other species is uncertain, but ranges between 80 and 82. (Courtesy, R. N. Shoffner, Department of Animal Science, University of Minnesota)

The modern breeder knows that the job of transmitting qualities from one generation to the next is performed by the germ cells—a sperm from the male and an ovum or egg from the female. All animals, therefore, are the result of the union of two such tiny cells, one from each of its parents. These two germ cells contain all the anatomical, physiological, and psychological characters that the offspring will inherit.

In the body cells of an animal, each of the chromosomes is duplicated; whereas in the formation of the sex cells, the egg and the sperm, a reduction division occurs and only one chromosome and one gene of each pair goes into a sex cell. This means that only half the number of chromosomes and genes present in the body cells of the animal go into each egg and sperm, but each sperm or egg cell has genes for every characteristic of its species. When mating and fertilization occur, the single chromosomes from the germ cell of each parent unite to form new pairs, and the genes are again present in duplicate in the body cells of the embryo.

■ **DNA (deoxyribonucleic acid)**—In recent years, the molecular basis of heredity has become much better understood. The most important genetic material in the nucleus of the cell is deoxyribonucleic acid (DNA), which serves as the genetic information source. It is composed of nucleotides containing adenine, guanine, cytosine, and thymine. The sequence of these four bases in DNA acts as a code in which messages can be transferred from one cell to another during the process of cell division. The code can be translated by cells (1) to make proteins and enzymes of specific structure that determine the basic morphology and functioning of a cell; (2) to control differentiation, which is the process by which a group of cells become an organ; or (3) to control whether an embryo will become a chicken, a cow, or a human.

But DNA is far more than a genetic information center, or master molecule. The recent development of the recombinant DNA technique ushered in a new era of genetic engineering—with all its promise and possible hazard.

Recombinant DNA techniques are of enormous help to scientists in mapping the positions of genes and learning their fundamental nature. It may lead to new scientific horizons—of allowing introduction of new genetic material directly into the cells of an individual to repair specific genetic defects or to transfer genes from one species to another. On the other hand, the opponents of tinkering with DNA raise the specter (1) of reengineered creatures proving dangerous and ravaging the earth, and (2) of moral responsibility in removing nature's *evolutionary barrier* between species that do not mate. Nevertheless, molecular biologists are working ceaselessly away in recombinant DNA studies; hence, the poultry geneticist should keep abreast of new developments in this exciting field.

FERTILIZATION

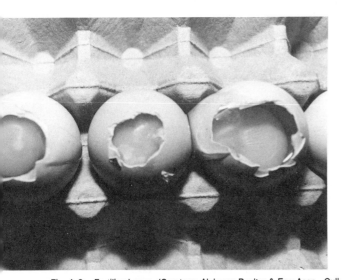

Fig. 4–3. Fertilized eggs. (Courtesy, Alabama Poultry & Egg Assn., Cullman, AL)

The act of copulation between male and female brings about the union of a spermatozoan and an ovum, which is known as fertilization.

Fertilization is the union of male and female gametes (sex cells) to form a new individual. The unfertilized egg has the haploid number of chromosomes, as do spermatozoa. Their

union in fertilization restores the diploid number of chromosomes in the fertilized egg.

The sperm ducts in the testes of the male produce spermatozoa. They combine with the secretion in the vas deferens to form semen, which is deposited in the oviduct of the female at the time of mating. Fertilization occurs after ovulation and before the egg reaches the magnum. Unlike mammals, the sperm of birds is relatively long-lived when stored in the female. In the vagina near the utero-vaginal junction, numerous cryptlike glands act as storage areas for sperm. Prior to ovulation, a certain number of sperm are released from these storage glands and are transported up the tract via peristaltic contractions. The mechanism by which sperm are retained and stored in these glands in not fully understood. However, this mechanism does enable birds to produce fertile eggs up to 3 weeks or longer after one insemination.

Once a swarm of spermatozoa are near the ovum, they tend to remain there, held by some chemical attraction which is not fully understood. A cone of ovum cytoplasm rises up to meet one of the sperm cells and draws it into the ovum; the tail of this sperm cell drops off as the sperm enters the ovum; and the surface covering of the ovum undergoes a chemical change and thickening which keeps out other sperm cells. The fertilized egg, known as a zygote, now carries the same number of chromosomes as a cell from either parent. It is now ready to undergo cell growth, mitosis, differentiation, and development into a new individual.

In higher animals, reproduction is possible only after fertilization. In most birds, fertilization is not a necessary preliminary to egg laying. Thus, the hen can lay eggs continuously without being mated or without being stimulated to lay by the presence of the male. This biological phenomenon has been advantageously utilized by poultry producers in producing infertile eggs for food. (Infertile eggs are of more economic value than fertile, because there is no danger of possible loss through the development of the embryo.)

In turkeys, and rarely in chickens, the egg may hatch without fertilization. This phenomenon is called *parthenogenesis.*

SEX DETERMINATION

In mammals, the male is the heterogametic sex (XY) and the female the homogametic sex (XX). In avian species, this condition is reversed, the female having the unlike pair and the male having the like pair.

The sex chromosomes in chickens are designated Z and W, which correspond to X and Y, respectively, in mammals.

Thus, when the fertilized egg contains two Z chromosomes, the resulting chick will be a male; and when the fertilized egg has ZW sex chromosomes, a female chick will result.

The Z chromosome seems to be especially important in chickens. It is the fifth largest of the chromosomes and comprises nearly 10% of the total DNA material. Of the six known linkage groups, the sex-linked group has the most known segregating loci.

SIMPLE GENE INHERITANCE
(QUALITATIVE TRAITS)

In the simplest type of inheritance, only one pair of genes is involved. Thus, a pair of genes is responsible for comb type, plumage color, and skin color.

Fig. 4–4 shows the inheritance of a pair of characters; the result of mating a rose comb White Wyandotte male and a single comb White Plymouth hen, comb types commonly found in chickens. A capital "R" is used as the symbol for the rose comb gene because it is dominant to the single comb gene; and a lower case "r" is used as a symbol for the recessive single comb.

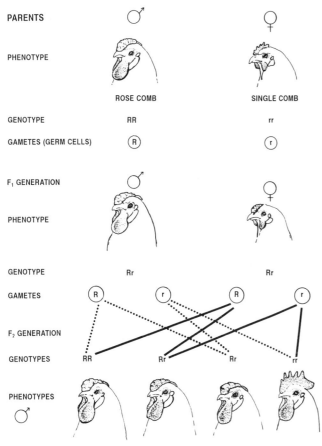

PARENTS

PHENOTYPE

ROSE COMB   SINGLE COMB

GENOTYPE   RR   rr

GAMETES (GERM CELLS)   R   r

F₁ GENERATION

PHENOTYPE

GENOTYPE   Rr   Rr

GAMETES   R   r   R   r

F₂ GENERATION

GENOTYPES   RR   Rr   Rr   rr

PHENOTYPES

Fig. 4–4. Inheritance of a pair of characters. When a rose comb (R) male is crossed with a single comb (r) female, all the first generation birds will have rose combs (dominant character). When the F₁ birds are mated among themselves, about 75% of the F₂ generation will have rose combs and the remainder will have single combs. Only the male offspring are illustrated in the F₂ generation.

MULTIPLE GENE INHERITANCE
(QUANTITATIVE TRAITS)

Relatively few characters of economic importance in poultry are inherited in as simple a manner as comb type, plumage color, and skin color. Commercially important characters—such as egg production, egg size, growth rate, and body conformation—are due to many genes; thus, they are called multiple-factor characters or multiple-gene characters.

Because such characters show all manner of gradation—from high to low production, for example—they are sometimes referred to as quantitative traits.

Quantitative traits are of particular interest to poultry breeders. They must devise breeding systems that will improve the average performance of a flock in several quantitative characters at the same time.

It is noteworthy that egg production and hatchability are lowly heritable; hence, they are sensitive to environmental influences and resistant to change by selection. On the other hand, body size and growth rate are highly heritable and readily improved by selection.

DOMINANT AND RECESSIVE FACTORS

Some genes have the ability to prevent or mask the expression of others, with the result that the genetic makeup of such animals cannot be recognized with perfect accuracy. This ability to cover up or mask the presence of one member of a set of genes is called *dominance*. The gene which masks the one is the dominant gene; the one which is masked is the recessive gene.

Both rose comb and pea comb are dominant to single comb. When rose- and pea-combed birds are crossed, the comb is somewhat intermediate and given the name walnut comb.

White in the White Leghorn is dominant to colored plumage. Thus, if a White Leghorn is mated to a colored bird, the F₁ generation will all be white, but some of the birds may have colored feathers. The dominance of white in the Leghorn is due to the presence of a gene which inhibits color; otherwise, the bird would be barred, for it carries genes for barring, color, and black pigment. Some strains of White Plymouth Rocks are also dominant.

It is noteworthy that white is not always dominant—that it depends on the breed. While white is the dominant color in the Leghorn, it may be recessive to color in the Wyandotte.

An example of the inheritance of two pairs of characters may be obtained by crossing a Black Wyandotte with a White Plymouth Rock. In this case, rose comb (R) is dominant to single (r), and black (B) is dominant to white (b). All of the first generation crosses from such a mating will be rose combed and black. However, the F₂ generation will consist of nine rose comb–black, three rose comb–white, three single comb–black, and one single comb–white; or a 9:3:3:1 ratio.

It is clear, therefore, that a dominant character will cover up a recessive. Hence, a bird's breeding performance cannot be recognized by its phenotype (how it looks), a fact which is of great significance in practical breeding.

As can be readily understood, dominance often makes the task of identifying and discarding all animals carrying an undesirable recessive factor a difficult one. Recessive genes can be passed on from generation to generation, appearing only when two animals, both of which carry a recessive factor, happen to mate. Even then, only one out of four offspring produced will, on the average, be homozygous for the recessive factor and show it.

Some dominant and recessive characters in chickens are listed in Table 4–1.

TABLE 4–1
SOME DOMINANT AND RECESSIVE CHARACTERS IN CHICKENS[1]

Character	Dominant or Recessive	Sex-Linked
Barred plumage	In Plymouth Rocks, dominant to nonbarring	Yes
Black Plumage	Dominant to recessive white	
Broodiness	Dominant to nonbroodiness	Yes
Buff plumage	Dominant to recessive white	
Close feathering	Dominant to loose feathering	
Early sexual maturity	Dominant to late sexual maturity	Yes
Feathered shanks	Dominant to nonfeathered shanks	
Rose comb	Dominant to single comb	
Side sprigs	Dominant to normal comb	
Silver plumage	Dominant to gold plumage	Yes
Slow feathering	Dominant to rapid feathering	Yes
White plumage	In White Leghorns, dominant to color	
	In Wyandottes, recessive to color	
White skin and shank color	Dominant to yellow skin and shank color	
Winter pause	Dominant to continuous laying	

[1]Adapted from Winter, A. R., and E. M. Funk, *Poultry Science and Practice*, 5th ed., J. B. Lippincott Company, 1960, p. 77, Table 4–2.

INCOMPLETE OR PARTIAL DOMINANCE

The traits listed and discussed in the previous section have been clear-cut dominant and recessive traits. Many traits exhibit incomplete or partial dominance so that it is possible to distinguish some or all of the heterozygous individuals as well as both types of homozygote with respect to that trait. Sometimes the heterozygote is preferred and on occasion has led to the recognition of a breed or variety. For example, a cross between a black and a white chicken gives the typical blue andalusian pattern. These hybrids never breed true because their color is due to the interaction of two genes, but they can transmit only one gene or the other to any one offspring.

The preceding discussion also indicates that there are varying degrees of dominance—from complete dominance to an entire lack of dominance. In the vast majority of cases, however, dominance is neither complete nor absent, but incomplete or partial. Also, it is now known that dominance is not the result of single-factor pairs, but that the degree of dominance depends upon the animal's whole genetic makeup together with the environment to which it is exposed, and the various interactions between the genetic complex (genotype) and the environment.

INHERITANCE OF SOME ECONOMICALLY IMPORTANT CHARACTERS IN POULTRY

The hereditary material of a chicken is located in 39 pairs of chromosomes. In the turkey, it is in 41 pairs of chromosomes.

Some of the Mendelian principles were first applied to poultry following the rediscovery of Mendel's laws about 1900.

Basically, all characteristics of poultry are genetically determined to some degree. The size, shape, color, behavior, or tissue enzyme content of a bird are all under genetic control. It is recognized, however, that the expression of hereditary characters may be modified by the environment to which an individual is exposed, and that some characters are affected by environment more than others. For example, a hen may have genes which allow her to lay 300 eggs a year, but if she is not fed properly, given good housing, or protected from disease, these genes affecting egg production may not be expressed. On the other hand, if the hen carries genes for white plumage, the environment will not affect this character.

The inheritance of some economically important characters in poultry follows:

1. **Plumage color.** White, or light-colored, feathers have become an important factor in the breeding of broilers because they are easier to pick clean than chickens with dark-colored feathers. Colored chickens often have pigmented pinfeathers at broiler age which have not broken through the skin and, therefore, cannot be readily removed with the result that they detract from the appearance of the carcass and cut-up parts. Hence, White Rock and specially developed *dominant white* meat-type male lines are preferred for broiler production. Also, white turkeys and white ducks are preferred for the same reason. Most bronze turkeys have been replaced with white turkeys; and most ducks produced commercially are of the White Pekin breed.

Fig. 4–5. Red hen for producing brown eggs and red-feathered broilers. Not all consumers want white chickens. (Courtesy, Hy-Line International, West Des Moines, IA)

Geneticists are well aware of the inheritance of plumage color and make use of it in the production of broilers. Also, in some cases, they make use of it where sexing of chicks is desired.

2. **Skin and shank color.** The several different shank colors found in fowl result from different combinations of pigments in the upper and lower layers of skin. Yellow shanks are due to the presence of carotenoid pigments in the epidermis, and the absence of melanic pigment. Black shanks are due to the presence of melanic pigments in the epidermis. White shanks are the result of complete absence of both types of pigments.

Blue, or slaty blue, shanks occur when melanic pigment is present in the dermis, but neither type of pigment is present in the epidermis. Green shanks occur when there is black in the dermis and yellow in the overlying epidermis.

3. **Rate of feather development.** Early feathering is essential for minimizing pinfeathers on the dressed carcass of broilers. To meet this requisite, all modern broiler strains now carry the sex-linked early feathering gene. This is a recessive gene. Early feathering chicks can be identified at hatching by the length of the covert feathers of the wing in proportion to the length of the primary feathers. Also, at about ten days of age, rapid-feathering chicks show well-developed tail feathers, whereas slow-feathering chicks show no tails.

In order to identify the sexes at hatching, the sire must have had early wing feathering and the dam late wing feathering. At hatching, their sons are late like their mother and

the daughters are early like their father. To utilize sex-linked inheritance, it is important that the sire have the recessive trait, in this case, early, and the mother the dominant trait. This type of sex identification is important in broiler chicks in order to separate the more rapidly growing broiler cockerels from their slower growing sisters.

Since the early feathering gene is sex-linked, it may be used in determining sex at hatching time, thereby alleviating the necessity of vent sexing. An early-feathering male mated to a late-feathering female produces slow-feathering male progeny and early feathering pullets. This makes it possible to select pullets at hatching time.

It is noteworthy that, in Leghorns, where the early-feathering gene is fixed, some breeders are now introducing the slow-feathering gene into one of their strains. The objective is to produce baby chicks which can be autosexed at hatching. The unwanted slow-feathering males can then be discarded.

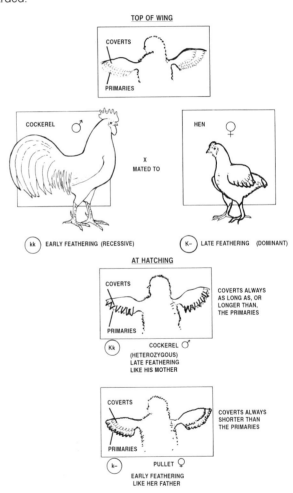

Fig. 4–6. Sex-linked feather (wing) sexing by the primary and covert feathers, viewing the outstretched wing from the top. Note that the coverts emerge from the top surface of the wing, whereas the primaries emerge from the lower edge of the wing. Note, too, that the relative length of the primaries and coverts is more important than the overall length of the feathers, since overall length depends upon the length of time that the chick has been out of the shell. This type of sex identification is increasing in broiler chicks because it makes it possible to separate the more rapidly growing broiler cockerels from their slower growing sisters.

4. **Color sexing.** Sex-linked matings to identify the sex of baby chicks are most effective if the silver (white) vs gold (red) down colors are characteristic of the two strains being crossed. In this case the gold color is recessive and used as the male parent mated with the female which was white (silver) at hatching. Such a mating produces white male and gold female baby chicks. This type of mating is used in both egg and broiler types when the red color is not objectionable to the chick purchaser.

5. **Blood groups.** In all species, blood-group information is useful in detecting pedigree errors. Also, red-cell groups and other biochemical variants of the blood of animals are useful tools in evaluating an animal's economic productivity. Some commercial breeders are now selecting chickens with specific blood-group genotypes in an attempt to develop strains with higher egg production and lower adult mortality. Although biochemical genetics will not likely replace conventional methods of selecting poultry, it could augment them.

About 12 different red-cell blood groups are recognized in chickens.

6. **Egg production.** Perhaps it is fair to say that geneticists have been more successful in increasing broiler (meat) production than they have in egg production. Yet, greater efficiency of egg production can be achieved through selecting layers for the economically important characters that follow:

a. **Feed efficiency.** Most of the income from layers is from the sale of eggs for food or for hatching purposes. Thus, the higher the egg production secured, the lower the feed and the total cost per dozen eggs. Consequently, producers are interested in high egg production. Egg production is affected by feeding. Likewise, management and environment exert considerable influence on egg production; for example, a bird will lay more eggs than it would otherwise if the eggs are removed from the nest daily.

Egg yield is the product of two forces: rate of laying and length of the laying period before molting.

b. **Sexual maturity** (approximately 30% heritability). A pullet is said to be sexually mature when she lays her first egg. The earlier a pullet commences laying, the longer will be her laying year, with the result that there is more possibility of her producing more eggs. In turn, this means lower cost per egg.

Age at maturity is hereditary. On the average, Leghorns become sexually mature at between 170 and 185 days of age; the dual-purpose breeds, such as the White Wyandotte, reach sexual maturity about 2 weeks later. However, there are differences between strains, and selection can be made accordingly.

Sexual maturity can also be greatly advanced or delayed by environment, especially by the lighting program followed during rearing. Feeding programs and disease can also delay age at first egg.

If birds of the light breeds, such as Leghorns, are to lay approximately 250 eggs during the pullet year, they should come into production when about 5 months old. The general-purpose breeds, such as Rhode Island Reds, should start laying when they are about 5½ to 6 months old.

In a flock of birds of about the same age, the first 75% to come into production will be the best layers.

c. **Egg size** (approximately 50% heritability). Size of egg is correlated with a number of factors, among them (1) body size—the larger breeds generally produce larger eggs, (2) age of pullets—egg size increases from the time pullets start to lay until about 6 months later, (3) weather—the size of eggs declines during the hot summer months, (4) second year—the eggs produced during the second year are larger than those produced the first year, (5) period of time within the clutch—those laid at the beginning of the clutch are larger than those laid at the end, and (6) total eggs laid—there is a tendency toward a decline in egg size with the total number of eggs laid in a year.

d. **Egg quality.** Breeders are concerned with several egg quality traits; among them (1) exterior traits, including shell color, texture, strength (resistance to breakage), and shape; and (2) interior traits, including freedom from blood and meat spots, amount of thick white, and proportion of white to yolk.

Shell color is also heritable. As is well known, some breeds produce white eggs, others produce brown eggs. Varying shades of color may be expected among brown eggs, but tinted shells should be avoided among white eggs, as these are not desired by consumers. For this reason, white eggs that have tinted shells should not be set.

WHITE   CREAM   BUFF   LIGHT BROWN   BROWN   DARK BROWN

Fig. 4–7. Eggshell color can be varied by breeding. (Courtesy, DEKALB AgResearch, Inc., DeKalb, IL)

e. **Intensity** (approximately 10% heritability). Intensity, or rate of production, which refers to the number of eggs laid by a hen during a given period of time, is most important from a profit standpoint. Thus, if a flock peaks at 85% production, the average hen in the flock must be laying an egg, on the average, more than 8 out of every 10 days. Such a flock is said to have a high intensity of lay.

f. **Hatchability** (approximately 12% heritability). Hatchability refers to the percentage of fertile eggs which hatch under artificial incubation. It is largely influenced by feeding and management. Even so, the poultry breeder should select against lethal genes which either reduce or prevent hatching, and against such indirect effects as large and small egg size and poor shell texture.

g. **Broodiness.** When the birds are brooding, they aren't laying. Hence, a minimum of broodiness is desired.

Broodiness is partially determined by a dominant sex-linked character. Breed differences exist; the light breeds, such as the Leghorn, are less broody than general-purpose breeds, such as the White Wyandotte. However, within a given breed or variety, there are strain differences in broodiness.

There is also evidence that broodiness is determined by complementary effects of genes. For this reason, when certain breeds are crossed, the progeny generally show more broodiness than that shown by either of the parent breeds.

Of course, broodiness was nature's way of aiding propagation. After laying a certain number of eggs, wild birds set on them until they hatched out. But, since this type of behavior is no longer needed in modern poultry production, it has been practically eliminated in today's egg-laying strains.

7. **Body weight and growth rate** (approximately 60% and 35% heritability, respectively). It is generally recognized that there are wide differences in body size between breeds—the extremes being illustrated by the difference in body size between bantams and Brahmas. Large body size is of importance to broiler and turkey breeders, chiefly because mature body size is correlated with rate of growth and efficiency of feed utilization.

Because weight of broilers is highly heritable, progress has been excellent through the selection of the larger individuals at broiler age. Further, there is some question whether heterosis, as obtained through hybrid breeding, contributes substantially to broiler weight; for example, the best New Hampshire strains generally equal the most rapid-gaining crosses. Despite the latter fact, crossbreeding is usually practiced in broiler production for, among other reasons, more uniformity of growth and fewer runts and culls are found in crossbreds than in standardbred stocks. Also, the crossbreds are generally more viable.

Growth rate and feed efficiency are highly correlated; hence, growth rate is of great importance in the breeding of both meat chickens and turkeys. Rapid growth makes for a saving in time, labor, feed consumption, and overhead in the production of meat.

8. **Body conformation.** Body conformation is especially important in turkeys, because, in comparison with broilers, turkeys are marketed at higher weights and the carcass may be marketed whole rather than cut up. With broiler producers, on the other hand, conformation is of secondary importance because of the increasing practice of marketing broiler meat in cut-up and packaged form. However, the poultry breeder has produced more desirable broiler carcasses through the infusion of broad-breasted, heavily-muscled, Cornish breeding.

Body conformation is of little consequence in layers, for the reason that after they have finished their usefulness to produce eggs, their carcasses are usually manufactured into chicken soup and other prepared foods.

9. **Sex-linked dwarfism.** Dwarfism, which causes an animal (or plant) to be subnormal in size, is well known in a variety of species, including chickens. Both egg producers and broiler producers are interested in sex-linked dwarfism.

Hens that carry the sex-linked dwarf gene are about one-third smaller than normal body size, but their egg size is only about 8% smaller. The objective in these "mini" layers is to lower body maintenance requirements and thereby increase the efficiency of feed utilization.

Fig. 4–8.   A normal and a mini meat-type breeder hen. (Courtesy, Euribrid B. V., Boxmeer, Holland)

Another application of sex-linked dwarfism is in broiler production, where the persistent problem is the high cost of producing hatching eggs. Broiler breeder hens are rather poor layers, and necessarily large-bodied to ensure that their broiler progeny have high growth rate. As a consequence, feed costs for hatching eggs are high. However, when the dwarf gene is introduced into the female parent line, body size is reduced about 30%, hens require less feed, and the cost of hatching eggs and broiler chicks is correspondingly reduced. When dwarf female breeders are mated to a normal nondwarf male line, the progeny are normal because the dwarf gene is recessive. Broiler progeny from dwarf hens show a slight reduction (2 to 3%) in growth to market age, because they are hatched from smaller eggs.

10. **Viability.** Viability, or livability, is influenced greatly by feeding and management practices. Experimental evidence also shows that there are family differences in susceptibility to and resistance against pullorum, fowl typhoid, range paralysis, roundworm infestation, crooked keels, and reproductive troubles. Hence, poultry geneticists have concentrated, and will continue to concentrate, on developing strains of higher livability.

The possibility of developing genetically resistant strains to certain diseases was clearly demonstrated by Cornell University. In their studies, resistant strains of White Leghorns showed only 2 to 3% mortality from leukosis, whereas susceptible strains showed a mortality of 25%.

As has already been pointed out, crossbreeding generally results in increased vigor and higher livability. On the other hand, inbreeding usually results in reduced vigor and increased mortality.

MUTATIONS

Gene changes are technically known as mutations. *A mutation may be defined as a sudden variation which is later passed on through inheritance and that results from changes in a gene or genes.* Mutations are not only rare, but they are prevailingly harmful.

Mutations are commonly called "sports." The White Plymouth Rock originated as a "sport" from the barred variety.

Gene changes can be accelerated by exposure to x-rays, radium, mustard gas, and ultraviolet light.

LETHALS AND ABNORMALITIES OF DEVELOPMENT

Lethals may be defined as congenital abnormalities which result in the death of an animal, either at birth or later in life. Other defects occur which are not sufficiently severe to cause death but which do impair the usefulness of the affected animals.

The embryological development—the development of the young from the time that the egg and the sperm unite until the chick is hatched—is very complicated. Thus, the oddity probably is that so many chicks develop normally rather than that a few develop abnormally.

Many such abnormalities are hereditary, being caused by certain "bad" genes. Moreover, the bulk of such lethals are recessive and may, therefore, remain hidden for many generations. The prevention of such genetic abnormalities requires that the germ plasm be purged of the "bad" genes. This means that where recessive lethals are involved, the breeder must be aware of the fact that both parents carry the gene.

More than 30 lethals have been identified in chickens. Fortunately for the breeder, these lethal genes are relatively rare. Nevertheless, they do contribute to the "genetic load" of chicken populations.

INBREEDING

As practiced by the poultry breeder, *inbreeding is the mating of closely related individuals, such as brother to sister, for several successive generations.* This approach to poultry improvement began to take shape during the years 1940 to 1945.

Because the procedure of developing inbred lines, then crossing them to form hybrids was borrowed from hybrid corn breeders, it is natural that two large U.S. commercial seed corn companies—Pioneer Seed Corn Company, Des Moines, Iowa; and DEKALB AgResearch, Inc., of DeKalb, Illinois—were the first in the field with chicken hybrids.

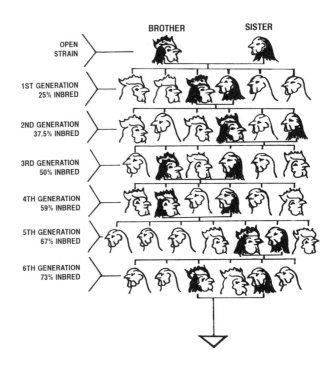

Fig. 4-9. Inbreds are the "building blocks" of the modern inbred-crossed chicken. The technique for making an inbred is illustrated in this figure. Of course, selections are made during each generation. (Courtesy, DEKALB AgResearch, Inc., DeKalb, IL)

Inbreeding is used in poultry to develop lines which when crossed will give progeny with high egg-production, large eggs, good meat type, good livability, etc.

The steps in developing inbred hybrids are:

1. The formation of inbred lines by intensive inbreeding.

2. The screening of the inbred lines by comparing the performance of test combinations.

3. After enough consistent field-test comparisons with an established commercial variety have been made, the breeder decides whether the new experimental variety is sufficiently superior to replace the established commercial variety.

So far, the inbred-hybrid scheme has not proved to be commercially feasible for broiler and turkey production.

HETEROSIS

Heterosis, or hybrid vigor, is a name given to the biological phenomenon which causes crossbreds to outproduce the average of their parents. For numerous traits, the performance of the cross is superior to the average of the parental breeds. This phenomenon has been well known for years and has been used in many breeding programs. The production of hybrid seed corn by developing inbred lines and then crossing them is probably the most important attempt by geneticists to take advantage of hybrid vigor. Also, heterosis is being used extensively in commercial poultry production today.

The genetic explanation for the hybrid's extra vigor is basically the same, whether it be layers, broilers, hybrid corn, or whatnot. Heterosis is produced by the fact that the dominant gene of a parent is usually more favorable than its recessive partner. When the genetic groups differ in the frequency of genes they have and dominance exists, then heterosis will be produced.

Heterosis is measured by the amount the crossbred offspring exceeds the average of the two parent breeds or inbred lines for a particular trait, using the following formula for any one trait:

$$\frac{\text{Crossbred average} \; (\text{minus}) \; \text{Purebred average}}{\text{Purebred average}} \times 100 = \begin{matrix} \text{Percent} \\ \text{Hybrid} \\ \text{Vigor} \end{matrix}$$

The level of hybrid vigor for all traits depends on the breeds crossed. The greater the genetic difference between two breeds, the greater the hybrid vigor expected.

NOT ALL HYBRIDS EXCEL PUREBREDS

The vast majority of chickens in America today are hybrids of one form or another—they are either strain crosses, breed crosses, or crosses between inbred lines. But there are exceptions!

Despite the fact that hybrids are widely used as commercial layers, it is noteworthy that egg-laying tests show that purebred Single Comb White Leghorns compete on even terms with hybrids under test conditions. Certainly, the hybrids are equal to the purebreds, but the point is that they do not excel them.

In broiler production, the main objective is the improvement of growth rate to market weight, although improvement in other economic factors is sought. Generally, growth rate and hybrid vigor are obtained by systematic matings that may involve crossing different breeds, different strains of the same breed, or the crossing of inbred lines. Most of the strains used as sires trace their ancestry to the broad-breasted Cornish breed. But there is some question whether heterosis, as obtained through hybrid breeding, contributes substantially to broiler weight. It is noteworthy, for example, that the best purebred New Hampshire strains generally equal the most rapid gaining crosses.

Body conformation is especially important in turkeys, because they are marketed at heavier weights than broilers and their carcasses may be left whole rather than cut up. Since conformation, size, and color of turkeys are highly heritable, they have responded well to simple methods of breeding and selection. As a result, most turkeys are bred as purebreds, rather than crossbreds. Also, it is noteworthy that in turkey breeding programs (1) selections are largely based on physical appearances (phenotype), and (2) mass matings (in which a number of males are allowed to run with the entire flock of hens) are the common practice.

BREEDING CHICKENS

It is improbable that all of the present-day breeds and varieties of chickens sprang from a common origin. The habits of the varieties in the Asiatic class indicate an ancestry which roosted on the ground and nested on a mound of earth. Such breeds as the Leghorn probably had tree roosting ancestors.

Authorities generally agree that the red jungle fowl, *Gallus gallus*, was one of the ancestors of domestic chickens (see Chapter 9, Fig. 9–2). But more recent investigations suggest that at least four species of jungle fowl may have contributed to the development of domestic fowl.

TYPES AND CLASSES OF CHICKENS

Type refers to the general shape and form, without regard to breed. Commercially speaking, chickens are of three types: (1) exhibition chickens, bred for competition in exhibitions; (2) the egg type, which is bred for egg production; and (3) the meat type, which is bred for meat production.

The term *class* is used to designate groups of breeds which have been developed in certain regions; hence, the class names—American, English, Asiatic, etc.

EXHIBITION CHICKENS

Exhibition chickens are produced both in the normal body sizes and in miniature (bantams). Breeding is restricted to members of the same breed and variety; hence, they are often called *purebreds*. The primary purpose for which exhibition chickens are reared is competition in exhibitions. The specific standards of excellence of each variety are detailed in a book entitled, *Standard of Perfection*, published by the American Poultry Association (A.P.A.). Anyone contemplating breeding of exhibition poultry (chickens, ducks, turkeys, or geese) should obtain a copy of the latest edition of this book, which contains pictures of ideal specimens of many varieties.

Exhibition chickens may be seen at poultry shows. Breeding for exhibition quality is both an art and a hobby of many people. The aim in breeding standard (exhibition) chickens is to approach closely the ideal for each variety. Breeding methods consist of selecting mature males and females which resemble the ideal or standard. Within each variety, individual birds have weak and strong points. Most of the weaknesses of color, size, and shape of the live bird may be improved by mating only the superior specimens. Success in breeding is partly dependent upon recognizing the superior characteristics peculiar to each variety.

Since there are so few specimens of some of the rarer varieties, problems of fertility, poor egg production, and hatchability often arise from inbreeding. Mating of relatives, is therefore, a dangerous practice although it is occasionally used to emphasize a particular color or shape characteristic observed only within a family group. Lack of vigor is also a characteristic of inbred poultry, although chickens may lack vigor for many other reasons such as poor nutrition and disease.

Fig. 4–11. White-egg layer. (Courtesy, Shaver Poultry Breeding Farm, Ltd., Cambridge, Ontario, Canada)

BROILER PRODUCTION CHICKENS

The major source of poultry meat is from broiler chickens which are slaughtered at about 6 to 7 weeks of age. In order to grow rapidly enough to reach a live body weight of about

Fig. 4–10.   Buff-Laced Polish. (Courtesy, *Poultry Tribune*, Mount Morris, IL)

EGG PRODUCTION CHICKENS

Chickens bred for egg production must be small bodied in order to consume the least feed while producing a large number of eggs. The most efficient egg-type hens weigh between 4 and 4½ lb at the end of the laying year. Excellence in egg-type chickens is judged by (1) the average egg yield per bird in large flocks containing several thousand layers, and (2) Random Sample Egg Laying tests.

Most of the egg laying chickens are white in color, tracing part of their inheritance to the White Leghorn. Likewise, most of the commercial layers produce white eggs. Colored birds and brown eggs have become less common in recent years. Some of the superiority in egg production is obtained by crossing synthetic strains developed by each breeder. Therefore, egg production chickens are sold under a breeder's trademark name, often followed by a number.

Fig. 4–12.   Broiler breeder hen. Note the large size. (Courtesy H & N, Inc., Redmond, WA)

4 lb, the parents of broiler chickens have become excessively large. These large hens are relatively poor egg layers and are retained as hatching egg producers for only about a 9-month laying period as contrasted to 12 to 16 months for egg-type chickens.

Broiler chickens are white or reddish-white in color. They are produced by crossing large broad-breasted males with females of a different origin. Also, crossbreeding is used to improve egg production of the mothers as well as the growth rate of the broiler chick. Both the sires and dams are selected for their ability to grow rapidly. While most egg-type chickens have single combs, the majority of broiler chickens have small, pea combs. However, pea comb inheritance is not necessary for good broilers. The Cornish breed, characterized by both a plump body and pea comb, has been retained in the synthetic sire strains, which are often called *Cornish-type* sire strains. Most of the hens have the single comb. Pea comb is inherited as a dominant trait, causing most of the progeny to have the pea comb.

Mothers of broiler chicks lay large brown-shelled eggs. Like the pea comb, the breeds from which the broiler stocks arose laid eggs with brown shells. Shell color has no influence on the growth rate of the chick or food quality of the egg.

Some breeders produce only the strain of chickens used as mothers while others produce only sire strains. For this reason there is a tendency to merchandise broilers under the trademark names of both breeders. When a breeder produces both parents, only one trade name may be used.

BREEDS AND VARIETIES OF CHICKENS

The term breed *refers to an established group of fowls, related by breeding, possessing a distinctive shape and the same general weight.*

A variety is a subdivision of a breed, distinguished either by color, color and pattern, or comb. Hence, a breed may embrace a number of varieties, distinguished by different colors—white, black, buff, etc.; or by color and markings—as Light Brahma, Dark Brahma, etc.; or by different combs—as single comb, rose comb, etc.

Standard of Perfection lists nearly 200 varieties of chickens. However, only two breeds (White Leghorn and White Plymouth Rock) are really important today, with three other breeds (Rhode Island Red, Barred Rock, and New Hampshire) of negligible importance.

Table 4–2 lists some representative breeds and varieties of chickens and gives their more important characteristics.

TABLE 4–2
SOME BREEDS AND VARIETIES OF POULTRY AND THEIR CHARACTERISTICS[1]

Breed and Variety	Plumage Color	Standard Weight		Type of Comb	Color of Ear Lobe	Color of Skin	Color of Shank	Shanks Feathered?	Color of Egg
		Cock	Hen						
		(lb)	(lb)						
American:									
Jersey Black Giant	Black	13	10	Single	Red	Yellow	Black	No	Brown
New Hampshire	Red	8½	6½	Single	Red	Yellow	Yellow	No	Brown
Rhode Island Red	Red	8½	6½	Single and rose	Red	Yellow	Yellow	No	Brown
White Plymouth Rock	White	9½	7½	Single	Red	Yellow	Yellow	No	Brown
White Wyandotte	White	8½	6½	Rose	Red	Yellow	Yellow	No	Brown
Asiatic:									
Black Langshan	Greenish-black	9½	7½	Single	Red	White	Bluish-black	Yes	Brown
Brahma (light)	Columbian pattern	12	9½	Pea	Red	Yellow	Yellow	Yes	Brown
Cochin (buff)	Buff	11	8½	Single	Red	Yellow	Yellow	Yes	Brown
English:									
Australorp	Black	8½	6½	Single	Red	White	Dark Slate	No	Brown
Buff Corpington	Buff	10	8	Single	Red	White	White	No	Brown
Silver-Gray Dorking	Black body with silver white markings	9	7	Single	Red	White	White	No	White
Sussex	Speckled/Red/Light	9	7	Single	Red	White	White	No	Brown
White Cornish	White	10½	8	Pea	Red	Yellow	Yellow	No	Brown
Mediterranean:									
Andalusian (blue)	Bluish slate (laced)	7	5½	Single	White	White	Slaty-blue	No	White
Anocona	Black, may be white tipped	6	4½	Single and rose	White	Yellow	Yellow	No	White
Minorca (S.C. Black)	Black	9	7½	Single	White	White	Dark slate	No	White
White Leghorn	White	6	4½	Single and rose	White	Yellow	Yellow	No	White

[1]To convert pounds to kilograms, multiply by 0.454.

A total of 342 breeds and varieties of domesticated land fowl and waterfowl are listed in *Standard of Perfection*, published by the American Poultry Association, Inc. The latter named association was first organized in 1873 by representatives of different sections of the United States and Canada. Its primary objective was to standardize the breeds and varieties of domestic fowl shown in exhibitions.

Fig. 4–14. Single Comb White Leghorns. (Courtesy, *Poultry Tribune*, Mount Morris, IL)

Fig. 4–13. White Wyandotte chickens. (Courtesy, J. C. Allen & Son, West Lafayette, IN)

Fig. 4–15. Barred Plymouth Rocks. (Courtesy, *Poultry Tribune*, Mount Morris, IL)

In the poultry industry, there are no breed registry associations like those for four-footed farm animals. Hence, in a sense, *Standard of Perfection* takes the place of such a registry association by recognizing as purebred only those individuals that show characteristics conforming to those given in *Standard of Perfection*.

Not all commercially important breeds of chickens have been officially recognized by *Standard of Perfection*. One such breed is the California Grey, an egg-laying strain which was once widely used by hatchery producers for crossing. Without doubt, the California Grey would have been eligible for recognition by the American Poultry Association. However, the developers of the strain did not request such recognition. This is indicative of the lack of interest in purebreds on the part of poultry breeders.

Fig. 4–16. White Plymouth Rocks. (Courtesy, *Poultry Tribune*, Mount Morris, IL)

Fig. 4–17. New Hampshires. (Courtesy, *Poultry Tribune*, Mount Morris, IL)

Fig. 4–18. White Cornish. (Courtesy, *Poultry Tribune*, Mount Morris, IL)

BUSINESS ASPECTS OF POULTRY BREEDING

Broadly speaking, commercial chicken producers are of two kinds; they are either egg producers or broiler producers. In either event, it is net return that counts. This means that the commercial producer is interested in increasing product output per bird, increasing the efficiency of producing the product, and improving the quality of the product produced. Improvements in fertility, hatchability, growth rate, body conformation, egg yield, meat yield, feed conversion, egg quality, meat quality, and viability (for example, pleuropneumonialike organism-free, which is usually abbreviated PPLO-free) are all important in that they contribute to the main goal—increased net returns.

No industry has done a better job than the poultry industry in combining science and technology. Literally speaking, these two have upped the ounce to the pound in broiler production, and upped the dozen to the gross in egg production. Hand in hand with the transformation of the poultry industry, there has come a very high degree of specialization.

From a breeding standpoint, today's poultry breeding is centered in two types of business enterprises: (1) the foundation breeder, and (2) the hatchery (or multiplier).

FOUNDATION BREEDERS

Most modern foundation breeders are large, well financed, and well managed. In fact, many of them have the characteristics of any other large business. Generally they are incorporated, departmentalized—breeding and development, sales, advertising, office management, purchases, etc. Many of them have sales representatives abroad, as well as throughout the United States. Most of them recognize the importance of and use copyrighted trade names. The larger ones have well-trained geneticists, physiologists, pathologists, and veterinarians on their staffs. Also, they make use of computers to handle the thousands of records which they must process.

The foundation breeder furnishes eggs to hatcheries.

Fig. 4–19. Progeny testing. Different combinations of lines and families within lines being tested by collecting data on progeny—egg numbers, egg size, body size, livability, etc. (Courtesy, Colonial Poultry Farms, Inc., Pleasant Hill, MO)

HATCHERIES

The hatcheries multiply the stock supplied by the foundation breeders, through hatching eggs. In turn, they sell chicks or poults to farmers or producers, who then grow them out as commercial egg layers, broilers, or market turkeys.

Until about 1950, foundation breeders and hatcheries operated independently of each other. Today, most of them are associated together through a franchise. A contract, signed by the two parties, specifies that the breeder will provide the hatchery with the breeding stock for his/her hatchery supply flock. Such supply flocks are commonly referred to as parent flocks, because the hatchery uses the chickens as parents of the commercial chicks that he/she sells.

Such franchise arrangements give the breeder virtual

Fig. 4–20. Hatchery trays full of baby chicks. (Courtesy, Indian River International, Division of Hy-Line Indian River Company, Nacogdoches, TX)

Fig. 4–21. Flock mating or mass mating, in which a number of males are allowed to run with an entire flock of hens. (Courtesy, The Cobb Breeding Corporation, Concord, MA)

Fertility can be improved by using cockerels with older hens, and by using older males with pullets.

2. **Pen mating.** In pen mating, a pen of hens is mated with one male. If the birds are trapnested and each hen's leg-band number recorded on its egg, this system makes it possible to know the parents of every chick hatched from a pen mating.

About the same number of hens are mated with one male in pen mating as in flock mating. However, fertility is generally not so good in pen mating as in flock mating be-

control over the stock sold by the associated hatcheries. On the other hand, hatcheries find the franchise agreement desirable because they do not have to spread themselves so thinly; they can leave the breeding problems to the foundation breeders and concentrate their efforts on hatching and selling chicks.

METHODS OF MATING

The two most common methods of mating are flock and pen mating, although stud mating and artificial insemination are sometimes used.

1. **Flock mating.** Flock or mass mating means that a number of males are allowed to run with the entire flock of hens. Other things being equal, better fertility is obtained from flock mating than from pen mating.

The number of hens per male will vary with the size and age of the birds. With the light breeds, such as Leghorns, it is customary to use one male for 15 to 20 hens. In the case of the general-purpose breeds, such as the White Wyandotte, it is customary to use one male with 10 to 15 hens. With the heavy breeds, one male is placed with each 8 to 12 hens. Also, age is a factor; young males are more active than older ones. Males past 3 years old are not too satisfactory as breeders.

Fig. 4–22. Pen mating, in which a pen of hens is mated to one male. (Photo by J. C. Allen & Son, Inc., West Lafayette, IN)

cause (a) there is no opportunity for the birds to mate with the ones they choose, and (b) there is no competition between males.

3. **Stud mating.** Stud mating is comparable to what is called hand mating among four-footed farm animals. In this method, the females are mated individually with a male that is kept by himself in a coop or pen. This system makes it possible to mate more females to a male than can be accomplished in either pen or flock mating. However, stud mating involves more labor than the other two systems because birds should be mated at least once each week in order to maintain good fertility.

Sometimes stud mating is used when a very valuable male is being used as a breeder, and it is desired to use him to the maximum.

ARTIFICIAL INSEMINATION

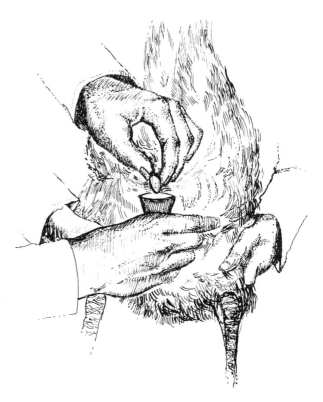

Fig. 4–24. Collection of semen from a cockerel. The easiest method to collect semen involves two people—one person to hold the bird and the other to collect the semen. The collection occurs when the operator places a hand against the tail pushing the feathers out of the way. The thumb and index finger should be placed on either side of the cloaca. A rapid, continuous massaging action (milking) is carried out by one hand, and the other hand is used to hold the collection vessel.

Fig. 4–23. Individually caged males at breeding farm, used in artificial insemination. (Courtesy, Euribrid B. V., Boxmeer, Holland)

Artificial insemination can be credited as a primary factor that has led to the rapid improvement of the genetic profile of poultry. It has served a threefold purpose.

1. A single ejaculate can be collected and used to inseminate several females, whereas natural mating only allows the ejaculate to be used for a single female.

2. Artificial insemination has permitted the geneticist to select for heavier, meatier birds. This is especially true in the turkey industry where many males are so large that natural mating is impossible.

3. Artificial insemination facilitates the use of cage systems. Natural mating is highly inefficient, or impossible, where breeders are housed in small cages.

In chickens, it is generally recommended that hens be inseminated with about 0.05 ml of whole semen. To attain maximum fertility, hens should be inseminated at 7-day intervals.

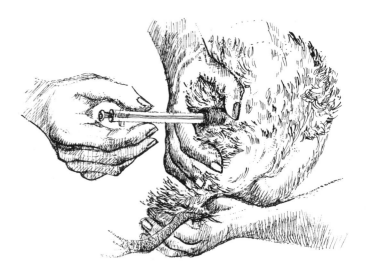

Fig. 4–25. Insemination of a hen. To inseminate the hen, one person holds the hen close to his/her body and applies pressure around the cloaca to cause the vagina to evert. A second person inserts a small syringe up the *left* oviduct about 1 in. where the semen is then deposited.

Turkey hens should be inseminated with 0.025 ml of whole semen per insemination. A fertility level in excess of 85% may be attained at an interval of 4 weeks between the first and second insemination, but most producers inseminate more frequently. Two-week intervals should maintain maximum fertility, but the insemination program should be tailored to the particular operation.

Artificial insemination is seldom used in waterfowl in the United States. However, it is being used more and more in ducks in Asia.

SYSTEMS OF BREEDING

Fig. 4–26.   Pedigreed eggs grouped by pen and hen leg band numbers. (Courtesy, Colonial Poultry Farm, Inc., Pleasant Hills, MO)

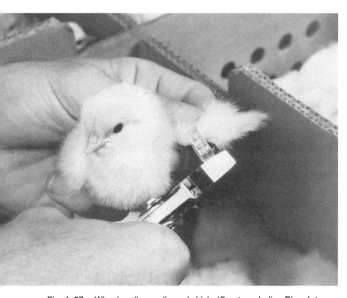

Fig. 4–27.   Wing banding pedigreed chick. (Courtesy, Indian River International, Nacogdoches, TX)

The breeding of chickens has passed through the total presently known systems of breeding; in fact, each method has been, and still is being, used successfully. But the vast majority of chickens in America today are hybrids of one form or another—either strain crosses, breed crosses, or crosses between inbred lines. A discussion of each method of breeding follows:

1. **Standard breed and variety.** This system of poultry breeding is similar to purebred breeding of four-footed farm animals. It consists in either (a) mass mating of phenotypically selected individuals of the same standard breed and variety, or (b) mating birds based on some measurement of performance, either individual, parental, sib, or progeny.

2. **Pure strains.** So-called pure strains, which are produced commercially today, are no more pure than the standard breeds and varieties (Point 1 above). A purebred strain of chickens generally takes the name of the breeder who developed it. It is called pure strain for the reason that the developer has closed the flock to outside blood for several years. Pure strains are now used extensively as parents of commercial crosses. Generally speaking, such pure strains are not for sale; they are the exclusive property of the developer who promotes them.

3. **Strain crosses.** These are crosses of two or more different strains within the same breed; hence, they are still purebreds. The crossing of two inbred strains may result in some favorable hybrid vigor in egg production, or in other economically valuable characteristics. Thus, this method of breeding has some advantages over pure-line breeding for the production of commercial stock. Most of the chickens produced today for commercial egg production are Leghorn strain crosses.

4. **Breed crosses or crossbreds.** This consists in crossing different breeds or varieties that combine well, based on the performance of the progeny. Such crossbreds usually show hybrid vigor or heterosis for egg or meat production. Among the more important commercial types of crossbreds are the following:

a. **Sex-linked cross.** When we speak of sex-linked traits, we refer to those traits that are determined by genes carried (linked) on one of the two sex chromosomes. A common sex-linked cross in chickens consists in the use of either a Rhode Island Red or a New Hampshire male on a Barred Plymouth Rock female. The male progeny from such a cross are barred like their mother, but the females are nonbarred like their father. The chicks from such a cross can be easily distinguished at hatching, with the result that the cockerels are raised for meat production, and the pullets are kept as layers.

It is noteworthy that the reciprocal cross—that is, a Barred Rock male mated on a Rhode Island Red or a New Hampshire female—produces all barred crossbred progeny.

b. **Leghorn-Red cross.** The offspring of a Leghorn male and a Rhode Island Red female is known as a Leghorn-Red cross. This cross, which was formerly used rather extensively in the Midwest in layer production, results in medium-weight layers which produce eggs of an intermediate color between the chalk white of the Leghorn and the brown of the Rhode Island Red. In addition to being good layers, when the hens are marketed as meat they usually command a better price than Leghorns.

The reciprocal cross — that is, a Rhode Island Red male mated to a Leghorn female — is quite different; the offspring mature more slowly, and are more inclined to broodiness, but they do have a lower adult mortality rate.

c. **Austra-White.** This cross, which results from mating and Australorp male and a White Leghorn female, was formerly used as layers in the Midwest and in southern California. Although they are good layers, they do tend toward excessive broodiness; and sometimes the dark shanks of the pullets and the tinted eggs are discriminated against on the market.

The reciprocal cross — Leghorn male on an Australorp female — results in a cross known as the White-Austra. These crossbreds are less inclined to broodiness than their counterparts, the Austra-White, yet they are apparently quite resistant to respiratory diseases.

5. **Inbred hybrids.** The use of inbreeding and hybridization to produce commercial hybrids is more or less patterned after the methods of the corn breeder. The actual details of commercially inbred hybrids are trade secrets of the companies producing them. Nevertheless, the general principles used by all are the same — the development and testing of many small inbred lines. Three or more generations of brother-sister matings, or sufficient generations of less intense inbreeding to produce an equivalent degree of inbreeding, may be required to produce a genetically stable stock. Some lines fall by the wayside in the process, due to reduction in hatchability, egg production, and viability under intense inbreeding, with the result that they must be discarded. The more viable and productive lines are mated with one another, and the cross-line progeny tested. Then, if the cross-line progeny is superior, further tests are conducted to find four-line (or more) combinations which will produce outstanding, uniform progeny in commercial volume. Because the inbred populations are, within themselves, not generally outstanding, only sufficient numbers of these inbred lines are maintained to produce the cross-line progeny which are mated to produce four-line combinations for commercial distribution.

In recent years, as a result of performance entries in random sample tests, inbred hybridization has lost some of its popularity, because strain-cross entries frequently prove superior to some of the inbred-hybrid entries. Thus, at this time, there is considerable debate as to the virtues of inbred hybrids. Eventually, the issue will be settled by commercial producers for, over a period of time, they will follow that system of breeding which is most profitable to them.

It is generally recognized that hybrid stocks show quicker recovery from disease outbreaks and greater resistance to unfavorable environment than standardbred stocks. On the other hand, some hybrids have suffered higher mortality than standardbred stocks from the avian leukosis complex. But hybrids with average or better resistance to leukosis have been developed.

6. **Recurrent reciprocal selection.** The most interesting development in systems of mating since inbred hybrids has been the breeding for cross-line performance by using the system of recurrent reciprocal selection. This system is much like strain crossing but differs from it in the way the pure strains are multiplied by using those individuals which cross best with the other line, rather than on the basis of their own performance.

Fig. 4–28. H & N "nick chick" Leghorn, produced by recurrent reciprocal selection. (Courtesy, Heisdorf & Nelson, Inc., Kirkland, WA)

SELECTING AND CULLING CHICKENS

Fig. 4–29. Hy-Line White Leghorn, bred and selected for white egg production. (Courtesy, Hy-Line International, West Des Moines, IA)

The terms *selection* and *culling* carry opposite connotations. Selection aims at progress; it deals with retaining the best in the flock, seldom more than the top 20 to 25%, and generally not more than 10 to 15%, for carrying forward. On the other hand, culling refers to the removal of the least productive part of the flock. It is aimed at prevention of retrogression rather than making progress.

METHODS OF SELECTION

Except for the poultry fancier, who is interested in breeds and varieties from the standpoint of *Standard of Perfection*, poultry is seldom exhibited today. Hence, only exhibition chickens are selected on the basis of show winnings.

The following methods of selection are used in poultry breeding:

1. **Individual or mass selection.** In this, selection of individuals is based on physical appearance. Many geneticists and producers maintain that this system has little or no value. Yet, two facts must be recognized: (a) A great deal of the poultry improvement to date has been the result of widely applied mass selection practices, particularly from the standpoint of lessening undesirable individuals that are slow gainers, poor layers, and which detract from the uniformity of the flock; and (b) mass selection may be effective in those cases where the flock is considerably below average in some trait— for example, in a flock where the average egg production is only 150 to 160 during the pullet year.

On the other hand, where a flock is average or above average in productivity, and for those characters of economic importance which are influenced by many different genes, individual or mass selection is ineffective.

2. **Pedigree selection.** Pedigree selection is of special importance when production data are not available, or when selection is being made between two males that are comparable in all other respects. For example, pedigree selection is of value where one is selecting breeding cockerels whose dams have very different production records, as between 250 eggs and 175 eggs. In making use of pedigree selections, however, it must be remembered that ancestors close up in the pedigree are much more important than those many generations removed.

3. **Family selection.** Family selection refers to the performance or appearance of the rest of the members of the family, particularly the bird's sibs (sisters and brothers). Sons of a 250-egg hen which was 1 of 10 sisters laying between 240 and 270 eggs each are more likely to transmit desirable genes for egg production to their daughters than are the sons of a 290-egg hen whose sisters finished the year with records of 180 to 240 eggs. Family testing has been the key to maximizing genetic progress in producing high-laying strains of chickens.

Indeed, individuality, pedigree, and family are important, and all should be used as tools in selection, but the only really sure basis concerning the ability of an individual to transmit genes for the desired characters to most of its progeny is based on a breeding test or progeny test.

4. **Blood group system.** Research work in genetics and animal breeding has demonstrated that in a population there may be more than two alternative genes that can occupy the loci on the chromosomes. Such genes are called multiple alleles. The best known theory of multiple alleles is that pertaining to blood types in humans. Three different genes are known to be involved in humans, called genes A, B, and O. So far, 12 blood group systems have been identified in chickens, and each group is controlled by a group of genes forming an allelic series.

To date, blood typing has not proved to be a simple and quick way for poultry improvement; someday it may. However, blood typing is valuable as a tool for obtaining greater insight concerning genetic mechanisms and for determining parentage.

MEASURING EGG PRODUCTION

With layers, annual egg production is, without doubt, the most important single criterion to invoke when it comes to selection. Yet, it is not the only trait of importance to the poultry breeder; he/she must select for body size and general appearance, livability, rapid growth and proper feathering, and egg size, shape, color, shell texture, and interior quality. It must be remembered, however, that selection for two or more characters automatically cuts down the effectiveness of selection for any one of them. Also, annual egg production is of low heritability, probably about 20%. All of this means that improvement in egg production through selection is relatively slow, a fact which is confirmed by egg records of the past.

1. **Trapnests.** Trapnests differ from regular nests in that they are provided with trapdoors by which birds shut themselves in when they enter. This was the invention of Professor James E. Rice, of Cornell University, about 1895.

There are two primary shortcomings to trapnesting chickens: (a) The female is in her second year of production before she is used as a breeder—hence, it lengthens the time between generations; and (b) the males are themselves untested—usually they are the sons of phenotypically superior females. Additionally, trapnesting requires more labor, particularly if it is done 7 days a week and each week of the year. However, studies have shown that a satisfactory measure of the egg-producing ability of a hen is possible by trapnesting 3 days a week. This reduces labor costs.

2. **Banding.** Some poultry producers give each pullet a numbered band, or make use of a series of colored bands of different color, in lieu of a complete quantitative trapnest record. By establishing a "key of colors," and using both the right and the left leg, it is possible to mark the birds so as to indicate such important characteristics as (a) time of starting to lay, (b) winter pause and molt, (c) broodiness, and (d) time of stopping lay in the fall.

3. **Examining the birds.** If one cares to do so, the hens may be examined early in the morning, and those which are going to lay on that day can be detected by feeling the egg in the shell gland. By going through this procedure on three successive days each month, a highly accurate record of the relative laying ability of the hens that make up the flock can be obtained, as well as an estimate of the actual number of eggs laid by each hen.

MEASURING BROILER PRODUCTION

Growth rate of broilers to market weight has an estimated heritability of 35 to 40%. This trait is of major importance in the breeding of meat chickens as well as turkeys, because rapid growth means a saving in time, labor, feed consumption, and overhead in production costs. Also, it is known that feed efficiency is highly correlated with growth rate.

Because of the high heritability of growth rate, progress in breed improvement for weight at broiler age is an excellent way in which to make progress.

Feed efficiency is also highly correlated with growth rate, and it is known that feathering is highly heritable. Because these traits are highly heritable, rapid progress in broiler improvement has been made through mass selection. Also, carcass quality and uniformity have been improved through crossbreeding.

RANDOM SAMPLING PERFORMANCE TEST; MULTIPLE UNIT POULTRY TEST

The idea for random sample testing was first proposed by Hagedoorn, a geneticist of The Netherlands, at the 1927 World's Poultry Congress held in Ottawa, Canada. Twenty years later, the first U.S. random sample laying test was conducted at Pomona, California.

Random Sample Performance Tests are designed to provide information on the performance of commercial chicks and poultry under uniform testing conditions. In these centralized tests, stocks from several growers are hatched at the test, the chicks raised, and the resultant pullets maintained to a fixed age (500 days or older) to determine egg production. Broiler tests are conducted for the purpose of evaluating growth, quality, and pounds of feed required to produce a pound of liveweight to the completion of the test. Entries are kept on the performance test for a period determined by the test management—usually 9 to 12 weeks.

Theoretically, centralized random performance tests of this kind should give a true comparative evaluation of the stocks tested. However, the following serious criticisms have been leveled at it: (1) that stocks sampled for testing are not truly random—which biases comparisons, (2) that such tests lead to monopolistic tendencies and to a serious reduction of the world's supply of potentially valuable genetic stock by the elimination of small breeders, and (3) that such tests are mainly a tool of breeders used to promote sales. Whatever the merits or demerits of the Random Sampling Performance Test, it is noteworthy that it has declined in popularity. In 1965, there were 21 egg-laying tests (17 in the United States and four in Canada), two meat production tests for broilers, and four turkey performance tests. By 1975, only seven egg-laying tests remained (six in the United States and one in Canada), and no broiler or turkey tests were conducted in the United States.

In order to alleviate some of the disadvantages and criticisms of the Random Sample Performance Test, the Multiple Unit Poultry Test evolved. In this program, as implied by the name, more than one type of test is carried out at more than one farm or location.

Even the most vociferous critics of random and multiple testing agree that performance tests of this kinds have exerted a powerful influence in developing high genetic merit of commercially sold poultry.

NATIONAL POULTRY IMPROVEMENT PLAN

The National Poultry Improvement Plan was established in 1935, with the approval of the Secretary of Agriculture and under the authority of an appropriation made by Congress for the United States Department of Agriculture, for the improvement of poultry, poultry products, and hatcheries. The National Turkey Improvement Plan became operative in 1943. Then in 1970, the two plans were consolidated into one National Poultry Improvement Plan with general provisions applicable to all classes of domesticated fowl and with special provisions applicable to problems and conditions peculiar to particular classes of such fowl.

The objective of the National Poultry Improvement Plan is to provide a cooperative state-federal program through which new technology can be effectively applied to the improvement of poultry and poultry products throughout the country. The provisions of the plan, developed jointly by industry members and state and federal officials, establish standards for the evaluation of poultry breeding stock and hatchery products with respect to production qualities and freedom from hatchery disseminated diseases. Products conforming to specific standards are identified by authorized terms that are uniformly applicable to all parts of the country.

The provisions of the plan are changed from time to time to conform with the development of the industry and to utilize new information as it becomes available.

Initially, this program aided in identifying the bacterial disease transmitted from the hen to the chick through the egg caused by the bacterium *Salmonella pullorum*. Today, breeders are not only carefully screened for this disease, but they are also screened or vaccinated against a host of other diseases, including pullorum, Arizona infection, fowl typhoid, infectious bronchitis, Newcastle, *Mycoplasma gallisepticum*, and *Mycoplasma synoviae*.

Acceptance of the plan is optional with the states and individual members of the industry within the states. Table 4–3 shows the participation of chickens and turkeys in the National Poultry and Turkey Improvement Plans, by leading states.

Information relative to the plan is available in *The National Poultry Improvement Plan*, a copy of which may be secured from the U.S. Department of Agriculture, Agriculture Research Service, BARC—East Building 265, Beltsville, MD 20705.

Proponents of the plan point out that it has been highly effective in increasing egg production and lowering pullorum disease. By 1980, pullorum disease had practically ceased to exist in the United States. Of the 35.4 million chickens tested that year, only 0.000002% were reactors; and of the 3.2 million turkeys tested in 1980, there were no reactors. This is no small achievement when it is realized that, in 1920, when pullorum testing was first started in the United States, there were 11% pullorum reactors in chicken flocks.

TABLE 4-3
PARTICIPATION OF CHICKENS AND TURKEYS IN THE NATIONAL
POULTRY AND TURKEY IMPROVEMENT PLANS, BY LEADING STATES[1]

National Poultry Improvement Plan				National Turkey Improvement Plan				
State	Hatcheries		Supply Flocks		State	Hatcheries		Supply Flocks	
	(number)	(egg capacity)	(flocks)	(birds)		(number)	(egg capacity)	(flocks)	(birds)
Alabama	57	72,316,000	417	4,602,000	Minnesota	7	8,124,000	118	808,000
Georgia	37	61,838,000	677	7,685,000	North Carolina	8	4,875,000	174	1,078,000
North Carolina	34	48,212,000	890	8,610,000	Missouri	4	2,520,000	15	188,000
Arkansas	65	42,311,000	477	3,951,000	Ohio	3	2,148,000	48	249,000
Mississippi	88	31,074,000	317	3,047,000	Virginia	3	1,788,000	43	213,000
Pennsylvania	33	15,684,000	130	804,000	Arkansas	2	1,204,000	20	122,000
Florida	35	15,561,000	146	1,330,000	Wisconsin	3	976,000	18	98,000
U.S. Total	925	423,633,000	6,181	37,324,000	U.S. Total	73	29,789,000	660	3,918,000

[1]*Agricultural Statistics 1988*, USDA, p. 355. Data for 1986.

CULLING

The attitude toward culling has changed as American laying flocks have increased in size. Today, extensive culling is no longer practiced in many high egg-producing commercial flocks. Because only healthy and well-developed birds are placed in commercial laying housing, these flocks normally contain a very small number of poor layers. Hence, it is not considered practical to disturb high producers and reduce egg production merely to identify a few culls. On the other hand, proper culling can increase efficiency of egg production in most farm flocks; and the poorer the flock, the greater the need to cull.

Culling is particularly valuable if the producer keeps hens for a second year of egg production. However, it is questionable whether or not it pays to keep a flock longer than the pullet year, since a hen lays 20 to 25% fewer eggs in her second year than in her first year.

Identifying and removing nonlaying and low-producing birds from the flock accomplishes the following: (1) keeps the egg production rate of the flock high; (2) saves the cost of feeding unproductive birds (and hens eat about 7 lb of feed per month whether laying or not); (3) reduces the spread of disease from hens to young birds; and (4) provides more space for the remaining birds.

Culling should take place throughout the year. When chicks are started, all obviously weak birds should be culled. As the flock gets older, runty and slow-growing pullets should be eliminated. When the flock is put into the laying house, slow-maturing birds should be removed. During the laying year, flock owners should remove the sick, lame, or injured birds. Generally speaking, heavy culling should not be necessary during the first 8 to 9 months of lay. However, if egg production slumps badly, it may be desirable to locate the cause and remedy it.

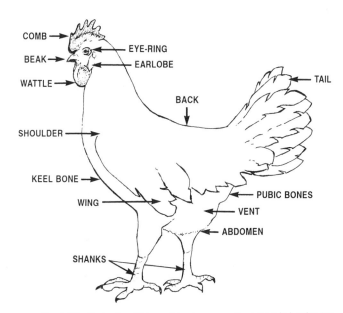

Fig. 4-30. Parts of a chicken, giving common names used in judging for production.

Fig. 4-31. Proper way to hold a bird (side view).

It is possible to secure an indication of a hen's present, past, and future egg production by physical examination of the bird. These characteristics, and their significance, are summarized in Table 4–4, Handy Culling Chart.

TABLE 4–4
HANDY CULLING CHART[1]

Separating Layers and Nonlayers

Character	Layer	Nonlayer
Comb	Large, smooth, bright red, glossy	Dull, dry, shriveled, scaly
Face	Bright red	Yellowish tint
Vent	Large, smooth, moist	Shrunken, puckered, dry
Pubic bones	Thin, pliable, spread apart	Blunt, rigid, close together
Abdomen	Full, soft, pliable	Contracted, hard, fleshy
Skin	Soft, loose	Thick, underlaid with fat

Separating High and Low Producers

Character	High Producer (continuous laying)	Low Producer (brief laying)
Vent	Bluish white	Yellow or flesh color
Eye-ring	White	Yellow
Earlobe	White	Yellow
Beak	White	Yellow
Shanks	White, flattened	Yellow, round
Plumage	Worn, soiled	Not much worn
Molting	Late, rapid	Early, slow

Characteristics of Desirable Producers

Time of maturity	Leghorns begin to lay at 5 to 5½ months; Rhode Island Reds, Plymouth Rocks, and similar breeds, at 5 to 6½ months.
Rate of production	Hens lay at least 220 eggs a year.
Broodiness	Birds are seldom broody.
Persistence of production	Good producers lay consistently for 12 to 15 months.

[1]*Cutting Hens*, USDA Farmers' Bull. No. 2216, p. 10.

BREEDS, AND BREEDING TURKEYS

Turkeys are native to America. They were found in great numbers by the pioneer settlers; and a limited number of wild turkeys still exist in certain remote areas of the United States and Mexico.

It is reported that turkeys were taken from this continent to Spain in 1498, and to England soon thereafter. Later, some of these European stocks were brought back to this country where, along with the native wild turkey, they were used in developing our present varieties.

Only one breed of turkeys is recognized by *Standard of Perfection*; hence, correctly speaking, we should refer to varieties rather than breeds.

Fig. 4–32. Bronze gobbler. (Courtesy, National Turkey Federation, Mount Morris, IL)

The varieties of turkeys of historical importance are listed in Table 4–5. Today, only the Broad-Breasted White is commercially important to the United States.

TABLE 4–5
VARIETIES OF TURKEYS AND THEIR CHARACTERISTICS[1]

Variety	Standard Weights		Plumage Color	Beak	Color of Throat Wattle	Beard	Shank and Toes	Comments
	Adult Tom	Adult Hen						
	(lb)	(lb)						
Black	33	18	Metallic black.	Slaty black.	Red, changeable to bluish white.	Black.	Pink in adults.	Black color evolved from selecting darker birds in the population. Black turkeys were popular in Spain, France, and Italy.
Bourbon Red	33	18	Brownish red, with white wing and tail markings.	Light horn at tip, dark at base.	Red, changeable to bluish white.	Black.	Reddish pink in adults.	Developed in Bourbon County, Kentucky. Very attractive.
Broad-Breasted White	33	18	Pure white.	Light pinkish horn.	Red, changeable to pinkish white.	Deep black.	Pinkish white.	Developed from crossing Broad-Breasted Bronze with white feathered variety. Very similar to Bronze; only white, and slightly higher fertility.
Bronze	36	20	Black; with an iridescent sheen of red, green, and bronze.	Light horn at tip, dark at base.	Red, changeable to bluish white.	Black.	Dull black in young; smoky pink in mature birds.	The Broad-Breasted Bronze is a sub-variety. Of all meat animals, the Broad-Breasted Bronze most uniformly produces a well-fleshed carcass.
Narragansett	33	18	Dull black, with white markings.	Horn.	Red, changeable to bluish white.	Black.	In adults, deep salmon.	Developed in Narragansett Bay area of Rhode Island, by crossing domestic stock on wild turkeys.
White Holland	33	18	Pure white.	Light pinkish horn.	Red, changeable to pinkish white.	Black.	Pinkish white.	White Holland is thought to be a sport from bronze turkeys.

[1]To convert pounds to kilograms, multiply by 0.454.

The Broad-Breasted White strain listed in Table 4–5 refers to strains that have been developed since 1950 by first crossing Bronze and White Holland varieties and then backcrossing the second generation white progeny to Bronze males. This procedure is repeated for several generations so that the resulting Broad-Breasted White is essentially a Bronze turkey with white feathering and broad-breasted in conformation. These birds have been developed in recent years in response to processors' objections to the dark pins of the Broad-Breasted Bronze. This accounts for the fact that Broad-Breasted Bronze turkeys have decreased in number since 1950, while the Broad-Breasted White variety has increased. In 1934, the USDA started the development of a small white turkey variety, which was released under the name of Beltsville Small White in 1941. However, turkey growers did not like the Small White, because the tonnage of meat produced was low. One person can take care of about the same number of big turkeys as of small turkeys, so labor cost per pound is higher on the smaller varieties. As a result, the Beltsville Small White has practically ceased to exist.

Fig. 4–33. Broad-Breasted White turkeys; male (left) and female. (Courtesy, *Turkey World*, Mount Morris, IL)

Fig. 4–34. Turkey hens in a breeder house showing trapnests on the left. Note ramp for easy access of the turkeys to the upper level nests. Breeders and facilities of Cuddy Farms, Ltd., North Carolina. (Photo courtesy James Strawser, University of Georgia, Athens)

Most turkeys are bred as standardbreds; that is, bred pure rather than hybridized. Also, individual selections and mass matings are the usual practices. Color, size, and conformation are highly heritable, with the result that they have responded well to these simple methods.

A few breeders have trapnested their birds and selected for egg production and hatchability.

In the case of turkeys, it appears that the breeding accomplishments in developing fast-growing, broad-breasted birds has left the breeder with a good market bird but with breeding populations seriously lacking in reproductive qualities. Part of this is attributed to the fact that heavily fleshed males are clumsy in mating. For this reason, many breeders are using artificial insemination.

BREEDS, AND BREEDING DUCKS

The wild mallard duck, *Anas boschas*, is the ancestor of all domestic breeds of ducks. Ducks must have been domesticated a long time, because the Romans referred to them as early as 2,000 years ago. Also, it is believed that commercial duck raising has been practiced longer in China than in any other country.

In 1985, 21.6 million ducks were marketed in the United States. Duck production was once centered on Long Island, New York. Although Long Island still produces many ducks, large numbers are now raised in Indiana, Michigan, North Carolina, and Minnesota.

The choice of a breed of ducks depends upon the market that is to be supplied. White Pekin, Aylesbury, and Muscovy ducks are excellent meat producers. Rouen, Cayuga, Swedish, and Call ducks reach market weights that make them valuable as meat producers; but poor egg production, and to some extent colored plumage, make them unsatisfactory for mass commercial production.

Khaki Campbells and Indian Runners are excellent egg-laying breeds. Accordingly, where special duck egg markets exist, the choice of either of these breeds would be wise.

Since the vast majority of ducks that are raised commercially in the United States are of the White Pekin variety, only this variety will be described. White Pekins reach market weight (7 lb) in 8 weeks. The breed originated in China and was introduced in the United States in the late 1870s. White Pekins are large, white-feathered birds, with orange-yellow bills, reddish-yellow shanks and feet, and yellow skin. Their eggs are tinted white. Adult drakes weigh 9 lb and adult females weigh 8 lb.

Fig. 4–35. White Pekin duck. (Courtesy, USDA)

White Pekins average approximately 160 eggs per year, but they are not good setters and they seldom bother to raise a brood. They are nervous, with the result that they should be treated gently.

In the United States, duck breeding has received less attention than chicken and turkey breeding, because of their lesser economic importance. Some pertinent facts about breeding ducks are: (1) breeders are usually selected from among the young ducks at market time; (2) hens should not be brought into full production before 7 months of age, because of small eggs and low hatchability; (3) one male may be mated to 6 to 7 females; (4) hens lay from 160 to more than 300 eggs per year, depending on the breed; (5) duck eggs are generally hatched in incubators; (6) the incubation period of most breeds is 28 days, although Muscovy duck eggs require 33 to 35 days; and (7) ducklings are easier to brood than chicks or poults—they require less warmth, for a shorter period of time.

(Also see Chapter 15, Ducks, Geese, and Miscellaneous Poultry.)

BREEDS, AND BREEDING GEESE

The goose was first domesticated 4,000 years ago in Egypt, where it was regarded as a sacred bird. History also records that the Romans valued goose liver as a delicacy, with the result that they placed large numbers of geese in pens and fattened them to increase the size of the liver. We are also told that they learned to use the feathers for filling mattresses, cushions, etc. Geese became well distributed over Europe during the Christian era, and even today goose raising is an important enterprise throughout eastern and western Europe.

BREEDING OTHER POULTRY

There are many other species, breeds, and varieties of poultry, some bred for fancy and show, some bred for game, others bred for racing, and still others bred to fight to the death. A few of these will be discussed:

1. **Game birds.** Game birds are raised by and for those who like to hunt, and those who like to eat them. Among such game birds are pheasants, quail, grouse, chukars (or Chukar partridge), wild ducks, and game birds bred for cockfighting.

2. **Guinea fowl.** Guinea fowl are native to Africa, but

Fig. 4–36. White Chinese geese. (Courtesy, USDA)

In the United States, geese, like ducks, are raised primarily for meat production. However, several varieties of geese are bred by poultry fanciers. Also, a considerable number of geese are used for weeding crops.

The principal meat-producing varieties of geese in the United States are the Toulouse, Emden, and African. As is true in chickens, turkeys, and ducks, a white or near-white goose can be dressed more easily and attractively.

Geese differ somewhat from ducks in their mating habits. The large breeds of geese mate best in pairs or trios, although ganders of some lighter breeds will mate satisfactorily with as many as five females. Canada wild geese are largely monogamous and will usually mate that way for life. Other facts pertinent to breeding geese are: (1) matings should not be changed from year to year, unless the birds will not mate; (2) Toulouse and Emden breeds will reproduce some in the second year, but they do not mature or give best results until the third year; (3) females may be kept until 8 to 10 years old, and ganders until 6 to 7 years old; (4) the incubation period of Canada and Egyptian geese is 35 days, for other breeds it is 29 to 31 days; and (5) when brooded in the spring, goslings need heat for only a week or so, but they should be kept dry and out of water.

(Also see Chapter 15, Ducks, Geese, and Miscellaneous Poultry.)

they were brought to Europe during the Middle Ages. They are sometimes used as a substitute for game birds. It is thought that they might be more popular were it not for their harsh and seemingly never-ending cry, and their nervous disposition.

There are three domesticated varieties of Guineas; the Pearl, the White, and the Lavender. The Pearl is by far the most popular. It has a purplish-gray plumage, dotted or "pearled" with white.

Like quail and most other wild birds, Guinea fowls have a tendency to mate in pairs. However, one male may be mated with three or four hens.

3. **Ornamental birds.** Peafowls and swans are kept chiefly for ornamental purposes.

Peafowls are native to India. They like the habitat of shrubbery or trees. Four or five hens may be mated with one cock bird.

Swans are more common in Europe than in the United States. They live in pairs and remain faithful to each other until death. Swans live to be very old; the females will breed for 30 years, and males have been known to live for more than 60 years.

4. **Ostriches.** The ostrich is the largest bird in the world. At maturity, it may stand 10 ft tall and weigh more than 330 lb. Young ostriches grow very rapidly; they reach full size in

Fig. 4–37. Ostrich. (Courtesy, Jo Schepers, Clingan's Junction, CA)

about 6 months, but they do not attain sexual maturity until 3 to 4 years of age. They may live to 70 years of age. The ostrich is the only bird that eliminates its urine and feces separately.

Ostriches are valued primarily for their skins, which are made into fine quality leather. Spasmodically, the plumes are popular for decorations and accessories. The eggs may be used for human food, but the meat is seldom consumed because it is tough and has an unpleasant taste.

Ostrich farms in the United States are relatively new; so, it remains to be seen whether ostriches in this country will be a passing fad or an infant industry.

5. **Pigeons.** Pigeons are kept in all parts of the United States for squab production, for racing, for messengers, and for exhibit.

There is a demand for squabs, especially in large cities, to take the place of game. Pigeons are the most rapid growing of all kinds of poultry. Squabs exceed the normal adult weight at the time they leave the nest, when 30 to 35 days of age. Flight and activity soon slim them down, however.

There are many varieties of pigeon, but the Homer, White King, and Swiss Mondaines are the most popular.

Pigeons mate in pairs and usually remain with their mates throughout life, although the mating may be changed if desired by placing the male and female in a coop together and leaving them there for 6 to 14 days, or until such time as they become settled.

QUESTIONS FOR STUDY AND DISCUSSION

1. Why is poultry breeding more flexible in the hands of breeders than the breeding of four-footed animals?

2. What is the essence of Mendelism?

3. Why is it said that DNA as a type of genetic engineering presents both promise and possible peril?

4. How does fertilization in poultry differ from fertilization in mammals?

5. Define (a) simple gene inheritance and (b) multiple gene inheritance, and give examples of each.

6. Define (a) dominant and (b) recessive, and give an example of each.

7. Discuss the economic importance of the inheritance of each of the following traits in poultry: (a) plumage color, (b) skin and shank color, (c) rate of feather development, (d) color sexing, (e) blood groups, (f) egg production, (g) body weight and growth, (h) body conformation, (i) sex-linked dwarfism, and (j) viability.

8. Define (a) mutations, and (b) lethals and abnormalities of development.

9. Define (a) inbreeding, and (b) heterosis.

10. Give the steps in developing inbred hybrids.

11. What is the genetic explanation for the hybrid's extra vigor?

12. Records indicate that (a) Single Comb White Leghorns compete on even terms with hybrids, and (b) the best New Hampshire strains generally equal the most rapid-growing crosses in broiler weights. Why crossbreed?

13. Will the standardbred breeds of chickens be completely eliminated in the United States?

14. Why has the inbred-hybrid scheme not proved to be commercially practical for broiler and turkey production?

15. Is there any economic justification for exhibition chickens?

16. Define (a) breed, (b) variety, and (c) *Standard of Perfection*.

17. What is the difference between foundation breeders and hatcheries?

18. Describe (a) flock mating, (b) pen mating, and (c) stud mating.

19. List the threefold purposes that artificial insemination has served.

20. Discuss each of the following methods of breeding: (a) pure strains, (b) inbred hybrids, and (c) recurrent reciprocal selection.

21. Discuss each of the following methods of selection: (a) mass selection, (b) pedigree selection, (c) family selection, and (d) blood-group system.

22. Should poultry breeders who are producing egg laying strains trapnest all of their birds?

23. Discuss each of the following: (a) Random Sampling Performance Test, (b) Multiple Unit Poultry Test, and (c) National Poultry Improvement Plan.

24. Is it necessary that modern chicken producers, of either layers or broilers, know anything about judging and culling chickens?

25. Why have not light breeds of turkeys such as the Beltsville Small White increased in importance?

26. The modern turkey breeder still makes most selections on the basis of physical appearance, whereas the modern chicken producer has largely discarded this basis of selection. Why the difference?

27. Give pertinent differences between the breeding of ducks and geese and the breeding of turkeys.

SELECTED REFERENCES

Title of Publication	Author(s)	Publisher
Animal Agriculture: The Biology of Domestic Animals and Their Use by Man	Edited by H. H. Cole M. Ronning	W. H. Freeman and Company, San Francisco, CA, 1974
Livestock and Livestock Products	T. C. Byerly	Prentice-Hall, Inc., Englewood Cliffs, NJ, 1964
Poultry Breeding and Genetics	Edited by R. D. Crawford	Elsevier Science Publishers, Amsterdam, The Netherlands, 1990
Poultry Meat and Egg Production	C. R. Parkhurst G. J. Mountney	Van Nostrand Reinhold Co., New York, NY, 1988
Poultry Production, Twelfth Edition	M. C. Nesheim R. E. Austic L. E. Card	Lea & Febiger, Philadelphia, PA, 1979
Poultry Science and Practice, Fifth Edition	A. R. Winter E. M. Funk	J. B. Lippincott Company, Chicago, IL, 1960
Standard of Perfection, The, Fifth Edition	Edited by M. C. Wallace	American Poultry Association, Inc., Crete, NE, 1966
Turkey Production: breeding and husbandry, Reference Book 242	G. A. Clayton, *et al.*	Ministry of Agriculture, Fisheries and Food, United Kingdom, 1985

Fig. 4–38.   Broiler breeder house, with part slat and part litter floor. Note nest boxes. (Courtesy, Dr. R. A. Ernst, University of California, Davis)

Fig. 4–39.   Turkey breeder females in feed conversion studies. (Courtesy, Nicholas Turkey Farms, Sonoma, CA)

Fig. 4–40.   Test lots of ducks at Cherry Valley Farms, in England, the largest duck breeding farm in the world. (Courtesy, Cherry Valley Farms, Ltd., Rothwell, England)

Fig. 5–1. By using a computerized control panel like this, the feed mill operator insures the proper mixture of ingredients for the poultry diet. (Courtesy, Delmarva Poultry Industry, Inc., Georgetown, DE)

FEEDS AND ADDITIVES

CHAPTER

5

Contents   Page

The economic importance of poultry feeding becomes apparent when it is realized that 55 to 75% of the total production cost of poultry is from feed, with the production of eggs toward the lower side of this range and production of broilers and turkeys toward the upper side. For this reason, the efficient use of feed is extremely important to poultry producers.

Fig. 5–2. This modern feed mill combines (1) a grain receiving operation, (2) an ingredient storage area, (3) a computer controlled mixing system, and (4) a pellet making operation. This feed mill provides the nutritionally balanced diets to Delmarva's nearly one-half billion chickens a year. (Courtesy, Delmarva Poultry Industry, Inc., Georgetown, DE)

The major objective of poultry feeding is the conversion of feedstuffs into human food. In this respect, the domestic fowl is quite efficient.

FEEDS FOR POULTRY

A wide variety of feedstuffs can be, and is, used in poultry rations. Broadly speaking, these may be classed as energy feedstuffs, protein supplements, mineral supplements, vitamin supplements, and nonnutritive additives.

ENERGY FEEDSTUFFS

The major energy sources of poultry feeds are the cereal grains and their by-products and fats.

GRAINS

Corn is the most important grain used by poultry, supplying about one-third of the total feed which they consume.

Fig. 5–3. Corn, the leading grain used for poultry. (Courtesy, USDA)

Oats, barley, and the sorghum grains are also used extensively in poultry rations. Oats are lower in energy than corn and are generally too expensive for broiler and layer rations. But oats can be used very effectively in feeds for replacement birds. Barley is less palatable than corn and is lower in vitamin A and energy. The sorghum grains can be readily substituted in place of corn as an energy feed, and they are being used extensively for this purpose in the southern states.

Fig. 5–3a. Barley is often used as a poultry feed.

Wheat may replace corn when available and the price is right. It is slightly lower in energy than corn, but higher in protein. Because wheat is gelatinous and has a tendency to "paste" on the beaks, it should be coarsely ground and pelleted when fed at high levels.

Surplus rice along with broken and low-grade rice, can be effectively used in poultry rations. Milled rice is quite comparable to corn in feeding value, except that it is lacking in vitamin A activity and pigmenting qualities.

Triticale, an intergeneric cross between wheat and rye, may be used as a replacement for corn or wheat with no adverse effect on growth or feed efficiency. Although millet is seldom used in poultry rations, it can be freely substituted for corn.

Rye and buckwheat are seldom used in poultry rations; rye because of being impalatable and producing sticky, pasty droppings when fed at moderate to high levels, and buckwheat because of scarcity and unpalatability. When used, rye should not replace over $\frac{1}{4}$ of the cereal grains in the ration of young chicks or more than $\frac{1}{3}$ of the cereal grains of older birds; and buckwheat should be limited to 15% of the cereal grains of the ration.

HIGH-LYSINE CORN (OPAQUE–2 OR O_2)

It has long been known that corn is nutritionally inadequate. In 1914, researchers at the Connecticut Agricultural Experiment Station induced starvation in laboratory rats by feeding them generous helpings of corn. Further, it was found that rats could be restored to health by supplementing the high-corn diet with two protein fractions — the amino acids lysine and tryptophan.

Although normal corn contains about 10% protein, half of the protein consists of zein, which is especially poor in lysine and tryptophan, essential amino acids that poultry cannot manufacture and must get from feed.

For years, plant scientists assayed the world's corn varieties one by one, looking for a strain with more nutritionally balanced protein. Finally, in 1963, a Purdue University team headed by biochemist Edwin T. Mertz analyzed an odd group of corns characterized by soft, floury endosperm inside an opaque, chalk-white kernel. The Purdue scientists found that the opaque characteristic of corn, which had been noted for years without exciting much scientific interest, is associated with a recessive gene that replaces some of the kernel's humanly useless zein with needed lysine and tryptophan. The mutant — routinely labeled opaque–2, or O_2 for short — had a lysine content of 3.4%, compared to 2.0% for normal corn. Additionally, opaque–2 showed higher levels of tryptophan and other amino acids.

Although the nutritional value of the high-lysine corn is recognized, a major hurdle between discovery and application must yet be overcome. The mutant gene is linked to opaque–2's soft, floury kernel, which is both light in weight and vulnerable to pest attacks, and which produces lower yields for farmers. But the need is great. So, plant breeders are crossing the opaque–2 gene on corn varieties that better meet the needs of poultry producers and others.

■ **High protein/high amino acid corn**—In 1989, University of Minnesota researchers announced the discovery of a gene in corn that controls the level of protein produced, which can be used to produce corn with 3% more protein and 20% more methionine and lysine than normal corn, without lowering the yields or producing soft kernels. A patent application on the process is pending.

CEREAL MILLING BY-PRODUCT FEEDS

Numerous by-products result from the milling and processing of grain. Many contain large amounts of protein as well as energy. Table 5-1 gives the cereal sources and lists the milling by-products that are commonly used in poultry feeds, whereas Fig. 5-4 shows the by-products obtained from the milling of wheat (see Chapter 18 for the feed compositions of these by-products).

TABLE 5–1
CEREAL MILLING BY-PRODUCT FEEDS

Cereal Source	Milling By-product[1]	Description	Comments
Corn, rice, wheat	Bran	Outer, coarse coat (pericarp) of grain separated during processing.	Laxative in action. Rice bran must have at least 14% crude protein and less than 14% crude fiber.
Corn, sorghum, wheat	Flour	Soft, finely ground and bolted meal from the milling of cereal grains and other seeds. Consists primarily of gluten and starch from endosperm.	Sorghum grain flour must have less than 1% crude fiber. Wheat flour must have less than 1.5% crude fiber.
Corn, sorghum, wheat	Germ meal	Embryo of the seed.	Wheat germ meal must contain at least 25% crude protein and 7% crude fat.
Corn, sorghum	Gluten (feed and meal)	Tough, viscid, nitrogenous substance remaining when flour is washed to remove the starch.	As currently manufactured, corn gluten meal contains approximately 60% protein.
All grains	Grain screenings	Small imperfect grains, weed seeds, and other foreign material of value as a feed that is separated through the cleaning of grain with a screen.	Quality varies according to percentage of weed seeds and other foreign material. Should be finely ground in order to kill noxious weed seeds.
Corn	Hominy feed	This is the by-product from producing pearl hominy.	Good hominy feed should contain a minimum of 1,350 kcal ME/lb *(2,970 kcal ME/kg)*.
Oats, rice	Groats	Grain from which the hulls have been removed.	A high-grade feed, but usually expensive. Sometimes used in chick mashes
Corn, oats	Meal	Feed ingredient in which the particle size is larger than flour.	Must contain at least 4% fat. Oatmeal must contain less than 4% fiber.
Rye, wheat	Middlings	A by-product of the flour milling industry comprising several grades of granular particles containing different proportions of endosperm, bran, and germ, each of which contains different percentages of crude fiber.	Deficient in calcium, carotene, and vitamin D. Wheat middlings cannot contain more than 9.5% crude fiber. Rye middlings cannot contain more than 8.5% crude fiber.
Barley, oats, sorghum, rice, rye, wheat	Mill-run (mill by-product or pollards)	State in which a grain product comes from the mill, usually ungraded and having no definite specifications.	Grain sorghum mill feed must contain more than 5% crude fat and less than 6% crude fiber. Oat mill by-product cannot contain more than 22% crude fiber. Rice mill by-product cannot contain more than 32% crude fiber. Rye mill-run cannot contain more than 9.5% crude fiber. Wheat mill-run cannot contain more than 9.5% crude fiber.
Rice	Polishings	By-product of rice, consisting of a fine residue that accumulates as rice kernels are polished (often hulls and bran have been removed).	High in fat; hence, rancidity can pose problems. Good source of thiamin.
Wheat	Red dog	By-product of milling spring wheat. Consists primarily of the aleurone with small amounts of flour and fine bran particles.	Cannot contain more than 4% crude fiber.
Wheat	Shorts	A by-product of flour milling consisting of a mixture of small particles of bran and germ, the aleurone layer and coarse fiber.	Cannot contain more than 7% crude fiber.

[1]Cereal by-products recognized by the Association of American Feed Control Officials. See Chapter 18 of this book for feed compositions.

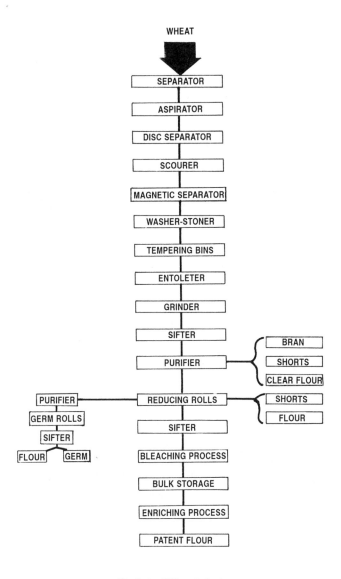

Fig. 5–4.   Milling of wheat.

BREWERS' AND DISTILLERS' GRAINS

Considerable quantities of grains are used in the brewing of beers and ales and in the distilling of liquors. After processing, the remaining by-product can be readily adapted to many feeding programs. In 1989, 125,000 tons of brewers' dried grains and 934,000 tons of distillers' dried grains were fed to livestock. In addition to those feeds, solubles and yeast products from these industries are used in livestock feeds, although to a much lesser extent.

When available at competitive prices, the mash from the brewing and distilling industries, as well as some of the pharmaceutical fermentations, can be used in poultry rations. Usually they contain some of the B vitamins and other nutrients which are supplied by fermentation.

FATS

Animal and vegetable fats are now used extensively in poultry feed. They serve the following functions: (1) increase the caloric density of the ration; (2) control dust; (3) lessen the wear and tear on feed mixing equipment; (4) facilitate pelleting of feeds; (5) increase palatability; and (6) help to homogenize and stabilize certain feed additives, especially those of a very fine particle size.

However, the use of fats in poultry feeds requires good mixing equipment. Also, it is necessary that the fat be properly stabilized in order to prevent rancidity.

Chickens are capable of tolerating high levels of fats, but costs usually permit only a maximum of 4 to 5% fat added to the ration.

MOLASSES

Molasses can be used effectively as an energy feed in poultry rations provided the level of usage is closely monitored. Excessive amounts cause wet droppings. Although molasses occasionally constitutes as much as 10% of poultry rations, levels are generally restricted to 2 to 5%.

In addition to its use as an energy feed, molasses is used in the following ways in poultry feeds: (1) to reduce the dustiness of a ration, and (2) as a binder for pelleting.

Cane and beet molasses are by-products of the manufacture of sugar from sugarcane and sugar beets, respectively. Citrus molasses is produced from the juice of citrus waste. Wood molasses is a by-product of the manufacture of paper, fiberboard, and pure cellulose from wood; it is an extract from the more soluble carbohydrates and minerals of the wood material. Starch molasses is a by-product of the manufacture of dextrose from starch derived from corn or grain sorghums in which the starch is hydrolyzed by use of enzymes and/or acid. Cane or blackstrap is, by far, the most extensively used type of molasses. The different types of molasses are available in both liquid and dehydrated forms.

PROTEIN SUPPLEMENTS

The usefulness of a protein feedstuff for poultry depends upon its ability to furnish the essential amino acids required by the bird, the digestibility of the protein, and the presence or absence of toxic substances. As a general rule, several different sources of protein produce better results than single protein sources. Both vegetable and animal protein supplements are used for poultry. Most of the protein supplements of animal origin contribute minerals and vitamins which significantly affect their value in poultry rations, but they are generally more variable in composition than the vegetable protein supplements.

PLANT PROTEINS

Even though they are not especially high in protein by comparison with other feedstuffs, the vegetative portions of many plants supply an extremely large portion of the protein in the total ration of poultry, simply because these portions of feeds are consumed in large quantities. Needed protein not provided in these feeds is commonly obtained from one or more of the oilseed by-products. The protein content and feeding value of the oilseed meals vary according to the seed from which they are produced, the geographic area in which they are grown, the amount of hull and/or seed coat included, and the method of oil extraction used.

Additional plant proteins are obtained as by-products from grain milling, brewing and distilling, and starch production. Most of these industries use the starch in grains and seeds, then dispose of the residue, which contains a large portion of the protein of the original plant seed.

OILSEED MEALS

Several rich oil-bearing seeds are produced for vegetable oils for human food (oleomargarine, shortening, and

salad oil), and for paints and other industrial purposes. In processing these seeds, protein-rich products of great value as poultry feeds are obtained. Among such high-protein feeds are soybean meal, coconut meal, cottonseed meal, linseed meal, peanut meal, rapeseed meal (canola meal), safflower meal, sesame meal, and sunflower seed meal.

Oil is extracted from these seeds by one of the following basic processes or modifications thereof: solvent extraction, hydraulic extraction, or expeller extraction.

Soybean Meal

Soybean meal has the highest nutritive value of any plant protein source and is the most widely used protein supplement in poultry rations.

In the past, oil was extracted by the solvent, hydraulic, and expeller processes (see Fig. 5–7). Today, almost all soybeans are solvent extracted. Soybean meal normally con-

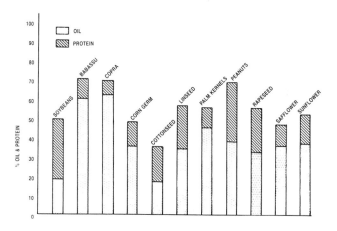

Fig. 5–5. The oil and protein content of commonly used protein supplements. (Adapted by the author from USDA data)

Fig. 5–6. A field of soybeans. (Courtesy, American Soybean Assn.)

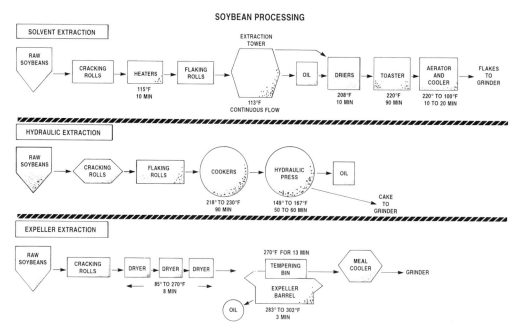

Fig. 5–7. Processing soybeans, showing the three methods.

tains 41, 44, 48, or 50% protein, depending on the amount of hull removed. Because of its well-balanced amino acid profile, the protein of soybean meal is of better quality than other protein-rich supplements of plant origin. However, it is low in calcium, phosphorus, carotene, and vitamin D.

Because raw soybeans contain several antinutritional factors, they should never be fed to poultry. However, heat processed soybean products are very acceptable poultry feeds.

Coconut Meal (Copra Meal)

This is the by-product from the production of oil from the dried meats of coconuts. The oil is generally extracted by either (1) the hydraulic process, or (2) the expeller process. Coconut meal averages about 21% protein content. The quality of the protein is not high; hence, its use should be restricted in poultry rations. If copra meal is fed to poultry, it should be supplemented with the amino acids, lysine and methionine. When properly supplemented with amino acids, copra meal can compose up to 20% of poultry rations. Light-colored meal is more digestible than the dark meal.

Cottonseed Meal

The protein content of cottonseed meal varies from about 22% in meal made from undecorticated seed to 95% in flour made from seed from which the hulls have been removed completely. Thus, by screening out the residual hulls, which are low in protein and high in fiber, the processor is able to make a cottonseed meal of the protein content desired—usually 36, 41, 44, or 48%.

Cottonseed meal is low in lysine and tryptophan and deficient in vitamin D, carotene (vitamin A value), and calcium. Also, it contains toxic substances, known as gossypol and cyclopropenoic fatty acids, varying in amounts with the seed and the processing. Gossypol can cause discoloration of egg yolks, and cyclopropenoic fatty acids impart pink color to egg whites. But, cottonseed meal is rich in phosphorus. It can be fed to growing chickens provided free gossypol levels are below 0.03% and supplemental lysine is provided.

Today, glandless cottonseed, free of gossypol, is being grown and improved. Someday, meal made from it may replace conventional cottonseed meal in poultry rations and alleviate (1) many of the restrictions as to levels of meal, and (2) the need to add iron to tie up free gossypol.

Linseed Meal

Linseed meal is the finely ground residue (known as cake, chips, or flakes) remaining after the oil extraction of flax seed. It averages about 35% protein content (33 to 37%). Linseed meal is lacking in carotene and vitamin D, and is only fair in calcium and the B vitamins. It is laxative, and can depress growth in poultry. Hence, linseed meal levels should be no higher than 5% of the poultry ration.

Peanut Meal

Peanut meal, a by-product of the peanut industry, is ground peanut cake, the product which remains after the extraction of part of the oil from peanuts by pressure or solvents. It is a palatable, high-quality vegetable protein supplement used extensively in poultry feeds. Peanut meal ranges from 41 to 50% protein and from 4.5 to 8% fat. It is low in methionine, lysine, and tryptophan, and low in calcium, carotene, and vitamin D.

Since peanut meal tends to become rancid when held too long—especially in warm, moist climates—it should not be stored longer than 6 weeks in the summer or 2 to 3 months in the winter.

Rapeseed Meal (Canola Meal)

Rape is grown extensively throughout the world. However, meal made from unimproved rape is rather unpalatable. Also, it contains several goitrogenic compounds that can pose potential hazards to poultry.

Canola was created from specially selected rapeseed by Canadian plant scientists in the 1970s. The new canola is low in glucosinolates in the meal, and low in erucic acid (a long-chain fatty acid) in the oil. Canola is grown mainly in Canada, but it is increasing in the United States.

Canola meal averages about 36% crude protein and its amino acids compare favorably with soybean meal. When the price is favorable, canola meal may be used as a protein supplement for all classes of livestock and poultry, but the amount in the total ration should be limited to about 10% for layers and breeding ducks, and 15% for breeding turkeys.

CAUTION: The lowering of the erucic acid and glucosinolate in rapeseed (canola) has proceeded at different rates in different countries. But the change is nearly complete in Canada, the leading nation, which produces about 20% of the world crop.

Safflower Meal

A large proportion of the safflower seed is composed of hull—about 40%. Once the oil is removed from the seeds, the resulting product contains about 60% hulls and 18 to 22% protein. Various means have been tried to reduce this high-hull content. Most meals contain seeds with part of the hull removed, thereby yielding a product of about 15% fiber and 40% protein. Safflower meal is deficient in lysine and methionine. Decorticated meal may be used successfully in layer rations at levels up to 15%.

Sesame Meal

Little sesame is grown in the United States despite the fact that it is one of the oldest cultivated oilseeds. The oil meal is produced from the entire seed. Solvent extraction yields higher protein (45%) but lower fat levels (1%) than either the screw press or hydraulic methods which produce meals containing about 38% protein and 5 to 11% oil. Sesame meal is extremely deficient in lysine. It is generally recommended that some soybean meal, along with added lysine, be fed with sesame meal to achieve optimum utilization.

Sunflower Seed Meal

Sunflower meal—a newcomer to the oilseed industry in the United States—is rapidly gaining acceptance as a high-quality source of plant protein. In 1989, about 1.9 million acres of sunflowers were cultivated in the United States.

Sunflower oil meal varies considerably, depending on the extraction process and whether the seeds are dehulled. Meal from prepressed solvent extraction of dehulled seeds contains about 44% protein, as opposed to 28% for whole seeds. Screw pressed sunflower meal ranges from 28 to 45% protein. When sunflower meal is used in poultry feeds, it should be combined with high-lysine supplements, such as meat scrap or fish meal, and/or lysine should be added. When sunflower meal constitutes all, or most, of the protein supplement, the ration should be pelleted or crumbled to prevent stickiness and necrosis of the beak.

ANIMAL PROTEINS

Protein supplements of animal origin are derived from (1) meat packing and rendering operations, (2) poultry and poultry processing, (3) milk and milk processing, and (4) fish and fish processing. Before the discovery of vitamin B–12, it was generally considered necessary to include one or more of these protein supplements in the rations of chickens. With the discovery and increased availability of synthetic vitamin B–12, high-protein feeds of animal origin have become less essential, although they are still included to some extent in rations for most monogastric animals.

With further improvement in the protein quality of plants, such as the development of high-lysine corn, the use of animal proteins may decline in the future. The cost of such proteins will need to be more competitive than they are at the present time if they are to be included in any major quantity in animal rations. Blending several proteins with complementary balances of amino acids and supplying more concentrated sources of individual amino acids may also be a factor affecting the future role of animal proteins in poultry rations.

Many protein supplements of animal origin are difficult to process and store without some spoilage and nutrient loss. If they cannot be dried, they must usually be refrigerated. If not heated to destroy disease-producing (pathogenic) bacteria, they may be a source of infection. On the other hand, protein availability will be reduced and some nutrients lost if the feed is heated excessively.

MEAT PACKING BY-PRODUCTS

Although the meat or flesh of animals is the primary object of slaughtering, modern meat-packing plants process numerous and valuable by-products, including protein-rich poultry feeds.

In the early 1890s, some of the selected meat residues from lard rendering—known as cracklings—were sold locally for chicken feed. Finally, about 1900, poultry feeders on the Pacific Coast, principally near Petaluma, California, started using dried tankage in poultry rations. Almost simultaneously, this new feed was utilized in hog feeding, with experiment stations leading the way. Thus was born a new era in the utilization of packinghouse by-products, and a new era in livestock feeding.

Tankage and Meat Meal

In 1989, about 2.5 million tons of tankage and meat meal were fed to livestock. Tankage and meat meal are made from the trimmings that originate on the killing floor, inedible parts and organs, cleaned entrails, fetuses, residues from the production of fats, and certain condemned carcasses and parts of carcasses.

The end products, and the methods of processing, are:

1. **Tankage (or digester tankage, meat meal tankage, feeding tankage)** is produced by the older wet-rendering method, in which all of the material is cooked by steam under pressure in large closed tanks; hence, the derivation of the name *tankage*. Tankage may also be made by the dry-rendering method (in which case it is known as *meat meal*), or by mixing products containing both wet-rendered and dry-rendered materials.

The level of protein of tankage (generally 60%) is often standardized during manufacturing by the addition of enough blood to raise the total protein to the desired level.

2. **Meat meal (or meat scrap)** is produced by the newer and more efficient dry-rendering method, in which all of the material is cooked in its own grease by dry heat in open steam-jacketed drums until the moisture has evaporated.

The level of protein of meat meal (generally 50 to 55%) was originally established because the normal proportions of raw materials available for rendering resulted in a product which, after being pressed and ground, contained approximately 50% protein. The protein content of meat meal is adjusted up or down by raising or lowering the quantity of bone and fat in the raw material.

Protein content alone is not an infallible criterion of the feeding value of tankage or meat meal, for it varies considerably according to the kind of raw material from which it is produced. For this reason, the poultry producer should (1) purchase tankage or meat meal from a reputable source, and (2) mix these products with other protein supplements, especially when the birds are confined.

Meat and Bone Meal Tankage, and Meat and Bone Meal

When, because of added bone, tankage or meat meal contains more than 4.4% phosphorus (P), the word *bone* must be inserted in the name; and they must be designated, according to the method of processing, as either (1) meat and bone meal tankage, or (2) meat and bone meal. Thus, when such high-phosphorus products are prepared by the older wet-rendering method, they are known as *meat and bone meal tankage*. Likewise, when products in excess of 4.4% phosphorus are prepared by the newer dry-rendering method, they are designated as *meat and bone meal*.

Meat and bone meal tankage and meat and bone meal, or other similar bone-containing products, contain less protein than tankage or meat meal, usually 45 to 50%.

Liver Meal, and Liver and Glandular Meal

These products are produced in limited quantities by U.S. meat-packing plants. Most of our supply comes from South America and Australia. *Liver meal* is made entirely by drying and grinding livers; whereas *liver and glandular meal* is made by drying and grinding liver and other glandular tissue, but at least 50% of the dry weight of the product must be derived from liver. These products contain proteins of high quality and are rich sources of both the fat-soluble and water-soluble vitamins.

Blood Meal

Blood meal is ground, dried blood. When prepared by a special process and reduced to a fine powder, the end product is called *blood flour*. It contains 80 to 82% protein, more than any other packinghouse product, and it is an excellent source of the amino acid lysine, of which about two-thirds is available to the bird. However, due to the high temperature of processing, the protein is less digestible and of lower quality than that in high-grade tankage or meat meal, being especially low in the essential amino acid isoleucine. Also, blood meal and blood flour differ from tankage and meat meal in that they are low in calcium and phosphorus.

In recent years, new methods of drying blood, such as flash drying, have shown considerable promise in producing a uniform high-lysine product.

POULTRY WASTES

By-product feedstuffs are derived from all segments of the poultry industry—from hatching all the way through processing for market; and they come from the broiler and turkey segments of the industry as well as from egg production. Centralization of these industries into large units with enough volume of wastes to make it feasible to process the potential feeds have opened new markets for what was previously a disposal problem. Certain precautions have had to be included to make the products most useful and safe, but considerable amounts of poultry products are currently being used in poultry rations.

The three most extensively used high-protein by-products of the poultry industry are hatchery by-products, poultry by-products, and poultry feathers. Cull birds, unsalable eggs, eggshells, and slaughter wastes are also used in poultry feeds.

Hatchery By-products

Of the various by-products from poultry, the most valuable is *hatchery by-products* consisting of infertile eggs, eggs with dead embryos, and unsalable sexed male chicks. However, these products deteriorate quite rapidly if not cooled promptly—a factor which is true of most poultry, fish, and meat products.

Poultry By-product Meal

Poultry by-products consist of nonrendered clean parts of carcasses of slaughtered poultry—such as heads, feet, and viscera—free from fecal content and foreign matter except for such trace amounts as are unavoidable in good factory practice. The meal is the dried, rendered, and ground poultry by-product.

Because of the heads and feet, poultry by-products are lower in nutritional value than the flesh of animals, including poultry. The biological value of the proteins is lower than the other animal proteins and the better plant proteins. They may be successfully used in animal rations, however, provided they are not the sole source of proteins.

Hydrolyzed poultry by-products aggregate is the product resulting from treatment under pressure of all by-products of slaughtered poultry, clean and undecomposed, including such parts as heads, feet, undeveloped eggs, intestines, feathers, and blood.

Feather Meal

Feathers, a by-product that is nearly all protein, can be used in rations after they are hydrolyzed with heat and pressure to make the proteins available. *Hydrolyzed feather meal is defined as the product resulting from the treatment under pressure of clean, undecomposed feathers from slaughtered poultry, free of additives, and/or accelerators*. Not less than 75% of its crude protein content must be digestible by the pepsin digestibility method. Although hydrolyzed feather meal is high in protein (from 85 to 90%), it is rather low in nutritional value, being low in the amino acids histidine, lysine, methionine, and tryptophan. Those amino acids which are present are readily available.

Because of the deficiencies of several amino acids, care must be used when incorporating feather meal into poultry feed. The addition of fish meal or meat meal tends to complement feather meal and facilitates its use. In practice, feather meal rarely exceeds 5% of poultry rations.

Poultry Manure

Nearly 100 million tons of poultry wastes (from layers, broilers, and turkeys) are produced annually. Because poultry production is highly intensive, with many birds in a small area, waste disposal is a major problem. Most cage-layer operations produce manure free of litter as the primary form of waste. Broiler operations, generally, produce litter.

On a moisture-free basis, cage-layer manure generally contains 25 to 35% crude protein and minimal fiber, while broiler litter contains somewhat less protein—about 18 to 30%—and substantially more fiber due to the presence of absorbent materials.

Poultry litter is the most collectable and the most nutritious of all animal wastes. It follows that many experiments have been conducted with it, involving feeding trials with different species. The results of numerous experiments are summarized in Table 5-2. The mean values for waste-fed animals reported therein were obtained by averaging all levels of feeding poultry wastes in the respective categories, though some of the levels were excessive. As shown in Table 5–2, the performance of animals fed wastes was generally slightly lower than that of the controls that were fed traditional feed ingredients. But, on a dry-matter basis, animal wastes generally make for least cost rations and highest net returns.

Dried Skimmed Milk and Dried Buttermilk

As the names indicate, these products are dehydrated skimmed milk and buttermilk, respectively. They contain less than 8% moisture and average 32 to 35% protein. One pound of dried skimmed milk or dried buttermilk has about the same

TABLE 5–2
PERFORMANCE OF ANIMALS FED RATIONS CONTAINING POULTRY WASTES[1]

Species of Experimental Animal Used	Kind of Poultry Waste Studied	Performance of Experimental Animals		
		Criteria	Control Group	Waste-Fed Group
Cattle	Dehydrated layer waste	Daily gain, lb Daily feed dry-matter intake, lb Feed/gain ratio	2.35 (*1.07 kg*) 15.82 (*7.19 kg*) 7.81	2.31 (*1.05 kg*) 15.44 (*7.02 kg*) 7.72
Lactating cows	Dehydrated layer waste	Milk yield, lb/day Milk fat, % Milk total solids, %	41.8 (*19.0 kg*) 3.51 12.04	38.94 (*17.7 kg*) 3.63 12.01
Sheep	Dehydrated layer waste	Daily gain, lb Feed/gain ratio	0.42 (*0.19 kg*) 5.52	0.40 (*0.18 kg*) 6.66
Swine	Dehydrated layer waste	Daily gain, lb Feed/gain ratio	1.32 (*0.60 kg*) 4.12	1.14 (*0.52 kg*) 4.82
Growing chicks	Dehydrated layer waste	Daily gain, g Feed/gain ratio	16.1 2.36	15.7 2.60
Laying hens	Dehydrated layer waste	Egg production, % lay Feed/dozen eggs, lb	71.9 4.18 (*1.90 kg*)	72.8 4.18 (*1.90 kg*)
Cattle	Poultry litter[2]	Daily gain, lb Feed/gain ratio	2.2 (*1.0 kg*) 10.18	1.91 (*0.87 kg*) 11.58

[1]Adapted by the author from *Unidentified Resources as Animal Feedstuffs*, NRC, National Academy Press, Washington, DC, 1983, pp. 132–144, Tables 35–41.

[2]Also, dried poultry litter has been fed successfully to dry and lactating dairy cows, to growing and breeding sheep, to growing swine, and to broilers.

Several methods have been used to process animal wastes for feed; among them, dehydration, ensiling, pelleting, preparation for liquid feeding, oxidation-ditch aerobic processing, commercial (patented) systems, and the use of wastes as substrates for single-cell protein production.

(Also see Chapter 8, Poultry Management, section headed "Poultry Manure.")

DAIRY PRODUCTS

Along about 1910, processes were developed for drying buttermilk. Soon thereafter, special plants were built for dehydrating buttermilk, and the process was extended to skimmed milk and whey. Beginning about 1915, dried milk by-products were incorporated in commercial poultry feeds. In 1989, 441,000 tons of dried dairy products (dried milk, dried skimmed milk, and dried whey) were fed to livestock.

The superior nutritive values of milk by-products are due to their high-quality proteins, vitamins, and good mineral balance, and the beneficial effect of the milk sugar, lactose. In addition, these products are palatable and highly digestible. They are an ideal supplement for balancing out the deficiencies of the cereal grains. The chief limitation to their wider use is price.

composition and feeding value as 10 lb of their respective liquid forms. Although dried skimmed milk and dried buttermilk are excellent poultry feeds, they are generally too high priced to be economical for this purpose.

Prior to World War II, dried skimmed milk was the most widely used dried milk product included in feeds. During and since the war, however, much of it has been marketed as a human food.

Whey

Whey is a by-product of cheese making. Practically all of the casein and most of the fat go into the cheese, leaving the whey which is high in lactose (milk sugar) and ash, but low in protein (0.6 to 0.9%) and fat (0.1 to 0.3%). However, its protein is of high quality.

Numerous whey products are commercially available. Those used in poultry rations when the price is right are:

1. **Dried hydrolyzed whey.** This product is the residue obtained by drying lactose hydrolyzed whey. It contains at least 30% glucose and galactose.

2. **Dried whey.** This product is derived from drying whey from cheese manufacture. It is high in lactose (milk sugar), containing at least 65%, and rich in riboflavin, pantothenic acid, and some of the important unidentified factors.

One pound of dried whey contains about the same nutrients as 13 to 14 lb of liquid whey.

3. **Dried whey product.** When a portion of the lactose (milk sugar) which normally occurs in whey is removed, the resulting dried residue is called dried whey product. According to the Association of American Feed Control Officials, the minimum percentage of lactose must be prominently declared on the label of the dried whey product. Dried whey product is a rich source of the water-soluble vitamins.

4. **Dried whey solubles.** This product is obtained by drying the whey residue from the manufacture of lactose following the removal of milk albumin and the partial removal of lactose. The minimum percentage of lactose must be prominently declared on the label.

Milk Protein Products

The four milk protein products currently available are:

1. **Dried milk protein.** This feed is obtained by drying the coagulated protein residue resulting from the controlled coprecipitation of casein, lactalbumin, and minor milk proteins from defatted milk.

2. **Dried milk albumin.** This product is produced by drying the coagulated protein residue separated from whey. It contains at least 75% crude protein on a moisture-free basis.

3. **Casein.** Casein is the solid residue that remains after the acid or rennet coagulation of defatted milk. It contains at least 80% crude protein.

4. **Dried hydrolyzed casein.** This is the residue obtained by drying the water-soluble product resulting from the enzymatic digestion of casein. It contains at least 74% crude protein.

Although the quality and quantity of protein from milk protein products are excellent, these sources of protein are generally too expensive to be used routinely as poultry feeds.

MARINE BY-PRODUCTS

Marine by-products are generally considered by poultry producers to be excellent sources of nutrients. Proteins, vitamins, and minerals are all readily available in most fish products.

Fish Meal

Fish meal — a by-product of the fisheries industry — consists of dried, ground whole fish or fish cuttings — either or both — with or without extraction of part of the oil. If it contains more than 3% salt, the salt content must be a part of the brand name. In no case shall the salt content exceed 7%. In 1989, 386,000 tons of fish meal were processed in the United States.

The feeding value of fish meal varies somewhat, according to:

1. **The method of drying.** It may be either vacuum, steam, or flame dried. The older flame drying method exposes the product to a higher temperature. This makes the proteins less digestible and destroys some of the vitamins.

2. **The type of raw material used.** It may be made from the offal produced in fish packing or canning factories, or from the whole fish with or without extraction of part of the oil.

Fish meal made from offal containing a large proportion of heads is less desirable because of the lower quality and digestibility of the proteins. Although few feeding comparisons have been made between the different kinds of fish meals, it is apparent that all of them are satisfactory when properly processed raw materials of good quality and moderate fat content are used.

3. **The amount of oxidation.** Much of the variation in the efficiencies of fish meal is due to the oxidation. Today, ethoxyquin is being added to many fish meals to prevent oxidation.

It is of interest to poultry producers to know the sources of the commonly used fish meals. These are:

1. **Menhaden fish meal.** This is the most common kind of fish meal used in the eastern states. It is made from menhaden herring (a very fat fish not suited for human food) caught primarily for their body oil. The meal is the dried residue after most of the oil has been extracted.

Fig. 5–8. Menhaden herring are a fat fish considered to be unsuitable for human consumption. They are caught primarily for their body oil, but the dried residue after oil extraction provides an excellent high-quality protein feed. (Courtesy, National Fisheries Institute, Inc., Washington, DC)

2. **Sardine meal or pilchard meal.** This is made from sardine canning waste and from the whole fish, principally on the West Coast.

3. **Herring meal.** This is a high-grade product produced in the Pacific Northwest and Alaska.

4. **Salmon meal.** This is a by-product of the salmon canning industry in the Pacific Northwest and in Alaska.

5. **White fish meal.** This is a by-product from fisheries making cod and haddock products for human food. Its proteins are of very high quality.

Fish meal should be purchased from a reputable company on the basis of protein content. It varies in protein content from 57 to 77%, depending on the kind of fish from which it is made. When of comparable quality, fish meal is superior to tankage or meat meal as a protein supplement for poultry. The protein of a good-quality fish meal is 92 to 95% digestible. If fish meal is poorly processed or improperly stored, the digestibility of protein decreases dramatically. Since fish meals are cooked, there is danger that certain amino acids — notably lysine, cystine, tryptophan, and histidine — will be denatured, but these losses are minimized when proper processing techniques are used.

Fish meals containing high levels of fat are considered to be low quality. If they are incorporated into poultry feeds, they tend to impart a fishy flavor to poultry products. Also, problems of rancidity are greater in high-fat fish meals.

Fish meal is an excellent source of minerals. Calcium and phosphorus are especially abundant, being present in the amounts of 3 to 6% and 1.5 to 3.0%, respectively. Many of the trace minerals required by poultry, especially iodine, can be supplied in part by fish meal.

Fish meal is not a particularly good source of vitamins. Most of the fat-soluble vitamins are lost during the extraction of oil, but a fair amount of the B vitamins remain. However, fish meal is one of the richest sources of vitamin B-12 and unidentified growth factors.

Fish Residue Meal

This is the dried residue from the manufacture of glue from nonoily fish. If the product contains more than 3% salt, the amount of salt must be included in the product name. However, salt content must not exceed 7%.

Condensed Fish Solubles

This is a semisolid by-product obtained by evaporating the liquid remaining from the steam rendering of fish, chiefly sardines, menhaden, and redfish. The water that comes from processing contains about 5% total solids. After this liquid is evaporated or condensed, it contains about 50% total solids. Condensed fish solubles, containing approximately 30% protein, are a rich source of the B vitamins and unknown factors. They are particularly rich in pantothenic acid, niacin, and vitamin B-12.

Dried Fish Solubles

Dried fish solubles are obtained by dehydrating the glue water of fish processing. It contains at least 60% crude protein.

Fish Liver and Glandular Meal

This marine by-product is obtained by drying the entire viscera of fish. The Association of American Feed Control Officials (AAFCO) specifies that it must contain at least 18 mg of riboflavin per pound, and that at least 50% of dry weight must consist of livers.

Fish Protein Concentrate

Fish protein concentrate is prepared through the solvent extraction processing of clean, undecomposed whole fish or fish cuttings. The product cannot contain more than 10% moisture and must contain at least 70% protein, according to AAFCO. The solvent residues must conform to the rules as set forth by the Food Additive Regulations.

Shrimp Meal and Crab Meal

Shrimp meal and crab meal are high-quality protein feeds for poultry.

Shrimp meal is the ground, dried waste of shrimp processing. It may consist of the head, hull (or shell), and/or whole shrimp. The provisions relative to salt content are the same as those for fish meal. It is either steam dried or sun dried, with the former method being preferable. On the average, it contains 32% protein and 18% mineral.

Crab meal, the by-product of the crab industry, is composed of the shell, viscera, and flesh. Mineral content is exceedingly high, about 40%, and protein content is above 25% (generally about 30%). A rule of thumb for feeding crab meal is that 1.6 lb of crab meal can replace 1 lb of fish meal.

Whale Meal

Whale meal is the clean, dried, undecomposed flesh of the whale after part of the oil has been extracted. If it contains more than 3% sodium chloride, the amount of salt must constitute a part of the product name. But the amount of salt cannot exceed 7%.

YEASTS

(**Note:** Herein, yeasts are listed as protein supplements. But they may also be classed as vitamin supplements.)

Three forms of dried yeast are used in poultry feeding: (1) dried yeast which is a by-product of the brewing industry; (2) torula yeast, resulting from the fermentation of wood residue and other cellulose sources; and (3) irradiated yeast which is used because of its vitamin D_2 content. Of these, only brewers' dried yeast, which must contain not less than 35% crude protein, is used to any extent in poultry feeds.

A wide variety of materials can be used as substrates for the growth of yeasts. Current research deals with the use of industrial by-products which otherwise would have little or no economic value. By-products from the chemical, wood and paper, and food industries have shown considerable promise as sources of nutrients for single-cell organisms; among them, (1) crude and refined petroleum products, (2) methane, (3) alcohols, (4) sulfite waste liquor, (5) starch, (6) molasses, and (7) cellulose.

Yeast proteins are generally of good quality, although they are deficient in methionine. They are good sources of lysine. Yeasts are excellent sources of the B complex vitamins.

MINERAL SUPPLEMENTS

Mineral supplements are required by poultry for skeletal development of growing birds, for eggshell formation of laying hens, and for certain other regulatory processes in the body.

Of the macrominerals demonstrated to be required by poultry, salt (sodium chloride), calcium, magnesium, potassium, and phosphorus, are routinely considered for poultry rations.

The producer should add trace minerals that are believed to be in short supply. This is usually accomplished through the use of a commercial trace mineral mix.

Several products offering minerals in chelated form are on the market. Those selling chelated minerals generally recommend a smaller quantity of them (but at a higher price per pound) and extoll their *fenced-in* properties. When it comes to synthetic chelating agents, much needs to be learned about their selectivity toward minerals, the kind and quantity most effective, their mode of action, and their behavior with different species of animals and with varying rations.

Salt is added to most poultry rations at a 0.2 to 0.5% level. Too much salt will result in increased water consumption and wet droppings.

(For further information on minerals, also see Chapter 6, Fundamentals of Poultry Nutrition; and Chapter 7, Poultry Feeding Standards, Ration Formulation, and Feeding Programs.)

GUIDELINES FOR MINERAL SUPPLEMENTATION

No single plan can be proposed as being the best for mineral supplementation. Rather, the producers must tailor their supplement regimens to encompass the following considerations:

1. **Needs of the birds.** Age, sex, weight, and production parameters must all be considered.
2. **Types of feed.** A high-energy ration will require a different mineral supplement than a low-energy ration.
3. **Region from which the feeds were obtained.** The mineral content of the feed will reflect the mineral composition of the soil on which it was grown.

CALCIUM AND PHOSPHORUS SUPPLEMENTS

The common calcium supplements used in poultry feeding are ground limestone, crushed oystershells or oyster flour, bone meal, calcite, chalk, and marble.

Most of the phosphorus in plant products is in organic form and not well utilized by young chicks or turkey poults. Hence, for poultry, emphasis is placed upon inorganic phosphorus sources in feed formulation. The most important phosphorus sources in poultry feeding are dicalcium phosphate, defluorinated rock phosphate, and steamed bone meal. All these are calcium phosphates which can supply both calcium and phosphorus.

Bone meal, dicalcium phosphate, defluorinated phosphate, colloidal phosphate, and raw rock phosphate are used where both calcium and phosphorus are needed in the ration.

PRECAUTIONS RELATIVE TO CALCIUM AND PHOSPHORUS SUPPLEMENTS

Phytin phosphorus is unavailable to poultry, and other simple-stomach animals, because of lack of the enzyme, phytase, in the gastrointestinal tract. In plants, approximately one-third of the phosphorus has been found to be present as nonphytin phosphorus and is available to chickens. In calculating the available phosphorus content of feed, the phosphorus from inorganic supplements and animal feedstuffs is considered to be 100% available while that from plants is assumed to be 30% available.

During World War II, the shortage of phosphorus feed supplements led to the development of defluorinated phosphates for feeding purposes. Raw, unprocessed rock phosphate usually contains from 3.25 to 4.0% fluorine, whereas steamed bone meal normally contains only 0.05 to 0.10%. Fortunately, through heating at high temperatures under conditions suitable for elimination of fluorine, the excess fluorine of raw rock phosphate can be removed. Such a product is known as defluorinated rock phosphate.

The Association of American Feed Control Officials has established maximum fluorine content for (1) mineral substances, and (2) total ration (see Table 5–3).

TABLE 5–3
MAXIMUM FLUORINE CONTENT FOR
(1) MINERAL SUBSTANCES AND (2) TOTAL RATION[1]

Class of Animals	Maximum Fluorine Content of Any Mineral or Mineral Mixture Which Is to Be Used Directly for the Feeding of Animals Shall Not Exceed—	Fluorine Content of Rock Phosphate (or other Ingredients) Shall Be Such That the Maximum Fluorine Content of the Total Ration Shall Not Exceed—
	(%)	(%)
Poultry	0.60	0.035
Cattle	0.30	0.009
Sheep	0.35	0.01
Swine	0.45	0.014

[1]*Feed Control*, official publication, Association of American Feed Control Officials, 1975, p. 46.

VITAMIN SUPPLEMENTS

As with mineral supplements, careful consideration must be given to the vitamin supplementation of poultry feeds. While the requirements of vitamins are extremely small in comparison with energy and protein, the omission of a single vitamin from the diet of a species that requires it will produce specific deficiency symptoms, thereby reducing production. Moreover, the cost for vitamin supplementation constitutes a very small fraction of the total feed bill.

It is to be emphasized that subacute deficiencies can exist although the actual deficiency symptoms do not appear. Such borderline deficiencies are both the most costly and the most difficult with which to cope, going unnoticed and unrectified; yet they may result in poor and expensive production. Also, under practical conditions one will usually not find a vitamin deficiency which involves only a single vitamin. Instead, deficiencies usually represent a combination of factors, and usually the deficiency symptoms will not be clear-cut.

Formerly, a wide variety of feed ingredients were added to poultry rations for their vitamin content. But it was found that the vitamin concentration of feedstuffs varied tremendously, being affected by plant species and part (leaf, stalk, or seed), harvesting, storing, and processing. Generally speaking, vitamins are easily destroyed by heat, sunlight, oxidation, and mold growth. So, today, nutritionists rely on vitamin supplements, which in many cases are chemically pure sources that need to be used only in very minute amounts. In modern feed formulation, premixes often represent the common-sense approach to providing vitamins.

(For further information on vitamins, also see Chapter 6, Fundamentals of Poultry Nutrition; and Chapter 7, Poultry Feeding Standards, Ration Formulation, and Feeding Programs.)

UNIDENTIFIED FACTORS

In addition to the vitamins as such, certain unidentified factors are important in poultry nutrition. They are referred to as *unidentified* because they have not yet been isolated or synthesized in the laboratory.

It is a common practice to include sources of unidentified factors in rations for starting chicks, broilers, and breeding hens. These sources include distillers' dry solubles, condensed fish solubles, various fermentation by-products, dried whey, dried yeast, and alfalfa meal. Most of the unidentified factor sources are added to the diet at a level of 1 to 3%.

ALFALFA PRODUCTS

Alfalfa hay or leaf meal is included in some poultry rations to supply vitamin A, riboflavin, vitamin E, vitamin K, unidentified factors, minerals, and protein. Dehydrated alfalfa meal is superior to sun-cured meal because of its higher vitamin content. Leaf meal is better than hay because of its lower fiber and greater nutrient content.

Poultry nutritionists interested in high-energy rations, are reducing or eliminating alfalfa and substituting synthetic forms of vitamins, because alfalfa meal is fibrous and unpalatable, and its vitamins are not stable.

A good dehydrated alfalfa meal should contain 100,000 I.U. of vitamin A activity per pound.

Alfalfa processors are improving alfalfa products by the addition of antioxidants to preserve the fat-soluble vitamins and carotenoid pigment, and pelleting alfalfa meal to prevent dustiness and help preserve nutritional values.

NONNUTRITIVE ADDITIVES

Modern poultry feeds commonly contain one or more nonnutritive additives. These additives are used for a variety of reasons. They are not nutrients, but some of them improve production under certain circumstances. Others prevent rancidity in the feed. There is no evidence of a nutritional deficiency when they are omitted from a ration.

When considering the use of a feed additive, the following questions should be asked:

1. **What are the specific needs of the birds?** As intensity of production increases, the need for production stimulants increases.

2. **Does the additive have a withdrawal period or a product discard period?** Quite often a drug will have a required withdrawal period before the birds can be slaughtered or before the eggs can be marketed. If a withdrawal period is required, the producer must have separate holding facilities for unmedicated and medicated feeds. Additionally, feed mixing facilities must be thoroughly cleaned after mixing medicated feed. If residues are found in the market products, the producer is subject to prosecution.

The ultimate guide to using any drug properly is the label or tag. These should be read and followed carefully.

3. **Can the additive be used in combination with other additives?** The FDA has strict regulations as to which additives can be used in combination. If there is any doubt, the producer should contact the county agent or a reputable feed company.

4. **What is the best form of the product to be used?** Likewise, the active ingredient of one product may be more stable or readily available than that of a competing brand.

5. **Must method of mixing and storing be considered?** The properties of various additives are such that methods of mixing and storing should be considered.

A long list of production-promoting additives is approved by FDA for incorporation in poultry rations. But this list is being constantly revised as each product is reviewed on the basis of (1) its safety to the birds, and (2) its safety to humans who consume the poultry products. Thus, the status of each of these products is subject to change. Hence, the user is urged to keep abreast of current recommendations. The label of each additive contains information concerning the restrictions of use in combination with other drugs, level of incorporation, and withdrawal periods for meat-type birds. Additional information pertaining to additives appears in the *Feed Additive Compendium* and in the Federal Register. These instructions must be followed carefully as misuse of feed additives can lead to costly condemnations and fines.

ANTIOXIDANTS

All feeds are susceptible to spoilage, but those which are high in fat content are especially prone to autoxidation and subsequent rancidity. To curb the oxidation of feeds, antioxidants are routinely added to many poultry feeds.

Antioxidants are compounds that prevent oxidative rancidity of polyunsaturated fats. It is important that rancidity of feeds be prevented because it may cause destruction of vitamins A, D, and E, and several of the B complex vitamins.

Also, the breakdown products of rancidity may react with the epsilon amino groups of lysine and thereby decrease the protein and energy value of the diet.

The antioxidants which are presently approved for addition to fat in poultry feeds are butylated hydroxyanisole (BHA), butylated hydroxytoluene (BHT), and ethoxyquin. These antioxidants may be used to prolong the induction period in fats and to prevent oxidation in mixed feeds. They are used as a level of 1/4 lb per ton. Antioxidants are capable of temporarily inhibiting the destructive effects of oxygen on sensitive feeding ingredients — the unsaturated fats, fat-soluble vitamins, and other constituents. They are normally incorporated in the vitamin-trace mineral premix to prevent vitamins A and E from oxidative destruction. Some are added to feed fats to stabilize them against rancidity.

FLAVORING AGENTS

Flavoring agents are feed additives that are supposed to increase palatability and feed intake.

Chickens have limited ability to differentiate between sucrose solutions, for which they show preference, and saccharine solutions which they avoid. Other studies have shown that chickens possess a sense of taste, but very limited ability to smell. Numerous chemical agents are so objectionable to chickens that they cause a decrease in normal feed consumption, but no flavoring agent has yet been found which will increase feed consumption above that normally obtained from well-balanced poultry rations composed largely of good-quality corn and soybean meal. However, there is evidence that certain natural feedstuffs are relatively unpalatable to chickens; among them, barley, rye, and buckwheat. Whether chickens avoid certain feedstuffs on the basis of taste, lack of eye appeal, or because of adverse effects upon metabolism or *sense of well-being* is unknown.

There is need for flavoring agents that will help to keep up feed intake (1) when highly unpalatable medicants are being administered, (2) during attacks of diseases, (3) when birds are under stress, or (4) when a less palatable feedstuff is being incorporated in the ration.

PELLET BINDERS AND ADDITIVES THAT ALTER FEED TEXTURE

Pellet binders are products that enhance the firmness of pellets. Several feed additives are known to produce a marked increase in the firmness of pellets; among them, (1) sodium bentonite (clay), (2) liquid or solid by-products of the wood pulp industry, consisting mainly of hemicelluloses, or combinations of hemicelluloses and lignins, and (3) guar meal, an annual legume produced in Asia. Although bentonites have no nutritive value, several reports indicate that at the level of common usage (2 to 2.5% of the ration) they may even improve the growth and/or feed utilization of animals. Hemicellulose preparations at levels up to 2.5% may serve as good energy sources, but lignin has practically no nutritive value. In addition to its binding properties, guar meal is a satisfactory source of protein and energy when limited to 2.5 to 5.0% of well-balanced diets.

Molasses or fat are sometimes added to feed as an aid in pelleting, as well as a concentrated source of energy.

ADDITIVES THAT ENHANCE THE COLOR OR QUALITY OF POULTRY PRODUCTS

Feeds that contain large amounts of xanthophylls and carotenoids produce a deep yellow color in the beak, skin, and shanks of yellow-skinned breeds of chickens. The consumer associates this pigmentation with quality and in many cases is willing to pay a premium price for a bird of this type. Also, processors of egg yolks are frequently interested in producing dark-colored yolks to maximize coloration of egg noodles and other food products. The latter can be accomplished by adding about 60 mg of xanthophyll per kilogram of diet. In recognition of these consumer preferences, many producers add ingredients that contain xanthophylls to poultry rations.

It is not necessary to incorporate high levels of xanthophyll in starter and grower rations. Low levels can be maintained through these periods of feeding, but finishing rations should contain high levels of these pigment-producing compounds.

Only a few natural products contain significant quantities of xanthophyll. The xanthophyll content of some feedstuffs follows:

Feedstuff	Xanthophyll Content (mg/kg)
Marigold petal meal	7,000
Algae, common, dried	2,000
Alfalfa juice protein, 40% protein	800
Alfalfa meal, 20% protein	240
Corn gluten meal, 41% protein	132
Corn, yellow	22

Although the xanthophylls in corn and alfalfa are utilized efficiently, both ingredients lose xanthophyll, or lutein, rather quickly when stored.

Today, certain synthetic xanthophylls are produced for use to supplement the natural xanthophylls of feedstuffs. Their use should be in keeping with Food and Drug Administration approval and manufacturer's directions.

GRIT

Since poultry do not have teeth to facilitate grinding of feed, most grinding takes place in the thick-muscled gizzard. The more thoroughly feed is ground, the more surface area is created for digestion and subsequent absorption. Hence, when hard, coarse, or fibrous feeds are fed to poultry, grit is sometimes added to supply additional surface for grinding within the gizzard. Additionally, grit serves to break down ingested feathers and litter which can sometimes lead to gizzard impaction. When mash or finely ground feeds are used, the value of grit is greatly diminished.

Oyster, clam, and coquina shells and limestone are sometimes used for grit. Being relatively soft and calcareous, they provide a source of calcium as they, too, are ground in the process. Gravel and pebbles have been used successfully as long-lasting sources of grit. Several granite products are available commercially.

CHELATES

Chelating agents, such as EDTA, are sometimes used to increase the availability and absorption of certain minerals. For example, in chicks zinc absorption has been demonstrated to be enhanced through the addition of EDTA.

Chelated forms of various trace minerals, such as iron, copper, and cobalt, are now being marketed as mineral supplements.

Some chelating agents perform useful and vital functions in the animal body. Others cause drastic interference with metabolism. These effects depend upon the stability constants, the absorbability, and the release ease with which the trace elements may be released from the chelating agents. Until we learn what chelates are useful and beneficial, and which are harmful and toxic, many unexpected results may be produced from attempts to make practical use of them.

ENZYMES

Enzymes are complex protein compounds produced in living cells which cause changes in other substances without being changed themselves. They are organic catalysts.

Normally, the enzymatic output of the digestive system of birds is adequate for maximum digestion of the starches, fats, and proteins. However, repeated experiments in many laboratories with western barley have shown that the metabolizable energy value of barley for poultry produced in the West under semiarid conditions is improved either by soaking in water before being fed or by the addition of enzyme preparations derived from fungal fermentations. Yet, little or no improvement in metabolizable energy has been shown from the use of the same fungal enzymes with other common poultry feedstuffs, or even with barley produced in the eastern part of the United States.

ANTIFUNGAL ADDITIVES (MOLD INHIBITORS)

Antifungals are agents that destroy fungi.

Feeds provide an excellent environment for the growth of fungi, such as *Aspergillus flavus, Fusarium,* and *Candida albicans,* which are detrimental to the health of poultry. *A. flavus* produces a potent toxin which is referred to as aflatoxin. *C. albicans* is the causative agent of a condition in poultry called *thrush* or *moniliasis.*

In recent years, nutritionists have been giving increasing attention to the effects of fungal infestations of feeds. It has been speculated that perhaps many nutritional problems of the past (for example, suspected nutrient deficiencies) were, in fact, caused by feeds contaminated with fungi.

Fungi can affect feed intake and subsequent production through contamination at one or more of four stages in the feeding chain: (1) in the field (preharvest), (2) during storage, (3) at mixing, and (4) in the bird itself. Fungal contamination can pose problems through the production of toxins, alterations of the chemical composition of the feed, or alterations of the metabolic functioning of the bird ingesting or harboring the fungus.

Certain fungi, most notably the organism *Aspergillus flavus,* produce a toxin. The toxin produced by *A. flavus* is known as *aflatoxin.* Aflatoxin, which has clearly been shown to be a carcinogen (tumor producing), causes much trouble in animals. But it is not the only mycotoxin to be feared. Mycotoxins affect all species, especially the young. Generally, ruminants appear to tolerate higher levels of mycotoxins and longer periods of intake than simple-stomached animals. Growing chickens are markedly less susceptible to aflatoxins than ducklings, goslings, pheasants, or turkey poults. Fish are one of the most susceptible animal species to aflatoxin poisoning.

The primary condition conducive to the growth of *A. flavus,* and therefore the production of aflatoxin, is moisture; hence, proper harvesting, drying, and storage will minimize mold growth and subsequent toxin production. All feedstuffs should be dried below the critical moisture content which permits the growth of molds—approximately 12%. Additionally, mold inhibitors should be added to high-moisture feeds that are exposed to air during storage. Propionic acid, acetic acid, and sodium propionate are used in high-moisture feeds to inhibit mold growth. Many feed manufacturers add such antifungals as nystatin and copper sulfate to concentrate feeds to prevent further growth by molds. It should be noted, however, that these compounds do not improve the value of moldy feeds.

The toxicity of aflatoxin-contaminated feed can be reduced by irradiation with ultraviolet light or exposure to anhydrous ammonia under pressure.

DRUGS

Poultry rations frequently contain drugs designed to promote production or prevent a specific disease. For example, a wide variety of chemical substances, sold under various trade names, is available for use in the prevention of coccidiosis. These drugs are known as *coccidiostats.*

Turkey rations are frequently formulated with drugs for the prevention of blackhead. This class of drugs, known as *histomonostats,* also contains a wide variety of chemical substances sold under various trade names.

When any drug is incorporated into poultry feed, it is mandatory that the producer obey all restrictions concerning (1) the length of usage, (2) specific types of birds the drug may be used for, and (3) withdrawal periods prior to marketing.

PRODUCTION-PROMOTING DRUGS

The list of drugs that has been shown to exert production-promoting effects is long and diverse. Nonetheless, most of these drugs can be classified as either an antibiotic, an arsenical, or a nitrofuran.

Antibiotics

Antibiotics are substances which are produced by living organisms (molds, bacteria, or green plants) and which have bacteriostatic or bactericidal properties.

In 1949, quite by accident, while conducting nutrition studies with poultry, Jukes of Lederle Laboratory and McGinnis of Washington State University, obtained startling growth responses from feeding a residue from Aureomycin production. Later experiments revealed that the supplement used by Jukes and McGinnis—the residue from Aureomycin production—supplied the antibiotic chlortetracycline. Such was the birth of feeding antibiotics to livestock.

The primary reason for using antibiotics in poultry feeds is for their growth-stimulating effect, for which purpose they are generally used in both broiler and market turkey rations. The reasons for the beneficial effects of antibiotics still remain obscure, but the best explanation for their growth-stimulating activity is the disease level theory, based on the fact that antibiotics have failed to show any measurable effect on birds maintained under germ-free conditions.

In addition to their use as growth stimulators, antibiotics are used to increase egg production, hatchability, and shell quality in poultry. They are also added to feed in substantially higher quantities to remedy pathological conditions.

Antibiotics are generally fed to poultry at levels of 5 to 50 g per ton of feed, depending upon the particular antibiotic used. Higher levels of antibiotics (100 to 400 g per ton of feed) are used for disease control purposes. The antibiotics most commonly used in poultry rations are bacitracin, virginiamycin, bambermycin, and lincomycin.

High levels of calcium in a laying mash will inhibit assimilation of certain tetracycline-type antibiotics to the bloodstream and reduce their effectiveness.

In all probability, antibiotics will always be used as feed additives to control and treat health problems in poultry. But the status of many commonly used antibiotics as production stimulators is, at the present time, tenuous. Pressure from consumer groups and medical people may result in banning many of the antibiotics that are primarily used for medicinal purposes in humans from the list of approved production promoters.

However, in the future, an increasing number of antibiotics will likely be developed specifically for the purpose of improving poultry performance. One example is that of bambermycin. This antibiotic was developed solely for use as a production promoter, serving to increase rate of gain and feed efficiency in chickens and swine. It has no medicinal applications, and, therefore, poses no health hazard with regards to bacteria becoming resistant to it.

Arsenicals and Nitrofurans

The use of arsenic and its compounds dates back to antiquity. Hippocrates (460–377 B.C.) used realgar (arsenic sulfide) in the treatment of ulcers. In the Middle Ages, its poisonous properties became known and widely used, particularly by women, to dispose of their adversaries or of unwanted suitors.

Arsenicals and nitrofurans exert many of the same effects as the antibiotics; hence, they are often added to poultry feeds to improve performance. It would appear that the action of arsenicals and that of antibiotics are very similar, since the effects of the two are not considered to be additive. For broilers, arsanilic acid (or sodium arsanilate) is used at 45 to 90 g per ton, and 3–nitro is used at 22.5 to 45.0 g per ton of ration; but, in keeping with the recommendation made relative to the use of any drug, the manufacturer's directions should be followed.

ANTIPARASITIC DRUGS

To combat parasitic infestations, many drugs in the form of feed additives are currently available. These additives can be separated into two general categories—anticoccidials and anthelmintics.

Anticoccidials (Coccidiostats)

Anticoccidial drugs are used to control coccidial infection—a parasitic disease caused by microscopic protozoan organisms known as *Coccidia*, which live in the cells of the intestinal lining of animals. Each class of domestic livestock harbors its own species of *Coccidia*, thus there is no cross-infection between species. However, within species, coccidiosis can spread very rapidly among animals.

This disease affects poultry, cattle, sheep, goats, and rabbits. However, poultry are most affected by coccidiosis. Hence, anticoccidials are routinely added to poultry feed. Each coccidiostat should be used at the level designated by the manufacturer.

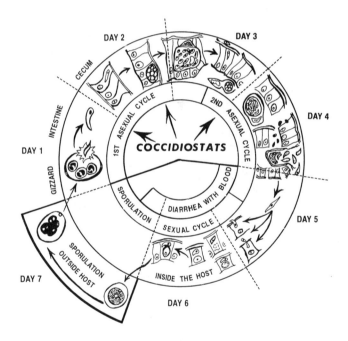

Fig. 5–9. Anticoccidial drugs are designed to kill the organisms at specific stages of development. There are currently a large number of anticoccidials on the market, many of which differ in mode of action—that is, the particular stage of development which is blocked.

Anthelmintic Drugs

Helminths are many-celled worm parasites varying greatly in size, shape, structure, and physiology. They may be classified as (1) flukes (or trematodes), (2) tapeworms (or cestodes), (3) roundworms (or nematodes), or (4) thorny-headed worms (or acanthocephala). With a few exceptions, the eggs or larvae must leave the host animal in which they originate to undergo further development on the ground, elsewhere in the open, or in intermediate hosts. Anthelmintics (wormers; vermifuges) generally require more than one administration. The first administration kills those worms which are present in the body, and subsequent wormings kill those worms which hatched from eggs after the initial dose. Some anthelmintics can be fed continuously.

The prevention and control of parasites is one of the quickest, cheapest, and most dependable methods for increasing production with no extra birds, no additional feed, and little more labor. This is important, for, after all, the producer bears the brunt of this reduced production and wasted feed.

From time to time, new vermifuges, or wormers, are approved and old ones are banned or dropped. Where parasitism is encountered, therefore, it is suggested that the producer obtain from local authorities the current recommendation relative to the choice and concentration of the vermifuge (wormer) to use. This information can be obtained from the county agent, entomologist, veterinarian, or agricultural consultant.

EVALUATING POULTRY FEEDS

It is not expected that poultry producers conduct experiments to determine the nutrient requirements of poultry or that they evaluate the different feeds that they use; that is, unless they are very large operators. It is important, however, that they have a working knowledge of the value of different poultry feeds from the standpoint of purchasing and utilizing them.

The nutritive requirements for a specific substance are determined by finding the minimum amount of that particular nutrient or substance that will permit maximum development of the physiological function or economic characteristics of concern. In general, the economic characteristics of importance in poultry are growth, efficiency of feed utilization, egg production, and hatchability. For example, if the need of a certain nutrient for growth is being determined, groups of birds must be fed on an experimental ration containing different levels of the nutrient in question until it is known that increasing the quantity of the test nutrient beyond a particular level will not result in further increases in growth. If the test ration is complete in all other respects, then the nutrient requirement will be equal to the minimum supplemental level found to give maximum growth.

Some feeds are more valuable than others; hence, measures of their relative usefulness are important. Among such methods of evaluating the usefulness of poultry feeds are: (1) physical evaluation, (2) chemical analysis, (3) biological tests, and (4) cost factor.

PHYSICAL EVALUATION OF FEEDSTUFFS

In order to produce or buy superior feeds, poultry producers need to know what constitutes feed quality, and how to recognized it. They need to be familiar with those recognizable characteristics of feeds which indicate high palatability and nutrient content. If in doubt, observation of the birds consuming the feed will tell them, for birds prefer and thrive on high-quality feed.

The easily recognizable characteristics of good grains and other concentrates are:

1. Seeds are not split or cracked.
2. Seeds are of low-moisture content—generally containing about 88% dry matter.
3. Seeds have a good color.
4. Concentrates and seeds are free from mold.
5. Concentrates and seeds are free from rodent and insect damage.
6. Concentrates and seeds are free from foreign material, such as iron filings.
7. Concentrates and seeds are free from rancid odor.

CHEMICAL ANALYSIS

While the biological response of animals (feeding trials) is the ultimate indicator of nutritive adequacy in a ration, tests of this type are difficult to perform, require extended periods of time, and are usually expensive. Thus, certain chemical analyses have been developed which are rough indicators of the value of a feedstuff or ration with regard to specific nutrient substances. The usual chemical analysis of feeds includes crude protein, ether extract or crude fat, crude fiber, ash or mineral, and moisture. It is recognized, however, that such proximate analysis of poultry feeds leaves much to be desired because in many cases the protein and nitrogen-free extract indicated may not be available to poultry.

In addition to the so-called proximate analysis, specific chemical and microbiological determinations can be made from many of the vitamins and individual mineral elements.

Fig. 5–10. The technician is analyzing the protein content of feed ingredients to insure a proper feed formulation for poultry. (Courtesy, Delmarva Poultry Industry, Inc., Georgetown, DE)

Feed composition tables serve as a basis for ration formulation and for feed purchasing and merchandising. Commercially prepared feeds are required by state law to be labeled with a list of ingredients and a guaranteed analysis. Although state laws vary slightly, most of them require that the feed label (tag) show in percent the minimum crude protein and fat; and maximum crude fiber and ash. Some feed labels also include maximum salt, and/or minimum calcium and phosphorus. These figures are the buyer's assurance that the feed contains the minimal amounts of the higher cost items – protein and fat; and not more than the stipulated amounts of the lower cost, and less valuable, items – the crude fiber and ash.

BOMB CALORIMETRY

When compounds are burned completely in the presence of oxygen, the resulting heat is referred to as gross energy or the heat of combustion. The bomb calorimeter is used to determine the gross energy of feed, waste products from feed (for example, feces and urine), and tissues.

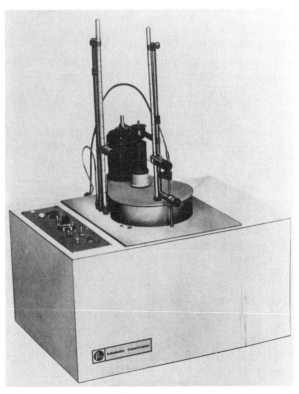

Fig. 5–11. Bomb calorimeter for the determination of gross energy. (Courtesy, Parr Instrument Company, Moline, IL)

The calorie is defined as the amount of heat required to raise the temperature of 1 g of water 1°C (precisely from 14.5° to 15.5°C). With this fact in mind, we can readily see how the bomb calorimeter works.

Briefly stated, the bomb calorimeter procedure is as follows: An electric wire is attached to the material being tested, so that it can be ignited by remote control; 2,000 g of water are poured around the bomb; 25 to 30 atmospheres of oxygen are added to the bomb; the material is ignited; the heat given off from the burned material warms the water; and a thermometer registers the change in temperature of the water. For example, if 1 g of feed is burned and the temperature of the water increases 1°C, 2,000 calories/gram are given off.

It is noteworthy that the determination of the heat of combustion with a bomb calorimeter is not as difficult or time consuming as the chemical analyses used in arriving at TDN values.

BIOLOGICAL TESTS

Most chemical and microbiological tests for nutrient substances give information about the total amount of nutrient present in a particular feedstuff or ration. However, these tests do not tell anything about the digestibility and utilization of the feedstuff or ration in the digestive tract of the animal. Hence, biological tests directly involving the bird are required to establish the true usefulness of feed supplying the nutrient needs of the bird. These biological tests are particularly important in evaluating protein and energy-yielding nutrients like carbohydrates and fats.

1. **Biological measure of protein utilization.** The amount of protein or nitrogen digested by the bird can be determined by a balance experiment in which a measured intake of protein is compared to the measured undigested protein in the feces of the bird. The biological value of a protein source is defined as the amount of protein retained in the body expressed as a percentage of the digestible protein available. Thus, this expression is a reflection of the kinds and amounts of amino acids available to the bird after digestion. If the amino acids available to the bird closely match those needed for body protein formation, the biological value of the protein is high. If, on the other hand, there are excesses of certain amino acids and deficiencies of other amino acids as a result of digestion, the biological value of the protein is low because of the increased number of amino acids which must be excreted via the kidney.

2. **Biological measure of energy utilization.** The total energy content of a feed can be measured by completely burning the feed in an apparatus known as a bomb calorimeter (see Fig. 5–11). Birds, like other animals, are not able to extract all of the energy present in feeds. Hence, the term *digestible energy* is used to describe the total energy of the feed minus that which remains undigested. Metabolizable energy is the total energy in the feed minus both fecal and urinary energy; it represents all the available energy for any use in the animal. The net energy value of a feed is the metabolizable energy content minus the energy employed in utilizing it; thus, net energy may be used for body storage or the production of heat and muscular activity. Metabolizable energy values are used to describe the energy content of poultry feedstuffs and rations. Metabolizable energy values are relatively easy to measure in poultry where the feces and urine are voided together and are little affected by various physiological conditions.

COST FACTOR

From the standpoint of a poultry producer the most important measurement of a feed's usefulness is in terms of net returns. Cost per pound or per ton of feed, and pounds of feed required to produce a pound of broiler or a dozen eggs, are important only as they reflect or affect the cost per unit of poultry products produced. For example, if the cost of a broiler ration is 9¢ a pound and 1.9 lb of the ration are required to produce 1 lb body weight, than the feed cost per pound of body weight can be arrived at by multiplying the above figures (9 × 1.9), which gives a feed cost of 17.1¢ per pound. Obviously when rations are compared, the ration that produces a unit of poultry product at the lowest total feed cost is the most desirable from an economic point of view.

BUYING FEEDS

Poultry producers who have familiarized themselves with the various types of feeds available are in a position to make reasonable and responsible decisions about what feeds to include in their rations. In order to maximize production and profit, they must choose the feeds that are most economical for the particular demands of the birds to be fed. A high-energy, high-protein feed that is fed to low-producing layers is unnecessarily expensive. Conversely, a low-cost but low-energy feed that is fed to birds at a high production level will depress potential for production and should be considered an expensive feed. It is important, therefore, that poultry producers know and follow good feed buying practices. Likewise, they should consider the possibility of using commercial feeds, all or in part, and be knowledgeable about feed laws.

WHAT PRODUCERS/FEED BUYERS SHOULD KNOW

Buying feeds is an integral part of modern poultry production. Moreover, the trend to purchase feeds, rather than grow them on location, will continue. In a broad sense, modern sophisticated buying involves knowledgeable buyers, futures trading, consideration of feed substitutions, volume buying, storage, capital outlay, and how to determine the best buy in feeds.

Most feeds used in poultry operations are purchased by producers—by practical operators who subsequently feed them. So, no one has a greater incentive to purchase wisely and well than these producer/buyers. However, altogether too often they think only in terms of cost per bushel, per hundredweight, or per ton. Too often they merely consider the feed analysis or the guarantee based on averages, oblivious to such quality-affecting factors as variety, soil, and weather. Too often buying means haggling with the seller over the price of a feed, rather than considering the net returns. Too often when buying feeds these practical operators are competing with highly trained professionals with great expertise in buying—they are competing with buyers representing commercial feed companies or brokers. For these reasons, more and more poultry producers are purchasing formula feeds, rather than separate ingredients. By so doing, they are acquiring a service—ration formulation—as well as a feed.

Of course, price of feed is very important. For this reason, a section in this chapter is devoted to "Best Buy in Feeds." Additionally, a number of complex and interrelated factors should be considered by the producer/buyer when purchasing feeds. Successful feed buying necessitates knowledge of all the factors that affect net returns, from the time a deal is made to buy the feed until the end product is marketed. Today, sophisticated poultry producers/feed buyers need to know the following:

1. They need to know the nutritive requirements of their birds.

2. They need to know feed terms and feed processing methods.

3. They need to know production and economic trends.

4. They need to know business aspects, such as sources of credit, interest rates, contracts, futures, and possible tax savings to accrue from purchasing feeds before the end of the year.

5. They need to know the different feed grade and quality classifications.

6. They need to know the restrictive use of certain feedstuffs, such as cottonseed meal.

7. They need to know the associative, or additive, effects of certain feedstuffs, especially of protein feeds.

8. They need to know the origin of the feed ingredients because soils in different areas produce different levels of minerals in plants.

9. They need to know the local potential to grow certain feeds.

10. They need to know the long term availability of feed.

11. They need to know the moisture content of the feed ingredients.

12. They need to know transportation costs.

13. They need to know the storage capabilities of the feed.

14. They need to know feed shrinkage.

15. They need to know the risks, such as where a perishable product is involved.

16. They need to know what processing will be involved.

17. They need to know that certain feedstuffs affect the product produced, such as alfalfa meal which can impart favorable color characteristics to poultry products.

18. They need to know about toxic residues.

19. They need to know the government regulations, especially as to levels, drug combinations, and withdrawals.

20. They need to know the impact of foreign feed purchases.

SOME FEED REQUISITES

In addition to the considerations already noted, it is important that all feeds—both bought and homegrown—meet the following requisites:

1. **Palatability.** If they don't eat it, they won't produce; and if they don't eat enough, feed efficiency will be poor. The relationship of feed consumption to feed efficiency becomes clear when it is realized that the maintenance requirement of a bird producing at a low rate represents a much greater percentage of the total feed required than for a bird producing at a more rapid rate.

Palatability is the result of the following factors sensed by the bird in locating and consuming feed: appearance, odor, taste, texture, and temperature. These factors are affected by the physical and chemical nature of the feed.

Fig. 5–12. A worker is shown checking the feed on the fryer ranch of Foster Farms, Livingston, California. (Courtesy, Foster Farms, Livingston, CA)

2. **Variety.** Some variety in the ration is desirable, particularly from the standpoint of assuring balance of nutrients – for example, all the essential amino acids.

3. **Digestive disturbances.** The choice of feeds can give a big assist in minimizing such disturbances.

4. **Cost.** Cost is important, but net return is even more important; hence, it may well be said that it is net return rather than cost per ton, or per bag, that counts.

BEST BUY IN FEEDS

Feed prices vary widely. For profitable production, therefore, feeds with similar nutritive properties should be interchanged as price relationships warrant.

In buying feeds, the poultry producer should check prices against value received.

Most poultry establishments and feed companies now use computers to formulate least-cost rations. The latter is described in Chapter 7, Poultry Feeding Standards, Ration Formulation, and Feeding Programs, in the section headed "Example of a Least-Cost Ration."

The hand method of determining cost per unit of nutrients is detailed in the section that follows.

COST PER UNIT OF NUTRIENTS

One method of arriving at the best buy in feeds is to compute and compare the cost per unit of nutrients, based on feed composition. Where a chemical analysis of a specific feed is not available, feed composition tables may be used as good indicators. Thus, feed composition tables may serve as a basis of feed purchasing and merchandising, as well as for ration formulation.

The use of the cost per unit of nutrients method can best be illustrated by the examples that follow:

■ **Cost per pound of protein** – If 44% protein (crude) soybean meal is selling at $13.93 per 100 lb whereas 60% protein (crude) fish meal sells for $17.55 per 100 lb, which is the better buy? Divide $13.93 by 44 to get 31.7¢ per pound of crude protein for the soybean meal. Then divide $17.55 by 60 to get 29.3¢ per pound of crude protein for the fish meal. Thus, at these prices fish meal is the better buy – by 2.4¢ (31.7 – 29.3 = 2.4) per pound of crude protein.

■ **Cost per unit of energy** – If wheat (ME_n = 1,334 kcal/lb) is selling at $7.90 per 100 lb and oats (ME_n = 1,157 kcal/lb) are selling at $6.25 per 100 lb, which is the best buy on cost per unit energy? In order to determine the best buy, we must base the comparison on a cost per kcal basis. Thus, the cost of wheat on a per pound basis, 7.9¢, must be divided by 1,334 kcal/lb and the cost of oats on a per pound basis, 6.25¢, must be divided by 1,157 kcal/lb. The analysis breaks down as follows:

	Cost/lb		kcal/lb		Cost/kcal
Wheat	7.90¢	÷	1,334	=	.0059¢
Oats	6.25¢	÷	1,157	=	.0054¢

Oats is the best buy since 1 kcal of metabolizable energy of oats costs .0005¢ less than 1 kcal of metabolizable energy of wheat.

Of course, it is recognized that many other factors affect the actual feeding value of each feed, such as (1) palatability, (2) grade of feed, (3) preparation of feed, (4) ingredients with which each feed is combined, and (5) quantities of each feed fed. It follows that, from the standpoint of the poultry producer, the important measurement of a feed's usefulness is in terms of *net returns*, rather than cost per bag or cost per ton.

BUYING COMMERCIAL FEEDS

Numerous types of commercial feeds, ranging from additives to complete rations, are on the market, with most of them designed for a specific species, age, or need. Among them are complete rations, concentrates, protein supplements (with or without reinforcements of minerals and/or vitamins), mineral and/or vitamin supplements, additives (antibiotics, hormones, etc.), different levels of production from the idle to the forced producers, and medicated feeds.

So, it may be said that there exist two good alternative sources of most feeds and rations – home-mixed or commercial – and the able manager will choose wisely between them.

HOME-MIXED VS COMPLETE COMMERCIAL FEEDS

The ultimate criterion for choosing between home-mixed and complete commercial feeds is which program will make for maximum returns to the producer for labor, management, and capital. Generally speaking, the use of complete commercial feeds makes it possible for the producer to have more birds and concentrate on production, whereas home-mixing restricts bird numbers and necessitates that part of the time and capital be devoted to feed formulating and manufacturing.

The poultry producer has the following options from which to choose for home-mixing feeds:

1. Purchase of a commercially prepared protein supplement (likely reinforced with minerals and vitamins), which may be blended with local or homegrown grain.

2. Purchase of a commercially prepared mineral-vitamin premix, which may be mixed with a protein supplement, then blended with local or homegrown grain.

3. Purchase of individual ingredients (including minerals and vitamins) and mixing the feed from the ground up.

Fig. 5–12a. Chicken feed. (Courtesy, Hy-Line International, Johnston, IA)

HOW TO SELECT COMMERCIAL FEEDS

There is a difference in commercial feeds! That is, there is a difference from the standpoint of what poultry producers can purchase with their feed dollars. Smart operators will know how to determine what constitutes the best in commercial feeds for their specific needs. They will not rely solely on the appearance or aroma of the feed, nor on the salesperson.

REPUTATION OF THE MANUFACTURER

The reputation of the manufacturer can be determined by (1) conferring with other producers who have used the particular products, and (2) checking on whether or not the commercial feed under consideration has a good record for meeting its guarantees. The latter can be determined by reading the bulletins and reports prepared by the respective state department in charge of monitoring feed quality and enforcing feed laws.

SPECIFIC NEEDS OF THE BIRDS

Feed requirements vary according to (1) the class, age, and productivity; and (2) whether the birds are fed primarily for maintenance, growth, commercial egg production, or hatching. The wise producer will buy different formula feeds for different needs.

Feeding poultry has become a sophisticated and complicated process. Feed manufacturers have extensive resources with which to formulate and test rations for different needs. As a result, most manufacturers have a large selection of feeds—one of which should be applicable to the needs of the producer. It is essential that the producer make clear to the feed company the kind of birds to be fed.

FEED LAWS

The U.S. Food and Drug Act was passed in 1906, giving the Federal Government authority to regulate and inspect feeds shipped in interstate commerce. Additional controls were authorized in the Food, Drug, and Cosmetic Act of 1938. In addition to the federal laws, nearly all states have laws regulating the sale of commercial feeds. These benefit both feeders and reputable feed manufacturers. In most states the laws require that every brand of commercial feed sold in the state be licensed, and that the chemical composition be guaranteed.

Since most commercial feeds are closed formulas, it is necessary to sample and analyze the feed periodically to ensure that it is fulfilling its guarantees. Generally, samples of each commercial feed are taken periodically and analyzed chemically in the state's laboratory. Additionally, skilled microscopists examine the sample to ascertain that the ingredients present are the same as those guaranteed. Flagrant violations on the latter point may be prosecuted as they represent willful mislabeling.

FEED SUBSTITUTIONS

Successful poultry producers are keen students of values. They know that feed conversion rate alone does not determine success. The cost of producing each pound of broiler, turkey, or each dozen eggs is the determinant of success. For this reason, producers want to obtain a balanced ration that will make for the highest net returns. They recognize that feeds of similar nutritive properties can and should be interchanged in the ration as price relationships warrant, thus making it possible at all times to obtain a balanced ration at the lowest cost. Thus, (1) the cereal grains may consist of corn, barley, wheat, oats, and/or sorghum; and (2) the protein supplements may consist of soybean, cottonseed, peanut, rapeseed, sunflower, and/or linseed meal.

Table 5–4, Feed Substitution Table for Poultry, is a summary of the comparative values of the most common U.S. poultry feeds. In arriving at these values, two primary factors besides chemical composition and feeding value have been considered—namely, palatability and product quality.

It is emphasized that the comparative values of feeds shown in the feed substitution table are not absolute. Rather, they are reasonably accurate approximations based on average-quality feeds, together with experiences and experiments.

TABLE 5–4
FEED SUBSTITUTION TABLE FOR POULTRY (As-Fed)

Feedstuff	Relative Feeding Value (lb for lb) in Comparison with the Designated (underlined) Base Feed Which = 100	Maximum Percentage of Base Feed (or comparable feed or feeds) Which It Can Replace for Best Results	Remarks
Grains, By-product Feeds:			
Corn, No. 2	100	100	Corn is the most widely used grain in poultry feed. It is a good source of vitamin A, xanthophyll pigments, and linoleic acid.
Bakery wastes	75	50	Bakery wastes are very similar to cereal grains in composition.
Barley	80–85	50	Barley is very low in vitamin A; less palatable than corn; western grown barley may be improved through enzyme treatment.
Beans (cull)	90	5	
Buckwheat	90	15	Unpalatable.
Cassava	85	20	Extremely low in protein.
Hominy	95	50	A good source of linoleic acid.
Millet	95–100	65	Best when used as a 50:50 mix with barley, corn, or oats; can be used as a scratch feed.
Molasses	70	5	High levels of molasses will produce wet droppings.
Oats	70–80	50	Usually too bulky for broilers and too expensive for layers, but used extensively in replacement rations.
Peas, dried	90–100	5–10	
Rice, rough	80–85	20–50	Deficient in vitamin A.
Rice, bran	50	5–10	Rice bran is high in fat and susceptible to oxidation.
Rice polishings	85–90	5–10	High in fat; good source of thiamin.
Rye	90	25–30	Unpalatable; will cause sticky, pasty droppings if excessive amounts are fed.
Sorghum	100	100	May replace corn, but lacks xanthophyll pigments and is lower in linoleic acid.
Triticale	80–90	30	Triticale is somewhat lower in feeding value than either wheat or corn.
Wheat	90–95	100	Wheat is the most variable of the cereal grains in protein; very low in xanthophyll; should be processed to increase palatability.
Wheat bran	75	10–15	
Protein Supplements:			
Soybean meal (48%)	100	100	Well balanced in amino acids. Best quality of all plant protein supplements. Very palatable.
Babassu oil meal	50	20	Similar to copra meal.
Blood meal, flash- or ring-dried ...	120	5–20	Excellent source of lysine.
Copra meal (coconut meal)	50	25	Copra meal should be supplemented with lysine and methionine.
Corn gluten meal	50–75	25	Used extensively to supply xanthophyll pigments in broiler finishing rations. Low in lysine and tryptophan.
Cottonseed meal	85	80[1]	Gossypol, a compound found in cottonseed meal, can cause discoloration of egg yolks. The addition of iron in the ration may minimize the dangers of gossypol poisoning. Cyclopropenoic fatty acids in cottonseed meal can cause discoloration of egg whites. Glandless cottonseed meal is recommended.
Feather meal	50	5	Feather protein is deficient in methionine, lysine, histidine, and tryptophan. Poor quality protein.

(Continued)

TABLE 5–4 *(Continued)*

Feedstuff	Relative Feeding Value (lb for lb) in Comparison with the Designated (underlined) Base Feed Which = 100	Maximum Percentage of Base Feed (or comparable feed or feeds) Which It Can Replace for Best Results	Remarks
Fish meal	115	50–65	Excellent balance of amino acids and good source of calcium and phosphorus. Most poultry rations incorporate some fish meal at levels of about 2–5% of the ration. Fish meal can impart fishy flavors if fed at levels in excess of 10%.
Linseed meal	80	10	Linseed meal depresses growth and can cause diarrhea. Levels should be restricted to a maximum of 5% of the ration.
Meat and bone meal	100	20–50	Variable protein quality; good source of calcium and phosphorus; deficient in methionine and cystine.
Meat meal (50–55% protein)	100	50–65	Meat meal is high in phosphorus and low in methionine and cystine. It is recommended that the maximum level of usage not exceed 10% of the ration.
Peanut meal (41% protein)	95	75–100	If peanut meal it to be substituted for soybean meal, lysine must be added.
Poultry by-product meal	100	20–50	Excellent protein if properly prepared.
Rapeseed meal (canola meal) ...	80	30	Not recommended for poultry starter ration. It may be used at levels of up to 10% of the total ration for fattening birds or layers. Meal from new low glucosinolate rapeseed may constitute 20% of layer rations and 10% of starter feeds.
Safflower seed meal (decorticated)	95	50–100	Safflower seed meal can be incorporated at levels up to 15% of the layer ration. Deficient in lysine and methionine.
Sesame meal	95–100	100	Extremely deficient in lysine, but good plant source of methionine.
Sunflower seed meal	95–100	100	Low in lysine and methionine.
Yeast, torula	100	60	Yeast is a good source of the B complex vitamins.

[1]When cottonseed meal is substituted for soybean meal at this level, it must be degossypolized.

FEED PREPARATION

Grains for poultry are prepared in the following three forms:

1. **Mash.** This is a ground feed and the usual end product resulting from mixing poultry feedstuffs.

2. **Pellets.** Composed of mash feeds that are pelleted. Birds usually consume more of a pelleted ration than the same ration in mash form. Usually there is more cannibalism with pellets than with mash or crumbles.

3. **Crumbles.** Produced by rolling pellets.

Pellets or crumbles cost slightly more than the same ration in mash form. Yet, they are used for broilers and turkeys because of improved feed efficiency (fewer pounds of feed to produce a pound of gain).

When alfalfa is to be included in poultry feeds, it should be ground.

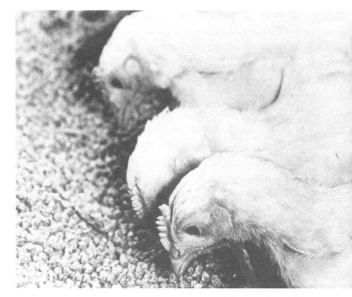

Fig. 5–13. Six-week-old White Rock broilers eating pellets. (Courtesy, Ralston Purina Company, St. Louis, MO)

PRESENCE OF SUBSTANCES AFFECTING PRODUCT QUALITY

The composition of the feed can affect the product. The color of the skin or shanks of a broiler or of the yolk of an egg is primarily due to the carotenoid pigments consumed in the feed. Corn, alfalfa meal, and corn gluten meal are the main feeds used to contribute these pigments.

Screw process cottonseed meal, which is high in gossypol, when fed to laying hens may cause egg yolk discoloration in stored eggs. Some fish products may impart off-flavors to poultry meat or eggs. Thus, certain feedstuffs may be undesirable simply because of the effect they produce on the end product.

TOXIC LEVELS OF INORGANIC ELEMENTS FOR POULTRY

Poultry are susceptible to a number of toxins, any one of which may prove disastrous in a flock. Among them are a number of inorganic elements. It is important, therefore, that the poultry producer guard against toxic levels of inorganic elements.

Toxicity is influenced by (1) the form of the element (for example, methyl mercury is considerably more toxic than inorganic mercury); and (2) the composition of the diet (particularly with respect to the content of other mineral elements).

QUESTIONS FOR STUDY AND DISCUSSION

1. Discuss the economic importance of poultry feeding.

2. How would you classify poultry feeds, broadly speaking?

3. What grains are commonly fed to poultry? Why are not more wheat and rye fed to poultry?

4. Should poultry producers be interested in buying high-lysine corn? If so, why?

5. Name the leading cereal grains and list the milling by-products of each that are used in poultry feeds.

6. Why are fats used in poultry feeds? Which is preferable — animal or vegetable fat?

7. Discuss the role of molasses in poultry feeds.

8. What determines the usefulness of a protein feedstuff for poultry?

9. Discuss the relative importance of plant and animal protein sources. Which sources have become increasingly more important in recent years?

10. List the commonly used oilseed meals. Which oilseed meal is most widely used for poultry?

11. What substances in cottonseed meal may affect egg color?

12. Until about 1940, it was commonly believed that animal protein had to be incorporated into nonruminant feeds. Why? Why has the use of animal products generally declined in recent years?

13. Define (a) tankage, and (b) meat meal.

14. Blood is the medium of nutrient transport in the body. One is tempted to conclude, therefore, that blood meal would be a better feed than tankage or meat meal. Often, this is not the case. Explain why.

15. What by-products and residues are included in hatchery wastes?

16. What are the limitations to the use of dairy products?

17. List and discuss the three major factors which affect the feeding value of fish meal.

18. What are condensed fish solubles? In what vitamins are they especially rich?

19. Give some guidelines for mineral supplementation of poultry.

20. Why must phytin phosphorus and fluorine be considered in phosphorus supplementation?

21. Why do today's nutritionists and poultry producers rely on vitamin supplements, many of which are chemically pure sources, rather than on feed ingredients as such?

22. Why isn't more alfalfa meal incorporated in poultry rations?

23. What is an *additive*?

24. Why are antioxidants routinely added to poultry rations?

25. Name three commonly used pellet binders.

26. How may a poultry producer enhance (a) the yellow color of the skin of broilers, and (b) the dark color of egg yolks?

27. What is the purpose of grit?

28. Discuss the role of each of the following additives in poultry feeding: chelates, enzymes, and antifungals.

29. Define antibiotic. Why are antibiotics used in poultry feeds? What is the reason for the beneficial effects of antibiotics?

(Continued)

QUESTIONS FOR STUDY AND DISCUSSION *(Continued)*

30. What are the two general categories of antiparasitic drugs?

31. Poultry feed analysis may be (a) physical, (b) chemical, or (c) biological. Which method would you recommend that a poultry producer use?

32. Enumerate the important things that a sophisticated poultry producer/feed buyer should know.

33. If a 48% protein (crude) soybean meal is selling at

$140 per ton whereas a 40% peanut meal is selling at $130 per ton, which is the better buy on the basis of cost per pound of protein?

34. On what bases should a poultry producer select commercial feeds?

35. Of what importance is a feed substitution table?

36. In what way is feed usually prepared for poultry?

SELECTED REFERENCES

Title of Publication	Author(s)	Publisher
Animal Feeding and Nutrition, Fifth Edition	M. H. Jurgens	Kendall/Hunt Publishing Company, Dubuque, IA, 1982
Animal Nutrition, Seventh Edition	L. A. Maynard, *et al.*	McGraw-Hill Book Company, Inc., New York, NY, 1979
Animal Nutrition, Second Edition	P. McDonald R. A. Edwards J. F. D. Greenhalgh	Oliver & Boyd, Edinburgh, Scotland, 1973
Animal Science, Ninth Edition	M. E. Ensminger	Interstate Publishers, Inc., Danville, IL, 1991
Association of American Feed Control Officials Incorporated, Official Publication	Association of American Feed Control Officials, Inc.	Association of American Feed Control Officials, Inc., Annual
Basic Animal Nutrition and Feeding, Second Edition	D. C. Church W. G. Pond	John Wiley & Sons, Inc., Salt Lake City, UT, 1984
Calcium in Broiler, Layer, and Turkey Nutrition	R. H. Harms, *et al.*	National Feed Ingredients Association, West Des Moines, IA, 1976
Cereal Processing and Digestion	U.S. Feed Grains Council, USDA Foreign Agricultural Service	U.S. Feed Grains Council, London, England, 1972
Commercial Chicken Production Manual, Fourth Edition	M. O. North D. D. Bell	Van Nostrand Reinhold Co., New York, NY, 1990
Feed Additive Compendium	Edited by D. Natz	The Miller Publishing Company, Minnetonka, MN, Annual
Feed Flavor and Animal Nutrition	T. B. Tribble	Agriads, Inc., Chicago, IL, 1962
Feed Formulation, Second Edition	T. W. Perry	The Interstate Printers & Publishers, Inc., Danville, IL, 1975
Feeds and Feeding, Third Edition	A. Cullison	Reston Publishing Company, Inc., Reston, VA, 1982
Feeds & Nutrition, Second Edition	M. E. Ensminger J. E. Oldfield W. W. Heinemann	The Ensminger Publishing Company, Clovis, CA, 1990
Feeds & Nutrition Digest	M. E. Ensminger J. E. Oldfield W. W. Heinemann	The Ensminger Publishing Company, Clovis, CA, 1990
Feeds for Livestock, Poultry, and Pets	M. H. Gutcho	Noyes Data Corporation, Park Ridge, NJ, 1973
Manual of Poultry Production in the Tropics	R. R. Say	CAB International, Wallingford, Oxon, United Kingdom, 1987

SELECTED REFERENCES (Continued)

Title of Publication	Author(s)	Publisher
Nutrient Requirements of Poultry and Nutritional Research	C. Fisher K. N. Boorman	Butterworths, Kent, England, 1986
Nutrient Requirements of Poultry, Eighth Revised Edition	National Research Council	National Academy of Science, Washington, DC, 1984
Nutrition of the Chicken, Third Edition	M. L. Scott M. C. Nesheim R. J. Young	M. L. Scott and Associates, Ithaca, NY, 1982
Nutrition of the Turkey	M. L. Scott	M. L. Scott of Ithaca, Ithaca, NY, 1987
Phosphorus in Poultry and Game Bird Nutrition	P. E. Waibel, *et al.*	National Feed Ingredients Association, West Des Moines, IA, 1977
Poultry: Feeds and Nutrition, Second Edition	H. Patrick P. J. Schaible	The Avi Publishing Company, Inc., Westport, CT, 1980
Poultry Meat and Egg Production	C. R. Parkhurst C. J. Mountney	Van Nostrand Reinhold Co., New York, NY, 1988
Poultry Nutrition, Fifth Edition	W. R. Ewing	The Ray Ewing Company, Pasadena, CA, 1963
Poultry Production, Twelfth Edition	M. C. Nesheim R. E. Austic L. E. Card	Lea & Febiger, Philadelphia, PA, 1979
Processing and Utilization of Animal By-Products	I. Mann	Food and Agriculture Organization of the United Nations, Rome, Italy, 1975
Raising Poultry Successfully	W. Graves	Williamson Publishing, Charlotte, VT, 1985
Scientific Feeding of Chickens, The, Fifth Edition	H. W. Titus J. C. Fritz	The Interstate Printers & Publishers, Inc., Danville, IL, 1971
Tropical Feeds		Food and Agriculture Organization of the United Nations, Rome, Italy, 1975

Fig. 5–14. Truck at feed mill. (Courtesy, Monsanto Agriculture Co., St. Louis, MO)

Fig. 5–15.   Caged layers feeding. (Courtesy, Euribrid B.V., Boxmeer, Holland)

Fig. 5–16.   Range turkeys at feeder.

Fig. 6–1. Vitamin B–12 made the difference! The perky chick at right and his smaller companion are both 3½ weeks old. *Left:* The small chick, fed a ration deficient in vitamin B–12, weighed 157 g. *Right:* The larger chick, fed the same ration plus vitamin B–12, weighed 280 g. (Courtesy, Merck Chemical Division, Rahway, NJ)

FUNDA-MENTALS OF POULTRY NUTRITION

CHAPTER

The primary purpose of raising poultry is to transform feeds into meat and eggs. But the conversion of feed to these uses must be done efficiently and economically. To do this, the principles of nutrition must be applied; and they must be augmented by superior breeding, good health, and competent management.

Like other sciences, nutrition does not stand alone. It draws heavily on the basic findings of chemistry, biochemistry, physics, microbiology, physiology, medicine, genetics, mathematics, endocrinology, and, most recently, animal behavior and cellular biology. In turn, it also contributes richly to each of these fields of scientific investigation.

Poultry nutrition is more critical than that of other farm animals with regard to a number of factors. This is so because birds are quite different from four-footed animals; their digestion is more rapid, their respiration and circulation are faster, their body temperature is 8° to 10°F higher (about 106°F), they are more active, they are more sensitive to environmental influences, they grow at a more rapid rate, and they mature at an earlier age. Also, egg production is an all-or-none phenomenon — that is, birds must have enough nutrients to produce an egg; otherwise, no egg is produced.

A thorough knowledge of the anatomy and physiology of the avian digestive system is requisite to fully understanding the fundamentals of poultry nutrition. This subject is fully covered in Chapter 2, Avian Anatomy and Physiology; The Egg; hence, the reader is referred thereto.

NUTRIENT COMPOSITION OF POULTRY AND EGGS

Nutrition encompasses the various chemical and physiological reactions which change feed elements into body elements. It follows that knowledge of body and egg composition is useful in understanding poultry response to nutrition. Table 6–1 gives the body composition of poultry compared to that of some of their mammalian counterparts used in meat production.

Within each type of animal there is a wide range in body composition, depending on age and nutritional status. These ranges are evident in poultry, as shown in Table 6–1. However, the following conclusions relative to body composition may be drawn:

1. **Water.** On a percentage basis, the water content shows a marked decrease with advancing age, maturity, and fatness.

2. **Fat.** The percentage of fat normally increases with growth and fattening. It is recognized, of course, that the amount of fat is materially affected by the feed intake.

3. **Fat and water.** As the percentage of fat increases, the percentage of water decreases.

4. **Protein.** The percentage of protein remains rather constant during growth, but decreases as the animal fattens.

TABLE 6–1
BODY COMPOSITION OF (1) ANIMALS, AT DIFFERENT
WEIGHTS AND AGES, INGESTA-FREE (EMPTY) BASIS, AND (2) EGG[1]

Species	Age or Status	Weight		Water	Fat	Protein	Ash
		(lb)	(kg)	(%)	(%)	(%)	(%)
Poultry:							
Chick	Newly hatched	0.09	0.041	78.8	4.0	15.3	1.9
Broiler	Market	3.5	1.6	64.0	14.2	18.8	3.7
Hen	Layer	4.5	2.04	62.4	15.6	18.7	4.0
Turkey	8 weeks	5.1	2.3	70.7	4.5	20.7	4.1
Turkey	Market	18.0	8.2	60.2	18.7	19.4	3.1
Egg	Newly laid	0.136	0.062	66.0	10.0[2]	13.0	11.0[3]
Cattle:							
Steer	Choice grade	1,050	477.3	53.5	26	17	3.5
Sheep:							
Lamb	Market	100	45.4	53.2	29	15	2.8
Pig:							
Pig	Market	220	100	50	34.4	13	2.6

[1]Prepared by the author from numerous sources.

[2]Chiefly in the yolk.

[3]Nearly all is calcium in shell.

On the average, there are 3 to 4 lb of water per 1 lb of protein in the body.

5. **Ash.** The percentage of ash shows the least change. However, it decreases as animals fatten because fat tissue contains less mineral than lean tissue.

6. **Composition of gain.** The data presented in Table 6–1 clearly indicate that gain in weight may not provide an accurate measure of the actual gain in energy of the animal, because it tells nothing about the composition of gain. This is important because efficiency of feed utilization (pounds feed per pound body gain) is greatly influenced by the amount of fat produced.

7. **Species comparison.** The following species differences are noteworthy:

a. The bodies of poultry contain less fat, but more water, than cattle, sheep, and pigs in comparable condition.

b. Because of their smaller skeletons, the bodies of pigs contain less ash than those of cattle, sheep, or poultry.

c. At normal market stage, broilers are higher in protein than four-footed animals.

The chemical composition of the body varies widely between organs and tissues and is more or less localized according to function. Thus, water is an essential of every part of the body, but the percentage composition varies greatly in different body parts; blood plasma contains 90 to 92% water, muscle 72 to 78%, and bone 45%. Proteins are the principal constituents, other than water, of muscles, tendons, and connective tissues. Most of the fat is localized under the skin, near the kidneys, and around the intestines. But it is also present in the muscles, bones, and elsewhere.

Although carbohydrates are very important in nutrition, they account for less than 1% of the body composition, being concentrated primarily in the liver, muscles, and blood. Instead, they provide the basic building blocks for fat (energy reserves) and carbon skeletons for amino acids and are metabolized rapidly for the production of energy.

CLASSIFICATION OF NUTRIENTS

Birds do not utilize feeds as such. Rather, they use those portions of feeds called *nutrients* that are released by digestion, then absorbed into the body fluids and tissues.

Nutrients are those substances, usually obtained from feeds, which can be used by the animal when made available in a suitable form to its cells, organs, and tissues. They include carbohydrates, fats, proteins, minerals, vitamins, and water. (More correctly speaking, the term *nutrients* refers to the more than 40 nutrient chemicals, including amino acids, minerals, and vitamins.) Energy is frequently listed with nutrients, since it results from the metabolism of carbohydrates, proteins, and fats in the body. Knowledge of the basic functions of the nutrients in the body and of the interrelationships between various nutrients and other metabolites within the cells of the bird is necessary before one can make practical scientific use of the principles of nutrition.

FUNCTIONS OF NUTRIENTS

Of the feed consumed, a portion is digested and absorbed for use by the bird. The remaining undigested portion is excreted and constitutes the major portion of the feces. Nutrients from the digested feed are used for a number of different body processes, the exact usage varying with the species, class, age, and productivity of the bird. All birds use a portion of their absorbed nutrients to carry on essential functions, such as body metabolism and maintaining body temperature and the replacement and repair of body cells and tissues. These uses of nutrients are referred to as *maintenance*. That portion of digested feed used for growth, fattening, or the production of eggs is known as *production requirements*.

Based on the quantity of nutrients needed daily for different purposes, nutrient demands may be classed as high, low, variable, or intermediate. Requirements for egg production are considered *high-demand uses*, whereas molting is a *low-demand use*. Growth and fattening may be classed as intermediate in nutrient demands.

MAINTENANCE

Birds, unlike machines, are never idle. They use nutrients to keep their bodies functioning every hour of every day, even when they are not being used for production.

Maintenance requirements may be defined as the combination of nutrients which are needed by the bird to keep its body functioning without any gain or loss in body weight or any productive activity. Although these requirements are relatively simple, they are essential for life itself. A mature bird must have (1) heat to maintain body temperature, (2) sufficient energy to keep vital body processes functional, (3) energy for minimal movement, and (4) the necessary nutrients to repair damaged cells and tissues and to replace those which have become nonfunctional. Thus, energy is the primary nutritive need for maintenance. Even though the quantity of other nutrients required for maintenance is relatively small, it is necessary to have a balance of the essential proteins, minerals, and vitamins.

No matter how quiet a bird may be, it requires a certain amount of fuel and other nutrients. The least amount on which it can exist is called its *basal maintenance requirement*.

There are only a few times in the normal life of a bird when only the maintenance requirement needs to be met. Such a status is closely approached by mature males not in service and by mature, nonproducing females. Nevertheless, maintenance is the standard bench mark or reference point for evaluating nutritional needs.

Even though maintenance requirements might be considered as an expression of the nonproduction needs of a bird, there are many factors which affect the amount of nutrients necessary for this vital function; among them, (1) exercise, (2) weather, (3) stress, (4) health, (5) body size, (6) temperament, (7) level of production, and (8) individual variation. The first four are *external factors* — they are subject to control to some degree through management and facilities. The others are *internal factors* — they are part of the bird itself.

Both external and internal factors influence requirements according to their intensity. For example, the colder or hotter it gets from the most comfortable (optimum) temperature, the greater will be the maintenance requirements.

GROWTH

Growth may be defined as the increase in size of bones, muscles, internal organs, and other parts of the body. It is the normal process before hatching, and after hatching until the bird reaches its full, mature size. Growth is influenced primarily by nutrient intake. The nutritive requirements become increasingly acute when birds are under forced production, such as when poults are stimulated to production at an early age.

Growth is the very foundation of animal production. Poultry will not make the most economical finishing gains unless they have been raised to be thrifty and vigorous. Nor can one expect the most satisfactory production of eggs from laying hens unless they are well developed during their growing period.

Generally speaking, organs vital for the maintenance of life—e.g., the brain, which coordinates body activities, and the gut, upon which the rest of the postnatal growth depends—are early developing; and the commercially more valuable parts, such as muscle and fat, develop later.

The nutritive needs for growth vary with age, breed, sex, rate of growth, and disease.

AGE

In comparison with older birds, young birds generally (1) consume more feed per unit of body weight; (2) utilize feed more efficiently, in pounds of feed eaten per pound of body gain; (3) have a higher requirement for protein, energy, vitamins, and minerals per unit of body weight; (4) require a more concentrated and more easily digested diet; and (5) are more subject to nutritional deficiencies.

The digestive systems of newly hatched birds have been largely nonfunctional and are relatively undeveloped. For this reason, the type of ration is particularly important during this period of functional and physical development of the digestive tract and should include a balance of all available nutrients in a readily available form. Young birds (chicks, poults, ducklings, and goslings) should be fed a highly nutritious diet.

After an initial adjustment period (which may be prolonged if the feed, environment, or disease level are unfavorable), the rate of gain of birds is very rapid when measured as a percentage of body weight. Table 6–2 shows the growth rate of chickens, ducks, and turkeys as measured by the days needed to double birth weight and the number of months needed to reach 50% of mature body weight. As shown, the juvenile period of humans is extremely long by comparison.

TABLE 6–2
DAYS NEEDED TO DOUBLE BIRTH WEIGHT
AND MONTHS NEEDED TO REACH 50 PERCENT OF
MATURE BODY WEIGHT, FOR POULTRY AND HUMANS

Species	Days to Double Birth Weight	Months to Reach 50% Mature Weight
	(days)	(months)
Chicken	5	2
Duck	4	1½
Turkey	5	4
Human	150	115–145

BREED

Larger breeds of birds grow more rapidly than smaller breeds and have a higher nutrient requirement.

SEX

Growth studies involving young birds of both sexes, and of all species, reveal the following: (1) Males gain more rapidly than females and have a higher feed requirement; (2) uncastrated males use feed more efficiently for body weight gains than females, because of the higher water and protein content and the lower fat content of the increased body weight; (3) mature average size is larger in males than in females; and (4) females reach maturity faster than males.

RATE OF GROWTH

In recent years, the accent in broiler production has been on forced production and marketing at an early age. Achieving this goal has involved improved nutrition, and, generally speaking, rapid gains and profits have been on the same side of the ledger. Today, broilers generally reach market weight in 6 to 7 weeks, with efficiencies in feed utilization commonly being less than 2 lb of feed per 1 lb of bird.

Rapid gains call for more nutrients. In turn, this necessitates high-energy, well-balanced rations. For the most part, fast gains are efficient gains; when birds grow at maximum rates, they require fewer nutrients and fewer pounds of feed per pound of gain.

HEALTH

Ill health—diseases and parasites—results in lack of thrift and poor development in young birds. When the causative factor is severe, growth may be stunted. Feed is always too costly to waste. Besides, the full productive potential of birds is needed.

REPRODUCTION

Being hatched is the most important requisite of poultry production; for if birds fail to reproduce, the poultry producer is soon out of business. A "mating of the gods," involving the greatest genes in the world, is of no value unless these genes result in (1) the successful joining of the sperm and egg, and

(2) hatching. Certainly, there are many causes of reproductive failure, but scientists are agreed that nutritional inadequacies play a major role.

Egg production involves feeding for number of eggs, egg quality, hatchability, and control of molt and broodiness. The nutritive needs for commercial egg production include those for maintenance of the birds, growth of pullet layers, and the formation of eggs. The nutritive requirements are greater for birds with an inherited capacity for high egg production than for those that lay only a few eggs. The standard-weight egg contains about 95 calories of gross energy, 7.5 g of crude protein, and 2 g of calcium.

With poultry, the development and hatching of a fertile egg constitute reproduction. As with mammals, the nutritive requirements of poultry breeders (including chickens, turkeys, ducks, geese, etc.) are more rigorous than those for commercial laying. For high hatchability and good development of embryos, breeders require greater amounts of vitamins A, D, E, B-12, riboflavin, pantothenic acid, niacin, and of the mineral, manganese. Birds intended as breeders should be started on special breeder rations at least a month before hatching eggs are to be saved.

NUTRIENTS

Nutrients are utilized in one of two metabolic processes: (1) for anabolism, or (2) for catabolism. *Anabolism is the process by which nutrient molecules are used as building blocks for the synthesis of complex molecules.* Anabolic reactions are endergonic—that is, they require the input of energy into the system. *Catabolism is the oxidation of nutrients, liberating energy (exergonic reaction) which is used to fulfill the body's immediate demands.*

ENERGY (CARBOHYDRATES AND FATS)

The energy requirement may be defined as the amount of available energy that will provide for growth or egg production at a high enough level to permit maximal economic return for the production unit.

Energy is required for practically all life processes—for the action of the heart, maintenance of blood pressure and muscle tone, transmission of nerve impulses, ion transport across membranes, reabsorption in the kidneys, protein and fat synthesis, and the production of eggs.

A deficiency of energy is manifested by slow or stunted growth, body tissue losses, and/or lowered production of meat and eggs rather than by specific signs, such as those which characterize many mineral and vitamin deficiencies. For this reason, energy deficiencies often go undetected and unrectified for extended periods of time.

It is common knowledge that a ration must contain carbohydrates, fats, and proteins. Although each of these has specific functions in maintaining a normal body, all of them can be used to provide energy for maintenance, for growth, and for egg production. From the standpoint of supplying the normal energy needs of birds, however, the carbohydrates are by far the most important, more of them being consumed than any other compound, whereas the fats are next in importance for energy purposes. Carbohydrates are usually more abundant and cheaper, and most of them are very easily digested, absorbed, and transformed into body fat. Also, carbohydrate feeds may be more easily stored than fats in warm weather and for longer periods of time. Feeds high in fat content are likely to become rancid, and rancid feed is unpalatable, if not actually injurious in some instances.

Feed intake is largely governed by the energy concentration of the feed. Although the bulkiness of feed can alter feed intake, the bird, for the most part, will eat to satisfy its energy needs. Because of this, special attention must be given to nutrient ratios, especially the ratio of energy to various nutrients such as amino acids and minerals. For example, if the producer uses a high-energy feed, the protein content of the feed must be high if the bird is to ingest adequate amounts of protein. Conversely, if the energy content of a feed is low, the protein content should be low, also; otherwise, the bird will consume excessive amounts of expensive protein.

The metabolic rate of poultry is an indication of the energy needs of the bird. Several factors can affect this rate, including the following:

1. **Breed and strain.** The development of breeds and strains for specific purposes has resulted in genetic differences in the efficiency of energy utilization. For example, the University of Guelph found that Leghorn chickens obtained 3% more metabolizable energy from their diet than broiler strain chickens. In turkeys, separate lines have been developed for males and females.

2. **Activity.** Birds that have access to large areas have a higher metabolic rate than their counterparts which are confined to small cages which restrict movement.

3. **Diurnal rhythm.** Within individual birds, the metabolic rate will vary according to the time of day.

4. **Environmental temperature.** Poultry are extremely sensitive to environmental temperature, especially hot weather. Since there is little heat dissipated through sweating, much of body heat must be lost through panting.

5. **Diet.** The metabolic rate of birds is affected by the type of diet. A low-energy, high-protein diet will necessitate different digestive processes than a high-energy, low-protein diet.

6. **Level of production.** A bird that is in heavy production will have different energy demands than one that is not in production.

7. **Other factors.** Other factors affecting dietary requirements for energy, in addition to those already listed, are: stress, body size, and feather coverage.

CARBOHYDRATES

Carbohydrates are organic compounds composed of carbon, hydrogen, and oxygen, which constitute about 75% of the dry weight of plants and grain and make up a large part of the poultry ration. They serve as a source of heat and energy in the bird's body, and a surplus of them is transformed into fat and stored.

In poultry feeds, the term *nitrogen-free extract (NFE)* is often used to refer to the soluble and digestible portion of carbohydrates, whereas the term *fiber* is used to refer to the insoluble and indigestible carbohydrates that are the structural components of plants.

No appreciable amount of carbohydrate is found in the animal body at any one time, the blood supply being held rather constant at about 0.05 to 0.1% for most animals. However, this small quantity of glucose in the blood, which is constantly replenished by changing the glycogen of the liver back to glucose, serves as the chief source of fuel with which to maintain the body temperature and to furnish the energy needed for all body processes. The storage of glycogen (so-called animal starch) in the liver amounts to 3 to 7% of the weight of that organ.

FATS

Lipids (fat and fatlike substances), like carbohydrates, contain the three elements — carbon, hydrogen, and oxygen. Fats are soluble in such organic solvents as ether, chloroform, and benzene.

As feeds, fats function much like carbohydrates in that they serve as a source of heat and energy and for the formation of fat. Because of the larger proportion of carbon and hydrogen, however, fats liberate more heat than carbohydrates when digested, furnishing approximately 2.25 times as much heat or energy per pound on oxidation as do the carbohydrates. A smaller quantity of fat is required, therefore, to serve the same function.

Although fats are used primarily to supply energy in poultry diets, they also improve the physical consistency of rations and the dispersion of microingredients in feed mixtures, and serve as carriers of the fat-soluble vitamins. The fats used for feeding poultry are derived from three sources: animal or poultry fats obtained from the rendering industry, restaurant greases, acidulated soapstocks from the vegetable oil industry, and/or mixtures thereof. The nutritional value of fats for poultry feed is determined by moisture, impurities, unsaponifiables, free fatty acids, total fatty acids, and fatty acid composition. The polyunsaturated linoleic and arachidonic acids are considered to be *essential fatty acids*. They have specific functions in the body that are not related to energy production. Birds exhibit poor growth, fatty livers, reduced egg size, and poor hatchability without these essential fatty acids. Fats for poultry feed should be stabilized against oxidation.

The metabolizable energy (ME) contribution of fats may be influenced by their fatty acid composition, free fatty acid content, level of fat inclusion in the ration, ingredient composition of the ration, and age of poultry. Fats often increase the utilization of dietary energy by poultry in excess of the increase expected when the ME of the fat is added to the ME values of the other ration constituents. Supplemental fats may increase energy utilization in adult chickens due to (1) a decreased rate of food passage through the gastrointestinal tract, and (2) the heat increment of fat being less than that of carbohydrates.

Fats constitute about 17% of the dry weight of market broilers and about 40% of the dry weight of whole eggs.

Food fats affect body fats. Thus, poultry consuming soft fat, such as vegetable oils, may accumulate fat that is somewhat soft and oily.

MEASURING AND EXPRESSING ENERGY VALUE OF FEEDSTUFFS

One nutrient cannot be considered as more important than another, because all nutrients must be present in adequate amounts if efficient production is to be maintained. Yet, historically, feedstuffs have been compared or evaluated primarily on their ability to supply energy to animals. This is understandable because (1) energy is required in larger amounts than any other nutrient, (2) energy is most often the limiting factor in poultry production, and (3) energy is the major cost associated with feeding poultry.

Our understanding of energy metabolism has increased through the years. With this added knowledge, changes have come in both the methods and terms used to express the energy value of feeds.

ENERGY DEFINITIONS AND CONVERSIONS

Some pertinent definitions and conversions of energy terms follow:

■ **Calorie (cal)** — The amount of energy as heat required to raise the temperature of 1 g of water 1°C (precisely from 14.5° to 15.5°C). It is equivalent to 4.184 joules. Although *not preferred*, it is also called a *small calorie* and so designated by being spelled with a lower case "c." **Note well:** In popular writings, especially those concerned with human caloric requirements, the term *calorie* is frequently used erroneously for the kilocalorie.

■ **Kilocalorie (kcal)** — The amount of energy as heat required to raise the temperature of 1 kg of water 1°C (from 14.5° to 15.5°C). Equivalent to 1,000 calories. In human nutrition, it is referred to as a kilogram calorie or as a *large Calorie* and is so designated by being spelled with a capital "C" to distinguish it from the *small calorie*.

■ **Megacalorie (Mcal)** — Equivalent to 1,000 kcal or 1,000,000 calories. Also, referred to as a *therm*, but the term *megacalorie* is preferred.

■ **British Thermal Unit (BTU)** — The amount of energy as heat required to raise 1 lb of water 1°F; equivalent to 252 calories. This term is seldom used in animal nutrition.

■ **Joule** — A proposed international unit (4.184J = 1 calorie) for expressing mechanical, chemical, or electrical energy, as well as the concept of heat. In the future, energy requirements and feed values will likely be expressed by this unit.

■ **Converting TDN to Mcal** — One pound of TDN = 2.0 Mcal or 2,000 kcal. It is recognized, however, that the roughage component in a ration affects its energy value.

CALORIE SYSTEM OF ENERGY EVALUATION

Calories are used to express the energy value of feedstuffs. To measure this heat, an instrument known as the bomb calorimeter is used, in which the feed (or other substance) tested is placed and burned in the presence of oxygen.

Through various digestive and metabolic processes, numerous losses of the energy in feed occur as the feed passes through the bird's digestive system. These losses are illustrated in Fig. 6–2.

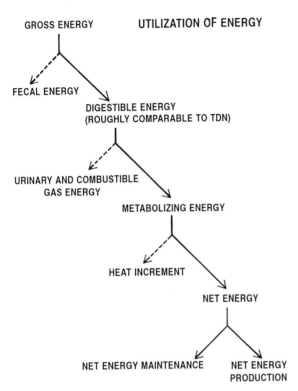

Fig. 6–2. Utilization of energy.

As shown in Fig. 6–2, energy losses occur in the digestion and metabolism of feed. Measures that are used to express energy requirements and the energy content of feeds differ primarily in the digestive and metabolic losses that are included in their determination. Thus, the following terms are used to express the energy value of feeds:

■ **Gross energy (GE)** — Gross energy represents the total combustible energy in a feedstuff. It does not differ greatly between feeds, except for those high in fat.

■ **Digestible energy (DE)** — Digestible energy is that portion of the GE in a feed that is not excreted in the feces.

■ **Metabolizable energy (ME)** — Metabolizable energy represents that portion of the GE that is not lost in the feces, urine, and gas. Although ME more accurately describes the useful energy in the feed than does GE or DE, it does not take into account the energy lost as heat.

■ **Net energy (NE)** — Net energy represents the energy fraction in a feed that is left after the fecal, urinary, gas, and heat losses are deducted from the GE. The net energy, because of its greater accuracy, is being used increasingly in ration formulations, especially in computerized formulations for large animals and large operations.

Of the various systems of expressing energy values of feeds and nutrient requirements — gross energy, digestible

energy, metabolizable energy, and net energy — the poultry industry has found metabolizable energy, commonly labeled ME_n, to be the most reliable expression of energy needs. In general, metabolizable energy represents 25 to 90% of gross energy.

The ME values given in the NRC tables (Tables 7–3 to 7–17) give perspective to the other nutrient requirement levels. Using the energy level for reference will enable the nutritionist to form a ratio of the amount of the nutrient per unit of energy, thereby keeping the nutrients in balance with available energy. In eating to satisfy its energy need, a bird will eat less of a high-energy diet and more of a low-energy diet. Having the nutrients in relation to dietary energy will ensure proper intake on a daily basis.

■ **True metabolizable energy (TME)** — *The TME for poultry is the gross energy of the feed minus the gross energy of the excreta of food origin.* A correction for nitrogen retention may be applied to give a TME_n value. In 1975, Dr. I. R. Sibbald of the Animal Research Center, Ottawa, Canada, suggested a new method of measuring the energy content of feed ingredients based on the precision feeding of a measured amount of a single ingredient to a mature rooster. The excreta was collected so that the amount of energy that had been digested could be measured. Because the excreta (feces and urine) contain metabolic and endogenous fractions from the body, Sibbald corrected for them (he deleted them) and called the resulting energy value *true metabolizable energy* (TME). To clearly differentiate between the TME values and ME values, Sibbald suggested using the term *apparent metabolizable energy* (AME) for the traditionally determined energy values which include metabolic and endogenous fractions.

In simple formulas AME and TME are: **AME = feed energy – (fecal + urinary + gaseous energy)**, whereas **TME = AME + (metabolic + endogenous energy)**. In poultry, the gaseous losses are negligible and usually ignored.

TME values are easier and much less expensive to determine than the traditional ME values. In recent years, the TME values have been corrected for the energy contained in nitrogen excretion; the objective being to provide an energy measure where the energy in the ingredient is used for only one purpose — energy in the animal. So, true metabolizable energy values are currently expressed either as (1) TME, or (2) TME_n when corrected for nitrogen excretion. Following the nitrogen correction, the values obtained by use of TME assay are virtually identical to those obtained by using the more traditional apparent metabolizable energy; hence, nitrogen corrected TME (TME_n) and apparent metabolizable energy (AME) values are interchangeable.

Some scientists question the Sibbald procedure for determining TME and AME because the assays are made with starved birds. Others are reluctant to adopt the TME system of feed evaluation because of the difficulty and cost in changing from current dietary standards. Still others feel that further research studies are needed. Nevertheless, the TME system is now in use worldwide, serving as the basis for feed formulation and specifications for purchasing. Initially, the energy part of the system attracted industrial attention. Now, more and more companies are using the amino acid part of the system.

PROTEINS

Proteins are complex organic compounds made up chiefly of amino acids, which are present in characteristic proportions for each specific protein. This nutrient always contains carbon, hydrogen, oxygen, and nitrogen; and in addition, it usually contains sulfur and frequently phosphorus. Proteins are essential in all plant and animal life as components of the active protoplasm of each living cell.

Crude protein refers to all the nitrogenous compounds in a feed. It is determined by finding the nitrogen content and multiplying the result by 6.25. The nitrogen content of protein averages about 16% (100 ÷ 16 = 6.25).

In plants, the protein is largely concentrated in the actively growing portions, especially the leaves and seeds. Plants also have the ability to synthesize their own proteins from such relatively simple soil and air compounds as carbon dioxide, water, nitrates, and sulfates, using energy from the sun. Thus, plants, together with some bacteria which are able to synthesize these products, are the original sources of all proteins.

Proteins are much more widely distributed in animals than in plants. Thus, the proteins of the animal body are primary constituents of many structural and protective tissues — such as bones, ligaments, feathers, skin, and the soft tissues which include the organs and muscles.

Birds of all ages require adequate amounts of protein of suitable quality for maintenance, growth, reproduction, and egg production. Of course, the protein requirements for growth are the greatest and most critical.

Typical broiler starter rations contain from 21 to 24% protein, and typical laying rations from 16 to 17% protein. Grain and millfeeds supply approximately one-half of the protein needs for most poultry rations. Additional protein is supplied from high-protein concentrates of either animal or vegetable origin.

From the standpoint of poultry nutrition, the amino acids that make up proteins are really the essential nutrients, rather than the protein molecule itself. Hence, protein content as a measure of the nutritional value of a feed is becoming less important, and each amino acid is being considered individually. The essential amino acid requirements of poultry are given in Chapter 7, Poultry Feeding Standards, Ration Formulation, and Feeding Programs, Tables 7–3, 7–5, 7–7, 7–10, 7–13, 7–15, 7–16, and 7–17.

The energy content of the diet must be considered in formulating to meet the desired intake of all essential nutrients other than the energy itself, including the intake of the essential amino acids. For example, if the producer uses a high-energy feed, the protein content of the feed must be high if the bird is to ingest adequate amounts of protein. Conversely, if the energy content of the feed is low, the protein content should be low, also; otherwise, the bird will consume excessive amounts of expensive protein.

Currently stated amino acid requirements have no reference to environmental conditions. Percentage requirements should probably be raised in warmer environments and lowered in colder environments in accordance with expected differences in feed or energy intakes.

The amino acid concentrations presented in the NRC tables in Chapter 7 are intended to promote maximum growth and production. Maximum economic returns may not, however, always be assured by maximum growth and production, particularly when protein prices are high. So, at times the dietary concentrations may be somewhat reduced, lowering growth rate to some degree, but maintaining economic returns.

It has been determined that the chick requires dietary sources of protein to furnish 13 different amino acids. These amino acids are referred to as *essential*, since the chicken cannot produce them in sufficient amounts for maximum growth or egg production, and because a dietary deficiency of any one of them interferes with body protein formation and affects growth or egg production. The primary object of protein feeding, therefore, is to furnish the bird with protein which, upon digestion, will yield sufficient quantities of the 13 essential amino acids needed for top performance. According to our present knowledge, the following division of amino acids as essential, semiessential, and nonessential for the chick seems proper:

Essential (indispensable)	Semiessential (semidispensable)	Nonessential (dispensable)
Arginine	Cystine	Alanine
Histidine	Glycine	Aspartic acid
Isoleucine	Tyrosine	Glutamic acid
Leucine		Hydroxyproline
Lysine		Proline
Methionine		Serine
Phenylalanine		
Threonine		
Tryptophan		
Valine		

When formulating poultry rations, they must be so designed as to supply all the essential amino acids in ample amounts. Special attention needs to be given to supplying the amino acids lysine, methionine and cystine, and tryptophan, which are sometimes referred to as the *critical amino acids* in poultry nutrition. Additionally, there must be sufficient total nitrogen for the chicken to synthesize the other amino acids needed.

In practice, the amino acid requirements of growing chickens and turkeys, and of laying hens, are met by proteins from plant and animal sources. Protein supplements that most nearly supply the essential amino acids of the bird are known as *high-quality* supplements. Usually it is necessary to choose more than one source of dietary protein, then combine them in such a way that the amino acid composition of the mixture meets the requirements of the bird.

Some high-protein feedstuffs may contain toxic compounds. For example, cottonseed meal may contain gossypol, which discolors the yolks in stored eggs. Certain strains of rapeseed meal contain high enough levels of goitrogenic compounds to be toxic. Even soybean meal, the most used protein supplement, contains harmful substances, such as trypsin inhibitor, but these are destroyed by proper heating.

Any excess protein consumed by the bird can be burned in the body to yield energy in somewhat the same manner as carbohydrates and fats. However, in practical feeding of poultry, it is seldom wise to use excessive protein because carbohydrates and fats are generally more economical sources of energy.

In addition to dietary energy concentration and ambient temperature, the following factors affect the amino acid requirements:

1. **The rate of growth or intensity of egg production.** The more rapid the growth and the higher the egg production, the higher the amino acid requirements.

2. **The strains.** Even within species of like body size, growth rate, or egg production, there may be differences in requirements among strains.

3. **The protein level.** The amino acid requirements tend to increase with dietary protein.

4. **The amino acid relationships,** specifically —

a. **Methionine-cystine.** The requirement for methionine can be met only by methionine, while the requirement for cystine may be met by cystine or methionine. This is because methionine is readily converted to cystine metabolically, while the reverse is not possible.

b. **Phenylalanine-tyrosine.** The requirement for phenylalanine may be met only by phenylalanine, while the requirement for tyrosine may be met by tyrosine or phenylalanine.

c. **Glycine-serine.** Glycine and serine can be used interchangeably in poultry diets.

5. **Antagonisms.** There are specific antagonisms among amino acids that may be structurally related, e.g., valine-leucine-isoleucine and arginine-lysine. Increasing one or two of such a group may raise the need for another of the same group.

6. **Imbalances.** In supplementing diets with limited amino acids, it is important to supplement first with the most limiting one, followed by the second most limiting one. Over-supplementation with only the second most limiting one may create an imbalance and accentuate the primary deficiency.

7. **Conversion of certain amino acids to vitamins.** High levels of methionine may partly compensate for a deficiency of choline or vitamin B–12 by providing needed methyl groups, and high levels of tryptophan may alleviate a niacin deficiency through metabolic conversion to niacin.

8. **Amino acid availability.** The usual assumption that amino acids are 80 to 90% available is not necessarily valid. For example, feathers or blood are either indigestible in native form or made indigestible by overheating in processing.

The consequences of a protein or amino acid deficiency vary with the degree of the deficiency. A *borderline deficiency* is characterized by poor growth and feathering, reduced egg size, poor egg production (but hatchability is not affected), tendency toward greater deposition of carcass and liver fat, poor feed conversion into eggs or meat, and lack of melanin pigment in black- or reddish-colored feathers with low lysine. A *severe protein deficiency* is marked by stopping of feed intake, stopping of egg production, loss of body weight, resorption of ova, a tongue deformity with leucine, isoleucine, and phenylalanine deficiency, stasis of the digestive tract, and death.

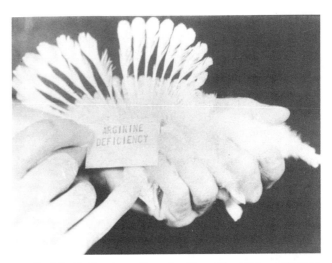

Fig. 6–3. Feather abnormality produced by arginine deficiency. (Courtesy, Department of Poultry Science, University of Wisconsin, Madison)

■ **True digestibility of amino acids**[1] — Recently, there has been much interest in determining amino acid availability, using some of the same techniques that were originally developed for determination of true metabolizable energy. This interest stemmed from the fact that, frequently, amino acids cannot be completely utilized due to inherent characteristics of the feedstuff. Details follow.

The available amino acids are the amino acids in the diet that are not combined with compounds interfering with their digestion, absorption, or utilization by the bird; they are actually supplied to the sites of protein synthesis.

The digestible amino acids are those that are absorbed from the gut lumen; they are calculated as the difference between the amount of amino acids in the feed and the excreta.

Differentiation must be made between apparent and true digestibility. In addition to nonabsorbed feed, the excreta also contain materials originating in the tissue of the bird, e.g., cells sloughed from the gut wall, mucus, bile, unabsorbed digestive juices, etc. Apparent digestibility makes no correction for these endogenous components. In simple formulas, *apparent digestibility coefficient* and *true digestibility coefficient* follow:

$$\text{Apparent Digestibility Coefficient} = \frac{\text{amino acid consumed} - \text{amino acid excreted}}{\text{amino acid consumed}} \times 100$$

$$\text{True Digestibility Coefficient} = \frac{\text{amino acid consumed} - \text{amino acid excreted} + \text{endogenous amino acid}}{\text{amino acid consumed}} \times 100$$

[1]In the preparation of this section, the author adapted selected material from the publication: *True Faecal Digestibility of Essential Amino Acids in Feedstuffs for Poultry*, published by Eurolysine, 16 Rue Ballu 75009 Paris, France, with the permission of Eurolysine's "sister company," Heartland Lysine, Inc., 8430 West Bryn Mawr, Suite 650, Chicago, IL 60631.

The availability of amino acids in feedstuffs is estimated by three main methods: (1) *in vitro* tests, (2) growth tests, and (3) digestibility tests. Most of the recent research has focused on digestibility determinations.

Digestibility trials measure the digestibility of all amino acids in a feedstuff in one run. Various experimental techniques may be applied to determine the digestibility of amino acids. Dr. I. R. Sibbald of the Animal Research Center, Ottawa, Canada, developed the cockerel precision feeding technique, which estimates the true digestibility of amino acids. It consists of force feeding a precise quantity of feed (1 to 2% of body weight) to a starved adult rooster, followed by a 48-hour collection period of the excreta. Other birds are fasted throughout the experiment to allow for collection of the endogenous excreta. It appears that true amino acid digestibility established in the chicken are valid for other bird species also.

Ration formulation on the basis of digestible amino acids has the following **advantages**:

1. It improves performance prediction under practical conditions.

2. It allows more by-product and alternate protein sources to be included in poultry feeds, along with the possibility of reducing costs while minimizing the risks of deterioration in performance.

3. It reduces the variability of poultry performance.

4. It evaluates feedstuff and rations more precisely.

5. It makes possible a more accurate supply of amino acids required for optimal performance.

The **disadvantages** or **reservations** relative (1) to the *rooster bioassay* technique for determining the true digestibility of amino acids in feedstuffs, and (2) to the use of such values in feed formulations are:

TABLE
POULTRY MINERAL

Minerals Which May Be Deficient Under Normal Conditions	Conditions Usually Prevailing Where Deficiencies are Reported	Function of Mineral	Some Deficiency Symptoms	Types of Poultry Rations Usually Requiring Supplementation			
				Starting	Growing	Laying	Breeding
Major or macrominerals:							
Salt (sodium and chlorine—NaCl)	Omitted at the mill.	Improves appetite, promotes growth, helps regulate body pH, and is essential for hydrochloric acid formation in the stomach.	Chloride-deficient chicks show poor growth, high mortality, nervous symptoms, and reduced blood chloride level. Sodium deficiency in layers results in decreased egg production, poor growth, and cannibalism.	Yes	Yes	Yes	Yes
Calcium (Ca)	Imbalance of Ca:P ratio. Presence of interfering elements.	Bone formation; eggshell formation; blood clotting; and neuromuscular function.	Anorexia, thin eggshells, rickets, or osteoporosis, tetany, abnormal walk.	Yes	Yes	Yes	Yes
Phosphorus (P)		Bone formation; metabolism of carbohydrates and fats; a component of all living cells; maintenance of the acid-base balance of the body; and calcium transport in egg formation.	Anorexia, weakness, rickets. Cage-layer fatigue, characterized by birds being paralyzed and unable to rise from a recumbent position. But there is evidence that this condition is not due to this factor alone.	Yes	Yes	Yes	Yes
Magnesium (Mg)	Diets containing high levels of Ca and P.	Essential for normal skeletal development, as a constituent of bone, enzyme activator, primarily in glycolytic system.	Decreased egg production, depressed growth and lethargy, convulsions.	No	No	No	No
Potassium (K)	Low plant protein level/high animal protein level.	Major cation of intracellular fluid where it is involved in osmotic pressure and acid-base balance. Muscle activity. Required in enzyme reaction involving phosphorylation of creatine. Influences carboydrate metabolism.	*Chicks:* Retarded growth and high mortality.	No	No	No	No

1. It does not consider the effects of microbes on dietary protein in the lower gut. Microbes both utilize and synthesize amino acids. Research has indicated that, under certain conditions, 20 to 25% of the amino acids found in the excreta are of microbial origin.

2. When conventional feed ingredients are used in poultry rations, consideration of amino acid digestibility is not particularly advantageous.

Crystalline amino acids, not being protein bound or enclosed in a feedstuff impairing their digestion, are 100% digestible and 100% available.

MINERALS

Minerals are inorganic elements, frequently found as salts with either inorganic elements or organic compounds.

Minerals are required for the formation of the skeleton, as parts of hormones or as activators of enzymes, and for the proper maintenance of necessary osmotic relationships within the body of the bird.

The minerals which have been shown to be essential for chickens and turkeys are calcium, phosphorus, magnesium, manganese, zinc, iron, copper, iodine, sodium, chlorine, potassium, sulfur, molybdenum, and selenium. Of these, calcium, phosphorus, manganese, sodium, chlorine, iodine, selenium, and zinc are considered to be of most practical importance since outside sources of them must be added to practical feed formulations for chickens and turkeys. Most of the pertinent facts relative to poultry minerals are summarized in Table 6–3.

Some minerals are required by poultry in relatively large amounts. They are referred to as *major* or *macrominerals*. Others are needed in very small amounts; they are referred to as *trace* or *microminerals*.

6–3
CHART (See footnote at end of table.)

Mineral Requirements[1] — Mineral Content (%) of Ration (Variable According to Class, Age, and Weight of Poultry) For . . . . See Table	Recommended Allowances[1]	Practical Sources of the Mineral	Comments
Layers . . . . 7–3 Broilers . . . . 7–5 Breeding hens . . . . 7–7 Turkeys . . . . 7–10 Ducks . . . . 7–13 Geese . . . . 7–15 Pheasants/Bobwhite . . . . . 7–16 Japanese Quail 7–17	0.2–0.5% of the diet.	Common table salt.	Sodium level is sometimes reduced to minimal to control the moisture level of the feces.
Layers . . . . 7–3 Broilers . . . . 7–5 Breeding hens . . . . 7–7 Turkeys . . . . 7–10 Ducks . . . . 7–13 Geese . . . . 7–15 Pheasants/Bobwhite . . . . . 7–16 Japanese Quail 7–17	The calcium allowance should vary with level of production and temperature. A minimum of 3.4% is believed to be optimum for layers in moderate climates. Growing rations should contain 0.8–0.9% of Ca and 0.4–0.7% of P.	Dicalcium phosphate. Limestone. Oystershell.	For young poultry, the Ca:P ratio should be about 1.2:1. However, ratios of 1:1 to 1.5:1 are well tolerated. An excess of calcium interferes with the utilization of magnesium, manganese, and zinc.
Layers . . . . 7–3 Broilers . . . . 7–5 Breeding hens . . . . 7–7 Turkeys . . . . 7–10 Ducks . . . . 7–13 Geese . . . . 7–15 Pheasants/Bobwhite . . . . . 7–16 Japanese Quail 7–17	Dependent on production. Laying hens require diets containing at least 3% P. Growing rations should contain 0.8–0.9% of Ca and 0.4–0.7% of P.	Defluorinated phosphate. Dicalcium phosphate. Monosodium phosphate. Phosphoric acid. Steamed bone meal.	Organic phosphorus (present in plants) is poorly utilized by growing birds, but is satisfactory for adult birds. Only about 30 to 40% of the phosphorus in plant products is available to the young chick, poult, or duckling.
Layers . . . . 7–3 Broilers . . . . 7–5 Turkeys . . . . 7–10 Ducks . . . . 7–13 Japanese Quail 7–17	600 ppm in the ration of broilers should suffice.	Magnesium oxide or magnesium sulfate.	Requirements are affected by Ca and P levels in the diet. Not normally deficient in poultry rations.
Layers . . . . 7–3 Broilers . . . . 7–5 Turkeys . . . . 7–10 Japanese Quail 7–17		Corn contains 0.33% potassium (as-fed), and other cereals contain 0.42–0.49% potassium.	Potassium is not deficient in normal rations, due to large amounts of plant products in poultry feeds.

(Continued)

TABLE 6–3

Minerals Which May Be Deficient Under Normal Conditions	Conditions Usually Prevailing Where Deficiencies are Reported	Function of Mineral	Some Deficiency Symptoms	Types of Poultry Rations Usually Requiring Supplementation			
				Starting	Growing	Laying	Breeding
Trace or microminerals:							
Copper (Cu)		Essential element in a number of enzyme systems and necessary for synthesizing hemoglobin and preventing nutritional anemia.	Anemia, depigmentation of feathers, digestive disorders. *Poults:* marked cardiac hypertrophy.	Yes	Yes	Yes	Yes
Iodine (I)	Feeds produced on iodine-deficient soils.	Needed by the thyroid gland for making thyroxin, an iodine-containing hormone which controls the rate of body metabolism or heat production.	Enlarged thyroid. Eggs produced from deficient breeder hens have a lowered hatchability, prolonged hatching time, and a subsequent retardation of yolk-sac absorption.	Yes	Yes	Yes	Yes
Iron (Fe)		Necessary for formation of hemoglobin, an iron-containing compound which enables the blood to carry oxygen. Iron is also important to certain enzyme systems.	*Chicks and poults:* Microcytic, hypochromic anemia. In red-feathered chickens, complete depigmentation of the feathers occurs.	Yes	Yes	Yes	Yes
Manganese (Mn)		Necesssary for growth, bone structure, and reproduction.	*Chicks and poults:* Perosis, shortened leg bones, skull deformation, parrot beak. Poor egg production, shell quality, and hatchability.	Yes	Yes	Yes	Yes
Selenium (Se)	Feeds that are grown in selenium deficient areas.	Involved in the destruction of peroxides within the cell as a constituent of glutathione peroxidase. Useful in preventing exudative diathesis.	Exudative diathesis. Pancreatic fibrosis. Steatitis. Muscular dystrophy. With a severe selenium deficiency, growth rate is reduced and mortality is increased. *Turkeys:* Myopathies of the gizzard and heart.	Yes	Yes	Yes	Yes
Zinc (Zn)		Zinc is a component of several enzyme systems, including peptidases and carbonic anhydrase. Also, zinc is required for normal protein synthesis and metabolism and is a component of insulin.	Bone problems, poor feathering (feather fraying occurs near the ends of the feathers), retarded growth, and loss of appetite. A zinc deficiency in breeder diets reduces egg production and hatchability.	Yes	Yes	Yes	Yes

[1]As used herein, the distinction between *mineral requirements* and *recommended allowances* is as follows: In *mineral requirements*, no margins of safety are included intentionally; whereas in *recommended allowances*, margins of safety are provided in order to compensate for variations in feed composition, environment, and possible losses during storage or processing.

■ **Trace minerals for a purified poultry diet**—For certain poultry experiments, it may be desirable to formulate a purified diet consisting of chemically defined ingredients, such as sugar, lard, casein, etc. When this is done, Table 6–4 may be used as a guide relative to the trace mineral elements and dietary concentrations to include in the ration.

TABLE 6–4
SUGGESTED TRACE MINERAL SUPPLEMENTS TO (CHEMICALLY DEFINED) RATIONS[1]

Element	Amount of Ration		Element	Amount of Ration	
	(mg/lb)	(mg/kg)		(mg/lb)	(mg/kg)
Boron (B)	0.91	2	Nickel (Ni)	0.05	0.1
Chromium (Cr)	1.36	3	Silicon (Si)	113.50	250
Fluorine (F)	9.08	20	Tin (Sn)	1.36	3
Inorganic sulfate	—[2]	—[2]	Vanadium (V)	0.09	0.2
Molybdenum (Mo)	0.45	1			

[1]Adapted by the author from *Nutrient Requirements of Poultry*, 8th rev. ed., NRC, National Academy Press, Washington, DC, 1984, p. 7, Table 1.

[2]Sulfur is supplied to the ration by methionine and cystine. There may be a response to inorganic sulfate if the ration is low in cystine.

(Continued)

Mineral Requirements[1] Mineral Content (%) of Ration (Variable According to Class, Age, and Weight of Poultry) For . . . . See Table	Recommended Allowances[1]	Practical Sources of the Mineral	Comments
Layers . . . . 7–3 Broilers . . . . 7–5 Turkeys . . . . 7–10 Ducks . . . . 7–13 Geese . . . . 7–15 Japanese Quail 7–17	Rations containing 6 to 8 ppm of copper should be adequate.	Copper sulfate and copper carbonate are about equally effective.	
Layers . . . . 7–3 Broilers . . . . 7–5 Turkeys . . . . 7–10 Pheasants/Bobwhite . . . . . 7–16 Japanese Quail 7–17	Laying hens require feed containing 300 ppb. Growing chicks require feed containing 350 ppb.	Stabilized iodized salt containing 0.007% iodine. Trace mineral mixes.	
Layers . . . . 7–3 Broilers . . . . 7–5 Turkeys . . . . 7–10 Japanese Quail 7–17	Rations should contain about 80 ppm of iron.	Alfalfa meal. Meat, liver, and fish meals.	Iron salts are used as a means of detoxifying gossypol from cottonseed meal.
Layers . . . . 7–3 Broilers . . . . 7–5 Turkeys . . . . 7–10 Ducks . . . . 7–13 Japanese Quail 7–17	Rations should contain at least 30 ppm for layers, and 60 ppm for broilers.	Alfalfa meal, distillers' solubles, or grain by-products. Manganese sulfate, manganous chloride, manganous carbonate, and manganous dioxide.	
Layers . . . . 7–3 Broilers . . . . 7–5 Turkeys . . . . 7–10 Ducks . . . . 7–13 Japanese Quail 7–17	0.2 to 0.3 ppm added selenium.	Fish meal and brewers' dried yeast. Sodium selenite or sodium selenate.	Selenium can pose toxicity problems. Hence, care should be taken when adding it to poultry rations. In 1987, FDA provided for an increase in the maximum allowance of selenium in complete feeds for chickens, turkeys, and ducks to 0.3 ppm. Added selenium is often put into the vitamin premix to avoid separation.
Layers . . . . 7–3 Broilers . . . . 7–5 Turkeys . . . . 7–10 Ducks . . . . 7–13 Japanese Quail 7–17		Zinc carbonate or zinc sulfate.	

MAJOR OR MACROMINERALS

The major or macrominerals of importance in poultry are salt (sodium chloride), calcium, phosphorus, magnesium, and potassium.

SALT (SODIUM CHLORIDE/NaCl)

Sodium and chlorine are essential for all animals, including poultry. Dietary proportions of sodium, potassium, and chlorine are important determinants of acid-base balance. The proper dietary balance of sodium, potassium, and chlorine is necessary for growth, bone development, eggshell quality, and amino acid utilization. It is generally recommended that 0.2 to 0.5% of the diet consist of salt.

As in most nutrient deficiencies, a lack of salt will reduce reproductive performance and retard growth. Also, cannibalism occurs in flocks when diets are deficient in salt.

CALCIUM (Ca) AND PHOSPHORUS (P)

The common calcium supplements used in poultry feeding are: ground limestone and oystershells. The most important phosphorus sources are: dicalcium phosphate, defluorinated rock phosphate, and steamed bone meal. Bone meal, dicalcium phosphate, defluorinated rock phosphate, colloidal phosphate, and raw rock phosphate are used where both calcium and phosphorus are needed.

The biological availability of calcium is high for most commonly used supplements, but the biological availability of phosphorus in supplements varies.

The calcium allowance of the laying hen is difficult to define. In Table 6–3, the listed requirement of 3.4% of the ration is believed to represent the mean dietary concentration for the quantities of feed likely to be consumed (110 g per hen per day) over a considerable range of environmental temperature. Most of the calcium in the diet of the mature laying fowl is used for eggshell formation, whereas most of the calcium in the diet of the growing bird is used for bone formation. Other functions of calcium include roles in blood clotting and neuromuscular function. High concentrations of calcium carbonate (limestone) and calcium phosphate make the ration unpalatable. Also, excess dietary calcium interferes with the availability of other minerals such as magnesium, manganese, and zinc.

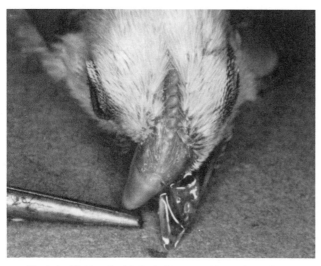

Fig. 6–4. "Rubbery beak" of a broiler, caused by feeding a high calcium layer ration which resulted in a severe phosphorus deficiency. (Courtesy, L. S. Jensen, Ph.D., Department of Poultry Science, The University of Georgia, Athens)

In addition to its function in bone formation, phosphorus is required in the metabolism of carbohydrates and fats and is a component of all living cells.

It is important that the minimum level of inorganic phosphorus be provided. The stated requirement is based on the

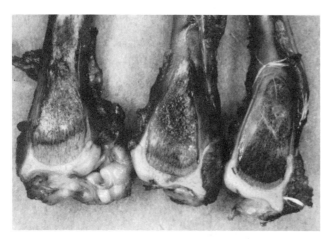

Fig. 6–5. Deficiencies of phosphorus, calcium, and/or vitamin D affect the normal calcification and development of the tibia in young chicks. The bone on the left was caused by a deficiency of phosphorus. The bone on the right was caused by a deficiency of calcium or vitamin D. The bone in the middle is from a chick that was fed a well-balanced diet. (Courtesy, L. S. Jensen, Ph.D., Department of Poultry Science, The University of Georgia, Athens)

generally greater availability of inorganic phosphorus than that of phytin phosphorus. Only about 30% of the phosphorus in plant products is available to the young chick, poult, or duckling, whereas the older bird has the ability to use most, if not all, of the phytin or organic phosphorus in plant products.

For growing chickens and turkeys, a calcium to phosphorus ratio of 1.2:1 is considered ideal. However, ratios of 1:1 to 1.5:1 are well tolerated.

Leg problems and thin eggshells occur when calcium and/or phosphorus are deficient. It is also essential, especially in poultry housed in confinement, that adequate amounts of vitamin D_3 be supplied in the ration to ensure proper absorption of these elements.

MAGNESIUM (Mg)

Most poultry feeds contain adequate amounts of magnesium, and supplementation generally is not necessary. A dietary level of 600 ppm of magnesium is required by broilers.

Excessive magnesium levels can interfere with the absorption of other elements—notably calcium and phosphorus. If it is known that a diet contains a high level of magnesium, levels of other elements should be increased.

Chicks that are fed magnesium-deficient rations, exhibit depressed growth rates and lethargy. When disturbed, magnesium-deficient chicks become hyperirritable and go into convulsions. If the condition is not corrected, chicks become comatose and eventually die.

POTASSIUM (K)

Potassium is found in large amounts in most plant feedstuffs, and generally supplementation is not required. It plays an essential role in cellular homeostasis, along with sodium and chlorine. In cases of potassium deficiency, generalized muscle weakness can be observed.

TRACE OR MICROMINERALS

Trace minerals are minerals that are required in small amounts. The trace minerals of special concern in poultry are copper, iron, iodine, manganese, selenium, and zinc.

As soils become leached, their content of trace minerals and the feedstuffs grown on them become borderline or deficient, thereby necessitating that more of certain minerals be added to poultry rations. Interactions between various trace minerals—such as copper and molybdenum, selenium and mercury, calcium and zinc, or calcium and manganese—are also important in poultry nutrition. Selenium is also metabolically involved with vitamin E and arsenic.

COPPER (Cu) AND IRON (Fe)

Both of these elements are necessary in the prevention of anemia. If a deficiency of either element exists, there is a reduction in the size of the red blood cells as well as a decreased oxygen-carrying capacity of the cells. Poultry rations should contain about 80 ppm of iron and 6 to 8 ppm of copper.

Iron salts may be used in poultry rations containing cottonseed meal to tie up gossypol—a compound that causes discoloration of yolks. However, iron salts do not prevent the pink discoloration of egg whites caused by cyclopropenoid fatty acids in cottonseed meal.

IODINE (I)

In certain areas of the United States (Northwest and Great Lakes regions), soils are deficient in iodine. In these deficient areas, it may be advisable to add iodine to the ration to prevent goiter. Feeds for laying hens should contain 300 ppb of iodine. For growing chicks, feed should contain 350 ppb of iodine.

MANGANESE (Mn)

This element is routinely added to poultry diets in such forms as manganous chloride, manganous carbonate, manganese dioxide, and manganese sulfate. Dietary levels of 30 ppm for layers and 60 ppm for broilers should be sufficient.

Chicks that are deficient in manganese characteristically display slipped tendons (perosis). A deficiency of either choline or biotin will produce similar symptoms. Hens suffering from a manganese deficiency will exhibit a reduction in egg production and hatchability. Many of the eggs that are laid are either thin-shelled or without shells.

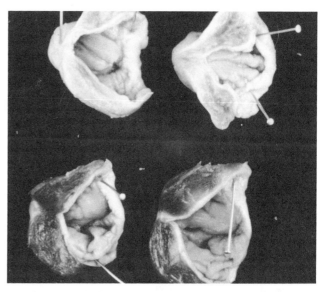

Fig. 6–7. White gizzard disease (muscular dystrophy) in turkey poults due to selenium deficiency. Note the whitish, underdeveloped gizzard at the top from a selenium-deficient poult compared to a normal gizzard below. (Courtesy, L. S. Jensen, Ph.D., Department of Poultry Science, The University of Georgia, Athens)

ZINC (Zn)

Chicks that are deficient in zinc exhibit bone problems, poor feathering, anorexia, and retarded growth. Zinc and calcium absorption are interrelated; hence, as the calcium level is increased, the zinc level should be increased, also.

Fig. 6–6. Perosis, or slipped tendon, due to manganese deficiency. (Courtesy, Department of Poultry Science, Cornell University)

SELENIUM (Se)

Chicks deficient in selenium develop exudative diathesis, steatitis, and pancreatic fibrosis. Selenium-deficient turkeys develop a condition commonly referred to as white gizzard disease—a form of muscular dystrophy. Care should be taken in the incorporation of selenium in poultry rations as it is toxic at a level of 10 ppm.

Fig. 6–8. Severe zinc deficiency. (Courtesy, Department of Poultry Science, University of Wisconsin, Madison)

VITAMINS

Vitamins are defined as complex organic compounds that are required in minute amounts by one or more animal species for normal growth, production, reproduction, and/or health.

The vitamins required by poultry, along with their deficiency symptoms and dietary sources, are shown in Table 6–5.

The column of Table 6–5 headed, "Types of Poultry Rations Usually Requiring Supplementation" indicates the types of poultry rations in which special attention must be paid to the inclusion of dietary sources of the vitamins. As shown, vitamin A, vitamin D, riboflavin, and vitamin B–12 are commonly low in most poultry rations. It is also to be emphasized that vitamin D_3, the animal form of vitamin D (made by the irradiation of 7–dehydrocholesterol), is more active for poul-

TABLE
POULTRY VITAMIN

Vitamins Which May Be Deficient Under Normal Conditions	Conditions Usually Prevailing Where Deficiencies are Reported	Function of Vitamin	Some Deficiency Symptoms	Types of Poultry Rations Usually Requiring Supplementation			
				Starting	Growing	Laying	Breeding
Fat-soluble vitamins							
Vitamin A	Old vitamin premix.	Essential for normal maintenance and functioning of the epithelial tissues, particularly of the eye and the respiratory, digestive, reproductive, nerve, and urinary systems.	*Chicks:* Depressed growth, weakness; loss of coordination; xerophthalmia; anorexia, lowered resistance to infection; alterations in mucous membranes. *Adults:* Depressed production; low hatchability; discharge from nose and eyes; lowered resistance to infection; alterations in mucous membranes.	Yes	Yes	Yes	Yes
Vitamin D_3	Birds that are in confinement.	Aids in assimilation and utilization of calcium and phosphorus, and necessary in normal bone development.	*Chicks:* Rickets, poor feathering, reduced growth. *Adults:* Weak bones, poor eggshell formation, reduced production and hatchability.	Yes	Yes	Yes	Yes
Vitamin E	Destruction by oxidation of the diet.	Antioxidant. Muscle structure. Reproduction.	*Chicks:* Encephalomalacia, exudative diathesis, muscular dystrophy. *Adults:* Poor reproductive performance; prolonged vitamin E deficiency results in permanent sterility in the male and reproductive failure in the female. *Poults:* Myopathy of the gizzard.	Yes	Yes	Yes	Yes
Vitamin K	Coccidiosis. When high levels of antibodies or sulfa drugs are fed. Newly hatched chicks from deficient females.	Essential for prothrombin formation and blood clotting.	Hemorrhaging. Increased clotting time.	Yes	Yes	Yes	Yes
Water-soluble vitamins							
Biotin	Broilers fed a milo, wheat, or wheat-barley based diet. Feeding avidin, a protein in uncooked egg white, which binds biotin and renders it unavailable nutritionally.	Involved in carbohydrate, lipid, and protein metabolism.	*Chicks:* Cracking and degeneration of skin on feet, around beak, and perosis (slipped tendon). *Adults:* Reduced hatchability. *Poults:* Broken flight feathers, bending of the metatarsus, and dermatitis of the footpads and toes, base of beak, eye ring and vent.	Yes	No	No	Yes

try, and should, therefore, be used instead of vitamin D_2, the plant form of the vitamin.

The fat-soluble vitamins (A, D, E, and K) can be stored and accumulated in the liver and other parts of the body, while only very limited amounts of the water-soluble vitamins (biotin, choline, folacin, niacin, pantothenic acid, riboflavin, thiamin, vitamin B-6, and vitamin B-12) are stored. For this reason, it is important that the water-soluble vitamins be fed regularly in the ration in adequate amounts.

Vitamin C is synthesized by poultry; hence, it is not considered as a required dietary nutrient. There is some evidence, nevertheless, of a favorable response to vitamin C by birds under stress.

Requirements for some of the vitamins may be met by the amounts occurring in natural feedstuffs. However, formulators of poultry feeds should be alert to the need for dietary supplementation with vitamins usually assumed to be supplied by the feedstuffs.

6–5

CHART (See footnote at end of table.)

Vitamin Requirements[1]	Recommended Allowances[1]	Practical Sources of the Vitamin	Comments
Vitamin Content (%) of Ration (Variable According to Class, Age, and Weight of Poultry) For . . . . See Table			
Layers . . . . 7–3 Broilers . . . . 7–5 Turkeys . . . . 7–10 Ducks . . . . 7–13 Geese . . . . 7–15 Japanese Quail 7–17	Variable according to class, age, and weight (see the suggested rations under each respective class).	Green forage, alfalfa meal, corn gluten meal, yellow corn, fish oils, synthetic vitamin A.	Toxicities can occur. The toxic level is on the order of 500 times the requirement. Symptoms of a vitamin A toxicity are weight loss, depressed feed intake, inflammation of epithelial tissue, and bone abnormalities. Requirements for vitamin A are expressed in either International Units (IU) or U.S. Pharmacopeia units (USP) per kilogram of diet.
Layers . . . . 7–3 Broilers . . . . 7–5 Turkeys . . . . 7–10 Ducks . . . . 7–13 Geese . . . . 7–15 Japanese Quail 7–17	Variable according to class, age, and weight (see the suggested rations under each respective class).	Irradiated animal sterols, fish liver oils, vitamin A and D feeding oils, synthetic vitamin D_3.	Vitamin D_3 is more than 30 times as efficient for preventing rickets in chickens as vitamin D_2. Hypervitaminosis can occur.
Layers . . . . 7–3 Broilers . . . . 7–5 Turkeys . . . . 7–10 Japanese Quail 7–17	Variable according to class, age, and weight (see the suggested rations under each respective class).	Alfalfa meal, vegetable oils, wheat germ, and pure vitamin concentrates such as alpha-tocopherol.	Vitamin E and selenium have a close interrelationship. In many cases, selenium can reduce the dietary requirement of vitamin E.
Layers . . . . 7–3 Broilers . . . . 7–5 Turkeys . . . . 7–10 Ducks . . . . 7–13 Japanese Quail 7–17	Variable according to class, age, and weight (see the suggested rations under each respective class).	Green pasture, alfalfa meal, synthetic vitamin K (menadione sodium bisulfite).	
Layers . . . . 7–3 Broilers . . . . 7–5 Turkeys . . . . 7–10 Japanese Quail 7–17	Variable according to class, age, and weight (see the suggested rations under each respective class).	Grains, soybean meal, alfalfa meal, dried yeast, milk products, green pasture.	Availability in wheat and barley is extremely low.

(Continued)

TABLE 6–5

Vitamins Which May Be Deficient Under Normal Conditions	Conditions Usually Prevailing Where Deficiencies are Reported	Function of Vitamin	Some Deficiency Symptoms	Types of Poultry Rations Usually Requiring Supplementation			
				Starting	Growing	Laying	Breeding
Water-soluble vitamins (Continued) Choline		Involved in nerve pulses. A component of phospholipids. Donor of methyl groups.	*Chicks, poults, ducklings:* Retarded growth and perosis (slipped tendon). *Adults:* Increased mortality, lowered egg production, and increased abortion of egg yolks from ovaries.	Yes	Yes	No	No
Folacin (Folic acid)		Related to B–12 metabolism. Metabolic reactions involving incorporation of single carbon units into larger molecules.	*Chicks:* Poor growth, poor feathering, perosis, and anemia. *Adults:* Reduced hatchability and egg production. *Turkey poults:* Nervousness, droopy wings, and a stiff extended neck. *Turkey breeder hens:* Normal egg production, but reduced hatchability.	Yes	Yes	Yes	Yes
Niacin (Nicotinic Acid, Nicotinamide)	A predominantly corn-soybean ration.	Required by all living cells, and an essential component of important metabolic enzyme systems involved in glycolysis and tissue respiration.	*Chicks:* Enlargement of hock joints and perosis, retarded growth, and inflammation of mouth and tongue ("black tongue"). *Adults:* No symptoms observed in hen except on protein-deficient diet. *Turkey poults:* A hock disorder similar to perosis.	Yes	Yes	Yes	Yes
Pantothenic Acid (Vitamin B–3)	Use of artificially dried corn (heating destroys the pantothenic acid) and the omission of milk by-products from the diet.	Part of coenzyme A, a necessary factor for intermediate metabolism.	*Chicks:* Poor growth, ragged feather development, degeneration of skin around beak, eyes, and vent, and liver damage. *Adults:* Reduced hatchability. Mortality is high in newly hatched chicks from pantothenic acid-deficient hens.	Yes	Yes	Yes	Yes
Riboflavin (Vitamin B–2)		A component of enzyme systems essential to normal metabolic processes.	*Chicks:* Curled toe paralysis, reduced growth, and diarrhea. *Adults:* Poor hatchability with many dying during 2nd week of incubation.	Yes	Yes	Yes	Yes
Thiamin (Vitamin B–1)		As a coenzyme in energy metabolism. Promotes appetite and growth, and required for normal carbohydrate metabolism.	*Chicks:* Anorexia, loss of coordination, poor feathering, polyneuritis. *Adults:* Blue comb, paralysis.	No	No	No	No
Vitamin B–6 (Pyridoxine, Pyridoxal, Pyridoxamine)		As coenzyme in protein and nitrogen metabolism. Involved in red blood cell formation. Important in endocrine systems.	*Chicks:* Poor growth, lack of coordination, and convulsions. *Adults:* Reduced body weight, egg production, and hatchability.	No	No	No	No
Vitamin B–12 (Cobalamins)		Numerous metabolic functions, and essential for normal growth and reproduction in poultry.	*Chicks:* Poor growth, perosis, mortality. *Adults:* Reduced hatchability, fatty heart, liver, and kidneys.	Yes	Yes	Yes	Yes

[1]As used herein, the distinction between *vitamin requirements* and *recommended allowances* is as follows: In *vitamin requirements*, no margins of safety are included intentionally; whereas in *recommended allowances*, margins of safety are provided in order to compensate for variations in feed compositions, environment, and possible losses during storage or processing.

(Continued)

Vitamin Requirements[1] — Vitamin Content (%) of Ration (Variable According to Class, Age, and Weight of Poultry) For . . . . See Table	Recommended Allowances[1]	Practical Sources of the Vitamin	Comments
Layers . . . . 7–3 Broilers . . . . 7–5 Turkeys . . . . 7–10 Pheasants/Bobwhite . . . . . 7–16 Japanese Quail 7–17	Variable according to class, age, and weight (see the suggested rations under each respective class).	Fish products and pure vitamin.	Recent evidence indicates that choline is synthesized by mature chickens in quantities adequate for egg production. Choline may be a factor in egg size of quail. Dietary requirements of growing quail appear to be higher than those for chicks or poults.
Layers . . . . 7–3 Broilers . . . . 7–5 Turkeys . . . . 7–10 Japanese Quail 7–17	Variable according to class, age, and weight (see the suggested rations under each respective class).	Alfalfa meal, wheat, soybean meal, and liver preparations.	
Layers . . . . 7–3 Broilers . . . . 7–5 Turkeys . . . . 7–10 Ducks . . . . 7–13 Geese . . . . 7–15 Pheasants/Bobwhites . . . . 7–16 Japanese Quail 7–17	Variable according to class, age, and weight (see the suggested rations under each respective class).	Chemically synthesized nicotinic acid, liver, yeast, and fermentation products, and most grasses.	Some niacin can be synthesized in the body through the conversion of tryptophan. The niacin in cereal grains and by-products is virtually unavailable and should not be included in the available niacin calculation.
Layers . . . . 7–3 Broilers . . . . 7–5 Turkeys . . . . 7–10 Ducks . . . . 7–13 Geese . . . . 7–15 Pheasants/Bobwhites . . . . 7–16 Japanese Quail 7–17	Variable according to class, age, and weight (see the suggested rations under each respective class).	Pure calcium pantothenate, alfalfa meal, dried milk products, and fermentation residues.	
Layers . . . . 7–3 Broilers . . . . 7–5 Turkeys . . . . 7–10 Ducks . . . . 7–13 Geese . . . . 7–15 Pheasants/Bobwhites . . . . 7–16 Japanese Quail 7–17		Alfalfa meal, milk products, distillers' solubles, fermentation products, and pure vitamin.	
Layers . . . . 7–3 Broilers . . . . 7–5 Turkeys . . . . 7–10 Japanese Quail 7–17	Variable according to class, age, and weight (see the suggested rations under each respective class).	Cereal grains and their by-products. Synthetic thiamin.	
Layers . . . . 7–3 Broilers . . . . 7–5 Turkeys . . . . 7–10 Ducks . . . . 7–13 Japanese Quail 7–17	Variable according to class, age, and weight (see the suggested rations under each respective class).	Milk products, meat and fish by-products, soybean meal.	
Layers . . . . 7–3 Broilers . . . . 7–5 Turkeys . . . . 7–10 Japanese Quail 7–17	Variable according to class, age, and weight (see the suggested rations under each respective class).	Fish meal, fish solubles, meat meal, liver preparations, fermentation products, and commercial vitamin B–12 concentrate.	Body reserves are rapidly depleted in hens that are fed high-protein diets.

FAT SOLUBLE VITAMINS

Vitamin A and vitamin D have long been routinely added to poultry feed. Today, vitamin E and vitamin K are also included in vitamin supplements for most poultry feeds.

VITAMIN A

The requirement of vitamin A in poultry is dependent upon the following factors:

1. **Individual bird variations.** Metabolic rates and subsequent nutrient requirements vary from bird to bird, even though they may be of the same strain.

2. **Type of bird.** There are genetic differences between different strains of birds with regards to their respective abilities to utilize vitamin A.

3. **Production and stress.** Vitamin A requirements vary according to the type and stage of production. The layer needs more vitamin A than the growing chicken, since vitamin A is passed into the egg. Likewise, birds that are subjected to environmental stress probably have different vitamin A requirements than birds under minimal stress.

TABLE 6–6
CONVERSION OF BETA-CAROTENE TO VITAMIN A FOR DIFFERENT SPECIES[1]

Species	Conversion of mg of Beta-Carotene to IU of Vitamin A		IU of Vitamin A Activity (calculated from carotene)
	(mg)	(IU)	(%)
Standard (rat)	1	1,667	100.0
Poultry	1	1,667	100.0
Beef cattle	1	400	24.0
Dairy cattle	1	400	24.0
Sheep	1	400–500	24.0–30.0
Swine	1	500	30.0
Horses:			
Growth	1	555	33.3
Pregnancy	1	333	20.0
Mink	Carotene not utilized		—
Human	1	556	33.3

[1]Adapted from the *Atlas of Nutritional Data on United States and Canadian Feeds*, NRC-National Academy of Sciences, p. XVI, Table 6.

Fig. 6–9. An advanced stage of vitamin A deficiency. Note the exudate from the eye and the general ruffled appearance. (Courtesy, Department of Poultry Science, Cornell University)

4. **Destruction of the vitamin.** Vitamin A in feed can be destroyed when fats in the feed become rancid. Likewise, certain processing methods can reduce the activity of the vitamin in the feed. Parasites and bacteria in the gut can also destroy vitamin A before it is absorbed.

5. **Absorbability.** Since vitamin A absorption is dependent on a lipoprotein in the blood, deficiencies of protein and/or fat can reduce absorption.

It is noteworthy that poultry are just as efficient as the rat in the conversion of β-carotene to vitamin A (see Table 6–6).

Young chicks are more susceptible to vitamin A deficiencies than adults as it takes a relatively long period for adult birds to deplete their body stores. Deficiency symptoms of vitamin A in chicks are characterized by retarded growth, emaciation, general weakness, staggered gait, ruffled plumage, lowered resistance to infection, xeropthalmia and disruption of mucosal membranes. In adults, eye problems are prevalent along with decreased egg production and hatchability.

Excessive amounts of vitamin A can be toxic, but these levels must be on the order of 500 times the recommended allowances. Symptoms of vitamin A hypervitaminosis are anorexia, emaciation, inflammation of epithelial tissues, abnormalities of bone, swelling of the eyelids, and mortality.

VITAMIN D

Cholecalciferol (D₃) is the form of vitamin D that has the highest activity in poultry feed. Today, most birds are reared in confinement where exposure to sunlight is insufficient for the conversion of 7–dehydrocholesterol – the precursor of vitamin D in the skin – to vitamin D_3. Thus, vitamin D_3 is routinely added to poultry feed. The cost of supplementation is small – especially when one considers the consequences of such deficiencies.

The dietary requirements of vitamin D depend on four factors: (1) exposure to sunlight; (2) Ca:P ratio; (3) levels of calcium and phosphorus in the feed; and (4) intensity of production.

Leg problems, poor growth, and poor feathering are the common deficiency symptoms in growing birds. Egg production, hatchability, and eggshell quality will be decreased in deficient hens.

Fig. 6-10. A chick deficient in vitamin D, showing ungainly manner of balancing body. The beak is also soft and rubbery. (Courtesy, Department of Poultry Science, Cornell University)

VITAMIN E

As with vitamins A and D, vitamin E is extremely susceptible to destruction from the oxidation of fats in the feed. To prevent this, antioxidants are commonly added to poultry feeds. Also, vitamin E is often added to feed in an esterified form to protect it from destruction.

The three classical symptoms of vitamin E deficiency in chicks are encephalomalacia, exudative diathesis, and nutritional muscular dystrophy. Encephalomalacia is a condition whereby there is a necrosis in the brain. Chicks exhibit an outstretching of the legs with toes curled; and the head is often in a retracted position. Prior to these symptoms of acute toxicity, chicks display a generalized lack of coordination. Exudative diathesis is a condition in which the walls of cap-

Fig. 6-11. A chick with nutritional encephalomalacia, due to a lack of vitamin E. Note head retraction and loss of control of legs. (Courtesy, Department of Poultry Science, Cornell University)

illaries become highly permeable. Nutritional muscular dystrophy in chicks is analogous to stiff lamb disease in sheep and white muscle disease in calves.

Reproduction is impaired in vitamin E-deficient adult birds. Degeneration of the testes is observed in deficient males—a condition that can lead to permanent sterility if not corrected in time. Layers suffering from a vitamin E deficiency do not show a dramatic drop in egg production, but hatchability is severely reduced.

In vitamin E-deficient turkey poults, a myopathy of the gizzard can be observed.

Selenium and vitamin E are closely related in physiological functions, but they cannot replace each other in the diet.

VITAMIN K

Vitamin K occurs in a number of naturally occurring and synthetic compounds with varying solubilities in fat and water. Menadione is a fat-soluble, synthetic compound that can be considered as the reference standard for vitamin K activity. Two naturally occurring forms are K_1 or phylloquinone and K_2 or minaquinone. Water-soluble forms include menadione sodium bisulfate (MSB), menadione sodium bisulfite complex (MSBC), and menadione dimethylpyrimidol (MPB). The theoretical activity of these compounds can be calculated on the basis of the proportion of menadione present in the molecule.

Vitamin K should be added to starter rations and rations that incorporate drugs, such as sulfaquinoxaline, which reduce the microbial population of the gut. When heavy parasitic infections occur, such as in coccidiosis, the vitamin K requirement is increased. Birds that are deficient in vitamin K have a greatly increased susceptibility to hemorrhaging and an increased clotting time. Newly hatched chicks from vitamin K-deficient females will have low storage of vitamin K and are generally susceptible to injury.

WATER-SOLUBLE VITAMINS

Of the water-soluble vitamins, biotin, niacin, pantothenic acid, riboflavin, and vitamin B-12 may be low in poultry feeds. Young chicks are most susceptible to vitamin deficiencies.

The requirements for the water-soluble vitamins are interrelated in some instances. They are also dependent upon the nature of the diet. The type of carbohydrate, protein concentration, and amino acid balance are major factors determining the dietary requirements for several vitamins.

BIOTIN

The availability of biotin in many of the cereal grains is very low. Hence, biotin levels and availability should be carefully monitored in poultry rations. Eggs are high in biotin. Therefore, careful consideration should be given to biotin in rations for layers. The symptoms of biotin deficiency in chicks are cracking and degeneration of skin on the feet and around the beak, and perosis. Reduced hatchability is observed in biotin-deficient hens.

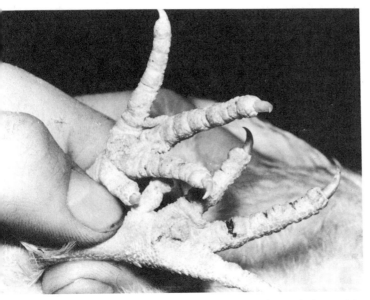

Fig. 6–12. Biotin deficiency. Note the severe lesions on the bottom of the feet. (Courtesy, Department of Poultry Science, University of Wisconsin, Madison)

FOLACIN (FOLIC ACID)

Because feeds rich in folacin — for example, alfalfa meal and liver meal — are used less frequently today in practical poultry rations, most poultry feeds are supplemented with folacin.

Deficiency symptoms of folacin in chicks are poor growth, abnormal coloration of feathers, perosis, and anemia. Hens that are deficient in folacin are observed to have reduced egg production and hatchability. Poults fed folacin-deficient rations may develop a paralysis of the neck.

NIACIN (NICOTINIC ACID/NICOTINAMIDE)

Synthetic niacin is routinely added to starter and breeder rations. Some niacin can be synthesized in the body from tryptophan; but since corn, which is notably low in tryptophan, is the most widely used feedstuff in poultry feed, this means of fulfilling the niacin requirement should be viewed with skepticism.

CHOLINE

Since choline levels can be marginal in starter and grower rations, choline is generally added to these types of feeds. Much like the deficiency symptoms of several of the other water-soluble vitamins, perosis (slipped tendon) can be observed in choline-deficient chicks. Since older birds can synthesize choline on the cellular level, it is difficult to produce a choline deficiency in laying hens.

Fig. 6–14. "Spectacled eye" in a niacin-deficient chick. Also, note the loss of feathers around the eye. (Courtesy, University of Wisconsin, Madison)

Chicks deficient in niacin show an enlargement of the hock joints, perosis, dermatitis, retarded growth, and an inflammation of the mouth and tongue. In addition to these symptoms, poor feathering and hyperirritability may be observed.

PANTOTHENIC ACID (VITAMIN B–3)

Pantothenic acid is widely distributed in nature. However, it is routinely added to starter and breeder rations — often in the form of calcium pantothenate — to ensure against any possibility of deficiency. Deficiency symptoms in chicks are rather nonspecific — poor growth, poor feathering, liver

Fig. 6–13. Perosis (slipped tendon) in a choline-deficient chick. (Courtesy, University of Arkansas, Fayetteville)

damage, and lesions around the beak, eyes, and vent. Lowered hatchability and a high mortality rate of newly hatched chicks can result from feeding hens a diet deficient in pantothenic acid.

Fig. 6–15. An advanced stage of pantothenic acid deficiency. Note the lesions at the corners of the mouth and on the eyelids and feet. (Courtesy, Department of Poultry Science, Cornell University)

RIBOFLAVIN (VITAMIN B–2)

All types of poultry rations should be supplemented with riboflavin. Chicks suffering from a riboflavin deficiency display a characteristic curled toe paralysis as well as depressed growth and diarrhea. In curled-toe paralysis, the brachial and sciatic nerves become greatly enlarged. Poor egg production and hatchability are observed in riboflavin-deficient hens.

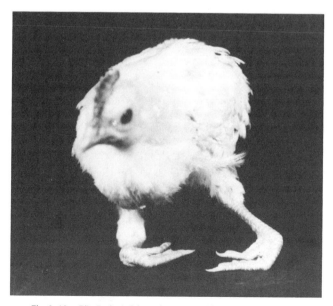

Fig. 6–16. Riboflavin deficiency in a young chick. Note the curled toes and tendency to squat on hocks. (Courtesy, Department of Poultry Science, Cornell University)

THIAMIN (VITAMIN B–1)

Although the requirement for thiamin is high in poultry, it is rarely added as a supplement to the ration because most feed ingredients contain high levels. Polyneuritis — a type of paralysis — is common in thiamin-deficient birds. Prior to this acute deficiency condition, anorexia, emaciation, ruffled feathers, and incoordination are observed in thiamin-deficient chicks. When polyneuritis sets in, a progressive paralysis is observed — beginning first in the toes and ultimately reaching the head, whereupon the head is retracted so that it lies on the back. Deficient adults frequently have a blue comb.

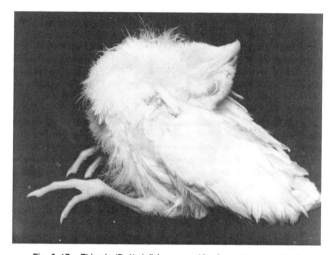

Fig. 6–17. Thiamin (B–1) deficiency, resulting in acute stage of polyneuritis. Note characteristic head retraction. (Courtesy, Department of Poultry Science, University of Wisconsin, Madison)

VITAMIN B–6 (PYRIDOXINE, PYRIDOXAL, PYRIDOXAMINE)

Deficiencies of vitamin B–6 are rare in poultry, since most feeds used in poultry rations contain relatively high levels of the vitamin. Deficiency symptoms in chicks are char-

Fig. 6–18. Pyridoxine (B–6) deficiency. Left: Normal, control chick. Right: Chick shows retarded growth and abnormal feathering due to vitamin B–6 deficiency. (Courtesy, University of Georgia, Athens)

acterized by depressed growth and neurological problems such as poor coordination and convulsions. Mature birds deficient in vitamin B–6 exhibit anorexia, rapid loss of weight, and lowered egg production and hatchability.

VITAMIN B–12 (COBALAMINS)

Vitamin B–12 is found only in animal products and bacterial fermentation products. Consequently, poultry feeds – which consists primarily of plant products – are supplemented with vitamin B–12.

Liver stores of vitamin B–12 may be high enough to sustain adults for several months, but high-protein diets can accelerate this depletion process.

Birds deficient in vitamin B–12 exhibit poor feed conversions, depressed growth, reduced hatchability, and in some cases perosis (slipped tendon). Fatty livers, kidneys, and hearts can be observed in some deficient birds.

UNIDENTIFIED GROWTH FACTORS (UGF)

In addition to the vitamins listed in Table 6–5 certain unidentified growth factors (UGF) are important in poultry nutrition. They are referred to as *unidentified* or *unknown* because they have not yet been isolated or synthesized in the laboratory. Nevertheless, rich sources of these factors and their effects have been well established. A diet that supplies the specific levels of all the known nutrients but which does not supply the unidentified factors is inadequate for best performance. There is evidence that these growth factors exist in fish solubles, dried whey, brewers' dried yeast, egg yolk, and corn distillers' dried solubles and other fermentation residues, packinghouse by-products, and green forages. Most nutritionists recognize the possible benefits of adding some UGF supplementation to the diets of broilers and breeding hens. Commonly, the UGF source is added to the diet at a level of 1 to 3%, although antibiotic fermentation residue products may be used at levels ranging from 2 to 8 lb per ton. This practice may have a twofold advantage: (1) providing possible unidentified growth factor responses, and (2) supplying additional amounts of some of the known vitamins.

WATER

Poultry should have free access to clean, fresh water at all times. It is needed as a solvent, a lubricant, and a temperature control device.

A general rule is that chickens drink approximately twice as much water by weight as the feed they consume.

The amount of water required by poultry varies considerably, as evidenced in Table 6–7. Several factors can affect the amount of water birds will consume: (1) age, (2) body weight, (3) production, (4) weather (heat and humidity), and (5) type of ration. The intensity of production dramatically affects the water requirement.

TABLE 6–7
DAILY WATER CONSUMPTION
BY CHICKENS AND TURKEYS OF DIFFERENT AGES[1, 2]

Age	Per 1,000 Birds					
	Leghorn-Type Pullets		Chicken Broilers[3]		Turkeys[3]	
(week)	(U.S. gal)	(liter)	(U.S. gal)	(liter)	(U.S. gal)	(liter)
1	5	19	5	19	10	38
2	10	38	13	50	20	76
3	12	45	24	90	30	114
4	17	64	37	140	40	151
5	22	83	53	200	50	189
6	25	95	69	260	60	227
7	28	106	85	320	75	284
8	30	114	100	380	95	360
9	35	132			115	435
10	38	144			125	473
12	40	151			150	568
15	42	158			160	606
20	45	170			200	757
Laying or Breeding						
35	50	189			M 240	908
					F 130	492

[1]Adapted by the author from *Nutrient Requirements of Poultry*, 8th rev. ed., NRC, National Academy Press, Washington, DC, 1984, p. 8, Table 2.

[2]Will vary considerably depending on temperature and ration composition.

[3]Mixed sexes.

Water is the largest single constituent of poultry tissue; it constitutes 85% of the body of a baby chick, 58% of the body of an adult bird, and 66% of an egg. Yet, water is often neglected. Birds can lose 98% of their body fat or 50% of their body protein and still survive. However, a 10% loss in body water causes serious physiological disorders, and 20% loss in body water will cause death.

In addition to being readily available, water quality is important. Water should be tested to determine that salts, pesticides, and microorganisms are at acceptable levels and that the water is palatable to poultry. Water that adversely affects growth, reproduction, or productivity should not be used.

QUESTIONS FOR STUDY AND DISCUSSION

1. What is the primary reason for keeping poultry?

2. Why is knowledge of composition of the body and the egg of poultry important to the nutritionist? How do each of the following affect body composition:
   a. Age
   b. Degree of fatness
   c. Species?

3. Discuss the nutrient needs of poultry for each of the following body functions:
   a. Maintenance
   b. Growth
   c. Reproduction

4. Discuss the effect of each of the following on the nutritive needs for growth: age, breed, sex, rate of growth, and health.

5. Define and discuss the differences between anabolism and catabolism.

6. Define energy. What are the chief sources of energy for poultry? How is a deficiency of energy manifested in poultry?

7. List and discuss the factors that can affect the metabolic rate of poultry.

8. How may vegetable oils affect the body fat of poultry?

9. Diagram and explain the utilization of energy.

10. Why has the poultry industry found metabolizable energy to be the most reliable expression of energy needs?

11. What is true metabolizable energy (TME)? Describe the Sibbald procedure for obtaining TME.

12. What are the original sources of all proteins?

13. From the standpoint of poultry nutrition, why are the amino acids that make up proteins really the essential nutrients, rather than the protein molecule itself?

14. Discuss amino acid relationships, antagonisms, imbalances, conversion of certain amino acids to vitamins, and amino acid availability.

15. How does a deficiency of amino acids affect poultry?

16. List the essential and semiessential amino acids of the chick.

17. What is meant by the *true digestibility of amino acids*? Describe the Sibbald method of obtaining these values.

18. Define macromineral. List the minerals which fit this definition. Define trace mineral, and list the minerals which fit this definition.

19. List the general functions of minerals.

20. List the physiological functions of calcium and phosphorus. What are the symptoms of their deficiencies? What ratio of these two minerals should be incorporated into poultry rations?

21. What are the hazards of (a) excessive magnesium, or (b) deficiency of magnesium?

22. Why is iron sometimes added to rations containing cottonseed meal?

23. What is perosis?

24. Discuss the role of selenium in poultry nutrition.

25. Describe the deficiency symptoms of zinc in poultry rations.

26. Define vitamins. List the fat-soluble vitamins and the water-soluble vitamins.

27. List and discuss the function of vitamin A.

28. Discuss the relative efficiency of poultry in comparison with other species in converting carotene to vitamin A.

29. How does vitamin E prevent peroxidation? How is this mechanism different from that of selenium?

30. List the symptoms of a vitamin E deficiency in chicks.

31. Normally dietary supplementation of vitamin K is of little concern. Why? Under what circumstances should the addition of vitamin K into the diet be of special concern?

32. List the water-soluble vitamins.

33. What are the symptoms of a biotin deficiency?

34. List the various functions of choline.

35. Describe niacin deficiency in the chick.

36. What is curled-toe paralysis in chickens?

37. List the physiological functions of water.

38. Discuss the factors that can affect the amount of water that birds can consume.

SELECTED REFERENCES

Title of Publication	Author(s)	Publisher
Animal Feeding and Nutrition, Fifth Edition	M. H. Jurgens	Kendall/Hunt Publishing Company, Dubuque, IA, 1982
Animal Growth and Nutrition	Edited by E. S. E. Hafez I. A. Dyer	Lea & Febiger, Philadelphia, PA, 1969
Animal Nutrition, Seventh Edition	L. A. Maynard, *et al.*	McGraw-Hill Book Company, Inc., New York, NY, 1979
Animal Nutrition, Second Edition	P. McDonald R. A. Edwards J. F. D. Greenhalgh	Oliver & Boyd, Edinburgh, Scotland, 1973
Animal Science, Ninth Edition	M. E. Ensminger	Interstate Publishers, Inc., Danville, IL, 1991
Basic Animal Nutrition and Feeding, Second Edition	D. C. Church W. G. Pond	John Wiley & Sons, Inc., Salt Lake City, UT, 1984
Bioenergetics and Growth	S. Brody	Reinhold Publishing Company, New York, NY, 1945
Calcium in Broiler, Layer, and Turkey Nutrition	R. H. Harms, *et al.*	National Feed Ingredients Association, West Des Moines, IA, 1976
Commercial Chicken Production Manual, Fourth Edition	M. O. North D. D. Bell	Van Nostrand Reinhold Co., New York, NY, 1990
Feeds and Feeding, Third Edition	A. Cullison	Reston Publishing Company, Inc., Reston, VA, 1982
Feeds & Nutrition, Second Edition	M. E. Ensminger J. E. Oldfield W. W. Heinemann	The Ensminger Publishing Company, Clovis, CA, 1990
Feeds & Nutrition Digest	M. E. Ensminger J. E. Oldfield W. W. Heinemann	The Ensminger Publishing Company, Clovis, CA, 1990
Feeds for Livestock, Poultry, and Pets	M. H. Gutcho	Noyes Data Corporation, Park Ridge, NJ, 1973
Manual of Poultry Production in the Tropics	R. R. Say	CAB International, Wallingford, Oxon, United Kingdom, 1987
Mineral Nutrition of Livestock, The	E. J. Underwood	Food and Agriculture Organization of the United Nations, Rome, Italy, 1975
Nutrient Requirements of Poultry and Nutritional Research	C. Fisher K. N. Boorman	Butterworths, Kent, England, 1986
Nutrient Requirements of Poultry, Eighth Revised Edition	National Research Council	National Academy of Science, Washington, DC, 1984
Nutrition of the Chicken, Third Edition	M. L. Scott M. C. Nesheim R. J. Young	M. L. Scott and Associates, Ithaca, NY, 1982
Nutrition of the Turkey	M. L. Scott	M. L. Scott of Ithaca, Ithaca, NY, 1987
Phosphorus in Poultry and Game Bird Nutrition	P. E. Waibel, *et al.*	National Feed Ingredients Association, West Des Moines, IA, 1977
Poultry: Feeds and Nutrition, Second Edition	H. Patrick P. J. Schaible	The AVI Publishing Company, Inc., Westport, CT, 1980
Poultry Meat and Egg Production	C. R. Parkhurst C. J. Mountney	Van Nostrand Reinhold Co., New York, NY, 1988
Poultry Nutrition, Fifth Edition	W. R. Ewing	The Ray Ewing Company, Pasadena, CA, 1963
Poultry Production, Twelfth Edition	M. C. Nesheim R. E. Austic L. E. Card	Lea & Febiger, Philadelphia, PA, 1979
Raising Poultry Successfully	W. Graves	Williamson Publishing, Charlotte, VT, 1985
Scientific Feeding of Chickens, The, Fifth Edition	H. W. Titus J. C. Fritz	The Interstate Printers & Publishers, Inc., Danville, IL, 1971
Trace Elements in Human and Animal Nutrition, Vols. 1 and 2, Fifth Edition	Edited by Walter Mertz	Academic Press, New York, NY, 1987

Fig. 7–1. Young turkeys at a typical feed pan of an automatic feeder line. (Courtesy, P. R. Ferket, North Carolina State University, Raleigh)

POULTRY FEEDING STANDARDS, RATION FORMULATION, AND FEEDING PROGRAMS

CHAPTER

7

Poultry feeding has changed more than the feeding of any other species — it has paced the entire livestock field. Today, the vast majority of commercial poultry is produced in large units wherein the maximum of science and technology exists. Confinement production is rather commonplace, and well-balanced rations containing adequate sources of all known nutrient materials are fed for maximum production. The current trend in poultry production is toward controlled environment, which usually results in lowered feed consumption. Under these conditions, the daily feed consumption is often taken into consideration and the nutrient content of feed (energy, amino acids, vitamins, and minerals) varied so as to compensate for the reduced feed intake and meet the requirements.

Table 7–1 shows the marked lowering of feed required to produce a unit of eggs, broilers, and turkeys since 1940.

TABLE 7–1
FEED UNITS REQUIRED PER UNIT OF EGGS AND POULTRY PRODUCED, SELECTED YEARS, 1940–90

Year Ending October	Per Dozen Eggs[1,2]	Per Pound Liveweight	
		Turkey[3]	Broiler[1,2]
	(feed units)	(feed units)	(feed units)
1940	7.4	4.50	4.7
1950	7.2	3.56	3.7
1960	6.4	3.37	3.0
1970	4.6	3.21	2.6
1980	4.14[4]	2.90[4]	2.1[4]
1990	3.75[4]	2.70[4]	1.9[4]

[1]Feed units used per dozen eggs or per pound of liveweight broiler produced. A feed unit is the economic equivalent of 1 lb of corn. 1940–1960 from *Handbook of Agricultural Charts 1965*, Ag. Hdbk. No. 300, USDA, Oct. 1965, p. 58.

[2]1970 from *Agricultural Statistics 1974*, USDA, p. 358, Table 518.

[3]The turkey data are based on estimates presented in *Efficiency in Animal Feeding with Particular Reference to Nonnutritive Feed Additives*, Council for Agricultural Science and Technology, Report No. 22, Jan. 18, 1974.

[4]Estimates by the author.

The most dramatic change has occurred in broilers. In 1940, it required 4.7 lb of feed to produce 1 lb of weight gain in broilers; in 1990, it took only 1.9 lb.

The nutrient composition of broiler and egg shown in Table 7–2 is indicative of the relative importance of these nutrients as broiler and egg constituents.

TABLE 7–2
NUTRIENT COMPOSITION OF BROILER AND EGG

Nutrient	Broiler[1]	Egg
	(%)	(%)
Water	64.0	66
Protein	18.8	13
Fat	14.2	10[2]
Minerals	3.7	11[3]

[1]Broiler data secured from the following source: Scott, M. L., "Nutrition in Reproduction—Direct Effects and Predictive Functions," *Breeding Biology of Birds*, National Academy of Sciences, 1973, pp. 51–52, Tables 1 to 4. Composition of broilers depends on age of bird; younger birds have a higher water content.

[2]Chiefly in the yolk.

[3]Nearly all is calcium in shell.

This chapter details two aspects of poultry feeding: (1) feeding standards and rations, and (2) feeding programs.

Part I—Feeding Standards and Rations

Poultry producers have the following alternatives for rations:

1. Purchase of a commercially prepared, complete feed.

2. Purchase of a commercially prepared protein supplement, reinforced with vitamins and minerals, which may be blended with local or homegrown grain.

3. Purchase of a commercially prepared vitamin-mineral premix which may be mixed with an oil meal, and then blended with local or homegrown grain.

4. Purchase of individual ingredients (including vitamins and minerals) and mixing the feed from the ground up.

Today, very few large poultry producers purchase a commercially prepared, complete mixed feed. Instead, most of them choose either option 3 or 4 from the above list.

FACTORS INVOLVED IN FORMULATING POULTRY RATIONS

Before anyone can formulate a poultry ration intelligently, it is necessary to know (1) the nutrient requirements of the particular birds to be fed, which calls for feeding standards; (2) the availability, nutrient content, and cost of feedstuffs; (3) the acceptability and physical condition of feedstuffs; (4) the average daily consumption of the birds to be fed; and (5) the presence of substances harmful to product quality.

NUTRIENT REQUIREMENT DETERMINATION

The nutritive requirements for a specific substance are determined by finding the minimum amount of that particular nutrient or substance that will permit maximum development of the physiological function or economic characteristic of concern. In general, the economic characteristics of importance in poultry are growth, efficiency of feed utilization, egg production, and hatchability. For example, if the need of a certain nutrient for growth is being determined, groups of birds must be fed on an experimental ration containing different levels of the nutrient in question until it is shown that increasing the quantity of the test nutrient beyond a particular level will not result in further increase in growth. If the test ration is complete in all other respects, then the nutrient requirement will be equal to the minimum supplemental level found to give maximum growth.

FEEDING STANDARDS

Feeding standards are tables listing the amounts of one or more nutrients required by different species of animals for specific productive functions, such as growth, fattening, and reproduction. Most feeding standards are expressed in either (1) quantities of nutrients required per day, and/or (2) concentration in the ration. The first type is used where animals are provided a given amount of a feed during a 24-hour period, and the second is used where animals are provided a ration without limitation on the time in which it is consumed.

Today, the most up-to-date feeding standards in the United States are those published by the National Research Council (NRC) of the National Academy of Sciences. Periodically, a specific committee, composed of outstanding researchers who have worked extensively with the class of animal whose requirements are being reviewed, revises the nutrient requirements of each species for different functions. Thus, the nutritive needs of each type of livestock are dealt with separately and in depth.

Feeding standards have been established for the various types of poultry, and through the use of these standards, producers can formulate rations tailored to meet the nutrient requirements of their birds.

Although feeding standards are excellent and needed guides, there are still many situations where nutrient needs cannot be specified with great accuracy. Also, in practical feeding operations, economy must be considered; for example, producers are interested in obtaining that level of egg production that will make for the largest net returns in light of current feed costs and the market price of eggs. Moreover, feeding standards tell nothing about the palatability, physical nature, or possible digestive disturbances of a ration. Neither do they give consideration to individual differences, management differences, and the effects of such stresses as weather, disease, parasitism, and surgery (debeaking, caponizing, etc.). Thus, there are many variables that alter the nutrient needs and utilization of birds—variables that are difficult to include quantitatively in feeding standards, even when feed quality is well known.

NATIONAL RESEARCH COUNCIL (NRC) REQUIREMENTS FOR POULTRY

The requirements for most of the nutrients needed by poultry have been established. These differ according to the kind and age of bird and the purpose for which it is being fed. A deficiency of a nutrient can be, and often is, a limiting factor in egg production or growth.

The National Research Council (NRC) nutritive requirements are given in the following tables:

For Chickens:

Table 7–3, Nutrient Requirements of Leghorn-Type Chickens as Percentages or as Milligrams or Units per Kilogram of Diet.

Table 7–4, Body Weights and Feed Requirements of Leghorn-Type Pullets and Hens.

Table 7–5, Nutrient Requirements of Broilers as Percentages or as Milligrams or Units per Kilogram of Diet.

Table 7–6, Body Weights and Feed Requirements of Broilers.

Table 7–7, Nutrient Requirements of Meat-Type Hens for Breeding Purposes.

Table 7–8, Typical Body Weights and Feed Allowances for Male and Female Meat-Type Chickens (Replacement Stock).

Table 7–9, Metabolizable Energy Required Daily by Chickens in Relation to Body Weight and Egg Production.

For Turkeys:

Table 7–10, Nutrient Requirements of Turkeys as Percentages or as Milligrams or Units per Kilogram of Diet.

Table 7–11, Growth Rate, Feed and Energy Consumption of Large-Type Turkeys.

Table 7–12, Body Weights and Feed Consumption of Large-Type Turkeys During Holding and Breeding Periods.

For Ducks:

Table 7–13, Nutrient Requirements of Pekin Ducks as Percentages or as Milligrams or Units per Kilogram of Diet.

Table 7–14, Typical Body Weights and Feed Consumption of Pekin Ducks to 8 Weeks of Age.

For Geese:

Table 7–15, Nutrient Requirements of Geese as Percentages or as Milligrams or Units per Kilogram of Diet.

For Pheasants and Bobwhite Quail:

Table 7–16, Nutrient Requirements of Pheasants and Bobwhite Quail as Percentages or as Milligrams or Units per Kilogram of Diet.

For Japanese Quail:

Table 7–17, Nutrient Requirements of Japanese Quail (Coturnix) as Percentages or as Milligrams or Units per Kilogram of Diet.

In establishing the NRC requirements for poultry, it was further assumed that the environmental temperature in which poultry of various species and ages are grown is ideal or as near optimum as possible for efficient growth and reproduction. Therefore, the energy level of the diet was first established for each species and age of poultry, then the other nutrients were determined based upon the established level of energy. If a higher level of energy than the NRC requirement is used in the diet, feed consumption will decrease; hence, the minimum level of the other nutrients should be increased in proportion to the energy content. Similarly, if a lower dietary energy level is used, then proportionately lower levels of other nutrients should be used in the diet.

TABLE 7–3
NUTRIENT REQUIREMENTS OF LEGHORN-TYPE CHICKENS AS PERCENTAGES OR AS MILLIGRAMS OR UNITS PER KILOGRAM OF DIET[1]

	Growing			Laying		Breeding
Energy Base: kcal ME/lb Diet[2] . . . . *kcal ME/kg Diet[2]*	0–6 Weeks 1,315 *2,900*	6–14 Weeks 1,315 *2,900*	14–20 Weeks 1,315 *2,900*	1,315 *2,900*	Daily Intake Per Hen (mg)[3]	1,315 *2,900*
Protein . . . . (%)	18	15	12	14.5	16,000	14.5
Amino acids:						
Arginine . . . . (%)	1.0	0.83	0.67	0.68	750	0.68
Glycine + serine . . . . (%)	0.7	0.58	0.47	0.5	550	0.5
Histidine . . . . (%)	0.26	0.22	0.17	0.16	180	0.16
Isoleucine . . . . (%)	0.6	0.5	0.4	0.5	550	0.5
Leucine . . . . (%)	1.0	0.83	0.67	0.73	800	0.73
Lysine . . . . (%)	0.85	0.6	0.45	0.64	700	0.64
Methionine + cystine . . . . (%)	0.6	0.5	0.4	0.55	600	0.55
Methionine . . . . (%)	0.3	0.25	0.2	0.32	350	0.32
Phenylalanine + tyrosine . . . . (%)	1.0	0.83	0.67	0.8	880	0.8
Phenylalanine . . . . (%)	0.54	0.45	0.36	0.4	440	0.4
Threonine . . . . (%)	0.68	0.57	0.37	0.45	500	0.45
Tryptophan . . . . (%)	0.17	0.14	0.11	0.14	150	0.14
Valine . . . . (%)	0.62	0.52	0.41	0.55	600	0.55
Linoleic acid . . . . (%)	1.0	1.0	1.0	1.0	1,100	1.0
Major or macrominerals:						
Calcium (Ca) . . . . (%)	0.8	0.7	0.6	3.4	3,750	3.4
Chlorine (Cl) . . . . (%)	0.15	0.12	0.12	0.15	165	0.15
Magnesium (Mg) . . . . (mg)	600	500	400	500	55	500
Phosphorus (P), available . . . . (%)	0.4	0.35	0.3	0.32	350	0.32
Potassium (K) . . . . (%)	0.4	0.3	0.25	0.15	165	0.15
Sodium (Na) . . . . (%)	0.15	0.15	0.15	0.15	165	0.15
Trace or microminerals:						
Copper (Cu) . . . . (mg)	8	6	6	6	0.88	8
Iodine (I) . . . . (mg)	0.35	0.35	0.35	0.3	0.03	0.3
Iron (Fe) . . . . (mg)	80	60	60	50	5.5	60
Manganese (Mn) . . . . (mg)	60	30	30	30	3.3	60
Selenium (Se) . . . . (mg)	0.15	0.1	0.1	0.1	0.01	0.1
Zinc (Zn) . . . . (mg)	40	35	35	50	5.5	65
Fat-soluble vitamins:						
Vitamin A . . . . (IU)	1,500	1,500	1,500	4,000	440	4,000
Vitamin D . . . . (ICU)	200	200	200	500	55	500
Vitamin E . . . . (IU)	10	5	5	5	0.55	10
Vitamin K . . . . (mg)	0.5	0.5	0.5	0.5	0.055	0.5
Water-soluble vitamins:						
Biotin . . . . (mg)	0.15	0.1	0.1	0.1	0.011	0.15
Choline . . . . (mg)	1,300	900	500	?	?	?
Folacin (Folic Acid) . . . . (mg)	0.55	0.25	0.25	0.25	0.0275	0.35
Niacin (Nicotinic Acid, Nicotinamide) . . . . (mg)	27	11	11	10	1.1	10
Pantothenic Acid (Vitamin B–3) . . . . (mg)	10	10	10	2.2	0.242	10
Riboflavin (Vitamin B–2) . . . . (mg)	3.6	1.8	1.8	2.2	0.242	3.8
Thiamin (Vitamin B–1) . . . . (mg)	1.8	1.3	1.3	0.8	0.088	0.8
Vitamin B–6 (Pyridoxine, Pyridoxal, Pyridoxamine) . . (mg)	3	3	3	3	0.33	4.5
Vitamin B–12 (Cobalamins) . . . . (mg)	0.009	0.003	0.003	0.004	0.00044	0.004

[1]Adapted by the author from *Nutrient Requirements of Poultry*, 8th rev. ed., NRC, National Academy Press, Washington, DC, 1984, p. 12, Table 4.

[2]These are typical dietary energy concentrations.

[3]Assumes an average daily intake of 110 g of feed/hen daily.

TABLE 7–4
BODY WEIGHTS AND FEED REQUIREMENTS OF LEGHORN-TYPE PULLETS AND HENS[1]

Age	Body Weight[2]		Feed Consumption[3]		Typical Egg Production (Hen/Day)
(weeks)	(lb)	(kg)	(lb/week)	(kg/week)	(%)
0	0.07	0.03	0.11	0.05	—
2	0.31	0.14	0.20	0.09	—
4	0.60	0.27	0.40	0.18	—
6	0.99	0.45	0.57	0.26	—
8	1.37	0.62	0.73	0.33	—
10	1.74	0.79	0.86	0.39	—
12	2.09	0.95	0.95	0.43	—
14	2.34	1.06	1.01	0.46	—
16	2.56	1.16	1.01	0.46	—
18	2.78	1.26	1.01	0.46	—
20	3.00	1.36	1.01	0.46	—
22	3.13	1.42	1.17	0.53	10
24	3.31	1.50	1.30	0.59	38
26	3.48	1.58	1.48	0.67	64
30	3.81	1.73	1.70	0.77	88
40	4.01	1.82	1.70	0.77	80
50	4.12	1.87	1.70	0.77	74
60	4.19	1.90	1.65	0.75	68
70	4.19	1.90	1.63	0.74	62

[1]Adapted by the author from *Nutrient Requirements of Poultry*, 8th rev. ed., NRC, National Academy Press, Washington, DC, 1984, p. 13, Table 5.

[2]Pullets and hens of Leghorn-type strains are generally fed *ad libitum* but are occasionally control-fed to limit body weights. Values shown are typical but will vary with strain differences, season, and lighting. Specific breeder guidelines should be consulted for desired schedules of weights and feed consumption.

[3]Based on rations containing 1,315 ME kcal/lb *(2,900 ME kcal/kg)*. Consumption will vary depending upon the caloric density of the ration, environmental temperature, and rate of production (see Table 7–8).

Fig. 7–2. Feed truck filling a feed tank outside a poultry house. (Courtesy, North Carolina State University, Raleigh)

TABLE 7–5
NUTRIENT REQUIREMENTS OF BROILERS AS PERCENTAGES OR AS MILLIGRAMS OR UNITS PER KILOGRAM OF DIET[1]

Energy Base:		Weeks 0–3	Weeks 3–6	Weeks 6–8
kcal ME/lb Diet[2]		1,452	1,452	1,452
kcal ME/kg Diet[2]		*3,200*	*3,200*	*3,200*
Protein	(%)	23	20	18
Amino acids:				
Arginine	(%)	1.44	1.2	1.0
Glycine + serine	(%)	1.5	1.0	0.7
Histidine	(%)	0.35	0.3	0.26
Isoleucine	(%)	0.8	0.7	0.6
Leucine	(%)	1.35	1.18	1.0
Lysine	(%)	1.2	1.0	0.85
Methionine + cystine	(%)	0.93	0.72	0.6
Methionine	(%)	0.5	0.38	0.32
Phenylalanine + tyrosine	(%)	1.34	1.17	1.0
Phenylalanine	(%)	0.72	0.63	0.54
Threonine	(%)	0.8	0.74	0.68
Tryptophan	(%)	0.23	0.18	0.17
Valine	(%)	0.82	0.72	0.62
Linoleic acid	(%)	1.0	1.0	1.0
Major or macrominerals:				
Calcium (Ca)	(%)	1.0	0.9	0.8
Chlorine (Cl)	(%)	0.15	0.15	0.15
Magnesium (Mg)	(mg)	600	600	600
Phosphorus (P), available	(%)	0.45	0.4	0.35
Potassium (K)	(%)	0.4	0.35	0.3
Sodium (Na)	(%)	0.15	0.15	0.15
Trace or microminerals:				
Copper (Cu)	(mg)	8	8	8
Iodine (I)	(mg)	0.35	0.35	0.35
Iron (Fe)	(mg)	80	80	80
Manganese (Mn)	(mg)	60	60	60
Selenium (Se)	(mg)	0.15	0.15	0.15
Zinc (Zn)	(mg)	40	40	40
Fat-soluble vitamins:				
Vitamin A	(IU)	1,500	1,500	1,500
Vitamin D	(ICU)	200	200	200
Vitamin E	(IU)	10	10	10
Vitamin K	(mg)	0.5	0.5	0.5
Water-soluble vitamins:				
Biotin	(mg)	0.15	0.15	0.1
Choline	(mg)	1,300	850	500
Folacin (Folic Acid)	(mg)	0.55	0.55	0.25
Niacin (Nicotinic Acid, Nicotinamide)	(mg)	27	27	11
Pantothenic Acid (Vitamin B–3)	(mg)	10	10	10
Riboflavin (Vitamin B–2)	(mg)	3.6	3.6	3.6
Thiamin (Vitamin B–1)	(mg)	1.8	1.8	1.8
Vitamin B–6 (Pyridoxine, Pyridoxal, Pyridoxamine)	(mg)	3	3	2.5
Vitamin B–12 (Cobalamins)	(mg)	0.009	0.009	0.003

[1]Adapted by the author from *Nutrient Requirements of Poultry*, 8th rev. ed., NRC, National Academy Press, Washington, DC, 1984, p. 13, Table 6.

[2]These are typical dietary energy concentrations.

TABLE 7–6
BODY WEIGHTS AND FEED REQUIREMENTS OF BROILERS[1, 2]

Age	Body Weights				Weekly Feed Consumption				Cumulative Feed Consumption				Weekly Energy Consumption		Cumulative Energy Consumption	
	Male		Female		Male		Female		Male		Female		Male	Female	Male	Female
(weeks)	(lb)	(g)	(lb)	(g)	(lb)	(g)	(lb)	(g)	(lb)	(g)	(lb)	(g)	(ME kcal/ bird)	(ME kcal/ bird)	(ME kcal/ bird)	(ME kcal/ bird)
1	0.29	130	0.26	120	0.26	120	0.24	110	0.26	120	0.24	110	385	350	385	350
2	0.71	320	0.66	300	0.57	260	0.53	240	0.84	380	0.77	350	830	770	1,215	1,120
3	1.23	560	1.14	515	0.86	390	0.78	355	1.70	770	1.55	705	1,250	1,135	2,465	2,255
4	1.90	860	1.74	790	1.18	535	1.10	500	2.88	1,305	2.66	1,205	1,710	1,600	4,175	3,855
5	2.76	1,250	2.45	1,110	1.63	740	1.42	645	4.51	2,045	4.08	1,850	2,370	2,065	6,545	5,920
6	3.73	1,690	3.15	1,430	2.16	980	1.76	800	6.67	3,025	5.84	2,650	3,135	2,560	9,680	8,480
7	4.63	2,100	3.85	1,745	2.41	1,095	2.01	910	9.08	4,120	7.85	3,560	3,505	2,910	13,185	11,390
8	5.56	2,520	4.54	2,060	2.67	1,210	2.14	970	11.75	5,330	9.99	4,530	3,870	3,105	17,055	14,495
9	6.45	2,925	5.18	2,350	2.91	1,320	2.23	1,010	14.66	6,650	12.21	5,540	4,225	3,230	21,280	17,725

[1]Adapted by the author from *Nutrient Requirements of Poultry*, 8th rev. ed., NRC, National Academy Press, Washington, DC, 1984, p. 14, Table 7.

[2]Typical for broilers fed well-balanced rations containing 1,452 ME kcal/lb *(3,200 ME kcal/kg)*.

Fig. 7–3. Broilers with automatic feed and water facilities. (Photo by J. C. Allen & Son, Inc., West Lafayette, IN)

TABLE 7–7
NUTRIENT REQUIREMENTS
OF MEAT-TYPE HENS FOR BREEDING PURPOSES[1, 2]

Energy Base: kcal ME/lb Diet[2] . . . . / kcal ME/kg Diet[2]	1,293[3] / 2,850[3]	Daily Intake Per Hen (mg)
Protein . . . . (%)	14.5	22,000
Amino acids:		
Arginine . . . . (%)	0.74	1,110
Glycine + serine . . . . (%)	0.62	932
Histidine . . . . (%)	0.14	205
Isoleucine . . . . (%)	0.57	850
Leucine . . . . (%)	0.83	1,250
Lysine . . . . (%)	0.51	765
Methionine + cystine . . . . . . (%)	0.55	820
Methionine . . . . (%)	0.35	520
Phenylalanine + tyrosine . . . . . (%)	0.75	1,112
Phenylalanine . . . . (%)	0.41	610
Threonine . . . . (%)	0.48	720
Tryptophan . . . . (%)	0.13	190
Valine . . . . (%)	0.63	950
Major or macrominerals:		
Calcium (Ca) . . . . (%)	2.75	4,125
Phosphorus (P), available . . . . (%)	0.25	375
Sodium (Na) . . . . (%)	0.10	150

[1]Adapted by the author from *Nutrient Requirements of Poultry*, 8th rev. ed., NRC, National Academy Press, Washington, DC, 1984, p. 14, Table 8.

[2]Rations are generally fed on a limited intake basis to control body weight gains. Adjust quantity of feed offered based on desired body weights and egg production levels for specific breed or strain.

[3]Rations for laying hens generally are fed to provide daily energy intakes of 375 to 450 ME kcal/day based on body weight, environmental temperature, and rate of egg production. Percentage of nutrients shown is typical of hens given 425 ME kcal/day.

TABLE 7–8
TYPICAL BODY WEIGHTS AND FEED ALLOWANCES FOR MALE AND FEMALE MEAT-TYPE CHICKENS (REPLACEMENT STOCK)[1,2]

Age	Male Body Weight[3]		Male Feed Consumption[4]		Female Body Weight[3]		Female Feed Consumption[4]		Typical Egg Production
(weeks)	(lb)	(g)	(lb/week)	(g/week)	(lb)	(g)	(lb/week)	(g/week)	(hen/day %)
0	0.09	40	0.22	100	0.09	40	0.17	75	—
2	0.55	250	0.55	250	0.50	225	0.56	225	—
4	1.20	545	0.77–0.85	350–385	1.00	455	0.69–0.73	315–330	—
6	1.75	795	0.86–0.94	390–425	1.46	660	0.73–0.77	330–350	—
8	2.25	1,020	0.89–1.03	405–475	1.85	840	0.77–0.88	350–400	—
10	2.98	1,250	1.03–1.21	475–550	2.20	1,000	0.85–0.98	385–445	—
12	3.26	1,480	1.19–1.38	540–625	2.60	1,180	0.94–1.06	425–480	—
14	3.75	1,700	1.27–1.54	575–700	3.00	1,360	1.01–1.21	460–550	—
16	4.25	1,930	1.38–1.69	625–765	3.42	1,550	1.09–1.32	495–600	—
18	4.74	2,150	1.47–1.82	665–825	3.81	1,730	1.16–1.48	525–670	—
20	5.29	2,400	—[5]	—[5]	4.25	1,930	1.26–1.61	570–730	—
22	5.82	2,640	—	—	4.65	2,110	1.40–1.75	635–795	10
24	7.05	3,200	—	—	5.40	2,450	1.76–2.04	800–925	15
26	7.80	3,540	—	—	6.02	2,730	2.09–2.31	950–1,050	30
28	8.27	3,750	—	—	6.35	2,880	2.38–2.52	1,078–1,141	56
30	8.60	3,900	—	—	6.61	3,000	2.38–2.52	1,078–1,141	75
32	9.02	4,090	—	—	6.81	3,090	2.38–2.52	1,078–1,141	80
34	9.30	4,220	—	—	6.90	3,130	2.38–2.52	1,078–1,141	78
36	9.57	4,340	—	—	6.97	3,160	2.38–2.52	1,078–1,141	76
38	9.81	4,450	—	—	7.01	3,180	2.36–2.50	1,071–1,134	73
40	10.01	4,540	—	—	7.01	3,180	2.35–2.48	1,064–1,127	72

[1]Adapted by the author from *Nutrient Requirements of Poultry*, 8th rev. ed., NRC, National Academy Press, Washington, DC, 1984, p. 15, Table 9.

[2]Broiler-breeder strains must be grown on a controlled feeding program to limit weight. Values shown are typical but will vary according to strain. Specific breeder guidelines should be consulted for desired schedule of weights and feed allotments.

[3]Values are typical for fall-hatched chicks. Spring-hatched chicks will have decreasing natural daylight during the time of sexual maturity and usually need to be heavier to attain sexual maturity at the desired age.

[4]Adjust as required to maintain desired body weight.

[5]Males and females intermingled.

Fig. 7–4.   Feed tanks on scales to measure daily feed consumption. (Courtesy, University of California, Davis)

TABLE 7–9
METABOLIZABLE ENERGY REQUIRED DAILY BY CHICKENS IN RELATION TO BODY WEIGHT AND EGG PRODUCTION[1,2]

Body Weight		Rate of Egg Production (%)					
		0	50	60	70	80	90
(lb)	(kg)	Metabolizable Energy/Hen Daily (kcal)[3]					
2.2	1.0	130	192	205	217	229	242
3.3	1.5	177	239	251	264	276	289
4.4	2.0	218	280	292	305	317	330
5.5	2.5	259	321	333	346	358	371
6.6	3.0	296	358	370	383	395	408
7.7	3.5	333	395	408	420	432	445

[1]Adapted by the author from *Nutrient Requirements of Poultry*, 8th rev. ed., NRC, National Academy Press, Washington, DC, 1984, p. 15, Table 10.

[2]A number of formulas have been suggested for prediction of the daily energy requirements of chickens. The formula used here was derived from that in *Effect of Environment on Nutrient Requirements of Domestic Animals* (NRC, 1981).

$$ME/\text{hen daily} = W^{0.75}(173 - 1.95T) + 5.5\,\Delta W + 2.07EE$$

where: W = body weight (kg), T = ambient temperature (°C), ΔW = change in body weight in g/day, and EE = daily egg mass (g).

[3]Temperature of 22°, egg weight of 60 g, and no change in body weight were used in calculations.

TABLE 7–10
NUTRIENT REQUIREMENTS OF TURKEYS AS PERCENTAGES OR AS MILLIGRAMS OR UNITS PER KILOGRAM OF DIET[1]

				Age (Weeks)				
Male	0–4	4–8	8–12	12–16	16–20	20–24		**Breeding**
Female	0–4	4–8	8–11	11–14	14–17	17–20	**Holding**	**Hens**
Energy Base:								
kcal ME/lb Diet[2]	1,270	1,315	1,361	1,406	1,452	1,497	1,315	1,315
kcal ME/kg Diet[2]	*2,800*	*2,900*	*3,000*	*3,100*	*3,200*	*3,300*	*2,900*	*2,900*
Protein . . . . (%)	28	26	22	19	16.5	14	12	14
Amino acids:								
Arginine . . . . (%)	1.6	1.5	1.25	1.1	0.95	0.8	0.6	0.6
Glycine + serine . . . . (%)	1.0	0.9	0.8	0.7	0.6	0.5	0.4	0.5
Histidine . . . . (%)	0.58	0.54	0.46	0.39	0.35	0.29	0.25	0.3
Isoleucine . . . . (%)	1.1	1.0	0.85	0.75	0.65	0.55	0.45	0.5
Leucine . . . . (%)	1.9	1.75	1.5	1.3	1.1	0.95	0.5	0.5
Lysine . . . . (%)	1.6	1.5	1.3	1.0	0.8	0.65	0.5	0.6
Methionine + cystine . . . . (%)	1.05	0.9	0.75	0.65	0.55	0.45	0.4	0.4
Methionine . . . . (%)	0.53	0.45	0.38	0.33	0.28	0.23	0.2	0.2
Phenylalanine + tyrosine . . . . (%)	1.8	1.65	1.4	1.2	1.05	0.9	0.8	1.0
Phenylalanine . . . . (%)	1.0	0.9	0.8	0.7	0.6	0.5	0.4	0.55
Threonine . . . . (%)	1.0	0.93	0.79	0.68	0.59	0.5	0.4	0.45
Tryptophan . . . . (%)	0.26	0.24	0.2	0.18	0.15	0.13	0.1	0.13
Valine . . . . (%)	1.2	1.1	0.94	0.8	0.7	0.6	0.5	0.58
Linoleic acid . . . . (%)	1.0	1.0	0.8	0.8	0.8	0.8	0.8	1.0
Major or macrominerals:								
Calcium (Ca) . . . . (%)	1.2	1.0	0.85	0.75	0.65	0.55	0.5	2.25
Chlorine (Cl) . . . . (%)	0.15	0.14	0.14	0.12	0.12	0.12	0.12	0.12
Magnesium (Mg) . . . . (mg)	600	600	600	600	600	600	600	600
Phosphorus (P), available . . . . (%)	0.6	0.5	0.42	0.38	0.32	0.28	0.25	0.35
Potassium (K) . . . . (%)	0.7	0.6	0.5	0.5	0.4	0.4	0.4	0.6
Sodium (Na) . . . . (%)	0.17	0.15	0.12	0.12	0.12	0.12	0.12	0.15
Trace or microminerals:								
Copper (Cu) . . . . (mg)	8	8	6	6	6	6	6	8
Iodine (I) . . . . (mg)	0.4	0.4	0.4	0.4	0.4	0.4	0.4	0.4
Iron (Fe) . . . . (mg)	80	60	60	60	50	50	50	60
Manganese (Mn) . . . . (mg)	60	60	60	60	60	60	60	60
Selenium (Se) . . . . (mg)	0.2	0.2	0.2	0.2	0.2	0.2	0.2	0.2
Zinc (Zn) . . . . (mg)	75	65	50	40	40	40	40	65
Fat-soluble vitamins:								
Vitamin A . . . . (IU)	4,000	4,000	4,000	4,000	4,000	4,000	4,000	4,000
Vitamin D[3] . . . . (ICU)	900	900	900	900	900	900	900	900
Vitamin E . . . . (IU)	12	12	10	10	10	10	10	25
Vitamin K . . . . (mg)	1.0	1.0	0.8	0.8	0.8	0.8	0.8	1.0
Water-soluble vitamins:								
Biotin . . . . (mg)	0.2	0.2	0.15	0.125	0.1	0.1	0.1	0.15
Choline . . . . (mg)	1,900	1,600	1,300	1,100	950	800	800	1,000
Folacin (Folic Acid) . . . . (mg)	1.0	1.0	0.8	0.8	0.7	0.7	0.7	1.0
Niacin (Nicotinic Acid, Nicotinamide) . . . . (mg)	70	70	50	50	40	40	40	30
Pantothenic Acid (Vitamin B–3) . . . . (mg)	11	11	9	9	9	9	9	16
Riboflavin (Vitamin B–2) . . . . (mg)	3.6	3.6	3	3	2.5	2.5	2.5	4
Thiamin (Vitamin B–1) . . . . (mg)	2	2	2	2	2	2	2	2
Vitamin B–6 (Pyridoxine, Pyridoxal, Pyridoxamine) . . (mg)	4.5	4.5	3.5	3.5	3	3	3	4
Vitamin B–12 (Cobalamins) . . . . (mg)	0.003	0.003	0.003	0.003	0.003	0.003	0.003	0.003

[1]Adapted by the author from *Nutrient Requirements of Poultry*, 8th rev. ed., NRC, National Academy Press, Washington, DC, 1984, p. 17, Table 11.

[2]These are typical ME concentrations for corn-soya rations. Different ME values may be appropriate if other ingredients predominate.

[3]These concentrations of vitamin D are satisfactory when the dietary concentrations of calcium and available phosphorus conform with those in this table.

TABLE 7–11
GROWTH RATE, FEED AND ENERGY CONSUMPTION OF LARGE-TYPE TURKEYS[1]

Age	Body Weight				Feed Consumption				Cumulative Feed Consumption				ME Consumption	
	Male		Female		Male		Female		Male		Female		Male	Female
(wks)	(lb)	(kg)	(lb)	(kg)	(lb/week)	(kg/week)	(lb/week)	(kg/week)	(lb)	(kg)	(lb)	(kg)	(Mcal/wk)	Mcal/wk
1	0.24	0.11	0.24	0.11	0.22	0.10	0.22	0.10	0.22	0.10	0.22	0.10	0.3	0.3
2	0.60	0.27	0.53	0.24	0.44	0.20	0.37	0.17	0.66	0.30	0.60	0.27	0.6	0.5
3	1.28	0.58	1.04	0.47	0.99	0.45	0.86	0.39	1.65	0.75	1.46	0.66	1.1	0.8
4	2.21	1.0	1.54	0.7	1.35	0.61	1.01	0.46	3.00	1.36	2.47	1.12	1.7	1.2
5	3.31	1.5	2.43	1.1	1.54	0.70	1.32	0.60	4.54	2.06	3.79	1.72	2.3	1.6
6	4.41	2.0	3.53	1.6	1.90	0.86	1.68	0.76	6.44	2.92	5.47	2.48	2.9	2.1
7	5.73	2.6	4.63	2.1	2.38	1.08	1.96	0.89	8.82	4.00	7.43	3.37	3.5	2.6
8	7.28	3.3	5.73	2.6	2.87	1.30	2.29	1.04	11.69	5.30	9.72	4.41	4.1	3.1
9	8.82	4.0	6.84	3.1	3.33	1.51	2.60	1.18	15.02	6.81	12.33	5.59	4.8	3.6
10	10.36	4.7	8.16	3.7	3.92	1.78	2.95	1.34	18.94	8.59	15.28	6.93	5.2	4.1
11	12.13	5.5	9.48	4.3	4.39	1.99	3.24	1.47	23.33	10.58	18.52	8.40	5.7	4.6
12	13.89	6.3	10.58	4.8	4.96	2.25	3.51	1.59	28.29	12.83	22.03	9.99	6.3	5.1
13	15.66	7.1	11.69	5.3	5.53	2.51	3.75	1.70	33.82	15.34	25.78	11.69	7.1	5.5
14	17.64	8.0	12.79	5.8	5.87	2.66	3.86	1.75	39.69	18.00	29.64	13.44	7.8	5.8
15	19.40	8.8	13.89	6.3	6.37	2.89	4.01	1.82	46.06	20.89	33.65	15.26	8.4	6.1
16	21.39	9.7	14.77	6.7	6.73	3.05	4.23	1.92	52.79	23.94	37.88	17.18	8.8	6.4
17	23.15	10.5	15.66	7.1	6.90	3.13	4.48	2.03	59.60	27.03	42.36	19.21	9.6	6.7
18	24.92	11.3	16.54	7.5	7.21	3.27	4.56	2.07	66.90	30.34	46.92	21.28	10.2	6.9
19	26.68	12.1	17.20	7.8	7.56	3.43	4.74	2.15	74.46	33.77	51.66	23.43	10.9	7.1
20	28.22	12.8	17.86	8.1	7.94	3.60	4.92	2.23	82.40	37.37	56.58	25.66	11.6	7.3
21	29.77	13.5	—	—	8.18	3.71	—	—	90.58	41.08	—	—	12.5	—
22	31.31	14.2	—	—	8.42	3.82	—	—	99.00	44.90	—	—	12.9	—
23	32.63	14.8	—	—	8.69	3.94	—	—	107.69	48.84	—	—	13.2	—
24	33.96	15.4	—	—	8.93	4.05	—	—	116.62	52.89	—	—	13.5	—

[1]Adapted by the author from *Nutrient Requirements of Poultry*, 8th rev. ed., NRC, National Academy Press, Washington, DC, 1984, p. 18, Table 12.

TABLE 7–12
BODY WEIGHTS AND FEED CONSUMPTION OF LARGE-TYPE TURKEYS DURING HOLDING AND BREEDING PERIODS[1,2]

Age	Hens					Toms			
	Weight		Egg Production	Feed		Weight		Feed	
(weeks)	(lb)	(kg)	(%)	(lb/day)	(g/day)	(lb)	(kg)	(lb/day)	(g/day)
20	15.4	7.0	—	0.44	200	26.5	12.0	0.88	400
25	17.6	8.0	—	0.47	215	29.8	13.5	0.93	420
30	19.8	9.0	Start light Stimulation	0.51	230	35.3	16.0	0.97	440
35	20.9	9.5	66	0.57	260	37.5	17.0	0.99	450
40	20.5	9.3	63	0.56	255	39.7	18.0	1.01	460
45	20.1	9.1	60	0.55	250	40.1	18.2	1.06	480
50	19.8	9.0	50	0.53	240	40.8	18.5	1.10	500
55	19.8	9.0	40	0.51	230	41.5	18.8	1.12	510
60	19.8	9.0	35	0.49	220	41.9	19.0	1.15	520

[1]Adapted by the author from *Nutrient Requirements of Poultry*, 8th rev. ed., NRC, National Academy Press, Washington, DC, 1984, p. 18, Table 13.

[2]These values are based on experimental data involving "in season" egg production (*i.e.*, November through July) of commercial stock. It is estimated that summer breeders would produce 70–90% as many eggs and comsume 60–80% as much feed, respectively, as "in season" breeders.

Fig. 7–4a. Turkeys on the range. (Courtesy, J. C. Allen & Son, Inc., West Lafayette, IN)

Fig. 7–5. White Pekin ducks. (Courtesy, California Polytechnic State University, San Luis Obispo)

TABLE 7–13
NUTRIENT REQUIREMENTS OF PEKIN DUCKS AS PERCENTAGES OR AS MILLIGRAMS OR UNITS PER KILOGRAM OF DIET[1, 2]

	Starting (0–2 Weeks)	Growing (2–7 Weeks)	Breeding
Energy Base:			
kcal ME/lb Diet[3]	1,315	1,315	1,315
kcal ME/kg Diet[3]	*2,900*	*2,900*	*2,900*
Protein . . . . (%)	22	16	15
Amino acids:			
Arginine . . . . (%)	1.1	1.0	—
Lysine . . . . (%)	1.1	0.9	0.7
Methionine + cystine . . . . . . (%)	0.8	0.6	0.55
Major or macrominerals:			
Calcium (Ca) . . . . (%)	0.65	0.6	2.75
Chlorine (Cl) . . . . (%)	0.12	0.12	0.12
Magnesium (Mg) . . . . (mg)	500	500	500
Phosphorus (P), available . . . (%)	0.4	0.35	0.35
Sodium (Na) . . . . (%)	0.15	0.15	0.15
Trace or microminerals:			
Manganese (Mn) . . . . (mg)	40	40	25
Selenium (Se) . . . . (mg)	0.14	0.14	0.14
Zinc (Zn) . . . . (mg)	60	60	60
Fat-soluble vitamins:			
Vitamin A . . . . (IU)	4,000	4,000	4,000
Vitamin D . . . . (ICU)	220	220	500
Vitamin K . . . . (mg)	0.4	0.4	0.4
Water-soluble vitamins:			
Niacin (Nicotinic Acid, Nicotinamide) . . . . (mg)	55	55	40
Pantothenic Acid (Vitamin B–3) . . . . (mg)	11	11	10
Riboflavin (Vitamin B–2) . . . . (mg)	4	4	4
Vitamin B–6 (Pyridoxine, Pyridoxal, Pyridoxamine) . . . (mg)	2.6	2.6	3

[1]Adapted by the author from *Nutrient Requirements of Poultry*, 8th rev. ed., NRC, National Academy Press, Washington, DC, 1984, p. 20, Table 15.

[2]For nutrients not listed, see requirements for chickens as a guide.

[3]These are typical dietary energy concentrations.

TABLE 7–14
TYPICAL BODY WEIGHTS AND FEED CONSUMPTION OF PEKIN DUCKS TO 8 WEEKS OF AGE[1]

Age	Body Weight				Feed Consumption By 1–Week Periods				Cumulative Feed Consumption			
	Male		Female		Male		Female		Male		Female	
(weeks)	(lb)	*(kg)*	(lb)	*(kg)*	(lb)	*(kg)*	(lb)	*(kg)*	(lb)	*(kg)*	(lb)	*(kg)*
0	0.11	*0.05*	0.11	*0.05*	—	—	—	—	—	—	—	—
1	0.60	*0.27*	0.60	*0.27*	0.49	*0.22*	0.49	*0.22*	0.49	*0.22*	0.49	*0.22*
2	1.72	*0.78*	1.63	*0.74*	1.70	*0.77*	1.61	*0.73*	2.18	*0.99*	2.09	*0.95*
3	3.04	*1.38*	2.82	*1.28*	2.47	*1.12*	2.45	*1.11*	4.65	*2.11*	4.52	*2.05*
4	4.32	*1.96*	4.01	*1.82*	2.82	*1.28*	2.82	*1.28*	7.50	*3.40*	7.34	*3.33*
5	5.49	*2.49*	5.07	*2.30*	3.26	*1.48*	3.15	*1.43*	10.74	*4.87*	10.50	*4.76*
6	6.53	*2.96*	6.02	*2.73*	3.59	*1.63*	3.51	*1.59*	14.33	*6.50*	14.00	*6.35*
7	7.36	*3.34*	6.75	*3.06*	3.70	*1.68*	3.59	*1.63*	18.04	*8.18*	17.60	*7.98*
8	7.96	*3.61*	7.25	*3.29*	3.70	*1.68*	3.59	*1.63*	21.74	*9.86*	21.19	*9.61*

[1]Adapted by the author from *Nutrient Requirements of Poultry*, 8th rev. ed., NRC, National Academy Press, Washington, DC, 1984, p. 20, Table 16.

TABLE 7–15
NUTRIENT REQUIREMENTS OF GEESE AS PERCENTAGES OR AS MILLIGRAMS OR UNITS PER KILOGRAM OF DIET[1, 2]

	Starting (0–6 Weeks)	Growing (After 6 Weeks)	Breeding
Energy Base:			
kcal ME/lb Diet[3]	1,315	1,315	1,315
kcal ME/kg Diet[3]	*2,900*	*2,900*	*2,900*
Protein .... (%)	22	15	15
Amino acids:			
Lysine .... (%)	0.9	0.6	0.6
Methionine + cystine ...... (%)	0.75	—	—
Major or macrominerals:			
Calcium (Ca) .... (%)	0.8	0.6	2.25
Phosphorus (P), available .... (%)	0.4	0.3	0.3
Fat-soluble vitamins:			
Vitamin A .... (IU)	1,500	1,500	4,000
Vitamin D .... (ICU)	200	200	200
Water-soluble vitamins:			
Niacin (Nicotinic Acid, Nicotinamide) .... (mg)	55	35	20
Pantothenic Acid (Vitamin B–3) .... (mg)	15	—	—
Riboflavin (Vitamin B–2) .... (mg)	4	2.5	4

[1]Adapted by the author from *Nutrient Requirements of Poultry*, 8th rev. ed., NRC, National Academy Press, Washington, DC, 1984, p. 19, Table 14.

[2]For nutrients not listed, see requirements for chickens as a guide.

[3]These are typical dietary energy concentrations.

Fig. 7–6. Geese. (Photo by J. C. Allen & Son, Inc., West Lafayette, IN)

Fig. 7–6a. Ringneck cock pheasant. (Courtesy, MacFarlane Pheasant Farm, Inc., Janesville, WI)

TABLE 7–16
NUTRIENT REQUIREMENTS OF PHEASANTS AND BOBWHITE QUAIL
AS PERCENTAGES OR AS MILLIGRAMS OR UNITS PER KILOGRAM OF DIET[1, 2, 3]

Energy Base:	Pheasant			Bobwhite Quail		
	Starting	Growing	Breeding	Starting	Growing	Breeding
kcal ME/lb Diet[4]	1,270	1,225	1,270	1,270	1,270	1,270
kcal ME/kg Diet[4]	*2,800*	*2,700*	*2,800*	*2,800*	*2,800*	*2,800*
Protein . . . . (%)	30	16	18	28	20	24
Amino acids:						
Glycine + serine . . . . (%)	1.8	1.0	—	—	—	—
Lysine . . . . (%)	1.5	0.8	—	—	—	—
Methionine + cystine . . . . (%)	1.1	0.6	0.6	—	—	—
Linoleic acid . . . . (%)	1.0	1.0	1.0	1.0	1.0	1.0
Major or macrominerals:						
Calcium (Ca) . . . . (%)	1.0	0.7	2.5	0.65	0.65	2.3
Chlorine (Cl) . . . . (%)	0.11	0.11	0.11	0.11	0.11	0.11
Phosphorus (P), available . . . . (%)	0.55	0.45	0.4	0.55	0.45	0.5
Sodium (Na) . . . . (%)	0.15	0.15	0.15	0.15	0.15	0.15
Trace or microminerals:						
Iodine (I) . . . . (mg)	0.3	0.3	0.3	0.3	0.3	0.3
Water-soluble vitamins:						
Choline . . . . (mg)	1,500	1,000	—	1,500	—	1,000
Niacin (Nicotinic Acid, Nicotinamide) . . . . (mg)	60	40	—	30	—	20
Pantothenic Acid (Vitamin B–3) . . . . (mg)	10	10	—	13	—	15
Riboflavin (Vitamin B–2) . . . . (mg)	3.5	3	—	3.8	—	4

[1]Adapted by the author from *Nutrient Requirements of Poultry*, 8th rev. ed., NRC, National Academy Press, Washington, DC, 1984, p. 21, Table 17.

[2]For pheasant values not listed see requirements for turkeys as a guide.

[3]For Bobwhite quail values not listed see requirements for Leghorn-type chickens as a guide.

[4]These are typical dietary energy concentrations.

The NRC standards presented in this chapter do not provide for margins of safety. Rather, the values reported represent adequacy, using as criteria growth, health, reproduction, feed efficiency, and quality of products produced.

It is recognized that a number of forces may affect the nutritive requirements of poultry; among them, the following:

1. **Temperature and humidity.** When temperature and humidity conditions deviate much from 60° to 75°F and 40 to 60%, respectively, adjustments in nutrients levels should be made to compensate for changes in feed intake.

2. **Genetic differences.** Genetic differences among strains affect nutrient requirements; hence, in this chapter consideration has been given to differences in the requirements between broiler-type and egg-type strains of chickens.

Also, the nutrient composition of feedstuffs is variable. In order to compensate for these conditions, the nutritionist usually adds a margin of safety to the stated requirements in arriving at *nutritive allowances* to be used in ration formulation.

TABLE 7–17
NUTRIENT REQUIREMENTS OF JAPANESE QUAIL (COTURNIX) AS PERCENTAGES OR AS MILLIGRAMS OR UNITS PER KILOGRAM OF DIET[1]

	Starting and Growing	Breeding
Energy Base: kcal ME/lb Diet[2] . . . . *kcal ME/kg Diet[2]*	1,361 *3,000*	1,361 *3,000*
Protein . . . . (%)	24	20
Amino acids:		
Arginine . . . . (%)	1.25	1.26
Glycine + serine . . . . (%)	1.2	1.17
Histidine . . . . (%)	0.36	0.42
Isoleucine . . . . (%)	0.98	0.9
Leucine . . . . (%)	1.69	1.42
Lysine . . . . (%)	1.3	1.15
Methionine + cystine . . . . . . (%)	0.75	0.76
Methionine . . . . (%)	0.5	0.45
Phenylalanine + tyrosine . . . . (%)	1.8	1.4
Phenylalanine . . . . (%)	0.96	0.78
Threonine . . . . (%)	1.02	0.74
Tryptophan . . . . (%)	0.22	0.19
Valine . . . . (%)	0.95	0.92
Linoleic acid . . . . (%)	1.0	1.0
Major or macrominerals:		
Calcium (Ca) . . . . (%)	0.8	2.5
Chlorine (Cl) . . . . (%)	0.2	0.15
Magnesium (Mg) . . . . (mg)	300	500
Phosphorus (P), available . . . . (%)	0.45	0.55
Potassium (K) . . . . (%)	0.4	0.4
Sodium (Na) . . . . (%)	0.15	0.15
Trace or microminerals:		
Copper (Cu) . . . . (mg)	6	6
Iodine (I) . . . . (mg)	0.3	0.3
Iron (Fe) . . . . (mg)	100	60
Manganese (Mn) . . . . (mg)	90	70
Selenium (Se) . . . . (mg)	0.2	0.2
Zinc (Zn) . . . . (mg)	25	50
Fat-soluble vitamins:		
Vitamin A . . . . (IU)	5,000	5,000
Vitamin D . . . . (ICU)	1,200	1,200
Vitamin E . . . . (IU)	12	25
Vitamin K . . . . (mg)	1.0	1.0
Water-soluble vitamins:		
Biotin . . . . (mg)	0.3	0.15
Choline . . . . (mg)	2,000	1,500
Folacin (Folic Acid) . . . . (mg)	1.0	1.0
Niacin (Nicotinic Acid, Nicotinamide) . . . . (mg)	40	20
Pantothenic Acid (Vitamin B–3) . (mg)	10	15
Riboflavin (Vitamin B–2) . . . . (mg)	4	4
Thiamin (Vitamin B–1) . . . . . (mg)	2	2
Vitamin B–6 (Pyridoxine, Pyridoxal, Pyridoxamine) . . . (mg)	3	3
Vitamin B–12 (Cobalamins) . . . (mg)	0.003	0.003

[1]Adapted by the author from *Nutrient Requirements of Poultry*, 8th rev. ed., NRC, National Academy Press, Washington, DC, 1984, p. 22, Table 18.

[2]These are typical dietary energy concentrations.

HOW TO BALANCE RATIONS

The increasing complexity of poultry rations, along with larger and larger enterprises, makes it imperative that the producers who choose to mix feed be absolutely sure that they will have a nutritionally balanced and adequate ration.

When fed free-choice, birds tend to eat to satisfy their energy requirements. Consequently, it is possible, within limits, to regulate the intake of all nutrients, except water, by including them in the diet in specific ratios to available energy. Thus, the energy content of the diet must be considered in formulating to meet a desired intake of all essential nutrients other than energy itself.

The larger commercial feed companies, and the larger poultry producers who do their own mixing or formulating, generally rely on the services of a nutritionist and the use of a computer in formulating their rations. Even though it is more time-consuming, and fewer factors can be considered simultaneously, a good job can be done in formulating rations by the hand method.

Good poultry producers should know how to balance rations. They should be able to select and buy feeds with informed appraisal; to check on how well their manufacturer, dealer, or consultant is meeting their needs; and to evaluate the results.

Ration formulation consists of combining feeds that will be eaten in the amount needed to supply the daily nutrient requirements of the bird. This may be accomplished by the methods presented later in this chapter, but first the following pointers are necessary:

1. In computing poultry rations, more than simple arithmetic should be considered, for no set of figures can substitute for experience and intuition. Formulating rations is both an art and a science—the art comes from bird know-how, experience, and keen observation; the science is largely founded on mathematics, chemistry, physiology, and bacteriology. Both are essential for success.

2. Before attempting to balance a ration, the following major points should be considered:

a. **Availability and cost of the different feed ingredients.** The first step in ration formulation is to determine what feeds are available and which feeds are the best buy. Preferably, cost of ingredients should be based on delivery after processing—because delivery and processing costs are quite variable.

b. **Moisture content.** When considering costs and balancing rations, feeds should be placed on a comparable moisture basis; usually, either *as-fed* or *moisture-free*.

c. **Composition of the feeds under consideration.** Feed composition tables ("book values"), or average analysis, should be considered only as guides, because of wide variations in the composition of feeds. For example, the protein and moisture contents of sorghum can be quite variable. Whenever possible, especially with large operations, it is best to take a representative sample of each major feed ingredient and have a chemical analysis made of it for the more common constituents — protein, fat, fiber, nitrogen-free extract, and moisture; and often calcium, phosphorus, and carotene. Such ingredients as oil meals and prepared supplements, which must meet specific standards, need not be analyzed so often, except as quality-control measures.

d. **Quality of feed.** Numerous factors determine the quality of feed, including:

(1) **Stage of harvesting.** For example, early-harvested grains are higher in moisture than those that are mature.

(2) **Freedom from contamination.** Contamination from foreign substances such as dirt, sticks, and rocks can reduce feed quality, as can aflatoxins, pesticide residues, and a variety of chemicals.

(3) **Uniformity.** Does the feed come from one particular area or does it represent a conglomerate of several sources?

(4) **Length of storage.** When feed is stored for extended periods, some of its quality is lost due to its exposure to the elements.

e. **Degree of processing of the feed.** Often, the value of feed can be either increased or decreased by processing. For example, grinding some types of grains makes them more readily digestible to poultry and increases their feeding value.

f. **Soil analysis.** If the origin of a given feed ingredient is known, a soil analysis or knowledge of the soils of the area can be very helpful; for example, (1) the phosphorus content of soils affects plant composition, (2) soils high in molybdenum and selenium affect the composition of the feeds produced, (3) iodine- and cobalt-deficient areas are important in animal nutrition, and (4) other similar soil-plant-animal relationships exist.

g. **The nutrient requirements and allowances.** These should be known for the particular class of poultry for which a ration is being tailored. Also, it must be recognized that nutrient requirements and allowances must be changed from time to time, as a result of new experimental findings.

3. In addition to providing a proper quantity of feed and to meeting the nutritive requirements, a well-balanced and satisfactory ration should be:

a. **Palatable and digestible.**

b. **Economical.** Generally speaking, this calls for the maximum use of feeds available in the area.

c. **One that will enhance,** rather than impair, the quality of the product (meat or eggs) produced.

4. In addition to considering changes in availability of feeds and feed prices, ration formulation should be altered at stages to correspond to changes in weight and productivity of birds.

STEPS IN RATION FORMULATION

The ideal ration is one that will maximize production at the lowest cost. A costly ration may produce phenomenal gains in poultry, but the cost per unit of production may make the ration economically infeasible. Likewise, the cheapest ration is not always the best since it may not allow for maximum production.

Fig. 7–7. Layers feeding. The final determinant of a successfully formulated layer ration is the cost of feed per dozen eggs. (Courtesy, University of Georgia, Athens)

Therefore, the cost per unit of production is the ultimate determinant of what constitutes the best ration. Awareness of this fact separates successful producers from marginal or unsuccessful ones.

The following four steps should be taken in an orderly fashion in order to formulate an economical ration:

1. Find and list the nutrient requirements and/or allowances for the specific bird to be fed.

2. Determine what feeds are available and list their respective nutrient compositions.

3. Determine the cost of the feed ingredients under consideration.

4. Consider the limitations of the various feed ingredients and formulate the most economical ration.

ADJUSTING MOISTURE CONTENT

The majority of feed composition tables are listed on an *as-fed* basis, while most of the National Research Council nutrient requirement tables are on either an *approximate 90% dry matter* or *moisture-free* basis. Since feeds contain varying amounts of dry matter, it would be much simpler, and more accurate, if both feed composition and nutrition requirement tables were on a dry basis. For information on how to adjust moisture content, the reader is referred to *Feeds & Nutrition*, by Ensminger, *et al.*

METHODS OF FORMULATING RATIONS

In the sections that follow, five different methods of ration formulation are presented: (1) the square method, (2) the trial-and-error method, (3) the simultaneous equation method, (4) the 2 × 2 matrix method, and (5) the computer method. Despite the sometimes confusing mechanics of each system, if done properly the end result of all five methods is the same — a ration that provides the desired allowance of nutrients in correct proportions economically (or at least cost), but, more important, so as to achieve the greatest net returns — for it is net profit, rather than cost, that counts. Since feed represents by far the greatest cost item in livestock production, the importance of balanced rations is evident.

An exercise in ration formulation follows for purposes of illustrating the application of the five methods.

It should be emphasized that each method of ration formulation may be used in balancing rations for all classes of livestock.

SQUARE (OR PEARSON SQUARE) METHOD

The square method is simple, direct, and easy. Also, it permits quick substitution of feed ingredients in keeping with market fluctuations, without disturbing the protein content.

In balancing rations by the square method, it is recognized that one specific nutrient alone — protein — receives major consideration. Correctly speaking, therefore, it is a method of balancing the protein requirement, with no consideration given to the vitamin, mineral, and other nutritive requirements.

To compute rations by the square method, or by any other method, it is first necessary to have available both feeding standards and feed composition tables.

The following example will show how to use the square method in formulating a ration:

Example: *A poultry producer desires to feed a 16% protein ration. Corn containing 9.5% protein is on hand. A 36% protein supplement, which is reinforced with minerals and vitamins is available. What percentage of the ration should consist of corn and what percentage of the 36% protein supplement?*

Step by step, the procedure in balancing this ration is as follows:

1. Draw a square, and place the number 16 (desired protein level) in the center thereof.
2. At the upper left-hand corner of the square, write *protein supplement* and its protein content (36); at the lower left-hand corner, write *corn* and its protein content (9.5).
3. Subtract diagonally across the square (the small number from the larger number), and record the difference at the corners on the right-hand side (36 − 16 = 20; 16 − 9.5

= 6.5). The number at the upper right-hand corner gives the parts of supplement by weight, and the number at the lower right-hand corner gives the parts of corn by weight to make a ration with 16% protein.

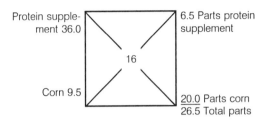

4. To determine what percent of the ration would be corn, divide the parts of corn by the total parts: 20 ÷ 26.5 = 75% corn. The remainder, 25%, would be supplement.

TRIAL-AND-ERROR METHOD

The trial-and-error method is exactly what the term implies. Feeds are interchanged by trial-and-error until the right combination is found.

Example: *A poultry producer desires to formulate a starting ration for Bobwhite quail. The ration will consist of fat, corn, wheat, soybean oil meal (49% protein), alfalfa meal (22% protein), and supplement (calcium-phosphorus, iodized salt, and micronutrients).*

Before the ration is formulated, the producer places the following restrictions on the feeds:

Feed Ingredient	Percent Incorporated into the Final Ration (%)
Fat	4.0
Alfalfa meal	3.0
Supplement (calcium-phosphorus iodized salt, and micronutrients) . . .	3.0

The feed compositions used in this sample are as follows:

	ME$_n$ (kcal/lb)	Crude Protein (%)
Corn	1,556	8.7
Wheat	1,334	14.9
Soybean meal (49% protein)	1,112	48.1
Alfalfa meal (22% protein)	754	22.0
Fat	3,583	0

Here are the steps in balancing this ration:

Step 1: The nutrient requirements for Bobwhite quail taken from Table 7–16 are as follows:

The starting ration for Bobwhite quail should contain 1,270 kcal of ME$_n$ per pound and 28% crude protein; per ton of feed, this calls for 2,540,000 kcal of ME$_n$ and 560 lb of protein.

Step 2: We must now calculate the amounts of energy and protein contributed to the ration by the feeds that have predetermined levels of incorporation.

	Amount/Ton (lb)	ME_n (kcal)	Crude Protein (lb)
Alfalfa meal	60	45,240	13.2
Fat	80	286,640	0
Supplements (calcium-phosphorus, iodized salt, micronutrients) ...	60	0	0
	200	331,880	13.2

Step 3: Remainder to be supplied by corn, wheat, and soybean meal—

1,800	2,208,120	546.8

Step 4: Let's try out (that's why it is called the *trial-and-error method*) a mixture of 600 lb of corn, 600 lb of wheat, and 600 lb of soybean meal; and see how much ME_n and crude protein is contributed by the mix.

	ME_n (kcal)	Crude Protein (lb)
Corn, 600 lb	933,600	52.2
Wheat, 600 lb	800,400	89.4
Soybean meal, 600 lb	667,200	288.6
	2,401,200	430.2

Step 5: The energy value of the corn-wheat-soybean mix is more than needed (2,401,200 ME_n supplied vs 2,208,120 needed), but the protein value is much less than needed (430.2 lb supplied vs 546.8 needed); hence, we must add soybean meal at the expense of corn and/or wheat. In this example, we shall substitute soybean meal for corn. To get a rough estimate of how much corn to remove, we must first find out how much protein is needed. This is found by determining the amount of additional protein needed in the mix (546.8 lb − 430.2 lb = 116.6 lb of protein needed). The protein content of soybean meal is 48.1% and of corn, 8.7%. Since the protein content of corn was included in the mix in Step 4, we will subtract 8.7% from 48.1% to find out how much additional protein is provided by soybean meal (39.4%). Now we divide 116.6 lb of protein needed by 0.394 and obtain a value of 295.9 lb; so, let's change our corn-wheat-soybean mix by subtracting 300 lb of corn, adding 300 lb of soybean meal, and rounding off wheat at 600 lb.

	ME_n (kcal)	Crude Protein (lb)
Corn, 300 lb	466,800	26.1
Wheat, 600 lb	800,400	89.4
Soybean meal, 900 lb	1,000,800	432.9
	2,268,000	548.4

Step 6: From the two groups of feeds, the ME_n and crude protein are:

	Amount/Ton (lb)	ME_n (kcal)	Crude Protein (lb)
Alfalfa meal, fat, and supplement	200	331,880	13.2
Corn, wheat, and soybean meal	1,800	2,268,000	548.4
	2,000	2,599,880	561.6

Step 7: Now our ration slightly exceeds the nutrient requirements of a starting ration for Bobwhite quail as given in Step 1.

Thus, for 1 ton of Bobwhite quail starter feed, we should include the following:

	Amount/Ton (lb)	ME_n (kcal)	Crude Protein (lb)
Wheat	600	800,400	89.4
Corn	300	466,800	26.1
Soybean meal	900	1,000,800	432.9
Alfalfa meal	60	45,240	13.2
Fat	80	286,640	0
Supplement (calcium-phosphorus, iodized salt, micronutrients ...	60	0	0
	2,000	2,599,880	561.6
		(1,300 kcal/lb)	*(28.1%)*

SIMULTANEOUS EQUATIONS METHOD

It is possible to formulate rations involving two sources and one nutrient quickly through the solving of simultaneous equations:

Example 1: *A poultry producer has on hand corn containing about 9% protein. A 50% protein supplement, which is reinforced with minerals and vitamins, can be bought at a reasonable price. A ration containing 22% protein is desired.*

Step by step, the procedure in balancing this ration is as follows:

Step 1: Let X = amount of corn to be used in 100 lb of mixed feed, and Y = amount of 50% protein supplement to be used in 100 lb of mixed feed. We know that the corn contains 9% protein and the protein supplement 50% and that the ration should be 22% protein. Therefore, the equation we must solve is as follows:

$$.09 + .50Y = 22 \text{ (lb of protein in 100 lb of feed)}$$

Step 2: In order to solve for two unknowns (X and Y), we must create a "dummy equation." This can be done in the following manner:

$$X + Y = 100 \text{ lb of feed}$$

Step 3: We must now multiply our dummy equation by .09 in order that our X term will cancel out with the original equation. Therefore:

$$X + Y = 100 \text{ becomes } .09X + .09Y = 9$$

Step 4: Subtracting our new dummy equation from the original equation, we can solve for Y as shown below:

$$
\begin{array}{ll}
.09X + .50Y = 22 & \text{original equation} \\
-.09X - .09Y = -9.0 & \text{dummy equation} \\
\hline
.00X + .41Y = 13 &
\end{array}
$$

$Y = \dfrac{13}{.41}$ or 31.71 lb of 50% protein supplement per 100 lb of feed or 31.71%

Step 5: We can now substitute our newly acquired value for Y in the original equation and solve for X as follows:

X = 100 − 31.71
= 68.29 lb of corn per 100 lb of feed or 68.29%

It is also possible to use simultaneous equations in the formulation of rations involving two sources and two nutrients. Since many formulations involve solving for more than one nutritional parameter, this method has many advantages.

Example 2: *A poultry feed manufacturer desires enough calcium and phosphorus to supply a ton of layer ration with an additional 40 lb of calcium and 5 lb of phosphorus using two sources, (1) oystershells and (2) monocalcium phosphate. The oystershells contain 37.0% calcium and 0.1% phosphorus. The monocalcium phosphate contains 15.9% calcium and 20.9% phosphorus. How much of each compound is needed to supplement the ration?*

Step by step, the procedure is as follows:

Step 1: Let X = amount of oystershells to be added to 1 ton of feed, and
Y = amount of monocalcium phosphate to be added to 1 ton of feed.

Our equations are:

.370X + .159Y = 40 calcium equation
.001X + .209Y = 5 phosphorus equation

Step 2: If we compare the coefficients for the oystershells in both equations, we derive an adjustment factor of 370 $\left(\frac{.370}{.001}\right)$ which is needed to balance the two equations. Thus, if we multiply the phosphorus equation by 370, the X terms will cancel out as follows:

```
     .370X +   0.159Y  =      40
 − (.370X + 77.330Y  =   1850)
          − 77.17Y  =  −1810
                 Y  =   23.45
```

Step 3: Now that we know the value of Y, we can substitute it in either equation to find X.

```
.370X + .159(23.45) = 40
    .370 + 3.729 = 40
         .370X = 36.27
             X = 98.03
```

Thus, the manufacturer must add 98.03 lb of oystershells and 23.45 lb of monocalcium phosphate to supply the supplementary calcium and phosphorus.

To check our calculations, we can substitute X and Y in our original equations as follows:

.370(98.03) + .159(23.45) = 40
.001(98.03) + .209(23.45) = 5

THE 2 × 2 MATRIX METHOD

In addition to the traditional algebraic method of solving simultaneous equations, matrix algebra provides an alternative which some people find easier and quicker. The 2 × 2 matrix provides a quick and accurate way of solving for two nutritional parameters — such as energy and protein — through the use of two ingredients.

A matrix is a mathematical array which allows for the solution of unknowns through the use of a series of equations. Consider the two equations:

$$a_1X + b_1Y = C_1$$
$$a_2X + b_2Y = C_2$$

Let us assume that X represents one type of feed, and that Y represents another type. In order to solve for X and Y, we can set up a 2 × 2 matrix using their respective coefficients. C_1 and C_2 could represent two nutrient levels that we want (for example, energy and protein). The 2 × 2 matrix would then be:

$$\begin{vmatrix} a_1 & b_1 \\ a_2 & b_2 \end{vmatrix}$$

The matrix would consist of two rows and two columns. In order to solve for X and Y, we must find the determinant of the matrix. The determinant is established as follows:

$$\begin{vmatrix} a_1 & b_1 \\ a_2 & b_2 \end{vmatrix} = a_1b_2 - a_2b_1$$

If the matrix is $\begin{vmatrix} 1 & 2 \\ 3 & 4 \end{vmatrix}$, the determinant would be

$$\begin{vmatrix} 1 & 2 \\ 3 & 4 \end{vmatrix} = (1 \times 4) + (-3 \times 2) = 4 - 6 = -2$$

Note that a determinant of a square is enclosed by straight vertical lines, and a square matrix is enclosed by curved lines. Through a series of derivations from the original two equations, our unknowns can be solved in the following manner:

$$X = \frac{\begin{vmatrix} c_1 & b_1 \\ c_2 & b_2 \end{vmatrix}}{\begin{vmatrix} a_1 & b_1 \\ a_2 & b_2 \end{vmatrix}} \text{ or } \frac{(c_1b_2 - c_2b_1)}{(a_1b_2 - a_2b_1)} \qquad Y = \frac{\begin{vmatrix} a_1 & c_1 \\ a_2 & c_2 \end{vmatrix}}{\begin{vmatrix} a_1 & b_1 \\ a_2 & b_2 \end{vmatrix}} \text{ or } \frac{(a_1c_2 - a_2c_1)}{(a_1b_2 - a_2b_1)}$$

Using the same example as in Example 2 of the simultaneous equation section, the 2 × 2 matrix method should arrive at the same answer.

Example: *A poultry feed manufacturer desires to formulate enough calcium and phosphorus to supplement a ton of layer ration with an additional 40 lb of calcium and 5 lb of phosphorus using two sources: (1) oystershells, and (2) monocalcium phosphate. The oystershells contain 37.0% calcium and 0.1% phosphorus. The monocalcium phosphate contains 15.9% calcium and 20.9% phosphorus. How much of each compound is needed to supplement the ration?*

Step by step, the procedure is as follows:

Step 1: Let X = amount of oystershells to be added to 1 ton of feed and
Y = amount of monocalcium phosphate to be added to 1 ton of feed.

Our equations are:

1370X + .159Y = 40 calcium equation
0.001X + .209Y = 5 phosphorus equation

Step 2: From these equations, we can set up the following 2 × 2 matrix:

$$\begin{pmatrix} a_1 & b_1 \\ a_2 & b_2 \end{pmatrix} = \begin{pmatrix} .370 & .159 \\ .001 & .209 \end{pmatrix} \text{ and } \begin{pmatrix} c_1 \\ c_2 \end{pmatrix} = \begin{pmatrix} 40 \\ 5 \end{pmatrix}$$

Step 3: Once we have set up our matrices, we can solve for X and Y by calculating the determinants as shown below:

$$X = \frac{40(.209) - 5(.159)}{.370(.209) - .001(.159)} = \frac{7.565}{.077} = 98.02 \text{ lb}$$

$$Y = \frac{.370(5) - .001(40)}{.370(.209) - .001(.159)} = \frac{1.81}{.077} = 23.51$$

Our values for X and Y agree with those from the same example using simultaneous equations. The slight difference is due to rounding error.

In least-cost formulations, matrix algebra is used by the computer, but the matrices that are used are much larger than 2 × 2 and are far more complicated. The 2 × 2 matrix offers a rapid means of calculating a simple ration using two feeds to fulfill two nutrients.

COMPUTER METHODS

Most large poultry establishments and feed companies now use computers in ration formulation. Also, many of the state universities, through their Federal-State Extension Services, are offering ration balancing computer services to farmers within their respective states on a charge basis. Consulting nutritionists are available throughout the United States and provide computer services, as well as other services. With the recent advent of the low cost personal computer, this powerful technology is available to almost everyone.

Fig. 7–8. Today, most poultry rations, like the ration for this layer operation in Maryland, are formulated with the use of a computer. (Courtesy, University of Maryland, College Park)

Despite their sophistication, there is nothing magical or mysterious about the use of computers in ration balancing. Their primary advantages are accuracy and speed of computation. In addition, computer programs (software) used in ration balancing provide a means of organizing needed information in a logical and systematic manner. The computer should be viewed as an extension of the knowledge and skills of the formulator.

At this time, there is no "pushbutton system" of feed formulation available. The degree of success realized is very dependent on the management of data put into the computer, and on the evaluation of the resulting formulations that the computer generates. In the hands of experienced users, the computer enables the producer and nutritionist to be more precise in carrying out ration formulation.

Two basic approaches to ration formulation are practiced with computers:

1. Trial-and-error formulation.
2. Linear programming (LP).

Trial-and-Error Formulation with the Computer

For a discussion of the trial-and-error method of ration balancing, see the earlier section in this chapter headed "Trial-and-Error Method." Many ration balancing software programs written for the computer allow for trial-and-error ration balancing. Feed mill nutritionists frequently use this technique to enter into the computer rations that are given to them by other nutritionists or by a producer. The objective in this case is to confirm the nutrient values for the ration based on the specific ingredients used by the feed manufacturer. In many cases, these rations are not to be altered without permission. In other cases, the number of ingredients for a specific ration may be limited so that the trial-and-error technique is just as fast as using linear programming to arrive at the desired nutrient levels in the ration.

NOTE: It does not take specialized computer software to use the trial-and-error method. Spreadsheet (or Financial Spreadsheet) programs, for instance, organize data into rows and columns. Information, such as nutrient values for a feedstuff, may be entered into data cells (see Fig. 7–9). Simple and complex arithmetic operations can be controlled by the user to the extent that rather large trial-and-error method rations can be programmed and run.

Spreadsheets have been developed with specific microcomputers in mind; and there are a great number of them on the market.

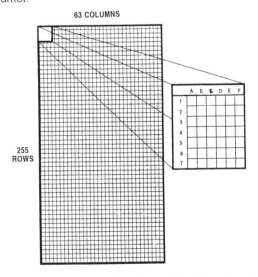

Fig. 7–9. Graphic representation of a spreadsheet (from Lane, R. J. and T. L. Cross, *Spreadsheet Applications for Animal Nutrition and Feeding*, Reston Publishing Co., Inc., Reston, VA, 1985).

Linear Programming (LP)

The most common technique for computer formulation of rations is the linear programming (LP) technique. At times, this is referred to as *least cost* ration formulation. This designation results from the fact that most LP techniques for ration formulation have as their objective *minimization of cost*. A few LP programs are in use that solve for *maximization of income over feed costs*. Regardless, the poultry producer and nutritionist should always keep in mind that maximizing net profit is the only true objective of most ration formulations. A skilled user of the LP system will control ration quality by writing specifications that lead to rations that will maximize profit.

Briefly described, the LP program is a mathematical technique in which a large number of simultaneous equations are solved in such a way as to meet the minimum and maximum levels of nutrients and levels of feedstuffs specified by the user at the lowest possible cost. It is not necessary to understand the inner workings of the computer program to use LP, though it does take experience to use it to good advantage and to avoid certain pitfalls. The most common pitfalls are incorrectly entered or missing data and the specification of minimums and maximums that cannot be met with the feedstuffs available. The latter is called an *infeasible solution*. When an infeasible solution is encountered, the user must determine (1) if this is due to incorrect or missing data, or (2) if the specifications must be relaxed.

Procedure for Use of Linear Programming (LP)

Before using the LP approach to ration formulation, the user should become familiar with the specific software package to be used. (See later section on "Selection of Computer Software and Hardware for Ration Formulation.") It is also desirable to study the LP technique as applied to feed formulation. After users are familiar with LP and their computer software, they are ready to begin using the computer for ration formulation by LP. It must first be understood that all data entered into the computer is directed to files. In most cases, these files are located on disks, or perhaps on tapes. Currently, most computers use keyboards and CRT (cathode ray tube) displays for entry of data. The necessary data files are generally created in steps as follows:

1. **Enter names of available feed ingredients, and the cost of each.** It is necessary that all of the available feeds be listed along with the unit cost. It matters little if the formulator uses cost per ton, cost per hundred weight (cwt), or cost per pound, but the same method of cost input must be used for all feeds. The computer software may call for a specific form for entering costs.

2. **Enter nutrient values for feeds.** Tables of feed composition using average or typical values, like those in this book, may be used, but, because of the wide variation in the composition of feeds, chemical analysis of a representative sample of each lot of feed is more precise and should be used if available.

3. **Enter ration specifications.** Ration specifications are generally broken into two parts: (1) nutrient limits and (2) ingredient limits. In each case, the formulator specifies either a lower limit and/or an upper limit for each item. If no specification for the particular item is desired, it may be specified

as zero (0) or left blank, depending on the circumstances. It is also appropriate to list feedstuffs available, but not currently on hand (with an upper limit of zero). Most LP solutions will then tell the user the highest cost at which such feeds would enter the solution if allowed to do so. Ratios between nutrients (such as a calcium/phosphorus ratio) or feedstuffs (corn/barley ratio) may also be specified in most LP software packages. The experienced formulator usually deals with palatability or feedstuff quality consideration by setting an upper limit on the amounts of problem feeds or a lower limit on feeds that contribute a positive quality to the ration. Nonnutritive attributes, such as bulk density, may also be programmed into the LP system. The LP technique is a very flexible and powerful ration balancing tool.

NOTE: Important additional items to consider when creating ration specifications are upper limits on the use of nonprotein nitrogen (or urea) and limits on the usage of feed additives like drugs, feed flavors, and the like.

Fig. 7–10 illustrates, by means of a worksheet, a logical method of organizing the restrictions for a ration.

LEAST-COST FORMULATION WORKSHEET

Specifications	Ingredient A	Ingredient B	Ingredient C	Restrictions
Cost				Minimize
Total weight				1,000 lb
Crude protein				133 lb
Digestible protein				100 lb
Ether extract				25 to 80 lb
Calcium				5 to 10 lb
Phosphorus				7 lb
Vitamin A equivalent				35,000
Vitamin D				60,000
Limits on ingredients				
Minimum				
Maximum				

Fig. 7–10. Sample worksheet for a least-cost formulation. The first column lists the specifications. The various feedstuffs to be considered are listed in the succeeding columns with their respective costs and nutritive values. The last column lists the restrictions desired on the final formulation.

4. **Submit all of the above information to the matrix building and solving portion of the LP software package.** Matrix building and solving are generally accomplished automatically by the computer software once the specifications have been entered into the computer. Mathematically, the procedure involves the solution of a complex algebraic problem, with an answer being derived in seconds or minutes. Using the LP program, the computer produces a mix that will meet the desired specifications at the lowest possible cost.

5. **Examine the solution provided by the computer software.** The end result should be feasible, both from a mathematical standpoint and from a nutritional standpoint.

The feedstuff mixture should be acceptable to the birds for which it is intended. In most cases the first solution provided to the user is not acceptable. Repeat runs may be necessary to obtain the best solution.

Figure 7–11 is a computer printout of an LP solution. The various columns of the report have been numbered for identification. Similar columns have been given the same number. The three sections of the report are each identified with a Roman numeral.

An explanation of the information contained in each column of Fig. 7–11 follows:

Column (1) – Ingredient and nutrient numbers.

Column (2) – Ingredient and nutrient names.

Column (3) – Solution *amounts* given in percentage for feed ingredients (Section I) and nutrients (Section II).

Column (4) – The percentage solution for ingredients has been converted to a *ton batch* using prespecified rounding factors for each ingredient. (The batch totals 1,992.73 lb, rather than an exact 2,000 lb because of the rounding requirements.)

Columns (5 & 6) – *Lower and upper limits* specified for each ingredient (Section I) and each nutrient (Section II).

Column (7) – Ingredient *costs* in dollars per hundred weight (cwt).

Columns (8 & 9) – The *stable cost range (lower)* gives the feed cost below which the present optimal solution would no longer be valid. Similarly, the *stable upper* gives the feed cost above which the present solution would no longer be valid. The *stable cost* figures let the LP user know when it is desirable to reprocess the ration.

```
ANYCO GRAIN AND MILLING
P. C. BCX 1234
  ANYTOWN, USA 90909    (123) 456-7890
* FEASIBLE* RATION: L507 LAYER MASH, 17 PCT.

                                    16-JAN-39 ( 7) ID. NO. 29706
                        COSTS....  $/CWT    $/TON
                        BASE LP    8.069    161.379
                        BATCH      8.03     161.60
```

Section I

(1) #	(2) INGREDIENT	(3) AMOUNT	(4) BATCH	(5) LIMITS LOWER	(6) UPPER	(7) COST	(8) STABLE COST RANGE LOWER	(9) UPPER	(10) COST PER UNIT DECREASE	(11) INCREASE	(12) EFFECTIVE RANGE DECREASE	(13) INCREASE
4	GROUND CORN	33.333	665.00	10.000		6.40	6.163	6.470	0.0007	0.0024	25.050	64.496
6	GROUND MILO	33.848	675.00			6.00	5.933	6.218	0.0022	0.0006	0.000	43.256
10	WHEAT MILLRUN	2.085	40.00		15.000	6.00	4.339	6.280	0.0028	0.0116	0.000	28.642
19	SOYBEAN MEAL,47.5	16.117	320.00			14.95	11.036	26.372	0.1142	0.0391	14.804	18.805
20	MEAT SCRAP,50	6.000	120.00		6.000	15.03		20.133	0.0510	-0.0510	3.938	7.540
21	DICAL PHOS(22CA/18P)	0.396	10.00			17.00	0.412	139.266	1.2227	0.1659	0.273	2.033
23	LIMESTONE	7.584	150.00			1.45	0.000	12.895	0.1144	0.0309	7.533	7.843
24	SALT, PLAIN	0.250	5.00	0.250	0.250	2.35	0.000		-0.0393	0.0393	0.000	0.930
27	POULTRY PREMIX	0.250	5.00	0.250	0.250	56.37	0.000		-0.5795	0.5795	0.000	0.930
33	DL-METHIONINE,99	0.086	1.73			157.00	76.279	540.275	3.8327	0.8072	0.067	0.094
37	SELENIUM, 90.6	0.050	1.00	0.050	0.050	17.50	0.000		-0.1908	0.1908	0.000	0.730

```
               TOTALS  100.000  1992.73   REQUESTED BATCH WEIGHT IS  2000.00 POUNDS
```

Section II

(1) #	(2) NUTRIENT	(3) AMOUNT	(5) LIMITS LOWER	(6) UPPER	(10) COST PER UNIT DECREASE	(11) INCREASE	(12) EFFECTIVE RANGE DECREASE	(13) INCREASE
1	WEIGHT	100.000	100.000	100.000	0.0158	-0.0158	99.320	104.209
8	CRUDE PROTEIN	17.000	17.000		-0.2428	0.2428	15.864	18.136
12	CRUDE FAT	2.930	0.000		0.0540	0.0532	2.822	5.748
13	CRUDE FIBER	2.308		4.000	0.1370	0.0969	2.250	2.368
14	ASH	11.963	0.000		0.6566	0.0333	11.907	12.203
15	CALCIUM	3.600	3.600	3.700	-0.0798	0.0799	2.001	3.858
16	PHOSPHORUS	0.671	0.650		0.5568	0.9380	0.661	0.961
17	AVAIL. PHOSPHORUS	0.470	0.470		-0.9095	0.9095	0.448	0.769
31	M. E. (POULTRY)/LB	1270.000	1270.000		-0.0035	0.0035	1203.931	1280.669
47	LYSINE	0.798	0.700		20.0361	1.6941	0.686	0.801
48	METHIONINE	0.350	0.350		-1.3590	1.3590	0.331	1.798
49	METHIONINE + CYSTINE	0.619	0.600		3.5320	1.3303	0.617	2.034
61	LINOLEIC ACID	1.134	1.000		0.0609	0.1073	1.038	2.531
62	XANTHOPHYLL /LB	3.000	3.000		-0.0264	0.0264	1.670	5.805

**Section III — ** NOT USED **

(1) #	(2) INGREDIENT	(3) COST	(14) RELATIVE WORTH	(6) UPPER LIMIT	(11) COST/UNIT INCREASE	(13) INCREASE RANGE
8	GROUND BARLEY	6.500	5.782		0.0072	6.531
12	ALFALFA, DEHY.,17	8.000	7.374	2.500	0.0063	0.932
31	VEGETABLE FAT	17.200	11.914		0.0529	2.835
32	CANE MOLASSES	4.450	2.384	0.000	0.0207	1.738

Fig. 7–11. Example Leghorn layer ration processed by computer linear programming. See text for explanation of numbered columns. (Courtesy, Nutri-Systems, Fresno, CA)

Columns (10 & 11)—The *cost per unit decrease* and *increase* values indicate how much the cost of the ration would be changed if either an ingredient or nutrient is increased or decreased by one unit in the percent solution. A positive value means that cost would be increased and a negative value means that cost would be decreased.

Columns (12 & 13)—The *effective range decrease* and *increase* values are related to the *stable cost range* columns and the *cost per unit decrease/increase* columns. The values delineate the limits over which the *stable cost* and *unit decrease/increase* columns are applicable. (An example: If the cost of ground corn decreases to $6.163/cwt [Column 8], then the usage of corn would increase to 64.496% [Column 13]. Of course, there would be changes in the usage of other ingredients as corn increases in amount.)

Column (14)—Section III contains information about the *ingredients not used* in the solution. The *relative worth* column indicates the cost at which each of these ingredients would enter the solution.

6. **Reformulate with LP at periodic intervals.** Changes in ingredient costs, in ingredient availability, and in the needs of birds dictate the need for reprocessing the ration. The good formulator monitors all these items on a regular basis. It is also critical to evaluate the feeding results to confirm that production goals and cost objectives are being met with the ration. Computers don't feed birds—people do!

Selection of Computer Software and Hardware for Ration Formulation

Numerous companies market computer software for ration formulation. The software varies from the very simple and straight-forward to very complex packages intended for large feed manufacturers. The latter packages include applications for formula costing, inventory control, control of usage of ingredients in limited availability, production of feed tags, etc. Most software is intended for use on a single computer model or at least a certain family of computers (IBM-PC, for example). It is therefore most desirable to select the software desired before purchasing the computer hardware. Computer type and size of memory and disk drive storage capacity must meet the criteria of the software developer or the software may not be usable.

Directories which list software by application are a good place to start looking. Other sources of information are feed and livestock trade publication advertisements, university personnel, and nutritionists who use feed blending software. Nutritionists are a good source of information as to how well a certain software package performs.

Ration formulation software may be generalized so that it can be made applicable to all species of animals or it may be designed with the unique requirements of specific species such as poultry, dairy cattle, etc. When the software has been designed for a certain species, it may incorporate tables of nutrient requirements and tables of typical feedstuffs and their nutrient values. This can save the user time, but it does not mean that software will run itself without the judgment of the user. No one has yet developed a software that will anticipate all the conditions under which poultry will be fed. Computers are not able to assess all aspects of ingredient quality, environment, and management. The judgment of the producer and formulator must be imposed on the computer software. Look for the freedom to make changes as needed. When in doubt, seek advice from those with experience.

USE OF THE FLEXIBLE FORMULA

The flexible formula is a ration formulation that allows for the substitution of various feeds on the basis of price and availability. The overall formula does not change. The only changes that take place are substitutions of feeds—for example oats for corn—within the formula to supply the same nutrient levels.

The following procedure may be used in setting up a flexible formula:

1. The nutrient requirement for the birds to be fed should be obtained from feeding standards, and should be listed.

2. A chart should be set up with four divisions: (1) energy feeds, (2) protein feeds, (3) mineral supplement, and (4) vitamin supplement. Within each division, subdivisions can be listed—for example, under protein feeds, two types of protein feeds, plant and animal, can be listed (see Table 7–18).

TABLE 7–18
EXAMPLE OF A FLEXIBLE FORMULA CHART

Nutrient Classi-fication of Feed	Subdivisions	Feeds for Utilization	Restric-tions for Use (Min.) (Max.)	Amount to be Incorporated (lb/100 lb feed)
Energy (Average 10% protein)	Grains	Corn Wheat Milo		86.7
	By-products	Wheat mill-run Corn gluten feed		
Protein (Average 45% protein)	Plant (Average 40% protein)	Soybean meal Cottonseed meal Linseed meal		11.8 Total (7.06 plant and 4.73 animal)
	Animal (Average 60% protein)	Fish meal Tankage Meat meal rendered		
Vitamin		Premix		0.01
Mineral		Premix		1.5
			Total	100.0

3. The proportion of the nutrient divisions must be established. This can be accomplished by balancing the various divisions for protein or any other feed parameter that is to be balanced. For example, a feed manufacturer wants to establish a flexible formula for a 14% protein feed. The manufacturer has premixes for vitamins and minerals and wants to incorporate them at levels of 0.01% and 1.5%, respectively. This then means that about 98.5% of the ration can be interchanged to fit protein and price specifications. If the energy sources average 10% protein and the protein feeds average 45% protein, the manufacturer can balance the classes of feeds using the square method. Since 98.5% of the 14% protein ration is to contain all of the proteins, this portion of the ration must contain 14.2% $\left[\frac{14}{.985}\right]$ protein.

Step 1.

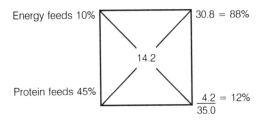

Step 2. Energy feeds to be incorporated in the ration.

100 × .88 × .985 = 86.7%

Protein feeds to be incorporated in the ration.

100 × .12 × .985 = 11.8%

4. Once the main nutrient classifications have been established, it is possible to subdivide the amounts of the various feedstuffs to be used. For example, we can divide the protein portions into plant and animal sources using the square method again.

Step 1. We know that 11.8% of the ration will be protein feeds and that the average protein level of this fraction will be 45%. Therefore, 100 × .118 × .45 = 5.31 lb of protein will be supplied by this factor for every 100 lb of feed.

Step 2.

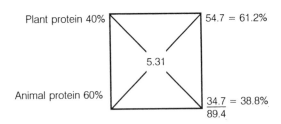

Step 3. Plant proteins to be incorporated into the ration.

100 × .612 × .118 = 7.22%

Animal proteins to be incorporated into the ration.

100 × .388 × .118 = 4.58%

5. Once the proportions of the nutrient classifications have been established, the feed manufacturer can then place minimum or maximum restrictions on the individual feeds within each classification. As the price fluctuates, feeds can be readily substituted for each other within their own restrictions for use.

Flexible formulas are extremely useful in ration formulation. No one ration is best and, if substitutions are made wisely, the prices of feed can be minimized and the feeder will continually obtain good results. Common sense is an invaluable tool when using flexible formulas because the feed manufacturer must always keep in mind the factors which cannot always be quantitated, such as palatability and bulkiness of feed.

FORMULATION WORKSHEET

When formulating rations, it is advisable to record the ration on a worksheet similar to that in Fig. 7–12. This worksheet is merely an example of the format that should be used. A similar sheet can be developed for micronutrient composition of premixes for minerals and vitamins as well as for amino acids. In modern feed formulation practice, computer printouts provide convenient worksheets. The worksheet serves three purposes:

1. It provides a means of reviewing and double checking the calculations used to formulate the ration. If there is a gross error, it will become obvious when listed on the worksheet.

2. It can be used to organize mixing procedures. It is vital that the person mixing feed be able to refer to a worksheet on which can be recorded what has been mixed and what mixing order should be followed.

3. The worksheet can be filed for future reference. If any questions should arise when the feed is fed, the worksheet provides an orderly record of the content of the feed and its mixing.

Each ration should be assigned a number for future reference, and the date of formulation and/or mixing should be recorded. In addition to listing the feed ingredients and their respective amounts, the nutrient requirements to be fulfilled by the ration should be listed on the worksheet immediately below the totals of various components of the feed. By subtracting the totals contained in the feed from the nutrient requirements, the person formulating the ration can then determine if there are any severe excesses or deficiencies in the ration.

COMMERCIAL POULTRY FEEDS

Commercial feeds are just what the term implies — instead of being farm mixed, these feeds are mixed by commercial feed manufacturers who specialize in the business. In 1988, a total of 103.1 million tons of primary feeds (complete feeds) were manufactured in the United States, and an additional 10 million tons of secondary feeds (supplements) were produced; making for a total of 113.1 million tons of commercial feeds. Primary feed is that which is mixed from individual ingredients, sometimes with the addition of a premix at a rate of less than 100 lb per ton of finished feed. Secondary feed is that which is mixed with one or more ingredients and a formula feed supplement (which is a pri-

MACRONUTRIENT WORKSHEET

Ration Number: **Date:**

Ingredient	✔ if Mixed	Amount	Proximate Analysis				Energy	Minerals and Vitamins		
			Crude Fiber	Ether Extract (Fat)	N-Free Extract	Crude Protein		Calcium (Ca)	Phos-phorus (P)	Vitamin A
		(lb)	(lb)	(lb)	(lb)	(lb)	TDN = lb ME = Mcal	(lb)	(lb)	(IU)
TOTAL										
NUTRIENT REQUIREMENTS										
NUTRIENT BALANCE (Total–Nutrient Requirements)										

Fig. 7–12. Formulation worksheet for macronutrients.

Fig. 7–13. Poultry feed mill. (Courtesy, University of California, Davis)

mary feed); normally, the supplement is used at a rate of 300 lb or more per ton of finished feed, depending upon the protein content of the supplement and the percentage of protein content desired in the finished feed. The breakdown, percentagewise, by classes of livestock for which primary (complete) commercial feeds were used in 1988 follows: poultry, 44.5%; beef and sheep, 17.6%; dairy, 16.9%; hogs, 14.2%; and all other, 6.8%.[1]

The commercial feed manufacturer has the distinct advantage of (1) purchasing feed in quantity lots, making possible price advantages; (2) economical and controlled mixing; (3) the hiring of scientifically trained personnel for use in determining the rations; and (4) quality control. Most producers have neither the know-how nor the quantity of business to provide these services on their own. Because of these several advantages, commercial feeds are finding a place of increasing importance in feeding. Also, it is to the everlasting credit of reputable feed dealers that they have been good teachers, often getting producers started in feeding balanced rations.

Numerous types of commercial feeds, ranging from additives to complete rations, are on the market, with most of

[1]Source: American Feed Manufacturer's Association.

them designed for specific species, age, or need. Among them, are complete rations, concentrates, protein supplements (with or without reinforcements of minerals and/or vitamins), mineral and/or vitamin supplements, additives (antibiotics, hormones, etc.), starters, rations for different levels of production, and medicated feeds.

Part II—Feeding Programs

The nutritive requirements of poultry vary according to species (between chickens, turkeys, ducks, and geese), according to age, and according to the type of production—whether the birds are kept for meat production, layers, or breeders. For this reason, many different rations are required.

To be successful, rations must meet the nutritive requirements of the birds to which they are fed. The National Research Council nutritive requirements are given in Tables 7–3 through 7–17. In using these tables to formulate practical rations, it must be remembered that they are minimum requirements; which means that they do not provide for any margins of safety. Further, the protein-energy relationships shown therein should be retained.

Birds eat primarily to satisfy their energy needs. Also, the temperature of the environment has an important influence on feed intake—the warmer the environment, the less the feed intake. Therefore, the requirement of all nutrients, expressed as a percentage of the diet, is dependent upon the environmental temperature. Other factors affecting feed intake are health, genetics, form of feed, nutritional balance, stress, body size, and rate of egg production or growth.

It is believed that feed intake is, in part, controlled by the amount of glucose in the blood. It has been observed that the addition of fat to the diet results in overconsumption on the part of the bird. As a result, some variation in the protein:energy ratio may be tolerated. In general, when free-choice feeding dietary protein levels that are low in relation to energy, fat deposition is markedly increased; with higher levels of protein, less fat is deposited. Increasing the protein level above that required for maximum growth rate reduces fat deposition still further.

FEEDING CHICKENS

Today, most commercial chickens are provided complete rations with all the needed nutrients available in the quantities necessary, and with the feed and water available on an *ad libitum* or free-choice basis. Formulations are varied according to the type of production—whether the birds are bred and kept for egg production (layers), hatchery production (breeders), or meat production (broilers). Also, consideration is given to age; sex; stage and level of production; temperature, disease level, and other stresses; management; and other factors.

FEEDING LAYERS

Chickens kept for the production of eggs for human consumption (Leghorn-type) have small body size and are prolific layers. They are generally fed *ad libitum* during the laying period. Occasionally, layers will consume excess feed during the latter phases of egg production (following peak production) with resultant obesity, reduced feed efficiency, and higher incidence of fatty liver syndrome. When this situation is detected, limiting feed intake to 85 to 95% of full feed consumption is desirable. Data on feed consumption in a particular flock, together with information on body weight, ambient temperature, and rate of egg production, may be used to determine the degree of feed restriction.

Fig. 7–14. A modern cage facility in a commercial laying house. Note the three tiers of cages, with a feed trough and an automatic egg collecting belt in front of each row of cages. (Courtesy, California Polytechnic State University, San Luis Obispo)

These additional pointers are pertinent to feeding layers:

1. The largest item of cost in the production of eggs is feed. It normally constitutes 50 to 70% of the total cost; though, in exceptional cases, it can run as low as 45% or as high as 75%.

2. In the final analysis, the objective of feeding laying hens is to produce a dozen eggs of good quality at the lowest possible feed cost. Thus, the actual cost of the feed that a layer eats in producing a dozen eggs—not the price per pound of feed—determines the economy of the ration.

3. Feed consumption per bird varies primarily with egg production and body size. It is also influenced by the health of the birds and the environment, especially the temperature.

4. Normally, a mature Leghorn, or other lightweight bird, eats about 82.5 lb of feed per year and produces about 22 doz eggs in that same period of time. Hence, it requires about 3.75 lb of feed to produce 1 doz eggs. A bird of the heavier breeds eats 95 to 115 lb of feed per year; hence, they are not as efficient egg producers. With lightweight layers, the producer should aim for a feed efficiency of 3.5 to 4.0 lb of feed per dozen eggs.

5. Feed may affect egg quality. Deficiencies of calcium, phosphorus, manganese, and vitamin D_3 lead to poor shell quality. Yolk color is almost entirely dependent on the bird's diet. Low vitamin A levels may increase the incidence of blood spots.

Example rations for *layers* producing eggs for human consumption are given in Table 7–19a.

TABLE 7–19a
EXAMPLE LAYER RATIONS[A, B, 1]

Ingredient	Protein Level of Rations[C]											
	15%		16%		17%		18%		19%		20%	
	(lb)	(kg)	(lb)	(kg)	(lb)	(kg)	(lb)	(kg)	(lb)	(kg)	(lb)	(kg)
Ground yellow corn[2, 3]	1,457	662.3	1,403	637.7	1,339	608.6	1,242	564.5	1,177	535	1,120	509.1
Alfalfa meal, 17%	25	11.3	25	11.3	25	11.3	25	11.3	25	11.3	25	11.3
Soybean meal, dehulled	292.2	132.8	340.6	155	393.6	179	451.6	205.3	504.6	229.4	554	251.8
Meat and bone meal, 47%[5]	50	23	50	23	50	23	50	23	50	23	50	23
DL-Methionine or equivalent	1.0	0.5	1.0	0.5	1.0	0.5	1.0	0.5	1.0	0.5	1.0	0.5
Dicalcium phosphate[6]	9	4.1	8	3.6	8	3.6	7	3.1	7	3.1	7	3.1
Ground limestone[7]	159	72.3	159	72.3	159	72.3	174	79.1	174	79.1	174	79.1
Iodized salt[4]	7	3.1	7	3.1	7	3.1	7	3.1	7	3.1	7	3.1
Stabilized yellow grease or equivalent	—	—	7	3.1	18	8.2	43	19.5	55	25	62	28.2
Antioxidant	9	9	9	9	9	9	9	9	9	9	9	9
Mineral and vitamin supplements:[12]												
Calcium pantothenate . . . . (mg)	5,000		4,500		4,500		4,500		4,000		4,000	
Manganese[11] (Mn) . . . . (g)	52		52		52		52		52		52	
Selenium[25] (Se) . . . . (mg)	90.8		90.8		90.8		90.8		90.8		90.8	
Zinc[17] (Zn) . . . . (g)	16		16		16		16		16		16	
Vitamin A . . . . (IU)	6,000,000		6,000,000		6,000,000		6,000,000		6,000,000		6,000,000	
Vitamin D₃ . . . . (IU)	2,000,000		2,000,000		2,000,000		2,000,000		2,000,000		2,000,000	
Vitamin K[20]	—		—		—		—		—		—	
Choline . . . . (mg)	274,000		231,000		184,000		140,000		94,000		50,000	
Niacin (Nicotinic Acid, Nicotinamide) (mg)	12,000		12,000		12,000		12,000		12,000		12,000	
Riboflavin (Vitamin B–2) . . . . (mg)	2,000		2,000		2,000		2,000		2,000		2,000	
Vitamin B–12 (Cobalamins) . . . . . (mg)	6		6		6		6		6		6	
Totals[21]	2,000	909.4	2,000	909.3	2,000	909.3	2,000	909.4	2,000	909.4	2,000	909.1
Calculated analysis:[27]												
Metabolizable energy . . . . (kcal)	1,306.2	2,873.6	1,303.9	2,868.6	1,303.4	2,867.5	1,304.1	2,869	1,304.5	2,870	1,301	2,862.2
Protein . . . . (%)	15.08		16.02		17.03		18.01		19.03		20.00	
Lysine . . . . (%)	0.68		0.75		0.83		0.91		0.98		1.06	
Methionine . . . . (%)	0.31		0.32		0.33		0.34		0.35		0.36	
Methionine + cystine . . . . (%)	0.54		0.57		0.59		0.62		0.64		0.67	
Fat . . . . (%)	3.29		3.54		3.98		5.05		5.54		5.76	
Fiber . . . . (%)	2.20		2.20		2.21		2.18		2.18		2.19	
Calcium (Ca) . . . . (%)	3.25		3.24		3.24		3.50		3.50		3.51	
Total phosphorus (P) . . . . (%)	0.52		0.52		0.53		0.52		0.53		0.54	
Available phosphorus[13] (P) . . . . . (%)	0.45		0.45		0.45		0.45		0.45		0.45	
Vitamins (units or mg/lb or kg):												
Vitamin A activity . . . . (IU)	5,904	12,988.8	5,842	12,852.4	5,770	12,694	5,660	12,452	5,586	12,239.2	5,522	12,148.4
Vitamin D₃, added . . . . (IU)	1,000	2,200	1,000	2,200	1,000	2,200	1,000	2,200	1,000	2,200	1,000	2,200
Choline . . . . (mg)	500.13	1,100.3	500.34	1,100.7	500.05	1,100.1	500.39	1,100.9	500.48	1,101.1	500.66	1,101.5
Niacin (Nicotinic Acid, Nicotinamide) (mg)	15.40	33.9	15.48	34.1	15.55	34.2	15.5	34.1	15.56	34.2	15.64	34.4
Pantothenic Acid (Vitamin B–3) . . (mg)	5.01	11	4.88	10.7	4.99	10.9	5.07	11.2	4.95	10.8	5.01	11
Riboflavin (Vitamin B–2) . . . . . . (mg)	1.84	4	1.85	4.1	1.86	4.1	1.86	4.1	1.87	4.1	1.88	4.1

[A]Adapted by the author from *NECC Chicken and Turkey Rations*, prepared by the New England College Poultry Conference Board, by poultry specialists from, and distribution by, the New England Land-Grant Universities: University of Connecticut, Storrs; University of Maine, Orono; University of Massachusetts, Amherst; University of New Hampshire, Durham; University of Rhode Island, Kingston; and University of Vermont, Burlington.

[B]See footnotes following Table 7–23, p. 180.

[C]Six rations varying in protein levels are presented. The ration which best meets the needs of a particular flock may be determined from the factors presented below.

See "Factors to Consider in Determining Which Layer Ration to Feed," which follows:

FACTORS TO CONSIDER IN DETERMINING WHICH LAYER RATION TO FEED

A. Feed consumption of the hens must be known to select the appropriate layer ration. Feed consumption may vary depending on type of bird, bird weight, pen temperature, rate of production, energy content of ration, disease problems, and many other factors.

B. **Table 7–19b.** SUGGESTED MINIMUM DAILY PROTEIN INTAKE.

Production Status	Daily Protein Required*
	(g)
Coming into production	19–20
90% hen day egg production	18–19
80% hen day egg production	17–18
70% hen day egg production	16–17
60% and under egg production	15–16

*Under conditions of higher production, disease, or stress, add 2 g protein to the above.

D. When feed consumption is not known, not less than 17% protein should be used in the diet.

WEIGHT CONVERSION TABLE

1 pound	= 453.57 grams or	. . . .	0.4536 kilogram
1 ounce	= 28.349 grams	1 kilogram = 2.2046 pounds	
1 kilogram (kg)	= 1,000 grams	1 gram (g) = 1,000 milligrams	
1 milligram (mg)	= 100 micrograms (gammas)		
1 part per million (ppm) . .	= 1 milligram/kilogram or 1 microgram/gram		
1 part per million (ppm) . .	= 0.454 mg/lb or 0.907 g/ton		

C. **Table 7–19c.** A GUIDE FOR ESTIMATING DIETARY PROTEIN LEVEL BASED ON FEED INTAKE.

Lb Feed Consumed Per lb/100 birds/Day	% Protein in Laying Ration					
	15	16	17	18	19	20
	Protein Intake					
	(g)	(g)	(g)	(g)	(g)	(g)
18–19	12	13	14	15	16	17
20–21	14	15	15	16	17	18
22–23	15	16	17	18	19	20
24–25	16	17	18	20	21	22
26–27	18	19	20	21	22	24
28–29	19	20	22	23	24	26
30–31	20	22	23	25	26	27

Example—Your birds are consuming 24 lb per 100 birds per day and laying at a rate of 80% hen day egg production. Their need for protein is 17–18 g (Table 17–19b). Therefore to supply 18 g of protein you should use the 17% protein ration (Table 7–19c).

PHASE FEEDING

Phase feeding refers to changes in the diet of the laying hen to adjust for (1) age and level of egg laying, (2) temperature and climatic changes, (3) differences in body weight and nutrient requirements of different strains of birds, and (4) changes in nutrients as feeds are changed for economic reasons.

(See Chapter 12, section on "Phase Feeding," for detailed information on the subject of *phase feeding*.)

MOLTING

Molting is the process of shedding and renewing of feathers. It is a normal process of chickens and other feathered species; and it occurs in both sexes.

In a natural molt, a chicken loses feathers from various sections of its body in the following order: head; neck; feather tracts of the breast, thighs, and back; and wing and tail feathers. Some birds molt earlier than others; and some molt more slowly than others. A high-producing flock generally molts late and rapidly.

(See Chapter 12, section on "Molting," for detailed information on the subject of *molting*.)

FEEDING BREEDERS

The following pointers are pertinent to feeding breeders:

1. The nutritive requirements for breeding flocks are more rigorous than those for commercial laying flocks. Breeders require greater amounts of vitamins A, D, E, and B–1, and of riboflavin, pantothenic acid, niacin, and manganese than do laying flocks. Rations with these added ingredients in the right proportions give high hatchability and good development of chicks. Such rations cost more than normal layer rations.

Fig. 7–15. Broiler breeder birds for the production of hatching eggs. Note the plastic, slat-raised floors. Separate feeders for males and females are now common in breeder houses better to control the body weight of the two sexes. (Courtesy, L. S. Jensen, Ph.D., Department of Poultry Science, University of Georgia, Athens)

Fig. 7–16. Single sire pedigree pens. (Courtesy, Indian River Company, Nacogdoches, TX)

Fig. 7–17. Broiler breeders, showing nesting. (Courtesy, California Polytechnic State University, San Luis Obispo)

2. Broiler breeder replacement pullets should receive low energy diets, in the range of 1,090 to 1,135 kcal/lb and/or the feed intake should be restricted, to avoid excess fat accumulation at the time they reach sexual maturity.

3. Broiler breeder hens, which are heavy and have a high energy requirement for maintenance, require approximately 400 to 450 kcal ME per hen per day, for maximum egg production. Since these hens tend to become over-fat when fed high energy diets, it appears best to limit the energy content of their diets to approximately 1,200 to 1,250 kcal/pound and/or to restrict their feed intake in some way so that they do not obtain much more than about 420 kcal ME per hen per day.

4. Male breeders require slightly less energy than females during growth. The lower fat deposition in the male compared with the female is offset by the energy needs for more rapid growth. Being larger, the adult male cock requires considerably more energy than the hen for maintenance, but this is largely offset by the hen's need for energy for egg production.

Example rations for the chicken breeders producing eggs for hatching are given in Table 7–20.

TABLE 7–20
EXAMPLE CHICKEN BREEDER RATIONS[A, B, 1, 14]

Ingredient	Body Weight			
	Egg-Type Breeders $3\frac{1}{2}$–5 Lb		Meat-Type Breeders 5–8 Lb	
	(lb)	(kg)	(lb)	(kg)
Ground yellow corn[2, 3]	1,305	593.2	1,379	627
Wheat middlings	—	—	70	32
Alfalfa meal, 17%	50	23	50	23
Soybean meal, dehulled	324	147.3	248	113
Fish meal, herring, 65%[4, 5]	50	23	50	23
Meat and bone meal, 47%[5]	50	23	50	23
Dicalcium phosphate[6]	6	2.7	4	1.8
Ground limestone[7]	157	71.4	142	64.5
DL-Methionine or equivalent	0.5	0.2	0.4	0.2
Stabilized yellow grease or equivalent	51	23	16	16
Iodized salt[4]	7	3.1	7	3.1
Antioxidant	9	9	9	9
Mineral and vitamin supplements:[12]				
Calcium pantothenate (mg)	6,000		6,000	
Manganese[11] (Mn) (g)	52		52	
Selenium[25] (Se) (mg)	90.8		90.8	
Zinc[17] (Zn) (g)	16		16	
Vitamin A (IU)	4,000,000		4,000,000	
Vitamin D$_3$ (IU)	2,000,000		2,000,000	
Vitamin E (IU)	2,000		2,000	
Vitamin K[20]	—		—	
Choline (mg)	172,000		208,000	
Niacin (Nicotinic Acid, Nicotinamide) (mg)	10,000		10,000	
Riboflavin (Vitamin B–2) (mg)	3,000		3,000	
Vitamin B–12 (Cobalamins) (mg)	6		6	
Totals[21]	2,000.5	909.9	2,000.4	910.6
Calculated analysis:[27]				
Metabolizable energy (kcal)	1,337	2,941	1,295	2,849
Protein (%)	17.01		16.03	
Lysine (%)	0.87		0.78	
Methionine (%)	0.33		0.31	
Methionine + cystine (%)	0.59		0.56	
Fat (%)	5.81		3.57	
Fiber (%)	2.42		2.67	
Calcium (%)	3.27		2.98	
Total phosphorus (P) (%)	0.54		0.54	
Available phosphorus[13] (P) (%)	0.46		0.46	
Vitamins (units or mg/lb or kg):				
Vitamin A activity (IU)	5,983	13,163	6,067	13,347
Vitamin D$_3$, added (IU)	1,000	2,200	1,000	2,200
Choline (mg)	500.4	1,101	500.57	1,101
Niacin (Nicotinic Acid, Nicotinamide) (mg)	15.21	33.5	17	37.4
Pantothenic Acid (Vitamin B–3) (mg)	5.67	12.5	5.74	12.6
Riboflavin (Vitamin B–2) (mg)	2.49	5.5	2.51	5.5

[A]Adapted by the author from *NECC Chicken and Turkey Rations*, prepared by the New England College Poultry Conference Board, by poultry specialists from, and distribution by, the New England Land-Grant Universities: University of Connecticut, Storrs; University of Maine, Orono; University of Massachusetts, Amherst; University of New Hampshire, Durham; University of Rhode Island, Kingston; and University of Vermont, Burlington.

[B]See footnotes following Table 7–23, p. 180.

FEEDING REPLACEMENT PULLETS

Pullets generally perform well during their laying year when their nutrient requirements have been met during the growing period.

Fig. 7–18. Newly hatched baby chicks—future replacement pullets.

Leghorn-type pullets are seldom restricted-fed during the growing period because varying the lighting during growth from 6 to 20 weeks of age can be used to control feed consumption and sexual development.

However, pullets of heavy breeds tend to accumulate excessive amounts of body fat; so, it is common practice to restrict the feed intake of these birds to produce pullets with leaner bodies at the time of sexual maturity. This is beneficial because: (1) it produces healthier pullets, and (2) it reduces feed costs during the pullet rearing period.

Also, some producers restrict the feed intake of light breeds, based on research which shows that, even among the light breeds, restricted feed intake results (1) in slightly higher mortality during the rearing period, but (2) in lower mortality and higher egg production during the laying year.

Most feed restriction programs are started at 9 to 12 weeks of age. The methods commonly employed follow:

1. **Skip-a-day method.** This involves feeding pullets on alternate days only, from 9 weeks to sexual maturity. When fed every other day, pullets consume more feed on the days that feed is available than they would normally consume on a daily basis. They are unable, however, to consume enough feed in one day to satisfy their total energy requirements for 2 days. Thus, growth and body fat content are reduced.

2. **Daily restriction of feed.** Under this system, the producer determines the amount of feed that would normally be consumed by the pullets each day, then provides them with a fraction of that amount on a daily basis. Often this fraction is 75 to 85% of the amount of feed that would be consumed on a free-choice basis. Typical body weight increases and feed requirements of pullets are shown in Table 7–8.

3. **Bulky, low-energy or low-protein and/or amino acid imbalanced rations.** Another form of restriction involves (a) the feeding of bulky, low-energy rations, or (b) the feeding of low-protein and/or amino acid imbalanced rations on a free-choice basis during the period 12 to 20 weeks of age. The rations are formulated to be adequate in all nutrients, but are sufficiently low in energy or low in protein and/or imbalanced in amino acids that pullets cannot consume enough to satisfy their energy or protein needs for maximum growth. Under such a program, it is possible to restrict the growth of young pullets by 10 to 15%, an amount comparable to the growth depression with the skip-a-day and daily restriction methods.

These further pointers are pertinent to feeding replacement chicks (pullets):

1. Feed accounts for approximately 60% of the cost of raising replacement pullets.

2. Replacement chicks are usually fed a diet lower in energy than broiler chicks. Also, feed and daily light periods may be restricted, so as to permit the pullets to reach larger body size before they start to lay than would be the case were they full fed, and fully lighted.

3. Always use complete starter feeds for chicks, and give chicks starter feeds without grain supplement until they are 5 weeks old.

4. When chicks are 5 weeks old, change to the growing ration.

The NRC requirements for replacement pullets are presented in tables in the earlier section of this chapter headed "National Research Council (NRC) Requirements."

Example *starter* and *grower* rations (1) for egg-type and meat-type strains, from days 1 to 35, and (2) for egg- and meat-type strains, from day 36 until egg production begins are given in Table 7–21.

TABLE 7–21
EXAMPLE REPLACEMENT CHICKEN RATIONS[A, B, C, 1, 14, 26]

Ingredient	Starter (For egg- and meat-type strains, from days 1 to 35)				Grower (For egg- and meat-type strains, from day 36 until egg production begins)					
	20% Protein		18% Protein		15% Protein		14% Protein		12% Protein	
	(lb)	(kg)	(lb)	(kg)	(lb)	(kg)	(lb)	(kg)	(lb)	(kg)
Ground yellow corn[2, 3]	1,267	576	1,310	595.4	1,412	641.8	1,438	653.6	1,481	673.2
Wheat middlings	130	59.1	200	90.9	254	115.5	254	115.5	323	146.8
Alfalfa meal, 17%	25	11.3	25	11.3	25	11.3	25	11.3	25	11.3
Soybean meal, dehulled	422	192	309	140.4	217	98.6	217	98.6	104.8	47.6
Fish meal, herring, 65%[4, 5]	50	23	50	23	—	—	—	—	—	—
Meat and bone meal, 47%[5]	50	23	50	23	50	23	—	—	—	—
Lysine	—	—	—	—	—	—	1.0	0.5	1.2	0.5
Dicalcium phosphate[6]	10	4.5	9	4.1	14	6.4	30	13.6	29	13.2
Ground limestone[7]	19	8.6	20	9.1	21	9.5	28	12.7	29	13.2
Stabilized yellow grease or equivalent	20	9.1	20	9.1	[16]	[16]	[16]	[16]	[16]	[16]
Iodized salt[4]	7	3.1	7	3.1	7	3.1	7	3.1	7	3.1
Antibiotic supplement	[8]	[8]	—	—	—	—	—	—	—	—
Antioxidant	[9]	[9]	[9]	[9]	[9]	[9]	[9]	[9]	[9]	[9]
Coccidiostat	[10]	[10]	[10]	[10]	[10]	[10]	[10]	[10]	[10]	[10]
Mineral and vitamin supplements:[12]										
Calcium pantothenate (mg)	4,000		4,000		3,000		3,000		3,000	
Manganese[11] (Mn) (g)	52		52		52		52		52	
Selenium[25] (Se) (mg)	90.8		90.8		90.8		90.8		90.8	
Vitamin A (IU)	3,000,000		3,000,000		3,000,000		3,000,000		3,000,000	
Vitamin D_3 (IU)	1,000,000		1,000,000		1,000,000		1,000,000		1,000,000	
Vitamin K[20]	—		—		—		—		—	
Choline (mg)	213,000		298,000		84,000		125,000		209,000	
Niacin (Nicotinic Acid, Nicotinamide) (mg)	10,000		10,000		10,000		10,000		10,000	
Riboflavin (Vitamin B–2) (mg)	1,500		1,500		1,500		1,500		1,500	
Vitamin B–12 (Cobalamins) (mg)	6		6		6		6		6	
Totals[21]	2,000	909.7	2,000	909.4	2,000	909.2	2,000	908.9	2,000	908.9
Calculated analysis:[27]										
Metabolizable energy (kcal)	1,361	2,994	1,362	2,996	1,343	2,955	1,341	2,950	1,342	2,952
Protein (%)	20.03		18.01		15.03		14.01		12.01	
Lysine (%)	1.04		0.89		0.63		0.63		0.49	
Methionine (%)	0.34		0.32		0.25		0.24		0.21	
Methionine + cystine (%)	0.64		0.59		0.48		0.46		0.41	
Fat (%)	4.48		4.70		3.76		3.54		3.74	
Fiber (%)	2.67		2.83		3.01		3.00		3.15	
Calcium (%)	0.90		0.90		0.89		0.90		0.90	
Total phosphorus (P) (%)	0.66		0.66		0.66		0.66		0.65	
Available phosphorus[13] (P) (%)	0.41		0.41		0.40		0.40		0.40	
Vitamins (units or mg/lb or kg):										
Vitamin A activity (IU)	4,188	9,214	4,237	9,321	4,354	9,579	4,381	9,638	4,430	9,746
Vitamin D_3, added (IU)	500	1,100	500	1,100	500	1,100	500	1,100	500	1,100
Choline (mg)	600.2	1,320	600.11	1,320	420.48	925	420.2	924	419.76	923
Niacin (Nicotinic Acid, Nicotinamide) (mg)	19.08	42	20.49	45	20.87	45.9	20.53	45.2	21.84	48
Pantothenic Acid (Vitamin B–3) (mg)	5.34	11.7	5.27	11.6	4.69	10.3	4.72	10.4	4.64	10.2
Riboflavin (Vitamin B–2) (mg)	1.78	3.9	1.76	3.9	1.68	3.7	1.64	3.6	1.62	3.6

[A]Adapted by the author from *NECC Chicken and Turkey Rations*, prepared by the New England College Poultry Conference Board, by poultry specialists from, and distribution by, the New England Land-Grant Universities: University of Connecticut, Storrs; University of Maine, Orono; University of Massachusetts, Amherst; University of New Hampshire, Durham; University of Rhode Island, Kingston; and University of Vermont, Burlington.

[B]See footnotes following Table 7–23, p. 180.

[C]Equivalent to 14% protein + 1 lb lysine.

FEEDING BROILERS

Fig. 7–19. Typical broiler farm, showing two houses and two bulk feed bins. (Courtesy, University of Georgia, Athens)

Fig. 7–21. Nutrition research has played a major role in the development of the modern broiler industry. Battery brooders such as these are extensively used in research to evaluate feeds and additives. (Courtesy, University of Georgia, Athens)

The following pointers pertaining to feeding broilers, roasters, and capons are pertinent:

1. Feed is the largest cost item in broiler production, representing 60 to 75% of the total cost.

2. Producers aim for broilers with an average weight of over 4.0 lb at 6 to 7 weeks of age, feed conversion of less than 2.0, and mortality under 3.0%. Many good producers are achieving feed conversion of about 1.9.

3. Some operations use a 2-stage ration program (starter and finisher) for broilers, but most are using at least 3 stages in their feeding programs (starter, grower, and finisher) to reduce costs and make more efficient use of the nutrients. In a 3-stage program the starter feed should be used for 3 to 4 weeks, the grower for about 2 weeks, and the finisher for the remainder of the feeding period.

Fig. 7–22. Broiler rations are routinely computed by least cost formulation. The computer determines the combination of feedstuffs that will meet the nutrient specifications set by the nutritionist at the lowest cost per unit of feed. (Courtesy, University of Georgia, Athens)

Fig. 7–20. Six-week-old broilers. Note two continuous pan feeders and four rows of waterers. (Courtesy, California Polytechnic State University, San Luis Obispo)

The NRC requirements for broilers are presented in the tables in the earlier section of this chapter headed, "National Research Council (NRC) Requirements."

Example broiler *starter* and *finisher* rations are given in Table 7–22.

TABLE 7–22
EXAMPLE BROILER RATIONS[A, B, 1]

Ingredient	Starter[18] (For Broiler Chicks Until 24 Days of Age)		Finisher[C] (For Broilers From Day 25 to Market)	
	(lb)	(kg)	(lb)	(kg)
Ground yellow corn[3]	1,106	503	1,235	561
Alfalfa meal, 17%	—	—	25	11
Soybean meal, dehulled	605	275	420	191
Corn gluten meal, 60%	50	23	75	34
Fish meal, herring, 65%[4, 5]	50	23	50	23
Meat and bone meal, 47%[5]	50	23	50	23
Dicalcium phosphate[6]	10	4	9	4
Ground limestone[7]	16	7	14	6.3
DL-Methionine or equivalent	0.8	0.3	—	—
Stabilized yellow grease or equivalent	106	48	115	52.2
Iodized salt[4]	7	3	7	3
Antibiotic supplement	[8]	[8]	[8]	[8]
Antioxidant	[9]	[9]	[9]	[9]
Coccidiostat	[10]	[10]	[10]	[10]
Mineral and vitamin supplements:[12]				
Calcium pantothenate (mg)	5,000		5,000	
Manganese[15] (Mn) (g)	75		75	
Organic arsenical supplement[19] (lb)	0.1		0.1	
Selenium[25] (Se) (mg)	90.8		90.8	
Zinc[24] (Zn) (g)	30		30	
Vitamin A (IU)	4,000,000		4,000,000	
Vitamin D$_3$ (IU)	1,000,000		1,000,000	
Vitamin E (IU)	2,000		2,000	
Vitamin K[20] (mg)	2,000		1,000	
Choline (mg)	503,000		672,000	
Niacin (Nicotinic Acid, Nicotinamide) (mg)	20,000		20,000	
Riboflavin (Vitamin B–2) (mg)	3,000		3,000	
Vitamin B–12 (Cobalamins) (mg)	12		12	
Totals[21]	2,000.9	909.3	2,000.1	908.5
Calculated analysis:[27]				
Metabolizable energy (kcal)	1,399	3,078	1,494	3,287
Protein (%)	24.08		21.00	
Lysine (%)	1.30		1.05	
Methionine (%)	0.45		0.38	
Methionine + cystine (%)	0.81		0.71	
Fat (%)	8.20		8.92	
Fiber (%)	1.97		2.22	
Calcium (%)	0.84		0.80	
Total phosphorus (P) (%)	0.64		0.60	
Available phosphorus[13] (P) (%)	0.40		0.38	
Vitamins (units or mg/lb or kg):				
Vitamin A activity (IU)	3,769	8,292	5,424	11,933
Vitamin D$_3$, added (IU)	500	1,100	500	1,100
Choline (mg)	800.03	1,760	800.48	1,761
Niacin (Nicotinic Acid, Nicotinamide) (mg)	21.36	47	21.33	46.9
Pantothenic Acid (Vitamin B–3) (mg)	5.69	12.5	5.51	12.1
Riboflavin (Vitamin B–2) (mg)	2.44	5.4	2.49	5.5
Xanthophyll[28] (mg)	9.5	20.9	14.05	30.9

[A]Adapted by the author from *NECC Chicken and Turkey Rations*, prepared by the New England College Poultry Conference Board, by poultry specialists from, and distribution by, the New England Land-Grant Universities: University of Connecticut, Storrs; University of Maine, Orono; University of Massachusetts, Amherst; University of New Hampshire, Durham; University of Rhode Island, Kingston; and University of Vermont, Burlington.

[B]See footnotes following Table 7–23, p. 180.

[C]For preslaughter drug withdrawal times see section headed "Regulations Regarding Feed Additives," following "Footnotes."

FEEDING TURKEYS

Fig. 7–23. Turkeys grown in confinement. (Courtesy, California Polytechnic State University, San Luis Obispo)

Feeding turkeys involves two distinct areas of emphasis: feeding market turkeys, and feeding turkey breeders.

■ **Feeding market turkeys**—Most market turkeys are of the large type. Males (toms) are usually marketed at 19 to 25 weeks of age and at liveweights of 23 to 35 lb. Younger toms are often sold as oven-ready dressed birds; older toms are generally further processed or used in the restaurant trade. Females (hens) are commonly marketed at 16 to 17 weeks of age and at about 14 lb weight. Medium- and small-type turkeys (roasters/fryers) are often sold at younger ages and lighter liveweights—normally, about 10 lb.

The formulations of the rations fed to market turkeys should be changed as the birds grow. Thus, the nutrient requirements shown in Table 7–10 provide for such changes at 3- to 4-week intervals; in practice, however, the changes may occur more or less frequently than indicated in Table 7–10. Nutritional adjustments are often made for expected ambient temperature variations in order to assure that the birds consume the necessary amount of protein, minerals, and vitamins, regardless of changes in feed consumption.

■ **Feeding turkey breeders**—The feeding programs for breeder stock are usually divided into prebreeding (holding) and breeding periods. The prebreeding, or holding, rations may be fed from the time the breeders are selected, at about 16 weeks of age. Holding rations are usually formulated at medium energy levels in order to stabilize development and weight gains after market age. Hens are fed the holding ration until the time of light stimulation, at about 30 weeks of age; thereafter, breeder rations are fed. Toms may be fed a nutritionally balanced holding ration from the time of breeder selection throughout the breeding season. In some programs, the body weight of toms is controlled by limited feeding. The light stimulation of toms is normally initiated at about 26 weeks of age.

It is not necessary to feed low-energy rations or to restrict feed intake of turkey hens in the prebreeding, or holding, period. Corn-soybean meal type rations may be fed *ad libitum*. Growth restriction does not result in any consistent improvement in reproductive performance. Nevertheless, the

use of holding rations for turkey breeders is common practice. These rations usually contain medium energy concentrations so as to stabilize development and weight gains after mature body weight is attained. Care should be taken to provide turkey breeder hens adequate intake of minerals and vitamins during the holding period so that they are not depleted of these nutrients prior to the onset of lay.

These further pointers are pertinent to feeding turkeys:

1. Prevent poult "starve out." Upon arrival, poults should be encouraged to consume feed and water as soon as possible. It may be necessary to dip the beaks of some of them in feed and water to start them eating and drinking.

2. Turkeys grow faster than chickens; hence, they have relatively higher feed and protein requirements.

3. Young turkeys use feed efficiently. Small White turkeys raised to a liveweight of 6 to 8 lb at 14 to 16 weeks of age require 3 lb of feed per pound of live turkey produced. Large White turkeys require about 3.5 to 3.75 lb of feed to produce 1 lb of liveweight, when grown to a market weight of about 30 lb and 24 weeks of age.

4. A high-fiber, low-energy holding ration retards sexual maturity and may result in some desirable effects upon later reproductive performance. The holding ration limits energy intake, but should not limit protein, vitamins, and minerals. When a holding ration is used, the birds should be switched to the breeder ration 2 weeks prior to egg production.

Fig. 7–24. Turkey eggs and newly hatched poult. Approximately 10% of all incubated turkey eggs are infertile. The B vitamins, vitamin E, and selenium have been shown to affect the quality of the semen. In the diet of the turkey hen, these same nutrients, along with magnesium, have been shown to affect egg fertility. (Courtesy, Nicholas Turkey Breeding Farms, Sonoma, CA)

5. Good range provides green feed and tends to reduce feed costs. However, it may make for higher losses from blackhead and other diseases, and predators; and range turkey operations may make the neighbors unhappy because of dust, odors, and noise.

6. As they approach maturity, turkeys fed for market purposes should be fed rations that are quite different from those that are fed as turkey breeders.

Note well: The NRC requirements for turkeys are presented in Tables 7–10, 7–11, and 7–12 in the earlier section of this chapter headed, "National Research Council (NRC) Requirements." Example rations for market turkeys and breeders are given in Table 7–23, Example Turkey Rations, including the footnotes on page 180.

TABLE 7–23
EXAMPLE TURKEY RATIONS[A, B, C, D, 1]

Ingredient	Starter (0–8 Weeks) (lb)	(kg)	Grower (8–16 Weeks) (lb)	(kg)	Finisher[C, 22] (16–Market) (lb)	(kg)	Breeder (lb)	(kg)
Ground yellow corn[2, 3]	929	422.3	1,199	545	1,490	677.2	1,218	554
Wheat middlings	100	45.4	50	23	—	—	250	114
Alfalfa meal, 17%	25	11.3	25	11.3	25	11.3	60	27.3
Soybean meal, dehulled	675	307	570	259	335	152.3	190	86.4
Fish meal, herring, 65%[4, 5]	100	45.4	—	—	—	—	100	45.4
Meat and bone meal, 47%[5]	100	45.4	50	23	50	23	50	23
Dicalcium phosphate[6]	13	6	32	14.5	23	10.5	10	4.5
Ground limestone[7]	10	4.5	14	6	17	8	92	42
DL-Methionine or equivalent	0.6	0.3	—	—	—	—	—	—
Stabilized yellow grease or equivalent	40	18.2	50	23	50	23	20	9
Iodized salt[4]	8	3.6	10	4.5	10	4.5	10	4.5
Antibiotic supplement	8	8	—	—	—	—	—	—
Coccidiostat or antihistomonal	10	10	10	10	10	10	—	—
Antioxidant	9	9	9	9	9	9	9	9
Mineral and vitamin supplements:[D, 12]								
Calcium pantothenate (mg)	4,500		4,500		6,000		10,000	
Manganese[23] (Mn) (g)	30		30		30		30	
Selenium[25] (Se) (mg)	181.6		181.6		181.6		181.6	
Zinc[24] (Zn) (g)	30		30		30		30	
Vitamin A (IU)	7,500,000		7,500,000		7,500,000		4,000,000	
Vitamin D_3 (IU)	1,700,000		1,700,000		1,700,000		1,700,000	
Vitamin E (IU)	10,000		5,000		—		30,000	
Biotin (mg)	100		100		100		100	
Choline (mg)	674,000		388,000		417,000		427,000	
Niacin (Nicotinic Acid, Nicotinamide) (mg)	42,000		46,000		48,000		50,000	
Riboflavin (Vitamin B–2) (mg)	4,000		5,000		5,000		5,000	
Vitamin B–12 (Cobalamins) (mg)	6		6		6		6	
Totals[21]	2,000.6	909.4	2,000	909.3	2,000	909.8	2,000	910.1
Calculated analysis:[27]								
Metabolizable energy (kcal)	1,322	2,908	1,371	3,016	1,440	3,168	1,291	2,840
Protein (%)	27.31		21.09		16.22		17.01	
Lysine (%)	1.60		1.10		0.75		0.87	
Methionine (%)	0.48		0.33		0.27		0.32	
Methionine + cystine (%)	0.88		0.65		0.53		0.57	
Fat (%)	5.31		5.43		5.84		4.88	
Fiber (%)	2.60		2.47		2.28		3.25	
Calcium (%)	1.17		1.00		0.93		2.26	
Total phosphorus (P) (%)	0.91		0.80		0.69		0.70	
Available phosphorus[13] (P) (%)	0.65		0.55		0.46		0.61	
Vitamins (units or mg/lb or kg):								
Vitamin A activity (IU)	6,054	13,319	6,361	13,994	6,691	14,720	6,382	14,040
Vitamin D_3, added (IU)	850	1,870	850	1,870	850	1,870	850	1,870
Choline (mg)	1,000.07	2,200	700.08	1,540	600	1,320	650.24	1,430
Niacin (Nicotinic Acid, Nicotinamide) (mg)	35.94	79.1	34.51	75.9	33.84	74.4	42.07	92.6
Pantothenic Acid (Vitamin B–3) (mg)	6.07	13.4	5.59	12.3	5.63	12.4	8.11	17.8
Riboflavin (Vitamin B–2) (mg)	3.19	7	3.44	7.6	3.38	7.4	3.65	8

[A]Adapted by the author from *NECC Chicken and Turkey Rations*, prepared by the New England College Poultry Conference Board, by poultry specialists from, and distribution by, the New England Land-Grant Universities: University of Connecticut, Storrs; University of Maine, Orono; University of Massachusetts, Amherst; University of New Hampshire, Durham; University of Rhode Island, Kingston; and University of Vermont, Burlington.

[B]See footnotes following Table 7–23, p. 180.

[C]For preslaughter drug withdrawal times, see section following footnotes entitled "Regulations Regarding Feed Additives."

[D]Folacin may be required under certain conditions. It may be added at the rate of 1,000 mg per ton of feed.

FOOTNOTES FOR TABLES 7–19a, b, & c, 7–20, 7–21, 7–22, AND 7–23

[1]Wherever substitutions are made in the rations, the total nutrient content should be adjusted to meet established requirements.

[2]Two to four hundred pounds of coarsely ground wheat or yellow hominy may be used to replace an equal amount of corn. If wheat is used, add 200,000 IU of vitamin A for each 100 lb of corn removed.

[3]There is usually some loss of provitamin A activity in corn and alfalfa meal during storage. If stored ingredients are used, it may be advisable to increase the added vitamin A level of the ration by 1,000 or 2,000 IU/lb. This can be accomplished by increasing the recommended supplement by 2,000,000 or 4,000,000 IU/ton of feed.

[4]The added salt level should be reduced by the amount supplied by the fish meal and other by-product ingredients.

[5]Poultry by-product meal may be substituted for all of the meat and bone scrap and up to 50% of the fish meal. Correct for calcium and phosphorus loss due to substitutions of poultry by-product meal.

[6]Based on an 18.5% phosphorus product, steamed bone meal or defluorinated rock phosphate may replace the dicalcium phosphate on a phosphorus basis.

[7]Based on 35% calcium, low magnesium limestone.

[8]An antibiotic may be used in these rations at the level recommended by the manufacturer (see section following footnotes entitled "Regulations Regarding Feed Additives").

[9]1,2-dihydro-6-ethoxy-2,2,4-trimethylquinoline (ethoxyquin) is recommended in the chick starter, broiler and breeder rations at the 0.0125% level to help prevent the appearance of encephalomalacia (crazy chick disease). If desired, it, or an equivalent antioxidant, may be added to help prevent the oxidation of dietary components. Total ethoxyquin from all sources must not exceed 0.25 lb per ton.

[10]A coccidiostat or antihistomonal drug may be used in these rations, as required, at levels recommended by the manufacturer (see section following footnotes entitled "Regulations Regarding Feed Additives").

[11]This amount of manganese will be furnished by 0.5 lb of manganese sulfate or 0.21 lb manganous oxide (70% feeding grades). An equivalent amount of manganese may be added from other acceptable sources.

[12]Caution should be used when high potency vitamin mixes are involved. It is recommended that 10 lb be the minimum amount of any item added to a ton of feed to insure proper mixing. Thus, high potency vitamin, mineral, or drug mixes should be premixed with a carrier (such as corn meal) to such a dilution that 10 lb of the final mix will be added for each ton of feed mixed. Minerals and vitamins should not be premixed together.

[13]Available phosphorus has been taken as 30% of total phosphorus from plant sources for chicks, and 75% of total phosphorus from plant sources for adult birds. Phosphorus from other than plant sources is considered to be 100% utilized.

[14]For those persons wanting a specific restricted feeding program, specific programs are available from individual breeders or Extension specialists.

[15]This amount of manganese will be furnished by 0.7 lb manganese sulfate or 0.3 lb manganous oxide (70% feeding grades). An equivalent amount of manganese may be added from other acceptable sources.

[16]Stabilized fats may replace an equal amount of cereal grains to provide a higher energy level, control dust, and aid pelleting. Where maintaining body weight in layers is a problem, increase fat by 1 or 2% during the winter by replacing an equal amount of cereal grains.

[17]Approximately this amount of zinc will be furnished by 29 g of zinc carbonate or 20 g of zinc oxide. An equivalent amount of zinc may be used from other acceptable sources.

[18]Feed starting ration until birds are 35 days old.

[19]Based on 3-nitro-4 hydroxyphenylarsonic acid at a level of 45 g (0.1 lb) per ton. Other compounds that may be used at a level recommended by the manufacturer are sodium arsanilate or arsanilic acid (see section following footnotes entitled "Regulations Regarding Feed Additives").

[20]In the absence of alfalfa or if the birds are raised on wire, 2 g of vitamin K activity should be added. Values in the broiler rations are based on menadione. Other compounds supplying equivalent levels of vitamin K may be used.

[21]If an even 2,000 lb is desired, adjust by removing or adding ground yellow corn.

[22]May be fed with grain after 20 weeks.

[23]This amount of manganese will be furnished by approximately 0.3 lb of manganese sulfate or 0.13 lb of manganous oxide (70% feeding grades). An equivalent amount of manganese may be added from other acceptable sources.

[24]This amount of zinc will be furnished by approximately 53 g of zinc carbonate or 37 g of zinc oxide. An equivalent amount of zinc may be used from other acceptable sources.

[25]Federal law, which strictly regulates the addition of selenium to poultry rations, should be consulted. Selenium, as sodium selenite or sodium selenate, may be added to complete feed for chickens, turkeys, and ducks at a level 0.3 ppm.

[26]For heavy caged layer pullets we suggest feeding 18% protein 0–6 weeks, 14% protein 7–12 weeks, and 12% protein 13–20 weeks of age.

[27]Any discrepancies in calculated analysis that occur in the decimal part of the figures are due to rounding errors.

[28]These are not highly pigmented diets. The xanthophyll of natural ingredients is variable, so if more pigment is desired use a high potency source of xanthophyll.

REGULATIONS REGARDING FEED ADDITIVES

The Food and Drug Administration of the U.S. Department of Health and Human Services has published a series of regulations concerning the use of additives, such as arsenicals, antibiotics, coccidiostats and other drugs, in animal feeds. For information concerning the use of any additive, consult the feed tag or label. If you still have questions about proper use of the additive, especially in conjunction with other additives, see your veterinarian, feed dealer or drug supplier.

FEEDING DUCKS

Until 1975, the annual production of ducks in the United States was approximately 10 million. Then, between 1975 and 1985, annual production doubled. In 1985, 21.6 million ducks were marketed in the United States. In 1989, U.S. per capita consumption of duck was estimated to be 0.66 lb.

Ducks are grown successfully in two different types of environments: (1) in an open rearing system in which the growing house opens onto an exercise yard with water for wading or swimming; and (2) in a confinement system in which they are raised in environmentally controlled houses, with litter or with a combination of litter and wire floors.

Typically, ducks are provided with two or three feeds during the growing period: when only two feeds are provided during the growing period, a 22% protein starter ration is usually fed the first 2 weeks, followed by a grower-finisher ration. When three feeds are provided during the growing period, they consist of a 22% protein starter ration, an 18% protein grower ration, and a 16% protein finisher ration.

Fig. 7–26. Filling tray feeders for young ducklings. (Courtesy, Cherry Valley Farms, Ltd., Rothwell, England)

Fig. 7–25. White Pekin ducks in an open rearing system—on the range, at Ward Duck Co., La Puenta, California. (Courtesy, California Polytechnic State University, San Luis Obispo)

Fig. 7–27. White Pekin ducks swimming in a creek. (Photo by J. C. Allen & Son, Inc., West Lafayette, IN)

The following pointers are pertinent to feeding ducks:

1. Ducks should be fed pellets rather than mash. Use 1/8 in. pellets for starter rations, and 3/16 in. pellets for older ducks. Pellets will make for a saving of 15 to 20% in the feed required to produce a market duck.

2. Ducks are very susceptible to aflatoxicosis; so, monitoring feeds for aflatoxin is important.

3. Ducks are nearly as good foragers as geese.

4. Ducks should be ready for market between 7½ and 8 weeks of age.

5. When used, holding rations are designed to maintain breeding ducks from about 8 weeks of age until the breeding season commences, without their getting too fat. It is recommended that birds fed holding rations be limited to about ½ lb per bird per day.

6. When a holding ration is used, the breeder diet should be substituted for it about 4 weeks before eggs are desired for hatching purposes.

7. When feeding ducks, pellet quality, proper feather development, and limiting carcass fat disposition are concerns, in addition to proper growth, and satisfactory feed conversion.

8. Commercial ducks grow as rapidly and efficiently as commercial broilers.

The NRC requirements for ducks are presented in Tables 7–13 and 7–14, in the earlier section of this chapter headed, "National Research Council (NRC) Requirements."

Examples of three-phase grower rations, and of a breeder ration, for ducks are given in Table 7–24a, b, and c.

TABLE 7–24a
EXAMPLE MARKET DUCK AND BREEDER DUCK RATIONS[1, 2]

Ingredients and Analysis	Starter (0–2 weeks)		Grower (2–4 weeks)		Finisher (4–8 weeks)		Breeder	
	(lb)	(kg)	(lb)	(kg)	(lb)	(kg)	(lb)	(kg)
Ingredients:								
Yellow corn	1,209	548.9	1,420	644.8	1,489	676	1,309.5	594.5
Soybean meal, 48.5%	510	231.5	320	145.3	260	118.1	318	144.4
Meat and bone meal, 50%	80	36.3	80	36.3	80	36.3	76	34.5
Fish meal, 60%	56	25.4	65	29.5	50	22.7	60	27.3
Dried whey, delactosed	—	—	—	—	—	—	45	20.4
Animal-vegetable fat	50	22.7	30	13.6	40	18.2	—	—
Dicalcium phosphate	—	—	—	—	3	1.4	—	—
Limestone	13	5.9	10	4.5	8	3.6	112	50.8
Salt	7	3.2	6	2.7	6	2.7	6	2.7
Trace mineral mix (Table 7–24b)	2	0.9	2	0.9	2	0.9	2	0.9
Vitamin mix (Table 7–24c)	20	9.1	15	6.8	10	4.5	20	9.1
Methionine, hydroxy analogue	3	1.4	2	0.9	2	0.9	1.5	0.7
Pellet binder	50	22.7	50	22.7	50	22.7	50	22.7
Total	2,000	908	2,000	908	2,000	908	2,000	908
Calculated analysis:								
Metabolizable energy . . . . (cal./lb)	1,400		1,425		1,450		1,300	
Metabolizable energy . . . . *(cal./kg)*	*3,087*		*3,142*		*3,197*		*2,867*	
Protein . . . . (%)	21.4		18		16.4		17.6	
Fat . . . . (%)	5.6		5		5.5		3.3	
Fiber . . . . (%)	2.6		2.6		2.6		2.4	
Calcium . . . . (%)	0.85		0.8		0.75		2.75	
Available phosphorus . . . . (%)	0.4		0.4		0.4		0.4	
Total phosphorus . . . . (%)	0.6		0.58		0.57		0.57	
Ash . . . . (%)	5.3		4.7		4.4		10.1	

Footnotes Table 7–24c.

TABLE 7–24b
THE TRACE MINERAL MIX[1]

Mineral	Percent per Kg of Mix
	(%)
Trace or microminerals:	
Copper (Cu)	0.30
Iodine (I)	0.06
Iron (Fe)	3.00
Manganese (Mn)	6.50
Zinc (Zn)	6.50

Footnotes Table 7–24c.

TABLE 7–24c
THE VITAMIN MIX[1]

Vitamin	Amount per Lb or Kg of Mix	
	(lb)	(kg)
Fat-soluble vitamins:		
Vitamin A . . . . (IU)	400,000	880,000
Vitamin D₃ . . . . (ICU)	60,000	132,000
Vitamin E . . . . (IU)	500	1,100
Vitamin K (msb)[3] . . . . (mg)	200	440
Water-soluble vitamins:		
Choline . . . . (mg)	13,018	28,639.6
Niacin (Nicotinic Acid, Nicotinamide) . . (mg)	2,500	5,500
Pantothenic Acid (Vitamin B–3) . . . . . (mg)	300	660
Riboflavin (Vitamin B–2) . . . . (mg)	300	660
Vitamin B–12 (Cobalamins) . . . . (mg)	0.4	0.88

Footnotes for Tables 7–24a, 7–24b, and 7–24c.

[1]From: *Complete Duck Grower and Breeder Rations*, Purdue University, West Lafayette, IN.

[2]For best results, all rations should be pelleted.

[3]Menadione sodium bisulfite.

FEEDING GEESE

Geese are very hardy, highly disease resistant, are the closest grazers known, and can live almost entirely on good pasture. Yet, the production of geese for meat purposes has never enjoyed the popularity in the United States that it has in some European countries.

Geese are raised under the following variety of feeding programs:

1. The production of *farm geese*, with the goslings given a starter feed for about 2 weeks, followed by foraging the farm for a variety of pasture and grain feedstuffs; then, marketed at about 18 weeks of age after liberal grain feeding for the last 2 or 3 weeks.

2. The goslings are limit-fed prepared feed throughout the growing period, but are allowed considerable foraging in addition; then, marketed at about 14 weeks of age following liberal feeding of a high-energy finishing ration.

Fig. 7–28. Goose. (Courtesy, USDA)

3. The goslings are full-fed in confinement and marketed as *junior green geese* at about 10 weeks of age.

4. The raising and use of geese for weeding purposes. Weeder geese are used with great success to control and eradicate troublesome grass and certain weeds in a great variety of crops and plantings, including cotton, hops, mint, onions, garlic, strawberries, nurseries, corn, orchards, groves, and vineyards. The geese eat grass and young weeds as quickly as they appear, but they do not touch certain cultivated plants. They will work continuously from daylight to dark, 7 days a week (even on bright moonlit nights) nipping off the grass and weeds as promptly as new growth appears.

At the end of the weeding season, geese are generally brought from the field and placed in pens for fattening for 3 or 4 weeks, until they weigh 10 to 12 lb or more. Markets are highest during the 4 to 6 weeks prior to Thanksgiving and Christmas.

The carrying of geese over from one season to the next for weeding purposes is not recommended, because older geese are less active in hot weather than young birds.

5. The production in some European countries of goose liver for *pate de foie gras*. In this program, the geese are grown to about 12 weeks of age, following which they are force-fed a high-grain ration for the production of livers of high-fat content.

For breeding purposes, geese are fed a prebreeding (holding) ration beginning 6 to 8 weeks before the breeding season, followed by a breeding ration formulated for the intensive production of fertile eggs.

The following additional pointers are pertinent to feeding geese:

1. Rations for geese should be pelleted, with ³/₃₂- or ³/₁₆-in. pellets preferred. Mash and crumbles cause too much feed wastage and should not be used.

2. Although all rations may be home-mixed, a commercially prepared ration is recommended for young goslings and breeders during the laying season.

3. Succulent green feed should provide the bulk of the ration for young growing geese.

The NRC requirements for geese are presented in Table 7–15, in the earlier section of this chapter headed, "National Research Council (NRC) Requirements."

Example rations for geese are given in Table 7–25.

TABLE 7–25
EXAMPLE GEESE RATIONS[1]

Ingredient	Starter 0–3 Weeks		Grower 3 Weeks to Market		Breeder Layer	
	(lb)	(kg)	(lb)	(kg)	(lb)	(kg)
Ground yellow corn	600	272.3	600	272.3	300	136.2
Ground oats	400	181.5	500	227	700	317.8
Ground barley	375	170.2	375	170.2	210	95.3
Dehydrated alfalfa	72	32.6	72	32.6	250	113.5
Soybean oil meal, 44% protein	500	227	400	181.5	450	204.3
Limestone	20	9.1	20	9.1	50	22.7
Dicalcium phosphate	20	9.1	20	9.1	25	11.4
Iodized salt	10	4.5	10	4.5	10	4.5
Vitamin premix	3	1.7	3	1.7	5	2.3
Total	2,000	908	2,000	908	2,000	908

[1]From: *Goose Production in North Dakota*, North Dakota State University, Fargo.

FEEDING PIGEONS

Pigeons are raised primarily for sport and hobby — being widely used in racing, shows, and training to perform tricks. But many people consider pigeon (squab) to be an epicurean delight, and a limited demand for pigeon meat exists. Squabs are marketable as early as 28 days of age, at which time their dressed weight is about 1 lb. At this age, squabs are tender and self-basting due to the fat under the skin.

Pigeons grow more rapidly than most birds during the first 20 days of life. They receive their first nourishment from "pigeon milk" regurgitated from the parent pigeon's crop. Pigeon milk is a thick, creamy, semi-digested substance high in protein and fat, but low in carbohydrates. When 20 to 40

Fig. 7–29. Pigeons at feed. (Courtesy, Ralston Purina Company, St. Louis, MO)

days of age, squabs may be fed a pigeon feed. Unlike other forms of poultry, pigeons will not eat mash, so pigeon feed either consists of whole or cracked grains or commercially prepared pellets.

Most pigeon producers feed (1) a complete pelleted ration, or (2) a complete pelleted feed plus whole or cracked grain. The most common grains are corn, wheat, sorghum, and peas. Grains can be offered in an open trough or cafeteria style where the self-feeder has individual compartments for each type of grain. If an open trough is used, it is recommended that pigeons be fed twice daily. Only enough feed should be offered at each feeding period as will be eaten in 1 hour.

Commercial pigeon feeds are available. But the fancier may prepare a suitable ration of grains, plus a free-choice mineral mixture, similar to the ration shown in Table 7–26.

TABLE 7–26
EXAMPLE RATION FOR PIGEONS[1]

Grain Mix (Whole Grain)	Amount	
	(lb)	(kg)
Yellow corn	35	15.8
Grain sorghum	20	9.1
Cowpeas or field peas	20	9.1
Wheat	15	6.8
Oat groats	5	2.3
Hempseed	5	2.3
Total	100	45.4
Mineral mix (fed in separate hopper free-choice):		
Medium-sized ground oyster shells	50	22.7
Grit (appropriate size)	25	11.3
Bone meal or dicalcium phosphate	20	9.1
Salt	5	2.3
Total	100	45.4

[1]Adapted by the author from *Managing Game Birds*, p. 6, Table 2, published by Michigan State University, East Lansing.

FEEDING GAME BIRDS: PHEASANTS, BOBWHITE QUAIL, JAPANESE QUAIL, CHUKARS, PARTRIDGES, GROUSE, AND DOVES

Game birds may be propagated for many different purposes. Fanciers may keep game birds as pets. Some growers produce dressed birds, especially pheasants and Bobwhite quail, for specialty markets in stores and restaurants. Other growers produce pheasants, Bobwhite quail, chukars and various other types of partridges and/or grouse for game release farms (shooting preserves). Still others produce birds for release to the wild.

Japanese quail are fed for egg production. They mature at 5 to 6 weeks of age and lay up to 300 eggs per year. As is true of other game birds, Japanese quail must be fed higher protein rations than chickens so as to achieve fast early growth.

The NRC requirements for pheasants and Bobwhite quail are presented in Table 7–16, and the NRC requirements for Japanese quail are given in Table 7–17, in an earlier section of this chapter headed, "National Research Council (NRC) Requirements."

Commercial game bird feeds are available in most areas. Generally, these are of the following types: *starter ration* containing 28% protein, for the first 6 weeks; *grower ration* containing 20% protein, for 7 to 14 weeks of age; and, depending on whether they are being marketed for game-release or for dressed game birds, *finisher ration* or *flight conditioner ration* containing 15% protein, from 15 weeks until market. Commercial game bird feeds may be in the form of mash, crumbles, or pellets. **Note:** Game birds require a higher level of protein in early life than chickens.

Example rations for home-mixing of game bird rations are given in Table 7–27. Grit should be available in a separate feeder, with the size grit determined by the size of the birds.

TABLE 7–27
EXAMPLE RATIONS FOR GAME BIRDS

Ingredient and Analysis	Starter	Grower	Breeder
	(%)	(%)	(%)
Alfalfa meal, sun-cured	7.5	5.5	5
Corn, yellow, ground	14	44	56.7
Meat and bone meal	7	0	0
Fish meal	7	0	0
Sorghum, ground	8	0	0
Soybean meal, 44% protein	41	35	14.7
Wheat, ground	12	0	0
Wheat middlings	2	0	0
Wheat bran	0	12	16.8
Limestone, ground	0	1.0	4.1
CaHPO$_4$•2H$_2$O	0	1.5	1.5
Salt, iodized	0.7	0.4	0.5
DL-methionine	0.3	0.1	0.2
Premix[1]	0.5	0.5	0.5
Calculated Analysis:			
Protein .... (%)	28.1	20.8	15.1
Metabolizable energy (kcal/kg) or (kcal/2.2 lb)	2,720	2,660	2,570
Calcium .... (%)	1.0	0.94	2.15
Total phosphorus .... (%)	0.76	0.76	0.74
Available phosphorus .... (%)	0.52	0.45	0.44

[1]Premix should contain: in mg per kg (or per 2.2 lb) diet—MnSO$_4$• H$_2$O, 40; ZnO, 60; vitamin B–12, (cobalamins) 0.005; menadione sodium bisulfite, 2; riboflavin (vitamin B–2), 6; niacin (nicotinic acid, nicotinamide), 40; calcium pantothenate, 20; folacin (folic acid), 0.5; antioxidant, 100; antibiotic, 10; in IU—vitamin A, 5,000; vitamin D$_3$, 1,500; vitamin E, 20. An equivalent commercial premix can be used, but follow the directions of the supplier.

Specific information relative to feeding pheasants, Bobwhite quail, and Japanese quail follows:

■ **Feeding pheasants** — Following the starter phase, a grain supplement consisting of equal portions of (1) corn or grain sorghum, and (2) oats, barley, or wheat can be effectively used in limited amounts; but the grain supplement should not constitute more than one-fourth of the entire ration. Pheasants can be fed and managed to produce fertile eggs at any time of the year. Egg production requires proper feeds and light stimulation.

■ **Feeding Bobwhite quail** — Wild quail survive on native grass seeds and insects. Confinement-reared quail require nutritionally balanced rations to promote growth and health. After 6 weeks of age, Bobwhites should be fed according to whether they will be utilized as breeders, flight conditioned for shooting preserves, or processed for meat purposes.

■ **Feeding Japanese quail** — Laying rations for Japanese quail should contain about 25% protein; additionally, a free-choice supply of calcium (limestone or oystershell) should be available. Some growers may wish to supplement commercial feeds with small seeds or cracked grain. When Japanese quail are fed whole seeds, fine grit should be provided.

FEEDING GUINEA FOWL

In recent years, there has been a growing interest, worldwide, in the production of guinea fowl. In France and Italy, the commercial production of guinea fowl is a highly profitable industry.

Hens lay for 35 weeks and produce 175 eggs. Baby guineas are called *guinea poults* or *keets*. At 86 days of age, the keets weigh about 3.3 lb, made with a feed conversion of 2.7 to 3.2.

A three-phase feeding program is normally followed, consisting of a 24% protein starter, a 19–21% grower, and a 17–18% finisher.

FEEDING OSTRICHES

Feeding ostriches is different! The ostrich is the largest bird in the world. At maturity, it may stand 10 ft tall and weigh more than 330 lb. Young ostriches grow very rapidly; they reach full size in about 6 months, but they do not attain sexual maturity until 2 years of age. They may live to 70 years of age. The ostrich is the only bird that eliminates urine and feces separately.

Ostriches are valued primarily for their skins, which are made into fine quality leather. Spasmodically, the plumes are popular for decorations and accessories. The eggs may be used for human food, but the meat is seldom consumed.

In their native habitat, ostriches feed on succulent plants, fruit, and leaves, as well as on occasional insects, lizards, birds, mice, and turtles. They also eat much sand and gravel to aid in grinding food for digestion. Ostriches drink water when they can find it, but they can survive for long periods without drinking if the plants they eat are green and moist.

Ostrich farms in the United States are relatively new; so, it remains to be seen whether ostriches in this country will be a passing fad or an infant industry.

Good nutrition is essential to the success of ostrich production. Because of the rapid growth of ostrich *chicks* (they reach full size at about 6 months of age), they must receive a diet that is high in protein, and that contains the essential amino acids, along with adequate energy, minerals, and vitamins. Also, the nutrition of breeders affects egg production and hatchability.

The nutritive requirements of ostriches and turkeys appear to be very similar. So, the rations presented in Table 7–23, Example Turkey Rations, may be used for ostriches.

Because of the very rapid growth of ostrich chicks, however, it is suggested that they be continued on the high-protein turkey *starter ration* given in Table 7–23 longer than the 8 weeks suggested for turkey poults.

FEEDING PEAFOWL

The peafowl belongs to the same family as pheasants and chickens, differing primarily in plumage. Although they are prized as ornamental birds (blue, white, black, or green—with blue most common), peafowl are edible and are regarded as a delicacy on special occasions, perhaps more for rareness than for taste.

The care and management of peafowl is similar to that of turkeys. Peachicks may be fed a high-protein (28 to 30%) turkey or game-bird starter feed, preferably crumbles. At about 6 to 8 weeks of age, a game bird grower diet may be substituted for the starter, and small amounts of cracked corn, wild bird seed, or chopped green grass (lawn clippings) may be added to the ration.

When roaming free, adult peafowl eat a variety of seeds, insects, and plants. Additionally, they should be provided some turkey or game-bird feed, bird seed, or grain, with the allowance increased in cold weather.

SALMONELLA/TOXIC INORGANIC ELEMENTS

Salmonella and/or toxic inorganic elements may be problems in poultry.

■ **Salmonella**—The bacteria that cause the disease Salmonellosis in animals and humans have been around for a very long time. The name *Salmonella* is after the American bacteriologist and veterinarian, Daniel E. Salmon, who first isolated the bacteria in 1885. In recent years, *Salmonella* has been much in the news, primarily because new and sophisticated equipment has enabled scientists more minutely to screen the microbe field, and more intense effort has been made to look for microbes.

Salmonella is a family of bacteria that consists of more than 2,000 different strains, which may be found in feeds—especially animal by-products (meat meal, poultry by-product meal, and fish meal), and in human foods—including broilers, turkeys, and eggs. The disease, known as Salmonellosis, may occur if foods contaminated with *Salmonella* are eaten raw, not properly cooked, or mishandled after cooking. The symptoms of the illness are diarrhea, nausea, vomiting, and sometimes fever. The illness may occur within 6 to 72 hours after eating the contaminated food and may last 2 to 6 days.

Salmonella in human foods can be destroyed by proper cooking at a temperature of 160°F, or more, at the centermost part of the thickest item being cooked. The bacteria may also be inactivated by treating dressed poultry with 1% lactic acid, but it will color the meat slightly; and the bacteria may be completely destroyed by irradiation, but such treatment is not approved by FDA, presently. So, consumers should be urged (1) to refrigerate all animal products, (2) to fully cook all foods of animal origin, and (3) to avoid the recontamination of all foods after cooking. Additionally, producers should use all animal drugs and medications in compliance with FDA regulations and in accord with the directions on the label; and all producers and processors should continue their vigilance in reducing *Salmonella* in all feeds and facilities by rigid sanitation and heat treatment, and by preventing humanly edible animal products from becoming contaminated by *Salmonella*.

Despite all the scare stories, however, American consumers have the blessed assurance that they enjoy the safest and most abundant high-quality animal products in the world.

■ **Toxic inorganic elements**—Poultry are susceptible to a number of toxins, any one of which may prove disastrous in a flock. Among them are a number of inorganic elements such as arsenic, lead, and selenium. It is important, therefore, that the poultry producer guard against toxic levels of inorganic elements.

Fig. 7–30. Turkeys on the range. Note self-feeders on the right. (Courtesy, *Turkey World*, Mount Morris, IL)

QUESTIONS FOR STUDY AND DISCUSSION

1. How do you account for the fact that poultry feeding has changed more than the feeding of any other species?

2. Discuss the impact of lowering the feed required to produce eggs, broilers, and turkeys from the standpoint of (a) the poultry producer and (b) the consumer.

3. List the factors involved in formulating poultry rations.

4. How are the nutritive requirements for a specific substance determined for growth, efficiency of feed utilization, egg production, and hatchability?

5. What are feeding standards? How are they expressed?

6. What are the NRC requirements? How are they used?

7. What is the difference between (a) nutrient requirements and (b) nutrient allowances?

8. Explain how the following forces may affect the nutritive requirements of poultry: (a) temperature and humidity, and (b) genetic differences.

9. When fed free-choice, birds tend to eat to satisfy their energy requirements. How does this affect balancing of poultry rations?

10. Before attempting to balance a ration, what major points should be considered?

11. List the four steps which should be taken in an orderly fashion in order to formulate an economical poultry ration.

12. List and discuss the five different methods of ration formulation.

13. More than 40% of the primary (complete) commercial feed used in the United States is fed to poultry. Why is so much U.S. commercially-prepared feed fed to poultry?

14. Describe *phase feeding* and *molting* of layers. Why do so many poultry producers apply these practices?

15. How do the nutritive requirements for breeder chickens differ from those for commercial layers?

16. List and discuss three methods of restricted (limited) feeding commonly employed in feeding pullets of the heavy breeds.

17. In modern broiler production, what is considered average (a) market weight, (b) market age, (c) feed conversion, and (d) mortality?

18. What are the primary differences between feeding market turkeys and feeding turkey breeders?

19. Why do the protein requirements of turkeys from one day of age to marketing differ so widely?

20. What forces caused the production of ducks in the United States to double from 1975 to 1985?

21. Discuss the two major types of environments in which ducks are grown.

22. List and discuss the variety of programs under which geese are raised in the United States.

23. What is pigeon milk?

24. Outline a feeding program for a game bird farm raising pheasants, Bobwhite quail, Japanese quail, and other game birds.

25. Discuss the feeding of guinea fowl, ostriches, and peafowl.

26. Discuss the importance of *toxic inorganic elements* and *Salmonella*, from the standpoints of (a) poultry and (b) people consuming poultry products.

27. Is the farm poultry flock a thing of the past? If so, why?

SELECTED REFERENCES

Title of Publication	Author(s)	Publisher
Animal Feeding and Nutrition, Fifth Edition	M. H. Jurgens	Kendall/Hunt Publishing Company, Dubuque, IA, 1982
Animal Nutrition, Seventh Edition	L. A. Maynard, *et al.*	McGraw-Hill Book Company, Inc., New York, NY, 1979
Animal Nutrition, Second Edition	P. McDonald R. A. Edwards J. F. D. Greenhalgh	Oliver & Boyd, Edinburgh, Scotland, 1973
Animal Science, Ninth Edition	M. E. Ensminger	Interstate Publishers, Inc., Danville, IL, 1991
Basic Animal Nutrition and Feeding, Second Edition	D. C. Church W. G. Pond	John Wiley & Sons, Salt Lake City, UT, 1984
Calcium in Broiler, Layer, and Turkey Nutrition	R. H. Harms, *et al.*	National Feed Ingredients Association, West Des Moines, IA, 1976
Commercial Chicken Production Manual, Fourth Edition	M. O. North D. D. Bell	Van Nostrand Reinhold Co., New York, NY, 1990
Feed Formulations, Second Edition	T. W. Perry	The Interstate Printers & Publishers, Inc., Danville, IL, 1975
Feeds and Feeding, Third Edition	A. Cullison	Reston Publishing Company, Inc., Reston, VA, 1982
Feeds & Nutrition, Second Edition	M. E. Ensminger J. E. Oldfield W. W. Heinemann	The Ensminger Publishing Company, Clovis, CA, 1990
Feeds & Nutrition Digest	M. E. Ensminger J. E. Oldfield W. W. Heinemann	The Ensminger Publishing Company, Clovis, CA, 1990
Feeds for Livestock, Poultry, and Pets	M. H. Gutcho	Noyes Data Corporation, Park Ridge, NJ, 1973
Linear Programming Application to Agriculture	R. R. Beneke R. Winterboer	Iowa State University Press, Ames, IA, 1973
Manual of Poultry Production in the Tropics	R. R. Say	CAB International, Wallingford, Oxon, United Kingdom, 1987
Nutrient Requirements of Poultry and Nutritional Research	C. Fisher K. N. Boorman	Butterworths, Kent, England, 1986
Nutrient Requirements of Poultry, Eighth Revised Edition	National Research Council	National Academy of Science, Washington, DC, 1984
Nutrition of the Chicken, Third Edition	M. L. Scott M. C. Nesheim R. J. Young	M. L. Scott and Associates, Ithaca, NY, 1982
Nutrition of the Turkey	M. L. Scott	M. L. Scott of Ithaca, Ithaca, NY, 1987
Phosphorus in Poultry and Game Bird Nutrition	P. E. Waibel, *et al.*	National Feed Ingredients Association, West Des Moines, IA, 1977
Poultry: Feeds and Nutrition, Second Edition	H. Patrick P. J. Schaible	The AVI Publishing Company, Inc., Westport, CT, 1980
Poultry Meat and Egg Production	C. R. Parkhurst C. J. Mountney	Van Nostrand Reinhold Company, New York, NY, 1988
Poultry Nutrition, Fifth Edition	W. R. Ewing	The Ray Ewing Company, Pasadena, CA, 1963
Poultry Production, Twelfth Edition	M. C. Nesheim R. E. Austic L. E. Card	Lea & Febiger, Philadelphia, PA, 1979
Raising Poultry Successfully	W. Graves	Williamson Publishing, Charlotte, VT, 1985
Scientific Feeding of Chickens, The, Fifth Edition	H. W. Titus J. C. Fritz	The Interstate Printers & Publishers, Inc., Danville, IL, 1971

Fig. 8–1. Debeaking and vaccination line. (Courtesy, Foster Farms, Livingston, CA)

POULTRY MANAGEMENT

CHAPTER

Management is the art of caring for, handling, or controlling. In a poultry operation, it gives point and purpose to everything else.

Pertinent facts relative to, along with methods of accomplishing, some important poultry management practices are covered in this chapter.

THE POULTRY ENTERPRISE

Generally speaking, when launching a poultry enterprise the owner must (1) choose between species and between specializing or integrating, (2) decide on the location, and (3) determine the size.

CHOICE OF SPECIES, AND SPECIALIZED VS INTEGRATED

Producers may choose among layers, broilers, turkeys, ducks, geese, or game birds.

Additionally, the large commercial operators usually have several options; and wise owners/managers will choose among them. They may choose (1) between breeding, hatchery, market eggs or meat production, or processing; or (2) between specializing or integrating—for example, they may choose between (a) specializing in either growing broilers or processing them, or (b) integrating the two types of operations.

LOCATION

Since people usually do those things which are most profitable to them, and since poultry producers are people, relative profitableness in comparison with other types of enterprises must be the chief reason why, geographically, the following states, or areas, lead in poultry production, by a considerable margin: (1) why California is the leading egg-producing state of the nation, followed by Indiana, Pennsylvania, Ohio, Georgia, and Arkansas; (2) why Arkansas, Georgia, Alabama, North Carolina, and Mississippi completely dominate broiler production; and (3) why North Carolina, Minnesota, and California hold a sizeable lead on number of turkeys produced.

Usually a combination of economic conditions, and not net returns alone, suggests the location best adapted to a specific type of poultry enterprise; among them, (1) availability of birds, (2) markets, (3) labor, (4) feed, (5) climate, and (6) the educational and promotional value gained from being in an area where others are engaged in similar enterprises.

So, when launching a commercial enterprise, generally owners select what they consider to be the most desirable location.

SIZE

Many poultry operations are less profitable than they should be because they are too small. An owner who is trying to make a living from a poultry enterprise alone must have a sizable flock and assume considerable risk. Economic studies have shown, almost without exception, that net returns are in direct proportion to the size of flock. Of course, in a bad year the bigger the flock, the greater the potential losses.

Also, it should be recognized that the bigger and the more complicated the operation (integrated operations are more complicated than specialized operations, for example), the more competent the management required. This point merits emphasis because, currently, (1) bigness is a sign of the times in the poultry business, and (2) the most common method of attempting to "bail out" of an unprofitable poultry business is to increase its size. Although it is easier to achieve efficiency of equipment, labor, purchases, and marketing in big operations, bigness alone will not make for greater efficiency. Management is still the key to success.

So, the eventual size of the operation is usually limited only by the ability of the owner/manager and its profitableness.

MANAGER

Four major ingredients are essential to success in the poultry business: (1) good birds, (2) good feeding, (3) good management, and (4) good records. A manager can make or break any poultry enterprise. Unfortunately, this fact is often overlooked in the present era, primarily because the accent is on scientific findings and automation.

In manufacturing and commerce, the importance and scarcity of top managers are generally recognized and reflected in the salaries paid to persons in such positions. Unfortunately, agriculture as a whole has lagged; and altogether too many owners still subscribe to the philosophy that the way to make money out of the poultry business is to hire a manager cheap, with the result that they usually get what they pay for—a "cheap" manager.

TRAITS OF A GOOD MANAGER

There are established bases for evaluating many articles of trade, including hay and grain. They are graded according to well-defined standards. Additionally, they are chemically analyzed and evaluated in feeding trials. But no such standard or system of evaluation has evolved for poultry managers, despite their acknowledged importance.

The author has prepared the "Poultry Manager Checklist," given in Table 8–1, which (1) students may use for guidance as they prepare themselves for managerial positions, (2) employers may find useful when selecting or evaluating a manager, and (3) managers may apply to themselves for self-improvement purposes. No attempt has been made to assign a percentage score to each trait, because this will vary among poultry establishments. Rather, it is hoped that this checklist will serve as a useful guide (1) to the traits of a good manager and (2) to what the boss wants.

TABLE 8–1
POULTRY MANAGER CHECKLIST

☐ CHARACTER—
Absolute sincerity, honesty, integrity, and loyalty; ethical.

☐ INDUSTRY—
Work, work, work; enthusiasm, initiative, and aggressiveness.

☐ ABILITY—
Poultry know-how and experience; business acumen—including ability systematically to arrive at the financial aspects and convert this information into sound and timely management decisions; knowledge of how to automate and cut costs; common sense; organization; imagination; growth potential.

☐ PLANS—
Sets goals, prepares organization chart and job description, plans work, and works plans.

☐ ANALYZES—
Identifies the problem, determines pros and cons, then comes to a decision.

☐ COURAGE—
To accept responsibility, to innovate, and to keep on keeping on.

☐ PROMPTNESS AND DEPENDABILITY—
A self-starter; has "T.N.T.," which means that the work is done "today, not tomorrow."

☐ LEADERSHIP—
Stimulates subordinants and delegates responsibility.

☐ PERSONALITY—
Cheerful, not a complainer.

JOB DESCRIPTIONS OF NEW-LAY RANCH[1]

Owner	General Manager
Responsible for:	Responsible for:
1. Making policy decisions. 2. Borrowing capital. 3. Preparing proposed long-term plan. 4. (List others.)	1. Supervising all staff. 2. Budgets. 3. Buying chicks (replacement pullets), feed, equipment, and supplies. 4. Marketing eggs and cull hens. 5. (List others.)
Replacement Pullet Manager	**Layer Manager**
Responsible for:	Responsible for:
1. Taking delivery on baby chicks. 2. Feeding, care, and management of replacement pullets. 3. Sanitation. 4. Vaccination. 5. Lighting. 6. Moving the pullets.	1. Health of the birds. 2. Culling. 3. Gathering, processing, and storing eggs. 4. Spraying. 5. Litter, manure disposal. 6. Ventilation. 7. Meeting egg production and feed conversion goals, and keeping mortality down.

[1]Job description columns, similar to those above, should be added for the Marketing Manager and Office Manager, and for such other major positions as exist or are planned.

ORGANIZATION CHART AND JOB DESCRIPTION

It is important that all workers know to whom they are responsible and for what they are responsible; and the bigger and more complex the operation, the more important this becomes. This should be written down in an organization chart and a job description. Samples of an organization chart and a job description follow.

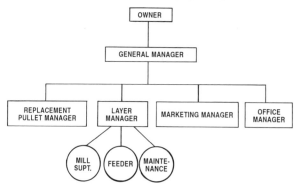

ORGANIZATION CHART OF NEW-LAY RANCH

AN INCENTIVE BASIS FOR THE HELP

Big poultry establishments must rely on hired labor, all or in part. Good help—the kind that everyone wants—is hard to come by; it is scarce, in strong demand, and difficult to keep. And the agricultural labor situation is going to become more difficult in the years ahead. There is need, therefore, for some system that will (1) give a big assist in getting and holding top-flight help, and (2) cut costs and boost profits. An incentive basis that makes hired help partners in profit is the answer.

The various plans in Table 8–2 (next page) are intended as guides only.

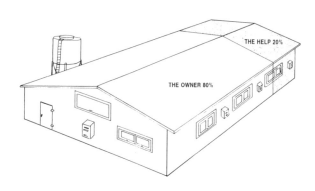

Fig. 8–2. A good incentive basis makes hired help partners in profit.

TABLE 8–2
INCENTIVE PLANS FOR POULTRY ESTABLISHMENTS

Types of Incentives	Pertinent Provisions of Some Known Incentive Systems in Use	Advantages	Disadvantages	Comments
1. **Bonuses**	A flat, arbitrary bonus; at Christmastime, year-end, quarterly, or other intervals. A tenure bonus such as (1) 5 to 10% of the base wage or 2 to 4 weeks additional salary paid at Christmastime or year-end, (2) 2 to 4 weeks vacation with pay, depending on length and quality of service, or (3) a certain sum of money/week set aside and paid if employee stays on the job a specified time.	It's simple and direct.	Not very effective in increasing production and profits.	
2. **Equity-building plan**	Employee is allowed to own a certain number of birds, which are usually fed without charge.	It imparts pride of ownership to the employee.	The hazard that the owner may feel that employees accord their birds preferential treatment; suspicioned if not true.	
3. **Production sharing**	Specified compensation per: (1) 100 dozen eggs produced, (2) 100 lb of broiler, or (3) 100 lb of market turkey.	It's an effective way to achieve higher production.	Net returns may suffer. For example, faster gains and heavier weights in broilers than is economical may be achieved by feeding more concentrated and expensive feeds than are practical. This can be alleviated by (1) specifying the ration and (2) setting an upper limit on the weight to which the incentive will apply. If a high performance level already exists, further gains or improvements may be hard to come by.	Incentive payments for production above certain levels are more effective than paying for all units produced.
4. **Profit sharing:** a. Percent of gross income b. Percent of net income	1 to 2% of the gross. 10 to 20% of net after deducting all costs.	It's a good plan for a hustler. Net income sharing works better for managers and supervisors than for common laborers because fewer hazards are involved in opening up the books to them. It's an effective way to get hired help to cut costs.	Percent of gross does not impart cost of production consciousness. Both (1) percent of gross income and (2) percent of net income exposes the books and accounts to workers, who may not understand accounting principles. This can lead to suspicion and distrust. Controversy may arise (1) over accounting procedures—for example, from the standpoint of the owner a fast tax write-off may be desirable on new equipment but this reduces the net shared with the worker and (2) because some owners are prone to overbuild and overequip, thereby decreasing net.	There must be prior agreement on what constitutes gross or net receipts, as the case may be, and how it is figured.
5. **Production sharing and prevailing price**	Establish break-even point(s), then split profit beyond this point on the basis of (1) 80% (owner) and 20% (help), or (2) an escalator arrangement, giving help greater percentage as profits rise. For example, the break-even points, above which profits would be split, might be— 1. Layers: 234 eggs/hen/year; feed conversion less than 4.25 lb/doz eggs; and mortality under 10%/year. 2. Broilers: weight production of 4.0 lb in 7 weeks; feed conversion less than 2.0 lb; and mortality under 1.5%.	It embraces the best features of both production sharing and profit sharing, without the major disadvantages of each. It (1) encourages high productivity and likely profits, (2) is tied in with prevailing prices, (3) does not necessitate opening the books, and (4) is flexible—it can be split between the owner and employee on any basis desired, and the production part can be adapted to a sliding scale or escalator arrangement.	It is a bit more complicated than some other plans, and it requires more complete records.	When properly done, and all factors considered, this is the most satisfactory incentive basis for a poultry enterprise.

Many manufacturers have long had an incentive basis. Executives are frequently accorded stock option privileges, through which they prosper as the business prospers. Common laborers may receive bonuses based on piecework or quotas (number of units, pounds produced). Also, most factory workers get overtime pay and have group insurance and a retirement plan. A few industries have a true profit-sharing arrangement based on net profit as such, a specified percentage of which is divided among employees. No two systems are alike. Yet, each is designed to pay more for labor, provided labor improves production and efficiency. In this way, both owners and laborers benefit from better performance.

Family-owned and family-operated poultry farms have a built-in incentive basis; there is pride of ownership, and all members of the family are fully cognizant that they prosper as the business prospers.

Many different incentive plans can be, and are, used. There is no best one for all operations.

The incentive basis chosen should be tailored to fit the specific operation; with consideration given to kind and size of operation, extent of owner's supervision, present and projected productivity levels, mechanization, and other factors.

For most poultry operations, the author favors a "production sharing and prevailing price" type of incentive.

INDIRECT INCENTIVES

Normally, we think of incentives as monetary in nature — as direct payments or bonuses for extra production or efficiency. However, there are other ways of encouraging employees to do a better job. The latter are known as indirect incentives. Among them are: (1) good wages; (2) good labor relations; (3) adequate house plus such privileges as the use of the farm truck or car, payment of electric bill, use of a swimming pool, hunting and fishing, use of a horse, and furnishing eggs, meat, and milk; (4) good buildings and equipment; (5) vacation time with pay, time off, and sick leave; (6) group health insurance; (7) security; (8) the opportunity for self-improvement that can accrue from working for a top person; (9) the right to invest in the business; and (10) an all-expense-paid trip to a short course, show or convention. These indirect incentives will be accorded to the employees of more and more establishments, especially the big ones.

HOW MUCH INCENTIVE PAY?

After (1) reaching a decision to go on an incentive basis, and (2) deciding on the kind of incentive, it is necessary to arrive at how much to pay. Here are some guidelines that may be helpful in determining this:

1. Pay the going base, or guaranteed, salary; then add the incentive pay above this.
2. Determine the total stipend (the base salary plus incentive) to which you are willing to go.
3. Before making any offers, always check the plan on paper to see (a) how it would have worked out in past years based on your records, and (b) how it will work out as you achieve the future projected production.

REQUISITES OF AN INCENTIVE BASIS

Owners who have not previously had experience with an incentive basis are admonished not to start with any plan until they are sure of both their plan and their help. Also, it is well to start with a simple plan; then a change can be made to a more inclusive and sophisticated plan after experience is acquired.

Regardless of the incentive plan adopted for a specific operation, it should encompass the following essential features:

1. A good owner (or manager) and good workers. No incentive basis can overcome poor managers. They must be good supervisors and fair to their help. Also, on big establishments, they must prepare a written organization chart and job description so the employees know (a) to whom they are responsible, and (b) for what they are responsible. Likewise, no incentive basis can spur employees who are not able, interested, and/or willing. This necessitates that they be selected with special care where they will be on an incentive basis. Hence, the three — good owner (manager), good employees, and good incentive — go hand in hand.
2. It must be fair to both employer and employees.
3. It must be based on and make for mutual trust and esteem.
4. It must compensate for extra performance, rather than substitute for a reasonable base salary and other considerations (house, utilities, and certain provisions).
5. It must be as simple, direct, and easily understood as possible.
6. It should compensate all members of the team.
7. It must be put in writing, so there will be no misunderstanding. For example, in a production-sharing plan involving a market turkey operation, it should stipulate the ration (or who is responsible for ration formulation).
8. It is preferable, although not essential, that workers receive incentive payments (a) at rather frequent intervals, rather than annually, and (b) immediately after accomplishing the extra performance.
9. It should give the hired help a certain amount of responsibility, from the wise exercise of which they will benefit through the incentive arrangement.
10. It must be backed up by good records; otherwise, there is nothing on which to base incentive payments.
11. It should be a "two-way street." If employees are compensated for superior performance, they should be penalized (or, under most circumstances, fired) for poor performance. It serves no useful purpose to reward the unwilling, the incompetent, and the stupid. For example, no overtime pay should be given to employees who must work longer because of slowness or in order to correct mistakes of their making. Likewise, if the reasonable break-even point on a layer operation is 220 eggs per bird per year, and this production level is not reached because of obvious neglect, the employee(s) should be penalized (or fired).

CUTTING COSTS

There are two ways in which to increase profits in any business, including the poultry business: (1) increase returns, and/or (2) cut costs.

Generally speaking, an individual poultry producer can do very little to increase returns, because supply and demand determine prices. However, a producer can cut costs. Cutting on those items accounting for the largest percentage of cost of production offers the greatest possibility to lower costs. For example, cutting feed costs is extremely important because feed accounts for 55 to 75% of the total cost of production, with the production of eggs toward the lower side of the range and the production of broilers and turkeys toward the upper side. There are other species differences in the relative importance of cost of production items; hence, the major production costs by species follow:

■ **Major production costs of layers**—The highest layer cost items are feed, pullets, house and equipment, and labor. Normally, these four items account for 96% of the cost of producing eggs, with feed alone accounting for 55 to 75%.

■ **Major replacement pullet costs**—The principal items of cost in raising replacement pullets, ranked in order, are feed, chicks, and labor; with feed running about twice the cost of the chicks.

■ **Major broiler production costs**—The chief broiler production costs, ranked in order, are: feed, chicks, and labor; with feed accounting for about 70% of the total cost of production.

■ **Major turkey production costs**—The chief turkey production costs, ranked in order, are: feed, poults, and fixed costs (land, buildings and equipment, maintenance, taxes, and insurance).

SAVING BY AUTOMATING

Fig. 8–3. Automatic egg gathering system for caged layers—gentle and versatile. (Courtesy, A. R. Wood Mfg. Co., Luverne, MN)

Before adding expensive equipment, owners/managers should determine if it will pay. They should compare the cost of mechanization with its saving in labor. Here are some useful guidelines:

1. **Guideline No. 1.** *The break-even point on how much you can afford to invest in equipment to replace hired labor can be arrived at by the following formula:*

$$\frac{\text{Annual saving in hired labor from new equipment}}{\text{(divide by) .15}} = \begin{array}{l}\text{amount you can}\\\text{afford to invest}\end{array}$$

Example:
If hired labor costs $6,000 per year, this becomes—

$$\frac{\$6,000}{.15} = \begin{array}{l}\$40,000\text{, the break-even point}\\\text{on new equipment}\end{array}$$

Since labor costs are going up faster than machinery and equipment costs, it may be good business to exceed this limitation under some circumstances. Nevertheless, the break-even point, $40,000 in this case, is probably the maximum expenditure that can be economically justified at the time.

2. **Guideline No. 2.** Assuming an annual cost plus operation of equipment equal to 20% of new cost, the break-even point to justify replacement of one hired worker is as follows:

Example:

If annual cost of one hired worker is[1]	The break-even point on new investment is
$5,000 (20%) × 5	$25,000
6,000 (20%) × 5	30,000
7,000 (20%) × 5	35,000

Assume that the new cost of added equipment comes to $6,000, that the annual cost is 20% of this amount, and that the new equipment would save one hour of labor per day for 6 months of the year. Here is how to figure the value of labor to justify an expenditure of $6,000 for this item:

$6,000 (new cost) × 20% = $1,200
  1,200 (annual ownership use cost) ÷ 180 hours (labor saved) = $6.67 per hour.

So, if labor costs less than $6.67 per hour, you probably shouldn't buy the new item.

[1]This is assuming that the productivity of workers at different salaries

FORCED MOLTING

Molting is the act or process of shedding and renewing feathers. Hens usually molt in the following order: head, neck, body (including breast, back, and abdomen), wing, and tail.

Birds inherit the tendency to shed their plumage annually. Under normal conditions, an early molter is a poor layer; and a later molter is a good layer.

Hens seldom lay and shed feathers at the same time. A high-producing bird may, for a time, molt and lay simultaneously, but usually she is in declining production when molting begins.

For many years, forced molting has been used as a method of recycling hens for an extra season of production. In recent years, there has been increased interest in forced molting.

Forced or induced molting can be achieved by a variety of techniques. However, all methods aim at shutting off the production of eggs as rapidly as possible, giving the hens a suitable rest period, and having the hens return to maximum production. In most cases, molting is induced at the end of a production season. However, it can also be used at the beginning or in the middle of a reproductive season to delay or stop production. For example, the procedure can be used in the middle of a season when hens show very poor production due to disease or some other stress. The procedure may also have application in situations where marketing of eggs has been restricted due to accidental dietary contamination which requires a depletion of the hen for several weeks.

Molting is normally done to reduce replacement cost on a per dozen egg basis. In general, higher net replacement costs (initial cost less salvage) and lower egg prices tend to favor the use of forced molting.

(Also see Chapter 12, Layers, section on "Molting.")

CAPONIZING

A capon is a male bird with the reproductive organs removed. It bears the same relation to a cockerel as a steer does to a bull, a wether to a ram, or a gelding to a stallion. The purpose of caponizing is to improve the quality of the poultry meat; research shows that it is more tender, juicier, and flavorful than meat from comparable birds that haven't been caponized.

The art of caponizing is very old, dating back to the pre-Christian era. As early as 37 B.C., Cato and Varro referred to "the altered males called capons" in their book *Roman Farm Management*. Reaumur, in his book entitled *The Art of Hatching and Bringing Up Domestic Fowls*, published in 1750, referred to the fact that capons can be trained to care for young chicks.

The most satisfactory way to learn to caponize chickens is to get instruction and experience by working with an experienced operator. Birds weighing ¾ to 2 lb are about the correct size for caponizing. The operation is relatively simple and can be learned quickly by almost anyone. After the operation, capons should be placed in clean quarters, separated from other chickens for 2 to 3 days. After recovering from the initial shock of the operation, capons do not need special care.

(Also see Chapter 13, Broilers [Fryers], Roasters, and Capons, section headed "Capons.")

CANNIBALISM

Cannibalism is the eating of others of its kind. It can be encountered in birds of all ages. Among baby chicks, the trouble is usually confined to toe and tail picking. With mature birds, the vent, tail, and comb are the regions most frequently picked.

The cause of cannibalism is not fully understood. It is known that it is more frequent under confined conditions. The stresses of overcrowding, too much light, and an unbalanced diet are among the factors that can lead to cannibalism in chickens and other domestic fowl. Also, poor ventilation, heat, and injury or sickness can cause birds to peck and eat at each other.

The best time to deal with the problem is before it starts. Producers should be sure that birds have enough space, especially at feed and watering facilities, and when nesting. Lighting should neither be too bright nor kept on too long; red lights are a temporary solution where cannibalism is already started (with such lighting, the birds are not able to see the blood or wounds). Low intensity lights are the best permanent solution; and they should not be on longer than 17 hours per day.

Good ventilation of the poultry house can lower heat and remove foul air, lessening the stress in the birds. Any diet in which there is too much high-energy feed and too little fiber can also lead to cannibalism. The birds should be watched closely and any injured, sick, or crippled individuals should be removed immediately.

Debeaking should be a normal practice where the birds are confined.

DEBEAKING

Debeaking is the practice of removing the pinching and tearing part of a bird's beak in order to lessen cannibalism and feather picking. It is a widely accepted practice in the poultry industry.

Beak trimming is a precision operation and experience is a great asset in doing it properly. Moreover, there are advantages and disadvantages to beak trimming, but the advantages far outweigh the disadvantages.

Machines for debeaking are in common use in most poultry areas. Most hatcheries and poultry equipment supply houses have them for sale or can get them. Hatcheries, veterinarians, and poultry service establishments will usually either rent the equipment or do the work at low cost.

Some producers recommend debeaking day-old chicks at the hatchery. They are more easily handled at this age, and hatchery debeaking is convenient. Other producers feel that stress is minimized if debeaking is delayed until chicks are 7 to 14 days of age. Debeaking at the latter stage is usually more uniform and effective for a longer period. At this age, a special attachment must be used with regular debeakers to prevent removal of too much beak.

Many producers regularly debeak a second time when pullets are between 12 and 17 weeks of age. When debeaked at this stage of development, pullets achieve complete recovery before the onset of sexual maturity.

Turkeys can also be debeaked. Young poults can be debeaked at between 1 and 3 weeks of age by removing about ⅝ to ¾ of the upper beak. Growing turkeys can be debeaked by removing ½ to ¾ of the upper beak. Mature birds should lose ½ or slightly more of the upper beak in the debeaking process.

There are several methods and several ages of effectively trimming beaks. The conventional method of debeaking chickens and turkeys is shown in Fig. 8–4.

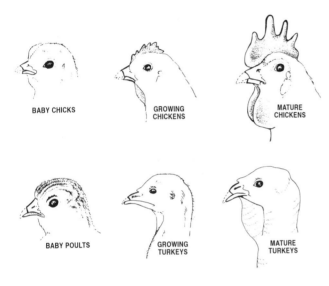

Fig. 8–4. Conventional method of debeaking chickens (three upper pictures) and turkeys (three lower pictures).

DUBBING

Dubbing refers to the removal of all or part of the comb. In certain instances, it may also mean the removal of wattles and possibly the earlobes. Although research reports vary, many producers feel that dubbing will increase egg production by 1 to 2%, for the following reasons:

1. Many of today's egg-laying strains have combs that are large enough to interfere with eating and drinking from automatic equipment. This can become a problem in today's high density manner of handling layers.

2. Dubbing reduces the chances of injury to the birds, either from catching their combs on sharp objects such as feeders, waterers, or exposed ends of wire cages, or from fighting.

3. Dubbing puts the hens on a more equal social basis. It has been reported that the social order, or peck order as it is often called, depends upon the recognition of individual birds by other birds. One of the chief means of recognition among chickens is their comb. It follows that when the social order is equalized, fighting and "bossing" are reduced.

4. A dubbed flock is usually a quieter and tamer flock.

5. Dubbing eliminates frozen comb problems in poultry houses where the temperature drops below freezing.

6. With breeding males, dubbing greatly reduces the chances of injury from fighting.

The best time to dub is at 1 day of age. At that time, a pair of curved manicure scissors can be used to snip off the comb easily with a minimum of pain or discomfort and practically no bleeding. In the case of males, dubbing should be done at a later date if the wattles are to be clipped.

TROUBLE SHOOTING

Producers should know the signs of bird well-being. Likewise, they should be able to (1) recognize trouble once it strikes, (2) diagnose the cause, and (3) institute corrective measures; thereby holding mortality and economic losses to a minimum. This means that they must know what to look for when trouble shooting.

CHICK TROUBLE SHOOTING

Fig. 8–5. When brooding chicks, the caretaker should (1) know the signs of chick well-being, and (2) be able to spot trouble. (Courtesy, Alabama Poultry & Egg Assn., Cullman, AL)

Producers should look for the following when chick trouble shooting:

1. Are the chicks spread out over the floor area and at the feeding and watering equipment? If so, that is good.

2. Are they crowded along the walls? If so, they are likely too hot.

3. Are they panting and gasping? If so, they may (a) be too hot or (b) have a disease problem.

4. Are they grouped close to the stove? If so, they are too cold.

5. Are they huddling and piling in small groups or in the corners? If so, they are too cold.

6. Are feeders and waterers crowded? If so, there is not enough space.

7. Do they seek sun spots? If so, that is good.

8. Is the litter packed and wet? If so, improve the ventilation and/or put down new litter.

9. Are ammonia odors strong? If so, improve ventilation and/or change litter.

LAYER TROUBLE SHOOTING

Fig. 8–6. The egg producer should (1) know the signs of layer well-being, and (2) be able to recognize trouble. (Courtesy, DEKALB AgResearch, Inc., DeKalb, IL)

When layer trouble shooting, the caretaker should look for the following:

1. Signs that all is well, which include: alert birds; red heads.

2. Well-bleached beaks and shanks characterize long-time layers. Also, white shanks and roughened feathers go together.

3. New plumage, and often yellow shanks, characterize hens that (a) have molted or (b) were poor layers.

4. Signs of molting, shrunken head parts, and perhaps yellow shanks, characterize hens that are out of condition.

5. An unsteady gait, and thin, shrunken head parts are indicative of hens in poor health.

6. Check egg records for rate of production.

7. Check length of the laying period if there is concern relative to (a) egg size, (b) shell texture, or (c) profitable lay.

8. Pullets that started laying at an early age or when undersized lay small eggs longer than pullets that were more mature and larger when they started laying.

9. Shell texture of hen's egg decreases in quality toward the end of the laying period due to fatigue of shell gland.

10. Make certain that there is plenty of shell; check shell hopper, sprinkle shell on top of mash, and mix shell with scratch grain. Provide coarsely cracked steamed bone as a source of extra phosphorus.

11. Use pen records to figure rate of production.

12. Use pen records to figure mortality rate.

13. If production is down, check on the amount of feed and water space.

14. Review recent feed consumption records.

15. Check causes for recent changes in production.

16. Sudden cold snaps may cut feed intake. When this happens, (a) moisten the mash, (b) stir the mash, (c) feed pellets, and/or (d) give frequent, small feedings of oats or scratch grain to stimulate the birds and warm their bodies.

17. If there is a shell problem, check (a) number of nests, (b) use of nests, (c) condition of nest litter, and (d) methods of gathering eggs.

18. If there are soiled eggs, check for (a) nesting problems and (b) dirty hen's feet caused by soiled litter. Soiled litter may result from poor insulation, overcrowding, poor litter management, or inadequate ventilation.

19. When heads appear blue, suspicion (a) lack of fresh air or (b) presence of blue comb disease.

20. Soiled fluff feathers may be caused by (a) kidneys not cleaning urates out of the bloodstream (in which case extra vitamin A or green feed may help), (b) vent picking, (c) albumen leakage from oviduct, or vent gleet.

21. Signs or symptoms of diseases and parasites.

EGG TROUBLE SHOOTING

The most common egg troubles, and their causes, are:

1. **Weak shells,** which may be caused by—
   a. Disease.
   b. Age of bird.
   c. Strain of chickens.
   d. Hot weather.
   e. Lack of shell-forming material in ration.
   f. Deficiency of vitamin D.

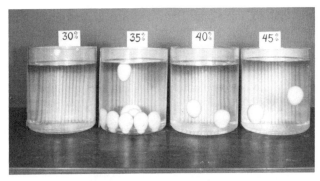

Fig. 8–7. Determining shell strength through the specific gravity test. (Courtesy, DEKALB AgResearch, Inc., DeKalb, IL)

2. **Too many cracks,** which may be caused by—
   a. Weak shells.
   b. Insufficient nests or nests not in good condition.
   c. Eggs not gathered often enough.
   d. Weak collection container or too many eggs in one container.
   e. Rough handling of eggs and container.
   f. Faulty washing of eggs and container.

3. **Mottled shells,** which may be due to—
   a. Eggs not gathered often enough.
   b. Egg-holding room too dry—below 75% relative humidity.
   c. Egg-packing material and cases too dry.

4. **Weak whites,** which may result from—
   a. Eggs not gathered often enough.
   b. Eggs not cooled quickly.
   c. Eggs not stored at cool temperature and high humidity.
   d. Eggs not packed daily as soon as cooled.
   e. Egg-packing material too warm or too dry.
   f. Disease.
   g. Age of birds. Pack eggs from each group separately.
   h. Strain of chickens. Pack eggs from each strain separately.

5. **Cooked whites,** generally caused by—
   a. The egg wash water being too hot. (Use thermometer to check temperature.)
   b. Eggs left in water too long.

6. **Yolk color,** which may be due to—
   a. Feed (yellow corn or green feed).
   b. Heat.
   c. Age.
   d. Drugs (mottling).

7. **Odors,** which generally result from—
   a. Moldy packing material.
   b. Mold in egg room.
   c. Fruit, vegetables, or other material stored in egg room.

COMPUTERS

Increasingly, the poultry industry is using computers as a management tool.

The vast majority of feed companies and large poultry establishments now use computers in ration formulation. This use is detailed in Chapter 7, Poultry Feeding Standards, Ration Formulation, and Feeding Programs, under the heading "Computer Methods."

Computers are also being used by poultry businesses in linear programming—as a way in which to analyze a great mass of data and consider many alternatives (see Chapter 17, Business Aspects of Poultry Production, section headed "Computers in the Poultry Business").

COMMUNICATION SYSTEM

Most visitors to poultry ranches are not interested in seeing the birds. They usually come on business—to talk to the manager or owner. Unless there is a ranch communication system, these visitors have no way of letting anyone know they are there. They often have to walk around the ranch or poultry houses looking for the manager or one of the employees.

In order to prevent the spread of poultry diseases, only those who need to come into the establishment should be allowed to enter. This necessitates fencing and locked gates, as well as a communication system so that the manager can screen all visitors. Then those who need to come in can be asked to take the necessary sanitary precautions first.

Also, to keep diseases from spreading within large poultry operations, it is important to limit all movement from one part to another. This applies to the owner, manager, and employees, as well as to visitors. A system that makes communication possible between workers in different parts of the ranch and between the office and the workers minimizes such movement.

The requirements for a workable communication system are: fences, a locked gate(s), a communication device, and a sign at the gate telling visitors how to reach the person they want to see.

In the ideal layout, the establishment is designed so that the office is located at the main gate to allow office personnel to control all entry. This is common on larger operations but it means that someone must be in the office at all times.

Five different types of communication systems are commonly employed. They are:

1. Horns, buzzers, or bells that are activated by a push button at the gate. Of all devices, these are the simplest and least costly. With this type of system, someone inside the ranch must go to the gate in order to see who is there.

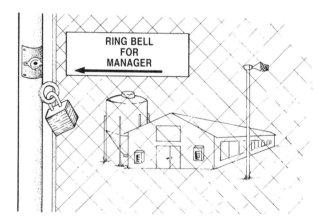

Fig. 8–8. The simplest and least costly devices are horns, buzzers, or bells that are activated by a push button at the gate. Such devices can be placed on posts at several locations around the ranch and installed inside poultry houses.

2. Two-way speaker system (squawk box). It may consist of a multiple-set, with units in various buildings and a central unit in the office. This arrangement is versatile. It can be set up so that managers can screen visitors without leav-

ing their work and so that workers at different locations on the ranch can talk to each other.

3. Belt-mounted "beeper." This device may signal the wearer to the phone, to the office, or to the front gate.

4. Hand-carried or vehicle-mounted two-way radio (a walkie-talkie). This equipment is expensive, but it is very useful if the manager has to spend a lot of time in a vehicle or if the poultry operation has several ranches.

5. A combination communication device and gate lock deactivator. Visitors can let the manager or office personnel know they are at the entrance. The gate lock can then be deactivated from the office or other location on the ranch.

The producer should decide what system best suits the particular operation. The important thing is that there be an effective communication system as a requisite of poultry disease prevention.

In addition to installing a communication device, the ranch owner or manager should establish a certain time during the day to be in the office to receive telephone calls. Also, salespersons should be encouraged to visit only by appointment. Most business should be conducted in the office or by telephone.

MANAGEMENT OF LIGHT

It has long been known that light stimulates egg production in chickens and other birds. Records show that the ancient Chinese made their canaries sing more by placing a lighted candle by the cage at night. Much later, early in the 1900s, poultry farmers in the state of Washington found that they could increase winter egg production by placing a lighted lantern in the chicken house for a few hours each evening. At that time, however, it was thought that the role of light was primarily a matter of increasing the "work day" of the bird. Today, the action of light is considered physiological; light enters the eye of the bird and stimulates the pituitary gland. In turn, the pituitary gland releases certain hormones which cause ovulation. Because of this phenomenon, the management of lighting systems is an integral part of modern poultry production.

Several lighting systems have been designed, differing primarily in windowless vs open houses, in light-to-dark ratio, and in continuous vs intermittent light. The author's recommendations for lighting regimens for poultry are given in Table 8–3.

It is noteworthy that chickens do not respond to all wave-

TABLE 8–3
RECOMMENDED LIGHTING SYSTEMS FOR POULTRY

Age of Birds	Minimum Light	Light-to-Dark Ratio (hours)		
		Light-tight Houses		Open-Sided House
		System 1	System 2	System 3
	(footcandles)			
Replacement pullets for egg production:				
0–3 weeks	1.0	20:4 hours.	20:4 hours.	20:4 hours.
3–12 weeks	0.1–0.5	Seasonal daylight.	16:8 hours.	Decreasing at rate of 15–30 minutes/week or natural light for March-September hatches.
12–21 weeks	0.5	Seasonal daylight, if decreasing; otherwise, short day length 6:18 or 8:16 hours.	Continue 16:8 to 16 weeks of age, then give 8:16 until 21 weeks.	Continue with simulated decreasing day lengths.
Commercial chicken layers:				
21 weeks or over	1.0	Seasonal light if increasing; or 16:8, 15:9, or 14:10 hours, depending on latitude.	16:8 hours.	Increasing light at rate of 15 minutes/week until 17 hours then maintain.
Chicken breeders:				
24–26 weeks or over	1.0	As above.	16:8 hours.	As above.
Broilers:				
0–2 days	3.5			Continous light.
After 2 days to market	0.5			23:1 hours. Lights off 1 hour at night to prevent panic in case of power failure.
0–5 days	3.5	Continuous light.	Continuous light.	
6 days to market	0.35	23:1 hours. 23 hours of continuous dim light and 1 hour of darkness. Lights off 1 hour at night to prevent panic.	1:3. Intermittent dim light. Provide 1 hour of dim light (feeding time) to 3 hours of darkness (resting time). This provides 6 hours of eating time each 24 hours.	

(Continued)

TABLE 8–3 *(Continued)*

Age of Birds	Minimum Light	Light-to-Dark Ratio (hours)		
		Light-tight Houses		Open-Sided House
		System 1	System 2	System 3
	(footcandles)			
Market turkeys:				
0–2 weeks	10.0–15.0	Continous light.	Continuous light.	Continuous light. Dim light at night only.
2 weeks to market	0.5–1.5	16:8. 1.0 footcandle during 16-hour day, and 0.5 footcandle during 8-hour night.		
Turkey breeders:				
(same lighting as market turkeys to 29 weeks of age)				
29 weeks and over	5.0–7.0	14:10 in spring and summer. 8:16 in fall and winter (for turkeys hatched August 1 to April 1).		
Duck breeders:	1.0	14:10.		

lengths of light. Orange and red lights are most effective. Shorter wavelengths are not effective. Normal white incandescent bulbs give sufficient light at effective wavelengths. It is noteworthy, too, that light stimulus seems to have a threshold level of intensity beyond which further increases in brightness of light have no effect. Thus, increased egg production cannot be brought about by using very bright lights. A level of 0.5 to 1.0 footcandle of light should be provided at the darkest points of exposure of the hens. Excessive light is both unnecessary and uneconomical.

Automatic time switches are available at moderate costs and should be installed in poultry houses for pullets or layers.

SAVE ENERGY

The energy shortage appears to be here to stay, and it will likely worsen. It is important, therefore, that poultry producers instigate energy conservation practices that will stretch fuel supplies and lower production costs.

It is estimated that the poultry industry can save 20 to 25% of the energy that it has been using to produce eggs and meat. Saving energy is a matter of attention to details and modification of existing practices. Little expense, if any, is needed to instigate energy-saving methods. Some ways in which to save energy follow:

1. **Brooding and growing.** Brooding uses the most energy of any phase of poultry production. Fuel for this purpose can be decreased 20 to 50% by practicing energy conservation and good management, including (a) partial house brooding; (b) clustering three or four brooders together and encircling the cluster with a single solid chick guard; (c) keeping the recommended number of chicks under each hover (undercapacity increases fuel cost per bird); (d) adjusting brooding temperatures to conserve heat—for example, starting the chicks at 88°F, rather than 95°F, but exercise good management by keeping the house free from drafts, having accurate thermostatic control, observing the birds carefully to see that they are comfortable, and lowering brooding temperature every three or four days instead of weekly; and (e) repairing or replacing torn curtains.

2. **Housing.** Good management of existing housing, along with modifying it if necessary (such as by adding insulation), will save much energy. Regular checking and sealing air leaks around doors, air intakes, and fans will reduce heat loss. Adding insulation to existing houses is expensive. But if the current insulation is inadequate, it will likely be cheaper to buy and install more insulation than to buy feed or fuel. Always make sure that wall and ceiling insulations have the recommended "R" values for the particular area, and make sure that there are adequate vapor barriers.

3. **Ventilation.** Much energy can be saved by a good, well-managed ventilation system. One way is to reduce ventilation rates. Temperature within the house can be raised to 70° to 75°F, provided air quality is kept acceptable and if ammonia does not exceed 50 ppm. This saves electricity and reduces feed usage, too.

Other ways of saving energy in ventilation include selecting the most efficient fans, having adequate electrical service for the house, along with proper wire size to fan motors, and keeping fans clean.

4. **Lighting.** Energy can be saved by careful management of the lighting program, including (a) installing reflectors (a 25-watt bulb with a reflector gives as much light as a 40-watt lamp without a reflector); (b) reducing light intensity, which can cut electricity use by 25% to 50% (CAUTION: never decrease the intensity of light of flocks in production); (c)

using intermittent lighting in windowless houses (CAUTION: birds already in production should not be switched from some other lighting program); and (d) going through the house every 2 to 3 weeks and replacing light bulbs which do not emit enough light.

5. **Feeding, watering, and management.** The energy requirements for watering and feeding are relatively small. Nevertheless, some savings can be effected in most operations. The number of times automatic feeders run each day should be reduced to the minimum needed for the desired level of feed consumption—usually, 3 to 4 runs a day will suffice.

Energy may be conserved by keeping the watering system in good repair to prevent leakage and spilling. Also, consideration should be given to switching from continuous-flow troughs to a valve-controlled or discontinuous system to reduce pumping in the water used. If continuous-flow troughs are used, turn the water off at night and operate the system on an intermittent schedule in the daytime.

Among the management programs which may be instituted to save energy are: keeping all moving equipment in good repair, cleaning manure-removal equipment after each use, and using only the recommended size of motors on feeders, waterers, egg-collection equipment, and other automation, as well as maintaining and adjusting such equipment regularly.

6. **Processing eggs.** Significant saving of energy can be made in processing eggs. Washing, sizing, and cartoning equipment should be operated at full capacity. Before turning the equipment on, the operator should make sure that sufficient volume of eggs is available for a full period of operation. Longer and fewer runs use less energy than frequent, short runs. Processing equipment should be shut off during rest breaks and at other times when it is not in use.

Egg-washing equipment should be set to maintain the wash-water temperature at 100° to 120°F. Higher temperatures waste energy, do not improve cleaning efficiency, and may harm interior and shell quality, depending on the conditions.

For short-term egg storage, keep the cooler temperature at 55° to 60°F. Lower temperatures require substantially more energy. Also, energy saving may be effected in egg storage through checking the temperature daily, keeping seals on the cooler door in good repair, using automatic or mechanical door closers on coolers, minimizing travel in and out of the cooler, cleaning the cooling unit filters and condensers regularly, and turning off the lights in the work and egg rooms when they are not in use.

CONTROL PESTS

The necessity of getting the greatest production per bird and the most return per dollar invested requires the timely use of pesticides to control insects (flies, lice, and mites), rodents, birds, and other pests. At the same time, the producer must be increasingly concerned about timing, choice, and dosage of registered pesticides, along with the method of application, in order to stay within the residue tolerances or limits allowed in eggs and meat by the Food and Drug Administration. Government agencies are continually sampling eggs and fowl for pesticide residues, feed additives, and antibiotics. Misuse of these chemicals can result in financial loss through confiscation and reflect discredit on the poultry industry. To the end that pesticides be used properly and efficiently, the following precautions and suggestions should be observed:

1. **Good sanitation and regular inspection.** The use of pesticides should be accompanied by good sanitation and regular inspection of buildings and birds.

2. **Start using pesticides early.** Pest control methods should be instituted before the problem builds up and reduces production efficiency.

3. **Use only approved pesticides.** Only registered, approved pesticides should be used on birds, in and around poultry houses, in egg rooms, or in storage areas. New products are developed, and sometimes old products are banned. Accordingly, it is recommended that the producer follow the current recommendations of the Cooperative Extension Service, or other recognized specialists, for the control of pests.

4. **Read and follow label directions.** When using any pesticides, the operator should always read and follow the label directions.

5. **Mix and use pesticides with care.** Pesticides should be mixed where birds cannot get to containers, equipment, or spillage. Likewise, pesticides should be kept away from eggs, feed, feeders, water, and watering equipment. Good ventilation should be provided in confined areas.

6. **Store pesticides and application equipment properly.** Pesticides and application equipment should be stored in a separate, marked, locked building or storage area away from children, poultry, feed, and water sources.

7. **Dispose of empty containers properly.** Empty paper and plastic containers should be burned in an approved manner, then the ashes should be buried.

LITTER

Litter is used primarily for the purposes of keeping the birds clean and comfortable. It absorbs moisture from the droppings and then gives this moisture to the air brought in by ventilation.

A good litter is highly absorbent and fairly coarse, so as to prevent packing. It should be free from mold and contain a minimum amount of dust. Availability and cost will determine the type of litter used. Table 8–4 (next page) lists some common litter materials and gives the average water absorptive capacity of each.

Naturally the availability and price per ton of various litter materials vary from area to area, and from year to year. Thus, in the New England states shavings and sawdust are available, whereas other forms of litter are scarce, and straws are more plentiful in the central and western states.

TABLE 8–4
WATER ABSORPTION OF LITTER MATERIALS

Material	Lb of Water Absorbed/Cwt of Air-Dry Bedding
Barley straw	210
Cocoa shells.	270
Corn stover (shredded)	250
Corn cobs (crushed or ground)	210
Cottonseed hulls	250
Flax straw	260
Hay (mature, chopped)	300
Leaves (broadleaf)	200
(pine needles)	100
Oat hulls	200
Oat straw (long)	280
(chopped)	375
Peanut hulls	250
Peat moss	1,000
Rye straw	210
Sand	25
Sawdust (top-quality pine)	250
(run-of-the-mill hardwood)	150
Sugarcane bagasse	220
Tree bark (dry, fine)	250
(from tanneries)	400
Vermiculite[1]	350
Wheat straw (long)	220
(chopped)	295
Wood chips (top-quality pine)	300
(run-of-the-mill hardwood)	150
Wood shavings (top-quality pine)	200
(run-of-the-mill hardwood)	150

[1]This is a micalike mineral mined chiefly in South Carolina and Montana.

Other facts of importance relative to certain litter materials and litter uses follow:

1. **Wood products (sawdust, shavings, tree bark, chips, etc.).** The suspicion that wood products will hurt the land is rather widespread but unfounded. It is true that shavings and sawdust decompose slowly, but this process can be expedited by the addition of nitrogen fertilizers. Also, when plowed under, they increase soil acidity, but the change is both small and temporary.

Softwood (on a weight basis) is about twice as absorptive as hardwood, and green wood has only 50% the absorptive capacity of dried wool.

2. **Cut straw.** Cut straw will absorb more liquid than long straw. But there are disadvantages to chopping; chopped straws may be dusty.

3. **Fertility value.** From the standpoint of the value of plant food nutrients per ton of air dry material, peat moss is the most valuable bedding and wood products the least valuable.

The minimum desirable amount of bedding to use is the amount necessary to absorb completely the moisture in the droppings.

REDUCING LITTER NEEDS

In most areas, litter materials are becoming scarcer and higher in price, primarily because (1) geneticists are breeding plants with shorter straws and stalks, (2) of more competitive and remunerative uses for some of the materials, and (3) the current trend toward more confinement rearing of all livestock requires more bedding.

Poultry producers may reduce litter needs and costs as follows:

1. **Ventilate quarters properly.** Proper ventilation lowers the humidity and keeps the litter dry. The condition of the litter and the amount of the fumes (ammonia fumes) are good indicators of the adequacy of ventilation.

2. **Chop litter.** Chopped straw, waste hay, fodder, or cobs will go further and do a better job of keeping the birds dry than long materials. Chopped straw, for example, will soak up approximately 25% more moisture than long straw.

3. **Use deep-litter system for layers.** In this system, built-up litter on the floor is used on about 60% of the area, with raised slats or wire floor over the remaining 40% of the area. This raised floor provides a place for feed, water, and roosts. About 70% of the droppings collect in the area below the raised floor.

4. **Consider slotted floors or cages.** Slotted floors and cages, for which no litter is needed, are extensively used for layers.

POULTRY MANURE

The term manure refers to a mixture of animal excrements (consisting of undigested feeds plus certain body wastes), with or without bedding.

No doubt, the manure pollution problem, suspicioned or real, will persist. However, the energy crisis, accompanied by high chemical fertilizer prices, and the recycling of animal waste as a livestock feed, has caused manure to be looked upon as a resource. Therefore, planned manure management is an important part of modern poultry management. The collection, transport, storage, and use of manure must meet sanitary and pollution control regulations.

Modern poultry buildings and equipment should be designed to handle the manure produced by the birds that they serve; and this should be done efficiently, with a minimum of labor and pollution, so as to retrieve the maximum value of the manure and make for maximum animal sanitation and comfort.

AMOUNT, COMPOSITION, AND VALUE OF MANURE PRODUCED

The quantity, composition, and value of poultry manure produced vary according to kind of poultry, weight, kind and amount of feed, and kind and amount of litter. Poultry manure, in pure form without added litter, is produced in about the quantities shown in Table 8–5.

TABLE 8–5
QUANTITY AND VALUE OF PURE MANURE (FREE OF LITTER) FROM VARIOUS FLOCKS[1]

Birds	Type of Flock	Average Bird Weight		Quantity Manure Produced (dry basis)		Time Period
		(lb)	(kg)	(lb)	(kg)	
100	Laying hens	4.5	2.0	2,400	1,091	12 months
1,000	Broilers (chickens)	4.0	1.8	2,700	1,227	9 weeks
1,000	Broilers (turkeys)	8.0	3.6	4,320	1,964	16 weeks
1,000	Heavy turkeys	20.0	9.1	35,000	15,909	24 weeks

[1]University of Maryland Extension Service, Fact Sheet 39.

Fig. 8–9 shows the comparative amounts of manure produced per year under confinement by poultry and other animals, with each class of animal on a 1,000 lb liveweight basis.

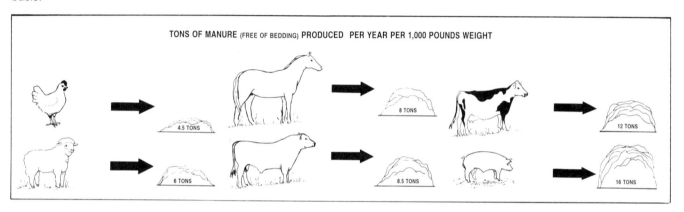

TONS OF MANURE (FREE OF BEDDING) PRODUCED PER YEAR PER 1,000 POUNDS WEIGHT

4.5 TONS 8 TONS 12 TONS
6 TONS 8.5 TONS 16 TONS

Fig. 8–9. On the average, each class of confined animals produces per year per 1,000 lb weight the tonnages of manure free of bedding shown above.

An egg farm with 100,000 layers will produce 12 tons of manure per day, or over 4,000 tons per year. In 1987, U.S. poultry produced the following tonnage of manure:

Species	Total Manure Production (tons)
Chickens, layers	5,672,835
Broilers	73,179,414
Turkeys	20,514,384
Total	99,366,633

It is noteworthy that, in 1987, U.S. livestock (poultry, cattle, sheep, swine, and horses) excreted an estimated total of 1,351,220,633 tons of manure, with an estimated fertilizer value of $10,268,628,485 (computed on the following basis per pound: nitrogen (N), 29¢; P_2O_5, 30¢; and K_2O, 12¢).

If properly conserved, poultry manure is rich in nutrients. Table 8–6 shows the composition and value of chicken manure, fresh vs different degrees of dryness.

TABLE 8–6
COMPOSITION AND VALUE OF CHICKEN MANURE, FRESH VS DIFFERENT DEGREES OF DRYNESS[1]

Kind of Manure	Water Content	Solid Material	Weight per Cubic Yard		Approximate Percentage of Plant Food by Weight (Multiply by 20 to get pounds per tons)					Estimated Value of Manure/Ton[2]
					Nitrogen (N)	Phosphorus		Potassium		
						(P)	(P_2O_5)	(K)	(K_2O)	
	(%)	(%)	(lb)	(kg)	(%)	(%)	(%)	(%)	(%)	
Fresh	75	25	1,750	795	1.0	0.6	1.3	0.6	0.7	15.28
Partially dried	50	50	—	—	2.0	1.1	2.6	1.1	1.4	30.56
Dry manure	25	75	800[3]	364	3.0	1.7	3.9	1.7	2.1	45.84
Completely dry	0	100	—	—	4.0	2.3	5.2	2.3	2.8	61.12

[1]Adapted by the author from data reported by California Agricultural Extension Service.

[2]Calculated on the following retail price per pound basis: nitrogen (N), 29¢; P_2O_5, 30¢; and K_2O, 12¢.

[3]The volume-weight relationship may vary considerably at the lower moisture content due to handling, compaction, and biological decomposition.

Since the manure of birds contains both the feces and the excretion from the kidneys, it is much richer in nitrogen than the manure of four-footed animals. As much as 80% of the urinary nitrogen is present as uric acid. So, unless the manure is properly handled, to prevent putrefaction, much of the uric acid will be changed to ammonium carbonate, with consequent loss of fertilizing value.

MANURE USES

Historically, manure has been used as a fertilizer. But high feed prices and shortages of fossil fuels have made for new uses. Although it is expected that manure will continue to be used primarily as a fertilizer for many years to come, increasingly it will be recycled and used as a feed and converted into energy. Other manure-based products will continue to evolve, but it is expected that they will be of minor importance.

MANURE AS A FERTILIZER

The actual monetary value of manure can and should be based on (1) equivalent cost of a like amount of commercial fertilizer, (2) increased crop yields, or (3) its value as a feed or for other uses. Table 8–6 gives the equivalent cost of a like amount of commercial fertilizer (see column headed, "Estimated Value of Manure/Ton"). Numerous experiments and practical observations have shown the measurable monetary value of manure in increased crop yields; and the results of more and more feeding trials are becoming available.

Poultry producers sometimes fail to recognize the value of this barnyard crop because (1) it is produced whether or not it is wanted, and (2) it is available for only the cost of handling.

HOW MUCH MANURE CAN BE APPLIED TO THE LAND

With today's heavy poultry concentration, the question is being asked: How many tons of manure can be applied per acre without depressing crop yield, creating salt problems in the soil, creating nitrate problems in feed produced thereon, contributing excess nitrate to groundwater or surface streams, or violating state regulations?

Based on earlier studies in the Midwest, before the rise of commercial fertilizers, it would appear that one can apply from 5 to 20 tons of manure per acre, year after year with benefit.

Heavier than 20-ton applications can be made, but probably should not be repeated every year. With higher rates annually, there may be excess salt and nitrate buildup. Excess nitrate from manure can pollute streams and groundwater and result in toxic levels of nitrate in crops. Without doubt the maximum rate at which manure can be applied to the land will vary widely according to soil type, rainfall, and temperature.

State regulations differ in limiting the rate of manure application. Missouri permits up to 30 tons per acre on pasture, and 40 tons per acre on cropland. Indiana limits manure application according to the amount of nitrogen applied, with the maximum limit set at 225 lb per acre per year.

When poultry producers have sufficient land, they should use rates of manure which supply only the nutrients needed by the crop rather than the maximum possible amounts suggested for pollution control.

PRECAUTIONS WHEN USING MANURE AS A FERTILIZER

The following precautions should be observed when using manure as a fertilizer:

1. Avoid applying waste closer than 100 ft to waterways, streams, lakes, wells, springs, or ponds.
2. Do not apply where percolation of water down through the soil is not good, or where irrigation water is very salty or inadequate to move salts down.
3. Do not spread on frozen ground.
4. Distribute the waste as uniformly as possible on the area to be covered.
5. Incorporate (preferably by plowing or discing) manure into the soil as quickly as possible after application. This will maximize nutrient conservation, reduce odors, and minimize runoff pollution.
6. Minimize odor problems by—
   a. Spreading raw manure frequently, especially during the summer.
   b. Spreading early in the day as the air is warming up, rather than late in the day when the air is cooling.
   c. Spreading only on days when the wind is not blowing toward populated areas.

7. In irrigated areas, (a) irrigate thoroughly to leach excess salts below the root zone, and (b) allow about a month after irrigation before planting, to enable soil microorganisms to begin decomposition of manure.

MANURE AS A FEED

Recycling manure as a ruminant feed is the most promising of the nonfertilizer uses. Various processing methods are being employed; some are even feeding manure without processing. More and more poultry manure will be either (1) incorporated in a cattle grower ration, or (2) fed to beef breeding herds during periods when pasture supplementation is beneficial, with the residues distributed over grazing areas where they would have fertilizing value. Further experimentation will be required before the use of manure feeds becomes widespread, but some researchers predict that eventually wastes may supply up to 20% of the nation's livestock feed, thereby freeing an equivalent amount of grain for human consumption.

MANURE AS A SOURCE OF ENERGY

Manure may also serve as a source of energy, which, of course, is not new. The pioneers burned dried bison dung, which they dubbed "buffalo chips," to heat their sod shanties. In this century, methane from manure has been used for power in European farm hamlets when natural gas was hard to get. While the costs of constructing plants to produce energy from manure on a large-scale basis may be high, some energy specialists feel that a prolonged fuel shortage will make such plants economical. India now has about 10,000 anaerobic digestion plants in operation.

Methane, of course, is usable like natural gas. There is nothing new or mysterious about this process. Sanitary engineers have long known that a family of bacteria produces methane when they ferment organic material under strictly anaerobic conditions. (Grandad called it swamp gas; his city cousin called it sewer gas.) However, it should be added that, due to capital and technical resources needed, for some time to come, the production of methane by anaerobic digestion will likely be limited to municipal or corporate industries. If all animal manure were converted to energy, it has been esti- mated that it could produce energy equal to 10% of the petroleum requirements or $12\frac{1}{2}$% of our natural gas require- ments.

SUMMARY OF POULTRY MANURE

In the future, as fertilizer, feed, and energy become in- creasingly scarce and expensive, the economic value of poultry manure will increase, and it will be looked upon as a resource and not as a waste that presents a disposal prob- lem. More and more poultry manure will be used as a fertil- izer, feed, or source of energy.

QUESTIONS FOR STUDY AND DISCUSSION

1. Define *management*. Evaluate the relative import- ance of each of the following in a poultry operation: breeding, feeding, management.

2. Discuss the importance of, and the consideration which should be accorded to, each of the following when planning a poultry enterprise:
    a. Choice of species, and specialized vs inte- grated.
    b. Location.
    c. Size.

3. Discuss the relative importance of each of the main traits of a good manager. How may a beginner acquire these traits?

4. Of what importance is an organization chart and job description in any business, including the poultry business?

5. Do you feel that an incentive basis for the help is essential to the success of a modern commercial poultry enterprise? If so, what type of incentive basis would you recommend for each of the following types of operations: commercial layers, broilers, and market turkeys.

6. In attempting to cut costs, what items would you scrutinize with special care in each of the following types of operations, and why would you do so?
    a. Layers.
    b. Replacement pullets.
    c. Broilers.
    d. Market turkeys.

7. Apply one guideline to a given poultry operation to determine if mechanization should replace labor.

8. Under what circumstances would you recommend forced molting (recycling) rather than buying replacement pullets? In your area, detail the method of molting that you would recommend.

9. Why are not more cockerels caponized?

10. Occasionally, some people criticize the following management practices as being cruel: debeaking and dub- bing. How would you answer such criticism?

11. What should a poultry producer look for when—
    a. Chick trouble shooting?
    b. Layer trouble shooting?
    c. Egg trouble shooting?

12. Why are computers increasingly being used as a management tool?

13. Why is it important to have a communication system in a poultry establishment? What type of communications system would you recommend?

14. What are the main differences between lighting pro- grams for replacement pullets, commercial layers, and broil- ers?

15. In what ways may a poultry producer save energy?

16. List the precautions that should be observed to the end that pesticides may be used properly and efficiently.

17. In your area, what kind of litter would you use, and why; or would you eliminate litter entirely?

18. How may litter be lessened or eliminated?

19. Can you justify the expense of saving and spreading poultry manure, or would it be cheaper to buy a commercial fertilizer and either give away the manure or wash it into a lagoon?

20. What can and should poultry producers do to lessen pollution when using manure as a fertilizer?

21. Discuss the potential of poultry manure as a feed.

22. How much potential does manure have as a source of energy?

23. How would you recommend that each (a) a small operator, and (b) a large operator handle and dispose of poultry manure?

SELECTED REFERENCES

Title of Publication	Author(s)	Publisher
Commercial Chicken Production Manual, Fourth Edition	M. O. North D. D. Bell	Van Nostrand Reinhold Co., New York, NY, 1990
Manual of Poultry Production in the Tropics	R. R. Say	CAB International, Wallingford, Oxon, United Kingdom, 1987
Poultry: Feeds and Nutrition, Second Edition	H. Patrick P. J. Schaible	The AVI Publishing Company, Inc., Westport, CT, 1980
Poultry Meat and Egg Production	C. R. Parkhurst G. J. Mountney	Van Nostrand Reinhold Co., New York, NY, 1988
Poultry Production, Twelfth Edition	M. C. Nesheim R. E. Austic L. E. Card	Lea & Febiger, Philadelphia, PA, 1979
Raising Poultry Successfully	W. Graves	Williamson Publishing, Charlotte, VT, 1985

Fig. 8–10. Service room showing carousel conveyors on which poults are placed, where they are graded, sexed, beak trimmed, toe nails clipped, vaccinated or injected with an antibiotic-vitamin mixture. Photo of integrated operation of Cuddy Farms, North Carolina. (Photo courtesy James Strawser, University of Georgia, Athens)

Fig. 9–1. Geese exhibiting their gregarious or flocking nature as they walk single file, one behind the other. (Courtesy, J. C. Allen & Son, Inc., West Lafayette, IN)

POULTRY BEHAVIOR AND ENVIRONMENT

CHAPTER

Fig. 9–2. *Gallus gallus*, one of the ancestors of the domestic chicken. The male weighed about 2½ lb.

The habits of chickens of the Asiatic class indicate an ancestry that roosted on the ground and nested on a mound of earth. Such breeds as the Leghorn probably had tree-roosting ancestors. The sport of cockfighting, which was once one of the principal uses of male chickens, applied the law of survival of the fittest and did much to develop the vigor and vitality of the species.

Very early, primitive people learned that eggs and poultry meat were good—and good for them. So, with the domestication of fowl more than 5,000 years ago, they selected the better layers and the larger and better fleshed birds so that their food supply could be increased.

In the early 1900s, U.S. poultry breeding was dominated by exhibition chickens—by fanciers-exhibitors who selected birds for such characteristics as colors, feather patterns, comb characteristics, and body weight. Chicken breeding shifted to inbred lines and hybridization during the years 1940–1945, with separate breeding programs designed to

achieve maximum egg or broiler production. Hand in hand with the transition in breeding methods and objectives, came confinement production, reaching the ultimate in batteries and cages; mechanical incubation and brooding—replacing normal parental behavior; and improved nutrition.

There are many biological differences between the 2-lb jungle hen, *Gallus gallus*, which laid only a few eggs, and the modern hen, *Gallus domesticus*, that lays more than 300 eggs per year. Likewise, there is little resemblance between the fighting cock and the present-day 13-lb Jersey Giant.

Wild turkeys (*Melleagris gallopavo*) are forest dwellers, although they prefer woods with clearing to dense forests. Domestication, with selection for broad breast and body weight, has had a negative effect on walking ability and mating, with the result that most commercial broad-breasted strains of turkeys are reproduced by artificial insemination.

The behavior of domestic ducks and geese, selected for large size and reduced flight ability, has changed greatly from their wild ancestors. Domestic ducks and geese neither fly nor migrate.

For the most part, the discussion of the behavior of ducks will be limited to the Mallard. Few studies have been made of the behavioral changes accompanying the domestication of the Mallard (parent of most domestic ducks).

Important aspects of the behavioral repertoire of the Japanese quail (*Coturnix coturnix japonica*)—a hardy, rapid-growing, strong, sexually motivated bird—are presented at appropriate places in this chapter.

Space limitations in this chapter will not permit a discussion of the behavior of each of the wide variety of birds of the several species classed as *poultry*. So, under each subject heading, the behavior of poultry as a whole will first be discussed, followed by a presentation of available, pertinent information relative to selected species in the following order: chickens, turkeys, ducks, and quail.

With domestication, environments became increasingly artificial until poultry were viewed primarily as economic, biological conversion units for changing feed to eggs and meat. Behavioral problems arose because the breeding of adapted poultry did not keep pace with environments.

Ethologists (those who study animal behavior) are explaining scientifically that all is not well with poultry; and the situation is being righted. Today, keepers of flocks are giving increasing attention to poultry behavior and environmental

NORMAL STANCE   **WALKING POSITION**

Fig. 9–3. Japanese quail.

control, including space requirements, light, air temperature, relative humidity, air velocity, wet bedding, ammonia buildup, odors, and manure disposal.

DO IT EARLY!

Fig. 9–4. Morning comes early for chickens! Thirty percent of all eggs are laid before 8:00 a.m.

ANIMAL BEHAVIOR

Animal behavior is the reaction of the whole organism to certain stimuli, or the manner in which it reacts to its environment. Through the years, poultry behavior has received less attention than the quantity and quality of the eggs and meat produced. But modern breeding, feeding, and management have brought renewed interest in behavior, especially as a factor in obtaining maximum production and efficiency. With the restriction, or confinement, of flocks, many abnormal behaviors evolved to plague those who raise them, including cannibalism and a host of other behavior disorders. Confinement has not only limited space, but it has interfered with the habitat and social organization to which, through thousands of years of evolution, the species became adapted and best suited.

We now know that controlled environment must embrace far more than an air-conditioned chamber, along with ample feed and water. The poultry producer needs to be concerned more with the natural habitat of birds. Nature ordained that they do more than eat, sleep, and produce eggs and/or meat.

What can be done about it? Preventing cannibalism by merely debeaking poultry is not unlike trying to control malaria fever in humans by the use of drugs without getting rid of mosquitoes. Rather, we need to recognize such a disorder for what it is—a warning signal that conditions are not right. Correcting the cause of the disorder is the best solution. Unfortunately, this is not usually the easiest. Correcting the cause may involve trying to emulate the natural conditions of the species, such as altering space per bird and group size, regulating the intensity and duration of light, promoting exercise, and gradually changing rations. Over the long pull, selection provides a major answer to correcting confinement

and other behavioral problems; we need to breed poultry adapted to artificial environments.

This chapter is for the purpose of presenting some of the principles and applications of poultry behavior. Those who have grown up around poultry and dealt with them in practical ways have already accumulated substantial workaday knowledge about their behavior. Those with urban backgrounds may need to familiarize themselves with the behavior of poultry, in order to feed and care for them properly and to recognize the early signs of illness. To all, the principles and applications of poultry behavior depend on understanding, which is the intent of this chapter.

LEARNING

Studies of the learning ability of poultry have not been as extensive or systematic as in mammalian forms such as the rat, monkey, or human. The few comparative studies that have been made sometimes show learning as good as that in lower mammals; sometimes not.

■ **Chickens**—In chicks, learning improves with age from hatching (or earlier) up to 2 weeks to 2 months of age, depending upon what is being taught.

Learning in animals is of two types: (1) simple learning (training and experience), and (2) complex learning (intelligence).

SIMPLE LEARNING (TRAINING AND EXPERIENCE)

In general, the behavior of animals depends upon the particular reaction patterns with which they were born. These are called *instincts* and *reflexes*. They are unlearned forms of behavior.

Several types of simple learning are known in poultry; among them, those that follow.

HABITUATION

Habituation is getting used to, or ignoring, certain stimuli.

■ **Turkeys**—A classical example of habituation is the response of young turkeys (poults) to the danger call of the mother when there is impending danger—like the approach of a hawk. At first, they scamper for cover or to the corners of their pen, pile in clumps, and remain motionless (freeze) for several minutes. However, they gradually get used to a repetition of this call, with each response becoming less intense and producing a shorter freeze. By the time they are 2 months old, they practically ignore such an alarm—they merely come to attention.

CONDITIONING (OPERANT CONDITIONING)

Conditioning is the type of learning in which the animal responds to a certain stimulus.

■ **Quail**—Quail have been trained to display courting behavior at the sound of a buzzer, a previously neutral stimulus. The conditioning consisted in presenting a female to the male in association with the sound of a buzzer. The courting response began to appear as a conditioned response as early as the fifth pairing with the buzzer; and, following 32 pairings, courting behavior was fully elicited in all birds by the buzzer alone.

Operant conditioning, or operant learning, refers to animal operation of some aspect of the environment to obtain access to feed or other animals.

■ **Quail and pigeon**—Both quail and pigeons can be taught to peck at a response key to obtain food.

IMPRINTING (SOCIALIZATION)

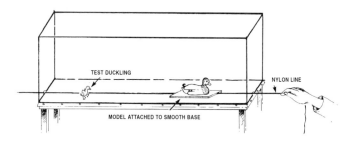

TEST DUCKLING

NYLON LINE

MODEL ATTACHED TO SMOOTH BASE

Fig. 9–5. Apparatus for studying imprinting behavior. Note that the movement of the decoy duck is controlled by the nylon line held by the operator. Note, too, that the duckling is following the decoy.

This is a form of early social learning which has been observed in some species. The pioneering work in this field was done by the Austrian zoologist Lorenz, with goslings. He found that if a baby gosling was exposed immediately after hatching to some moving object, especially if the object emitted sound, it would adopt that object as its parent-companion. Further studies revealed that goslings would adopt any other moving object in the same manner—dogs, cats, humans, and so forth. Also, it was found that the same principle applied to other fowl and to mammals.

Apparently, inheritance controls the time and the length of the critical period when an individual can be imprinted, the type of object to which it can be imprinted, the tendency to respond to the first object to which it is exposed, and the permanence of the attachment to the object following imprinting. For example, goslings can usually be imprinted only within the first 36 hours following hatching.

Although human-socialized animals make adorable pets, they are often a nuisance on the farm. People who have hand reared a duck have found that such animals seek human companionship and never fit in well with their own kind.

■ **Chickens**—Normally, the critical period for imprinting baby chicks begins soon after hatching and lasts about 3 days. Initially, a chick will follow any hen that crosses its path, but after a time it will follow only its mother.

MEMORY

Memory is the ability to remember or keep in mind; the capacity to retain or recall that which is learned or experienced.

The existence of a peck order is evidence that birds do remember (recognize) each other; otherwise, pecking would be promiscuous and continue without end.

■ **Chickens**—The memory of chickens is inferior to that of mammals. After only 2 weeks in isolation, hens are no longer able to recognize members of their flock. However, after 6 days, chicks still remembered the solution of a detour problem.

COMPLEX LEARNING (INTELLIGENCE)

Complex learning (intelligence) is the capacity to acquire and apply knowledge—the ability to learn from experience and to solve problems. It is the ability to solve complex problems by something more than simple trial-and-error, habit, or stimulus-response modifications. In people, we recognize this capacity as the ability to develop concepts, to behave according to general principles, and to put together elements from past experience into a new organization.

Animals learn to do some things, whereas they inherit the ability to do others. The latter is often called instinct. Thus, ducks do not have to learn to swim—instinctively, they take to water.

Some folks judge the intelligence of animals by the size of their brain in relation to body size. Others rank them according to their ability to solve a maze (a pathway complicated by at least one blind alley, used in learning experiments and intelligence tests) in order to get food.

Generally speaking, behavioral scientists are agreed that each species has its own special abilities and capacities, and that it should only be tested on these. For example, the dog, pig, and rat are more adept at solving a maze test than poultry. Hence, solving a maze in order to find food favors the scavengers (and the dog, the pig, and the rat are all scavengers)—they have connived for their food since the beginning of time.

Thus, each species is uniquely adapted to only one ecological niche. Moreover, a niche is filled by the particular species that can solve food finding therein, and that is best adapted under the conditions that prevail. It follows that intelligence comparisons between species are not meaningful, and that it is absurd to say that one species is smarter than another.

Of course, human intelligence is generally recognized.

■ **Chickens**—Chicks can learn to alternate turning responses in a temporal maze, which is a fairly difficult task for lower mammals. Also, chicks can learn by observing others learn.

■ **Quail**—In maze tests involving a single forced turn preceding a choice-point, it has been found that neither initial preference nor the direction of forced turn significantly affected choice behavior of quail. Also, quail show evidence of visual imitation; they have been trained to open a puzzle box by an appropriate pecking response merely by watching trained birds performing.

SENSES OF PERCEPTION

Vision and hearing, which are the most highly developed senses in the bird, play a crucial role in social behavior, communication, and responses to predators. However, taste, smell, touch, and temperature are involved in the well-being of poultry.

VISION

The eyes of birds are on the side of the head. The eyeballs are flatter than those of mammals and fit tightly in the orbits, with minimum eye movement. Moving objects are followed by moving the head and neck. Other visual characteristics include color vision (with preference for violet and orange), rapid accommodation, high visual acuity, a visual field of about 300°, and a limited binocular field.

- **Chickens** — Chickens possess natural color vision, with preferences for violet and orange and negative preference for green. Distance visual is normal or slightly farsighted. Because of the rather flat eyeball, there is little eye movement, but the position of the eyes in the orbits gives the chicken a visual field of approximately 300°.

- **Turkeys** — Turkeys have color vision, rapid accommodation, high visual acuity, a visual field of about 300°, and a limited binocular field.

HEARING

Except for the lack of an earlobe, the ears of poultry are similar to those of mammals. Hearing is well developed with frequency discrimination and the ability to determine the direction of sound comparable to that of humans.

- **Chickens** — Baby chicks are attracted by maternal calls and repetitive tapping.

- **Turkeys** — Hearing is well developed, with frequency discrimination and the ability to determine the direction of sound comparable to humans.

TASTE

Taste is involved in the well-being of both chickens and turkeys.

- **Chickens** — Chickens have taste buds, located on the base of the tongue and the floor of the pharynx. Though their taste buds resemble those of mammals, their taste recognition is quite different. In comparison with mammals, chicks are indifferent to the common sugars, evidence a wider range of tolerance for acidity and alkalinity, discriminate between carbohydrates, are sensitive to bitter-tasting substances, and will accept only up to about 0.9% salt solution.

- **Turkeys** — It has generally been assumed that the taste capabilities of chickens and turkeys are similar. However, wide differences between the species in the acceptance and rejection of certain chemical compounds suggest that specific studies are needed to determine the gustatory sensitivity in turkeys.

SMELL

Birds have an olfactory epithelium, but the sense of smell is poorly developed.

- **Chickens** — Chicks fail to show a preference when given a choice of feeds and waters, even though some of them are odorous to humans.

- **Turkeys** — Smell plays a negligible role in the behavior of turkeys.

TOUCH (TACTILE)

The sense of touch is well developed and of much practical importance in poultry, as evidenced by the following: foreign matter adhering to the feathers is readily located and removed; and contact with eggs when setting, and with young when brooding, involves a sense of touch.

- **Turkeys** — Tactile receptors appear to be highly developed in turkeys as evidenced by (1) sexual responses in the female being readily elicited by tactile stimulation over extensive areas of the body; and (2) their ability to detect distant explosions as a result of the sensory organs in the legs receiving vibrations through the ground.

TEMPERATURE

The ability to detect radiant heat is evident in sunbathing and preening in the sun.

- **Chickens** — Cold reduces activity and causes chickens (both chicks and adults) to huddle.

COMMUNICATION METHODS

Communication between animals involves giving by one individual some sign or signal which, on being received by another, influences its behavior. Among poultry, these stimuli are essentially auditory and visual.

VOCAL

Vocal communication is of special interest because it forms the fundamental basis of human language. The gift of language alone sets humans apart from the rest of the animals and gives them enormous advantages in their adaptation to their environment and in their social organization.

Sound is also an important means of communication among animals. They use sounds in many ways; among them, (1) feeding, in sounds of hunger by young, or food finding, and of hunting cries; (2) distress calls, which announce the approach or presence of an enemy and the all-clear signal following the departure of a predator; (3) sexual behavior, courting songs, and related fighting; (4) mother-young interrelations to establish contact and evoke care behavior; and (5) maintaining the group in its movements and assembly.

- **Chickens** — Vocalizations in chickens are especially important in mother-young relationships and during threat behavior.

The cock's crow functions as a means of recognition, since a given cock crows much the same and crows vary greatly between cocks. Also, the duration of crow differs between breeds and is highly heritable.

The duration and primary frequency of the adult distress call are genetically controlled.

■ **Turkeys**—Among the several vocalizations of turkeys related to social interactions are: the trill emitted during threat behavior and as the vocal component of the male strut; various yelps given by hens in attracting the poults; and the variety of vocalizations given by the poults, including a distress peep, a trilling contentment call, a high-pitched screamlike call when pecked, and the sleepy call just before going to sleep.

In older birds, the peep is replaced by the yelp, a lower-pitched call given frequently by adult females and less frequently by males.

The gobbling call of the male appears to serve as individual recognition within the flock. If one turkey gobbles, usually most of the males in the vicinity follow suit. At the height of the breeding season, a male turkey will respond with gobbling to a wide variety of stimuli.

■ **Ducks**—Ducks are very vocal. The familiar quack made by female Mallards is given with a variety of intensities and rhythms, producing several distinctive patterns and serving different functions; among them, (1) the series of quacks, of variable number, given by standing or swimming birds; (2) the going-away call, given just before flying; (3) alarm quacks, when alerted or alarmed; (4) two-syllable conversation calls; (5) the persistent quacking from mated females; (6) the inciting call associated with pairing and interactions between birds, indicating the female's preference for one male and rejection of another; and (7) the food call of females in feeding flocks, which is imitated by hunters to lure birds to the gun.

The male Mallard produces the following types of sounds: (1) a nasal sound, given in situations of mild alarm, after flight take-off when flushed, and in hostile situations; (2) a loud whistle, followed by a grunting sound, which is often the first major display to appear in a bout; and (3) a courtship whistle.

■ **Quail**—The vocal communication of quail is characterized by (1) clicking, a sharp and brief burst of sound that accompanies each air intake during the last stages prior to hatching—a significant and necessary stimulating factor in the synchronization of hatching; and (2) adult vocalization, including male crowing, the call of feeding and contentment, the danger call, and the invitation-to-mating call of the female.

VISUAL DISPLAYS

Among the important visual displays of poultry are raised hackles (feathers on the top of the neck) to denote aggressive intentions and make them look larger and more formidable; low crouch in a submissive response; and the waltz plus wing-flutter in a displacement reaction. Birds are especially noted for their sexual behavior in the act of courtship.

■ **Chickens**—In chickens, the male typically takes the initiative in sexual behavior. There are a variety of approaches by which a cock may evoke a sexual response in the hen.

The most spectacular of these is the wing-flutter, waltz, or dance. These initial sexual reactions of cocks have been called *courting*. A hen may be indifferent to courting, or she may respond either negatively or positively. As a negative reaction, she may step aside, walk or run away, or struggle if captured. These types of avoidance of the male by the female may be accompanied by vocalizations varying in intensity from faint screams to loud squawks. A positive reaction to courting takes the form of a crouch, often with head low and wings spread. This behavior has been called a *sex invitation*, or *crouching*. The sexual crouch is a strong stimulus for the cock to mount and tread, particularly when the rooster approaches from the rear. The male stands on the outstretched wings, grasps the comb or hackle of the hen, and moves the feet up and down in a treading manner. Subsequently, the male rears up, the hen moves the tail to one side, and each everts the cloaca as vents meet. The male usually steps off in a forward direction and the hen shakes herself vigorously as she gets to her feet. She may run in an arc and the cock may execute a waltz.

(Also see Fig. 9–6, p. 214.)

■ **Turkeys**—When attacked by a predator, young poults exhibit a frenzy of vocalization, dart about, freeze momentarily, and dash again in all directions.

■ **Ducks**—The hostile posture of ducks, characterized by a pronounced uptilting of the bill, is seen when two males are chasing, fighting, or threatening, or when males are competing in an attempt to rape a female.

The visual sexual displays of Mallard ducks are detailed under the heading "Sexual Behavior" and illustrated in Fig. 9–8, p. 215; hence, the reader is referred thereto.

HOMING; MIGRATION; AND TERRITORIAL BEHAVIOR

No single explanation is really satisfying relative to the homing and migrating of birds. Through sound, scent, or some sense of which we do not know, many birds, such as homing pigeons, find their way back home when moved to distant places. Each year, an estimated 10 billion American birds, ranging from tiny hummers to giant eagles, sweep in flood tide up the face of North America, sending their ripples into every field and woodlot. It is suspected that birds use the sun and stars as compasses to guide them, but there is no proof; it is one of nature's best-kept secrets, for which there is no complete answer.

■ **Quail**—The migrating behavior of quail is characterized by (1) the sexes not migrating together, with the males arriving first; (2) migrating flights occurring at night; and (3) the date of spring arrival varying in different latitudes from March to May, and departure dates from September to November.

Territoriality often results in wide spacing of individuals in natural species and elaborate behavioral patterns have evolved to maintain it. Territoriality in which individuals maintain minimal distances from each other is important for population control, food distribution, nesting, care of young, and inhibition of parasites and predators. When individuals are crowded together, as is common with domesticated species, the behaviors associated with territory must be modified.

Long-term modification of poultry can occur in display, threshold to fighting movements, or value of the birds. Short-term modification can occur through surgical procedure, such as debeaking. When natural species in captivity are subjected to crowding, feeding, sexual behavior, and care of young are adversely affected.

■ **Ducks**—The males of some dabbling ducks defend strict territories throughout the incubation period, whereas other species make little or no attempt to defend an area. Mallards are intermediate between these two extremes.

HOW POULTRY BEHAVE

Animals behave differently, according to species. Also, some behavioral systems or patterns are better developed in certain species than in others. Moreover, ingestive and sexual behavioral systems have been most extensively studied because of their importance commercially. Nevertheless, most animals, including poultry, exhibit the following eight general functions or behavioral systems, each of which will be discussed:

1. Ingestive behavior.
2. Eliminative behavior.
3. Sexual behavior.
4. Parental behavior (maternal behavior).
5. Agonistic behavior (combat).
6. Allelomimetic behavior (gregarious behavior).
7. Shelter-seeking behavior.
8. Investigative behavior.

INGESTIVE BEHAVIOR

This type of behavior includes eating and drinking; hence, it is characteristic of animals of all species and all ages. It is very important because animals cannot live without feed and water.

Each species has its own particular method of ingesting feed. Chickens and turkeys ingest their feed by pecking; ducks scoop their feed with their broad, soft bills. Except for geese (which graze on grass), poultry do not eat much forage.

The feed intake of poultry is governed by the energy concentration of the feed; that is, birds tend to eat to satisfy their energy requirements if fed free-choice. While the bulkiness of the feed can alter feed intake, for the most part the bird will eat to satisfy its energy needs. It follows that birds eat more when it is cold than when it is warm, because of the higher energy requirements.

■ **Chickens**—Chicks do not peck much until their second day after hatching, presumably due to ingestion of the yolk sac during hatching. Normal pecking experience requires some light. Initially, chicks peck and ingest both nutritive and nonnutritive substances.

Chickens prefer crumbles; scratch vigorously when grain is scattered in the litter; require increased feeding space with growth; peck and feed more in the presence of a feeding companion; and peck more when there are sounds associated with feeding, such as made by finger taps or a pecking model. After having been fed, a dominant bird may

return when its inferiors begin active feeding, thereby increasing its consumption and reducing the feed consumption of those in the lowest rank.

The number of birds feeding at any given time is influenced by dominance relations, hunger, and feeding space.

The water intake of chicks can be increased slightly by adding blue food coloring to the water supply.

■ **Turkeys**—Turkeys swallow without the necessity of raising the head. But drinking is accomplished differently; after getting a beak full of water, the bird raises its head, extends the beak upward and repeats several rapid closures of the beak.

Newly hatched poults peck indiscriminately at bright spots and small objects that contrast with the background. Thus, they may be encouraged to consume feed and water by the use of colored feed, or by placing brightly colored marbles in the feed and water. In a few days, this indiscriminate eating changes to selective eating and drinking.

Low-fiber, high-energy pelleted rations make for maximum feed efficiency and growth of turkeys. But, because such rations are consumed rapidly and the birds have more idle time, they predispose the flock to feather picking (denuded backs) and cannibalism.

■ **Ducks**—Mallard ducks employ several feeding methods, including (1) dabbling (as the birds swim or walk through mud puddles or along shorelines) of the bill along the water surface or in mud which achieves a straining of planktonic organisms and mud-dwelling invertebrates; (2) feeding from the bottom of ponds; and (3) snapping flying insects by rapid closing of the bill.

Drinking by ducks involves a distinctive movement sequence; the bill is dipped in water and then lifted slightly above the horizontal.

■ **Quail**—The ingestive behavior of the quail is characterized by (1) sharp increases in feed intake shortly after the onset of morning light and during the period of approximately 3 hours prior to darkness; (2) preference for sweet and sour solutions, and avoidance of salt and bitter solutions; and (3) the ability to tolerate moderate levels of salt in the drinking water for prolonged periods, although they normally avoid a salt solution.

ELIMINATIVE BEHAVIOR

In recent years, elimination has become a most important phenomenon, and pollution has become a dirty word. Nevertheless, nature ordained that if animals eat, they must eliminate.

A full understanding of the eliminative behavior will make for improved animal building design and give a big assist in handling manure. Right off, it should be recognized that the eliminative behavior in farm animals tends to follow the general pattern of their wild ancestors; but it can be influenced by the method of management.

In mammals, urine and feces are voided separately; but in poultry the end products coming from the urinary and gastrointestinal tracts are mixed together and voided by the cloaca. In poultry droppings, the fecal material prevents much of the nitrogen from leaching out, which is the primary reason that poultry manure has a higher nitrogen content than manure from four-footed animals.

■ **Chickens** — Except when on the roost at night, chickens deposit their excreta at random.

■ **Turkeys** — Like chickens, turkeys deposit their excreta at random.

SEXUAL BEHAVIOR

Reproduction is the first and most important requisite of livestock breeding. Without young being produced, the other economic traits are of academic interest only. Thus, it is important that all those who breed animals should have a working knowledge of sexual behavior.

Sexual behavior involves courtship and mating. It is largely controlled by hormones.

Each animal species has a special pattern of sexual behavior. As a result, interspecies matings do not often occur.

■ **Chickens** — In chickens, which are polygamous, sexual behavior is usually referred to as mating behavior. Mating in chickens is preceded by various behavior patterns known as displays or courting, which synchronize sexual activities of males and females (see Fig. 9–6).

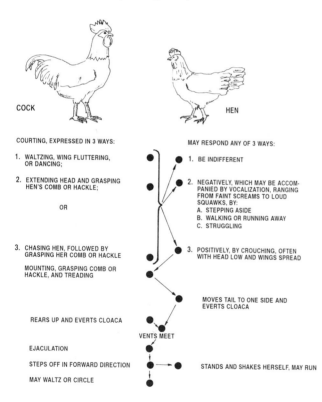

Fig. 9–6. Mating behavior in chickens, showing chain reaction between rooster (left) and hen (right).

Preferential or nonrandom mating has been observed in both cockerels and hens.

Additional pertinent information relative to the sexual behavior of chickens follows: (1) both early social experience and hormonal level influence the level of sexual behavior; (2) the earlier the separation of males from females, the less the

sexual performance of males; (3) the relation of sex hormones and sexual behavior is well established; (4) when several cocks are placed with a flock of hens, the dominant cock (the one ranking highest in the peck order) is usually most successful in mating, fertilizing a large number of eggs, and siring the most chicks; (5) dominance among hens interferes with mating, because such hens do not readily submit to the crouch; and (6) there is a difference in the mating frequency of males belonging to different sire families.

■ **Turkeys** — The mating behavior in turkeys tends to follow a chain reaction similar to chickens. That is, the behavior of one sex partner elicits a specific response from the other, and, in turn, that response elicits a further response from the first partner. However, the following differences exist between the sexual behavior of chickens and turkeys: (1) The movements of chickens during courtship and mating are more rapid and the feather display of the male is less elaborate than in turkeys; (2) typically, male chickens force matings with hens, but this behavior is not seen in turkeys; (3) turkey females sometimes follow the male and crouch near him, whereas female chickens seldom approach the cock before crouching; and (4) although males of both species move away if another male attempts to mount, cockerels may tread each other whereas male turkeys will not.

Fig. 9–7. Courtship display in the male turkey, commonly referred to as "strutting," is characterized by slow and restricted movements and elaborate feather display. (Courtesy, *Poultry Tribune*, Mount Morris, IL)

■ **Ducks** — Ducks form pair-bonds. Both males and females breed in their first year; and normally they raise only one brood annually.

Unlike most birds, drakes (and other male waterfowl) have a well-developed penis.

Social courtship displays and displays associated with pair formation occur only during the autumn, winter, and early spring months.

Female Mallard sexual displays include the following:

1. **Inciting**, a distinctive display associated with pairing and maintenance of the pair-bond, in which the female walks, swims, or flies beside or behind a chosen male, while making

threatening movements over one shoulder and uttering loud calls. Inciting appears to indicate the female's preference for one male and her rejection of others.

THE "RAB RAB" CHATTER OF A MALLARD PAIR (HEN LEFT; DRAKE RIGHT) CALMS BOTH MEMBERS.

A FEMALE (IN FOREGROUND) NOD-SWIMMING AMONG THE DRAKES.

REJECTION, GESTURE USED BY THE MALLARD FEMALE WHEN HARASSED BY DRAKES INTENT ON RAPE.

IN THIS POSTURE, THE MALLARD DRAKE UTTERS THE ATTRACTION AND WARNING CALL, "RAB."

HEAD-SHAKE OF THE DRAKE.

THE GRUNT-WHISTLE OF THE DRAKE.

THE HEAD-UP, TAIL-UP OF THE DRAKE.

Fig. 9–8. Social courtship displays of Mallard ducks.

2. **Nod-swimming**, associated with social courtship, in which the female swims rapidly among the drakes with her head stretched forward over the water surface. Head-nodding, in which the head is held high but moved forward and backward, occurs in similar situations. Both actions stimulate males to perform displays.

3. **Repulsion**, characteristic of incubating or brooding females when they are harassed by drakes intent on rape, in which the head is tucked back, the bill is opened, the plumage on the back and flanks is ruffled, the tail is fanned, and loud, harsh calls are emitted.

4. **Distraction displays**, given when incubating Mallards are flushed from the nest. The female flaps across the ground or water, thrashing with both wings and giving loud quacks.

Male Mallard sexual displays include the following:

1. **Preliminary displays**, which occur ahead of the burst of major display. The male adopts a characteristic posture with the head sunk in the shoulders and the head feathers ruffled, followed by head-shakes, head-flicks, prolonged tail-wags, and swimming-shakes.

2. **Major courtship displays**, characterized by the sudden occurrence of one or the other of the following social displays:

a. **Grunt-whistle**, characterized by lowering of the bill to the water surface; arching the body upward; flicking the bill to one side and sending a fine spray of water toward the female; and uttering a loud whistle followed by a grunt.

b. **Head-up, tail-up**, a complex display involving a sudden raising of the head along with simultaneous vertical cocking of the tail and raising of the closed wings so that the scapulars almost touch the back of the head. The movement is accompanied by a loud whistle. As the wings and tail are lowered, the male turns his head so as to point his bill toward a female (the mate in the case of paired birds.)

c. **Down-up**, involving a sudden complex of body actions in which the drake dips his breast deeply into the water; jerks his bill upward and outward, flipping up a column of water as he goes; raises his tail high out of the water, and emits a whistle when the head is at the highest level.

3. **Ritualized jump-flights**, in which, in rapid succession, the drake rises steeply with head extended horizontally, flies 3 to 10 ft, alights a short distance in front of the female, and drinks.

4. **Bridling**, in which the drake pulls his head back so that the breast is raised slightly from the water. This display normally follows copulation.

Ducks form pair-bonds each year, although the strength and duration of the pair-bond and the extent to which males try to steal matings vary between species. Generally speaking, the drakes of dabbling ducks are noted for being unfaithful. Most Mallard pair-bonds break while the female is incubating, the males moving off to molting areas. To succeed in forming a pair-bond, each male and female must attract and hold the attention of a potential mate, select and indicate a preference for one individual, and discourage rivals. Mated birds reaffirm their attachment to one another by staying close together, synchronizing their activities, performing displays together, and by threatening rivals.

Copulation may be normal or forced (rape). The female controls the frequency of normal copulations, but the male determines when rape will occur. Drakes attempt to rape females other than their mate frequently.

■ **Quail**—Sexual development of domestic quail is photic controlled. At all times other than spring, quail maintained at approximately 32.5° of latitude require supplemental light for rapid sexual maturation and maintenance of reproductive functions. Keeping birds in continuous dark results in almost complete inhibition of sexual behavior and egg production, whereas keeping birds in continuous light results in a very high rate of egg production during the first year, with production coming to a halt at the end of the second year.

Fig. 9–9. Male Japanese quail courting. The male approaches the female, walking stiff-legged on his toes, feathers erect, and the neck extended horizontally.

■ **Geese**—The larger breeds mate best in pairs or trios, although ganders of some lighter breeds will mate satisfactorily with an many as five females. Canadian wild geese are largely monogamous and will usually mate that way for life.

■ **Guinea fowls**—They have a tendency to mate in pairs; however, one male may be mated with three or four hens.

■ **Pigeons**—They mate in pairs and usually remain with their mates throughout life, although the mating may be changed if desired by placing the male and female in a coop together, leaving them there for 6 to 14 days or until they become settled.

■ **Peafowl**—The hens are usually grouped with four or five hens to one cock bird.

■ **Swans**—They live in pairs and remain faithful to each other until death.

PARENTAL BEHAVIOR (MATERNAL BEHAVIOR)

The parental behavior is largely confined to females among domestic poultry; and the care-seeking behavior is normal for young birds. Parental behavior and care-seeking vary widely among different species.

■ **Chickens**—Modern incubators have replaced the setting hen and precluded the need for broody hens. As a result, breeders have increased egg production by selecting against this trait. Also, with incubators, few hens and chicks are allowed to run together. Nevertheless, when this happens, the maternal behavior is intense. For example, when foraging for food, a mother hen will cluck to her chicks and call them each time a choice morsel is found. All the chicks will come running to participate in the find. Hens hover over their chicks by covering them with their spreading wings and nestling them close to their bodies during the night or at other periods when they rest or need protection. A hen with chicks exhibits a definite antagonistic behavior and will attack any enemy that bothers her young. To warn her chicks of danger, she emits a loud, shrill cry. The chicks react quickly and seek protection.

■ **Turkeys**—Parental care is usually absent in the male but assumed by the female.

Domestic hens adapt to laying in artificial nests, although some individuals must be trained to use them.

The termination of laying and the start of incubation are closely related. Decreased levels of broodiness are desirable under domestication. Fortunately, selection against broodiness is very effective.

Turkeys hatch at approximately 27 days after the eggs are set with pipping of the shell the first outward sign. Within about 8 hours after hatching, the young poults are dry and start to move about slowly. Poults imprint to the mother and are soon able to discriminate her individual calls. They crouch, rest, or sleep under the mother's wings or body as she hovers over them. The mother defends her brood against intruders. She may attack by hissing, running in a low crouch, and jumping at the intruder with extended claws and beating wings. At other times, the hen may hover over her brood and remain motionless; or, if disturbed, she may feign injury as she flies off in a wobbly flight—directing attention away from the poults. Also, the mother sounds various alarm calls as strange objects are sighted on the ground or in the air (see the section on "Habituation" for the responses of poults to the danger call).

■ **Ducks**—Natural behavior in nest-site selection, nest-building, and egg-laying do not exist under domestication; hence, the discussion that follows pertains to Mallards.

Mallards arrive on the breeding grounds in flocks of pairs, homing to the area which they used the previous year. Females inspect potential nesting sites—on the ground in dense grasses, under bushes, in open fields, in holes in trees, and a variety of other situations; and a nest, consisting of dead grasses and small twigs, along with pieces of down from the female's breast, is gradually built during the egg-laying and incubation periods.

Eggs are laid during the morning at the rate of one a day, with an occasional day missed. Once the clutch (consisting of 8 to 10 eggs) is complete, the female sets on the eggs constantly with the exception of one or two short periods off each day.

The incubation period of Mallard eggs averages 22 to 24 days, with extremes of 21 to 28 days. (Domestic birds of Khaki Campbell and Indian Runner breeds have a 28-day incubation period.)

When setting (incubating), Mallard hens spend most of the time sleeping. When approached while nesting, they often remain frozen in a crouched posture with feathers tightly sleeked and head lowered. Domestic hens nesting in close association with humans may refuse to leave the nest even when touched. Such birds will assume an intimidating posture with body feathers erected and tail fanned; hiss; and lunge and peck at the intruder. Incubating birds leave the nest 1 to 3 times each day to feed, drink, and bathe.

Hatching behavior embraces three stages: (1) pipping, (2) emergence and drying, and (3) brooding on the nest. Emergence of all ducklings normally takes 3 to 8 hours following the first clicking noises of hatching. Thereafter, the ducklings remain in the nest and are brooded for 12 to 24 hours. Ducklings have two common kinds of vocalizations: (1) contentment notes, and (2) distress calls.

Young ducklings are brooded by their mother, but as they grow larger they sleep beside her. The mother protects them from danger and guides them to feeding areas. She also leads them to favorite resting areas, where they bathe, oil, and sleep.

The phenomenon of imprinting is important in ducklings, with the optimum period being between 13 and 16 hours after hatching. When exposed to an object (a box, wooden duck decoy, a human being, and even flickering lights) during this period, they become attached to it, follow it when it moves, sit near it and utter contentment notes when it is stationary, and emit distress calls if it is removed. Sexual imprinting can result in undesirable fixations. When male Mallard ducklings are raised with a group of males only (without any females), they show no interest in females; instead, they direct their courtship displays to males only and form strong homosexual bonds.

■ **Quail** — Nest-building, which is usually not started until laying of the first egg, is done by the female alone; and normally she incubates the eggs. Clutches average about six eggs. Hens may produce two broods within a season.

AGONISTIC BEHAVIOR (COMBAT)

This type of behavior includes fighting, flight, and other related reactions associated with conflict. Among all species, males are more likely to fight than females. Nevertheless, females may exhibit fighting behavior under certain conditions. Castrated males are usually quite passive, which indicates that hormones, (especially testosterone) are involved in this type of behavior. Thus, farmers have for centuries used castration as a means of producing docile males.

The intensity of fighting depends upon the tenacity of the two combatants. Although fighting rarely results in death, it usually continues until one gives up.

■ **Chickens** — The Greeks regarded the cock as the symbol of pugnacity. Legend has it that the sight of two fighting cocks had emboldened the Athenians to take up the struggle with the Persians. In commemoration, annual cockfights were held in Athens and other Greek cities. The Romans added iron spurs to the sharp claws of the cocks, so that a fight became a lethal business. Henry VIII of England was passionately fond of cockfighting and drew up a set of rules for the sport.

In chickens, agonistic behavior includes attack, escape, avoiding, and submissive behavior.

When a number of strange hens are placed together in a pen, fights occur by twos until each bird has engaged all the others. The winner of each initial contest thereafter has the right to peck the loser, and the latter usually avoids the former. Some individuals give way without a fight, whereas others may challenge the winner again and again before dominance is settled. At subsequent meetings, one member of each pair pecks or threatens the other, definite dominance-subordination patterns become habitual and the peck order is established. Submission in hens is characterized by crouching. Combats between hens are much less serious than between roosters.

Where roosters are run with a flock of hens, fighting may continue for several days with both combatants covered with blood. Eventually, one gives up and escapes by running away.

A mixed flock of males and females has two peck orders, one for each sex.

Breed differences in the agonistic behavior of chickens are very great. Usually the meat breeds are quite placid, whereas in game chickens (cockfighting), special strains or breeds of chickens are selected because of their agonistic behavior. In passing, it is noteworthy that the sport of cockfighting exercised a tremendous influence in the domestication of wild chickens, and also in the subsequent distribution of chickens throughout the world. Much importance was attached to the pastime of cockfighting by many human races. The literature of various nations contains many references to this sport, and it would appear that cockfighting had as much to do with the domestication of the fowl as the demand for food and that the sport was chiefly instrumental in the widespread distribution of chickens that followed.

Factors other than breed differences which affect the peck order of chickens are: appearance, aggressiveness, level of gonadal hormones, and experience.

■ **Turkeys** — Fighting generally occurs when two strange birds meet for the first time, and the winner of the fight subsequently dominates the loser. Groups of young turkeys raised together establish their social hierarchies at 3 to 5 months of age, after which fighting is reduced. Males fight more vigorously than females. Also, some varieties of turkeys are better fighters than others.

Fig. 9–11. Termination of a male turkey fight. The defeated bird cowers on the ground while the victor walks threateningly around him, but no longer attacks him.

Fig. 9–10. Greek cockfight (3rd century B.C.).

Turkeys exhibit a variety of threat displays, ranging from mild threat (head raised, looking at the opponent) to strutting (against other males). During the threat display, the birds vocalize in a distinctive trill of relatively high pitch.

Actual fighting begins by both birds jumping at each other with their feet extended forward. If one bird lands a stroke on the other's back, the latter gives up. Following 1 to 20 jumps, combat shifts to a tugging battle, in which the head is darted forward to grasp the caruncles, snood, wattle, or beak of the opponent. Simultaneously, they tug and push in an attempt to force the opponent's head downward. A fight is terminated by submission, with the loser retracting its snood, lowering its head, attempting to hide under the victor's breast, and fleeing. The winner may chase and peck the defeated bird.

■ **Ducks**—Dabbling duck males defend strict territories throughout the incubation period, but other species make little or no attempt to defend an area. Mallards are intermediate in this respect. During this phase of the breeding cycle, males may chase each other and engage in fights, usually on the ground. Also males participate in "rape fights," in which they pursue a female over a long distance, then compete for an opportunity to rape her when she alights.

■ **Quail**—Agonistic behavior in quail consists of pecking and grabbing the neck and head, especially around the eyes. In some parts of Asia, quail are used in cockfighting.

ALLELOMIMETIC BEHAVIOR

Allelomimetic behavior is mutual mimicking behavior. Thus, when one member of a group does something, another tends to do the same thing; and because others are doing it, the original individual continues.

Under domestication, animals are usually protected from predators. Nevertheless, the allelomimetic behavior still has important consequences. By stimulating each other, animals produce the phenomenon known as *social facilitation.* Thus, the competition between animals makes for higher feed consumption.

■ **Chickens**—Pecking and feeding are facilitated by social stimulation; e.g., the presence of a feeding companion.

■ **Turkeys**—Social facilitation of the feeding and drinking of poults is provided by other poults or by the hen as she pecks the ground and calls.

GREGARIOUS BEHAVIOR

Gregarious behavior refers to the flocking or herding instinct of certain species. It is closely related to allelomimetic behavior. If animals imitate each other, they must stay together. If they stay together as a mobile group, they must use allelomimetic behavior to do so. All such behavior arises out of the process of social attachment.

Gregarious behavior differs among species.

Chickens, turkeys, ducks, and geese tend to flock together. Of course, under domestication, people have interfered with the normal flocking instinct of both chickens and turkeys. Even under domestication, however, ducks and geese exhibit their gregarious or flocking nature as they walk single file, one behind the other.

SHELTER-SEEKING BEHAVIOR

All species of animals seek shelter—protection from the sun, wind, rain and snow, insects, and predators.

In the wild, one of the ancestors of chickens (the red Burmese jungle fowl, ancestor of the Leghorn breed) and the ancestors of turkeys lived in forests where they had natural shelter and roosted in trees. Even today, turkeys show little tendency to seek shelter in a severe blizzard. It is not unusual, therefore, for turkey losses to be high in a severe storm.

■ **Chickens**—When cold, chicks and adults huddle together.

■ **Turkeys**—Poults seek shelter under the mother's wings or body as she hovers over them.

INVESTIGATIVE BEHAVIOR

All animals are curious and have a tendency to explore their environment. Investigation takes place through seeing, hearing, smelling, tasting, and touching. Whenever an animal is introduced into a new area, its first reaction is to explore it.

The investigative behavior exhibited by chickens and turkeys is more casual and subtle than with the four-footed animals. They will, however, walk slowly toward a new object, looking it over as they go. Usually, they stop a short distance from the object, then turn and go on their way.

BEHAVIORAL NORMS

The poultry producer needs to be familiar with behavioral norms of birds in order to detect and treat abnormal situations—especially illness.

Some of the signs of good health are:

1. Contentment.
2. Alertness.
3. Eating with relish.
4. Sleek feathers and pliable and elastic skin.
5. Large, smooth, bright red comb.
6. Soft, loose skin.
7. Normal feces.
8. Normal temperature, pulse rate, and breathing rate.

The above signs of good health, including normal temperature, pulse rate, and breathing rate of different species of farm animals, are detailed in Chapter 10 of this book. Other behavioral norms, knowledge of which is important in poultry production, are: locomotion, thermoregulation, grooming, and sleeping.

LOCOMOTION

Locomotion is observable in healthy animals from birth, and it is important throughout their lives. Prior to domestication, survival often depended upon all species being able to travel to their feed and water, which often involved great distances, and even migration.

■ **Chickens**—The ancestors of such breeds as the Leghorn probably roosted in the trees. But modern chickens are larger than their ancestors and poorly adapted to flying.

■ **Turkeys**—Wild turkeys prefer walking to flying. They will run to escape their predators, although they will fly if necessary.

Domestication of turkeys, with selection for heavier weights and broad breasts, has had a negative effect on locomotion, adversely affecting both mating and fertility.

■ **Ducks**—Wild Mallards are good walkers, swimmers, and fliers. Domestic ducks, with abundant food supply and protection from predators, along with increased size, have greatly reduced locomotion° abilities. The heaviest breeds waddle, rather than walk, and they have lost the power of flight.

THERMOREGULATION

The relatively high deep-body temperature, the absence of sweat glands, and the very effective insulation provided by the plumage distinguish the thermoregulatory physiology of birds.

Thermoregulation behavior usually involves movement of the entire bird or part of the bird, such as the foot, in response to either environmental or body temperature.

The most conspicuous thermoregulatory behavior of birds is migration to warmer or cooler areas.

Poultry also respond to thermal stress by increasing body temperature and respiration rates, decreasing feed consumption, increasing evaporation cooling of the air sac system and oral mucosa surfaces, spreading their wings as they crouch on the ground (so that air can circulate past the less insulated undersurface of the wings, and so that squashing of the breast feathers facilitates heat loss to the soil), and opening the mouth and panting.

■ **Chickens**—In a cold environment, the chicken reduces its surface area, and hence its heat loss, by *hunching*. An additional reduction in heat loss, amounting to 12%, may be achieved by tucking the head under the wing. A still further saving of 20 to 50% can be made if the chicken sits rather than stands, thereby reducing the heat loss from the unfeathered legs and feet.

In warm weather, the wings are held away from the body and the scapular feathers are elevated, both responses facilitating heat loss to the air.

GROOMING

When not confined, chickens and turkeys groom themselves by taking dust baths.

All waterfowl spend a considerable amount of time caring for their bodies. Among such activities are: shaking movements to remove water from the feathers; distributing oil to the feathers from the uropygial gland above the tail, and wetting the feathers during bathing. Apparently these activities are necessary to keep the feathers in good condition, and of special importance in preserving waterproofing and thermoregulatory properties.

■ **Ducks**—The very considerable grooming activities of ducks include the following: (1) several distinct shaking movements involving the head, body, tail, and wings; (2)

stretching movements; (3) cleaning movements of the head, feet, bill, and eyes; (4) oiling-preening, the sequence of which is (a) bathing and removal of water, (b) taking oil from the oil gland and distributing it over the feathers, (c) prolonged nibbling, (d) stretching movements, and (e) stopping preening and going to sleep.

BATHING-SOMERSAULTING

AFTER SWIMMING BODY SHAKE

Fig. 9–12. Two of the series of grooming movements executed by ducks.

SLEEPING

Normal behavior in sleep should be recognized, especially since it differs widely between species.

Chickens and turkeys wind their claws tightly around a pole, or roost, and snuggle closely together. They like to close their eyes and hide their heads in the feathers of their wings. Ducks and geese sleep on both land and water. On land they often drowse while standing on one leg; on water they paddle every now and then, in order not to drift ashore. The eyes frequently blink open; and sleeping birds can become fully alert instantly if disturbed.

ABNORMAL BEHAVIOR

Abnormal behavior in animals develops where there is a combination of confinement, excess stimulation, and forced production, along with a lack of opportunity to adapt to the situation. Also, it is recognized that confinement of animals makes for lack of space; this often leads to unfavorable changes in habitat and social interactions for which the species have become adapted and best suited over thousands of years of evolution.

Homosexual behavior is common where adult mammals of one sex are confined together.

Abnormal behavior often provides a way in which to recognize diseases early. Sick birds usually eat less, may be dull and inactive, and may isolate themselves from the rest of the flock. Layers produce fewer eggs, and fertility and hatchability of eggs decline. Specific signs of disease may be evident, such as diarrhea, paralysis, coughing, wheezing, sneezing, inflammatory exudates of the skin, or blood in the droppings.

■ **Chickens**—The most common abnormal behavior observed in chickens in confinement is cannibalism. This trait may be encountered among birds of all ages. Among baby chicks, the trouble is usually confined to toe and tail picking. With mature birds, the vent, tail, and comb are the regions most frequently picked.

The cause of cannibalism is not fully understood. It is known that it is more frequent under confined conditions. Without doubt, it may be accentuated by deficiencies in management and nutrition. Also, it may be brought on by just plain boredom or too much light.

The best way to control cannibalism is by debeaking. Many broiler producers, and some egg producers, have their chicks debeaked at the hatchery. Additionally, layers can be debeaked at the time of housing in the laying house.

Chickens confined to small cages in laying batteries will develop stereotyped head movements.

■ **Turkeys**—Feather picking (from the backs) and cannibalism increase with low-fiber, high-energy pelleted rations. This has been attributed to the birds having too much idle time. More than twice as much time is required to eat the same feed as a mash as in pelleted form; and birds eating mash spend far more time wiping their beaks through the feathers of other birds in a cleaning movement.

Desnooding, or the cutting off the snood of the young poults, is practiced by some growers to prevent injury among mature males. It is believed that this practice helps prevent the spread of erysipelas in the flock. Most hatcheries will desnood poults for a small fee upon request.

SOCIAL RELATIONSHIPS

Social behavior may be defined as any behavior caused by or affecting another animal, usually of the same species, but also, in some cases, of another species.

Social organization may be defined as an aggregation of individuals into a fairly well-integrated and self-consistent group in which the unity is based upon the interdependence of the separate organisms and upon their responses to one another.

When we restrict or confine birds and force them into spaces that bring them within the individual distance that has been established, we immediately create stress. Thereupon, the dominants have to pay more attention to maintaining their dominance and to protecting their own field of territory. They will have to be more aggressive in their reactions. The subservients become far more nervous, and their nervousness spreads throughout the group.

DOMINANCE

Within most groups of animals of the same species, there is a well-organized social rank. Animals observe this order in their relationships just as carefully as protocol demands that it be observed at a *State Dinner*.

In chickens, in which dominance was first observed, the social rank order is called the *peck order* (see earlier section headed, "Agonistic Behavior [Combat]").

A social rank order similar to the peck order in chickens exists in other species of poultry.

Once the social rank order is established, it results in a peaceful coexistence of the flock. Thereafter, when the dominant bird merely threatens, the subordinate one submits and avoids conflict. Of course, there are some pairs that fight every time they chance to meet. Also, if strange birds are introduced into such a group, social disorganization results in the outbreak of new fighting, as a new social rank order is established.

Of course, social rank becomes of importance when a group of birds is fed in confinement; and it becomes doubly important if limited feeding is practiced. Under such circumstances, the dominant individuals crowd the subordinate ones away from the feed hopper, with the result that they may go hungry.

Several factors influence social rank; among them, (1) age—both young birds and those that are senile rank toward the bottom; (2) early experience—once a subordinate in a particular flock, usually always a subordinate; (3) weight and size; and (4) aggressiveness or timidity. Also, it is noteworthy that social rank is influenced by hormones; for example, a capon (castrated male chicken) automatically goes to the bottom of the totem pole, whereas the injection of roosters and hens with the male sex hormone, testosterone, increases their social rank.

Dominance and subordination are not inherited as such, for these relations are developed by experience. Rather, the capacity to fight (agonistic behavior) is inherited, and, in turn, this determines dominance and subordination.

■ **Turkeys**—Both male and female turkeys develop a typical peck order with each sex forming an independent hierarchy in heterosexual groupings.

Dominant turkey hens peck the backs of the necks of subordinate birds; hence, the area of denudation of a hen's neck is a good indicator of the number of birds pecking that individual.

INTERSPECIES RELATIONSHIPS

Social relationships are normally formed between members of the same species. However, they can be developed between two different species. In domestication this tendency is important (1) because it permits several species to be kept together in the same area, and (2) because of the close relationship between people and animals. Such interspecies relationships can be produced artificially, generally by taking advantage of the maternal instinct of females and using them as foster mothers. It is not unusual, for example, to set duck eggs under a hen. All goes well until the young ducklings take to water for their first swim. Thereupon, the mother hen becomes quite excited, fearful that her babies will drown (little realizing that swimming comes naturally for young ducks).

FILIAL BEHAVIOR

Filial behavior, as used herein, refers to the relation of the young to the mother hen. Call plays an important roll in early socialization of poultry, especially in relation to the mother hen.

■ **Chickens**—In response to the mother hen's warning note, the chicks remain silent. Also, chicks give positive responses to the hen's purring sound when she settles down to brood, to sources of warmth, and to attractive moving objects (imprinting). Repeated exposure to the broody hen, accompanied by protection and food guidance, strengthens the filial bond.

■ **Turkeys**—Young poults respond to calls from the mother hen. Strange objects or animals on the ground elicit an alert call or cluck to which young turkeys respond by dispersing over a wide area in a creeping posture. Upon sighting a high-flying predator, the mother gives a "singing" note, which calls the poults to attention. As the predator comes close enough to spell danger, the mother sounds another call. To the latter, the poults respond by dashing violently to cover or the corners of a pen, pile in clumps, and remain motionless for several minutes.

SOCIAL STRESS

Social stress refers to those changes in social behavior and population density which may influence growth, reproductive performance, and many types of stimuli or the environment.

■ **Poultry**—All domestic poultry are subjected to a variety of stresses, including heat and cold, deficiencies in feed and water; number of birds together; lack of space; diseases; and social competition. Other reactions involve the endocrine glands, the rectifying of which is very complex.

APPLIED POULTRY BEHAVIOR

At the outset of this chapter, it was stated that the application of animal behavior depends upon understanding. The presentation to this point has been for the purpose of understanding. Let us next turn to some practical applications of poultry behavior.

It should be recognized that domestication of poultry relieved them of certain stresses, but introduced new ones. Among natural species, wide spacing of individuals is common, and elaborate behavioral patterns evolved to maintain it. In the wild, it was important that animals maintain minimal distances from each other for reasons of population control, feed availability, nesting, care of the young, and control of predators, diseases, and parasites. With domestication, people exerted control over all these—and more. They even provided mates. But captivity introduced new stresses—boredom, invasion of personal space, increased production, and even artificial mating. The challenge of the modern poultry producer is either to change the environment and/or to breed and select birds that are adapted to an artificial environment—birds that not only survive, but thrive, under the conditions that are imposed upon them.

BREEDING FOR ADAPTATION

The wide variety of livestock in different parts of the world reflects a continuous process of natural and artificial selection which has resulted in the survival of animals well adapted to climate and other environmental factors.

Animal behavior can be changed through heredity and selection as evidenced by the following changes in poultry.

1. The reproductive cycle of chickens has been changed, with egg laying occurring throughout much of the year.
2. Domestic cocks selected for fighting show elements of fighting behavior with minimal stimulation, with breed differences evident.
3. Females of some breeds of turkeys have such low thresholds for tactile stimulation that they refuse the full mating stimulation of the male.

COMPANIONSHIP

Companionship in animals is of great practical importance. Except for the cat, all domestic animals are highly social and have constant need for companionship.

If not too crowded, placing animals together sometimes accomplishes two things: (1) greater feed consumption, due to the competition between them (mutual facilitation); and (2) a quieting effect.

The best-known animal companionship of all pertains to high-strung racehorses and stallions, in which all sorts of companions are used—a chicken, a duck, a pony, a sheep, or a goat. Such companions are commonly referred to as *mascots*.

The great Stymie, Thoroughbred winner of $918,485, became attached to a hen of nondescript breeding that came to dinner one day and never left.

MAKING IT RIGHT WITH ANIMALS

People need animals, and animals need people. As primitive people adopted a more settled mode of life, there came the desire to safeguard their food supply for times when hunting was poor and to have their food close at hand; at this stage, nearly all our modern animals were tamed or confined, or, as we say, domesticated.

In domesticating animals, caretakers recognized the importance of behavior; they selected those species which could both be tamed and used to satisfy their needs. However, in the breeding, care, and management that followed, behavior received less attention than the quantity and quality of meat, milk, eggs, fiber, and power produced. The race was on for greater rate and efficiency of production. Animals in forced production were confined and automated. Then, suddenly, in sign language that spoke louder than words, animals told us that all was not well in the barnyard. They told us that something was missing—something as vital to them as an essential amino acid, mineral, or vitamin—something as important as disease prevention and environmental control. They told us that consideration of their habitat and social organization had not kept pace with advances in genetics, nutrition, environmental control, and other areas of animal care. They told us what was wrong and what they wanted through a whole host of abnormal behaviors, including cannibalism, loss of appetite, poor parental care, over-aggressiveness, dullness, and degenerate sexual behavior. These warning signals are being heeded. Today, there is renewed interest in the study and application of animal behavior; we are trying to make it right with animals by correcting the causes of the disorders. For the time being, this calls for emulating the natural conditions of the species, including their space requirements, social organization, and training and experience. Over the long pull, it calls for breeding and selecting animals better adapted to artificial environments. It is hoped that the principles and applications of animal behavior presented in this chapter will speed the process.

ANIMAL ENVIRONMENT

Environment may be defined as all the conditions, circumstances, and influences surrounding and affecting the growth, development, and production of animals. The most important influences in the environment are the feed and quarters (space and shelter).

The branch of science concerned with the relation of living things to their environment and to each other is known as ecology.

Poultry producers were little concerned with the effect of environment on birds so long as they were kept on the range. But rising feed, land, and labor costs, along with the concentration of birds into smaller spaces, changed all this. Today, most broilers and layers are maintained in confinement throughout life; many layers are kept in cages, and essentially all broilers are on litter floors. Turkeys are shifting rapidly from range to confinement. Water is important for ducks, but even with ducks the trend is toward higher population densities and more confinement.

In animals, environmental control involves space requirements, light, air temperature, relative humidity, air velocity, wet bedding, ammonia buildup, dust, odors, and manure disposal, along with proper feed and water. Control or modification of these factors offers possibilities for improving animal performance. Although there is still much to be learned about environmental control, the gap between awareness and application is becoming smaller. Research on poultry environment has lagged, primarily because it requires a melding of several disciplines—nutrition, physiology, genetics, engineering, and climatology. Those engaged in such studies are known as ecologists.

In the present era, pollution control is the first and most important requisite in locating a new poultry establishment, or in continuing an old one. The location should be such as to avoid (1) the neighbors complaining about odors and insects; and (2) pollution of surface and underground water. Without knowledge of poultry behavior, or without pollution control, no amount of capital, native intelligence, and sweat will make for a successful poultry enterprise.

HOW ENVIRONMENT AFFECTS POULTRY

No matter how good the genetics, a good environment is essential to obtain high production.

Also, selection of genetically superior individuals to be parents of the next generation is hampered by environmental factors that tend to mask the actual breeding values of indi-

viduals being selected. Thus, the contribution of these environmental factors to the total phenotypic variation should be minimized before estimating the genetic parameters.

The environmental factors affecting animals vary.

The following factors are of special importance in any discussion of poultry environment:

1. Feed and nutrition. 4. Health.
2. Light. 5. Stress.
3. Weather.

FEED AND NUTRITION

Poultry may be affected by either (1) too little feed, (2) rations that are too low in one or more nutrients, (3) an imbalance between certain nutrients, or (4) objection to the physical form of the ration.

Forced production and feeds which are often produced on leached and depleted soils have created many problems in nutrition. These conditions have been further aggravated through the increased confinement of poultry. Under these unnatural conditions, nutritional diseases and ailments have become increasingly common.

LIGHT

Light affects both the growing bird and egg production. It not only acts by way of the eye, but also penetrates the skull and appears to exert a direct effect on the diencephalon of the pituitary to increase gonadotrophin secretion. Under natural sunlight, hormonal secretions are activated once the total length of the light day reaches 11 to 12 hours, as in the spring months. During the winter, the length of the light day is not normally long enough to make for maximum egg production; hence, poultry producers must use artificial lighting. Added light can simulate lengthening days or compensate for decreasing amounts of natural light. Diminishing day lengths encourage growth of adolescent poultry, however.

Light greatly influences sexual maturity—egg production in females and semen production in males. Light also acts as a signal to govern the time of day eggs are laid. It is noteworthy, too, that birds are more stimulated by light near the red end of the spectrum than by light in the green, blue, or violet regions. Light from almost any commercially produced bulb or tube will be satisfactory for such stimulation, provided that the intensity and duration of light are controlled.

(Also see Table 8–3, Recommended Lighting Systems for Poultry.)

WEATHER

Extreme weather conditions can cause wide fluctuations in poultry performance. The difference in weather impact from one year to the next, and between areas of the country, causes difficulty in making a realistic analysis of buildings and management techniques used to reduce weather stress.

The research data clearly show that environmentally controlled buildings almost always improve production and feed efficiency. The issue is clouded only because the additional costs incurred by shelters have frequently exceeded the benefits gained by the improved performance, particularly in those areas with less severe weather and climate.

The temperature of the environment has a large influence on the feed intake. The warmer the environment, the less the feed intake; therefore, the requirements for all nutrients, expressed as a percent of the diet, is dependent upon the environmental temperature.

Turkey performance data indicate that growth, feed conversion, and fat deposition are greatest in the range of 60° to 70°F.

ENVIRONMENTALLY CONTROLLED BUILDINGS

With the shift of poultry to confinement structures and high-density production operations, building design and environmental control became more critical. Limited basic research has shown that birds are more efficient—that they produce and perform better, and require less feed—if raised under ideal conditions of temperature, humidity, and ventilation.

However, the per bird cost is much higher for environmentally controlled facilities. Thus, the decision on whether or not confinement and environmental control can be justified should be determined by economics. Will the birds in environmentally controlled quarters produce sufficiently more efficiently to justify the added cost? Of course, manure disposal and pollution control should also be considered.

There is still much to be learned about environmental control, but the gap between awareness and application is becoming smaller.

Environmentally controlled buildings are costly to construct, but they make for the ultimate in bird comfort, health, and efficiency of feed utilization. Also, like any confinement building, they lend themselves to automation, which results in a saving in labor; and, because of minimizing space requirements, they effect a saving in land cost. If they malfunction, however, they can cause birds to suffocate and result in large economic losses. Today, environmental control is rather common in poultry housing; and it is on the increase.

Before an environmental system can be designed for poultry, it is important to know their (1) heat production, (2) vapor production, and (3) space requirements. This information is as pertinent to designing poultry buildings as nutrient requirements are to balancing rations. This information is presented in Chapter 11, Poultry Houses and Equipment.

HEALTH

Diseases and parasites (external and internal) are ever present poultry environmental factors. Death takes a tremendous toll. Even greater economic losses result from lowered egg production, retarded growth, poor feed efficiency, carcass condemnations, decreases in meat quality, and in labor and drug costs. The signs of good health are summarized earlier in this chapter in the section headed, "Behavioral Norms"; hence, the reader is referred thereto.

Any departure from the signs of good health constitutes a warning of trouble. Most sicknesses are ushered in by one or more signs of poor health—by indicators that tell expert caretakers that all is not well—that tell them that their birds will go off feed tomorrow, and that prompt them to do something about it today.

STRESS

Stress is defined as physical or psychological tension or strain. Stress of any kind affects birds. Among the external forces which may stress birds are previous nutrition; abrupt ration changes; change of water, space, level of production, number of birds together, quarters, or mates; irregular care; transporting; excitement; presence of strangers; fatigue; illness; management; temperature; and abrupt weather changes.

Birds experience many periods of stress.

CONTROL POLLUTION

Pollution is a major issue. It matters little whether pollution is due to agriculture or factories. Everything that defiles, desecrates, or makes impure or unclean streams or atmosphere, or mars the landscape, must be controlled.

We must ever be mindful that life, beauty, wealth, and progress depend upon how wisely people use nature's gifts—the soil, the water, the air, the minerals, and the plant and animal life.

Poultry producers need to give particular attention to any pollution that may be caused by manure, fertilizer, insecticides, herbicides, and growth stimulants.

Today, there is worldwide awakening to the problem of pollution of the environment (air, water, and soil) and its effect on human health and on other forms of life. Much of this concern stems from the amount and concentration of manure produced by the sudden increase of animals in confinement. Certainly, there have been abuses of the environment (and it has not been limited to agriculture). There is no argument that such neglect should be rectified in a sound, orderly manner; but it should be done with a minimum disruption of the economy and lowering of the standard of living.

In altogether too many cases, extreme environmentalists advocate policy changes and legislation that may in the end be detrimental to agriculture, to our food production potential, and to society in general. Frequently, these activists have only used the data that support their theories about ecological doom. One of their favorite comparisons deals with the relative magnitude of the effect on the environment caused by animal manure, industrial waste, and municipal waste.

POLLUTION LAWS AND REGULATIONS

Invoking an old law (the Refuse Act of 1899, which gave the Corps of Engineers control over runoff or seepage into any stream which flows into navigable waters), the U.S. Environmental Protection Agency (EPA) launched a program to control water pollution by requiring that all cattle feedlots which had 1,000 head or more the previous year must apply for a permit by July 1, 1971. The states followed suit; although differing in their regulations, all of them increased legal pressures for clean water and air. Then followed the Federal Water Pollution Control Act Amendments, enacted by Congress in 1971, charging the EPA with developing a broad national program to eliminate water pollution.

Owners/operators of animal feeding facilities with more than 1,000 animal units must apply. Animal units are computed as follows: multiply number of slaughter and feeder cattle by 1.0; multiply number of mature dairy cattle by 1.4; multiply swine weighing over 55 lb by 0.4; multiply number of sheep by 0.1; and multiply number of horses by 2.0. (See Table 9–1, footnote 1, for what constitutes 1,000 animal units, including poultry.)

TABLE 9–1
SUMMARY OF REGULATIONS

Feedlots with 1,000 or More Animal Units[1]	Feedlots with Less than 1,000 but with 300 or More Animal Units[2]	Feedlots with Less than 300 Animal Units
Permit required for all feedlots with discharges[3] of pollutants.	Permit required if feedlot— 1. Discharges[3] pollutants through an unnatural conveyance, or 2. Discharges[3] pollutants into waters passing through or coming into direct contact with animals in the confined area. Feedlots subject to case-by-case designation requiring an individual permit only after on-site inspection and notice to the owner or operator.	No permit required unless— 1. Feedlot discharges pollutants through an unnatural conveyance, or 2. Feedlot discharges pollutants into waters passing through or coming into direct contact with the animals in the confined area, and 3. After on-site inspection, written notice is transmitted to the owner or operator.

[1]More than 1,000 feeder or slaughter cattle, 700 mature dairy cows (milked or dry), 2,500 swine weighing over 55 lb *(24.9 kg)*, 500 horses, 10,000 sheep or lambs, 55,000 turkeys, 100,000 laying hens or broilers with continuous overflow watering, 30,000 laying hens or broilers with liquid manure handling, 5,000 ducks; or any combination of these animals adding up to 1,000 animal units.

[2]More than 300 slaughter or feeder cattle, 200 mature dairy cows (milked or dry), 750 swine weighing over 55 lb *(24.9 kg)*, 150 horses, 3,000 sheep, 16,500 turkeys, 30,000 laying hens or broilers with continuous overflow watering, 9,000 laying hens or broilers with liquid manure handling, 1,500 ducks; or any combination of these animals adding up to 300 animal units.

[3]Feedlot not subject to requirement to obtain permit if discharge occurs only in the event of a 25-year, 24-hour storm event.

Before constructing a poultry facility, the owner should become familiar with both state and federal regulations. The state regulations can be secured from the state water board. They differ from state to state, but most states require a catch basin (detention pond) sufficient to contain the runoff from a storm of the magnitude of the largest rainfall during a 48-hour period of the most recent ten years.

MANURE

Poultry operations located near centers of population are having an increasing number of complaints lodged against them because of manure and odor. Lawsuits, based on the nuisance law, have been filed against some of them.

No doubt, the manure pollution problem, suspicioned or real, will persist. However, the energy crisis, accompanied by increased chemical fertilizer and feed prices, has caused manure to be looked upon as a resource and not a waste that presents a disposal problem. As a result, a growing number of American farmers are returning to organic farming—they are using more manure—the unwanted barnyard centerpiece of years gone by. They are discovering that they are just as good reapers of the land and far better stewards of the soil. Additionally, more and more poultry manure is being fed to ruminants.

In the future, as fertilizer and feed become increasingly scarce and expensive, the economic value of manure will increase.

From the standpoints of using manure as a fertilizer and pollution control, it is important that the poultry producer (1) exercise certain precautions, and (2) know how much manure can be applied to the land. These and other matters pertinent to the subject are fully discussed in Chapter 8, Poultry Management, under the heading "Poultry Manure."

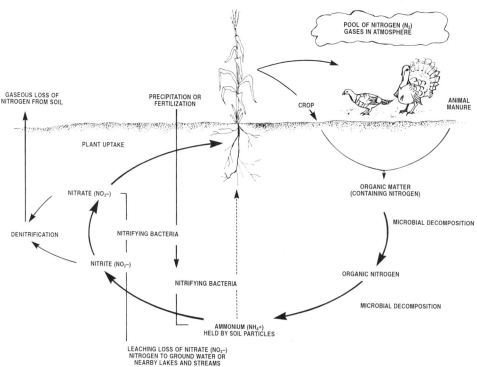

Fig. 9–13. The cycle of nature: manure applied to soils is decomposed by microbes; complex protein in manure is broken down to release nitrogen as ammonia (NH_4^+); aerobic microbes convert ammonium nitrogen to nitrite (NO_2^-), thence to nitrate (NO_3^-) nitrogen; nitrate is either (1) taken up by plants and built back into protein compounds; (2) leached downward when the soil is saturated—contaminating surface and groundwater if excessive nitrogen has been applied to the soil; and/or (3) released into the atmosphere when soils are wet for extended periods of time and the absence of air causes anaerobic microbes to convert the nitrate nitrogen to gaseous forms.

AGRICULTURAL CHEMICALS

In the everyday pursuit of modern agriculture, more and more chemicals and drugs are being used. Hand in hand with this development, there has been increased public concern over the use of these products, for fear of poisoning human food.

Chemicals and drugs must be used with discretion, especially those designed to kill some living organism. But sometimes choices must be made; for example, between malaria-carrying mosquitoes and some fish, or between hordes of locusts and grasshoppers and the crops that they devour. This merely underscores the need for (1) careful testing through properly designed experiments of all products prior to use, (2) conforming with federal and state laws, and (3) accurate labeling and use of products. Additionally, food producers need to relate the miracle of agriculture. They need to tell that back of food and clothing are agricultural chemicals and drugs—herbicides, insecticides, pesticides, disease control materials, feed additives, fertilizers, and many others. They need to show how these products are as indispensable to modern food production as tractors, trucks, hybrid seeds, and improved livestock.

In an era of food shortages, losses of feed, food, and fiber will increasingly concern all people, producers and consumers alike.

Poultry producers have an obligation to produce more foods efficiently, to reduce production costs, and to increase their income. In turn, consumers—all—benefit from agricultural chemicals and drugs which produce more products at less cost.

When properly used, agricultural chemicals are an important adjunct in providing feed for animals and food for people. However, improper use can result in toxicity. Moreover, certain chemicals can accumulate in animal products.

The vast majority of agricultural chemicals and drugs have been properly used. Of course, it should not be too surprising that a few have been improperly used when it is realized that there are approximately 300,000 trade name products on the market.

When chemical poisoning or drug misuse happens, it can be both devastating and perplexing. Usually, the causative agent can be diagnosed after an investigation of the environment and the feed. However, few poisons can be diagnosed with certainty by clinical symptoms alone. When trouble is encountered, the producer should promptly call a veterinarian if animals are involved, and a medical doctor if people are involved.

A voluminous amount of information is available on the deleterious effects of poisonous chemicals and drugs—both artificial and natural.

Poultry producers know that unless they follow state and federal regulations they risk having their products condemned and seized, or refused by food processors. Nevertheless, the economics dictate that new products be used as soon as they prove useful. On the other hand, food faddists may feel that they are being poisoned; wildlife conservationists may be concerned over possible damage to songbirds and other animals; beekeepers become unhappy if insecticides kill honeybees; and public health agencies are concerned about contamination of soil, water, and food supplies.

CONSERVE ENERGY

Currently, the global use of resources is increasing at an alarming rate—far faster than population.

When considering the energy required to produce food products, it is insufficient to focus entirely on energy use on the farm. Energy is consumed during the extraction of raw materials, in the manufacture of farm inputs (fertilizer, farm machinery, feed, pesticides, petroleum products), in processing, and in the distribution of food products. Fig. 9–14 and Table 9–2 illustrate basic energy utilization.

Table 9–2 points up the increasing drain that modern food production is putting on the energy supply. In 1990, U.S.

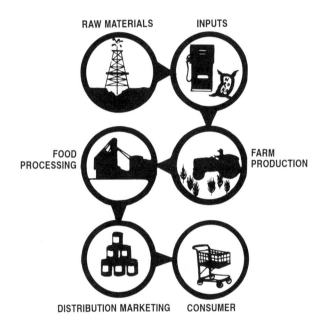

Fig. 9–14. Energy use in the food system.

TABLE 9–2
MODERN FOOD PRODUCTION IS INEFFICIENT IN ENERGY UTILIZATION—THE STORY FROM PRODUCER TO CONSUMER[1]

Year	On the Farm	Food Processing	Marketing and Home Cooking	Total/ Person/ Year
1940[2]				
Million kcal . . .	0.9	2.2	2.1	5.2
Percent	18.0	42.0	40.0	100.0
1990[3]				
Million kcal . . .	2.8	5.7	4.6	13.1[4]
Percent	21.4	43.5	35.1	100.0
Increase, times, 1940–1990 . . .	3.1	2.6	2.2	2.5

[1]Energy in million kcal used per capita to produce one million kcal of food in the U.S.

[2]Values from Borgstrom, G. "The Price of a Tractor," *Ceres*, FAO of the U.N., Rome, Italy, Nov.-Dec., 1974, p. 18, Table 3.

[3]Author's estimate based on several reports detailing trends in energy usage.

[4]This means that in 1990, it required 13.1 million kcal to produce 1 million kcal of food for each person, a daily consumption of 2,740 kcal (1,000,000 ÷ 365 = 2,740).

farms put in 2.8 calories of fuel per calorie of food grown, 3.1 times more than the on-farm energy input in 1940.

Table 9–2 also shows that, in the United States in 1990, a total of 13.1 calories were used in the production, food processing, and marketing-cooking for every calorie of food consumed, with a percentage distribution of the total cost of energy at each step from producer to consumer as follows: on the farm, 21.4%; food processing, 43.5%; and marketing and home cooking, 35.1%. In 1940, it took only 5.2 calories — about 40% of the 1990 figure — to get 1 calorie of food on the table. It is noteworthy, too, that more energy is required for food processing and marketing-home cooking than for growing the product; and that, from 1940 to 1990, the on-the-farm energy requirement increased by 3.1 times, in comparison with an increase of 2.6 and 2.2 times for each of the other steps — processing and marketing-home cooking.

Prior to the advent of machines and fuel in crop production, 1 calorie of energy input on the farm produced about 16 calories of food energy. Today, on the average, U.S. farms put in about 2.8 calories of fuel per calorie of food grown; hence, to produce a daily intake of 3,000 calories of edible food from cultivated crops may require 8,400 calories of energy from fossil fuels — an exhaustible source. It is more surprising yet — and thought-provoking — to know that, even today in the poorer or developing countries, it takes only 1 calorie to produce each 10 calories of food consumed. The Oriental wet rice peasant uses only 1 unit of energy to produce 50 units of food energy. This gives the Orientals a favorable position among the major powers as the energy crisis worsens.

It is noteworthy that U.S. agriculture, in all its phases from farm to consumer, accounts for about 15% of total U.S. energy consumption.[1] It is noteworthy, too, that the United States now uses about one-third of the world's energy.

We can, and we must, use energy more efficiently and wisely — and already much has been accomplished.

The type of energy used is very important, especially should substitutions become necessary. Presently, the energy inputs in agriculture include petroleum, natural gas, and electricity. About half of the energy consumed in the food system is petroleum, and another 30% is natural gas. Almost 90% of the energy used in the manufacture of fertilizer and

other inputs is natural gas. Gasoline and diesel fuels for tractors, combines, and other farm machinery comprise four-fifths of the direct energy requirements for crop and livestock production. Electricity, which is a secondary form of energy, has many applications throughout the food industry.

The following additional points are pertinent to any energy conservation program:

1. **Photosynthesis fixes energy**. Photosynthesis is by far the most important energy-producing process. But currently only about 1% of the solar energy falling on an area is fixed by photosynthesis; and only 5% of this captured energy is fixed in a form suitable as food for people. Thus, (a) manipulation of plants for increased efficiency of solar energy conversion, and (b) converting a greater percentage of total energy fixed as chemical energy in plants (the other 95%) into a form available to humans would appear to hold great promise in solving the future food problems of the world.

2. **Other potential energy sources**. Other potential energy sources that may be considered are the production of alcohol from either grain or crop residues, the production of methane gas from animal wastes, and the production of energy from direct burning of agricultural residues.

Alcohol can be produced from either crop residues or grain. High-cellulose residue materials can be converted to sugars, which can then be fermented to produce alcohol, or grain can be fermented directly to produce alcohol. However, the energy yield from either process is quite low.

Methane can be generated from animal wastes. But this does not appear to be a likely source of much energy.

Nor does the direct burning of agricultural residues as a source of energy appear promising. It would be costly because of the relatively low energy densities of these materials. Also, energy costs of residue transportation would be so high as to make net energy return very low.

Increasing consideration will be given to the conservation of energy by the methods listed above. Additionally, farmers will conserve energy by using minimum tillage techniques, and by switching to fuel-conserving diesel tractors which use approximately 73% as much fuel as a gasoline tractor in performing equivalent work. Also, up to a point, big farms utilize energy more efficiently than little farms; hence, energy shortages favor the trend to bigness.

We must be ever mindful that photosynthesis is by far the most important energy-producing process.

(Also see Chapter 8, Poultry Management, section entitled, "Save Energy.")

---

[1]Slick, W. J., Jr., Senior Vice President, Exxon Company, U.S.A., paper entitled, "Energy, Agriculture, and the Future," address before the National Council of Farmer Cooperatives, San Francisco, CA, Jan. 11, 1978.

QUESTIONS FOR STUDY AND DISCUSSION

1. Discuss the habitat and biological differences between (a) domestic chickens, turkeys, and ducks, and (b) their respective ancestors.

2. Why do we need to breed poultry better adapted to an artificial environment?

3. What is imprinting? Who did the pioneering work in this field, and with what species did he work?

4. Do chickens have "memory"? Justify your answer.

5. Discuss each of the following senses of perception in poultry: (a) vision, (b) hearing, (c) taste, (d) smell, (e) touch, and (f) temperature.

6. How can the natural ability of turkey poults to discriminate color be used when starting them on feed?

7. Discuss the (a) vocal and (b) visual displays of chickens.

(Continued)

8. Discuss the importance of homing, migration, and territorial behavior of domestic chickens, turkeys, and ducks.

9. How may a modern chicken breeder utilize each of the following behavioral systems:

a. Ingestive behavior
b. Eliminative behavior
c. Sexual behavior
d. Parental behavior
e. Agonistic behavior
f. Allelomimetic behavior
g. Shelter-seeking behavior
h. Investigative behavior?

10. What are the chief differences between chickens, turkeys, and ducks in each of the following behavioral systems:

a. Ingestive behavior
b. Sexual behavior
c. Parental behavior
d. Agonistic behavior?

11. How has the domestication and selection of turkeys affected locomotion, mating, and fertility?

12. Describe the thermoregulation of chickens.

13. What causes cannibalism in chickens and feather picking in turkeys? How can these abnormal behaviors be rectified?

14. How is the *peck order* in a flock of chickens established?

15. Has artificial incubation and brooding changed the filial (mother-young) behavior of chickens and turkeys?

16. Why was it important that animals maintain minimal distances from each other in the wild?

17. What new stresses came with animal domestication?

18. The challenge of the modern poultry breeder is either (a) to change the environment, or (b) to breed and select birds that are better adapted to an artificial environment. Which alternative would you recommend?

19. Define (a) environment and (b) ecology.

20. How do the following environmental factors affect each—chickens, turkeys, and ducks:

a. Feed and nutrition
b. Light
c. Weather
d. Health
e. Stress?

21. Why and how should pollution be controlled?

22. What are the pertinent federal guidelines relative to pollution control?

23. What is the highest and best use for poultry manure?

24. Discuss the pros and cons of using agricultural chemicals, including herbicides, insecticides, pesticides, and other chemicals.

25. In what ways can a poultry producer conserve energy?

SELECTED REFERENCES

Title of Publication	Author(s)	Publisher
Behavior of Domestic Animals, The, Third Edition	Edited by E. S. E. Hafez	The Williams and Wilkens Company, Baltimore, MD, 1975
Bibliography of Livestock Waste Management	J. R. Miner D. Bundy G. Christenbury	Office of Research and Monitoring, U.S. Environmental Protection Agency, Washington, DC, 1972
Biology of Stress In Farm Animals: an integrative approach	P. R. Wiepkema P. W. M. van Adrichem	Kluwer Academic Publishers, Hingham, MA, 1987
Brazil's Imperiled Rain Forest, Rondonia's Settlers Invade	W. S. Ellis	*National Geographic*, Vol. 174, No. 6, pp. 772–799, Dec., 1988
Concise Survey of Animal Behavior, A	E. K. Honore P. H. Klopfer	Academic Press, Inc., Harcourt Brace Jovanovich, Publishers, San Diego, CA, 1990
Development and Evolution of Behavior	Edited by L. R. Aronson, *et al.*	W.H. Freeman and Company, San Francisco, CA, 1970
Domestic Animal Behavior	J. V. Craig	Prentice-Hall, Inc., Englewood Cliffs, NJ, 1981
Effect of Environment on Nutrient Requirements of Animals	D. R. Ames, Chairman	NRC, National Academy Press, Washington, DC, 1981
Environmental and Functional Engineering of Agricultural Buildings	H. J. Barre L. L. Sammet G. L. Nelson	Van Nostrand Reinhold Co., New York, NY, 1988
Environmental Biology	P. L. Altman D. S. Dittmer	Federation of American Societies for Experimental Biology, Bethesda, MD, 1966
Environmental Control for Agricultural Buildings	M. L. Esmay J. E. Dixon	The AVI Publishing Company, Inc., Westport, CT, 1986

(Continued)

SELECTED REFERENCES *(Continued)*

Title of Publication	Author(s)	Publisher
Environmental Management in Animal Agriculture	S. E. Curtis	Animal Environment Services, Mahomet, IL, 1981
Ethology, The Biology of Behavior, Second Edition	I. Eibl-Eibesfeldt	Holt, Rinehart and Winston, New York, NY, 1975
Farm Animal Manures: an overview of their role in the agricultural environment	J. Azevedo P. R. Stout	Agricultural Publications, University of California, Berkeley, CA, 1974
Guide to Environmental Research on Animals, A	R. G. Yeck, Chairman	NRC, National Academy of Science, Washington, DC, 1971
Health Issues Related to Chemicals in the Environment: A Scientific Perspective	A. L. Craigmill, Chairman	Council for Agricultural Sciences and Technology, Ames, IA, 1987
Impact of Stress, The, Proceedings	Edited by R. E. Moreng J. R. Herbertson	Colorado State University, Ft. Collins, CO, 1986
Introduction to Animal Behavior, An,: ethology's first century, Second Edition	P. H. Klopfer	Prentice-Hall, Inc., Englewood Cliffs, NJ, 1974
Livestock Behavior, a practical guide	R. Kilgour C. Dalton	Westview Press, Boulder, CO, 1984
Livestock Environment, Proceedings, Second International Livestock Environment Symposium	D. S. Bundy, Planning Chairman	American Society of Agricultural Engineers, St. Joseph, MI, 1982
Mechanisms of Animal Behavior	P. Marler W. J. Hamilton	John Wiley & Sons, New York, NY, 1966
Organic Farming: current technology and its role in a sustainable agriculture	Edited by D. M. Kral	American Society of Agronomy, Madison, WI 1984
Our Friendly Animals and Whence They Came	K. P. Schmidt	M. A. Donohue & Co., Chicago, IL, 1938
Portraits in the Wild	C. Moss	Houghton Mifflin Company, Boston, MA, 1975
Poultry Welfare, Proceedings	Edited by R. M. Wegner	German Branch, The World Poultry Science Association, Federated Agricultural Research Centre, Braunschweig-Volkenrode, Germany, 1985
Principles of Animal Behavior	W. N. Tavolga	Harper & Row, New York, NY, 1969
Principles of Animal Environment	M. L. Esmay	The AVI Publishing Company, Inc., Westport, CT, 1978
Readings in Animal Behavior	Edited by T. E. McGill	Holt, Rinehart and Winston, New York, NY, 1973
Safe and Effective Use of Pesticides, The	P. J. Marer	University of California Publications, Oakland, CA, 1988
Scientific Aspects of the Welfare of Food Animals	F. H. Baker, Chairman	Council for Agricultural Science and Technology, Ames, IA, 1981
Social Hierarchy and Dominance	Edited by M. W. Schein	Dowden, Hutchinson & Ross, Inc., Stroudsburg, PA, 1975
Social Space for Domestic Animals	Edited by R. Zayan	Kluwer Academic Publishers, Hingham, MA, 1985
Social Structure in Farm Animals	G. J. Syme L. A. Syme	Elsevier Scientific Publishing Co., Amsterdam, The Netherlands, 1979
Stress Physiology In Livestock Vol. 1, *Basic Principles* Vol. 2, *Ungulates* Vol. 3, *Poultry*	M. K. Yousef	CRC Press, Inc., Boca Raton, FL, 1985
Structures and Environment Handbook		Midwest Plan Service, Iowa State University, Ames, IA, 1972
Utilization, Treatment, and Disposal of Waste on Land, Proceedings	E. C. A. Runge, President of Society	Soil Science Society of America, Inc., Madison, WI, 1986
Wild Animals in Captivity	H. Hediger	Dover Publications, Inc., New York, NY, 1964

Fig. 10–1. Pictures of health, evidenced by these "Golden Comets." (Courtesy, Hubbard Farms, Walpole, NH)

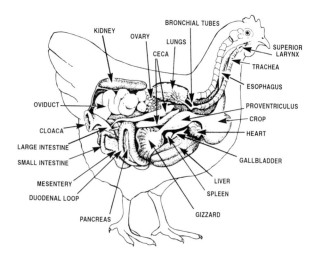

Fig. 10–2. Identification of poultry diseases calls for knowledge of the internal organs, as changes in them often provide the clues to the disease. (Adapted by the author from material provided by Salsbury Laboratories, Charles City, IA)

Fig. 10–3. When poultry are placed in close confinement, in great numbers, under forced production, and eat and sleep in close contact with their own body discharges, the control of diseases and parasites becomes of paramount importance. (Courtesy, DEKALB AgResearch, Inc., DeKalb,

Healthy birds are a requisite for profit; unhealthy birds cause financial losses. This fact is generally recognized by poultry producers.

Deaths, condemnations, and medications make for tremendous losses. But even greater economic losses may result from decreased egg production and growth, lowered feed efficiency, and downgrading.

Not all poultry losses can be prevented, but they can be reduced; and the more knowledgeable producers and those who counsel with them are, the more successful they will be in instituting and carrying out programs of poultry health, disease prevention, and parasite control. To the latter end, this chapter is presented.

NORMAL SICKNESSES AND LOSSES

Certain rules of thumb are helpful in evaluating the importance of a developing disease problem; among them, are the following:

1. **More than 1% of the birds sick at one time.** If the poultry producer finds that more than 1% of the flock is sick at any one time, it should be obvious that a disease problem is present and that it should be given immediate attention.

2. **First 3 weeks' losses; 2% chickens, 3% poults.** During the first 3 weeks of a chick's life, mortality losses generally average about 2%. For turkey poults, they run about 1% higher. If losses are greater than these figures, there may be cause for alarm.

3. **After 3 weeks, 1% per month.** The normal mortality after three weeks of age should not exceed 1% per month. However, a slight rise in mortality can be expected as adult flocks come into egg production.

SIGNS OF GOOD HEALTH

It is important to know the *normal* in order to recognize and evaluate the *abnormal*.

The signs of good health in poultry are:

1. Normal feed and water consumption, droppings, egg production, sounds, and actions. Good poultry producers sense immediately when any of these conditions are abnormal and interpret them as signs of abnormal health.
2. Normal rectal temperatures:
    Average, 106°F
    Range, 105° to 107°F
3. Normal pulse rate:
    200 to 400 per minute
4. Normal breathing rate:
    15 to 36 per minute

SYMPTOMS OR SIGNS OF DISEASE

Any departure from health is known as disease, the manifestations of which are known as symptoms or signs.

Many symptoms are general—manifestations that are seen in any diseased bird, such as droopiness, ruffled feathers, diarrhea, loss of appetite, drop in egg production or lowered gains, shell quality faults, and not sounding right. Other symptoms are more specific—they are usually seen only when certain diseases are present; examples of such specific symptoms are tremors associated with avian encephalomyelitis (epidemic tremors) and the flaccid paralysis associated with botulism.

Visible changes in the size, color, shape, or structure of an organ are known as lesions. Like symptoms, lesions may be general or specific. For example, enteritis is associated with many diseases, but the "gray eye" of ocular leukosis is specific.

CAUSES OF AVIAN DISEASES

The study of the causes of diseases is known as *etiology*.

A disease often results from a combination of two or more causes: (1) the indirect or predisposing factors which may lower the bird's resistance, and (2) the direct factors which produce the actual disease.

Predisposing causes of disease are frequently referred to as *stress* factors. Among such factors are chilling, poor ventilation, overcrowding, inadequate feeding and watering space, and overmedication. Also, one disease may predispose another; for example, infectious bronchitis may predispose *air sac* infection.

The direct causes of disease are either infectious or noninfectious. Infectious diseases may be contagious or noncontagious. All contagious diseases are infectious, but all infectious diseases are not contagious. *A contagious disease is one that is transmitted readily from one individual or flock to another. An infectious disease is one produced by living organisms.* Most infectious diseases of poultry are contagious; however, a few such as aspergillosis are not.

Contagious diseases are the greatest threat to poultry health. They are caused by bacteria, viruses, rickettsia, and fungi.

When living agents such as bacteria enter the body and multiply, they produce a disturbance of function, and infection occurs. Disease is caused by the chemical toxins (poisons) produced by invading organisms. But in some protozoan diseases, such as coccidiosis, mechanical damage to tissues is an important factor.

The ability of an organism to cause disease in a host is known as its virulence or pathogenicity. Many microorganisms unable to cause disease under most conditions may cause disease under certain conditions. Therefore, such microorganisms would be considered pathogenic in that particular host under the existing conditions. On the other hand, some organisms almost always are pathogenic and produce diseases when they enter the body of a susceptible host. Some will invade the body of only one species of birds and are said to be specific for that particular species. For example, infectious bronchitis virus will cause disease only in the chicken. Other organisms affect a large number of species. For example, some of the Salmonella organisms affect a large variety of species including reptiles, rodents, domestic animals, poultry, and humans.

The ability of an organism to cause disease is not a fixed characteristic. It depends upon many factors, such as ability to invade tissues and produce chemical toxin. Often pathogenicity can be altered intentionally. This characteristic has been used in developing some vaccines. Variation in pathogenicity of organisms also partially explains why the same disease may present different forms and degrees of severity.

HOW CONTAGIOUS DISEASES ARE SPREAD

Some common ways contagious diseases are introduced into and spread within poultry flocks are:

1. Introduction of diseased birds.
2. Introduction of healthy birds which have recovered from disease but which are still carriers.
3. Contact with inanimate objects that are contaminated with disease organisms, such as poultry crates, feeders, and waterers.
4. Carcasses of dead birds that have not been disposed of properly.
5. Impure water, such as surface drainage water.
6. Rodents and free-flying birds.
7. Insects; for example, fowl pox is commonly transmitted by mosquitoes.
8. Shoes and clothing of people who move from flock to flock.
9. Contaminated feed and feed bags.
10. Contaminated soil or litter.
11. Airborne; organisms do not spread far through the air, but this source of infection can be a strong factor in heavily populated poultry areas.
12. Egg transmission; a number of diseases such as pullorum and fowl typhoid are transmitted from the hen to the chick through the egg.

BODY DEFENSES AGAINST DISEASE

The body has a well-developed defense mechanism that must be understood and utilized in controlling infectious diseases. Immunity is the ability to resist infection, although this ability can be overcome under certain conditions. Resistance is used interchangeably with immunity.

An animal has two types of protective mechanisms: (1) those that hinder or prevent invasion of organisms, and (2) those that combat agents which invade the body.

Mechanisms which hinder or prevent invasion of organisms include the intact skin and mucous membranes which create a direct barrier, secretions such as mucous which tend to dilute and wash out invading organisms, and cilia (hairlike projections on some mucous membranes) which, with wavelike action, move foreign material out of such structures as the trachea (windpipe).

Mechanisms which combat agents that invade the body include the white blood cells and circulating antibodies.

Immunity (resistance) may be either (1) inherited, or (2) acquired; and the acquired immunity may be either (a) active or (b) passive.

Inherited resistance may be complete or partial; for example, turkeys are not susceptible to laryngotracheitis, and although chickens are more resistant than turkeys to blackhead, they may become infected under certain conditions. Inherited resistance or susceptibility to lymphoid leukosis is well established, but no completely resistant breed or strain of chickens has been developed. Individual resistance is apparent in practically every disease outbreak in a poultry flock. Some birds fail to develop evidence of the disease, even though exposed to the same chances of infection.

While inherited immunity is important, acquired immunity is a more controllable reaction that can be used by the poultry producer. Acquired immunity is the reaction stimulated by the application of vaccines. The purpose of vaccines is to stimulate an active production of antibodies by safe means. Active immunity depends upon the production of antibodies within the body of each individual. Antibodies are proteins associated with the globulin fraction of the blood serum. Antibody production is not understood completely, but antibodies apparently are produced in the liver, spleen, and bone marrow. In general, antibodies are specific for the organism which stimulated their production; thus, immunity to one disease ordinarily does not provide immunity to others.

Passive immunity is the transfer of antibodies from the individual in which they are produced to another individual. This may be done by the injection of serum collected from an immunized individual. Antibodies also are transferred from the hen to the offspring through the egg; thus, hens that have had Newcastle disease transfer antibodies through the yolk to their chicks. Such passive immunity is an important consideration in vaccination programs. Passive immunity is of short duration and there is usually a marked decline in the antibody level within 21 to 30 days. Passive protection against infection usually lasts no longer than 4 to 6 weeks.

DISEASE PREVENTION AND CONTROL

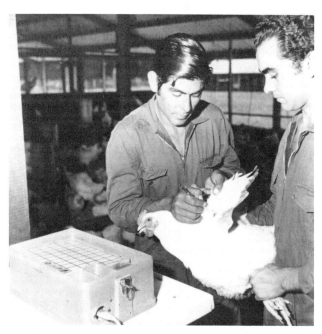

Fig. 10–4. Blood testing for MG (*Mycoplasma gallisepticum*) and MS (*Mycoplasma synovial*). Hatching eggs should be produced under MG- and MS-negative programs. (Courtesy, Arbor Acres Farm, Inc., Glastonbury, CT)

An effective program of disease prevention and control is important in today's intensive poultry operations. Although the exact program will and should vary according to the specific conditions existing on each individual poultry farm, the basic principles will remain the same. With this in mind, the following program is presented. Poultry producers may

use it (1) as a yardstick with which to compare existing programs, and (2) as a guide so that they and their veterinarian, and other advisors, may develop similar and more specific programs for their enterprises. Also, it must be recognized that cumulative changes in this program will need to be made from time to time, in light of new problems, new biologics, and new regulations. A modern and effective disease prevention and control program should embrace (1) a flock health program, (2) hatchery management and sanitation, (3) use of disinfectants, (4) vaccination, (5) use of a diagnostic laboratory, (6) treatment of disease outbreaks, and (7) consideration of Specific Pathogen Free (SPF) birds.

FLOCK HEALTH PROGRAM

There is no such thing as a standard disease prevention program that will apply to all poultry farms. Nevertheless, there are certain basic principles which should always be observed. These follow:

1. Select a well-known, reliable source from which to purchase chicks, poults, pullets, or hatching eggs—one that can supply healthy stock, inherently vigorous, and developed for a specific purpose.

2. Purchase only day-old chicks or hatching eggs. If necessary to purchase started birds, select the best possible source.

3. Regulate temperature, humidity, and ventilation during brooding to the comfort of the chicks and poults. Prevent drafts, overheating, and chilling.

4. Keep birds separate according to source and age groups. Mixing birds is an invitation to trouble.

5. Follow an "all-in, all-out" program.

6. Don't crowd birds. Crowding retards growth, reduces feed efficiency, and decreases production. Also, crowding increases cannibalism, feather picking, hysteria, and other stress-related problems.

7. Change litter and thoroughly clean and disinfect the house and equipment between each group of birds. While litter selection and management is a controversial subject, applying this recommendation as a general practice will prevent many disease and parasite problems.

8. Keep chickens and turkeys separate. Preferably, only chickens or only turkeys should be kept on the same premises.

9. Maintain hatchery supply flocks on separate premises from other birds.

10. Provide either a good commercial feed or a carefully formulated home-mixed feed.

11. Provide an adequate supply of clean, fresh water. Avoid watering from surface tanks, streams, or ponds.

12. Carry out a precise vaccination schedule for each flock. Develop the vaccination program in consultation with poultry disease authorities in each state or local area.

13. Control internal and external parasites. Feed is always too costly to give to parasites. Besides, controlling parasites keeps stress to a minimum and helps resist diseases.

14. Control vermin and screen out free-flying birds.

15. Discourage persons other than the caretaker or essential personnel from visiting the poultry house or yard.

16. If a disease problem develops, obtain an early, reliable diagnosis and apply the best treatment, control, and eradication measures for that specific disease.

17. Dispose of all dead birds as follows (listed in order of preference): incineration, pit, and deep burial.

18. Maintain good records relative to flock health. These should include vaccination history, disease problems, and medication employed.

HATCHERY MANAGEMENT AND SANITATION

The information contained in this section has been adapted from the recommended procedures of the National Poultry Improvement Plan.

■ **Hatching egg sanitation** — Hatching eggs should be collected from the nests at frequent intervals, and the following practices should be observed:

1. Use clean and disinfected containers in which to collect the eggs, and take precautions to prevent contamination from organisms that may be on the hands or clothing.

2. Maintain the identity of all eggs relative to the breeder flocks of origin.

3. Do not use dirty eggs for hatching. Collect them in a separate container from hatching eggs. Slightly soiled eggs may be dry cleaned by hand or by a motor-driven buffer.

4. Fumigate the eggs as described later in this section (under "Fumigation") as soon as possible after collection.

5. Store eggs in a cool place in properly cleaned and disinfected racks following fumigation. But hold the storage period to as short a time as possible before setting.

6. Use new or fumigated cases to transport eggs to the hatchery. Burn soiled egg case fillers.

■ **Hatchery sanitation** — An effective program for the prevention and control of Salmonella and other infections includes the following practices:

1. Arrange the hatchery building so that separate rooms, with separate ventilation, are provided for each of the four operations: (a) egg receiving, (b) incubation and hatching, (c) chick holding, and (d) disposal of offal and cleaning of trays. Place these rooms under isolation so that admission is permitted only to authorized personnel who have taken proper precautions to prevent introduction of diseases.

2. Fumigate all eggs prior to setting or within 12 hours after they are placed in the incubator. Also, fumigate them after transfer to the hatching compartment.

3. Clean thoroughly and disinfect frequently the hatchery rooms, tables, racks, and other equipment in them. Burn all hatchery wastes and offal or otherwise properly dispose of them. Clean and sterilize containers used to remove such materials after each use.

4. Clean thoroughly, then fumigate the hatching compartment of incubators, including hatching trays, after each hatch.

5. Distribute day-old chicks, poults or other newly hatched poultry in clean, new boxes or in disinfected plastic cartons. After each use, clean and disinfect all crates and vehicles used for transporting started or adult birds.

6. Maintain the identity of all chicks and poults relative to the breeder flock of origin.

7. Do not mix the progeny of different breeder flocks if it can be avoided.

Fig. 10–5. Thorough and frequent cleaning of the hatchery is important. (Courtesy, Foster Farms, Livingston, CA)

■ **Cleaning and disinfecting** — Poultry houses and hatcheries are quite different; hence, they involve different cleaning and disinfecting.

1. In poultry houses, cleaning and disinfecting should include these steps:

a. Settle dust by spraying lightly with disinfectant.

b. Remove all litter and droppings to an isolated area where there is no opportunity for dissemination of any infectious disease organisms that may be present.

c. Scrub the walls, floors, and equipment with a hot soapy water solution. Rinse to remove soap.

d. Spray with a suitable disinfectant, following the manufacturer's directions.

2. In the hatcheries, cleaning and disinfecting should include the following procedure:

a. Remove trays and all controls and fans for separate cleaning. Thoroughly wet the ceiling, walls and floors with a stream of water; then scrub with a hard bristle brush. Rinse until no deposits are on the walls, particularly near the fan opening.

b. Replace cleaned fans and controls. Replace trays, preferably while still wet from cleaning, and bring the incubator up to the normal operating temperature.

c. Fumigate the hatcher before inserting eggs.

d. If eggs are hatching and incubating in the same machine, clean the entire machine after each hatch. Use a vacuum cleaner to remove chick down from the egg trays.

■ **Fumigation**—Fumigation of eggs and incubators is an essential part of a hatchery sanitation program. When improperly done, fumigation can be a hazard; hence, it should always be done by, or under the supervision of, an experienced person.

Fig. 10–6. Hatching eggs in fumigation cabinet. (Courtesy, *Turkey World*, Mount Morris, IL)

Routine preincubation fumigation of hatching eggs on the farm is highly recommended for eliminating Salmonella infection from poultry flocks. Each egg entering the hatchery should have been subjected to preincubation fumigation as soon as possible after its collection from the nest.

High levels of formaldehyde gas will destroy Salmonella organisms on shell surfaces if used as soon as possible after the eggs are laid. An inexpensive cabinet for enclosing the eggs is required. Fans circulate the gas during the fumigation process and then exhaust it from the cabinet. Eggs for fumigation should be placed on racks so that the gas can reach the entire surface. Plastic trays used for washing market eggs are ideal for this purpose.

A high level of formaldehyde gas is provided by mixing 1.2 cubic centimeters (cc) of formalin (37% formaldehyde) with 0.6 g of potassium permanganate ($KMnO_4$) for each cu ft of space in the cabinet. An earthenware, galvanized, or enamelware container having a capacity at least 10 times the volume of the total ingredients should be used for mixing the chemicals. The gas should be circulated within the enclosure for 20 minutes, then expelled to the outside.

Humidity for this method of preincubation fumigation is not critical; but temperature should be maintained at approximately 70°F. Extra humidity may be provided in dry weather. Eggs should be set as soon as possible after fumigation and extra care taken to ensure that they are not exposed to new sources of contamination.

Eggs should be routinely refumigated after transfer to the hatchery to destroy organisms that may have been introduced as a result of handling. Recommendations for loaded-incubator fumigation vary widely, depending upon the make of the machine. Therefore the method, concentration, and duration of fumigation should be in accordance with the manufacturer's instructions.

DISINFECTANTS

A disinfectant is a bactericidal or microbicidal agent that frees from infection (usually a chemical agent which destroys disease germs or other microorganisms, or inactivates viruses).

The high concentration of birds and continuous use of modern poultry buildings often results in a condition referred to as disease buildup. As disease-producing organisms—viruses, bacteria, fungi, and parasite eggs—accumulate in the environment, disease problems can become more severe and be transmitted to each succeeding group of birds raised on the same premises. Under these circumstances, cleaning and disinfection become extremely important in breaking the life cycle. Also, in the case of a disease outbreak, the premises must be disinfected.

Under ordinary conditions, proper cleaning of poultry houses removes most of the microorganisms, along with the filth, thus eliminating the necessity of disinfection.

Effective disinfection depends on five things:

1. Thorough cleaning before application.

2. The phenol coefficient of the disinfectant, which indicates the killing strength of a disinfectant as compared to phenol (carbolic acid). It is determined by a standard laboratory test in which the typhoid fever germ often is used as the test organism.

3. The dilution at which the disinfectant is used.

4. The temperature; most disinfectants are much more effective if applied hot.

Fig. 10–7. Many disease problems can be avoided by thorough cleaning and disinfecting between broods. (Courtesy, DEKALB AgResearch, Inc., DeKalb, IL)

5. Thoroughness of application, and time of exposure.

In all cases, disinfection must be preceded by a very thorough cleaning, for organic matter serves to protect disease organisms and otherwise interferes with the activity of the disinfecting agent.

Sunlight possesses disinfecting properties, but it is variable and superficial in its action. Heat and some of the chemical disinfectants are more effective.

The application of heat by steam, by hot water, by burning, or by boiling is an effective method of disinfection. In many cases, however, it may not be practical to use heat.

A good disinfectant should (1) have the power to kill disease-producing organisms, (2) remain stable in the presence of organic matter (manure, feathers, soil), (3) dissolve readily in water and remain in solution, (4) be nontoxic to birds and humans, (5) penetrate organic matter rapidly, (6) remove dirt and grease, and (7) be economical to use.

The number of available disinfectants is large because the ideal universally applicable disinfectant does not exist. Table 10–1 gives a summary of the limitations, usefulness, and strength of some common disinfectants.

When using a disinfectant, *always read and follow the manufacturer's directions.*

Fig. 10–8. Worker scrubbing boots with a disinfectant before entering a facility, for the prevention of transmitting diseases. (Courtesy, Arbor Acres Farm, Inc., Glastonbury, CT)

TABLE 10–1
DISINFECTANT GUIDE[1]

Kind of Disinfectant	Usefulness	Strength	Limitations and Comments
Alcohol (ethyl-ethanol, isopropyl, methanol)	Primarily as skin disinfectants and for emergency purposes on instruments.	70% alcohol—the content usually found in rubbing alcohol.	They are too costly for general disinfection. They are ineffective against bacterial spores.
Boric acid[2]	As a wash for eyes, and other sensitive parts of the body.	1 oz in 1 pt water (about 6% solution).	It is a weak antiseptic. It may cause harm to the nervous system if absorbed into the body in large amounts. For this and other reasons, antibiotic solutions and saline solutions are fast replacing it.
Chlorines (sodium hypochlorite, chloramine-T)	Used (1) for egg dipping and washing, (2) in processing plants, (3) for sanitizing poultry drinking water, and (4) as a deodorant. Chlorines will kill all kinds of bacteria, fungi, and viruses, providing the concentration is sufficiently high.	Generally used at about 200 ppm for disinfection and 50 ppm for sanitizing.	They are corrosive to metals and neutralized by organic materials. Not effective against TB organisms and spores.
Cresols (many commercial products available)	Recommended for disinfecting houses, equipment, and footbaths. Cresols in fuel oil are the best disinfectants for dirt floors. Effective against tuberculosis and the red mite.	Cresol is usually used as a 2 to 4% solution (1 cup to 2 gal of water makes a 4% solution). Cresols can be incorporated in water, kerosene, or fuel oil.	Effective on organic material. Cannot be used where odor may be absorbed.
Formaldehyde (may be used as a gas or as a liquid)	Effective against viruses, bacteria, and fungi. It is often used to disinfect buildings following a disease outbreak. It is commonly used in fumigating hatchery eggs, prior to and during incubation.	As a liquid disinfectant, it is usually used as a 1 to 2% solution. As a gaseous disinfectant (fumigant), use 1½ lb potassium permanganate plus 3 pt of formaldehyde. Also, gas may be released by heating paraformaldehyde.	It has a disagreeable odor, destroys living tissue, and can be extremely poisonous. The bactericidal effectiveness of the gas is dependent upon having the proper relative humidity (above 75%) and temperature (about 86°F).
Heat (by steam, hot water, burning, or boiling)	In the burning of rubbish or articles of little value, and in disposing of infected body discharges. The steam "Jenny" is effective for disinfection (example: poultry equipment) if *properly employed*, particularly if used in conjunction with a phenolic germicide.	10 minutes' exposure to boiling water is usually sufficient.	Exposure to boiling water will destroy all ordinary disease germs but sometimes fails to kill the spores of such diseases as anthrax and tetanus. Moist heat is preferred to dry heat, and steam under pressure is the most effective. Heat may be impractical or too expensive.

(Continued)

TABLE 10–1 *(Continued)*

Kind of Disinfectant	Usefulness	Strength	Limitations and Comments
Iodine[2] (tincture)	Extensively used as skin disinfectant, for minor cuts and bruises.	Generally used as tincture of iodine, either 2% or 7%.	Never cover with a bandage. Clean skin before applying iodine. It is corrosive to metals.
Iodophor ("tamed" iodine)	Effective against all bacteria (both Gram-negative and Gram-positive), fungi, and most viruses. Used for (1) egg dipping, (2) hatchery or poultry house disinfection, and (3) sanitizing processing plants, footbaths, and poultry drinking water.	Usually used as disinfectants at concentrations of 50 to 75 ppm titratable iodine, and as sanitizers at levels of 12.5 to 25 ppm. At 12.5 ppm titratable iodine, they can be used as an antiseptic in drinking water.	They are inhibited in their activity by organic matter. They are quite expensive. They should not be used near heat. When the characterisitic iodine color fades, effectiveness is gone.
Lime (quicklime, burnt lime, calcium oxide)	As a deodorant when sprinkled on manure and animal discharges, or as a disinfectant when sprinkled on the floor or used as a newly made "milk of lime" or as a whitewash.	Use as a dust; as "milk of lime"; or as a whitewash, but *use fresh*.	Not effective against spores. Wear goggles when adding water to quicklime.
Lye (sodium hydroxide, caustic soda)	On concrete floors.	Lye is usually used as either a 2% or 5% solution. To prepare a 2% solution, add 1 can of lye to 5 gal of water. To prepare a 5% solution, add 1 can lye to 2 gal water.	Damages fabrics, aluminum, and painted surfaces. Be careful, for it will burn the hands and face. Not effective against organism of TB. Lye solutions are most effective when used hot. Diluted vinegar can be used to neutralize lye.
Lysol (the brand name of a product of cresol plus soap)	For disinfecting surgical instruments. Useful as a skin disinfectant and for use on the hands before surgery.	0.5 to 2.0%.	Has a disagreeable odor. Does not mix well with hard water. Less costly than phenol.
Phenol (carbolic acid): 1. Phenolics—coal tar derivatives 2. Synthetic phenols	Commonly used for (1) egg dipping, (2) disinfection of hatcheries, poultry houses, and equipment, and (3) footbath solutions. Effective against bacteria and fungi.	Both phenolics (coal tar) and synthetic phenols vary widely in efficacy from one compound to another. So, note and follow manufacturer's directions. Generally used at 100 ppm for disinfecting and 50 ppm for sanitizing.	Organic materials have a diluting effect, but do not inactivate them. Ineffective against viruses.
Quaternary ammonium compounds (QAC) "quats"	Effective against bacteria and fungi. Used for (1) egg washing and dipping, (2) disinfecting hatcheries, poultry houses, and equipment, and (3) sanitizing poultry drinking water.	For sanitizing, use 200 ppm. For disinfecting, use 400 to 800 ppm.	They can corrode metal. Adversely affected by organic matter. Not very potent in combatting viruses. Not effective against TB organisms and spores. Inactivated by soaps, calcium, magnesium, iron, and aluminum salts.
Sal soda	It may be used in place of lye against certain diseases.	10½% solution (13½ oz to 1 gal of water).	
Sal soda and soda ash (or sodium carbonate)	They may be used in place of lye.	4% solution (1 lb to 3 gal of water). Most effective in hot solution.	Commonly used as cleaning agents, but have disinfectant properties, especially when used as a hot solution.
Soap	Its power to kill germs is very limited. Greatest usefulness is in cleansing and dissolving coatings from various surfaces, including the skin, prior to application of a good disinfectant.	As commercially prepared.	Although indispensible for sanitizing surfaces, soaps should not be used as disinfectants. They are not regularly effective; staphylococci and the organisms which cause diarrheal diseases are resistant.

[1]For metric conversions, see the Appendix.

[2]Sometimes loosely classed as a disinfectant but actually an antiseptic and practically useful only on living tissue.

VACCINATION

Vaccination may be defined as the injection of some agent (such as vaccine) into an animal for the purpose of preventing disease.

Vaccines are products containing high numbers of the organism known to cause a particular disease. They are specific in that a separate vaccine must be used for immunization against each disease. Vaccines contain either live or killed microorganisms. *Live-virus* vaccines are more profi-cient because the vaccine virus will grow and reproduce in the host, whereas a *killed-virus* product is dependent upon the antigenic units (virus cells) present in the vaccine dose to stimulate antibody production. Most poultry vaccines are of the live-virus type, produced by growing laboratory strains of virus in embryonated chicken eggs or in cell culture systems. Bacterial vaccines, called *bacterins*, are killed or inactivated preparations of bacteria, produced by growing selected strains of bacterial organisms in artificial media.

Short-term protection against a particular disease can

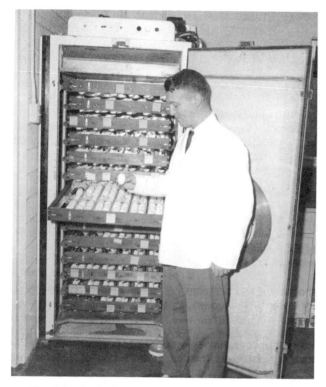

Fig. 10–9. A scientist at DEKALB's Poultry Pathology Department is shown examining a group of eggs which have been inoculated with the causative agent for epidemic tremors. After 18 days of incubation, the embryos will show whether the eggs came from a susceptible or resistant flock. (Courtesy, DEKALB AgResearch, Inc., DeKalb, IL)

be given by vaccination with an *antiserum*, containing antibodies against the virus from animals that have been exposed to that disease.

Reliable immunity can be obtained when vaccination is given under optimal conditions: that is, when the vaccine is handled, stored, and administered properly; the birds are susceptible, healthy, and vigorous; the nutrition is adequate; the host is parasite-free; and the environment is good.

So-called vaccination outbreaks do occur. The factors that influence vaccine-response in poultry are many, mainly depending on the host and environment. Ordinarily some protective immunity is conferred when birds are vaccinated, although the vaccine within itself cannot guarantee it.

A sound vaccination program is part of a good management program, but not a substitute for a sanitation program.

DISEASES FOR WHICH VACCINES ARE AVAILABLE

Of the products on the market for the control of virus infections, vaccines to prevent respiratory infections lead the list, but others are becoming increasingly available. Virus diseases which can be controlled by vaccination include:

1. Avian encephalomyelitis (epidemic tremors).
2. Fowl pox.
3. Gumboro (infectious bursal disease).
4. Infectious bronchitis.
5. Laryngotracheitis.

6. Marek's disease.
7. Newcastle disease.

Bacterins which are available commercially include:

1. Erysipelas bacterin.
2. Fowl cholera bacterin.
3. Mixed bacterins.

Coccidiosis control by vaccination has gained acceptance with producers as one approach toward controlling coccidiosis outbreaks in both light and heavy breed layers.

VACCINATION PROGRAMS FOR CHICKENS AND TURKEYS

In a concentrated poultry area, various diseases are a constant threat to the producer's profits. Vaccination is cheap insurance against heavy losses from some of these diseases.

The particular vaccination program and schedule followed should vary according to the area and from flock to flock. Thus, it had best be worked out in consultation with a poultry disease specialist and/or vaccine manufacturer. For guidance purposes only, a suggested vaccination program and schedule follows:

■ Replacement chicks (commercial egg or breeders) –

1. Purchase day-old chicks vaccinated at the hatchery against Marek's disease.
2. Vaccinate for Newcastle, with the type of vaccine and schedule tailored to suit the prevailing disease situation and other factors.
3. Vaccinate for infectious bronchitis, with the type of vaccine and schedule tailored to the prevailing situation.
4. Vaccinate for fowl pox, with the type of vaccine and schedule tailored to the prevailing situation.
5. Vaccinate breeder replacements against avian encephalomyelitis during the growing period to assure that they do not become infected after maturity, thereby preventing dissemination of the virus by the egg-borne route.

■ **Chicks for broiler production** – Broilers may be vaccinated against Newcastle disease, infectious bronchitis, and fowl pox, with the choice of each vaccine and the schedule tailored to suit the prevailing disease situation.

Undertake vaccination of chickens for diseases other than specifically outlined above only after getting expert advice.

■ **Turkeys** – Turkeys may be vaccinated against Newcastle disease, fowl pox, erysipelas, and fowl cholera. The decision on whether or not to vaccinate should be made on the basis of area, experience, and expert advice.

DIAGNOSTIC LABORATORY

Most states have one or more poultry pathology laboratories, operated by the state, available to their producers and the state's poultry industry. Additionally, many veterinarians in private practice provide limited laboratory services for areas adjacent to their practice.

Accurate identification of poultry diseases is often difficult for even the best pathologists if they must work without the aid of a properly equipped laboratory. A wrong diagnosis often results in faulty recommendation with improper medication and unsatisfactory results. This can prove to be quite costly due to ineffective medication and a continuance of high mortality and increased diseased birds.

When a disease outbreak is suspected, live birds showing typical symptoms should be immediately submitted to a poultry diagnostic laboratory for examination. Such laboratories are equipped to identify the disease problem and make recommendations for control. Practicing veterinarians, industry service people, and trained extension personnel working with producers and a diagnostic laboratory can bring about a reduction in losses due to disease.

TREATMENT OF DISEASE OUTBREAKS

Drug treatment of disease should be initiated only after a reliable diagnosis has been established. To do otherwise is costly and often produces serious bird losses. For example, an outbreak of erysipelas in turkeys does not respond to the usual treatments employed for fowl typhoid or fowl cholera. Misdiagnosis results in drug expense and a continued death loss. Hemorrhagic anemia syndrome of chickens is confused easily with coccidiosis. If a flock affected with the former condition is treated for coccidiosis, the existing problem is aggravated and severe losses may occur.

Once an accurate diagnosis has been established, follow the prescribed recommendations for treatment closely. Many drugs produce toxic effects if used improperly.

SPECIFIC PATHOGEN FREE (SPF) PROGRAM

The Specific Pathogen Free (SPF) program is a combination of breeding, testing, sanitation, and management practices designed to establish and maintain breeder flocks free of specific known infectious diseases. It has been shown that progeny of SPF breeders have a significantly better livability and gain efficiency than do chicks from non-SPF breeders. This program is directed toward egg-passed infection and diseases for which there is no other effective control, primarily pullorum, fowl typhoid, mycoplasma infections, lymphoid leukosis, inclusion-body hepatitis, and infectious bursal disease.

The term *SPF* implies varied meanings, but the operation of an SPF program generally involves the following:

1. No dirt floors in houses.

2. Houses must be screened against wild birds (maximum 1-in. mesh).

3. Houses must be locked at all times.

4. The farm shall have bulk-feed facilities with outside filler pipes; and handling of feed in bags is not permitted.

5. The farm shall have outside oil or fuel-filling facilities.

6. Houses must be at least 100 ft from public highways and at least 1,000 ft from other poultry houses on adjacent premises.

7. Houses must be at least 1,000 ft from poultry litter piles and other vermin-attracting debris.

8. All houses must have a footbath with an approved disinfectant and a stiff brush next to the door in the entry room. The disinfectant solution should be changed at least once daily and caretakers must clean their footwear upon entering or leaving.

9. Litter material should be delivered in clean trucks.

10. Each house must be thoroughly cleaned and disinfected in an approved manner prior to the introduction of baby chicks.

11. Only chicks from N.P.I.P. flocks (pullorum-typhoid free) may be used.

12. Medications such as coccidiostats (chemicals for the control of coccidiosis) as are currently used should be premixed in the feed.

13. All birds in the program will be of the same age; no pet birds or other poultry will be allowed on the farm.

14. Visitors, service persons, feed company representatives, sales persons, and neighbors will not be permitted in the houses at any time.

15. A logbook recording dates and times of all persons entering poultry houses will be kept.

16. Other animals (dogs, cats, etc.) shall not be permitted to enter the poultry house.

17. Baby chicks shall be delivered directly from the hatchery in new or disinfected shipping equipment, by attendants wearing disinfected shoes, freshly laundered coveralls, and caps.

18. Caretakers shall not visit premises where other poultry are kept or poultry products are processed.

19. Members of the caretaker's household should not work on other poultry farms or in poultry processing plants.

20. Poultry meat and eggs for home use should, preferably, be purchased at commercial outlets (inspected poultry meats and sanitized eggs).

21. Mortality records and reports of abnormalities in the health of the flocks shall be kept by the caretaker.

22. An incinerator or approved disposal pit for dead birds must be available.

GENETIC RESISTANCE TO DISEASE

Genetics as a tool for eliminating or controlling certain diseases holds promise. In this area, plant breeders have led the way. In 1905, it was discovered that certain varieties of wheat were more resistant to mycotic stem rust than others, thereby laying the foundation for important advances in the knowledge of genetic resistance to disease. Subsequently, scientists have evolved with many varieties of plants showing genetic resistance to disease. Evidence that similar genetic resistance to disease holds for animals has been demonstrated by experiences and experiments. For example, it has been demonstrated that selected lines of chickens vary in their resistance and susceptibility to fowl leukosis.

In the future, scientists may be able to genetically engineer poultry resistant to some of the most costly diseases. The goal is to improve the overall health of birds without compromising desirable production traits like egg lay, growth, or meat quality. But for this to be a reality, the genes responsible for disease resistance, or susceptibility, must be identified and understood.

Fig. 10–10. These White Leghorn roosters look alike, but their genes are different. The rooster on the left was injected with genes of avian leukosis virus when it was a 1-day-old embryo. The center and right roosters are of two succeeding generations, which directly inherited these virus genes. Viruses may someday be used as carriers for genes for leaner, tastier, bigger, and more profitable chickens. (Courtesy, USDA)

The application of genetics to disease control in poultry presents greater problems than in plants; it is more expensive and time-consuming. Also, to be of greatest practical value, it would be necessary to develop strains or breeds of birds that are genetically resistant to several diseases. Nevertheless, the stakes are high and this approach is worthy of greater attention than it has received in the past.

DISEASES

For purposes of convenience and discussion, the diseases of poultry are herein grouped as follows: (1) bacterial diseases, (2) respiratory diseases, (3) viral diseases (exclusive of respiratory diseases), (4) protozoan diseases, and (5) nutritional diseases. Pertinent facts relative to each of these groups are presented, followed by an alphabetically listed tabular summary of nonnutritional diseases.

BACTERIAL DISEASES

Bacteria are one of the smallest and simplest known forms of life. They are microscopic, possess just one cell, vary in shape, multiply by transverse fission, and possess no chlorophyll.

Bacteria are grouped into spherical forms, straight rods, curved or spiral rods, and filamentous forms. Bacteria, in common with other living organisms, have certain requirements as to environmental temperature, moisture, and nutrition for propagation.

All bacteria are not detrimental to animal health. In fact, many bacteria are necessary for such processes as food digestion. The classification of bacteria into species is done so disease-producing organisms may be separated from those that are harmless or beneficial.

The successful control of bacterial diseases entails isolating and identifying the disease-producing species, and preventing the multiplication and spread of these organisms within the animal body or to other birds. This should be done with a minimum detrimental effect upon the beneficial bacteria.

RESPIRATORY DISEASES

Poultry disease affecting the air passages, windpipe, lungs and air sacs are classified as *respiratory diseases*. They are among the major threats to poultry health.

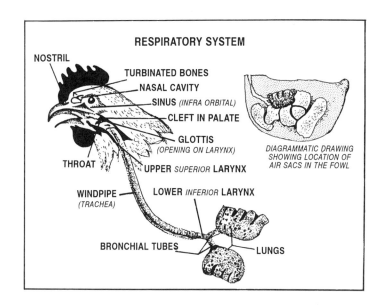

Fig. 10–11. Many poultry diseases affect the respiratory system, shown in this sketch. Starting with the nostrils and mouth, the air passes through the throat, windpipe, and bronchial tubes to the lungs. The nasal cavity and sinuses are also shown. (Courtesy, Salsbury Laboratories, Charles City, IA)

At first, all respiratory system diseases were known as *colds*; and often they were considered to be caused by environmental factors such as drafts and chilling. As more information accumulated, it became apparent that colds actually were a group of separate infectious diseases having many common characteristics.

Possibly all infectious agents causing respiratory symptoms have not been recognized, but many have been isolated and the diseases they produce well defined. Because the nature of the causative organism tells much about a disease, the following outline according to cause is useful in understanding respiratory diseases.

- **Caused by viruses:**
  Infectious bronchitis
  Laryngotracheitis
  Newcastle disease

- **Caused by bacteria:**
  Chronic respiratory disease of chickens
  Coryza (roup)
  Infectious sinusitis of turkeys
  Mycoplasmosis

- **Caused by molds:**
  Aspergillosis (brooder pneumonia)

VIRAL DISEASES (EXCLUSIVE OF RESPIRATORY DISEASES)

Viruses are disease-producing agents that (1) are so small that they cannot be seen through an ordinary microscope (they can be seen by using an electron microscope), (2) are capable of passing through the pores of special filters which retain ordinary bacteria, and (3) propagate only in living tissue.

A number of viral diseases of poultry produce symptoms and lesions, for the most part not affecting the respiratory system. Among them are some of the most devastating diseases of chickens and turkeys, including avian pox, leukosis, Marek's disease, avian encephalomyelitis (epidemic tremor), and bluecomb disease.

PROTOZOAN DISEASES

Protozoa are the simplest and most primitive form of animal life; they consist of only a single cell. Although many microscopic protozoan organisms are harmless, others produce severe disease. Five of the more common and serious poultry diseases are caused by these organisms; namely, coccidiosis, blackhead, trichomoniasis, hexamitiasis, and leucocytozoonosis.

NUTRITIONAL DISEASES

A summary of the most common nutritional diseases, the vast majority of which are brought about by the deficiency of one or more of the minerals or vitamins, is given in Chapter 6, Fundamentals of Poultry Nutrition, Tables 6–3 and 6–5.

TABULAR SUMMARY OF NONNUTRITIONAL DISEASES

Pictures showing the symptoms and signs of selected poultry diseases follow; and an alphabetically listed tabular summary of the common nonnutritional diseases of poultry is presented in Table 10–2, pp. 242–247.

Fig. 10–12. Bluecomb disease afflicted hen (left) vs normal hen (right). Note the dark and dried comb and wattles of the affected bird. This darkening starts at the outer edge, and progresses toward the base. (Courtesy, Salsbury Laboratories, Charles City, IA)

Fig. 10–13. Coryza (roup), showing swelling of tissue around the eye. (Courtesy, A. S. Rosenwalk, DVM, University of California, Davis)

Fig. 10–14. Ducklings with *duck viral hepatitis (DVH)*. Symptoms: Ducklings lie on side with heads drawn back and paddle feet spasmodically. (Courtesy, Salsbury Laboratories, Charles City, IA)

Fig. 10–15. Epidemic tremor (avian encephalomyelitis), a virus disease which can kill chicks and cause a lowering of egg output in laying flocks. Note that the chicks rest on their haunches or on their sides, and appear too weak to move about. The lack of muscle coordination prevents them from reaching feed and water. (Courtesy, Salsbury Laboratories, Charles City, IA)

Fig. 10–16. Fowl cholera, an advanced, localized case. The swelling of the wattle is caused by accumulation of cheesy pus. If opened, these swellings have a foul odor. (Courtesy, Salsbury Laboratories, Charles City, IA)

Fig. 10–17. Fowl cholera afflicted turkey. Symptoms: fever, purplish head, greenish-yellow droppings, ruffled feathers, and sudden death. (Courtesy, Virginia Poultry Federation, Harrisonburg, VA)

Fig. 10–18. Fowl pox—the "dry" or skin form. This common, virus-caused disease may produce brownish scabs on the comb, face, and wattles. It usually occurs in the fall or early winter and may appear in the same locality from year to year. Pox scabs are wartlike, deep seated, and won't peel off. (Courtesy, Salsbury Laboratories, Charles City, IA)

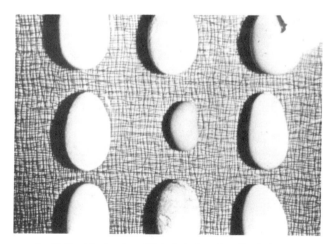

Fig. 10–19. Infectious bronchitis in the laying or breeding flock causes egg production to drop quickly; and the few eggs laid are often misshapen, rough, or soft-shelled, and may vary widely in size. Egg quality may never return to normal. (Courtesy, Salsbury Laboratories, Charles City, IA)

Fig. 10–20. Laryngotracheitis (gapes), a highly contagious virus disease. The gasping or cawing position is typical, and blood may fleck from the mouth. Birds die suddenly. (Courtesy, Salsbury Laboratories, Charles City, IA)

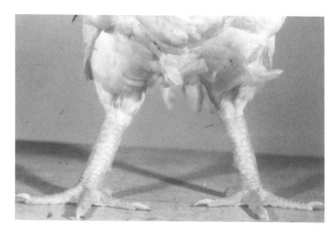

Fig. 10–21. Leukosis, showing a close-up view of leg bone enlargement caused by the bone form of leukosis. (Courtesy, Salsbury Laboratories, Charles City, IA)

TABLE 10–2

Disease	Species Affected	Cause	Symptoms and Signs	Postmortem
Arizonosis	Chiefly turkeys.	*Salmonella arizona.*	Poults unthrifty, and a good many develop eye opacity and blindness. Mortality usually confined to first 3 to 4 weeks of age.	Yolk sacs slowly absorbed. Liver enlarged and mottled. Infected intestinal tract.
Aspergillosis (brooder pneumonia)	Almost all birds and animals, including humans.	Mold *(Aspergillus).*	Fever, difficult breathing, nervous symptoms.	Nodules in lungs and air sacs, pus in air sacs.
Avian encephalomyelitis (epidemic tremor)	All birds.	Virus.	Unsteadiness, sitting on hocks, inability to move, muscular tremors of head, neck, and limbs.	No gross lesions of the nervous system seen.
Bluecomb disease (coronaviral enteritis)	Turkeys.	Virus.	Poults appear cold and seek heat, stop eating and lose weight, and have frothy or watery droppings. In growing turkeys, the appearance of the disease is sudden with a concurrent drop in feed and water consumption. Sick birds show darkening of the head and skin.	Birds show few lesions. The contents of the duodenum, jejunum, and ceca are watery and gaseous.
Botulism (limber-neck; food poisoning)	All birds except vultures.	Toxin produced by the anaerobic bacterium *Clostridium botulinum.*	Convulsions, paralysis, and sudden death.	Enteritis.
Chronic respiratory disease of chickens (also see mycoplasmosis)	Chickens. Turkeys.	*Mycoplasma gallisepticum.*	Coughing, gurgling, sneezing, nasal exudate, slow spread, loss of weight.	Mucus in trachea, air sacs thickened and containing yellow pus, thickened membrane over heart.
Coryza (roup)	Primarily chickens.	Bacterium *(Hemophilus paragallinarum).*	Gasping, swollen eyes, nasal discharge, offensive odor.	White to yellow pus in eyes and sinuses.
Duck plague (duck virus enteritis)	Ducks. Geese. Swans.	Virus.	Spreads rapidly. Can cause heavy mortality. Affected birds reluctant to walk.	Hemorrhages of internal organs.
Duck virus hepatitis	Ducks.	Virus.	Acute disease. Especially affects ducklings up to 3 weeks of age. Mortality ranges from 5 to 90%.	Characterisitic liver lesions.

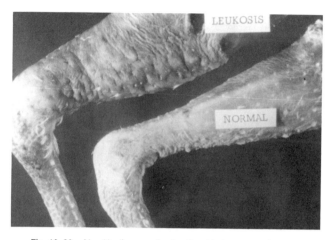

Fig. 10–22. Marek's disease, showing the skin form, sometimes referred to as skin leukosis. (Courtesy, Salsbury Laboratories, Charles City, IA)

Fig. 10–23. Newcastle disease in 12-week-old turkey. (Courtesy, Virginia Poultry Federation, Harrisonburg, VA)

NONNUTRITIONAL DISEASES OF POULTRY

Distribution and Losses Caused By	Treatment	Prevention	Remarks
Widespread.	Chemotherapy may reduce losses in acute outbreaks of avian arizonosis.	Elimination of infected breeder flocks. Hatchery fumigation and sanitation.	It is an egg-transmitted infection.
The incidence of the disease is not great.	No treatment. Remove source of infection.	Avoid musty and moldy feed and litter, provide good ventilation.	Young fowl most susceptible.
Worldwide.	Affected birds are usually destroyed.	Vaccination of breeder pullets at 10 to 15 weeks of age, repeated at molting if held second year.	Differentiated from vitamin E deficiency by history, signs, and histological study.
Heaviest losses are in condition and production, but death losses may be high in young turkey poults.	Antibiotics or nitrofurans according to directions may be helpful in reducing the mortality.	Depopulation and decontamination of turkey buildings and surrounding areas with a rest period before restocking.	Turkeys that recover from the disease are immune to challenge, but remain carriers for life.
Botulism occurs worldwide.	If isolated and provided with water and feed, many sick birds will recover. Move birds to clean facilities. Antitoxin may be used on valuable birds, but it is difficult to obtain and expensive.	Do not feed spoiled or decomposing feed. Prompt disposal of dead birds and rodents. Avoid wet spots in litter.	The toxin is very potent, being 17 times more deadly than cobra venom for the guinea pig.
Worldwide.	Antibiotics in feed or water according to directions. In severe outbreaks, inject birds with appropriate antibiotics.	Secure mycoplasma-free stock.	*Mycoplasma gallisepticum* (MG) infection is commonly designated as chronic respiratory disease (CRD) of chickens and infectious sinusitis of turkeys.
Worldwide.	Various sulfonamides or antibiotics used according to directions.	Keep age groups separate. Periodic complete depopulation.	Do not expose susceptible birds to recovered birds. Latter are lifetime carriers.
Worldwide.	No successful treatment.	Different kinds of vaccines give variable results.	This is a reportable disease.
Worldwide.	Administration of serum from immune ducks.	Vaccination of breeder ducks and vaccination of young ducklings. Strict isolation.	

(Continued)

TABLE 10–2

Disease	Species Affected	Cause	Symptoms and Signs	Postmortem
Epidemic tremor (avian encephalo-myelitis)	All birds, but primarily young chickens.	Virus.	Tremors of head and neck, muscular incoordination. Temporary drop in egg production of layers.	None.
Erysipelas	Turkeys primarily, but other fowl affected. Also humans and swine.	Bacterium *(Erysipelothrix)*.	Sudden losses, swollen snood, discoloration of parts of face, droopy.	Hemorrages in muscles, mucus in mouth; reddened intestines.
Fowl cholera	Chickens. Turkeys. Water fowl. Other birds.	Bacterium *(Pasteurella multocida)*.	Fever, purplish head, greenish-yellow droppings, sudden deaths.	Enlarged liver, hemorrhages in heart and in other organs.
Fowl pox (avian pox; avian diphtheria; contagious epithelioma; sore head)	Chickens. Turkeys. Other birds.	Virus.	Small clear to yellow blisters on comb and wattles that soon scab over; decreased egg production.	May have lesions in throat and trachea.
Fowl typhoid	Chickens. Turkeys. Ducks. Pigeons. Pheasants.	Bacterium *(Salmonella gallinarium)*.	Inactive, fever, greenish-yellow droppings.	Liver and spleen enlarged, bronze or greenish colored liver with some lesions.
Gumboro (infectious bursal disease)	Chickens.	Virus.	Sleepy. White, watery diarrhea.	Enlarged bursa, hemorrhages.
Hemorrhagic enteritis (enteritis)	Turkeys.	Unknown, probably a virus.	Usually the only sign is one or more dead birds.	Severe hemorrhagic inflammation of intestinal lining from gizzard to ceca.
Infectious bronchitis	Chickens only.	Virus.	**Young birds:** Gasping, wheezing, nasal discharge. **Older birds:** Sharp and prolonged drop in egg production, and soft-shelled eggs.	Yellowish mucus or plugs in lower trachea and air passages of lungs.
Infectious sinusitis of turkeys (also see Mycoplasmosis)	Turkeys.	Specific causative agent under investigation.	Nasal discharge, swollen sinuses, labored breathing, coughing.	Exudate in sinuses, cheesy material in air sacs.
Infectious synovitis (*Mycoplasma synovial* infection)	Chickens. Turkeys.	*Mycoplasma synoviae.*	The disease occurs primarily in growing birds from 4 to 12 weeks of age. Enlarged hocks, foot pads, lame. Breast blisters.	Exudate at joints. Enlarged liver, spleen.
Laryngotracheitis (gapes, chicken flu)	Chickens.	Virus.	Gasping, coughing, loss of egg production, soft-shelled eggs, extending of neck outward on inhalation and slumping on exhaling, weeping of eyes.	Blood-stained mucus in trachea.
Leukosis (big liver disease; lymphoid leukosis)	Chickens. Turkeys. Other fowl.	Virus.	Loss of weight, diarrhea, thickened bones, gray eyes.	Enlarged liver and spleen, tumors in various parts of body.
Marek's disease (range paralysis, acute leukosis)	Chickens. Other fowl.	Herpes virus.	Sudden death, loss of weight, diarrhea, paralysis of legs or wings. Skin lesions in young birds. May occur as early as 5 to 8 weeks of age.	Enlarged liver, spleen, kidney, ovary, and testicles; or nodular tumors on these organs. Enlarged nerves in wings or legs. Skin lesions in young birds.
Mycoplasmosis	All poultry.	Primarily the following three: *M. gallisepticum, M. meleagridis, M. synovial.*	Infection of the air passages. Coughing, sneezing, nasal discharge. *M. synovial* may affect the joints, producing an exudative tendonitis and bursitis.	Thickened air sacs filled with exudates.

(Continued)

Distribution and Losses Caused By	Treatment	Prevention	Remarks
Worldwide. Morbidity in affected flocks averages 5 to 10%	No satisfactory treatment is known.	Vaccination of breeders.	Vaccination of commercial laying flocks is of questionable value.
The disease is of economic concern to turkey growers throughout the world.	Use antibiotics according to recommendations.	Vaccinate.	Transmitted via wounds or skin abrasions.
Worldwide. At times it causes high mortality; at other times the losses are nominal.	Sulfonamides and antibiotics according to directions.	Sanitation, disposal of sick birds, isolation of new stock. Vaccination. Commercially produced bacterins and live vaccines are available.	When there is an outbreak, work fast in treatment.
Worldwide. Mortality is not high. Economic loss is in reduced feed efficiency and production.	Treatment is of little value.	Vaccination.	Control mosquitoes.
Worldwide. Mortality of affected birds ranges from 1 to 40% if treatment not instituted promptly. The cost of fowl typhoid is primarily in testing under the National Poultry Improvement Plan.	Nitrofurans or sulfa drugs according to directions. But every effort should be made to eradicate the disease.	Get stock from disease-free sources.	Egg and mechanical transmission.
Disease occurs in most of concentrated poultry-producing areas of the world. Heaviest losses in chicks up to 6 weeks.	None.	Live and dead vaccines.	The disease damages the birds' immune processes. Thus, vaccines for other diseases are less effective.
The disease has been reported in the U.S., Canada, Japan, Australia, India, and Israel.	Injection of convalescent antiserum, which is obtained from healthy flocks and usually collected at slaughter.	Vaccine.	
Worldwide. Economic loss is in lowered production and quality of eggs, and in mortality and lowered gains and feed efficiency of young chickens.	No specific treatment.	Strict isolation and repopulation with only day-old chicks following the cleaning and disinfecting of the poultry house. Inactivated and live vaccines.	
It may cause significant mortality in poults.	Drain sinuses and treat with an antibiotic, injected or administered in feed or water.	Secure poults from disease-free breeders and keep isolated from chickens.	
Probably worldwide. Mortality varies from 2 to 75% with 5 to 15% being the average.	Antibiotics.	Test breeders and purchase clean chicks and poults. Treatment of eggs with antibiotics; egg inocculation; or heat.	
Laryngotracheitis has been identified in most countries.	No drug treatment is effective.	Vaccination.	Farm eradication can be accomplished if security measures are superior.
With few exceptions, leukosis virus infection occurs in all chicken flocks; by sexual maturity most birds have been exposed.	None.	No vaccine. Buy birds from complement fixation avian leukosis (Cofal) flocks. Raise birds in isolation away from old or adult stock.	Now virtually eliminated from major breeding units.
Worldwide. Prior to development of a vaccine, losses ranged from 25 to 60%.	None.	Vaccination of day-old chicks at the hatchery.	Genetic resistance and isolation-rearing are important adjuncts to vaccination.
Worldwide.	Antibiotics.	Eradication from breeding flocks.	Mycoplasmosis infection is commonly designated as a chronic respiration disease of chickens and infectious sinusitis of turkeys.

(Continued)

TABLE 10–2

Disease	Species Affected	Cause	Symptoms and Signs	Postmortem
Mycosis of the digestive tract	Chickens. Turkeys. Other fowl.	Pathogenic fungi.	Signs are not particularly characteristic. Affected birds show poor growth, stunted appearance, listlessness, and roughness of feathers.	Cheesy scum on crop lining.
Mycotoxicoses	All poultry.	Ingestion of toxic substances caused by molds growing on feeds and possibly litter. *Aspergillus flavis*, which produces aflatoxins, is of most concern. The B–1 toxin is the most toxic and of greatest concern to the poultry industry.	Reduced growth and egg production and high mortality.	Hemorrhages. Pale, fatty liver, kidneys.
Newcastle disease	Chickens. Turkeys. Other fowl.	Virus.	Gasping, wheezing, twisting of neck, paralysis, severe drop in egg production, soft-shelled eggs.	Often none. Sometimes mucus in trachea and thickened air sacs containing yellow exudate.
Paratyphoid	Chickens. Turkeys. Waterfowl and other birds.	Bacteria; Salmonella species other than *S. pullorum* and *S. gallinarum*.	Seen mainly in poults.	Enteritis, nodules in wall of intestines.
Pullorum	Chickens. Turkeys. Other domestic and wild fowl.	Bacteria (*Salmonella pullorum*).	Sleepy, pasted up, inactive, high mortality in young birds.	Lesions on lungs, liver, and intestines. Unabsorbed egg yolks.
Tuberculosis	All poultry.	Bacterial (*Mycobacterium avium*).	Unthriftiness, lowered egg production, and finally death.	Characteristic grayish-white or yellowish nodules of varying sizes in the liver, spleen, and intestines.
Turkey coryza (bordetellosis)	Turkeys.	Bacterium (*Bordetella avium*).	Snicking, rales, and discharge of excessive nasal mucus.	Lesions and excessive mucus in the upper respiratory system.
Turkey veneral disease (*Mycoplasma meleagradis* infection)	Turkeys.	*Mycoplasma meleagridis*.	Cull poults, lowered fertility and hatchability.	Infected air sacs.
Ulcerative enteritis (quail disease)	Chickens. Turkeys. Game birds.	Bacterium (*Clostridium colinum*).	Sleepy, loss of appetite.	Ulcers on intestines, enteritis.
Vibrionic hepatitis	Chickens.	Bacterium.	Large number of culls. Decreased egg production.	Enlarged liver, spleen.

(Continued)

Distribution and Losses Caused By	Treatment	Prevention	Remarks
Mycosis probably occurs frequently, but most cases are mild.	Fungicidal drugs.	Sanitation. Do not overcrowd. Nystatin: 50 gm/ton continuously.	
The disease first became prominent in 1960 when 100,000 turkeys died, later found to be caused by a toxin in moldy peanut meal.	Remove source of aflatoxin from the diet.	Avoid feed spoilage. Treat high moisture grains with propionic or acetic acids.	Is on increase. Young are more susceptible than mature birds.
Worldwide. Mortality of affected chickens varies from 0 to nearly 100%.	There is no effective treatment.	Vaccination.	It was first recognized in England in 1926; and it is named after the town of Newcastle. Newcastle disease was first reported in the U.S. in 1944. Symptoms in turkeys are mild; reduction of egg production of turkey breeders is main economic loss. This is a notifiable disease.
Worldwide. Mortality in turkey poults usually 1 to 20%. Outbreaks in ducks (keel disease) often run very high.	Sulfonamides, antibiotics, and nitrofurans may be employed to reduce mortality.	Hatchery and flock sanitation are the most important factors in paratyphoid prevention.	Egg and mechanical transmission; and through "blow-up" of infected eggs during incubation, and through some feeds.
Presently, the main economic loss from pullorum in the U.S. is due to the necessity of testing breeding flocks of chickens and turkeys to ensure freedom from infection.	Sulfonamides, nitrofurans, and antibiotics may be used to check mortality, but there is no substitute for a sound eradication program.	Eggs from disease-free breeders, hatched in disease-free incubators. Eradication from breeding stock. Buy chicks and poults only from pullorum typhoid-clean breeding stock.	Primarily egg-transmitted, but transmission may be by other means.
Worldwide, but occurs most frequently in the North Temperate Zone. In 1972, tuberculosis was the cause for condemnation of 0.04% of the 186.9 million mature chickens slaughtered under federal inspection in the U.S.	None.	Sanitation; put disease-free birds in a clean house or on clean ground.	Avian tuberculosis is transmissible to swine, so keep swine and chickens separated.
Worldwide.	No treatment is entirely effective, although use of antibiotics in the early stages of the disease may be helpful.	Commercial vaccines are available.	Morbidity is generally 100%, while mortality ranges from 5 to 75%.
Worldwide.	Antibiotics.	Strict sanitation. Vaccines are not available.	
Death losses may be high in replacement pullets and quail.	Antibiotics.	Sanitation. Raising birds on wire is an effective preventive measure.	Increasing problem in chickens.
Vibrionic hepatitis has been identified in the U.S., Canada, Netherlands, Germany, Italy, and Switzerland. Egg production may drop as much as 35%. Mortality is usually low, but may be as high as 10 to 15%.	Nitrofurans, antibiotics, or sulfas.	There is no successful immunization.	Affects young and immature chickens.

PARASITES

Parasites are organisms living in, on, or at the expense of another living organism. A wide variety of external and internal parasites attack poultry.

The prevention and control of parasites is one of the quickest, cheapest, and most dependable methods of increasing egg and meat production with no extra birds, no additional feed, and little more labor.

EXTERNAL PARASITES

External parasites cause birds to look unsightly, reduce weight gains and egg production, and mar the skin. The result is downgrading of quality and lower market value. Heavy infestations cause high mortality among young poults. Studies have shown that poultry lice and mites can sometimes reduce weight gains and egg production from 2 to 25%, or more.

The most common external parasites of poultry are listed in Table 10–3, pp. 250–251. Many insecticides are available to help control external parasites of poultry, but few of them are suggested herein because the list of approved materials changes rapidly. Information about what is available and approved for a particular area should be obtained from the county agricultural agent or local veterinarian and used in keeping with manufacturer's directions.

INTERNAL PARASITES

In summary form, the most common internal parasites of poultry, including the treatment and prevention of each, are given in Table 10–3, pp. 250–251. In using any of the vermifuges, the poultry producer should use them according to the manufacturer's directions and in keeping with pure food and drug regulations.

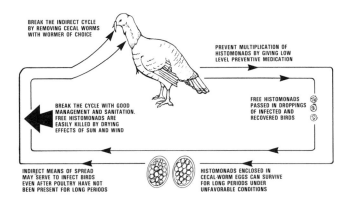

Fig. 10–26. Blackhead (Histomoniasis), to which turkeys are especially susceptible. Preventive measures are: (1) good sanitation; (2) control of cecal worms; (3) turkeys kept on wire, away from chickens; and (4) a histomonastat administered continuously in feeds to turkeys over 6 weeks old.

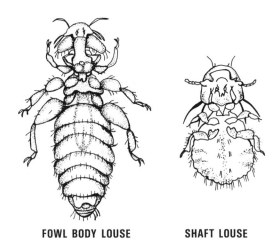

FOWL BODY LOUSE   **SHAFT LOUSE**

Fig. 10–24. Lice.

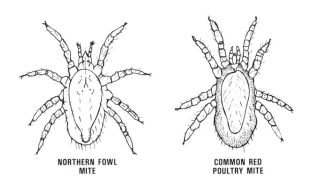

NORTHERN FOWL MITE   **COMMON RED POULTRY MITE**

Fig. 10–25. *Mites:* Northern fowl mite, *Arnithonyssus sylviarum* (left), and common red mite, *Dermanyssus gallinae* (right).

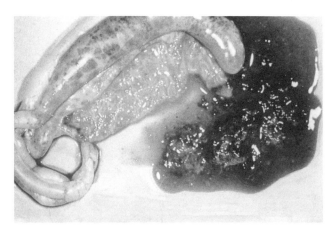

Fig. 10–27. Cecal, or bloody, coccidiosis in a chicken, showing accumulation of blood in ceca, caused by the protozoan organism, *Eimeria tenella*. (Courtesy, C. F. Hall, DVM, College of Veterinary Medicine, Texas A&M University, College Station)

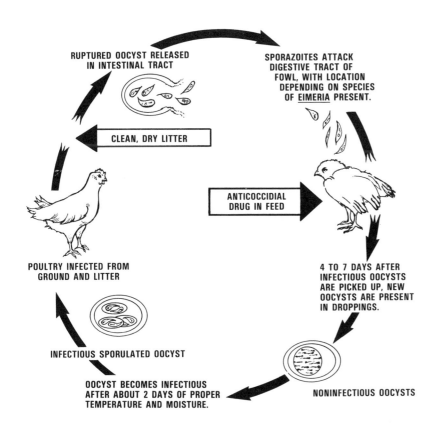

Fig. 10–28. Life cycle of *E. tenella*, one of the genus containing many species causing coccidiosis in poultry, showing (1) the typical 7- to 9-day cycle, and (2) the stage of development at which anticoccidial drugs kill the organism.

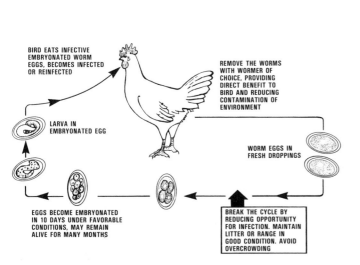

Fig. 10–29. Life cycle of the common, large roundworm, *Ascaridia galli*, the most prevalent of the worm parasites of chickens and the cause of heavy economic losses.

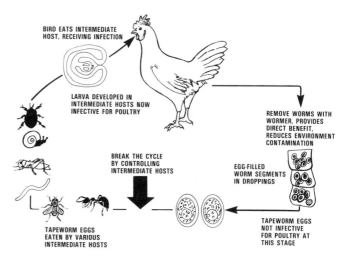

Fig. 10–30. Life cycle of tapeworms, several species of which infect chickens and turkeys. Poultry become infected by eating the intermediate host.

TABULAR SUMMARY OF PARASITES

An alphabetically arranged tabular summary of internal and external parasites of poultry is given in Table 10–3, Parasites of Poultry (next page).

TABLE 10–3

Parasite	Species Affected	Cause	Symptoms and Signs	Postmortem
Blackhead (histomoniasis)	Chickens, but more resistant than turkeys. Turkeys.	Protozoa (*Histomonas meleagridis*).	Droopiness, loss of weight, sulfur-colored diarrhea, and darkened heads.	Lesions in liver and ceca, ceca enlarged, liver enlarged and spotted with dark red or yellow circular areas.
Coccidiosis	Chickens. Turkeys. Ducks. Geese. Game birds.	Protozoa (*Eimeria* species). Six species of *Eimeria* cause the disease in chickens and three in turkeys. Other species cause the disease in ducks, geese, and game birds.	Bloody droppings (usually cecal-type only), sleepy, pale, ruffled feathers, unthrifty.	Bloody or cheesy plugs in ceca (cecal-type). Intestinal wall thickened with small white or reddish areas (intestinal-type).
Hexamitiasis (infectious catarrhal enteritis)	Turkeys.	The one-celled parasite *Hexamita meleagridis*.	Listlessness, foamy and watery diarrhea, and convulsions.	Dehydration, emaciation, thin and watery intestinal contents, and bulbous areas in intestines.
Large roundworms	Chickens. Turkeys.	Ascarid infestation (*Ascaridia galli*).	Droopiness, emaciation, and diarrhea.	Roundworms, 1½ to 3 in. long, in intestines.
Leucocytozoonosis	Turkeys. Ducks.	Protozoa (*Leucocytozoon*).	Symptoms are not apparent in older birds. Loss of appetite, droopiness, weakness, increased thirst and rapid, labored breathing in young birds.	
Lice	All poultry.	More than 40 species of lice.	Frequent picking, pale head and legs, loss of weight.	
Mite infestation (mites on body only during night)	Chickens are the commonest hosts, but these mites may occur on all poultry.	Common red, or roost, mites (*Dermanyssus gallinae*).	Reduced egg production, retarded growth, lowered vitality, damaged plumage, and even death.	
Mite infestation (mites always on body)	All poultry.	Northern fowl mites, the most common of which is *Ornithonyssus sylviarum*.	Droopy, pale condition and listlessness.	
Tapeworms	Chickens. Turkeys. Other birds.	Several species of tapeworms.	Pale head and legs, poor flesh.	Tapeworms in intestines.
Trichomoniasis	Chickens. Turkeys. Pigeons. Quail.	Protozoa (*Trichomonas gallinae*).	Loss of appetite, droopiness, loss of weight, and darkened head.	Lesions—necrotic ulcerations—in the upper digestive tract, affecting the crop in particular.

MISCELLANEOUS HEALTH PROBLEMS

Although not properly considered diseases, there are several other conditions which can cause serious problems, and even mortality, in a flock; among them, the following:

■ **Cage fatigue** — Cage fatigue is a paralytic condition observed in birds held in cages. The disease is most common among high-producing young pullets during the summer.

Fig. 10–31. Cage fatigue paralysis. As the condition progresses, the bird will lie on its side. (Courtesy, University of Wisconsin, Madison)

PARASITES OF POULTRY

Distribution and Losses Caused By	Treatment	Prevention	Remarks
Average annual losses in turkeys estimated to exceed $2 million.	A number of drugs are on the market, from which a selection may be made and used according to manufacturer's directions.	Sanitation, frequent range rotation. Do not crowd. Preventive medication in feed and water according to directions. Do not keep chickens and turkeys on the same premises.	Transmitted by droppings from infected birds. Control cecal worms.
Coccidia are found wherever poultry are raised. Estimates of annual losses in the U.S. range up to $200 million. More than $80 million spent on preventive medication, annually.	Anticoccidials in feed or soluble form of drugs in water according to directions.	Preventive medication in feed or water according to directions. Vaccination is useful in certain types of operations, but seek expert advice before using it.	Transmitted by droppings of infected birds.
Reported in the U.S., Canada, Scotland, England, and Germany. Hexamitiasis is not of major importance today.	The disease does not respond well to treatment. Furazolidone, aureomycin, and terramycin have been used with some success.	Segregation of age groups, and sanitation.	
	Several deworming drugs are on the market. Select and use according to manufacturer's directions.	Sanitation and rotation of range and yards. Careful use of old litter. Cage rearing and housing.	See Fig. 10–29 for life cycle.
The disease occurs most frequently in southern and southeastern U.S.	Drug treatment of leucocytozoonosis has had limited success. Clopidol may be used as a treatment or preventive.	Exterminate black fly population, and do not raise turkeys near running streams. Segregate breeding and brooding operations. Brooding in houses with cheesecloth over openings during black fly season. Sulfadimethoxine and sulfaquinoxaline prevent infection of certain types of the parasite.	
Worldwide. Heavy infestations affect bird performance.	Select an approved insecticide and use according to manufacturer's directions.	Buy louse-free birds and never add lousy birds to clean stock.	
Worldwide. Can cause reduced performance.	Select an approved insecticide and use according to manufacturer's directions.	Sanitation. Examine birds frequently for signs of mites. Preventive insecticide treatment of quarters. Control sparrows.	
The Northern fowl mite, *Ornithonyssus sylviarum*, is rated as the most important permanent parasite in all major producing areas of the U.S.	Select an approved insecticide and use according to manufacturer's directions.	Examine birds frequently for evidence of mite infestation. Preventive insecticide treatment of quarters. Control sparrows.	
	Butynorate is the most widely used product for treatment. Use according to manufacturer's directions.	Eliminate the intermediate insect hosts. Control snails, earthworms, beetles, and flies.	See Fig. 10–30 for life cycle.
Worldwide.	Sanitation; remove infected birds; and use one of the recommended drugs according to manufacturer's directions.	Sanitation, clean feed and water. Eliminate recovered or carrier birds.	

The exact cause is not understood; however, the disorder is considered to be a disturbance in mineral metabolism. Affected birds usually make a spontaneous recovery if placed on the floor or if the cage bottom is covered with newspaper or other such material. As a preventive measure, it is recommended that, during the summer months, caged pullets be limit-fed hen-size oystershell at the rate of 25 lb per 1,000 birds per week, divided into three equal feedings. Also, putting 2 to 3 birds in a single cage will reduce the incidence.

■ **Fatty liver syndrome**—The *fatty liver disease* continues to be a serious anomaly in commercial egg operations. It is characterized by deranged fat metabolism resulting in the deposition of excess fat in the liver and body cavities. It is seen most commonly in caged birds, but on occasion it may strike floor birds, particularly in the heavy breeds. The cause is unknown. However, factors which predispose the condition include reduced activity in a cage operation and the use of high-energy feeds.

Lipotropicagents such as choline, methionine, and inositol, combined with added vitamin B–12 and vitamin E₁, have been used to treat fatty liver syndrome. The adjustment of energy and protein levels in the ration may also be indicated. A nutritionist should be consulted before attempting to treat the disease.

■ **Pendulous crop**—This condition, sometimes known as *baggy crop*, or *drop crop*, may be found among chickens and among growing poults 2 to 3 months of age. The condition may be caused by irregular feeding and by overconsumption of feed or water at any given time. It also seems that there are fewer baggy crops where turkeys are in houses which are not overheated, and when there is adequate shade for poults on the range.

■ **Hysteria**—Sometimes, a condition of excessive fright has been observed in growing pullets or layers. With floor-housed birds, it may take the form of flight and piling up in the corners of the house, with the result that many birds may be suffocated. Caged layers may attempt to fly, resulting in injuries to wings and legs or broken necks. The causes of outbreaks of hysteria are unknown, but care must be taken not to frighten birds needlessly by such things as loud noise, quick movements, and change in light intensity.

BEHAVIORAL PROBLEMS

With the restriction, or confinement, of flocks, many abnormal behaviors evolved to plague those who raise them including cannibalism of many types and egg eating.

■ **Cannibalism**—Many types of cannibalism occur in domestic fowl and game birds reared in captivity; among them, the following:

1. **Vent picking.** Picking of the vent, or the area below the vent, is the severest form of cannibalism. This type is generally seen in pullet flocks in high production. Predisposing factors are prolapsus or tearing of the tissue caused by passage of a very large egg.
2. **Toe picking.** Toe picking is most commonly seen in chicks or young game birds. It may be brought on by hunger.
3. **Head picking.** Head picking usually follows injuries to the comb or wattles caused by freezing or by fighting between males.
4. **Blueback.** Blueback in turkeys describes the condition in which the backs of the affected birds turn blue or black. It is caused by feather picking, followed by exposure to sunlight. Blueback may result from overcrowding in the brooder, keeping the poults on the sun porch too long, and lack of sufficient fiber in the ration.

■ **Egg eating**—Egg eating is a costly vice. If one bird acquires the habit, it usually spreads quickly throughout the flock.

Egg eating is predisposed by factors favoring egg breakage, including insufficient nests, insufficient nesting material, not collecting egg frequently enough, and soft-shelled or thin-shelled eggs. Prevention consists in alleviating these conditions.

Once the egg eating habit has started, it is very difficult to stop. If the birds have not been debeaked, this should be done immediately. Also, nests should be darkened and eggs should be collected frequently.

GLOSSARY OF POULTRY HEALTH TERMS

Poultry health, like other subjects in the field of animal science, has a language somewhat of its own. Here is a list of some of the most frequently used animal health words and their definitions, most of which are applicable to poultry.

Active immunity—Immunity or resistance to disease that has been acquired by host response to a disease agent. It can be acquired by having a disease and recovering or by vaccination.

Acute—A disease which has a short and relatively severe course.

Allergy—A severe reaction, or sensitivity, which occurs in some individuals following the introduction of certain antigens into their bodies.

Anemia—A condition in which the blood is deficient in quantity or quality (lack of hemoglobin content or in number of red blood cells), characterized by paleness of skin and mucous membranes and loss of energy.

Anthelmintic (vermifuge)—A product which removes worm parasites.

Antibiotic—A chemical substance, produced by molds or bacteria, which has the ability to inhibit the growth of, or to destroy, other microorganisms.

Antibody—A substance that opposes the action of another substance.

Antigen—A foreign substance which, when introduced into the body, stimulates formation of protective antibodies.

Antiseptic—A compound that inhibits the growth of microorganisms, and which is usually applied to the skin.

Antiserum—A serum containing a specific antibody used to treat a specific disease.

Antitoxin—A specific kind of antibody that will neutralize toxin.

Atrophy—Wasting away or diminution in size.

Autopsy—Inspection, and partial dissection, of a dead body to determine the cause of death.

Bacteria—One-celled microorganisms, smallest of the non-green plants, which are chiefly parasitic.

Bactericide—A product which kills bacteria.

Bacterin—Killed suspension of bacterial organisms used as an immunizing agent.

Bacteriostat—A product which retards bacterial growth.

Breathing rate—Normal for poultry ranges from 15 to 36 per minute.

Broad-spectrum antibiotic—An antibiotic which attacks both Gram-positive and Gram-negative bacteria, and which may also show activity against other disease agents.

Carrier—An apparently healthy animal that harbors disease organisms and is capable of transmitting them to susceptible animals.

Catarrhal—Describes an inflammatory condition of the mucous membranes characterized by an increased flow of mucus.

Chronic—A disease of long duration.

Coccidiostat—A drug incorporated in feed at low levels and fed continuously to prevent coccidiosis.

Congestion—Excessive accumulation of blood in a part.

Contagious—An infectious disease that may be transmitted readily from one individual to another.

Culture—The propagation of microorganisms, or of living tissue cells, in special media conducive to their growth.

Cyanosis — Bluish discoloration of the skin, particularly the comb and wattles in birds.

Debilitating — Weakening.

Disease — Any departure from a normal state of health.

Disinfectant — A product which, at certain concentrations, will kill on contact a wide range of disease organisms.

Edema — Presence of abnormal amounts of fluid in tissues.

Emaciated — A severe loss of weight.

Enzootic — A disease confined to a certain locality.

Etiology — Study of the cause of disease.

Excreta — Excreted material; waste matter.

Exudate — A fluid oozing from tissue.

Filterable virus — An organism so small that it is capable of passing through filters which will retain the ordinary bacteria.

Fungi — Certain vegetable organisms such as molds, mushrooms, and toadstools.

Gram-negative bacteria — Those bacterial species which are decolorized by acetone or alcohol.

Gram-positive bacteria — Those bacterial species which retain crystal-violet colors even when exposed to alcohol or acetone.

Gross — A change in tissue which can be seen with the naked eye.

Hematocrit — Concentration of red blood cells in a given amount of blood.

Hemorrhage — Escape of blood from vessels; bleeding.

Hypersensitivity — A state in which the body reacts to a foreign agent more strongly than normal.

Immune — Resistant to a particular disease.

Immunity — Condition of being immune.

Infection — Invasion of the tissue by pathogenic organisms resulting in a disease state.

Infectious — A disease produced by living organisms.

Inflammation — Response of tissues to an injury or other irritant.

Ingestion — The taking in of food and drink.

Intradermal — Into, or between, the layers of the skin.

Intramuscular — Within the substance of a muscle.

Intraperitoneal — Within the peritoneal cavity.

Intrauterine — Within the uterus.

Intravenous — Within the vein or veins.

In vitro — Occurring in a test tube.

In vivo — Occurring in the living body.

"Itis" — Suffix denoting an inflammatory state, such as enteritis — an inflammation of the intestines.

Lesion — Visible change in size, shape, color, or structure of an organ.

Listless — Indifferent to surroundings.

Medium-spectrum antibiotic — An antibiotic which attacks a limited number of Gram-positive and Gram-negative bacteria.

Metabolic — Pertaining to the nature of metabolism.

Metabolism — Refers to the changes which take place in the nutrients after they are absorbed from the digestive tract including (1) the building-up processes in which the absorbed nutrients are used in the formation or repair of body tissues, and (2) the breaking-down processes in which nutrients are oxidized for the production of heat and work.

Microorganisms — Any organism of microscopic size, applied especially to bacteria and protozoa.

Microscopic — Invisible to the naked eye. Visible only by the aid of a microscope.

Morbidity — Sick rate.

Mortality — Death rate.

Narrow-spectrum antibiotic — An antibiotic whose activity is restricted to either Gram-negative or Gram-positive bacteria. For example, penicillin is active primarily against Gram-positive organisms, whereas streptomycin attacks only Gram-negative organisms.

Necrosis — Death or dying of local tissue.

Neoplasm — Abnormal growth such as a tumor.

Oral — Given by mouth.

Parasite — An animal form that lives on or within a bird to its detriment.

Parenteral — As applied to drug or vaccine administration, to inject subcutaneously, intramuscularly.

Pathogenic — Disease-producing.

Pathological — Diseased, or due to disease.

Postmortem — Examination after death.

Predispose — To confer a tendency toward disease.

Prophylaxis — Preventive treatment against disease.

Protozoa — One-celled animals which reproduce by splitting in half; found largely in water, and include many parasitic forms.

Pulse rate — Normal for poultry ranges from 200 to 400 per minute.

Rickettsial — A group of microorganisms intermediate between the bacteria and viruses, some of which are pathogenic to humans and animals.

Rigor mortis — Stiffening of the body after death.

Rod — As applied to bacteria, a cylindrical shaped organism.

Secondary invaders — Infective agents which attack after a primary organism has established an infection.

Serum, blood — The clear portion of blood separated from its more solid elements.

Sign — Discernable evidence of disease; symptoms or lesions.

Sporadic — A disease outbreak occurring here and there; not widely diffused.

Spore — Bacteria or fungi capable of resisting unfavorable environmental conditions.

Stress — Factors tending to lower resistance of an animal to disease, such as chilling, moving, etc.

Subcutaneous — Under the skin.

Sulfa drug (sulfonamide) — A synthetic organic drug which has the ability to inhibit growth of, or to destroy, microorganisms.

Supportive treatment — Treatment of individual symptoms of a disease where diagnosis is obscure or where a specific treatment has not been established.

Symptoms — Detectable signs of disease.

Temperature — Normal for poultry is 106°F, with a range of 105° to 107°F.

Therapy — Treating disease.

Toxemia — A condition produced by the presence of poisons (toxins) in the blood.

Toxin — Poison produced by microorganisms.

Vaccine — A suspension of live microorganisms (bacteria or virus) or microorganisms that have had their pathogenic properties removed but their antigenic properties retained.

Virulence — As applied to pathogenic, microscopic organisms, refers to its ability to overcome the body defenses of the host.

Virus — The smallest living microorganism, not visible under an ordinary microscope, which lives parasitically upon plants and animals, and sometimes causes disease.

QUESTIONS FOR STUDY AND DISCUSSION

1. What might be considered the normal (a) percentage of sick birds at one time, (b) percentage mortality losses during the first 3 weeks of life of chicks and of poults, and (c) mortality of chickens after 3 weeks?

2. Give the signs in poultry (a) of good health, and (b) of disease.

3. List and discuss the most important predisposing causes of poultry diseases.

4. Discuss the difference between infectious and contagious diseases.

5. List the major ways in which contagious diseases of poultry are spread.

6. List poultry body defenses against disease.

7. Outline a flock health program.

8. Outline a hatchery sanitation program.

9. What is a disinfectant? List the five things upon which an effective disinfectant depends.

10. Name five of the most common disinfectants used in poultry operations.

11. What is the difference between live-virus vaccines, killed-virus vaccines, and bacterins?

12. Name five virus diseases which can now be controlled by vaccination. Of the products on the market for the control of virus infections, why do vaccines to prevent respiratory infections lead the list?

13. When and by whom should chicks be vaccinated against Marek's disease?

14. Why and how should a poultry producer use the state diagnostic laboratory?

15. What is the purpose of the Specific Pathogen Free (SPF) program? List a few of the pertinent recommendations of the SPF program.

16. What characterizes each of the following disease groups from the standpoint of cause: (a) bacterial diseases, (b) respiratory diseases, (c) viral diseases, (d) protozoan diseases, and (e) nutritional diseases?

17. Name three external parasites of poultry and give the (a) cause, (b) symptoms, (c) treatment, and (d) prevention of each.

18. Name three internal parasites of poultry and give their (a) cause, (b) symptoms, (c) treatment, and (d) prevention.

19. Relate pertinent information to each of the following conditions: (a) cage fatigue, (b) fatty liver syndrome, and (c) hysteria.

20. It is generally believed that cannibalism was a rarity among wild chickens, and that it was accentuated by confinement. Should this transition be construed as cruelty to chickens? Justify your answer.

21. Define each of the following terms: (a) active immunity, (b) antibiotic, (c) carrier, (d) disease, (e) fungi, (f) morbidity, (g) sporadic, and (h) stress.

SELECTED REFERENCES

Title of Publication	Author(s)	Publisher
Diseases of Poultry, Ninth Edition	Edited by B. W. Calnek, *et al.*	Iowa State University Press, Ames, 1991
Merck Veterinary Manual, The, Sixth Edition	Edited by C. M. Fraser, *et al.*	Merck & Co., Inc., Rahway, NJ, 1986
Poultry Health Handbook	L. D. Schwartz	The Pennsylvania State University Press, University Park, 1972
Serviceman's Poultry Health Handbook	E. H. Peterson	Better Poultry Health Company, Fayetteville, AR, 1975

In addition to these selected references, valuable publications on different subjects pertaining to poultry diseases, parasites, disinfectants, and related matters can be obtained from the following sources: Division of Publications, Office of Information, U.S. Department of Agriculture, Washington, DC; your state agricultural college; and several biological, pharmaceutical, and chemical companies.

Fig. 11–1. Inside view of turkey grow-out facility. Integrated operation of Cuddy Farms, Ltd., North Carolina. (Photo courtesy James Strawser, University of Georgia, Athens)

POULTRY HOUSES AND EQUIPMENT

CHAPTER

11

Contents

Contents

Poultry housing and equipment are needed for the comfort, protection, and efficient production of eggs and meat.

The jungle fowl, ancestor of the modern chicken, sought comfort and safety on the high limb of a tree or in the thick underbrush. The jungle also provided protection from the sun and the wind.

The natural laying and mating season of birds is in the spring of the year, when the weather is comfortable and there is the added stimulus of gradually increasing length of day. It follows that for maximum egg production and growth during other seasons of the year, springlike conditions should be emulated. The latter is achieved by housing, which provides protection from the elements, plus the stimulus of artificial light when needed. Additionally, modern poultry houses and equipment make it possible to care for the flock in accordance with recommended practices, and with a minimum of time and effort.

No standard set of poultry buildings and equipment can be expected to be well adapted to such diverse conditions and systems of poultry production as exist in the United States. This is evidenced by Fig. 11–14, p. 267, a map showing three U.S. winter temperature zones as related to the amount of insulation needed in poultry houses. Extremes in temperature, and the duration of these extremes, along with other climatic factors that are closely associated with temperature, such as amount of sunlight, rainfall, latitude, and elevation, determine the type of housing and construction necessary. Thus, detailed plans and specifications for poultry houses in a given area should be obtained from a local architect or from the college of agriculture of the state.

LOCATION

It is important that poultry buildings be located and planned correctly for efficient, profitable, and pleasant operation. To this end, consideration should be given to the following:

1. **Water supply.** Water must be available and plentiful. The availability of electricity and automatically operated pumps makes it possible to locate the headquarters farther from the source of water supply when it is desirable to do so to obtain other advantages.

2. **Roads.** It is preferable that poultry buildings be located near an all-weather road or highway that is well maintained. Normally, a location along an all-weather road has

better access to electric and telephone lines, the school bus, mail, religious and recreational facilities, and other services.

3. **Telephone and electricity.** The headquarters should be near well-maintained telephone and electric lines. Poultry farming is a business, and it is difficult to conduct any kind of business without access to a telephone. Likewise, electricity is essential for the operation of most modern utilities and automated equipment.

4. **Service facilities.** The poultry farm should have convenient access to an established mail route, a school bus from a good school, delivery services (milk, laundry, bread, etc.), the church of preference, and recreational facilities of interest.

5. **Topography.** The topography should be high and level with no abrupt slopes. A relatively level area requires less site preparation, thereby lowering building costs.

6. **Drainage.** The soil should be porous and the slope gentle, for this makes for dryness.

7. **Layout of operations.** Prior to starting construction, poultry producers may avoid much subsequent difficulty and expense by first doing some paper and pencil planning. They should first decide on the specific kind, or kinds, of poultry and the size of the enterprise. Then, they should sketch out the buildings and equipment required to meet these needs in the most efficient and economical manner. In particular, the preliminary layout of poultry operations should include the following:

    a. **The management system.** The management system will greatly affect the kind, size, and amount of buildings and equipment. For example, commercial egg producers should decide (1) whether they are going to raise or buy replacement pullets, and (2) whether they are going to buy commercial, ready-mixed feeds, or buy ingredients and do their own mixing.

    b. **Plans for the flow of all materials.** Producers should develop detailed plans for the flow of all materials, with primary consideration given to maximum automation and disease prevention. These plans should include provision for (1) delivering the feed to the birds at the desired time and place, (2) providing a sanitary water supply, (3) delivering and distributing litter, (4) removing manure, and (5) marketing. All these considerations, and more, enter into the handling of materials to, within, and from buildings.

The above information, constituting the layout of operations, should first be put on paper in sketch form, by the producer. From this, the architect and/or engineer can design, or recommend for purchase, buildings and equipment which most effectively and economically meet the production requirements of the specific enterprise.

8. **Vegetation; windbreaks.** Natural shade, trees for windbreaks, and a well-sodded area are valuable attributes. If a natural windbreak (hills or trees) is not available, wind protection may be provided by planting trees, or by utilizing the buildings themselves as windbreaks to protect yards, lots, and other open areas.

9. **Orientation.** Fortunately, the poultry farm need not be oriented with the compass. Although in general the farmstead plan will be developed to present the front to the road, most buildings can be turned, quarter-turned, or reversed, as

may be necessary to take advantage of the prevailing winds, sunlight, view, etc. In general, poultry houses are placed with the long axis north and south.

10. **Fire protection.** Poultry buildings should be far enough apart so that fire will not spread easily from one building to another. In general, this means at least 100 ft apart in the case of large buildings. In acquiring added fire protection through spacing buildings farther apart, one should avoid extreme distances that will mean inefficiency in operation; fire insurance is probably cheaper than labor.

11. **Appearance.** Careful attention to the headquarters arrangement can add to the attractiveness of the entire unit. Manure piles and unsightly objects should not be visible from the main highway or house; shrubbery and trees should be planted to screen unsightly objects; fences and buildings should be repaired and painted regularly; and yards and driveways should be kept free of rubbish, scattered equipment, etc.

12. **Expansion.** Provision should be made for easy expansion. Many times buildings can be expanded in size by extending their length, provided no other buildings or utilities interfere.

SPACE REQUIREMENTS OF BUILDINGS AND EQUIPMENT

The intensification of animal agriculture, along with increased confinement, has created the need for scientific, yet practical, information relative to space requirements. Moreover, the recommendations have changed from time to time, as a result of new experiments and experiences.[1] In 1958, Siegel and Coles showed that with adequate control of other environmental factors, broilers could be reared in either 72, 108, 144, or 178 sq in. of floor space. The conclusions were questioned because routine space allowances in the field were 144 sq in. at the time. Yet, within a few years, allowances of 86 sq in. became routine.

One of the first and frequently one of the most difficult problems confronting the producer who wishes to construct a building or item of equipment is that of arriving at the proper size or dimensions. Suggested space requirements are given in Table 11–1.

---

[1]Siegel, P. B., and W. B. Gross, "Confinement, Behavior, and Performance with Examples from Poultry," *Journal of Animal Science*, Vol. 37, No. 2, August 1973, p. 612.

TABLE 11–1
SPACE REQUIREMENTS FOR POULTRY[1]

Type of Bird	Type of Facility	Age of Birds (wk)	Space/Bird (sq ft)	Space/Bird (sq cm)	Feeder Space/Bird Linear (in.)	Feeder Space/Bird Linear (cm)	Water Space/Bird Linear (in.)	Water Space/Bird Linear (cm)
Commercial layers (Leghorn-type hybrids)	Floor	0–4	0.30	279	1.0	2.5	0.2	0.5
	Floor	4–8	0.60	557	1.0	2.5	0.4	1.0
	Floor	9–16	1.25	1,161	1.5	3.8	0.6	1.5
	Floor	16 over	1.50	1,394	1.5	3.8	1.0	2.5
	Cage, individual	Adult	0.50	465	3.0	7.6	1.5	3.8
	Cage, colony	Adult	0.50	465	3.0	7.6	1.5	3.8
Broiler-breeder pullets	Floor	0–8	0.80	743	1.0	2.5	0.5	1.3
	Floor	9–16	1.30	1,208	3.0	7.6	0.6	1.5
	Floor	16 over	2.00	1,858	4.0	10.2	1.0	2.5
Broiler-breeder hens	Floor	Adult	2.50	2,322	4.0	10.2	2.0	5.0
	Slat floor	Adult	2.00	1,858	4.0	10.2	2.0	5.0
Broilers	Floor	0–4	0.30	279	1.0	2.5	0.2	0.5
	Floor	4–8	0.75	697	1.0	2.5	0.2	0.5
Roasters	Floor	0–8	0.80	743	1.0	2.5	0.5	1.3
	Floor	9–16	1.50	1,394	2.0	5.0	1.0	2.5
Turkey-breeders	Confinement on floor	Adult	4.50	4,181	3.0	7.6	1.0	2.5
	Confinement and range	Adult	2.50 in house plus range	2,322	3.0	7.6	1.0	2.5
Turkeys-market	Confinement on floor	0–4	1.25	1,161	1.0	2.5	0.5	1.3
	Confinement on floor	4–16	2.50	2,322	2.0	5.0	1.0	2.5
	Confinement on floor	16–29	4.00	3,716	2.5	6.4	1.0	2.5
Duck-breeders	Confinement-yard	Adult	2.50 in house plus yard	2,322	2.0	5.0	1.5	3.8
Ducklings-market	Wire	0–3	0.50	465	1.0	2.5	0.5	1.3
	Floor	0–3	1.00	929	1.0	2.5	0.5	1.3
	Floor	3–5	1.50	1,394	1.5	3.8	1.0	2.5
	Floor	After 5	2.00	1,858	2.0	5.0	1.5	3.8
Geese	Floor	0–2	1.25	1,161	1.5	3.8	1.0	2.5
	Floor	After 2	2.50	2,322	2.5	6.4	2.0	5.0

[1]"Report of the Committee on Avian Facilities," *Poultry Science*, 1974, 53(6):2257, with turkeys, ducks, and geese added by the author.

POULTRY HOUSES

Birds of good breeding, no matter how well fed, will not return maximum profit unless well housed. Good housing includes everything which contributes to the comfort and health of the birds and to the convenience and ease of management for the poultry producer at a reasonable cost.

In order to design any building properly it is necessary to know the purpose for which it is to be constructed. In addition, the designer must know the nature and number of units to be housed, their size and the space required for each, and the conditions required for best operation or production. The designer must also be aware of the normal weather conditions of the area and have a knowledge of available materials that are best suited for use under the given conditions.

REQUISITES OF POULTRY HOUSES

There are certain general requisites of all animal buildings, regardless of the species, that should always be considered; among them, reasonable construction and maintenance costs, minimum labor costs, and minimum utility costs. In the case of poultry, however, increased emphasis needs to be placed on the following features of buildings:

1. **Temperature.** Feathers give some protection against cold. However, the bird's efficiency in egg production, meat production, and feed utilization is greatly lowered when it must endure temperatures appreciably below the comfort zone.

Birds have a very poor defense against heat, and their cooling system is not very efficient because they do not have sweat glands. They attempt to adjust to heat (a) by panting or breathing rapidly with the mouth open, (b) by eating less and drinking more, (c) by holding their wings away from the body, and (d) by resting against a cool surface such as the damp earth or a concrete floor.

The optimum temperature for layers is 55° to 70°F and for broilers, 75°F.

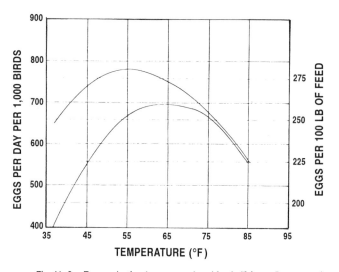

Fig. 11–2. Egg production (upper curve) and feed efficiency (lower curve) at various house temperatures. (Courtesy, USDA)

Tests conducted by the U.S. Department of Agriculture showed that hens kept at 55°F laid at a rate of 75% and consumed 3.5 lb of feed for each pound of eggs, whereas those maintained at 85°F laid at a rate of only 50% and consumed 4 lb of feed for each pound of eggs, while those kept at 23°F laid at a rate of only 26% and ate 12.3 lb of feed for each pound of eggs. Experiments conducted at the University of Connecticut showed a 12.5% increase in feed efficiency in broilers grown in a 75°F house compared to those grown at 45°F.

Humidity influences are tied closely to temperature effects. For laying houses, the relative humidity should be within the range of 50 to 75%.

2. **Insulation.** The term *insulation* refers to materials which have a high resistance to the flow of heat. Such materials are commonly used in the walls and ceilings of poultry houses. Proper insulation makes for a more uniform temperature—cooler houses in the summer and warmer houses in the winter, and for a substantial fuel saving in brooder houses.

Heat produced by layers varies with their weight and the environmental temperature. White Leghorn layers will produce 9 BTUs per pound per hour at a housing temperature of 55°F. This is approximately 40 BTUs per hour for a typical 4½-lb White Leghorn hen.

3. **Vapor barrier.** There is much moisture in poultry houses; it comes from open water fountains, wet litter, the respiration of the birds, and from the droppings. When the amount of water vapor in the house is greater than in the outside air, the vapor will tend to move from inside to outside. Since warm air holds more water vapor than cold air, the movement of vapor is most pronounced during the winter months. The effective way to combat this problem in a poultry house is to use a vapor barrier with the insulation. It should be placed on the warm side or the inside.

4. **Ventilation.** Ventilation refers to the changing of air — the replacement of foul air with fresh air. Poultry houses should be well ventilated, but care must be taken to avoid direct drafts and coldness. Good poultry house ventilation saves feed and helps make for maximum production.

Three factors are essential for good ventilation: (a) fresh air moving into the poultry house, (b) insulation to keep the house temperatures warm, and (c) removal of moist air.

In most poultry houses, easily controlled electric fans do the best job of putting air where it is needed. Gravity flow — air movement without fans — is suitable only for narrow houses and few birds.

A complete ventilation system has three parts: (a) a fan, or fans, to move fresh air through the house; (b) enough inlets to let plenty of fresh air in; and (c) enough outlets to let stale, moisture-laden air out. All these parts are necessary for success.

Exhaust-type ventilation systems are usually used. Wall fans are generally cheaper and easier to install than ceiling fans. However, ceiling fans provide gravity ventilation if electricity should fail.

The water-holding capacity of air increases with rising temperature (Fig. 11–3).

Fans are rated in cubic feet per minute (cfm) of air they

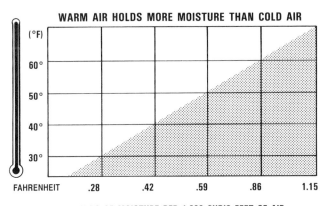

WARM AIR HOLDS MORE MOISTURE THAN COLD AIR

POUNDS OF MOISTURE PER 1,000 CUBIC FEET OF AIR

Fig. 11–3. Influence of air temperature on its water holding capacity. (From: Washington State University Ext. Bull. No. 529)

move. To arrive at fan capacity, allow (a) 3 cfm per bird for layers of the light breeds, (b) 4 cfm per bird for layers of the medium-weight breeds, (c) 5 cfm per bird for heavy breeds, and (d) ½ cfm per pound of body weight for broilers and turkeys.

5. **Lighting.** In the early 1900s, when lighting was first used to stimulate egg production, it was thought that the role of light was primarily a matter of increasing the "work day" of the bird. Today, the action of light is considered physiological; light enters the eye of the bird and stimulates the pituitary gland. In turn, the pituitary gland releases certain hormones which cause ovulation. Because of this phenomenon, artificial lighting in the poultry house is exceedingly important. For pullets, an increase in the day length during the growing period will stimulate early maturity, whereas a decrease in day length will delay the age of maturity. For mature layers, an increase in day length will stimulate egg production, whereas a decrease in day length will suppress egg production.

Automatic time switches are available at moderate cost and should be installed in poultry houses for pullets or layers.

6. **Manure management.** The handling of manure is probably the single most important problem confronting commercial poultry producers today. It must be removed for sanitary reasons, and there is a limit to how much of it can be left to accumulate in pits or other storage areas. Also, to the suburbanite and city dweller alike, manure odor is taboo. Hence, it has two primary outlets: (a) as a fertilizer, and (b) as a feed for ruminants. Its value for both purposes has increased in recent years – as a replacement for high-priced petroleum-based chemical fertilizer, and as a substitute for scarce and high-priced ruminant feed (see Chapter 8, Poultry Management, section on "Poultry Manure").

NATURALLY VENTILATED BUILDINGS

There is continuing interest among poultry producers in naturally ventilated buildings as opposed to environmentally controlled buildings, primarily because of their significantly lower construction and operating costs. Because no attempt is made to regulate temperature, the costs of heavy insulation, tight fitting doors and windows, and a mechanical ventilation system are averted. Of course, buildings which for management reasons must be maintained at temperatures above winter levels are not suited to natural ventilations (e.g., brooder houses).

Naturally ventilated buildings are mainly a shell to protect birds from rain and snow and act as a sunshade. Winter inside temperatures will often be within 3° to 10°F of outside temperatures. Thus, such buildings are often referred to as cold confinement buildings.

Typical naturally ventilated buildings can be divided into two types as follows:

1. **Open sides.** These buildings have long sides which are partially open.

Fig. 11–4. Naturally ventilated, curtain-sided broiler house in California. (Courtesy, D. [Bell, University of California, Riverside)

Fig. 11–5. Inside view of a naturally ventilated, curtain-sided turkey grow-out house (Courtesy, P. R. Ferket, North Carolina State University, Raleigh)

2. **Enclosed.** These buildings have all sides closed but have a continuous opening at the high point (normally the ridge) of the building for air exhaust and continuous openings or inlets along the long sidewalls of the building for fresh air. The size of these openings is based on rules of thumb or experience. Air entering along the sidewalls (normally under the eaves) of the building is warmed by the heat from the birds in the building and picks up moisture as it rises toward the ridge. The continuous open ridge allows this warm, moist air to escape, thus completing the air exchange process.

ENVIRONMENTALLY CONTROLLED BUILDINGS

Environment may be defined as all the conditions, circumstances, and influences surrounding and affecting the growth, development, and production of a living thing. In poultry, this includes the air temperature, relative humidity, air velocity, wet bedding, dust, light, ammonia buildup, odors, and space requirements.

Optimum environmental conditions must be provided for maximum production. Today, with genetically superior birds and scientifically balanced rations it is only with the optimum environment that their genetic potential and maximum feed conversion can be realized. There is still much to be learned about environmental control, but the gap between awareness and application is becoming smaller.

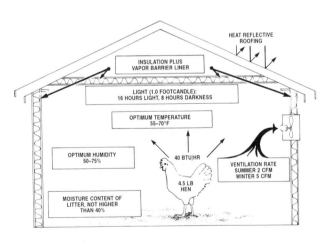

Fig. 11–6. A favorable environment for a laying hen.

The *critical temperature* is that temperature at which the heat created by digestion and body metabolism just equals that which the bird dissipates by convection, evaporation, radiation, and conduction. The *comfort zone* is the range in temperature within which the bird may perform with little or no discomfort. At temperatures below the comfort zone, additional nutrients need to be converted to heat to keep the

body warm; and at temperatures above the comfort zone, nutrients are needed to help keep the bird cool. The *optimum temperature* is the temperature at which the bird responds most favorably, as determined or measured by maximum rate of gains or production, feed efficiency, and/or production.

The critical temperature and comfort zone vary with different species, ages, breeds, and the physiological and productive status. The species differences result primarily from the kinds of thermoregulatory mechanism provided by nature, such as type of coat (feathers, hair, wool), sweat glands, etc.

The temperature varies according to age, too. For example, the comfort zone of baby chicks is 95°F, whereas the comfort zone of adult layers is 50° to 75°F.

Stresses at both high and low temperature are increased with high humidity. The respired air has less of a cooling effect. As humidity of the air increases, discomfort at any temperature increases, and nutrient utilization decreases.

Air movement (wind) results in body heat being removed at a more rapid rate than when there is no wind. In warm weather, air movement may make the birds more comfortable; but in cold weather, it adds to the stress of temperature. At low temperatures, the nutrients required to maintain body temperature are increased as the wind velocity increases. In addition to the wind, a drafty condition where the wind passes through small openings directly onto some portion or all of the body will usually be more detrimental to comfort and nutrient utilization than the wind itself.

Poultry producers were little concerned with the effect of environment on birds as long as they were out on the range. Space requirements, wet bedding, ammonia buildup, odors, and manure disposal were no problem. But the concentration of birds into smaller spaces changed all this. With the shift to confinement structures and high-density production operations, building design became more critical.

Environmentally controlled buildings are costly to construct, but they make for the ultimate in bird comfort, health, and efficiency of feed utilization. Also, they lend themselves to automation, which results in a saving in labor; and, because of minimizing space requirements, they effect a saving in land cost. Today, environmental control is rather common in poultry, and it will increase.

The decision on whether or not confinement and environmental control can be justified should be determined by economics. Will the birds in environmentally controlled quarters lay sufficiently more eggs, or gain sufficiently more rapidly and efficiently, to justify the added cost? Of course, manure disposal and pollution control should also be considered.

Before an environmental system can be designed for poultry, it is important to know the (1) heat production, (2) vapor production, and (3) space requirements (see Table 11–1, p. 257, for space requirements). This information is as pertinent to designing poultry buildings as nutrient requirements are to balancing rations.

HEAT PRODUCTION OF POULTRY

The heat production of a laying hen is given in Table 11–2.

TABLE 11–2
HEAT PRODUCTION OF LAYING HEN[1]

Heat Source	Unit		Heat Production, BTU/Hr			Heat Production, Kcal/Hr		
			Temperature	Total	Sensible	Temperature	Total	Sensible
	(lb)	(kg)	(°F)			(°C)		
Layer hen	4.5	2.04	55	40	28	12.8	10.1	7.1

[1]Adapted by the author from *Agricultural Engineers Yearbook*, St. Joseph, MI, ASAE Data Sheet D–249.2, p. 424.

Table 11–2 may be used as a guide, but in doing so, consideration should be given to the fact that heat production varies with age, body weight, ration, breed, activity, house temperature, and humidity at high temperatures. As noted, Table 11–2 gives both total heat production and sensible heat production. Total heat production includes both sensible heat and latent heat combined. *Latent heat refers to the energy involved in a change of state and cannot be measured with a thermometer; evaporation of water or respired moisture from the lungs are examples. Sensible heat is that portion of the total heat, measurable with a thermometer, that can be used for warming air, compensating for building losses, etc.* Heat is measured in British thermal units (BTUs). *One BTU is the amount of heat required to raise the temperature of 1 lb of water 1°F.*

Since ventilation also involves a transfer of heat, it is important to conserve heat in the building to maintain desired temperatures and reduce the need for supplemental heat. In a well-insulated building, mature birds may produce sufficient heat to provide a desirable balance between heat and moisture, but newly hatched poultry will usually require sup-

plemental heat. The major requirement of summer ventilation is temperature control, which requires moving more air than in the winter.

VAPOR PRODUCTION OF POULTRY

Birds give off moisture during normal respiration; and the higher the temperature the greater the moisture. This moisture should be removed from buildings through the ventilation system. Most building designers govern the amount of winter ventilation by the need for moisture removal. Also, cognizance is taken of the fact that moisture removal in the winter is lower than in the summer; hence, less air is needed. However, lack of heat makes moisture removal more difficult in the wintertime. Table 11–3 gives the information necessary for determining the approximate amount of moisture to be removed for layers.

TABLE 11–3
VAPOR PRODUCTION OF LAYING HEN[1]

Vapor Source	Unit		Temperature		Vapor Production		Vapor Production	
	(lb)	(kg)	(°F)	(°C)	(lb/hr)	(BTU/hr)	(kg/hr)	(kcal/hr)
Layer hen	4.5	2.04	50	10	0.012	12	0.005	3.0

[1]Adapted by the author from *Agricultural Engineers Yearbook*, St. Joseph, MI, ASAE Data Sheet D–249.2, p. 424.

RECOMMENDED ENVIRONMENTAL CONTROLS

The comfort of poultry is a function of temperature, humidity, and air movement. Likewise, the heat loss from birds is a function of these three items.

Temperature, humidity, and ventilation recommendations for layers, broilers, and turkeys are given in Table 11–4. This table will be helpful in obtaining a satisfactory environment in confinement poultry buildings, which require careful planning and design.

TABLE 11–4
RECOMMENDED ENVIRONMENTAL CONDITIONS FOR POULTRY

Kind of Poultry	Temperature				Acceptable Humidity	Commonly Used Ventilation Rates[1]			Drinking Water			
	Comfort Zone		Optimum			Basis	Winter	Summer	Winter		Summer	
	(°F)	(°C)	(°F)	(°C)	(%)		(cfm)	(cfm)	(°F)	(°C)	(°F)	(°C)
Layers	50–75	10–24	55–70	13–21	50–75	per bird	2	5	50	10	60–75	15–24
Broilers	95[2]	35	70[3]	21	50–75	per lb body weight	0.5	1	50	10	60–75	15–24
Turkeys	95–100[2]	35–38	70[3]	21	50–75	per lb body weight	0.5	1	50	10	60–75	15–24

[1]Generally two different ventilating systems are provided; one for winter, and an additional one for summer.

[2]First week after hatching.

[3]After sixth week.

Fig. 11–7. Environmentally controlled, three-tier caged layer house, with controlled temperature, humidity, air movement, and lighting, along with automatic feeding by feed conveyor chain, and automatic egg collection system on longitudinal belts. (Courtesy, Landwerk, Veghel, Holland)

The prime function of the winter ventilation system is to control moisture, whereas the summer ventilation system is primarily for temperature control. If air in poultry houses is supplied at a rate sufficient to control moisture—that is, to keep the inside relative humidity in winter below 75%—then this will usually provide the needed fresh air, help suppress odors, and prevent an ammonia buildup.

LIGHTING

The intensity of light required by poultry (about 1 footcandle) is much less than is required for plant growth or for an everyday activity of people such as reading.

■ **Layers**—Artificial light in the laying house should give a 16-hour day. If natural daylight is longer, artificial light should maintain the longest daylight period, although a light regimen in excess of 17 hours is of doubtful value. Use intensity of 1 footcandle at bird height. (Approximately 40-watt bulb with reflector every 12 ft, 7 ft above the floor in floor systems.) Never decrease day length or light intensity during the laying period. A photoelectric cell may be used in connection with a time clock to turn lights off at dawn and on at twilight.

■ **Replacement pullets**—During the growing period, pullets should not be exposed to increasing amounts of light prior to 21 weeks of age if best egg production is to be attained. Rather, short or decreasing light during the growing period is desirable.

The following lighting regimen is recommended for pullets: 0 to 3 weeks, 1.0 footcandle, 20 hours of light and 4 hours of darkness; 3 to 12 weeks of age, 0.1 to 0.5 footcandle, 16 hours of light and 8 hours of darkness; 12 to 21 weeks of age, 0.5 footcandle, continue 16 hours of light and 8 hours of darkness to 16 weeks of age, followed by decreasing light to 8 hours of light and 16 hours of darkness at 21 weeks of age.

Long or increasing light should begin at 21 weeks of age, with the light increased at the rate of 15 minutes per week until 17 hours, then maintained.

One 25-watt bulb with reflector every 12 ft, 7 ft above the floor will give approximately 0.5 footcandle intensity.

■ **Broilers**—Recommended broiler lighting in light-tight houses is: 0 to 5 days, 3.5 footcandle continuous light; 6 days to market, 0.35 footcandle, 23 hours of continuous light and 1 hour of darkness. Or from 6 days to market, intermittent light may be used; 1 hour of dim light (feeding time) to 3 hours of darkness (resting time).

Recommended lighting for broilers in open-sided houses is: 0 to 2 days, 3.5 footcandle of continuous light; 2 days to market, 0.5 footcandle, 23 hours of continous light and 1 hour of darkness.

■ **Turkeys**—Market turkeys: 0 to 2 weeks, 10 to 15 footcandle, continuous light; 2 weeks to market, 0.5 to 1.5 footcandle, 16 hours of continuous light and 8 hours of darkness.

Turkey breeders: same lighting as market turkeys to 29 weeks of age. After 29 weeks, 5 to 7 footcandle, 14 hours light and 10 hours darkness in spring and summer (or 8 hours of light and 16 hours of darkness in the fall).

■ **Ducks, breeders**—Mature ducks may be brought into production in the winter by providing a 14-hour day (14 hours of light and 10 hours of darkness).

■ **Other poultry**—Pheasants and chukars appear to have a short day requirement. Coturnix (quail) are unique, for they may be sexually mature by 6 weeks of age if exposed to day lengths in excess of 12 hours; in comparison with the 21 to 29 weeks for chickens and turkeys, respectively.

(Also see Chapter 8, Poultry Management, Table 8–3, Recommended Lighting Systems for Poultry.)

VENTILATION

A 4-lb Leghorn breathing 40 times per minute uses air at the rate of 0.019 cu ft per minute. Compared with each other, the air breathed in is 5% richer in oxygen and the air breathed out 5% richer in carbon dioxide. Fig. 11–8 shows what happens to respired air each minute.

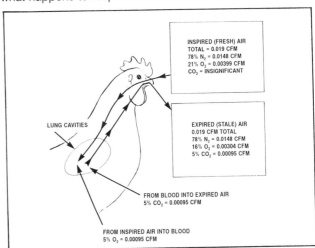

INSPIRED (FRESH) AIR
TOTAL = 0.019 CFM
78% N_2 = 0.0148 CFM
21% O_2 = 0.00399 CFM
CO_2 = INSIGNIFICANT

LUNG CAVITIES

EXPIRED (STALE) AIR
0.019 CFM TOTAL
78% N_2 = 0.0148 CFM
16% O_2 = 0.00304 CFM
5% CO_2 = 0.00095 CFM

FROM BLOOD INTO EXPIRED AIR
5% CO_2 = 0.00095 CFM

FROM INSPIRED AIR INTO BLOOD
5% O_2 = 0.00095 CFM

Fig. 11–8. Respiration air changes each minute.

But respiration needs are greatly overshadowed by the rate of air change necessary to remove moisture from a tightly built house. In most instances, if ventilation is adequate to keep the house dry, it will more than meet the air requirements of the birds.

The factors affecting and the procedure for calculating ventilation requirements of poultry are discussed later in this chapter in the several subsections under "Environmental Factors." Only the ventilation equipment is presented at this point.

Forced-air ventilation is used in most commercial poultry houses. Most systems use exhaust fans and air intakes. Poultry specialists should be consulted in planning and installing a fan system for ventilation.

DESIGN AND CONSTRUCTION

The types, uses, and plans of poultry houses are very diverse.

■ **Type**—Poultry housing may be one of the following types: (1) open shed, (2) semienclosed, or (3) enclosed with light and temperature control.

Open sheds are suited for breeders, especially turkeys, ducks, geese, and game birds.

Semienclosed houses give more protection to the birds and provide more environmental control.

Enclosed houses are environmentally controlled, with the light, temperature and ventilation regulated.

■ **Purposes**—Large and highly specialized poultry enterprises have buildings adapted to their specialized purposes. Thus, chicken houses are designed for the brooding of chicks, replacement pullets, layers, or broilers; and turkey houses are designed for breeder turkeys, brooding poults, or market turkeys.

■ **Plans**—Plans also vary. For example, layers may be housed (1) on solid floor, litter-type houses, (2) on slotted floors, or (3) in cages.

Although types, purposes, and plans of poultry houses differ widely, certain principles are similar; that is, certain principles are observed relative to roof types, floor types, cage types, and building materials.

ROOF TYPES

Fig. 11–9 illustrates various roof types. The shape, slope, and type of roof construction selected should be based upon the function to be served, the strength, the environment, the climate, and the economy of construction and appearance.

On permanent buildings, the roofing should last 15 to 20 years without replacement. Among the more durable roofings are: cedar shingle; cement-asbestos shingles; 250-lb asphalt shingles; 28-gauge, 2-oz coated steel sheets; and aluminum.

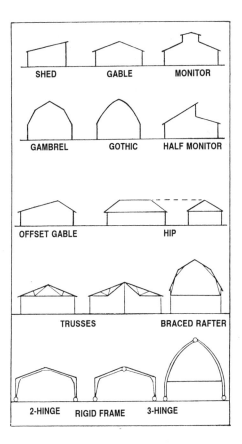

Fig. 11–9. Roof types. (*Structures and Environment Handbook*, 7th ed., Midwest Plan Service, Iowa State University, 1975, p. 98, Fig. 141–12)

FLOOR TYPES

Floors may be either solid or slotted.

SOLID FLOORS

Solid floors should be dry, easily cleaned, ratproof, and durable. They may be and are constructed of numerous materials—including clay, clay with a concrete border, plank, concrete, concrete with board surfacing, cork brick, creosoted wooden blocks, cinders, or various combinations of these materials. Regardless of the type of flooring material, for a good dry bed there should be a combination of surface and subsurface drainage, together with a cover provided by a suitable absorbent litter.

After considering both the advantages and the disadvantages of the many types of flooring material, most producers are agreed that, where a solid floor is desired, concrete is usually the most satisfactory. Concrete floors are sanitary, durable, and ratproof. If properly constructed, they are dry; and if bedded with litter, they are not cold.

Fig. 11–10. Solid floor layer house. (Courtesy, Babcock Industries, Ithaca, NY)

Dirt floors and deep litter are sometimes used in broiler houses. They are much less expensive than other types of floors, and many growers are well satisfied with them. In some cases, the entire accumulation of litter and manure is removed after each lot of broilers is sold; in other instances, the litter is used for three or four lots before changing.

SLOTTED FLOORS

Fig. 11–11. Slotted floor layer house. (Photo by J. C. Allen & Son, Inc., West Lafayette, IN)

Slotted floors are floors with slots through which the feces pass to a storage area below or nearby. Such floors are not new; they have been used in Europe for over 200 years.

The **advantages** of slotted floors are (1) less space per bird is needed, (2) bedding is eliminated, (3) manure handling is reduced, (4) increased sanitation, and (5) saving in labor.

The chief **disadvantages** of slotted floors are (1) higher initial cost than conventional solid floors, (2) less flexibility in the use of the building, (3) any spilled feed is lost through the slots, and (4) environmental conditions become more critical.

Slotted floors may be used for the entire floor area, or they may cover only about half the floor—usually a strip down the center of a long house. In both cases, a pit is provided below the slotted floor for the collection of manure. Such pits are often equipped with mechanical scrapers for periodic removal of manure. In some installations, the pits are deep enough to permit removal of manure by the use of a tractor with scraper blade.

CAGE TYPES

Fig. 11–12. Layers in cages. (Courtesy, Alabama Poultry and Egg Assn., Cullman, AL)

In the 1920s, the keeping of poultry in cages made of welded wire or hardware cloth was developed as a means of keeping the birds from fouling the feed and water with droppings. At that time, the system failed, primarily because of poor temperature and ventilation controls. Two decades later, environmental controls were perfected. Today, the vast majority of layers are raised in cages. Also, more and more replacement pullets are being raised in cages. With certain modifications, mainly in size, turkeys, ducks, and quail may be kept successfully in cages. Although some problems, such as breast blister, must be overcome, scientists, equipment manufacturers, and practical producers are working ceaselessly away to perfect cages for broilers.

BUILDING MATERIALS

Several types of building materials are used in poultry house construction; among them, wood, concrete blocks, hollow tile, and metal (steel or aluminum).

Wood houses are most common because of ease of construction and availability of material. The outside wall is generally drop siding. There is considerable heat loss through a single layer of wood; hence, the use of sheathing or insulation board on the inside, to provide double-wall construction, is recommended. Wood houses require frequent painting.

Concrete block houses are not widely used for poultry because they are costly and there is considerable heat loss through the walls. Some houses are built of blocks made of cinders and concrete, which are lighter and less costly than concrete.

Hollow tile houses are popular in areas where tile is produced and is cheaper than other construction material. The tile is durable, does not require paint, inside insulation is not necessary, the walls are ratproof, and the house is fireproof.

Metal houses (steel or aluminum) are durable and easy to keep clean, but expensive. Double-wall construction or insulation is recommended, because single-wall houses are difficult to keep warm during winter.

ENVIRONMENTAL FACTORS

Effective environmentally controlled poultry houses have a variety of features designed to provide an optimum environment of temperature, moisture, air movement, and gas content—features like tight construction, insulation, vapor barriers, and ventilation. To understand the basis of all the environmental factors, however, it is first necessary to know how heat is produced and eliminated by the bird.

HEAT PRODUCTION AND ELIMINATION BY THE BIRD

During the process of body metabolism, heat is liberated as a by-product. But the process is not constant; it is affected by (1) oxygen intake, (2) feed intake, and (3) activity of the bird. Increases in any of these three factors will lead to greater heat production in the deep body section of the chicken. This heat must be liberated; otherwise, the body temperature will rise. Because of the variation in the heat of metabolism, the "normal" body temperature of the chicken fluctuates between 105° and 107°F, with 106.5°F being average.

Deep body temperature varies between individuals; it is higher in small birds than large birds, higher in males than in females, higher in active birds than in inactive ones, and higher in nonbroody than in broody hens. Also, it rises with the presence of the food in the digestive tract, with increased activity, and with higher ambient temperature. Also, deep body temperature is highest in the morning and lowest in the evening (about three hours after darkness).

Heat is moved from the body of the bird to the air by the following four pathways:

1. **From the skin** (known as *sensible* or direct heat loss), by (a) radiation, (b) convection, and (c) conduction.
2. **By vaporization** of moisture in the respiratory tract (*insensible* or indirect heat loss).
3. **By fecal excretions** (not great).
4. **By production** of eggs (minor).

The amount of heat lost from the skin surface to the surrounding air by radiation depends on the spread between the two temperatures. When air temperature is low, heat loss by radiation is great; when air temperature is high, heat loss

by radiation is low. At ambient temperatures that are optimum for the bird's well-being, about 75% of the total heat lost from the skin is by radiation and is the major means of dissipation.

When cool air comes in contact with the surface of the bird, the air is warmed, expands (becomes lighter), and rises. As heat is carried away with the warm air, cooler air moves in; and the process is repeated. Heat lost in this manner is by convection.

Loss of heat from the surface of the bird by conduction is that type which occurs when the body surface of the bird comes in contact with any cooler surrounding object, such as the floor or soil. Generally, this type of heat loss is minimal.

As room temperature rises, less sensible heat is lost from the body and the bird resorts to vaporization—the changing of water from liquid to vapor in the respiratory tract—as a means of removing heat. Heat is necessary to change water from liquid to vapor. At ambient temperatures of about 86°F or above, the bird pants in an endeavor to cause more air to (1) move through the respiratory system, (2) become vaporized, and (3) carry off more heat; and the higher the ambient temperature, the faster the panting. If the inhaled air is completely saturated (has 100% relative humidity), it cannot absorb moisture in the respiratory tract; hence, there can be no vaporization, and the bird has lost its last chance to survive.

Unlike heat lost through the skin, heat lost by vaporization does not raise the room temperature. It does cause the birds to drink more water in order to prevent dehydration, often resulting in wet droppings.

The heat production of a 5-lb bird at a building temperature of 55°F is given in Table 11–5.

TABLE 11–5
POULTRY HEAT PRODUCTION[1]

Type of Heat	5–lb Bird at Building Temp. of 55°F
	(BTU/hr/bird)
Sensible heat	31.2
Insensible heat	13.0
Total heat	44.2

[1]*Structure and Environment Handbook*, 7th ed., Midwest Plan Service, Iowa State University, p. 165, Table 214–7.

INSULATION

Insulation is any material that reduces the rate at which heat is transferred from one area to another. Although all building materials have some insulation value, the term *insulation* is generally reserved for those products that provide this one service.

Insulation has several functions, the most important of which are:

1. **To reduce the rate of heat loss from buildings during cold weather.** Insulation helps conserve heat during periods of cold weather. Conserving bird heat helps maintain desirable housing conditions without the addition of great amounts of supplemental heat.

2. To reduce the rate at which heat passes into buildings during hot weather. Insulation helps reduce the rate of heat gain in hot weather, an important consideration in the South.

3. To control moisture condensation by making the wall and ceiling surfaces warmer. In a poorly insulated building, the inside ceiling and wall surfaces become cold in the winter, bringing discomfort to birds. When the surface temperature is low enough, the air next to the surface becomes saturated and moisture condenses. If the surface temperature is below freezing, frost forms. Well-insulated buildings reduce condensation by keeping the walls and ceilings relatively warm.

4. To reduce frost heaving. Correct foundation insulation (insulating the outside of the foundation wall if possible) reduces frost heaving, keeps floors warmer, and reduces heat losses.

Most insulation materials are bulky, porous, and lightweight, with countless air spaces. Generally speaking, the lighter the material, the better its insulating properties; and the more air pockets in the material, the better it insulates. Some building materials, like wood, are good insulators, while others, like concrete and metal, are poor insulators.

Fig. 11–13 shows how thick building materials must be to have the same insulating value as 1 in. glass wool batt. As shown, it takes 2.96 in. of plywood or 46.3 in. of concrete to equal the insulation value of 1 in. of glass wool.

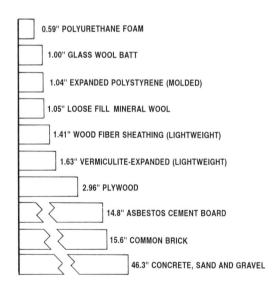

Fig. 11–13. Thickness of materials required to equal the insulation value of 1 in. glass wool batt. (*Structures and Environment Handbook*, 7th ed., Midwest Plan Service, Iowa State University, p. 162, Fig. 214–6. To convert inches to centimeters, multiply by 2.54)

Insulation, as well as other building materials, is rated according to its ability to resist the flow of heat. This is commonly referred to as the "R" value. The R, or insulation value, may be given per inch of thickness, or for the total thickness of a material. The relative insulation values of some common building materials are given in Table 11–6.

TABLE 11–6
INSULATION R VALUES OF SOME COMMON MATERIALS[1]

Material	Insulation Value[2, 3]	
	Thickness	For Thickness Listed
	(in.)	(in.)
Batt or blanket insulation:		
Glass wool, mineral wool or fiberglass	3.70	
Fill-type insulation:		
Glass or mineral wool	3.00 to 3.50	
Vermiculite (expanded)	2.13 to 2.27	
Shavings or sawdust	2.22	
Paper or wood pulp	3.70	
Rigid insulation:		
Wood fiber sheathing	2.27 to 2.63	
Expanded polystyrene, extruded	4.00 to 5.26	
Expanded polystryene, molded	3.57	
Expanded polyurethane (aged)	6.25	
Glass fiber	4.00	
Ordinary building materials:		
Concrete, poured	0.08	
Plywood, ⅜ in.	1.25	0.47
Plywood, ½ in.	1.25	0.63
Hardboard, ¼ in.	1.00 to 1.37	
Cement asbestos board, ⅛ in.		0.03
Lumber (fir, pine), ¾ in.	1.25	0.94
Wood beveled siding, ½' × 8"		0.81
Asphalt shingles		0.44
Wood shingles		0.94
Window glass, includes surface conditions:		
Single-glazed		0.89
Single-glazed + storm windows		1.79
Double-pane insulating glass		1.45 to 1.73
Air space (¾ in. or larger)		0.90
Surface conditons:		
Inside surface		0.68
Outside surface (15 mph wind)		0.17

[1]*Structure and Environment Handbook*, 7th ed., Midwest Plan Service, Iowa State University, p. 162, Table 214–1.

[2]Mean temperature of 76°F.

[3]To convert inches to centimeters, multiply by 2.54.

Table 11–7 shows insulation values for some common types of construction, including roofs and ceilings.

VAPOR BARRIER

In poultry buildings, and in other buildings where the relative humidity is high, insulation should be protected from moisture. Moisture, in the form of water vapor, tends to move from warmer moist areas to the cooler outside. So, moisture enters the wall, moves outward, then condenses when it reaches a cold enough area. Condensed water in the wall greatly reduces the value of the insulation and may damage the wall. To eliminate this flow of moisture, a vapor barrier should be placed near the warm side of the wall, preferably immediately beneath the interior lining.

Common vapor barriers are aluminum foil, 4 mil plastic film, and some of the asphalt-impregnated building papers.

TABLE 11–7
INSULATION R VALUES FOR SOME COMMON CONSTRUCTION[1]
(Surface Conditions Included)

Type of Construction	"R"
Roofs and Ceilings:	
Asphalt shingles, wood sheathing, vented attic space, ½ in. insulating board ceiling	4.95
Wood shingles, wood sheathing, vented attic space, ½ in. insulating board ceiling	5.45
Metal roofing on nailing girts, wood sheathing, vented attic space, ½ in. insulating board ceiling	3.53
Metal roofing on nailing girts, vented attic, 3 in. blanket insulation (mineral wool), ½ in. plywood ceiling	13.94
Metal roofing on nailing girts, vented attic, 4 in. fill insulation (glass or mineral wool), ½ in. plywood ceiling . . .	16.88
Metal roofinig on nailing girts, vented attic, 6 in. blanket insulation (mineral wool), ½ in. plywood ceiling	25.04
Metal roofing on nailing girts, hay mow, 12 in. of hay or straw	20 (approx.)
Doors:	
Wood siding, beveled, ¾" × 10"	1.90
Plywood, ¾ in. blanket insulation between two sheets ½ in. plywood	4.88
Plywood, 1½ in. blanket insulation between two sheets ½ in. plywood	7.65
Floor perimeter (per foot of length of exterior wall):	
Concrete, without perimeter insulation	1.23
Concrete, with 2" × 24" perimeter insulation	2.22

[1]Adapted by the author from *Structure and Environment Handbook*, 7th ed., Midwest Plan Service, Iowa State University, p. 163, Table 214–5.

INSULATION NEEDS

The amount of insulation needed in poultry houses varies according to winter temperature (see Fig. 11–14).

CLIMATE ZONE	TOTAL RESISTANCE (R)	
	WALLS	CEILINGS
MILD 9		12
MODERATE 9–14		16
COLD 14		23

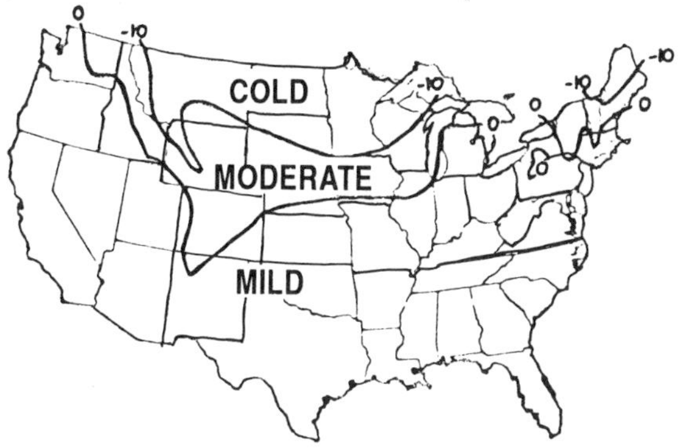

Fig. 11–14. Recommended insulation, showing three U.S. winter temperature zones as related to the amount of insulation needed in poultry houses. (*Structures and Environment Handbook*, 7th ed., Midwest Plan Service, Iowa State University, p. 163, Fig. 214–7)

AIRFLOW CALCULATIONS

When designing a large poultry house, most architects calculate the air exchange, step by step, as follows:

Step 1: Record the desired environmental conditions. For layers, these are:

Temperature, 55° to 70°F (see Table 11–4).
Relative humidity, 50 to 75% (see Table 11–4).
Difference between inside and outside temperature, maximum of 20°F.
Moisture content of litter, not higher than 40%.

Step 2: Record the amount of insulation needed for the particular climatic zone (see Fig. 11–14).

Step 3: Record the difference between the inside and outside temperature in degrees Fahrenheit.

Step 4: Determine the heat produced from the birds (see Table 11–2 for per bird).

Step 5: Determine the wall and ceiling area exposed to heat loss, with due allowance for glass area.

Step 6: Determine the insulation R value of the walls and ceilings (or roof) (see Tables 11–6 and 11–7.)

Step 7: Determine the building heat loss.

Step 8: Determine the volume of air change in cubic feet per hen per hour.

Step 9: Apply the following basic rule: Each 50 cu ft of air will remove 1 BTU of heat for each 1°F rise in temperature.

SIMPLIFIED DETERMINATION OF VENTILATION REQUIREMENTS [2]

The output of heat liberated from a bird is measured in British thermal units (BTU). One BTU is the amount of heat required to raise the temperature of 1 lb of water 1°F at or near 39.2°F.

When the ambient temperature is 70°F, the heat output for one bird at rest in still air will approximate that shown in Table 11–8 (next page). This table shows that, on a unit of body weight basis, a 4-lb bird produces only half as much heat as a 1-lb bird; hence, heat per unit of body weight decreases as a bird becomes larger and heavier.

[2]This section, including Tables 11–8, 11–9, and 11–10, adapted from North, M. O., "Basics of Poultry House Ventilation," *Poultry Tribune*, March 1978, p. 12.

TABLE 11–8
TOTAL HEAT OUTPUT BY A CHICKEN AT 70°F [1]

Average Body Weight		Total Heat Output per Hour		
		Per Lb Body Weight	Per Kg Body Weight	Per Bird
(lb)	(kg)	(BTU)[2]	(BTU)[2]	(BTU)[2]
1	0.5	20.0	44.0	20.0
2	0.9	14.5	31.9	29.0
3	1.4	11.5	25.3	34.5
4	1.8	10.0	22.0	40.0
5	2.3	9.0	19.8	45.0
6	2.7	8.2	18.0	49.2

[1]North, M. O., "Basics of Poultry House Ventilation," *Poultry Tribune*, March 1978, p. 12.

[2]One BTU is the amount of heat required to raise the temperature of 1 lb of water 1°F; equivalent to 252 calories.

But the ration of sensible heat to insensible heat loss varies according to house temperature; and it is sensible heat with which we are concerned in house ventilation. This relationship is shown in Table 11–9.

TABLE 11–9
SENSIBLE AND INSENSIBLE HEAT PRODUCTION BY A SINGLE, RESTING LEGHORN LAYER AS INFLUENCED BY AMBIENT TEMPERATURE[1]

Ambient Temperature		Sensible Heat	Insensible Heat	Output of Sensible Heat per Hour of Body Weight in BTU[2]	
(°F)	(°C)	(%)	(%)	(lb)	(kg)
40	4	90	10	9.0	19.8
60	16	80	20	7.9	17.4
80	27	60	40	6.1	13.4
100	38	40	60	4.3	9.5

[1]North, M. O., "Basics of Poultry House Ventilation," *Poultry Tribune*, March 1978, p. 12.

[2]One BTU is the amount of heat required to raise the temperature of 1 lb of water 1°F; equivalent to 252 calories.

In calculating the amount of air necessary to carry the heat given off by the bird, it is necessary to know the temperature of the outside air entering the house, where it is warmed and then exhausted from the room.

The rule of thumb is that each 50 cu ft of air will remove 1 BTU of heat for each 1°F rise in temperature. This means that if the air temperature were increased 5°F, 50 cu ft of air leaving the building would remove 5 BTUs of heat per hour, or it would require 10 cu ft to remove 1 BTU.

As a practical example, let's assume a house containing 10,000 Leghorn layers, with an outside temperature of 65°F and an in-house temperature of 70°F; hence, the rise is 5°F. Ten thousand hens each weighing 4 lb totals 40,000 lb of birds, with a sensible heat production per pound of 7 Btu (see Table 11–9: 70°F is midway between 7.9 and 6.1 BTU, which is 7 BTU), or 280,000 BTU per hour for the 10,000 birds

(40,000 × 7). Since 10 cu ft of air are necessary to remove each BTU, the ventilation requirement is 2,800,000 cu ft per hour, or 46,666 cu ft per minute.

The above figures are based on a single hen at rest, the only figures available. Obviously, this would not be the status of 10,000 layers, for birds are seldom still. Also, there is heat loss from the building, and the birds warm each other.

So, the following formula is a simpler method for determining the ventilation requirement in a poultry house:

Provide 0.012 cu ft of airflow per minute per pound of body weight of chickens in the house for each 1°F of house temperature. Typical examples are shown in Table 11–10.

TABLE 11–10
RECOMMENDED AIRFLOW THROUGH A CHICKEN HOUSE[1]

Air Temperature in House		Cu Ft of Air per Minute per Lb of Body Weight	M³ of Air per Minute per Kg of Body Weight
(°F)	(°C)		
40	4	0.48	0.03
60	16	0.72	0.04
80	27	0.96	0.06
100	38	1.20	0.07
110	43	1.32	0.08

[1]North, M. O., "Basics of Poultry House Ventilation," *Poultry Tribune*, March 1978, p. 12.

Although the formula and the Table 11–10 examples do not consider the temperature of the air coming into the poultry house, under actual conditions the temperature of the incoming air fluctuates very little; hence, the table is practical and adequate under most circumstances.

PREFABRICATED HOUSES

Preengineered and prefabricated poultry houses — complete with insulation, ventilating system, and all equipment — are becoming increasingly important. Fabricators of such buildings have the distinct advantages of (1) price savings due to purchase of materials in quantity lots, (2) economical and controlled fabricating, and (3) well-trained personnel for developing the best in plans and specifications.

From the standpoint of the poultry producer, a factory-built house is often appealing because (1) the full cost of such a building is known in advance, rather than a contractor's estimate; (2) financing may be easier to arrange as a result of the entire transaction being carried out with one supplier; and (3) construction will likely take less time than where a local contractor is involved.

Many poultry producers, however, still prefer to design and build their own houses and to select and install the necessary equipment. By so doing, they can give vent to their personal preferences; and they may be able to effect a saving.

Fig. 11–15. Prefabricated poultry house. (Courtesy, FACCO, Padova, Italy)

POULTRY EQUIPMENT

Poultry equipment refers to structures other than houses or shelters used in feeding, watering, and caring for poultry.

Much poultry equipment had best be purchased, rather than homemade. When buying, look for rugged construction, simple operations, and durable materials and finishes. When making equipment yourself, use a good plan and follow the recommendations as to types and grades of materials. Working drawings of many items of equipment have been prepared by the universities and the U.S. Department of Agriculture. These may be purchased from your extension agricultural engineer or from the following planning service:

Midwest Plan Service
122 Davidson Hall
Iowa State University
Ames, IA 50011

REQUISITES OF POULTRY EQUIPMENT

The size and design of poultry equipment may differ; that is, not all self-feeders, for example, are the same. Yet there are certain fundamentals of poultry equipment that are similar regardless of the kind of equipment, the design, or the size. These requisites are:

1. **Utility value.** Equipment should be useful, practical, and efficient.

2. **Simple construction.** If poultry equipment is homemade, simple construction is essential.

3. **Durable.** Poultry equipment receives hard and heavy use. Thus, it should be strongly and durably built.

4. **Dependable.** Poultry equipment should be dependable, so that it will function without getting out of order. Overly complicated equipment sometimes requires more of the operator's time than if it were not available.

5. **Low annual cost and upkeep.** Like poultry houses, poultry equipment must be paid for out of profits. It is important, therefore, that it have a low annual and upkeep cost. Because of lower maintenance cost and longer years of usefulness, it may be cheaper in the long run to pay a higher initial cost for more durable and substantial equipment than to purchase whatever is cheapest.

6. **Movable.** Much poultry equipment should be movable so that it may be shifted from one location to another.

7. **Accessible.** Stationary or less-portable equipment, such as nests, should be readily accessible.

8. **Save feed.** Much feed may be saved when fed in properly constructed equipment. When such equipment is used, birds eat their feed without throwing it out of the feeder.

9. **Reduce labor.** Modern equipment reduces the labor required in caring for the birds.

10. **Conserve manure and prevent pollution.** Manure is worth money, and manure will pollute; hence, it should be conserved.

AUTOMATION

Automation is a coined word meaning the mechanical handling of materials. Producers automate to lessen labor and cut costs.

Modern equipment has eliminated much hand labor. Such chores as feeding, watering, adding litter, cleaning, and handling eggs have been mechanized. Producers are using self-unloading trucks, self-feeders, feed augers and belts, laborsaving processing equipment, automatic waterers, and manure disposal units. Automation of the poultry industry will increase.

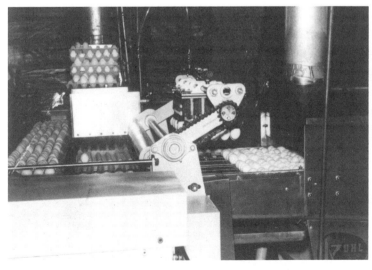

Fig. 11–16. Automatic-vacuum egg lift. (Courtesy, D. D. Bell, University of California, Riverside)

TYPES OF EQUIPMENT

Fig. 11–17. Broilers. Note the continuous feeders and waterers. (Courtesy, California Polytechnic State University, San Luis Obispo)

Good equipment is essential for satisfactory poultry production. It should be simple in construction; trouble-free; adequate for the intended function, but not excessive; movable and easily cleaned; and make for a minimum of labor.

There is hardly any limit to the types of equipment. However, the feeders, waterers, nests, and perches are of particular importance.

FEEDING EQUIPMENT

Since feed cost is the major item in poultry production, it is necessary that there be adequate feeder space (for feeder space requirements of different species and ages of poultry, see Table 11–1, p. 257). They should be easy to fill, easy to clean, built to avoid waste, arranged so that the birds cannot roost on them, high enough so that the birds cannot scratch litter into them, and so constructed that as long as they contain any feed, the birds can reach it.

Fig. 11–18. Broilers feeding at a metal feeder in a broiler house. (Courtesy, USDA)

Automated feeders are now standard equipment on large commercial layer and broiler farms. They save labor and keep feed fresh. It is important, however, that standby generating equipment be available for their operation in case of power failure.

WATERERS

The importance of water is evidenced by the fact that laying hens drink 2 to 3 lb of water for each pound of feed that they consume, and the recommendation that laying hens not be required to walk more than 15 ft to get a drink.

Watering devices should keep the water clean, be easily cleaned, prevent spillage of water around the vessels or containers, and keep the water cool in warm weather and from freezing in cold weather.

To keep the water clean, the container(s) should be high enough that litter will not be scratched into it, and located so that the birds cannot contaminate it with droppings.

Automatic watering devices are commonly used on commercial layer, broiler, and turkey operations. For large flocks, watering facilities running the entire length of the house are satisfactory. For cages, drip-type or small-cup waterers minimize cleaning and spillage.

(For water space requirements for different species and ages of poultry, see Table 11–1, p. 257.)

OTHER POULTRY FACILITIES AND EQUIPMENT

In modern large-scale poultry operations, two other types of facilities and equipment, which for the most part are used in all types of poultry production operations, merit mentioning. Namely, (1) facilities and equipment for handling manure, and (2) an emergency warning system.

FACILITIES AND EQUIPMENT FOR HANDLING MANURE

The facilities and equipment for handling poultry manure will vary according to the disposition made of it. Among the common manure disposal systems and equipment are the following:

1. **Dry spreading.** In most places, the spreading of dry manure on crop land is still the most common and economical way to dispose of it. The major problems or drawbacks to this method are the land acreage required and the odors produced. A manure spreader, a manure pit where storage is planned, and a field constitute the necessary facilities and equipment for this system.

2. **Wet spreading.** This refers to manure to which water has been added for the purpose of facilitating handling. Liquid manure is stored in large watertight storage pits or tanks (a 30,000 gal storage facility is required to store manure from 10,000 laying hens for 8 weeks), with 90 to 93% water where a sludge pump is used and 80 to 85% water for a vacuum pump. For conveying the manure away from the storage tank or pit, either a watertight manure spreader or an irrigation pipe must be available.

3. **Lagoon.** Outdoor lagoons should be between 3 and 5 ft deep, and 1 acre in size for each 1,000 to 2,000 laying hens. Indoor lagoons should have a minimum of 3.5 cu ft of water per bird, proper inlet and outlet mechanisms to control water depth, and provision for supplying oxygen.

4. **Dehydration.** Drying equipment and marketing require a sizable investment for this type of operation. Dehydration equipment to handle manure is expensive. Also, there is an odor problem, which precludes the possibility of locating a manure drying operation near any population center.

5. **Incinerators.** Incinerators are rather expensive, require the use of considerable fuel to consume manure, and create an odor problem; hence, the use of incinerators for manure disposal does not appear to be the answer.

6. **Processing for feed.** Poultry house litter may be used as either an energy or a protein source for beef cattle. As an energy source, it has 10 to 40% the feeding value of No. 2 corn, and it may replace 15 to 25% of the grain ration. As a protein supplement, it has 50 to 55% the value of soybean meal (41%), and it may replace 25% of the oil meal supplement of a ration.

Additional information pertaining to the subject of "Poultry Manure" is presented in Chapter 8, Poultry Management; hence, the reader is referred thereto.

EMERGENCY WARNING SYSTEMS

The confinement of large flocks of birds in a mechanically controlled environment entails considerable risk because of the possibility of (1) power or equipment failure or (2) fire or abnormal temperatures. To guard against such troubles, an emergency warning system should be installed. Such a warning system may save its cost during just one

Fig. 11–19. Acoustic and visual alarm for minimum and maximum temperature and electricity failure. (Courtesy, FACCO, Padova, Italy)

power interruption, fire, or undesirable temperature in an incubator, brooder house, layer house, broiler house, egg storage room, or furnace room. The continuation of the poultry enterprise as the result of a timely warning can be far more valuable than any monetary insurance settlement after the operation has failed.

CLEANING AND SPRAYING EQUIPMENT

Thorough cleaning should precede disinfecting. For big operations, a power sprayer should be available for use in applying both disinfectants and insecticides for fly control.

DISPOSAL PIT

A disposal pit is a 7-ft-deep, airtight, underground pit covered with a tight lid, with an opening through which dead birds can be dropped. Disposal pits should not be used where there is a hazard of polluting ground water. With the exception of the latter precaution, they are entirely satisfactory if properly constructed.

INCINERATOR

Cremation, or burning, is the most satisfactory form of flesh disposal. If done properly, there is no odor, disease, rodent, fly, or water pollution problem. Incinerators may be home-constructed, or a steel-jacketed commercial unit may be purchased. Propane, gas, or oil may be used as the fuel for the incinerator.

RANGE

Almost all poultry are now raised in confinement rather than on the range. Turkeys, ducks, and geese are sometimes kept on the range.

SPECIALIZED BUILDINGS AND EQUIPMENT FOR SPECIALIZED PURPOSES

With the decline in farm flocks, fewer and fewer general-purpose type poultry houses and equipment are seen. The large and highly specialized poultry enterprises have developed buildings and equipment adapted to their highly specialized purposes. Although there is hardly any limit to the number of different styles, sizes, and colors of poultry buildings and equipment, within specialty types of operations certain principles are similar; that is, certain principles are observed in the facilities and equipment for layers, for brooding of chicks, for replacement pullets, for broilers, for breeder turkeys, for turkey poults, for market turkeys, and for ducks and geese. Building and equipment requisites for each of these specialty areas follow.

HOUSING AND EQUIPMENT FOR LAYERS

The design and construction of houses and equipment for layers should be such as to provide for top performance of the layers, optimum environmental control, functional arrangement of equipment, maximum labor efficiency, satisfactory waste disposal, and minimum housing and care costs per dozen eggs produced. Also, futuristic housing and equipment must be big. It is estimated that the average egg farm had 150,000 layers in 1990. By the year 2000, it is projected that this figure will exceed 250,000 per farm.

HOUSES

Layer houses may be colony houses, multiple-unit houses, or multiple-story houses; some are permanent and others are movable.

The starting point in designing a layer house is the selection of the type of laying system. Presently, layer houses are being arranged in the following ways, or according to the following systems:

1. **Cage system.** Wild hens were caught and kept in cages by primitive people in order to facilitate egg gathering. Then, beginning in the 1930s, a limited number of commercial layers were kept in cages. Today, more than 90% of the layers in the United States are kept in cages.

Fig. 11–20. Cage system of housing layers, with three decks of cages. (Courtesy, DeKalb Agricultural Association, DeKalb, IL)

Concrete floors are widely used in caged layer houses. In the cage system, small wire cages placed side by side, in long rows, hold the birds. These cages stand in rows on each side of an aisle (usually about 30 to 36 in. wide), placed at a convenient working height for the operator. The cage arrangement may range from single-deck to five decks. Where multiple decks are used, the second row (and subsequent rows) of cages may be located directly above the lower row, or placed so that the bottom row projects forward about one-half the cage's depth, giving a stair-step effect.

Originally, one or two hens were kept in individual wire cages. To reduce the cost per bird, however, *colony cages* which hold up to 20 to 25 hens have been developed. Three or four hens per cage appears to be the most popular arrangement at the present time.

Cage sizes are not standardized, but Table 11–11 will serve as a useful guide.

TABLE 11–11
CAGE SIZES, BIRDS PER CAGE, AND FEED AND WATER SPACE[1]

Cage Size	No. Birds	Total Sq. Inches	Sq. Inches per Bird	Inches Feed and Water Space per Bird
8" × 16"	1	128	128	8
8" × 16"	2	128	64	4
12" × 16"	3	192	64	4
12" × 18"	3	216	72	4
12" × 18"	4	216	53	3
24" × 18"	7	432	62	3.5
24" × 18"	8	432	54	3
12" × 20"	4	240	60	3
30" × 24"	10	720	72	3
48" × 36"	20	1,728	86	2.4
48" × 36"	25	1,728	69	1.9

[1]To convert to metric, see the Appendix.

While the cage system (a) accommodates more birds in a given floor area than the litter floor system and (b) eliminates many internal parasite troubles, it does give rise to problems. High initial investment and high labor requirements have been experienced in the hen-per-cage system. With the colony-type cage arrangement, cannibalism and similar social problems have often been encountered.

Controlling flies and removing manure have been particularly difficult with the cage system.

2. **Slat floor.** This means just what the name indicates—the use of slats or wire over the entire floor. Slats, usually of metal or wood, are placed on edge. In slat or wire floors, the droppings collect in the space beneath the floors, and can be removed with a mechanical cleaner, or a floor section can be taken out once each year and the droppings removed with tractor-mounted equipment.

Fig. 11–21. New slatted floor layer house, with White Leghorn pullets. (Photo by J. C. Allen & Son, Inc., West Lafayette, IN)

The **advantages** of slat floors are: They require less floor space per bird than when the birds are kept on a litter floor, no litter is needed, there is better control of bacterial diseases, and it is not necessary to clean during the laying year — except to eliminate moisture.

The **disadvantages** of slat floors are: high humidity, no place where the birds can relax, birds appear more nervous, feather conditions become rougher, more egg breakage, more feather picking and cannibalism, and lower egg production.

3. **Combination slat and litter floor.** A slat-and-litter combination floor is popular for birds producing hatching eggs, particularly meat-type breeders. Such a system provides good fertility and keeps hatching eggs cleaner than other systems. In this system, slats are generally placed over two-thirds of the floor area, with the feeders and waterers located on the slats. The other third is covered with litter. Generally, the slats/litter are in either of two locations: (1) half the slats on each side of the house, with the litter area down the center; or (2) the slats across the center of the house, with half the litter area in front of the house and half in back of the house.

Fig. 11–22. Combination slat and litter floor, used for breeders. Note slatted floor on left and litter floor and nest boxes on right. (Courtesy, Dr. M. J. Wineland, North Carolina State University, Raleigh)

4. **Floor or litter-type house.** This is the oldest of the systems or arrangements, and at one time it was the exclusive type of layer house. It consists of litter covering the entire floor. Feeders and waterers are located on the litter, and nests usually line one or both sides of the house. This arrangement calls for a minimum amount of equipment. Also, fly control is simple. But it requires a well-insulated house with a good ventilation system.

Litter floor houses may be used in climates where slotted or wire floor houses or cages are unsatisfactory. They are expensive to construct, but maintenance costs are low. Also, they are more flexible; for example, additional brooding facilities are not required since the chicks may be brooded in floor-laying type houses which have been cleared of hens.

As would be expected, each of these systems has its strong advocates, and the ardent supporters of each system can cite experiences and experiments to substantiate their claims. After studying the overall results obtained in the form of net income per hen or per laborer by these four main types of systems, the author came to the conclusion that there is no clear-cut advantage for any one type — that among the many factors influencing profit in the egg business, the type or system of housing is not paramount. The most important things are to design the building to fit the local climate; to provide adequate protection, ventilation, and cooling for the birds; and to make for good working conditions for the operator. It is further recommended that anyone planning to construct a layer house should inspect all types or systems, and confer with those who have used them; then, reach a decision.

In addition to the layer house, a building for handling, cooling, and holding eggs under refrigeration on the egg farm is essential.

Most poultry feed is purchased in bulk and delivered in bulk tanks from which it is withdrawn as used, by gravity or mechanical conveyors. Where whole grain is used, it may be practical to have sufficient storage facilities to permit purchase of grain at harvest time, when prices are usually more favorable.

A service building for storing supplies and small equipment, and for making repairs, is also needed. It may be a separate building, or it may be a part of the garage or egg room.

EQUIPMENT

Good laying-house equipment is essential for satisfactory production. It should be simple in construction, movable, and easily cleaned. The nests, roosts, feeders, and waterers are of particular importance.

1. **Nests.** Nests should be roomy, movable, easily cleaned, cool and well ventilated, dark, and conveniently located. They are usually about 14 in. square, 6 in. deep, and with 15 in. head room. All-metal nests are preferred to wooden nests because of ease of cleaning and less chance of becoming infected by mites.

Nesting material should consist of small particles that are highly moisture absorbent, such as shavings, oat hulls, sawdust, or excelsior pads.

Roll-away nests, without nesting material, are becoming more common. They consist of plastic-covered wire bottoms, sloped to a covered egg tray.

Fig. 11–23.   Layer in plastic nest. The egg collector is in front. (Courtesy, FACCO, Padova, Italy)

Trapnests differ from regular nests in that they are provided with trap doors by which the birds shut themselves in when they enter. They are the accepted means of securing accurate individual egg records, and they are essential where pedigree breeding is practiced — that is, where more than one female is continuously mated with one male.

2. **Roosts.** Where roosts are used, they are commonly made of 2 in. × 3 in. or 2 in. × 4 in. lumber, placed sideways, with the edges rounded off.

3. **Feeders.** Since feed cost is the major item in egg production, it is necessary that there be adequate feeder space and that the feeders be good. They should be easy to

Fig. 11–24.   Layers at feeder pan of an automatic feeder line. (Courtesy, Hy-Line International, Division of Hy-Line Indian River Company, Johnston, IA)

fill, easy to clean, built to avoid waste, arranged so that the birds cannot roost on them, high enough so that the birds cannot scratch litter into them, and so constructed that as long as they contain any feed the birds can reach it.

Automatic (mechanical) feeders are standard equipment in large commercial egg operations.

4. **Waterers.** Laying hens drink 2 to 3 lb of water for each pound of feed they eat. Watering devices should keep the water clean, be easily cleaned, and prevent spillage of water around the vessels or containers. Also, it is important that the waterers be distributed throughout the laying house so that a hen never has to travel more than 15 ft for a drink.

Many cage installations for layers are equipped with either drip-type or small-cup waterers that minimize cleaning problems as well as spillage.

CARE OF THE HOUSE

Good husbandry and housekeeping are essential for optimum production and high egg quality. Also, it is necessary to minimize the spread of diseases. Accordingly, the following practices should be a part of the regular chores of the caretaker: inspect the birds daily, clean waterers daily, keep nests clean, keep light bulbs clean, clean windows regularly, control flies and rodents, and inspect all equipment routinely. Additionally, the following management and environmental factors should receive special attention.

1. **Litter.** Where litter is used, 6 to 8 in. of it should be provided. Litter absorbs moisture from droppings and then gives this moisture to the air brought in by ventilation. A good litter is highly absorbent and fairly coarse, so as to prevent packing. The litter should be free of mold and contain a minimum amount of dust. Availability and cost will determine the type of litter used. Common litter materials include shredded cane pulp, soft wood shavings, peanut hulls, ground corncobs, rice hulls, and peat moss.

2. **Cooling.** In warm areas, summer heat may cause retarded egg production and even result in death losses. Well-ventilated houses help, but when temperatures become extreme, artificial cooling is necessary. During extremely hot weather, the house can be kept more comfortable by increasing the movement of air, cleaning the fan blades and screens, painting the roof white, and sprinkling the roof. Foggers, controlled by thermostats, may also be used to produce a fine spray. In an emergency, sprinkling water over the hens with a garden hose, using a fine spray, will help cut down death due to heat prostration if enough breeze is available to evaporate the water and cool the birds.

3. **Manure removal.** A manure removal or holding system should be planned before the operation is started. The importance of this becomes apparent when it is realized that 100,000 layers will produce over 12 tons of manure a day, or well over 4,000 tons a year.

BROODER HOUSES AND EQUIPMENT FOR CHICKS

Wherever chicks are raised—whether for broilers or replacement layers, and whether for continued rearing in confinement or on the range—artificial brooding of some kind is necessary. No phase of the poultry business is so important as brooding—it is the part that makes for a proper start in life.

BROODER HOUSES

Until recently, brooder houses for chicks were, for the most part, either portable or stationary buildings that were used for other purposes the balance of the year—part of the laying house, or the garage, or perhaps one end of the machine shed was used for brooding purposes. Housing arrangements of this type are still common among farm flock owners. However, in large commercial installations, where it is not uncommon to find 30,000 to 50,000 or more chicks being brooded together as a unit, special brooder houses or arrangements are common.

Fig. 11–25. Brooder house, with baby chicks. (Courtesy, Gold Kist-Communications, Atlanta, GA)

BROODING EQUIPMENT

Heating, feeding, and watering equipment are the three main items of equipment needed for the brooding of chicks.

1. **Heating equipment.** The heat requirements for artificially brooding chicks may be supplied by a wide variety of devices and methods, among them the following:

a. **Portable brooders.** These units are, as indicated by the name, portable or movable. Although they come in a wide variety of styles and sizes, they generally consist of a central heating unit surrounded by a hover. They may be heated by gas, oil, or electricity. Portable brooders cost less to install than central heating systems, but they cost more per chick to operate. Also, they require more attention and labor, and there is more fire hazard from using them than central systems.

b. **Infrared lamps.** In this method, infrared lamps are suspended 18 to 27 in. above the floor litter. These lamps do not heat the surrounding air, but they warm the chicks in the same manner as direct rays from the sun. One 250-watt infrared bulb will suffice for 60 to 100 chicks.

c. **Battery brooders.** Commercial battery brooders may be either (1) unheated brooders made for use in warm rooms, or (2) those equipped with heating units and warm compartments for use in rooms held at 60° to 70°F. Most batteries of today have heating units warmed by electricity and equipped with thermostat regulators.

d. **Central heating.** Large commercial operations, which handle 5,000 to 25,000 or more chicks per house, need a central heating system. Several different heating systems have been developed for such use. Most of them are highly automated, thermostatically controlled, and fueled by oil, gas, or electricity. Central heating may provide warmth through either hot water (pipes or radiant heating) or hot air (direct or indirect).

2. **Feeders and waterers.** Feeding and watering equipment which is suited to 6-week-old birds is not satisfactory for day-old chicks. Hence, special feeding and watering equipment must be provided, and the chicks must learn to eat and drink when they are first placed in the brooder.

HOUSING AND EQUIPMENT FOR REPLACEMENT (STARTED) PULLETS

Regular replacement of the laying flock is one of the most important, and most expensive essentials of egg production. It ranks next to feed in the cost of egg production, and it exceeds labor, housing, equipment, interest, taxes, and other costs.

Fig. 11–26. Cage-reared replacement pullets. (Courtesy, D. D. Bell, University of California, Riverside)

Producers may choose between growing their own pullets or purchasing started pullets. In any event, the building and equipment requirements are the same, whether pullets are raised for sale or raised by the one who will retain them as layers.

Usually it is wise to provide pullets with levels of environment—housing, feed, and general management—considerably above minimum standards, but consistent with realistic cost consideration. This calls for housing that gives reasonable protection from heat and cold and provides good ventilation; and it means adequate feed and water space and facilities.

HOUSING

Effective isolation of growing stock from older birds is important for disease control, particularly during the early stages. Hence, separate housing should be provided for replacement pullets, completely separated from layers. Housing that is suitable for brooding chicks, or for housing a laying flock, will be satisfactory for rearing pullets. Usually, they are started out in the brooder house, then switched to the layer house.

FEEDERS AND WATERERS

Water space is less important than a good distribution of waterers through the house with a constant supply of fresh water. The same principle applies to some degree to feeders. Both feeders and waterers should be located to permit birds in any part of the house to feed or drink conveniently, without having to find their way around barriers or to travel more than 15 ft.

OTHER EQUIPMENT

Hoppers for grit should be provided for replacement pullets throughout the growing period, and, as they near maturity, additional hoppers should be provided for oystershell or other calcium supplement.

Roosts are not always used where pullets are reared, but they are a desirable addition to the equipment. When used over enclosed pits, roosts aid in sanitation. Also, they add to the comfort of the birds in hot weather and help to establish desirable roosting habits for the laying period that follows. Some producers also feel that flocks with good roosting habits are less likely to present severe problems of floor eggs.

Nesting equipment should be available by the time the pullets begin to lay. Of course, only a few nests are necessary if pullets are moved from rearing quarters to their laying quarters before production becomes heavy.

Except for automatic feeders and waterers, little mechanization exists in the vast majority of pullet rearing facilities, especially when compared to layer facilities.

HOUSES AND EQUIPMENT FOR BROILERS

No other segment of agriculture is as well suited to assembly-line production techniques as broiler production.

In modern commercial broiler production, the bird spends its entire life in one house; that is, it is not brooded in a special brooder house, then moved to a house for growing, for broiler raising is basically a brooding operation. Instead, brooder houses are thoroughly cleaned and disinfected between flocks, preferably with the quarters left idle one or two weeks before starting a new group.

Today, the vast majority of the nation's broilers are produced in large production units; and virtually 100% of them are grown under some type of vertical integration or contractual arrangement. Hence, in the discussion that follows, only the buildings and equipment common to these larger establishments will be described.

HOUSING

A broiler house should provide clean, dry, comfortable surroundings for birds throughout the year. The house should be kept warm enough, but not too warm; the litter should be kept reasonably dry; provisions should be made to modify the air circulation as broilers grow; and fresh air should be circulated, but the house should be free from drafts. In short, the broiler house should not be too cold, too hot, too wet, or too dry.

Most of the new broiler houses being constructed today are 24 to 40 ft wide, with gable-type roofs. Truss-type (of either wood or steel) construction is replacing pole-type due to lower labor costs in construction and greater ease of cleaning with a tractor. The length varies from 200 to 600 ft, with most of them averaging 300 to 400 ft. Capacity varies, but in the newer houses they generally range from 7,200 to 20,000 broilers. All of the birds may be in one large pen, but the newer trend is to pen units of 1,200 to 2,500 birds. There

Fig. 11–27.   Broiler house. Note insulation under trusses, which facilitates cleaning. (Courtesy, California Polytechnic State University, San Luis Obispo)

is increasing interest in controlled environment housing for broilers. When the broiler house is insulated and environmentally controlled, a centrally heated brooding system may be difficult to justify. However, where broiler houses are not environmentally controlled, a central heating system, of either hot or cold air or hot water, is preferred. Where a chick hover is used, large—1,000 capacity—units are most popular.

Although maintaining adequate temperature is of great importance, constant attention is needed to ventilate the broiler house properly. This calls for a building that is properly insulated and in which there is a forced ventilation system.

FEEDERS

Baby chicks should be started eating from new, cut-down chick boxes or box lids placed at floor level, allowing one feeder lid per 100 chicks. These feeders allow the chicks to find the feed easily.

Following the *box lid* stage, broilers may be fed from trough feeders, hanging tube-type feeders, or from mechanical feeders. Bulk feed bins and mechanical feeders are the most costly of the various types of feeding equipment, but they also make for more saving in labor. As a rule of thumb, installation of a mechanical feeder or feeders is worthwhile if the investment is no more than five times the labor saved per year.

WATERERS

Clean water at 55°F should be available at all times. Gallon fountains should be provided for baby chicks, but these should be replaced by automatic, hanging waterers as soon as the chicks have learned to drink from the latter.

All waterers should be cleaned and washed daily.

HOUSES AND EQUIPMENT FOR TURKEYS

Although there are some species differences between turkeys and chickens, primarily because of a difference in size, the general principles relative to buildings and equipment for turkeys and chicks are very similar.

FACILITIES FOR BREEDER TURKEYS

Breeding turkeys may be kept (1) on restricted range, (2) in confinement housing, or (3) in semiconfinement.

Open range without shelters should not be used unless the flock can be protected from cold winds by trees and/or sloping hillsides. Turkey range should be enclosed by a well-constructed, permanent fence that is at least 5 ft high, preferably with an electric wire around the outside further to discourage predators. A 4-acre area will accommodate about 600 breeders, with provision for some rotation of pens and two night pens. Night pens should be equipped with roosts and lights arranged to ensure the entire area is lighted.

Fig. 11–28.   Turkey shelters on range in Illinois. (Courtesy, Watt Publishing Co., Mount Morris, IL)

Day pens should be equipped with waterers, feeders, and nests.

Buildings for strict confinement are usually 40 to 50 ft in width and covered with wire on all sides. Most operators enclose three sides of the shelter with plastic to give protection during the winter months. Strict confinement lends itself to automatic feeding and watering.

Semiconfinement gives the protection advantage of complete confinement during the bad weather, with the added yard space for improved sanitation and mating. Buildings for semiconfinement are usually of lower cost construction than those used in strict confinement; generally they consist of an open-front, pole-type building, with a fenced yard.

Choice between the system—restricted range, strict confinement, and semiconfinement—depends to a very large extent upon the physical characteristics of the building site on the farm, the cost involved, and the weather.

FACILITIES FOR BROODING TURKEY POULTS

Turkey poults should be brooded and reared separately from all ages of chickens and adult turkeys. In large commercial turkey operations, they are usually brooded in permanent-type houses. Some growers prefer to brood in batteries for the first 2 or 3 weeks before placing the poults on the floor. Regardless of the method of brooding employed, at least 1 sq ft of floor space is required to 8 weeks of age. More floor space must be provided as the poults grow older.

Any of the types of eating equipment commonly used for baby chicks can be used for brooding poults. To begin with, poults should be maintained at a temperature of 95° to 100°F, with the temperature lessening following the first week. A guard should be placed around the brooder for the first two weeks to prevent crowding and smothering in the corners of the house.

The watering and feeding facilities used by turkey poults are very similar to that used by baby chicks, but generally they are somewhat larger and, of course, more space needs to be provided per bird.

Fig. 11–29. Turkey poults in a brooder house with access to feed, water, and heat from a gas brooder. Integrated operation of Cuddy Farms, Ltd., North Carolina. (Photo courtesy, James Strawser, University of Georgia, Athens)

FACILITIES FOR MARKET TURKEYS

For the first 8 to 10 weeks of life, the brooding of turkey poults is the same regardless of whether they are subsequently to be raised in confinement or moved to the range. Although a limited number of market turkeys are reared on the range, there is an increasing tendency to grow them in confinement in pole-type shelters, or even in environmentally controlled buildings. However, breeder turkeys are usually only partially confined.

Fig. 11–30. Turkey grow-out house in California. (Courtesy, University of California, Davis)

Where turkeys are to be range raised under a rotation system, portable range shelters, roosts, feeders, and waterers are necessary. These may be placed on runners or wheels and moved where and when needed.

When confinement reared, turkeys are generally provided with a pole-type shelter, although more costly environmentally controlled units are being used, particularly for turkey-fryer production. When raised in strict confinement, 3 to 5 sq ft of floor space should be provided to carry each bird to market age and weight, the amount depending on the size of the strain of turkeys. Automatic waterers and feeders, or bulk feeders, help reduce labor costs. Care should be taken to provide adequate ventilation and occasional additions to the litter to reduce dampness and dirty litter conditions.

HOUSES AND EQUIPMENT FOR DUCKS AND GEESE

Elaborate and expensive housing facilities are not necessary for ducks or geese. Except during storms, they prefer to be outdoors. A fenced-in area, in which there is a colony house or open shed to provide protection during inclement weather, is all that is needed.

Most commercial producers of ducks and geese allow breeders access to water for swimming. However, ducklings and goslings can be reared successfully without swimming facilities, provided they have a constant supply of readily available fresh drinking water.

Commercially produced ducks and geese are brooded much like chicks and poults. Brooding temperatures should be 85° to 90°F under the hover at the start, but reduced to 80°F by the end of the first week and 70°F by the end of the third week.

Fig. 11–31. Breeder ducks in breeding house on the world's largest duck farm. Note the nest boxes in the foreground. (Courtesy, Cherry Valley Farms, Ltd., Rothwell, England)

Many different types of feeders and waterers are used for ducks and geese. The main requisites are that feeders handle pellets without wastage, and that the waterers be of a type which birds cannot get into and splash.

Also see Chapter 15, "Ducks, Geese, and Miscellaneous Poultry."

QUESTIONS FOR STUDY AND DISCUSSION

1. What was the natural laying and mating season of chickens? How may this be emulated in modern housing?

2. After deciding on the type of enterprise and the area, what location factors should be considered?

3. What are the likely consequences of (a) allowing too much space for housing and equipment, and (b) allowing too little space for housing and equipment?

4. Why is poultry housing important?

5. What are the main requisites of modern poultry housing?

6. Under what circumstances, and for what kind of poultry, might naturally ventilated buildings be practical today?

7. What constitutes favorable environment in a layer house in (a) temperature, (b) humidity, and (c) ventilation rate for each summer and winter?

8. What lighting program would you recommend for replacement pullets?

9. Discuss the relative importance of ventilation for each (a) respiration air changes of the birds, and (b) removal of moisture from a tightly built house.

10. For a layer house, what is your preference for (a) roof type, and (b) solid floor, slotted floor, or cages? Justify your choices.

11. How is heat produced and eliminated by the bird? If the air inhaled by the bird is completely saturated (has 100% humidity), what will be the consequences?

12. What are the functions of insulation?

13. What is meant by the insulation "R" value?

14. What is a vapor barrier, and why is this important in a poultry house?

15. Would you recommend that beginning producers build their own houses or buy prefabricated houses? Justify your decision.

16. What are the chief requisites of poultry equipment?

17. What is meant by the term *automation*?

18. What main methods of using manure are being employed by large commercial poultry operators?

19. Why should there be an emergency warning system to guard against power failure or fire in an environmentally controlled poultry house?

20. Describe each of the following systems or arrangements of layer houses, and give the advantages and disadvantages of each: (a) cage system, (b) slat floor, (c) combination slat and litter floor, and (d) floor or litter-type house.

21. It is estimated that more than 90% of layers in the United States are kept in cages today. Why have cages become so popular?

22. Discuss the types, along with the importance, of each of the following kinds of laying-house equipment: (a) nests, (b) roosts, (c) feeders, and (d) waterers.

23. Discuss the special attention that should be accorded to each of the following management and environmental factors for optimum production and high egg quality: (a) litter, (b) cooling, and (c) manure removal.

24. For chicks, discuss the kind, along with the importance, of modern (a) brooder houses, (b) heating equipment, and (c) feeders and waterers.

25. For replacement pullets, discuss the kind, along with the importance, of modern (a) housing, (b) feeders and waterers, and (c) other equipment.

26. For broilers, discuss the kind, along with the importance, of modern (a) housing, (b) feeders, and (c) waterers.

27. For turkeys, discuss the kind, along with the importance, of modern (a) facilities for breeder turkeys, (b) facilities for breeding turkey poults, and (c) facilities for market turkeys.

28. Discuss the kind, along with the importance, of modern houses and equipment for ducks and geese.

SELECTED REFERENCES

Title of Publication	Author(s)	Publisher
Agricultural Engineers Yearbook	Edited by R. H. Hahn, Jr.	American Society of Agricultural Engineers, St. Joseph, MI, Annually
Commercial Chicken Production Manual, Fourth Edition	M. O. North D. D. Bell	Van Nostrand Reinhold Co., New York, NY, 1990
Farm Building Design	L. W. Neubauer H. B. Walker	Prentice-Hall, Inc., Englewood Cliffs, NJ, 1961
Farm Builder's Handbook, Second Edition	R. J. Lytle	Structures Publishing Company, Farmington, MI, 1973
Livestock Waste Management and Pollution Abatement		American Society of Agricultural Engineers, St. Joseph, MI, 1971
Poultry Handbook	L. D. Schwartz F. W. Hicks	The Pennsylvania State University Press, University Park, 1972
Poultry Meat and Egg Production	C. R. Parkhurst G. J. Mountney	Van Nostrand Reinhold Co., New York, NY, 1988
Poultry Production, Twelfth Edition	M. C. Nesheim R. E. Austic L. E. Card	Lea & Febiger, Philadelphia, PA, 1979
Principles of Animal Environment	M. L. Esmay	The AVI Publishing Company, Inc., Westport, CT, 1978
Structures and Environment Handbook, Seventh Edition		Midwest Plan Service, Iowa State University, Ames, 1975

Fig. 11–32.   Outside view of typical curtain-sided poultry grow-out house. (Courtesy, North Carolina State University, Raleigh)

Fig. 12–1. What it's all about! A layer and an egg. (Courtesy, Hy-Line International, Division of Hy-Line Indian River Company, Johnston, IA)

LAYERS

CHAPTER

12

The production of commercial eggs in the United States is big and important, and characterized by specialization and business methods, as evidenced by the following profile of the industry in 1989 (or in the year indicated):

■ **Number of layers; number of eggs** — A total of 269 million layers laid an average of 250 eggs per hen, for a total of 67.3 billion eggs.

■ **Price and income** — Farmers averaged 68.9¢ per dozen eggs, for a gross egg income of $3,861,270,833, which represented 2.42% of U.S. farm income and 26% of poultry income.

■ **Per capita consumption** — The per capita consumption of eggs has steadily declined for many years. In 1945, the U.S. per capita average exceeded 400 eggs, nearly double the rate of 236 in 1989.

■ **Top ten states** — The top ten egg-producing states of the nation were: California, Indiana, Pennsylvania, Ohio, Georgia, Arkansas, North Carolina, Texas, Florida, and Minnesota. Eggs are produced more generally across the United States than broilers or turkeys.

■ **Fewer but larger flocks** — The trend to fewer but larger flocks has been going on for many years. In 1910, more than 5½ million farms in the United States kept chickens; and the average size flock numbered 50 laying hens. In recent years, there has been a marked trend to fewer and bigger layer flocks. In 1989, the average egg farm had 154,000 hens. Also, it is noteworthy that, in 1989, 59 U.S. firms each owned a million or more commercial layers; and that the 10 largest U.S. egg companies owned 61.2 million layers, which represented 27% of the nation's laying flock; and that one firm — Cal-Maine Foods, Inc., Jackson, Mississippi — owned 15 million laying hens.

■ **Capital requirements** — Considerable capital is needed to start a commercial egg operation today. It is unusual for anyone to start an enterprise with fewer than 50,000 birds. Housing and equipment require an investment of $6.00 to $10.00 per bird. The 20-week-old started pullets may cost $2.50 to $3.00 each. Additionally, operating capital for feed and supplies is needed until the flock peaks in production.

Also, financing is needed for land and for housing for the manager.

■ **Integrated or contract egg production is increasing** — About 89% of the nation's eggs are produced under some kind of integrated or contract arrangement. Generally speaking, the contractor supplies the birds, feed, medication, market, and supervision, and the contractee provides the housing, equipment, and labor; and the contractee is paid so much per dozen eggs produced or on some kind of percentage basis.

■ **More than 90% of layers in cages or on wire** — More than 90% of the layers in the United States are in pampered cages or on wire; and more and more of all replacement pullets are brooded and reared in cages from day one. The main reasons for this trend to cages are reduced labor requirements and all around greater efficiency.

■ **More environmental control** — The trend is toward environmentally controlled housing, with light and temperature regulated. The major advantage of environmentally controlled housing is better feed conversion, especially during the winter months.

■ **Increased automation** — All new egg production units are highly automated — for feeding, watering, and collecting eggs. At many of the large production units, the eggs are moved directly from the laying houses by belts to the egg processing plant on the farm.

■ **Changes in feed efficiency** — In 1940, it required 7.4 lb of feed to produce a dozen eggs; in 1990, it took 3.75 lb.

■ **Mostly white eggs** — About 95% of the eggs are white-shelled eggs. Brown egg production is concentrated in the six New England states, with a scattering in a number of other states for specialty markets.

Fig. 12–2. Commercial brown-egg layer. Some consumers prefer brown eggs. (Courtesy, Babcock Industries, Ithaca, NY)

Economics and consumer preference are the main reasons for the high percentage of white eggs being produced in the United States. White eggs can be produced at a lower cost than brown eggs because of lower feed consumption of the white-egg layers. It is noteworthy, however, that egg breakers prefer brown eggs because of their stronger shells.

■ **Processed eggs take 18%** — About 18% of the eggs are processed — as liquid, frozen, or dried products (see Fig. 12–3). Most of these eggs are used by bakeries, confectioners, noodle manufacturers, and institutions, such as hospitals, schools, hotels, restaurants, and airlines.

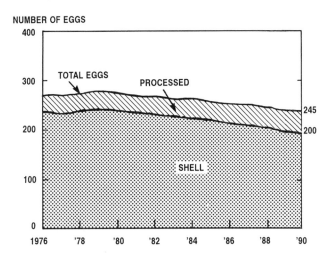

Fig. 12–3. Per capita consumption of processed eggs. (From: *1989 Agricultural Chartbook*, USDA, Agricultural Handbook No. 684, p. 91)

■ **Replacement pullets raised or bought** — Most egg producers either raise their own replacement pullets or make arrangements to buy started pullets just prior to housing time. In most states, started pullets are produced by the hatcheries, although in the Northwest, and in a few other areas, replacements are raised by professional started-pullet growers who purchase day-old chicks from hatcheries.

■ **Trend toward cage rearing of replacement pullets** — There is a trend toward cage rearing of replacement pullets. A number of producers brood their replacements on floors for the first 6 to 8 weeks, then move them to growing cages. Also, there is a definite trend toward rearing replacement pullets in cages from day old.

■ **Pullet layers** — Pullets are usually housed at about 21 weeks of age and kept in production 12 to 14 months.

■ **Bigger and fewer processing industries** — Egg processing industries have continued their trend toward greater concentration of the market.

■ **Marketing** — In the movement of eggs from the producer to the consumer, fewer agencies are being used and there is more direct marketing. An estimated 75% of the eggs are sold in supermarkets.

There is a tendency of egg producers to do the entire job of production, processing, and marketing; with their eggs trucked directly from their egg holding room facilities to warehouse or store-door delivery.

Most small, independent producers do their own processing and marketing; with their eggs marketed to smaller stores, restaurants, and other retail outlets.

Also, layer operations are diverse and ever changing with respect to business aspects, breeds and breeding, replacement programs, housing and equipment, feeding and watering, health programs, management, and marketing; all of which characterize ongoing egg production operations, and should be considered when planning a new egg production enterprise.

BUSINESS ASPECTS

Today's commercial egg production units are generally large, and intricate, too; and they are destined to get bigger and more complicated. Today, it is unusual for anyone to start a commercial egg enterprise with fewer than 50,000 birds. With the increase in size and complexity of units, the business aspects have become more sophisticated. It is important, therefore, that large and modern commercial egg producers be knowledgeable in financing, labor requirements, contracts, record keeping, and costs and returns. To this end, each of these business aspects is covered more completely in the sections that follow.

FINANCING

One of the first things to consider when thinking of becoming a commercial egg producer is how much money will it take, and where will the money come from.

As previously indicated, it is unusual to start a layer enterprise with fewer than 50,000 birds. For the house and equipment, this will call for an investment of $8.00 per bird, or an outlay of $400,000. The 20-week-old started pullets may cost as much as $2.75 each, or $137,500. Land is needed for the physical plant, with the price determined by location and prevailing price. Operating capital is needed, including 75 to 85 lb of feed per hen per year (4 to 4.5 lb per dozen eggs) and supplies. Additionally, provision must be made for family housing and living expenses. In total, a unit of 50,000 layers will require an investment of $750,000 to $1,000,000. Hence, commercial layer operations necessitate both money and knowledge of financing. (Also see Chapter 17, Business Aspects of Poultry Production.)

LABOR REQUIREMENTS

Caring for layers is a full-time job, 7 days a week, and 52 weeks a year. It calls for long hours and hard work.

In 1990, one worker cared for (1) 15,000 layers on litter floor, or (2) 40,000 layers in cages. But, with mechanical feeding and egg collection systems, one person can care for upwards of 100,000 hens. In addition to full-time help, homeworkers and part-time help will be needed at times.

EGG CONTRACTS

Narrowing profit margins, fluctuating egg prices, expanded flock size, large investment, and a desire for a stable income caused producers to consider contract production and marketing. In 1989, 89% of the nation's eggs were produced under contracts or in integrated operations; only 11% of the production remained largely independent.

The usual contract provisions for egg production cover the following: management of laying house (feeding, space, etc.); strain and number of birds; months of lay (production and quality); egg gathering and delivery; egg cleaning (who does it, methods used); cooling (temperature, humidity, and time); oiling specifications, if used; disease, parasite, and cannibalism control; quantity and quality of eggs to be delivered; restriction of other poultry and visitors; record keeping and maintenance; pullet replacement program; sale and removal of fowl at end of lay period; provision for disposal of manure; provision for use of eggs and fowl; obligation of assets to ensure the contract; credit (amount and repayment); labor (present and future); incurring additional expense; share of fixed cost during the down period.

The main kinds of egg contracts are:

1. Shell and/or breaker egg.
2. Hatching egg.
3. Hatching egg foundation stock, lease basis.
4. Pullet growing contracts for laying flock.

Many adaptations of methods and techniques are used in the various types of egg contracts — with producer contracts, credit contracts, marketing contracts, or a combination of these, available.

In egg contracts, producers usually own the buildings and equipment and furnish labor in return for a fixed rate per layer per week or per dozen top-grade eggs produced. Contractors usually furnish hens, feed, supervision, and a market for the eggs. Electricity, medication, litter, and fuel are usually furnished by the producer, but in some cases the contractor shares in these costs. In some areas, the contractors haul the eggs; in others the producers do the hauling. Generally, the producer only washes the dirty and badly stained eggs.

The terms of contracts vary among areas and are affected by the amount of competition and the level of commercial egg prices. The following types of egg production contracts are rather typical:

■ **A fixed fee per dozen eggs** — The producer furnishes housing, equipment, labor, utilities, and sometimes litter and receives a payment of 5 to 8¢ per dozen eggs produced. The contractor furnishes ready-to-lay hens, feed, medication, bookkeeping, and owns all eggs as well as the salvage hens at the end of the production period. Some contracts of this type also pay bonuses for good feed conversion, high percentage of salable eggs, or low mortality.

■ **A fixed fee per hen per month** — The producer furnishes the same items as in the above contract, but receives a payment of from 8 to 15¢ per hen per month. Bonuses or incentive payments may also be added.

■ **Guaranteed price agreement** — Producers agree to supply eggs of designated grade and in a given volume for a guaranteed price. In this type of contract, producers usually supply all inputs (housing, pullets, feed, etc.) and are protected against price declines, but they cannot benefit from market increases.

Generally, this type of contract does not endure for continued renewals because the guaranteed price is usually set slightly below market averages.

■ **Percentage of returns** — The producer furnishes the items listed in the first type of contract and receives a percentage of total egg returns (usually averages 15 to 18%). The feed supplier usually receives 50 to 55% of the return, while the supplier of ready-to-lay pullets gets 26 to 28% (usually retains ownership of birds). The percentage received by any party must be proportional to the contribution in the form of material and/or services.

This type of contract has the built-in incentive of increasing returns by doing a good job. This is the only type of contract that treats all parties proportionately as egg prices change.

In cases of dispute and disagreement over an egg contract that cannot be resolved by the producer and the contractor, the contract should provide for arbitration to be conducted by a committee of three — each party to the contract choosing a representative and these two then choosing a third party, to study the case and recommend settlement.

RECORDS

Keeping, analyzing, and using performance records are vital to success in egg production. The following types of records are important in a well-run commercial egg operation:

1. **Feed consumption.** A daily record of feed consumption should be kept. A drop in feed consumption is one of the first signs that precedes a drop in egg production.

2. **Number of eggs produced daily.** A second meaningful record for layers is the number of eggs produced daily. The last gathering of eggs should be scheduled to take place at the same time each day. If the final daily collection is done at different times, the number of eggs gathered will vary greatly; hence, a true picture of what is going on in the laying house may not be reflected.

3. **Egg quality and egg size.** The efficient operator produces 90% or more grade As, fewer than 5% small eggs, and fewer than 20% medium ones. Large grade As, or better in size and quality, are the ones that have the most profit potential.

4. **Mortality.** The producer should always know how many birds remain in the house and whether mortality is normal or excessive. A perpetual inventory of birds is not difficult.

COSTS AND RETURNS

Costs of egg production include both cash and noncash items. Feed purchased is by far the largest cash cost. Among the noncash items are labor, interest, and depreciation.

Eggs are the chief source of returns on a commercial egg farm. Cull (spent) hens sold for meat are of only minor importance.

As is true in any business, big or little, profits in egg production can be increased by either, or both, (1) lowering costs, or (2) increasing returns.

LOWERING COSTS

In 1989, it cost Southern California egg producers 51.7¢ to produce a dozen table eggs, with a breakdown as shown in Table 12-1.

TABLE 12-1
COST OF PRODUCING A DOZEN EGGS[1]

Item	Cost per Dozen	Percent
	(¢)	(%)
3.3 lb feed at $8.00/100 lb	30.4	58.8
Pullets, 2.50—32 dozen	8.2	15.9
Labor	3.5	6.8
Buildings and equipment depreciation . . .	2.5	4.8
Interest	2.1	4.1
Management	1.0	1.9
Miscellaneous	4.0	7.7
Total	51.7	100.0

[1]Adapted by the author from *Poultry Digest*, April 1990, Vol. 49 #4, pp. 65–66, estimates prepared by D. D. Bell, Poultry Specialist, University of California, Riverside.

With an estimated total return of 62.2¢ per dozen eggs in Southern California in 1989, the net return was 10.5¢ per dozen, or $2.10 net income per bird for the year. This is the highest net return that most Southern California egg producers have ever experienced. Of course, not all Southern California egg producers had a 51.7¢ per dozen production cost, and not all Southern California egg producers had a return of 62.2¢ per dozen eggs.

Perhaps some production costs can be lowered. The place to start is the highest cost items — feed, pullets, and labor. These accounted for 42.1¢, or 81%, of the cost of production.

■ **Lowering feed cost** — To lower feed cost, the producer should be located in an area where suitable major feed ingredients are produced and near a feed supplier who has a large volume, low operating costs, and a willingness to pass some of this efficiency on to the producer. For every $5.00 a ton that can be saved on feed, the cost of production will be lowered 1¢ a dozen.

■ **Lowering pullet cost** — Twenty-week-old started pullets may be purchased for less than $2.50 each, if the order is placed in sufficient time and one's payment record is good. However, bird quality should never be sacrificed for price. Some producers have been raising their own pullets for $2.00 a bird. Every 20¢ extra paid for pullets is equal to 1¢ a dozen increase in the cost of production. It may be profitable to raise, rather than purchase, pullets; at least the possibility should be investigated.

■ **Lowering labor cost** — The best way in which to lower labor costs is to maximize automation.

INCREASING RETURNS

Returns can be increased! Traditionally, egg prices have fluctuated widely and wildly! Therefore, labor income and profits have sometimes been either nonexistent or very high. Returns from commercial egg production can be improved by efficient business and poultry management, pullet rearing, feed mixing, more eggs per bird, egg processing, and egg marketing.

■ **Increase eggs per bird** — The one thing that will do more to increase returns than all else is to increase the number of eggs produced per bird. Birds eat little more feed, but use no more lights, ventilation, or any other cost item, when they produce more than 20 dozen eggs per bird in 12 months in the laying house, rather than fewer eggs.

Increased production does not just happen; it requires planning and work. It may mean adjusting the lights, feeding a level of protein commensurate with the age of the birds and the season, and other similar practices, as well as paying attention to details.

Increased production of 24 eggs per bird in a 30,000-bird unit will mean $37,320 (at 62.2¢ a dozen) more income per flock. Yet costs will be increased very little.

■ **Set high goals** — High goals spur high achievement and increase returns. Some realistic goals for a commercial egg operation are:

Eggs per hen, 12-month cycle 270
Feed per dozen eggs produced . . . . 3.6 lb
Laying house mortality, 12 months . . . . . . 10% or less

BREEDS AND BREEDING

Unless there is a premium market for brown eggs, it is recommended that White Leghorns be used for commercial production. Hens require about 8 lb of feed per year per pound of body weight just for maintenance. Heavy breeds average from 1 to 1½ lb more body weight than Leghorns and require 8 to 12 lb more feed per year. If feed cost 6¢ per pound, this means that it will cost 48 to 72¢ more to keep a heavy hen for a year than a Leghorn. Of course, the heavy hen brings more as a cull, but the difference does not compensate for the extra feed consumed. Egg production is about the same, or perhaps even a little lower for the heavy hen.

Hatcheries and breeders are now franchised and reproduce and sell "name lines." These may be either in-bred lines or strain crosses.

In-bred lines are developed then crossed to produce the commercial chick; or breeders may cross various strains until they find a combination that "nicks" well, embodying the greatest number of desired characteristics.

Each line or strain of a breeder carries a brand name or number as identifying as those for automobiles.

Fig. 12–4. Hisex-white egg layer. (Courtesy, Euribrid B. V., Boxmeer, Holland)

Selection of a top in-bred line or cross strain is extremely important. A check with successful poultry producers in the area regarding the lines or strains that have performed well for them will be helpful. Also, Random Sample Test summaries can be helpful in comparing characteristics of economic importance. Random Sample Test at various stations throughout the nation have been the proving grounds for most breeder's lines. There is some hazard in acting solely on results from a single year or test. But an average of several tests gives a fair appraisal of the anticipated performance of a single line. Continuous improvement is being made in breeding.

In recent years, several strains of "mini" chickens have evolved, in both meat and egg lines. Mini Leghorns are now becoming popular and taking their place in commercial production. In comparison with conventional Leghorns, the mini:

(1) is smaller — 15 to 25% lighter at 21 weeks of age; (2) requires less floor space in the poultry house or cage and allows more birds to be housed in a given area; (3) consumes 10 to 20% less feed; (4) requires 5 to 10% less feed per dozen eggs; and (5) lays eggs that are about 10% lighter.

REPLACEMENTS

A recent trend in poultry production involves specialization within a specialization. For example, some commercial egg producers are only concerned with producing market eggs — leaving the rearing of replacement pullets to other specialists. This means that the commercial chicken producer has two alternatives when specializing — raising replacement pullets or producing eggs. Likewise, the commercial egg producer has two alternatives in procuring replacement stock:

1. Day-old chicks.
2. Ready-to-lay pullets (usually 20 to 22 weeks of age).

Today, the majority of large commercial egg producers prefer to buy started pullets rather than raise their own. In the final analysis, the choice of the system should be based on the cost of the pullets of the desired quality, with added consideration given to available facilities and labor.

Planned disease prevention should start with the purchase of replacement stock. Economic losses can be prevented much more easily by not introducing a disease than by acting after the disease has gained a foothold.

The source of day-old chicks or ready-to-lay replacement pullets, and their quality when delivered, are extremely important to the economic success or failure of the poultry operation. Today's major commercial strains of egg production stock all have good genetic potential for high egg production, but their actual performance depends on the management of the breeder flocks that produced the hatching eggs, as well as the conditions under which the eggs were incubated and the chicks hatched and grown out.

DAY-OLD-CHICKS

Price should be one of the least important considerations when buying disease-free, high-quality, day-old chicks for raising replacement pullets. Some searching questions to which answers should be obtained, are:

■ Do the breeder and hatchery firms have reputations for honesty and fair dealings?

■ Will the hatching eggs come from healthy breeders certified free of pullorum-typhoid and maintained in a relatively disease-free area?

■ Are the parent flocks free of *Mycoplasma gallisepticum* (MG) and are the chicks also MG-free, having been hatched, sorted, and sexed with chicks from similar sources?

■ Do the chicks carry a high degree of passive immunity to infectious bronchitis and Newcastle disease as a result of the vaccination regime of the parent flock?

Fig. 12–5. A 20-week-old mini pullet. The mini layer produces more eggs per ton of feed in less housing space. (Courtesy, Colonial Poultry Farms, Pleasant Hill, MO)

■ What is the Marek's disease vaccination history of the parent flock? Were the chicks vaccinated against Marek's disease?

■ Will the entire order be filled by chicks of only one strain coming from just one hatchery and from eggs produced in a single location?

■ What will be the age spread, if any, between the youngest and oldest chicks delivered?

■ Is the strain of stock relatively free of such mortality-causing problems as flightiness, hysteria, and cannibalism?

■ Does the strain's body size permit housing at the cage density planned?

■ Do the egg size and quality characteristics of the strain meet the marketing requirements of the poultry producer?

■ Will the chicks be delivered in clean, uncontaminated boxes and equipment, without a prolonged period of transit? If new boxes are not used, how are the boxes cleaned and disinfected?

READY-TO-LAY PULLETS

Quality stock is essential regardless of the age at the time of purchase. However, when purchasing started pullets, more information is needed than when purchasing day-old chicks. With ready-to-lay pullets, it is important to make sure that they are well developed, healthy, and capable of expressing their full genetic potential as layers.

The choice of a supplier of started pullets is most critical. The supplier (1) should be a specialist — one whose business is raising pullets to sell, and who has the knowledge and experience to raise quality replacements; (2) should not have any adult birds on the same premises unless the area is large enough to keep the two operations well separated; and (3) should have adequate facilities for producing high-quality stock that will meet high standards.

When contracting for replacement pullets, the buyer and the pullet grower should agree on as many as possible of the following management practices that the grower will use:

■ Records to be kept.

■ Strain and source of stock.

■ Number of age-groups brooded on the premises.

■ Proximity to other poultry operations.

■ Number of birds to be housed in each unit (depending on size of unit).

■ Sanitation procedures and safeguards used to prevent introducing diseases.

■ Feeding and lighting programs.

■ Schedules and procedures for vaccinating and debeaking.

■ Health monitoring practices during brooding and grow out; diagnoses of losses.

■ Precautions in delivering the pullets.

Written instructions should specify that the buyer will receive: periodic weight and mortality information; data on feed consumption; the brand and serial number of each vaccine used, with vaccination dates, procedure followed, and postvaccination results; copies of reports from the monitoring veterinarian; and the right to visit the operation (obviously in clean protective clothing) during the growing period.

As a further precaution, the buyer may wish to have the pullet grower submit samples of birds or blood samples to a poultry disease laboratory for examination and testing. A copy of the laboratory report should be sent to the buyer before the birds are delivered. Prior agreement should be reached relative to who will pay for this work. The buyer should also reserve the right to reject individual birds or the entire lot if they do not meet the quality standards specified.

The manner in which the pullets are loaded, transported, and placed in the cages on the buyer's premises will significantly affect their subsequent performance. Thus, the contract should specify who is to deliver the birds, and that all equipment be thoroughly sanitized. Delivery instructions should be sufficiently detailed to include method and time of loading, type of hauling equipment to be used, number of birds per crate or cage, the route of travel, and arrival time. The owner or manager should be on hand to supervise the distribution of the pullets into their laying cages.

Fig. 12–6. Twenty-week-old ready-to-lay pullets being moved from grower to laying house in coops by a large transport truck. Moving is usually done at night to avoid summer heat and so that the pullets may become accustomed to the new quarters in daylight. (Courtesy, Colonial Poultry Farms, Pleasant Hill, MO)

ALTERNATIVE REPLACEMENT PROGRAMS

When the average rate of lay falls below 55%, most commercial egg producers either cull or molt.

The replacement program that egg producers choose is one of the most important management decisions. It should be based on the performance and marketing arrangement of the flock.

The choice may be made from among many alternative programs, but not all of them give equal financial returns. Forced molting may or may not be an integral part of the program chosen (see Chapter 8, Poultry Management, section headed "Forced Molting.") Also, what may be optimum for one producer and flock may cause failure for another.

Replacement represents the second largest cost of producing eggs (feed is first). Replacement cost is reduced primarily by increasing the number of dozens of eggs a given flock produces during its lifetime; and this is achieved by keeping the flock for a longer period of lay. When the savings exceed increases in other costs or reductions in income, the extended period of lay is more profitable.

A sound replacement program should be well thought out; and should be based on the flock's normal performance, realistic estimates of costs and prices, and marketing arrangements.

ALL-PULLET FLOCKS

An all-pullet flock is made up of birds that are less than 1 year old when the production year starts. In comparison with hens, pullets have the following advantages:

1. Pullets lay 20 to 40% more eggs than hens will lay during their second year of production.
2. Pullets require less feed to produce a dozen eggs.
3. Pullet eggs have a higher interior quality (a high percent of thick white compared to thin white).
4. Pullets lay eggs with stronger shells.
5. Pullets can be fed to laying age on about the same amount of feed (20 lb) that is required to feed a hen through a normal molt.

The all-pullet system is still the most popular throughout the United States, except in California, the leading egg-producing state of the nation, where molting is practiced extensively.

All-pullet flocks are easy to manage. Once a cut-off age is selected, new pullets can be ordered at routine intervals. The only real problem is to determine the selling age at which total profits will be maximized.

Egg producers choose a wide range of cut-off ages of all-pullet programs—from as short as 50 weeks of age at sale to as long as 108 weeks, with an average of about 82 weeks.

At the 82-week (20-month) age, most flocks are at 50 to 55% production; the number of undergrade eggs approaches 20%; and the flock size is approximately 80% of the number originally housed.

MOLTING

Molting is a normal process of chickens and other feathered species; and it occurs in both sexes. In the wild state, birds usually shed and renew old, worn plumage before the beginning of the cold weather and their migratory flights. Since undomesticated birds lay only a few eggs, molting and reproduction are not usually associated.

Chickens kept for commercial egg production have a different molting pattern. They have been bred for high performance, and their environment, with respect to temperature and light, is usually modified to remove major seasonal influences. A natural molt does not normally occur until after a period of 8 to 12 months of egg production. If nothing is done to alter the normal molting cycle, it requires about 4 months for a hen to drop her feathers and grow a new set. It is possible, however, to speed the process through a program of forced molting, thereby recycling the hens for another period of egg production and improved egg quality.

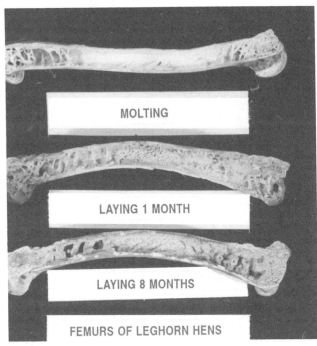

Fig. 12–7. Femurs (medullary bones) of Leghorn hens at three stages of egg laying. Note the lack of mineralization in the cross sections of femur bones due to calcium depletion for egg shell formation, with the calcium depletion increasing from (1) laying 1 month, (2) to laying 8 months, (3) to molting. (Courtesy, North Carolina State University, Raleigh)

Molting is controlled by the gonads and the thyroid gland and is associated with a drop in estrogen levels and a decreasing rate of egg production. Egg production is not greatly affected by the process of natural molting, but molting is prolonged when the birds are kept in production. High producers tend to molt late; but once production ceases, molting is rapid.

Several factors affect the onset and length of molting; among them, (1) weight and physical condition of the bird, (2) length of light exposure, (3) nutrition of the bird, and (4) environmental influences, such as temperature and humidity. Thus, if one drastically reduces the amount of light or starves the birds in such a way as to knock them out of production, molting can be induced or speeded up.

Decreasing day length is the normal trigger for molting. Therefore, lighting programs for layers should provide either constant or increasing day length. Minor stresses, such as temporary feed or water shortage, disease, cold temperature, or sudden changes in the lighting program, can also initiate a partial or premature molt, evidenced by a considerable number of feathers on the floor of the poultry house. In these cases, chickens lose some head and neck feathers. If the molt continues beyond this point, a more severe drop in performance can be expected.

■ **Forced (induced) molting**—Under forced molting, a layer flock is induced to shed and replace its feathers at a time selected by the flock manager. This may come near the end of a normal laying cycle, or a flock may be force molted earlier as part of a multiple molting program.

An induced molt causes all of the hens in a flock to go out of production for a period of time. During this period, regression and rejuvenation of the reproductive tract occur, accompanied by the loss and replacement of feathers. After a molt, the hen's reproductive rate usually peaks slightly below the previous peak rate, and egg quality is improved.

Forced molting has been practiced in a limited way since the turn of the century, but always as economic conditions dictated. In 1990, it was estimated that approximately 60% of the nation's laying flocks are recycled, with California leading in the percentage of flocks that are recycled. The first and most important reason for induced molting is that it usually improves profits and, therefore, is part of a planned replacement policy. The question of whether it pays to recycle depends primarily upon the relative performance of all-pullet vs recycled flocks.

■ **Types of recycling**—Egg producing hens may be molted one or more times, giving rise to the following types of recycling programs:

1. **Two-cycle molting program.** This involves one molt and two cycles of egg production. The hens are force molted after about 10 to 12 months of egg production, brought back into egg production for about 6 months, then sold. United States commercial egg producers commonly peak their first-cycle (all-pullet) flocks at 85 to 95% production. Second-cycle flocks commonly peak between 75 and 85%—about 10% lower than first-cycle flocks, but a sizeable number of flocks now peak in excess of 85%. If molting is done properly, it will result in a return of all egg quality characteristics equivalent to a 10- to 12-months-old pullet, thereby alleviating the deterioration of quality traits with extended laying periods; many plants will not carton eggs from layers beyond 70 weeks of age in the first cycle or from layers beyond 30 weeks in the second cycle.

The two-cycle molting program generally extends laying life by an additional 6 to 8 months beyond the standard age at which all-pullet flocks are sold; and each hen produces an extra 100 or more eggs, which reduces the overall hen replacement cost per dozen.

2. **Three-cycle molting program.** This involves two molts and three cycles of egg production. The hens are first molted after about 9 months of production, then held through another shorter production period, molted again, followed by an even shorter period of lay, then sold. This periodic molting program totals about 24 months; longer programs are seldom profitable. Commercial egg producers commonly peak three-cycle flocks at 75%, which is 10 to 20% below first-cycle (all-pullet) flocks, and 7% below two-cycle flocks.

■ **Methods of molting**—Many satisfactory molting programs are in use, but most of them are simply modifications of two basic concepts—feed withdrawal, followed by a low nutrient feed intake period. Modifications involve the number of days without feed, the composition of the feed, and the duration of the low-nutrient intake period. Other minor factors include the choice of lighting programs during the molt and whether water should be restricted. Of the many satisfactory molting programs, three that feature different basic procedures follow:

1. **Conventional forced molting program.** This procedure, which is sometimes referred to as an on-again/off-again program, is outlined in Table 12-2.

TABLE 12–2
CONVENTIONAL FORCED MOLTING PROGRAM
(ON-AGAIN/OFF-AGAIN PROGRAM)

Day	Feed	Water	Light
1	None	None	8 hours
2	None	None	8 hours
3	Egg-type layers: 10 lb/100 hens	Water	8 hours
4	None	None	8 hours
5	Egg-type layers: 10 lb/100 hens	Water	8 hours
6	None	None	8 hours
7	Egg-type layers: 10 lb/100 hens	Water	8 hours
8	None	None	8 hours
9	Egg-type layers: 10 lb/100 hens	Water	8 hours
10 through 55–60	Restricted feeding— about 75% of full feed intake	Water	8 hours
61	Full-feed layer ration	Water	14–16 hours

Additional pertinent feeding instructions relative to the conventional molting program follow:

a. Self-feed oystershell from the start of molting until 2 weeks after egg production is reestablished, then return to controlled shell feeding.

b. Do not use skip-a-day feeding programs until after 10 days following the start of the molting procedure.

2. **California forced molting program.** California has a higher percentage of force molting than any other state. For this reason, the California method is presented. It is characterized by simplicity, low cost and high subsequent performance, and with all birds getting equal treatment and having uniform recovery. It is presented in Table 12-3.

TABLE 12–3
CALIFORNIA FORCED MOLTING PROGRAM

Day	Feed	Water	Light
1 through 10 to 14	None	Water	Discontinue artificial light or limit to 8 hours
11 to 15 through 35	Full feed cracked grain or low-protein, low-calcium molt mash	Water	Discontinue artificial light or limit to 8 hours
36 through 68	Full-feed laying mash	Water	14–16 hours

Additional pertinent information about the California molting program follows:

a. The 10 to 14 days without feed will usually result in 25 to 30% loss in body weight, with less than 1% mortality.

b. The feed withholding period (from day 1 through up to day 10 to 14) should end when the accumulated mortality reaches 1.25%, when body weight reduction for a 3.6-lb hen reaches 30%, or when it has gone 14 days — whichever comes first.

3. **North Carolina forced molting program.** This program was developed at North Carolina State University. It consists of a premolt period of 7 days during which artificial lights are turned on at night to make for a 24-hour day, followed by removal of feed. This method places emphasis on body weight losses of 30 to 35%, with the loss in weight of lightweight hens being at the lower end of the range and the weight loss of heavy hens being at the upper end. The weight reduction is achieved in about 14 days. After the weight reduction is reached, the birds are given 0.1 lb of feed per hen per day for 2 days, followed by a 15 to 16% protein and 2% calcium molt ration until the 28th day, after which the flock is returned to the regular layer diet. Twelve hours of total light are provided for the first 3 weeks, with the light increased to normal on the 35th day.

4. **Other methods of forced molting.** Many other methods of forced molting have evolved, with variable results; among them, (a) high level of zinc, and (b) a low sodium diet. The zinc method consists of adding about 20,000 ppm of zinc to the feed, preferably as zinc oxide, for 5 days, followed by returning to the normal ration. Hens come back into production about 7 days after the high zinc diet is discontinued, and peak at 75 to 80%.

The low sodium method consists of reducing the sodium content of the ration to 0.04%, which is achieved by feeding an all-grain diet which may be supplemented with high-fiber feedstuffs. Additionally, the lights should be turned off in open-sided houses, or limited to an 8-hour light day in environmentally controlled houses. Application of this method will result in the birds molting in about 6 weeks, following which they may be returned to the regular layer ration and regular lighting program. It is noteworthy that the low-sodium method is used extensively in Europe where animal welfare codes prohibit lengthy fasting programs.

■ **Will forced molting pay?** — There is no simple answer to this question, since it depends on a variety of economic factors. However, the following general guidelines should be taken into consideration:

1. Recycling becomes less profitable as egg prices increase and as price differential between egg sizes decreases.

2. As the cost of replacement pullets increases, it becomes more advantageous to recycle.

3. The profitability of recycling increases as the price paid for *spent* hens decreases.

4. Using replacement pullets instead of recycling ties up additional capital.

(Also see Chapter 8, Poultry Management, section on "Forced Molting.")

HOUSING AND EQUIPMENT

Fig. 12–8. Open-sided layer house with adjustable curtains. (Courtesy, Arbor Acres Farm, Inc., Glastonbury, CT)

Layer houses and equipment come in many styles and sizes. Floor houses are generally 36 to 70 ft wide and 200 to 600 ft long, with a refrigerated area for storage of eggs. The most common type of manual cage house is 9 to 12 ft wide and 250 to 300 ft long, with a concrete aisle down the center and a refrigerated area at one end, in which the layers are fed by manually driven electric feed carts and eggs are collected by hand. Mechanical cage houses vary from 8 to 32 ft in width and 270 to 450 ft in length; and they are equipped with mechanical feeders and egg belts. The space requirements for houses and equipment are given in Chapter 11, Poultry Houses and Equipment, Table 11–1, Space Requirements for Poultry; hence, the reader is referred thereto.

HOUSING REQUISITES

Poultry housing should be reasonable in construction and maintenance costs, reduce labor, have utility value, and confine the birds. Additionally, more and more poultry houses are environmentally controlled, which involves temperature regulation, insulation, a vapor barrier, ventilation, lighting, and manure management.

(See Chapter 11, Poultry Houses and Equipment.)

HOUSING SYSTEMS

Housing systems for egg production include (1) cages, (2) floor, (3) slats, and (4) partial floor and partial slats. Each system can be varied as to size, construction material, and extent of mechanization. There is no one best system or arrangement. All of them must take into account:

Bird comfort
Operator efficiency
Operational costs
Egg handling
Durability
Initial cost
Service availability

Bird comfort must head the list. The hen must be reasonably comfortable if she is to produce a large number of eggs with a small amount of feed and stay healthy. Also, the house must be arranged efficiently for the operator. The operating costs of all fans, feeders, manure removal systems, and other machines must be held down or their use may not be justified. Provision must be made for handling eggs rapidly with minimum breakage. Also, total cost must be within the competitive levels established by egg producers in other areas.

■ **Cage system** — The cage system is the most popular method of producing table eggs; more than 90% of the layers in the United States are kept in pampered cages or on wire. The cage system involves many small wire compartments each equipped with its own feeder and waterer and sloping floors so the eggs roll out of the cage for easy gathering by hand or egg belt. Droppings fall through the cage floor into pits or onto dropping boards, thence they are scraped into pits. Dropping pits range from 6 in. to 8 or more feet deep. With shallow pits, the droppings are scraped or washed into a holding chamber or directly into manure spreaders every few days. In the deeper pits they can accumulate for prolonged periods of time (a year or more).

Fig. 12–9. Pampered layers in cages. (Courtesy, Ralston Purina Company, St. Louis, MO)

Feed may be supplied to cages by hand, eggs picked up by hand, and ventilation adjusted by the operator; or these jobs may be totally mechanized so that the feed, air, water, eggs, and manure are all moved mechanically.

Cages vary in size and bird capacity. Each will hold 2 to 15 or more birds. Up to the present time, the most successful cage has 4 birds or less. Each bird should have from 54 to 64 sq in. of cage floor space. The dimensions of the smallest cage (2 birds) are 8 in. wide, 16 in. deep, and 14 in. high. Cages for more than 2 birds are simply wider and/or deeper.

Many cages have been installed in a stair-stepped arrangement with three, four, and sometimes five rows of cages. There are service walks down each side of the house and between each row of cages. Walks are 26 in. to 30 in. wide, while cage rows may be up to 90 in. wide.

Some stair-stepped cages have been modified so that the bottom rows of cages are separated by approximately 8 in. This necessitates the use of a dropping board on the top of the bottom cages to prevent manure from the birds in the top cages from falling through the cages below. The dropping boards have to be scraped frequently.

Cages are suspended from the ceiling of a house. The bottom of the lower cage is usually 24 to 28 in. above the aisle walkway. This puts the top cage at shoulder height. The manure drops into the pit below the cages.

Recently, triple-decked, narrow-aisle cage systems have been developed to increase house capacity. The interior aisles are 12 to 14 in. wide, but the outside aisles are approximately 24 in. to provide work space. The bottom cage is level with the aisle. The top of the bottom cage is about knee high, the middle cage a little over waist high, and the top cage head high. This type of house is usually completely mechanized.

Presently, some four-deck systems are being installed. It appears that five, six, or more decks can be used, but several problems need to be solved to make them feasible. Inspection of birds and equipment in a "tall house" is more difficult since some birds are higher and some lower than eye level.

Another cage arrangement is known as the flat-deck system. Two arrangements are possible, one a wall-to-wall plan (without aisles) and the other with aisles. The difference between the two is that an overhead personnel carrier is needed to service the birds in the wall-to-wall system. The carrier enables the caretaker to move over the cages to inspect and care for the birds and equipment.

Cage-system equipment costs more per bird than floor-systems equipment. This is offset by housing more birds in a given size building making the overall cost of house and equipment quite similar. Cages offer the following **advantages** compared to floor systems:

1. Eliminates the need for caretakers to enter the pens with the birds (this often appeals to some workers).

2. Eliminates part of the social pressure on the hens by breaking the flock up into many small societies.

3. Eliminates the floor-egg problem.

4. Produces an egg that is more acceptable to egg receivers and processors.

5. Programs easily for the inexperienced operator.

6. Reduces some diseases.

Some of the **disadvantages** of cage systems compared to floor units include:

1. Hens tend to have a very rough and ragged appearance.

2. The entire house must have a very uniform and acceptable total environment since the hen cannot move to a more comfortable location.

3. Egg production (total eggs laid) may be somewhat lower.

4. There may be more odor because the droppings are not mixed with litter and because the hens are housed closer together.

The cage system appeals to many because it can be augmented by many sophisticated mechanical devices to take the place of human labor. However, a large number of hens are needed to justify the use of such mechanics.

■ **Floor system** — This is the oldest of the systems. It consists of litter, varying in depth from light to built-up, covering the entire floor. Feeders and waterers are located on the litter, and nests line one or two sides of the house. This arrangement takes a low investment, since it needs a minimum of equipment. But it requires a well-insulated house with a properly-designed and well-managed ventilation system.

■ **Slat or wire floor** — This involves slats or wire over the entire floor. Wire floors are subject to rust and mechanical damage, and sometimes they produce foot problems in the chickens. Slat floors are made by arranging strips of wood, metal, or concrete parallel to each other about ½ in. apart so that the droppings are pushed between the slats by the birds' feet.

Droppings then accumulate in a pit below. The depth of the pit may vary from 2 or 3 ft up to 8 or 10 ft. The floor must be removed at least once a year to clean shallow pits. With deep pits, droppings may accumulate for several years if no water is permitted to enter the dropping area.

Breeder flocks are frequently housed on slat floors.

Mechanical cleaning with a cable-type cleaner beneath slats or wire floors frequently presents problems because of the large area and volume of manure. The operator of slat or wire floor houses should maintain a constant program of rat and mouse control.

Feather picking and cannibalism may be a problem on slat or wire floors. Hysteria may affect chickens when pen size, bird concentration, equipment placement or noise levels become intolerable.

■ **Partial floor, partial slat** — It consists of litter covering about 60% of the floor area, and of a raised slat or wire floor over the remaining 40% — usually down the middle of the house. This raised floor provides a place for feed, water, and roosts. About 70% of the droppings collect below the raised floor and are removed by a mechanical cleaner about twice a week. Removal of droppings also gets rid of considerable moisture, thus reducing the ventilation problem. The floor litter is easier to manage since it contains fewer droppings and less moisture.

PULLET HOUSING AND EQUIPMENT

Regular replacement of the laying flock is one of the most important, and most expensive, essentials of egg production. It ranks next to feed in the cost of egg production, and it exceeds labor, housing, equipment, interest, taxes, and other costs.

Producers may choose between growing their own pullets or purchasing started pullets. In any event, the building and equipment requirements are the same, whether pullets are raised for sale or raised by the one who will retain them as layers.

Usually it is wise to provide pullets with levels of environment — housing, feed, and general management — consider-

ably above minimum standards, but consistent with realistic cost consideration. This calls for housing that gives reasonable protection from heat and cold and provides good ventilation; and it means adequate feed and water space and facilities.

HOUSING

Effective isolation of growing stock from older birds is important for disease control, particularly during the early stages. Hence, separate housing should be provided for replacement pullets, completely separated from layers. Housing that is suitable for brooding chicks, or for housing a laying flock, will be satisfactory for rearing pullets. Usually, they are started out in the brooder house, then switched to the layer house.

In the past, most replacement pullets were raised on solid floors covered with litter. But more and more replacements are being grown in cages.

Fig. 12–10. Replacement pullets on floor. (Courtesy D. D. Bell, Poultry Specialist, University of California, Riverside)

Fig. 12–11. Twenty-week-old pullets which were raised in cages. (Courtesy, D. D. Bell, Poultry Specialist, University of California, Riverside)

FEEDERS AND WATERERS

Water space is less important than a good distribution of waterers through the house with a constant supply of fresh water. The same principle applies to some degree to feeders. Both feeders and waterers should be located to permit birds in any part of the house to feed or drink conveniently, without having to find their way around barriers or to travel more than 15 ft.

OTHER EQUIPMENT

Hoppers for grit should be provided for replacement pullets throughout the growing period, and, as they near maturity, additional hoppers should be provided for oystershell or other calcium supplement.

Roosts are not always used where pullets are reared, but they are a desirable addition to the equipment, unless they are subsequently to be caged. When used over enclosed pits, roosts aid in sanitation. Also, they add to the comfort of the birds in hot weather and help to establish desirable roosting habits for the laying period that follows. Some poultry producers also feel that flocks with good roosting habits are less likely to present severe problems of floor eggs.

Nesting equipment should be available by the time the pullets begin to lay. Of course, only a few nests are necessary if pullets are moved from rearing quarters to their laying quarters before production becomes heavy.

Except for automatic feeders and waterers, little mechanization exists in the vast majority of pullet rearing facilities, especially when compared to layer facilities.

LAYER HOUSING AND EQUIPMENT

The design and construction of houses and equipment for layers should be such as to provide for top performance of the layers, optimum environmental control, functional arrangement of equipment, maximum labor efficiency, satisfactory waste disposal, and minimum housing and care costs per dozen eggs produced.

HOUSING

The starting point in designing a layer house is the selection of the type of laying system. Presently, layer houses are being arranged in the following ways, or according to the following systems: cage system, floor or litter-type house, slat or wire floor, or combination floor and slat.

The subject of housing for layers is detailed in Chapter 11, under the heading "Houses"; hence, the reader is referred thereto.

Fig. 12–12. Cage system of housing layers, with three decks of cages. (Courtesy, Salopian Industries, Ltd., Shrewsbury, England)

CARE OF THE HOUSE

The care of the layer house is fully covered in Chapter 11, under the heading "Care of the House"; hence, the reader is referred thereto.

LAYER EQUIPMENT

Layer equipment refers to structures other than housing as such used in feeding, watering, and caring for layers. It includes ventilation equipment, feeders, waterers, egg handling equipment, and refrigeration.

Because 90% of the layers in the United States are in cages, the feeders, waterers, and egg handling equipment described in the sections that follow are adapted to cages.

VENTILATION EQUIPMENT

Of the various ventilation systems, a mechanical system is best. There are two basic types of mechanical ventilation — (1) exhaust fans, and (2) pressure or intake fans.

In an exhaust system, fans force air out of the building. This creates a slight vacuum in the building. Air then comes in through intake openings to equalize pressure. Where fresh air inlets appear uniformly around the building, fresh air distribution should occur uniformly.

In a pressure fan system, fans draw fresh air from the outside and build up enough pressure to push stale, moisture-laden air out through exhaust ports and any other openings.

Fan capacity is an important factor. The correct capacity depends primarily on the total weight of poultry in the house. Approximately 1 cfm (cubic foot per minute) handles 1 lb of bird. This rule of thumb may be stated as 3 cfm per bird for light breeds, 4 cfm per bird for medium breeds, and 5 cfm per bird for heavier breeds. Others say 3 to 5 cfm per bird. Many commercial systems have a 7 cfm per bird rating for added safety.

For 1,000 4-lb birds, 4,000 cfm is needed. A fan or fans should then be selected to deliver this quantity of air at 1/8 in. static pressure.

Small houses (24 ft wide or less) with birds housed on the floor may be satisfactorily ventilated with the gravity system. This is based on the principle that warm air rises. A gravity system needs draft-free inlets and controlled outlets.

Adequate summer ventilation is extremely important. Egg production drops when temperatures exceed 84°F, due to decreased feed consumption.

During the winter, fans are operated at one-fourth or more cfm per bird, depending on the outside temperature. Individual thermostats on each fan are set to run at different temperatures to provide fresh air regularly while minimizing heat loss.

FEEDERS

Two types of feeding mechanisms are used to distribute feed to the feed troughs serving cages. The most commonly used one in small operations is the motorized feed cart. Feed is delivered to the farm in bulk feed trucks, thence put into bulk feed bins outside the house. From the bins, it is augered into a feed cart. Then, a motorized feed cart is driven by the caretaker to distribute feed to the multidecked cages.

Mechanical feeders are used almost universally in large enterprises. Feed is delivered by a chain, cable, or traveling feed hopper to the feed trough in each cage. Feed distribution is activated by a time clock.

WATERERS

Presently, there are three ways of delivering water to birds in cages. One is the continuous flow V-shaped water trough, in which water enters one end and flows down a slight slope or grade to the other end. In these, clean, fresh water is always available, but this system uses a lot of water and requires frequent cleaning, preferably daily.

The other types of waterers in use are nipple and cup drinkers. The nipple waterer is hung from the top of each cage. Birds drink by pushing the nipple up, which allows the water to flow into their mouths.

One cup drinker is placed in each cage or between every two cages. The hen depresses a small trigger with her beak which causes water to flow into the cup. Birds must be taught to use these drinkers. Cup and nipple drinkers usually do not have to be cleaned.

EGG HANDLING EQUIPMENT

Eggs are collected by hand in smaller houses. They are usually put into filler flats on a cart that is pushed or driven up and down the aisles. Next, they are cased and then held in a cooler while awaiting delivery to a processing plant.

Egg collection belts are used in larger units. In some installations, eggs are cross-conveyed to a packing station within the house. In very large production units, eggs may be conveyed through a tunnel from all of the laying houses to an egg processing room. In the processing room, they can be directed through an egg washer, dryer, candler (to check interior quality), grader (to determine size), and packer (to carton) without being handled except to remove eggs with quality defects.

REFRIGERATION

Eggs are a perishable product and should be handled with care. When they are being candled, washed, and sized, the room temperature should be 72°F or less. Eggs should be cooled to 60°F or lower as soon after production as possible, then held at that temperature until used. An egg cooler room should have sufficient capacity to hold at least one week's production.

FEEDING

Feed is by far the greatest expense item in producing eggs. Therefore, the efficient producer uses a feeding program that maximizes profit.

A layer ration must furnish an adequate supply of nutrients. The essential nutrients are: (1) adequate energy, furnished by starches and fats; (2) proper proteins (amino acids); (3) necessary minerals; and (4) sufficient vitamins. Hens kept for breeding need more minerals and vitamins than those retained for commercial egg production.

Many producers buy all their feed commercially prepared. Some producers mix locally grown grain either with commercial supplements or individual ingredients. Still other producers, especially the large ones, do on-farm mixing. However, it should be remembered that feed mixing is complex and exacting and should be attempted only by those persons who are knowledgeable regarding it. Also, regardless of whether poultry feed is commercially prepared or home-mixed, it should be formulated for the intended purpose.

(Also see Chapter 6, Fundamentals of Poultry Nutrition; and Chapter 7, Poultry Feeding Standards, Ration Formulation, and Feeding Programs.)

FEEDING PROGRAMS

Generally speaking, egg producers may choose from among three feeding programs: all mash, grain and mash, or cafeteria.

■ **All mash** — Commercial egg producers use this feeding program exclusively. Under this program the operator uses a complete feed, which is almost a must when a mechanical feeder is used. It may be fed as meal, as granules, or as crumbles of various sizes. An all-mash program offers the inexperienced or the average feeder the greatest assurance of good production. Also, it comes closest to ensuring uniform yolk color. A 14 to 17% protein mash is recommended for layers.

■ **Grain and mash** — This program combines a constant supply of a 20 to 26% protein mash with a light grain feeding scattered in the litter each morning and a feeding of corn each evening. Results depend on the feeding skill of the producer.

■ **Cafeteria**—The cafeteria program maintains separate hoppers of grain—corn and either wheat or oats (or both)—and of a 26 to 32% protein supplement or concentrate, and lets the flock do its own balancing. This program can be tricky with older hens. They may become grain feeders, resulting in lowered production. Two-thirds to three-fourths of the feeders should contain grain and one-third to one-fourth should supply the supplement. This system is more likely to work if layers are reared on a cafeteria feeding plan.

A general rule in the floor system of management is that feeders and waterers be within 10 to 15 ft of every bird in the house. A similar rule applies to the distribution of nests.

Generally, a 15 to 20 ft trough feeder will take care of 100 birds. Feeding from both sides doubles the space, allowing 30 to 40 linear feet per 100 birds.

To minimize waste, feeders and waterers should be adjusted to shoulder height of the birds, and hoppers should be filled only one-third full. In nonmechanized houses, hoppers should be placed on a 4 in. block or suspended at an equivalent height so that birds reach out rather than down to feed. Feeders on 18 or 20 in. legs should be avoided because their use predisposes the flock to cannibalism as a result of the birds standing at eye level to each other.

Allow feed hoppers to empty occasionally to avoid mold development. Palatable, fresh mash increases consumption.

A 4 ft water trough will accommodate 100 birds. A laying flock drinks about 2 lb of water per pound of feed consumed; more in warm weather. This amounts to about 50 gal per day per 1,000 hens.

The water supply must be kept clean. Some researchers suggest that blood spot development in eggs may be due to contaminated drinking water. The flock appreciates clean, fresh water. It is important for high rates of egg production.

FEED CONVERSION (EFFICIENCY)

Feed conversion refers to the pounds of feed required to produce a dozen eggs, without regard to size of eggs produced, i.e., medium, large, or jumbo.

Rate of production is a very important factor in determining efficiency. Divide the pounds of feed by the dozens of eggs to obtain the feed conversion, or pounds of feed required to produce a dozen eggs. This number should be about 4.3 for profitable production with laying strains, although effort should be made to reduce it to about 4.0. In some experimental groups, this figure has been reduced to 3.2 for a 10-month period. When feed conversion approaches 5, check on feed wastage, rate of production, size of the layer, and make-up of diet.

The strain of birds, season of year, management, disease level, and feed used influence actual feed consumption. Table 12–4 gives the estimated feed consumption of confinement reared replacement pullets from 1 to 20 weeks of age. Table 12–5 shows the feed consumption of laying chickens, including average amount of feed required per day and per dozen eggs by 100 hens of different weights and egg production.

TABLE 12–4
ESTIMATED FEED CONSUMPTION OF
CONFINEMENT REARED REPLACEMENT PULLETS[1]
(Commercial Egg-type Pullets)

Age	Feed per 1,000 Birds per Day		Feed per 1,000 Birds per Week		Accumulative Feed Consumption per 1,000 Birds	
(wk)	(lb)	(kg)	(lb)	(kg)	(lb)	(kg)
1	25.7	11.7	180	82		
2	35.7	16.2	250	114	430	195
3	51.4	23.4	360	164	790	359
4	71.4	32.5	500	227	1,290	586
5	78.6	35.7	550	250	1,840	836
6	85.7	39.0	600	273	2,440	1,109
7	97.1	44.1	680	309	3,120	1,418
8	108.6	49.4	760	345	3,880	1,764
9	117.1	53.2	820	373	4,700	2,136
10	125.7	57.1	880	400	5,580	2,536
11	135.7	61.7	950	432	6,530	2,968
12	150.0	68.2	1,050	477	7,580	3,445
13	158.5	72.0	1,110	505	8,690	3,950
14	162.8	74.0	1,140	518	9,830	4,468
15	171.4	77.9	1,200	545	11,030	5,014
16	171.4	77.9	1,200	545	12,230	5,559
17	172.8	78.5	1,210	550	13,440	6,109
18	174.3	79.2	1,220	555	14,660	6,664
19	175.7	79.9	1,230	559	15,890	7,223
20	182.8	83.1	1,280	582	17,170	7,805

[1]Adapted by the author from *Producing Commercial Eggs*, University of Arkansas Leaflet 188 (Rev.).

TABLE 12–5
FEED CONSUMPTION OF LAYING CHICKENS[1]

Eggs per 100 Hens per Day[2]	Feed for 4-lb Hens (1.8 kg)		Feed for 5-lb Hens (2 kg)		Feed for 6-lb Hens (2.7 kg)		Feed for 7-lb Hens (3.2 kg)	
	Per Day	Per Dozen Eggs	Per Day	Per Dozen Eggs	Per Day	Per Dozen Eggs	Per Day	Per Dozen Eggs
0	15.3	—	17.7	—	19.9	—	22.0	—
10	16.7	20.0	19.0	22.8	21.3	25.6	23.9	28.7
20	18.1	10.9	20.4	12.2	22.7	13.6	25.8	15.4
30	19.5	7.8	21.8	8.7	24.1	9.6	26.7	10.3
40	20.9	6.3	23.3	7.0	25.6	7.7	27.6	8.3
50	22.3	5.4	24.7	5.9	26.9	6.4	29.0	6.9
60	23.7	4.7	26.1	5.2	28.3	5.6	30.4	6.1
70	25.1	4.3	27.4	4.7	29.7	5.0	31.8	5.4
80	26.6	4.0	28.9	4.3	31.2	4.7	33.3	5.0
90	27.9	3.7	30.3	4.0	32.6	4.3	34.6	4.6
100	29.3	3.5	31.7	3.8	34.0	4.1	36.0	4.3

[1]Adapted by the author from *Producing Commercial Eggs*, University of Arkansas Leaflet 188 (Rev.).

[2]Two-oz eggs are assumed.

PHASE FEEDING

The trend is to phase feeding laying hens. Phase feeding refers to changes in the laying hen's diet (1) to adjust to age and state of production of the hen, (2) to adjust for season of the year and for temperature and climatic changes, (3) to account for differences in body weight and nutrient requirements of different strains of birds, and (4) to adjust one or more nutrients as other nutrients are changed for economic or availability reasons. Research has shown, for example, that a hen laying at the rate of 60% has different nutritional requirements than one laying at the rate of 80%; hens have different requirements in summer and in winter; a 24-week-old layer has different needs than one 54 weeks old. The main objective, therefore, of phase feeding is to reduce the waste of nutrients caused by feeding more than a bird actually needs under different sets of conditions. In this way, feed efficiency can be improved and the cost of producing a dozen eggs reduced.

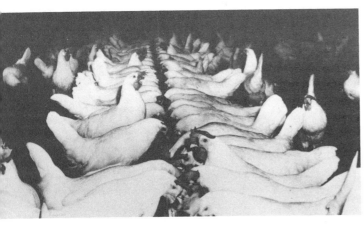

Fig. 12–13. Feeding layers went ultramodern with the development of *phase feeding*, which is the practice of changing ration formulations to meet changing nutritional requirements due to such factors as rate of lay, environmental temperature, etc. (Courtesy, DEKALB AgResearch, Inc., DeKalb, IL)

A phase feeding program for laying hens generally calls for use of a rather high-protein feed (usually 17 to 18%) from the onset of egg production through the peak production period. Thereafter, a lower level of protein (about 16%) is fed for the next 5 or 6 months, followed by still lower levels (usually 15%) until the laying period is completed. This general plan takes age into consideration, but for greatest benefits other factors will also need to be considered.

Although phase feeding has its advantages, it does present some problems: It is a complicated procedure, it necessitates a knowledgeable poultry producer, and it requires more bulk bins, closer check on feed deliveries, etc.

Phase feeding is practiced widely in commercial operations. Although it does not promise to bring about large increases in egg production, it can help production reach a higher peak and sustain it longer if other conditions are right. Most of all, phase feeding offers a good potential for lowering costs and increasing income. Like many other developments, it favors the larger operator.

FLOCK HEALTH

Good livability and health are requisites for layer profits. Deaths, medications, and condemnations make for tremendous losses. But even greater economic losses may result from decreased egg production.

More than 1% mortality per month is excessive. Also, any abnormal drop in egg production or in daily feed consumption indicates the need for a quick check of flock health and corrective treatment.

All chickens are susceptible to diseases, most of which result from a combination of two causes: (1) stress, and (2) infections. Stress factors — including such things as lack of sanitation, faulty nutrition, and poor environment — increase the chance of disease occurring and intensify any disease outbreak. Infections may occur in a single individual or be widespread, depending upon the infectious agent involved, the resistance of the flock, and the environment provided.

LAYER HEALTH PROGRAM

There is no such thing as a standard health program for all layer farms. Nevertheless, the following points are basic to any disease prevention program:

1. Purchase healthy, disease-free replacement stock (day-old chicks or ready-to-lay replacement pullets).
2. Minimize stress, including such things as lack of sanitation, faulty nutrition, and poor environment.
3. Develop and follow a precise vaccination program.
4. Keep visitors away from poultry buildings.
5. Do not mix birds of different ages.
6. Remove sick, injured, and dead birds as soon as noticed, and burn or bury dead birds immediately.
7. Screen or cover all poultry house openings so wild birds and rodents cannot enter.
8. Clean trough-type waterers daily. Check operation of individual cups and direct action valves or nipples. At the same time, check operation of feeders, fans, and lights.
9. Remove wet litter spots as they occur.
10. Never permit contaminated equipment (crates, tools, trucks, etc.) from other poultry farms in the buildings.
11. Obtain reliable diagnosis before administering drugs or biologics.
12. Check birds regularly for internal and external parasites. Treat promptly if parasites are found.
13. Maintain good records relative to flock health, including vaccination history, disease problems, and medication employed.

KEYS TO DISEASE PREVENTION

All the points of the above "Layer Health Program" are important, but the following five are keys in any disease prevention program; hence, they are elaborated upon in the sections that follow:

1. Sanitation.
2. Good nutrition.
3. Good environment.
4. A vaccination program.
5. A parasite control program.

SANITATION

There is no substitute for sanitation. This includes everything the birds or eggs come in contact with, either directly or indirectly.

Clean, disinfect, and air-out a poultry house before putting birds of any age in it. Disinfecting is not a substitute for cleaning. Disinfectants are only effective on clean surfaces. When preparing a previously occupied poultry house, follow these steps:

1. Remove dust, dirt, crusted manure, litter, feed, etc.

2. Clean the building and all equipment thoroughly, including air intakes, overhead ledges, and fans.

3. Disinfect; select the proper disinfectant and follow the manufacturer's directions. All disinfected surfaces must be dry and the building aired out before birds are put in.

Clean automatic waterers daily. Slime and decomposing feed can harbor disease organisms and molds.

GOOD NUTRITION

Nutrition plays a major role in the health of the flock and in its ability to resist disease. The nutritional needs of a healthy flock are normally satisfied by an adequate supply of clean fresh feed, easily accessible, and containing the necessary amounts of required ingredients.

Tonics, appetizers, and stimulants are not necessary. They should be used only under the direction of a qualified poultry specialist.

GOOD ENVIRONMENT

Environmental conditions contribute directly to flock health. A poor environment makes birds more susceptible to disease and contributes to the spread of many diseases.

Adequately lighted, well-ventilated, relatively dry quarters are desirable. A good environment provides:

1. Shelter, protection, and comfort.
2. Convenient and adequate supplies of proper feed, clean water, and fresh air.
3. Equipment and facilities that are conveniently arranged for both birds and caretaker.

Waterers should be easily reached by the birds, should be of adequate capacity, and should not overflow. Feeders which waste feed or become fouled aid the spread of disease.

If litter is used, it should be absorbent, relatively dust-free, and resistant to matting. Litter under waterers and feed troughs should be replaced when it becomes damp or matted. Moisture in litter fosters parasite development and growth of molds and bacteria.

With cage systems, the birds are unable to move to a more acceptable or healthy location; the manager must control the environment uniformly to get satisfactory performance.

A VACCINATION PROGRAM

Most of the serious diseases to which layers are susceptible can be prevented by vaccination of pullets during the growing period and by proper management of the birds during both the growing and laying periods. Diseases for which effective vaccines are available and may be used are Marek's disease, Newcastle, bronchitis, pox, epidemic tremors, and laryngotracheitis. (CAUTION: Laryngotracheitis vaccine should be used only in areas where the disease has been a problem.) While the birds are in the laying house, they can be given booster vaccinations with Newcastle and bronchitis vaccines every 3 months.

Advance planning of a vaccination program is important because immunity is not effective immediately, and because immunities developed by vaccinations may wear out in time—requiring a booster shot to raise protection to a satisfactory level. When planning a vaccination program, consideration should be given to the disease history of the farm, the community, and the state or region; and only healthy birds should be vaccinated. The vaccination program should be developed in consultation with a knowledgeable poultry specialist; and vaccines and drugs should be administered in keeping with the manufacturer's directions.

A PARASITE CONTROL PROGRAM

A wide variety of external and internal parasites attack poultry. The prevention and control of these parasites is one of the quickest, cheapest, and most dependable methods of increasing egg production with no extra birds, no additional feed, and little more labor.

All flocks should be checked periodically for the presence of worms. Worms are not usually a problem in flocks grown and maintained on slat or wire floors.

Lice and mites, of which there are several kinds, are the most common external parasites of layers. Heavy infestations cause irritation, discomfort, and general unthriftiness, and reduce egg production as much as 25%, or more.

Lice and mite control is simple if the right insecticide is used properly. Information about what insecticides are available and approved for a particular area should be obtained from the county agricultural agent, local veterinarian, or poultry specialist, and used according to the manufacturer's label.

Pertinent information relative to the common external and internal parasites of poultry is presented in Chapter 10, Poultry Health, Disease Prevention, and Parasite Control, in Table 10–3, Parasites of Poultry; hence, the reader is referred thereto.

STARTER PULLET MANAGEMENT

Well-bred and well-grown, healthy, vigorous started pullets are requisite to success in the laying house. Proper feeding, care, and management during the pullet starting period can develop the potential productive egg capacity of the birds, whereas poor practices and management may reduce or obliterate the genetic potential that has been bred

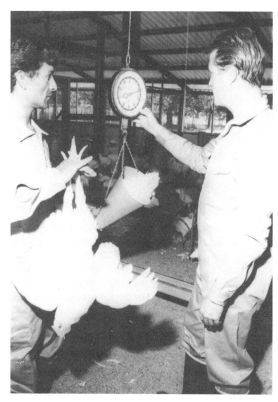

Fig. 12–14. Weighing pullets to monitor growth rate. (Courtesy, Arbor Acres Farm, Inc., Glastonbury, CT)

into them. The following management pointers in raising started pullets are recommended:

1. **Sanitation.** A strict sanitation program is a must in raising started pullets. Recommended rules are (a) do not mix ages or strains of birds in the same house; (b) do not allow visitors in the house; (c) screen out birds and control rats, mice, and other rodents; (d) do not permit contaminated equipment to come on the farm; (e) use incinerator, disposal pit, or plastic bags (which are sealed and removed from the premises) for dead birds; (f) keep birds free from external and internal parasites; and (g) post all dead birds.

2. **Vaccination.** The vaccination program should be in keeping with the needs of the area. Also, it should be completed far enough ahead of moving the pullets so that they will not be going through a vaccination reaction during or soon after moving.

3. **Debeaking and dubbing.** Pullets should be debeaked at 9 to 16 weeks of age, perhaps at the same time that they are vaccinated for fowl pox. This will make for a permanent job.

Dubbing (removal of the comb) should be done when the birds are a day old. Dubbing alleviates a source of injury, especially to caged birds, and results in 2 to 4% more eggs.

4. **Lights.** The lighting system used will depend on the type of building and the season of the hatch. It should be designed to avoid an increasing photo-period during the last one-half to three-fourths of the growing period.

5. **Moving the pullets.** The pullets should be moved in such manner that there will be a minimum of stress. This calls for clean (preferably steamed) coops and truck, not moving in inclement weather, providing fresh air but avoiding drafts during the move, and gentle handling.

LAYER MANAGEMENT

Management is decision making. Many of the decisions egg producers make are only made once; hence, they had better be right. Among the layer management decisions that must be made are:

1. Whether to go into egg production, and where.
2. Method of financing; contract vs independent production.
3. The breed and breeding; and whether to buy or raise replacements.
4. Type and size of house and equipment.
5. The ration to use; and whether to buy commercial feed or home-mix.
6. The flock health program.
7. How and where to market eggs.
8. What is ahead.

Additionally, the able manager sets high production goals, in recognition that high goals spur high achievement; follows certain guides; performs certain practices — known as *management practices*, routinely and on time; and is quick to sense when all is not well in the hen house — and to rectify the situation.

PRODUCTION GOALS

Producers of commercial eggs strive for the maximum of egg production at a minimum cost — yet, they recognize that net returns are more important than costs as such.

The largest item of cost in the production of eggs is feed. It normally constitutes 50 to 70% of the total cost; though, in exceptional cases, it may run as low as 45% or as high as 75%. Labor costs and cost of replacement pullets are the other two major cost items in production.

Generally speaking, higher egg production means lower costs per dozen eggs. This is so because the feed required for maintenance is constant for hens of any given weight and bears no relation to the number of eggs laid. Hence, it is important that commercial producers strive for high egg production.

The following production goals are suggested for the commercial egg producer:

1. Production of 270 eggs per hen per year.
2. Feed conversion of less than 3.6 lb of feed per dozen eggs.
3. A laying house mortality of less than 10%.
4. Seventy-five percent or more extra large and large Grade A eggs.
5. Ninety-five percent or more marketable eggs.
6. On-farm egg breakage under 2%.
7. Mortality of less than 5% from 1 day old to 5 months.

MANAGEMENT GUIDES

Bigness alone does not assure success and profit in commercial egg production. Rather, it makes it imperative that there be superior management. Among the management practices and guides followed by successful commercial egg producers are the following:

1. An average of 20 eggs per bird per month is a good standard for high-producing strains.

2. Mortality of about 1% per month for layers is normal.

3. Lights should be provided so as to make for about a 16-hour working day for the hens. Lights may also be used to help bring slow-maturing pullets into lay.

4. Culling should be done to remove diseased or low-vitality birds. With high-producing strains, little, if any, culling may be required. Accurate records should be kept on percentage of lay, feed intake, mortality, and other pertinent facts.

5. Eggs should be gathered 3 to 5 times daily, cleaned immediately, and stored at 55°F and 75 to 80% relative humidity.

6. Where the cage system is followed, it is important that there be a regular spray schedule, along with proper dropping management to control flies, that birds be debeaked if multiple caging is used, and that cages and waterers be cleaned once each week.

7. Ventilation should provide fresh air and remove moisture, without producing drafts. The condition of the litter and the amount of fumes (ammonia fumes) are good indicators of the adequacy of ventilation.

BEHAVIORAL PROBLEMS

With the restriction, or confinement, of flocks, many abnormal behaviors evolved to plague those who raise them, including cannibalism and egg eating.

CANNIBALISM

Many types of cannibalism occur in domestic fowl and game birds reared in captivity; among them, the following:

1. **Vent picking.** Picking of the vent, or the area below the vent, is the severest form of cannibalism. This type is generally seen in pullet flocks in high production. Predisposing factors are prolapsus or tearing of the tissue caused by passage of a very large egg.

2. **Toe picking.** Toe picking is most commonly seen in chicks. It may be brought on by hunger.

3. **Head picking.** Head picking usually follows injuries to the comb or wattle caused by freezing or by fighting between males.

EGG EATING

Egg eating is a costly vice. If one bird acquires the habit, it usually spreads quickly throughout the flock.

Egg eating is predisposed by factors favoring egg breakage, including insufficient nests, insufficient nesting material, not collecting eggs frequently enough, and soft-shelled or thin-shelled eggs. Prevention consists in alleviating these conditions.

Once the egg-eating habit has started, it is very difficult to stop. If the birds have not been debeaked, this should be done immediately. Also, nests should be darkened and eggs should be collected frequently.

MANAGEMENT PRACTICES

Even though considered routine, and even mundane, certain management practices are requisite to success in egg production; among them, culling, dubbing, debeaking, lighting, nesting, broodiness, and manure handling.

CULLING

The culling of commercial layers has changed with improvements in breeding. The present practice is to (1) discard the weak, crippled, and unhealthy pullets at the time of housing; (2) remove only the unthrifty and unhealthy birds during the first 7 or 8 months of production; and (3) cull later in the laying year, removing unhealthy birds, early and slow molters, and those with excessive fat deposits in the abdomen.

DUBBING

Dubbing is the removing of the comb and, in some instances, the wattles from chickens. It may be done to either males or females. Usually, dubbing is done when chicks are a day old, using curved manicure scissors. At this age very little bleeding or discomfort is noted.

The reasons for dubbing are:

1. It reduces the possibility of injury to the large tender comb surface.

2. It allows birds with large combs to eat and drink more easily from some types of feeders and waterers.

3. It prevents some birds with large combs from becoming frightened due to sight restriction from one side.

The comb evolved as a secondary sex characteristic functioning in temperature reduction (heat exchange) during periods of egg production.

Dubbing does create some stress; hence, it should be done only if it serves a definite need or purpose.

DEBEAKING

Debeaking is the removing of a portion of the upper (and often a lesser portion of the lower) beak of the fowl. It is the most effective way to control cannibalism.

Debeaking can be done at anytime from day old on. It is done with an electrically heated blade which both cuts and cauterizes the beak to prevent bleeding. The amount of beak removed and the heat of the blade determine the speed of regrowth. Chicks debeaked moderately (about one-fourth of the beak) at 6 to 10 days of age will usually not need further attention for about 1 year. Observe the flock closely and re-debeak any or all when regrowth makes it advisable. Remember that debeaked birds can no longer pick the last particles of feed from the trough or floor. Therefore, maintain at least $\frac{3}{8}$ in. of feed and water at all times for debeaked birds—more if the birds were severely debeaked.

LIGHTING

All birds, including laying hens, are sensitive to light. They usually seek or build nests, mate, and lay eggs as day length increases (spring). They cease egg production, molt, and regrow feathers as day length decreases (fall). Poultry producers have taken advantage of this phenomenon and, with electric lights, given the laying hen a condition of perpetual spring. The result is year-round egg production.

Windowless houses make exact control of day length a simple matter. Operators with windowed houses can get the same effect if they consider the present day length and what it will be at the various times in the bird's life, then make adjustments accordingly.

Many pullets are grown under short or decreasing daylight to 21 weeks of age. From this point, light increases of 15 minutes per week give the effect of increasing day length for much of the hen's productive life. It is doubtful if any stimulating effect can be obtained from more than 17 hours of light per day.

(Also see Chapter 8, Poultry Management, Table 8-3, Recommended Lighting Systems for Poultry.)

NESTING

When housed pullets come into production, encourage nesting by making frequent trips through the house. On each trip, pick up the floor eggs and place them in a nest. Level off the floor nests birds have made in the litter. Persistence will pay off — pretty soon the pullets will get the "idea."

Place nest pads in wire floor nests for the first 3 months or until the flock has reached peak production. Arrange nests in the pen convenient to the birds, and with consideration given to established social orders.

BROODINESS

Broodiness has been largely bred out of laying strains by selective breeding. In recent years, strain crossing has caused some of it to return. It is rarely seen in hens in cages. Some broodiness may be observed in floor managed flocks. Broodies should be removed as soon as observed on the nest, and placed in a slat or wire floored broody coop. Feed them a complete ration. They should be ready to return to the laying pen in five to seven days.

MANURE HANDLING

One of the major problems on a large egg farm is disposal of the manure. The best solution is to spread the manure on fields as fertilizer, although the possibility of using manure as a feed for ruminants should be considered. Some producers have attempted to handle the manure as a liquid. However, the odor and extra volume of materials (2 to 3 times more volume) to be handled tend to eliminate this method as a practical possibility.

In-house drying offers a possibility for those producers who encounter odor problems with neighbors, and for those who have limited land on which to spread manure. Manure that is fresh or that has been dried to 30 to 50% moisture has little odor. A combination of daily removal, hauling, and spreading is a good disposal method. Daily cleaning and spreading in the summer, storage in winter, and early spring clean out can be a good alternative.

(Also see Chapter 8, Poultry Management, section on "Poultry Manure.")

PRODUCING QUALITY EGGS

The quality of eggs produced by today's commercial egg-laying flocks is becoming increasingly superior through improved breeding, feeding, management, and marketing. However, eggs are a perishable product, and they should be handled as such. It is of utmost importance, therefore, that producers, jobbers, and retailers maintain superior quality.

Consumers want eggs with fresh-laid appearance, good flavor, and high nutritive value. The shells should be strong, regular, and clean; the white (or albumen) should be thick, clear, and firm; the yolk should be light-colored, well-centered, and free from blood and meat spots.

Contrary to popular belief, not all eggs gathered from the nest are necessarily of first-rate quality. However, a large percentage of top-quality eggs can be produced by adopting the following practices:

1. **Selecting a strain of birds noted for its ability to lay eggs of high quality.** Inherent capacity for producing high initial egg quality is important. Egg shape, shell color, shell strength, albumen quality, and incidence of blood and meat spots are quality factors which can be improved through selective breeding. Most breeders are giving considerable attention to this possibility. Data from Random Sample Tests are highly useful for this purpose.

Fig. 12–15. Shell thickness is an important factor considered by breeders. Thickness is correlated with shell strength. (Courtesy, DEKALB AgResearch, Inc., DeKalb, IL)

2. **Feeding well-balanced rations.** Deficiencies of calcium, phosphorus, manganese, and vitamin D_3 lead to poor shell quality. Yolk color is almost entirely dependent on the bird's diet. Low vitamin A levels may increase the incidence of blood spots.

3. **Keeping the flock disease-free.** Certain diseases, especially Newcastle and infectious bronchitis, often cause

birds returning to production to lay eggs of poor shape, poor shell quality, and low interior quality.

4. **Replacing birds in the laying flock when they are 18 to 20 months old.** The finest quality eggs are laid by pullets. Older hens lay eggs lacking in acceptable shell quality and albumen firmness.

5. **Producing infertile market eggs.** Keep males out of the laying flock.

HANDLING TABLE EGGS

Fig. 12–16. Properly handled table eggs—uniform, small ends down, in egg flat. (Courtesy, Morton & Associates, Gainesville, GA)

Observance of the following rules will aid in maintaining top quality all the way to the consumer:

1. **Gather eggs frequently.** Eggs should be gathered 3 to 4 times daily. Frequent gatherings reduce the amount of body heat to which eggs are exposed. Also, the number of broken and cracked eggs will be reduced.

2. **Produce clean eggs.** Eggs are usually cleanest when they are laid. If plenty of nonstaining, dry, nesting material is provided and changed when needed, and eggs are gathered three time a day, very few eggs will become dirty or stained.

3. **Clean soiled eggs.** To avoid excessive handling, clean all eggs rather than take time to separate the dirty eggs. The essential items for properly cleaning shells include water, detergent-sanitizer, and an egg-washing machine. Use clean water for washing eggs and maintain the water temperature between 110° and 120°F. Add detergent-sanitizer to water according to manufacturer's directions.

4. **Coat shell.** To preserve the high initial quality, coat the eggshell with a thin, odorless, colorless, and tasteless mineral oil to prevent the egg's carbon dioxide from escaping into the surrounding atmosphere. Apply oil, either automatically with in-line egg processing facilities or with a hand-type aerosol spray.

5. **Cool eggs properly.** Cooling eggs immediately after gathering removes the animal heat and retards any reaction which might be conducive to deterioration of quality. The egg cooler should be large enough to accommodate the daily production plus eggs held until they are marketed. An egg cooler operating at 55°F or lower, with a relative humidity of 75 to 80%, is considered adequate.

Eggs cooled prior to packing will sweat if removed from the cooler for candling or sizing. Therefore, eggs should be processed in coolers or in an adjoining air-conditioned room.

6. **Candle eggs.** Candling is the most practical way to determine interior quality of shell eggs. The object of candling is to discover and cull out eggs with blood spots and checks.

The shell should be sound and free from checks or cracks, and it should be of good texture.

7. **Separate eggs into weight classes.** After candling, eggs should be separated into weight classes. In large commercial operations, this is done by conveying the eggs over a series of scales to obtain uniform weight for a given pack. When so classified, eggs have eye appeal. Weight classes of eggs based on minimum net weight per dozen are given in Chapter 16, Table 16-7.

8. **Pack eggs properly.** Eggs should be packed with small ends down. It is possible for the air cell to break loose and move to the small end when the large end of the egg is packed down.

It is also important that cartons and cases be kept clean. This prevents the possibility of mold formation which may pass off-flavors to the eggs.

9. **Make frequent deliveries.** Eggs should be moved to market as quickly as possible — at least twice a week. Also, it is essential to keep all eggs cool en route to market.

After eggs are delivered to jobbers, or retailers, maintenance of quality is their responsibility. Generally speaking, they have adequate facilities to protect the high-quality product that the producer has delivered to them.

Refrigeration is a must for both short-time holding and displaying of eggs.

HANDLING HATCHING EGGS

The fertile egg is usually in a fairly advanced stage of development from an early embryological standpoint at the time of laying. Accordingly, ideal handling would consist of setting the egg at once, so that development could proceed without being checked. Obviously, this is not practical under most conditions, for hatching eggs must be held for varying lengths of time. The practical problem, therefore, is to hold these eggs in a suspended state of development without destroying the developing embryo. To this end, certain handling practices are essential. These are detailed in Chapter 3, Incubation and Brooding, under the heading "Handling Hatching Eggs."

MARKETING EGGS

Some eggs are marketed through processors who pick them up at the farm. They grade the eggs for size and interior quality and pay producers according to the quality and size. However, large producers are grading and packaging eggs on their establishments. An increasing volume of eggs will be processed at the production facility due to the lower costs involved.

Some smaller producers process and market their own

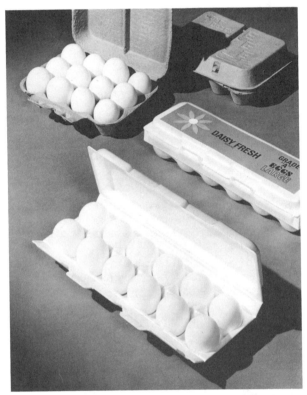

Fig. 12–17. Marketing eggs gives point and purpose to all that has gone before. (Courtesy, Hy-Line International, Division of Hy-Line Indian River Company, Johnston, IA)

eggs. But problems may exist in self-marketing programs. Many producers are not critical enough of their own eggs to package a really fine pack of eggs. Producers usually are not highly effective sales people nor good bill collectors. Frequently, they are not good labor supervisors in that they would rather do the work themselves than tell someone else how to do it.

Producers interested in marketing their own eggs should contact their State Department of Agriculture to learn the requirements, if any, for marketing eggs. The regulations are not difficult to meet.

Processing and marketing are time consuming, and costly, too. A representative cost of processing and marketing eggs would total approximately 15¢ a dozen, with a breakdown about as follows:

Farm pick-up cost	$ .01
Processing costs	.04
Cartons	.055
Delivery cost	.03
Administration	.015
Total	$ .15

An important, but often overlooked, cost is the loss of eggs by breakage between the time they are produced and sold. Cost should be figured on dozens sold, not dozens produced.

Egg prices will vary considerably at the different levels of marketing because of the various services performed from one level to another.

(Also see Chapter 16, Marketing Poultry and Eggs, section on "Marketing Eggs.")

WHAT'S AHEAD FOR LAYERS

Science and technology have made for great changes in the egg industry in recent years. Still further advances lie ahead.

1. **Concentrated areas.** The egg industry will be concentrated in areas with abundant grain production and major markets; cheap labor will no longer be of significance as a cost factor.

2. **Energy costs.** Energy costs will be higher in the decades to come, with the result that those areas having the lowest energy cost and/or requiring the least energy will be in favored positions from the standpoints of environmentally controlled buildings and automation.

3. **Pollution control.** Pollution control (due primarily to manure and dead bird disposal, odor, and flies) will be a critical factor in the site selection for an operation of large layer units in the years ahead, favoring remoteness from urban development and ease of compliance with state and federal regulations.

4. **Manure management will evolve.** Manure will no longer be an unwanted source of pollution. Rather, it will become a wanted product for fertilizer, feed, and other uses.

5. **Production units will be larger and more commercial.** Some flocks of layers now number 1,000,000; and units of 100,000 or more are not uncommon. There will be further consolidation and merger of existing operations.

6. **More integration will come.** In particular, it appears that large and well-financed commercial feed companies, processors, co-ops, and others will be doing more integrating, through ownership or contract in the future.

7. **More specialization will evolve.** An increasing number of egg producers will specialize in just one kind or phase of production only; for example, in the production of either started pullets or egg production. Fewer and fewer of them will diversify in any way whatsoever.

8. **Laborsaving devices and mechanization will increase.** Higher priced labor, along with more sophisticated equipment, will make for increased mechanization all the way along the line, from production of eggs through processing and marketing.

Fig. 12–18. Modern egg handling is highly automated. The operation pictured above shows automatic egg grading, sizing, and cartoning. (Courtesy, DEKALB AgResearch, Inc., DeKalb, IL)

9. **Improved housing and environmental control will come.** Physics, engineering, and physiology will be combined in such a way as to bring about improved houses and laborsaving equipment.

10. **Cages will increase.** Most layers are already in cages. More and more pullets will be reared in cages from day old.

11. **Bird density will increase.** With improved housing and more cages, along with better environmental control and mechanization, bird density will be increased.

12. **Egg production per layer will increase.** Some hens among the better strains now lay more than 300 eggs per year. There is increasing evidence that this trait (egg production) may have plateaued in high-producing strains. If this is true, perhaps the time has come when breeders of egg producers should pay more attention to body size and egg size. Without doubt most of the advances in egg production in the future will come in raising the level of the lower producers.

13. **Livability will increase.** Drugs and vaccines have helped increase livability, and they will be used in the future. However, it is expected that breeding stock will be selected for greater livability.

14. **"Mini" layers.** With further perfection, "mini" layers may provide a practical means of concentrating layers in less space and achieving greater feed efficiency.

15. **Quality of products will be improved.** No other country in the world produces as high-quality eggs as the United States. Yet, further improvements can, and will, be made.

16. **Production geared to consumption will improve.** Without doubt, the greatest problem facing the poultry producer today is that of overproducing at intervals. As a result, from time to time, the market is flooded and prices plummet, thereby making for severe losses. Without doubt, some solution—through marketing orders or quotas, or as a result of the controls exerted by fewer and larger owners—will evolve.

17. **Marketing costs will be lowered and efficiency will increase.** Lower marketing costs and greater efficiency will be achieved largely through increased grading and packaging on the farm and direct marketing from farm egg holding facilities to supermarkets and other large users.

18. **Egg consumption will increase.** There will be a gradual increase in per capita egg consumption as medical opinions and researchers allay consumer fears of cholesterol and heart failure from eating eggs.

19. **New products from spent layers will evolve.** New further processed products, made from spent chicken layers, will increase returns. Among such canned products being test marketed are: whole chicken, boned chicken, chicken and noodles, chicken à la king, chicken breasts and gravy, spaghetti sauce with chicken, and barbecued chicken.

QUESTIONS FOR STUDY AND DISCUSSION

1. Present a profile of the U.S. commercial egg industry in support of the assertion that it is big and important business.

2. What forces have caused more and more egg producers to enter into contracts of some sort? Is this trend good or bad?

3. What kind of records should a commercial egg producer keep?

4. How can profits in egg production be increased by (a) lowering costs, and/or (b) increasing returns?

5. What bases should a beginning commercial egg producer use in arriving at the choice of breed and breeding?

6. Will more or fewer egg producers grow their own started pullets in the future? Justify your answer.

7. List five searching questions to which a purchaser of day-old chicks should first obtain answers.

8. What factors favor a commercial egg producer keeping layers through one or two molts, rather than all-pullet flocks?

9. Why are cull (spent) hens worth so little on the market?

10. Describe in detail one method of molting a flock of layers.

11. List four housing systems for egg production, and give the advantages and disadvantages of each.

12. Describe how the housing and equipment of pullets differ from the facilities of layers.

13. What factors should be considered by a commercial egg producer in (a) determining whether to use commercial feed or home-mix, and (b) choosing between all mash, grain and mash, or cafeteria feeding programs?

14. List the major factors that influence the feed consumption of layers.

15. Define *phase feeding*. Why is phase feeding practiced widely in commercial layer operations?

16. List the two major causes of chicken diseases.

17. List the basic points that should be incorporated in any layer disease prevention program.

18. How should an egg producer determine (a) what wormer to use for roundworms, and what insecticide to use for lice; and (b) how to use each of them?

19. What are the major differences between started pullet management and layer management?

20. Discuss the importance and timing of each of the following management practices: (a) culling, (b) dubbing, (c) debeaking, (d) lighting, (e) nesting, (f) broodiness, and (g) manure handling.

21. In what ways may management give an assist in producing (a) quality market eggs, and (b) quality hatching eggs?

22. How are most commercial eggs marketed today?

23. What changes ahead do you see for layers?

SELECTED REFERENCES

Title of Publication	Author(s)	Publisher
Commercial Chicken Production Manual, Fourth Edition	M. O. North D. D. Bell	Van Nostrand Reinhold Co., New York, NY, 1990
Feeds & Nutrition, Second Edition	M. E. Ensminger J. E. Oldfield W. W. Heinemann	The Ensminger Publishing Company, Clovis, CA, 1990
Feeds & Nutrition Digest	M. E. Ensminger J. E. Oldfield W. W. Heinemann	The Ensminger Publishing Company, Clovis, CA, 1990
Poultry: Feeds & Nutrition, Second Edition	H. Patrick P. J. Schaible	The AVI Publishing Company, Inc., New York, NY, 1980
Poultry Husbandry I	C. J. Price	Food and Agriculture Organization of the United Nations, Rome, Italy, 1969
Poultry Husbandry II	C. J. Price J. E. Reed	Food and Agriculture Organization of the United Nations, Rome, Italy, 1971
Poultry Keeping in Tropical Areas	W. Thomann	Food and Agriculture Organization of the United Nations, Rome, Italy, 1968
Poultry Meat and Egg Production	C. R. Parkhurst G. J. Mountney	Van Nostrand Reinhold Co., New York, NY, 1988
Poultry Production, Twelfth Edition	M. C. Nesheim R. E. Austic L. E. Card	Lea & Febiger, Philadelphia, PA, 1979
Poultry Products Technology, Second Edition	G. J. Mountney	The AVI Publishing Company, Inc., Westport, CT, 1976
Poultry Science and Practice, Fifth Edition	A. R. Winter E. M. Funk	J. B. Lippincott Company, Chicago, IL, 1960

Fig. 13–1. Broiler. (Courtesy, Hubbard Farms, Walpole, NH)

BROILERS (FRYERS), ROASTERS, AND CAPONS

Contents

Fig. 13–2. Broilers are adapted to contract production and assembly line techniques. As a result, it has become a highly integrated, efficient industry, concentrated in relatively few areas. (Courtesy, National Broiler Council, Washington, DC)

A broiler is a young chicken (usually 6 to 8 weeks of age), of either sex, that weighs 3 to 5 lb, that is tender meated, and that has a soft, pliable, smooth-textured skin, and a flexible breastbone cartilage. Formerly, the term *fryer* implied a larger bird than a broiler; today, the terms *broilers, fryers,* and *young chickens* are interchangeable.

Roasters are older and heavier than broilers. At market time, straight-run flocks (both sexes) range from 9 to 11 weeks of age and weigh 6 to 8 lb.

Capons, which are surgically unsexed, are older and heavier than roasters—they range up to 20 to 24 weeks of age and weigh 12 to 14 lb.

Because much of the management of broilers, roasters, and capons is similar, and because of the dominant position of broilers, most of the discussion that follows is devoted to broilers. However, at the end of the chapter there are separate sections pertaining to roasters and capons, with the differences peculiar to each set forth therein.

Commercial broiler production was started in 1923, on the Delmarva Peninsula, by Mrs. Wilmer Steele of Ocean View, Sussex County, Delaware. Today, it is a highly specialized, complex, competitive business, characterized by a phenomenal growth. The total United States output increased from 34 million birds in 1934 to 5.5 billion in 1989; and per capita consumption increased from half a pound in 1934 to 65 lb in 1989. In 1989, broilers accounted for over 8.8 billion

dollars of the 15.3 billion dollars total annual income from poultry and eggs in the United States.

Much of the increased broiler production can be attributed to contract production and rapid advances made in breeding, nutrition, disease control, management, processing, and marketing. No other segment of agriculture is as well suited to assembly-line techniques as broiler production. As a result, since the thirties, broiler production has gradually changed from an industry characterized by small, independent farm flocks and small processors scattered across the country to a highly integrated, efficient industry, concentrated in relatively few areas.

A profile of the U.S. broiler industry of 1989 (or in the year indicated) follows:

■ **Production time**—Improvement in broiler strains, along with improved technology and production practices, have reduced the time needed to produce a broiler. In 1955, it took 12 to 14 weeks to produce a 3.5 lb bird; today, it takes only 6 to 8 weeks to produce a 3 to 5 lb bird.

■ **Feed efficiency**—The feed required to produce a pound of live bird has been reduced from 4.7 lb in 1940 to 1.9 lb in 1990.

■ **Number and pounds produced**—In 1989, a total of 5.5 billion broilers, averaging 4.3 lb per bird, produced 24 billion pounds of ready-to-cook meat.

■ **Price and income**—Farmers averaged 36.6¢ per pound liveweight for broilers; for a gross broiler income of 8.8 billion dollars, which represented 5.5% of the U.S. farm income.

■ **Per capita consumption**—The per capita consumption of chicken (broilers and other chickens) on a ready-to-cook basis totaled 68.9 lb.

■ **Broilers dominate chicken consumption**—Broiler meat as a share of total chicken consumption increased from 3.7% in 1934 to 97% in 1989.

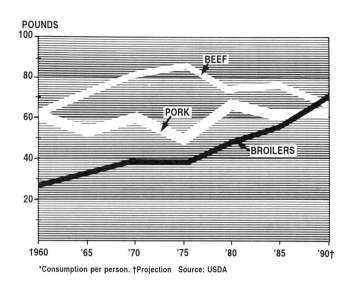

Fig. 13–3. Broilers changed the order of meat consumption. (Source: USDA)

■ **Top ten states** — The South dominates broiler production. The ten leading states, in order of numbers, are: Arkansas, Georgia, Alabama, North Carolina, Mississippi, Texas, Delaware, Maryland, California, and Virginia. In recent years, these states have accounted for about 83% of the total U.S. production.

■ **Fewer but larger broiler producers** — Output of broiler per farm increased from 33,600 birds in 1959 to 157,785 birds in 1989. One farm family, devoting full time to broiler production, can care for an average 5 to 6 broods per year. However, many broiler producers have only 1 to 2 houses; hence, their returns from contract broiler production are considered supplementary to their full-time job elsewhere.

■ **Capital requirements** — Considerable capital is needed to start a commercial broiler operation today. Housing and equipment require a minimum investment of about $3.75 per bird capacity; that is $300,000 for buildings to accommodate an 80,000 broiler operation. Also, financing is needed for land and for housing for the manager.

■ **Grower fees** — Broiler grower fees for labor and facilities range from 3¢ to 5¢ per pound of broiler marketed.

■ **Integrated** — Approximately 95% of all broilers are produced under contract, and integrator-owned farms produce another 5%.

■ **More environmental control** — The trend is toward environmentally controlled housing, with light and temperature regulated. The major advantage of environmentally controlled housing is better feed conversion, especially during the winter months.

■ **Increased automation** — All new broiler production units are highly automated for feeding and watering, with the result that a broiler producer family, devoting full time to the operation, can care for 80,000 to 120,000 broilers at one time.

■ **Trend toward cage rearing of broilers** — Although the vast majority of broilers are reared on the floor, growers and equipment manufacturers are striving to improve cages for broilers as a means of improving efficiency.

■ **Processors** — In 1989, the 20 leading broiler companies in the United States processed 80.28% of the broilers. One company, Tyson Foods, Inc., accounted for over 20% of the total industry volume.

■ **Federally inspected slaughter** — Today all commercial poultry slaughter is under federal or state inspection.

■ **Product form** — In 1989, processors sold about 18.3% of their broilers as whole carcasses, 50.4% as cut parts, 6.3% as further processed, and 25% in other forms. The fast-increasing further processed products included chicken franks, and boneless chicken in the form of patties, fillets, chunks, and nuggets.

■ **Primary market outlet** — Retail grocery stores continue to be the primary market outlet for distributors, with 51.2% of broilers marketed through this channel in 1989.

■ **Supporting enterprises created** — In addition to improving the production and marketing of poultry meat, broiler producers and processors have helped create many new supporting enterprises, including plants for manufacturing special processing equipment and packaging materials, specialty chicken shops and restaurants, food processing plants, and plants for manufacturing fertilizers and broiler by-products.

BUSINESS ASPECTS

A typical broiler unit might consist of a hatchery, a feed mill, a processing plant, a field service and management staff, and 150 to 300 growers. Fig. 13–4 illustrates the functional framework of an integrated broiler production unit, although not all firms are as fully integrated as the illustration suggests.

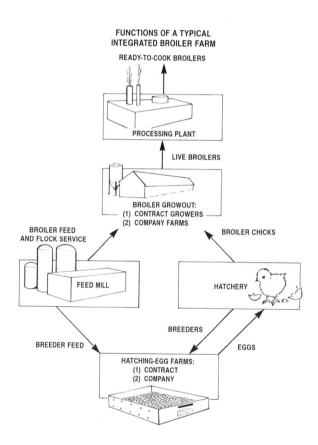

Fig. 13–4. Functions of a fully integrated broiler operation, which has its own hatchery, feed mill, and processing plant, and depends almost entirely on contract production.

With the increase in size and integration of broiler enterprises, the business aspects have become more important. More capital is required, knowledge of contracts has become important, competent management is in demand, records are essential, computers have come in, and such things as tax management, estate planning, and liability require more attention.

(Also see Chapter 17, Business Aspects of Poultry Production.)

FINANCING

Local banks and the Production Credit Association largely finance the broiler industry through credit extended to feed manufacturers, dressing-plant processors, hatcheries, and integrated operators. Much of this money is loaned at the local level to growers with mortgages as collateral.

Throughout the broiler industry, integration has increased. The principal reasons back of this trend have been (1) the need for more financing, and (2) the fact that more liberal credit is available to integrated operations than to individually owned operations.

Historically, bankers have had a "fear of feathers" — they have shied away from loans on poultry operations. This situation first changed in the South. It is easier to get financing on poultry operations there. Also, a dollar goes further in the South, especially in poultry house construction.

LABOR REQUIREMENTS

Efficiency of labor is an important factor in determining profits in the broiler business. Generally speaking, the more broilers raised per person the lower the production costs, and the bigger the operation the more feasible it is to automate. Greater labor efficiency has been a primary force back of increased number of broilers per farm.

In 1990, a farm family, devoting full time to the operation, could care for 80,000 to 120,000 broilers at one time, on one farm, in 5 to 6 houses; and grow out 5 to 6 broods during the year.

Many devices are available for reducing the number of hours required for each 1,000 broilers raised. Automatic feeding and watering systems may reduce labor costs by 60%. Additional savings in labor can be effected by automating manure disposal and by moving live birds from the farm to the processing plant by means of a suction system.

BROILER CONTRACTS

Fig. 13–5. Today, 95% of U.S. broilers are grown under contract; and integrated farms produce another 5%. (Courtesy, National Broiler Council, Washington, DC)

Broilers were one of the first animal industry commodities to shift from the conventional merchant-producer credit arrangement to a contractual-type program.

Today, 95% of all broilers produced in the United States are grown under contract. Integrated farms produce another 5%.

Typically, broiler production is carried out between an integrated firm (feed mixing, hatching, and processing plant) and a contract grower.

For maximum protection and minimum misunderstanding, all contracts should be in writing. Also, those involved should know and understand what constitutes a good contract.

■ **Major specifications to be spelled out** — In most broiler contracts, the following matters are spelled out: number and size of broods; disease, parasite, and cannibalism control; ownership, facilities, and labor; technical assistance and supervision; taxes and insurance; marketing method, price, age, weight, loading, weighing, and transportation; feeding program and its control; records and accounting; changing or adjusting contract; responsibility for and division of losses; credit arrangements, notes, and chattels; other poultry on premises and visitors; method of figuring costs; and method of settlement and payment for broilers at time of marketing, leftover supplies, and division of interest.

■ **Types of contracts** — There are numerous types of broiler contracts. But most of them fall into one of the following seven categories:

1. **Conventional-type credit plan or "open account" agreement.** The producer-owner retains title, makes all decisions, assumes all risks, takes any profit and/or loss (except where dealer provides a "no-loss" clause), and pays credit account at time of sale of birds. Credit may be secured by note or chattel.

2. **Share.** Under this type of contract, the producer furnishes the house, equipment, labor, and may or may not furnish litter and fuel; shares in the proceeds above costs as designated by terms of the contract; and shares in condemnation losses.

The contractor retains title, provides all production supplies, medication, supervision, makes necessary decisions, and usually assumes financial losses.

3. **Flat fee.** The grower receives a guaranteed fixed amount on a per bird or on a per pound basis. Some contracts include other incentives and provisions on feed conversion, gains, market price, etc.

4. **Feed conversion.** The grower is paid on specific rates based on feed conversion. Variations of incentives in addition to feed conversion are: share, market price and share, and production cost and share.

5. **Market price.** Payment to the grower is calculated on a specific schedule of rates based on the various market prices.

6. **Guaranteed price.** The dealer guarantees the grower a specific (prearranged) market price per pound for the broilers sold. The dealer may provide a "no-loss" clause as further protection for the grower.

7. **Salary.** Payment is made to the grower on a regular wage scale, paid on a weekly or monthly basis. If the grower owns buildings and equipment, income may be increased through rental or lease to contractors.

A typical contract is one in which the grower (usually a farmer) provides the housing and grow-out equipment — feeders, waterers, brooders — and such other items as water, electricity, fuel, litter, and labor. The contractor (the company) provides the chicks, feed, necessary medication, supervision, and labor and equipment for catching and hauling the birds to market.

The integrator retains title to the birds and the grower is paid a return for labor and facilities, usually plus a bonus or incentive arrangement. Some contractors pay growers on a per pound basis, whereas others pay on a per bird basis. Bonuses are earned by having lower mortality than the average grower, by producing more pounds of broiler on a given amount of feed, by having fewer birds condemned during processing, or by a combination of part of all of these accomplishments.

ANALYZING OUTCOME OF BROILER CONTRACT

Good management necessitates factual information about the performance of each flock. Growers should establish a system for determining their fixed investment and depreciation schedule. They should keep a record of variable or "out-of-pocket" expenses for each brood of broilers and recap these on a flock and annual basis. Then they should study the information provided by the integrator on each flock of broilers when contract payments are received. This provides the basis for estimating the grow-out earnings on future flocks. Table 13–1, Contract Broiler Result Summary, provides an outline for recapping and analyzing the performance factors related to the earnings of each flock.

BROILER FUTURES AND OPTIONS

Broiler futures are back again on the Chicago Mercantile Exchange! Chicken futures were traded at the "Merc" from 1962 to 1963 and broiler futures were traded from 1979 to 1982. But both of the earlier contracts were based on physical delivery of the commodity, which proved to be a hindrance. Also, there have been other changes. Broilers have become big business; the annual price volatility in percentage terms is over 40%; and the term *fast food* was unknown in 1960.

The basic unit of the broiler contract is the 40,000-lb load, and pricing is on a cents per pound basis. On both futures and options contracts the minimum offer is $10.00 per contract, or .025¢ per pound; maximum on the futures is $800.00 per contract, or 2¢ per pound. The option contract has no maximum. Contract months are February, April, May, June, July, August, October, and December, and hours of trading are 9:10 a.m. through 1 p.m. Chicago time. Settlement on the futures is cash on the USDA composite-weighted average price.

The subject of "Futures Trading" is more fully covered in Chapter 17, Business Aspects of Poultry Production; hence, the reader is referred thereto.

RECORDS

Good records are exceedingly important in order to determine economic progress, and to see how the flock compares with a standard or the average. Three types of broiler records should be kept: (1) those pertaining to growing the flock, (2) those pertaining to making the contract settlement, and (3) those pertaining to profit or loss.

COSTS AND RETURNS

Costs and returns in the broiler business are pertinent to decision making; when deciding whether or not to enter the business, and in determining how well an established enterprise is doing.

Costs and returns of broiler production differ from year to year and from area to area. Nevertheless, the figures presented in Table 13–2 will serve as useful guides.

Table 13–2 (next page) shows the estimated 1990 costs and returns on a typical contract broiler production unit in Georgia, the second ranking broiler state in the nation. Note that this unit consists of 16,000 sq ft, and can accommodate 21,600 birds. Note, too, that this building has a cement block foundation, insulated ceiling, side curtains, forced air furnaces, foggers, nipple fountains, and curtain controls.

TABLE 13–1
CONTRACT BROILER RESULT SUMMARY[1]

Lot number _____
House number _____

Costs (variable):	Total Value	Per Pound Sold
Litter	$_____	$_____
Fuel	_____	_____
Electricity	_____	_____
Hired labor	_____	_____
Repairs and maintenance	_____	_____
Other	_____	_____
Other	_____	_____
Total variable costs	$_____	$_____
Costs (fixed):		
Depreciation on buildings and equipment	$_____	$
Taxes	_____	_____
Insurance	_____	_____
Other	_____	_____
Total fixed costs	$_____	$_____
Total all costs	$_____	$_____
Income, contract payment	$_____	$
Earnings, grower and invested capital	$_____	$

Statistical:
Chicks hatched, date _____ , head delivered _____
Broilers marketed, date _____ , head _____ , net weight _____
Age of broilers marketed _____ days, livability _____ %, avg. weight _____
Feed used: _____ lb; grower _____ lb; total _____ lb
Pounds of feed to produce a pound of broiler _____

[1]Adapted by the author from *Keys to Profitable Broiler Production*, Texas Agricultural Extension Service Bull. L-962, Texas A&M University.

TABLE 13–2
ANNUAL COST AND RETURN ANALYSIS CONTRACT PRODUCTION UNIT WITH CEMENT BLOCK FOUNDATION, INSULATED CEILING, SIDE CURTAINS, FORCED AIR FURNACES, FOGGERS, NIPPLE FOUNTAINS, AND CURTAIN CONTROLS
(16,000 sq ft; 21,600 bird capacity; in North Georgia; 10 year amortization)[1]

Income Item	Quantity ×	Payment per Pound	Value	Your Value
Broilers[2] 4.00 lb	497,664 lb	$ 0.039	$19,409	$_____
(21,600 × 6 × .96 × 4.0 = 497,664 lb)				

Operating Expenses

Item	Quantity ×	Rate	Value	Your Value
Litter[3]	7 loads	$100	$ 700	$_____
Electricity	129,600 birds	$3.36/1,000	$ 435	$_____
Fuel (gas)[4]	129,600 birds	$22/1,000	$ 2,851	$_____
Miscellaneous[5]	129,600 birds	$2/1,000	$ 259	$_____
Repairs:				
Building	$41,500	1.0%	$ 415	$_____
Equipment (incl. pump)	$27,000	1.0%	$ 270	$_____
Total operating expenses			$ 4,930	$_____
Net returns to capital (labor, land & management) ($19,409 − 4,930 = $14,479)			$14,479	$_____
Annual fixed cost (see Table 13–3)			$11,989	$_____
Net return to grower (labor, land & management)[6] ($14,479 − 11,989 = $2,490)			$ 2,490	$_____

[1]Adapted by the author from *1990 Poultry Production Systems in Georiga, Costs and Returns Analysis*, by Dan L. Cunningham, Extension Poultry Scientist, University of Georgia, Athens.

[2]Assumed 96% of capacity marketed.

[3]Annual cleanout with some new litter added between flocks.

[4]Partial house brooding practiced.

[5]Miscellaneous expenses include: insecticides, disinfectants, rodent control, light bulbs, and telephone.

[6]The value of manure at cleanout time is assumed to equal the cost of removal.

As Table 13–2 shows, annual returns to contract poultry producers are modest (an estimated $2,490 for each house), at least until the unit is paid for. Contract broiler production requirements for labor are quite modest, but the capital requirements are great. One family, working full time, can care for 5 or 6 units at one time, on one farm; and they can grow 5 to 6 broods each year. The $2,490 per house per year income appears to be on the low side. A more often quoted income per house per year is $3,000 to $5,000. Thus, if a farm family owns and operates 6 broiler houses, they may expect an annual income of $18,000 to $30,000.

The fact that returns from contract broiler production are consistent and dependable, while labor requirements are low, makes this type of enterprise attractive to many farm families. Moreover, many contract broiler growers have only one to two houses, which they care for on a part-time basis, and which they consider as supplementary income to a full-time job off the farm.

Table 13–3 details the building and equipment costs for the Georgia broiler unit presented in Table 13–2. Note that a total building and equipment investment of $73,700 is required to house 21,600 broilers, or $3.41 per bird.

As is true in any business, profits in broiler production can be increased by either, or both (1) lowering costs, or (2) increasing returns. Based on the information presented in Tables 13–2 and 13–3, along with practical experience, it may be concluded that net returns depend on a satisfactory contract, reasonable broiler prices, efficiency of production, and volume.

TABLE 13–3
DETAILED BUILDING AND EQUIPMENT COSTS FOR THE GEORGIA BROILER UNIT PRESENTED IN TABLE 13–2
(Note that this building consists of 16,000 sq ft and will accommodate 21,600 birds)[1]

Item	Investment		Years of Life	Annual Fixed Cost			Total
	Purchase Price	Average Price		Dep.	Int.	Taxes & Ins.[2]	
Building[3]	$41,500	$20,750	10	$4,150	$2,075	$560	$ 6,785
Grading and gravel	$ 3,000	$ 1,500	10	$ 300	$ 150	$ 40	$ 425
Well	$ 1,700	$ 850	10	$ 170	$ 85	$ 23	$ 278
Well pump	$ 500	$ 250	10	$ 50	$ 30	$ 7	$ 87
Equipment[4]	$27,000	$13,500	10	$2,700	$1,350	$364	$ 4,414
Tractor w/front loader (used)	optional						
Manure spreader	optional						
Total investment	$73,700						
Total fixed cost							$11,989

[1]Adapted by the author from *1990 Poultry Production Systems in Georiga, Costs and Returns Analysis*, by Dan L. Cunningham, Extension Poultry Scientist, University of Georgia, Athens.

[2]Taxes determined at 40% of fair market value at 30 mills (1.2%); insurance estimated at 1.5% of current value; combined taxes and insurance rate = 2.7%. Interest = 10%.

[3]Building size: 40 ft × 400 ft = 16,000 sq ft (21,600 chicks started per flock).

[4]Equipment estimate—litter handling equipment may be available from other farm operations.

LOWERING COSTS

Most broilers are grown under some form of contract. Normally, the integrator supplies the feed, chicks, and vaccines/drugs; and the farmer supplies the labor, buildings and equipment, brooding fuel, litter, electricity, and miscellaneous items. Table 13–4 shows the cost of producing a 4-lb broiler, with a breakdown of the costs normally assumed by each the integrator and the owner.

TABLE 13–4
COSTS OF PRODUCING A 4-LB *(1.8 KG)* BROILER[1]

Integrator		Grower	
Item	Dollars	Item	Dollars
Feed	.1940	Labor	.0151
Chick	.0392	Building & Equipment	.0109
Vaccines & drugs . . .	.0039	Brooding fuel	.0029
		Litter	.0013
		Electricity	.0013
		Miscellaneous	.0003
Total	.2371		.0318

[1]Adapted by the author from Chapter V, *The Broiler Enterprise*, prepared by the University of Georgia, Athens.

■ **Lowering production costs** — Because most commercial growers raise broilers in large volume, small differences in production costs have a great impact on profit.

As shown in Table 13–4, the chief broiler production costs, ranked in order are: feed, chicks, and labor.

Because feed is the highest production cost item, any change in feed cost or feed efficiency will dramatically influence the cost of production.

Feed costs can be lowered through (1) locating the broiler operation in an area where major feed ingredients are produced, (2) using improved feed formulations, (3) selecting efficient broiler strains, (4) housing the birds in environmentally controlled buildings (it should be noted, however, that high energy costs have nullified, in part at least, the saving in feed costs effected by environmental control), and (5) lowering mortality. (High mortality makes for increased feed consumption per pound of broiler.)

■ **Lowering building and equipment costs** — Facility costs are high. Hence, they should be investigated with care. Comparison should be made between (1) package deals consisting of houses and equipment on an installed basis, (2) buying the houses and equipment from different suppliers, and (3) home-construction, especially of the houses. Also, growers should be aware that the design of houses and equipment changes. For this reason, they should visit successful growers and learn what is new and working well.

INCREASING RETURNS

Contract payments vary with the contractor and with such factors as mortality, feed conversion, average weights, and condemnations. Two types of broiler contracts are commonly used: (1) payment per bird raised, and (2) payment per pound market weight.

Growers have increased returns by turning out younger and heavier birds for market. Birds reach market weight one to two days earlier each year. This phenomenon has shortened the time each brood spends in grow out by nearly two weeks in the decade of the 1980s, permitting growers to raise one more brood per year than in the 1970s. Also, average time between broods has been shortened, so that many growers can now produce six or more broods per year.

The combination of an extra brood and heavier average weight increased grower output by about one-third in the 1980s. Growers can reduce per pound production costs by spreading fixed costs over more weight. The higher volume has increased their gross receipts.

BREEDS AND BREEDING

In the early years of the broiler industry, after selecting the females for egg production, the males of dual-purpose crosses were used for broiler production rather than being discarded. Although these day-old male chicks were obtainable at a low price, it was soon recognized that birds bred especially for broiler production were much superior in performance.

Today, parental broiler flocks consist of a female line and a male line. The female lines usually have been developed from birds of White Rock background and the male lines from birds of Cornish ancestry. Hybrid vigor is obtained by systematic matings that may involve crossing of different breeds, different strains of the same breed, or inbred lines. In addition to hybrid vigor, improvement in economic factors is also obtained, provided that matings include stocks having superior qualities of genetic origin.

The male lines for producing broilers usually have dominant white feathers and are selected for rapid growth; meat characteristics, such as breast width, body depth, live market grade, and dressing yield; and rapid feathering. Female lines used in producing broilers must also have outstanding growth rate, high hatchability, and good, but not outstanding, production of eggs of desirable size and texture. Considerable attention is also given to such additional factors as freedom from *Mycoplasma gallisepticum*, feed efficiency, skin texture, skin and shank pigmentation, and, in the case of large broilers, roasters, and capons, feathering on the breast or absence of breast blisters.

If male and female lines are produced by different breeders, as they often are, each line must cross well with other lines if it is to remain competitive. Progeny test and family selection have been effective in the development of broiler lines. The buyer must judge the ability of male and female lines to cross well, whether the lines are developed by separate breeders or the same breeder. The end product of all crosses should be a modern white, yellow-shanked broiler, produced efficiently and economically.

The selection of commercial broiler stock can be the determining factor between profit and loss. Usually, the county agricultural agent and the extension poultry specialist can provide information that will be helpful in making selection. Also, consideration should be given to the results obtained by other growers in the area from different stocks on such factors as growth, feed conversion, viability, and rate of condemnation.

Although the genetic makeup of parent broiler stocks is of primary importance, other factors can have a marked influence on the performance of broilers. Nutrition of the parent flock, size of the egg from which the broiler chick was hatched, and the lighting program used on the parent flock can affect progeny performance. It has also been demonstrated that storage time and method of storing hatching eggs can have an important effect on the growth rate of broilers hatched from stored eggs. Therefore, it is not only important to use stock of good breeding, but also to ensure that parental stocks are properly managed, with particular emphasis given to feeding, lighting, and handling of hatching eggs. Additionally, when selecting chicks for broiler production, growers should give consideration to the following factors if they are at liberty to do so (generally, the contract stipulates that the integrator is responsible for choosing the chicks, with the result that the growers must take the type and quality of chicks offered to them): (1) the vaccination program for the breeders, (2) the breeder and hatchery disease control programs, (3) the vaccination(s) at the hatchery, and (4) whether or not the chicks are sexed and debeaked at the hatchery.

In large broiler operations, a common practice is to test several stocks on one farm where suitable facilities are available and careful records are kept. For such tests, it is important that the housing, feeding, and management be uniform for all stocks in order to obtain a valid comparison of the factors affecting economic returns — rate of growth, pounds of chicken per hundred pounds of feed, mortality, and quality of broilers produced. Of course, the information will provide a more reliable basis for the selection of stocks if the tests are repeated on a second or third farm. The farm or farms selected for the test should follow production practices typical of the area.

Characteristics of economic important to which special attention should be given in selection and breeding programs include growth and feed efficiency, feathering, breast development, shank pigmentation, disease resistance, nutrient requirement inheritance, plumage color, and egg size. Pertinent information relative to each of these traits follows.

■ **Growth and feed efficiency** — Genetically, growth is inherited as a simple quantitative character. Also, growth is positively correlated with other important body factors — breast width, body depth, and fleshing.

Feed efficiency is inherited. Also, the more rapidly growing birds utilize feed more efficiently than the slower growing birds. Since growth rate and feed efficiency are highly correlated, commercial broiler breeders generally select for growth rate rather than feed efficiency as a specific factor.

■ **Feathering** — Processors demand broilers that are uniformly covered with feathers. Because the sex-linked factor for feathering is inherited in such a simple manner, it is highly improbable that any reputable strain does not already contain a large proportion of early-feathering stock.

Selection to obtain breeders that feather rapidly is most easily accomplished at the time of hatching. Fast feathering is a sex-linked recessive to slow feathering; hence, when mated together, male and female breeding birds selected for fast feathering will produce fast-feathering chicks.

Rate of feathering can be determined in day-old chicks by the length of the primary and secondary feather sheaths of the wing and the number of the secondary feather sheaths. The chick with the highest rate of feathering has well-developed primaries and secondaries, with six or more secondaries. The chick of the slow-feathering type has no secondaries, or less than six short ones, and no primaries or very short ones.

After fast-feathering chicks have been selected, they should either be raised by themselves or identified by some means.

■ **Breast development** — Since the breast meat is the most valued part of the broiler, this characteristic should be given considerable attention in any selection and breeding program. Observations of breast development should be made when birds are 6 weeks of age, but it may be made when they are from 6 to 12 weeks old. For these observations, each bird should be examined individually and held in a similar position. A good way is to hold the chicken by the legs in the left hand with the head downward, then with the right hand examine the width and length of the breast. Birds can be divided into at least four grades as shown in Fig. 13–6.

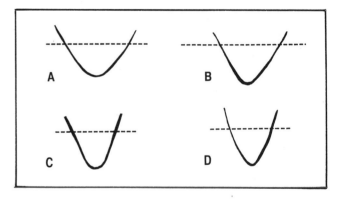

Fig. 13–6. Breast types and grades of broiler breeding stock: *A* and *B* show desirable fullness of breast; *C* and *D* are poorer types. (Observations made about three-quarters of an inch from edge of keel.)

At the time the breast is examined, observations may also be made on any imperfections of the breastbone or skin, such as curved, indented breastbones and breast blisters. Any individual with such imperfections should be rejected as a breeder.

■ **Shank pigmentation** — Shank pigmentation is inherited. Within the yellow-skinned varieties there are distinct differences — ranging from bright yellow and orange-yellow to green and sometimes bluish shades. It should be noted, however, that nutrition and parasites and/or diseases affect pigmentation much more than inheritance.

■ **Disease resistance** — Chickens differ genetically in their ability to resist invasion by protozoa, bacteria, fungi, viruses, and parasitic worms. It is feasible to develop strains comparatively resistant to *Eimeria tenella* coccidium, fowl typhoid, pullorum disease, visceral lymphomatosis, Newcastle disease, blue comb, and encephalomalacia. However, it has not been determined whether it is economically feasible to select completely resistant strains against these diseases and parasites.

■ **Nutrient requirement inheritance**—The nutrient requirement for certain amino acids—arginine and methionine—is inherited. Genetic differences have also been observed in the requirements (1) for vitamins A, B–1, D_3, E, and riboflavin, (2) for the unidentified growth factors, and (3) for manganese, calcium, and zinc. Basically, any individual with an abnormal requirement for any nutrient should be eliminated as a breeder.

■ **Plumage color**—Growth rate and other characteristics may be affected by the genes that determine plumage color of broilers. For example, for maximum growth the gene for silver apparently has to be associated with the gene that extends black and with recessive white. If white birds are desired, the dominant white gene needs to be combined with the gene causing barring. Processors want all birds to be white.

■ **Egg size**—Broiler chicks from large eggs (24 to 28 oz per dozen) are definitely larger at hatching than those from small eggs (18 to 22 oz). Each 1 oz per dozen increase in egg size makes for mature broilers that average about 0.04 oz heavier per bird. As broilers grow, the effect of this relationship is less, but it is still definite at 8 to 12 weeks of age for body weight and possibly for feed conversion. Also, it is noteworthy that mortality appears to be higher in broiler chicks hatched from small eggs.

When hatching eggs are scarce and more small eggs are set, broiler raisers observe that mortality tends to increase and growth and feed conversion suffer.

BROILER FLOCK PERFORMANCE STANDARD

When selecting breeding stock or analyzing the performance of a broiler flock, those less familiar with the general performance of a broiler flock may find it advantageous to have a standard for guidance or comparison. For this purpose, performance goals are presented in Table 13–5.

Other facts pertinent to an understanding of broiler performance are:

1. At any given age, males are heavier than females (see Table 13–5). Also see section headed "Sexing" for other differences in the performance of the sexes.

2. Weekly increases in weight are not uniform; gains increase each week until reaching a maximum at about the eighth week for straight-run (both sexes together) flocks.

3. Weekly feed consumption increases as weight increases; each week the birds eat more feed than they did the week before.

4. Generally the more feed consumed, the better the feed conversion at a given age.

5. Fast gains are efficient gains; as weekly gains increase, feed efficiency increases, also.

6. Healthy birds consume more feed and have better feed conversion than sick birds.

7. The greater the activity, the lower the feed efficiency.

8. Cannibalism results in lowered feed consumption, growth, and feed conversion.

9. Changes in temperature cause changes in feed consumption; broilers eat about 1% more feed for each 1°F

TABLE 13–5
BROILER PERFORMANCE GOALS[1]

Age in Weeks	Age in Days	Average Weight		F.C.R. Cum.[2]	Calorie Conversion	
		(lb)	(g)		(kcal/lb)	(kcal/kg)
Males						
1	7	0.38	172	0.94	1,321	2,906
2	14	0.93	422	1.21	1,700	3,740
3	21	1.69	767	1.43	2,009	4,420
4	28	2.60	1,179	1.60	2,288	5,034
5	35	3.65	1,656	1.72	2,460	5,411
6	42	4.74	2,150	1.85	2,646	5,820
7	49	5.83	2,644	1.98	2,871	6,316
8	56	6.90	3,130	2.12	3,074	6,763
9	63	7.94	3,602	2.26	3,277	7,209
Females						
1	7	0.36	163	0.96	1,349	2,967
2	14	0.84	381	1.28	1,798	3,956
3	21	1.51	685	1.50	2,108	4,637
4	28	2.32	1,052	1.67	2,388	5,254
5	35	3.10	1,406	1.80	2,574	5,563
6	42	3.88	1,760	1.94	2,774	6,103
7	49	4.75	2,155	2.09	3,031	6,667
8	56	5.54	2,513	2.24	3,248	7,146
9	63	6.23	2,826	2.40	3,480	7,656
Straight-run (as hatched)						
1	7	0.37	168	0.95	1,335	2,936
2	14	0.89	404	1.25	1,756	3,864
3	21	1.60	726	1.47	2,065	4,544
4	28	2.46	1,116	1.64	2,345	5,159
5	35	3.38	1,533	1.76	2,517	5,537
6	42	4.31	1,955	1.90	2,717	5,977
7	49	5.30	2,404	2.04	2,958	6,508
8	56	6.22	2,821	2.18	3,161	6,954
9	63	7.09	3,216	2.33	3,379	7,433

[1]Adapted by the author from *Hubbard Broiler Management Guide*, 1991–1992, Hubbard Farms, Walpole, New Hampshire.

[2]Feed conversion ratio.

decrease in temperature, and they eat about 1% less feed for each 1°F rise in temperature. Very high temperatures drastically reduce feed consumption and cause poor feed conversion. During very cold weather, growth and feed conversion are poorer because a greater portion of the feed is used to maintain body temperature.

10. Flocks are not uniform, with the result that all birds are not of the same weight at market time. The males are heavier than the females. But neither sex is uniform; there are large, medium, and small cockerels and pullets. When approximately 75% of the birds are within extremes of 10% of the average weight of each sex within a given flock, the flock is of acceptable uniformity.

11. Broiler growing efficiency is usually measured in the following three ways:
   a. Mature live body weight.
   b. Feed conversion over the life of the bird.
   c. Age to reach a desired weight, or growth rate.

Generally speaking, as the efficiency of a broiler operation increases, feed consumption is reduced, feed conversion is improved, growth rate is increased, and the length of time necessary to reach a certain weight is decreased. Above all, the measurement which best tells the kind of job being done is growth rate.

HOUSING AND EQUIPMENT

Fig. 13–7. Modern broiler house and equipment. (Courtesy, Foster Farms, Livingston, CA)

In modern commercial broiler production, the bird spends its entire life in one house; that is, it is not brooded in a special brooder house, then moved to a house for growing, for broiler raising is basically a brooding operation. Instead, brooder houses are thoroughly cleaned and disinfected between flocks, preferably with the quarters left idle 1 to 2 weeks before starting a new group.

Today, virtually all of the nation's broilers are produced in large production units and grown under some type of vertical integration or contractual arrangement. Hence, in the discussion that follows, only the buildings and equipment common to these larger establishments will be described.

There are, of course, many different styles and designs of houses, and even more variations in equipment. The important thing is that broiler houses and equipment provide comfortable conditions, including adequate feed and water, so that the birds can perform at the highest level of which they are genetically capable.

(Also see Chapter 11, Poultry Houses and Equipment.)

HOUSING

The broiler house should provide total confinement for the birds in clean, dry, comfortable surroundings throughout the year. It should be kept warm enough, but not too warm; the litter should be kept reasonably dry; provision should be made to modify the air circulation as broilers grow; and fresh air should be circulated, but the house should be free from drafts. In short, the broiler house should not be too cold, too hot, too wet, or too dry.

The importance of adequate housing to protect broilers from adverse conditions is apparent from the seasonal effects on broilers: mortality, condemnations, and medication costs are highest, and feed efficiency is usually poorest, in birds sold during the winter; growth is best in the fall; feed efficiency is best in birds sold in the early summer; medication costs are lowest in the early summer; and hot weather adversely affects both growth and feed efficiency.

The broiler house should be located where water drainage and air movement are good. There needs to be a good access road from the houses to the highway because of feed and bird hauling. The cost of getting utilities, such as water, gas, and electricity, should be considered. Houses should be convenient to the caretaker's house, and the prevailing winds should be away from it.

CONSTRUCTION

Broiler houses in various areas are of many types and dimensions, and are constructed of many kinds of materials.

Housing should be economical, yet substantial enough to last many years with a minimum of maintenance. The house should be sufficiently insulated to heat easily in winter and have sufficient ventilation for cooling birds in the summer. From $8/10$ to 1 sq ft of floor space should be provided per bird. Currently, most broiler houses that are being constructed have gable-type roofs, truss-type (either wood or steel) construction, and single floors; and they are 32 to 40 ft in width and of sufficient length to give the desired bird capacity. Most of them have a minimum capacity of 20,000 to 50,000 or more.

Most broiler houses have concrete floors. However, whenever the soil is dry, porous, and sandy, some houses have dirt floors.

Two basic house designs give satisfactory results — (1) environmentally controlled houses, which are light-tight and ventilation-controlled according to the requirements of the birds, and (2) open houses, usually with curtains or windows.

The trend in larger and better financed broiler operations is toward environmentally controlled housing. These are insulated, windowless, and fan-ventilated. With this type of arrangement, it is possible to provide a more uniform temperature, along with the proper supply of clean, fresh air without drafts. The final result is an improved market quality bird and higher income.

TEMPERATURE; BROODERS

Temperature is important! The recommended temperature at chick level at the edge of the hover is 95°F for the first week, with a reduction of 5°F weekly until 70°F is reached the sixth week. For best feed efficiency with good growth, and especially when there is any sickness present, a room temperature of 75°F is recommended. When chicks circle wide or leave the house, it is too hot. If they tend to crowd under the hover, it is too cool. In either case, the temperature is probably extremely high or extremely low and adjustment is needed immediately. Especially avoid temperatures below 80°F the first 2 weeks and over 85°F after 6 weeks. Avoid wide daily fluctuations of 20 to 30° in the first 6 weeks. Temperature fluctuations should be watched, especially when trying to brood without heat in warm weather.

During periods of stress, such as disease, it may be advisable to alter the environmental temperature. Likewise, it may be advantageous to use water-cooling devices during periods of extremely warm weather, to prevent losses and maintain maximum growth during the later stages of the rearing period.

Fig. 13–8. Broilers being brooded in the house in which they will spend their entire lives. (Courtesy, University of Georgia, Athens)

The type of brooders to be used varies with the source of fuel available. Natural gas, butane or propane gas, and electrical brooders are the types in widest use. Large – 1,000 capacity – hovers are most popular. Manufacturer's recommendations for number of chicks per brooder are usually based on day-old chicks. As the birds grow larger, they need additional space. For this reason, only about two-thirds of the rated chick capacity should be started under one brooder. In any case, never exceed the manufacturer's recommendations for the rated chick capacity. Follow the manufacturer's instruction for operating the brooder. Brooder guards to keep chicks confined to the brooding area are recommended by some companies. Corrugated cardboard 14 to 18 in. high may be used in winter not only to confine the chicks but to aid in preventing floor drafts. Poultry netting may be used to replace cardboard for summer brooding. The guards should extend out about 3 ft from the edge of the brooder to allow

chicks to move out from under the brooder if they become too hot. The guards may be removed after about the first week.

VENTILATION

The main functions of ventilation are to maintain oxygen, keep carbon dioxide levels low, remove dust or moisture and ammonia from the building, and maintain suitable temperatures.

Air movement requirements are best determined by observing bird comfort, litter condition, and odor build-up. If the house is fan-ventilated, air movement of 3 to 4 cu ft per bird per minute is sufficient under most conditions. However, total fan capacity of 7 cu ft per bird per minute should be installed to handle air on excessively hot, humid days.

Proper ventilation requires considerable management, because of large variations in exterior temperatures from time to time and increasing requirements of broiler flocks as they grow.

During the first week, excessive ventilation should be avoided. A rapid rate of air change at this time is neither necessary nor desirable, as there is a danger of chilling the young chick before it has developed enough to achieve good physiological control over its body temperature.

Generally, environmentally controlled broiler houses are constructed so that ventilation is controlled by the use of high-speed or multispeed fans operated both continuously and thermostatically. The ventilation system must let air into the building in a way that avoids drafts on the chicks, and yet promotes air mixing, air change, and dust removal. A smaller air intake is required in winter as cold air entering the building has a much larger expansion ratio than warmer air entering at other times of the year.

Most ventilation systems may be classified as either of two types: negative pressure or positive pressure. In the negative pressure system, exhaust fans expel air that has been drawn into the pen through intakes, usually located in the opposite wall. The positive pressure system uses fans to force air into the pen and air escapes out through ventilation openings; this makes it easy to filter the incoming air – a decided advantage from the standpoint of disease control. As previously mentioned, it is important to have a system that provides uniform air change without drafts.

In environmentally controlled buildings, some provision should be made for emergency ventilation in the event of an electric power failure. This can be done by either providing sufficient auxiliary electric power to operate fans, or installing doors in the sides of the building which can be opened to allow natural airflow and thus prevent smothering if the fan ventilation system fails.

RELATIVE HUMIDITY

It has been demonstrated that the performance of a broiler flock is affected by the humidity level. Optimum humidity levels reduce dust and promote better feathering and growth. A relative humidity of 60 to 70% appears optimum. Birds may show discomfort by huddling when the relative humidity is 45% or less and the room temperature is 60° to 70°F.

When dry, dusty conditions prevail, it may be advantageous to spray the walls and ceiling with a fine mist, using fogging nozzles to bring relative humidity up to acceptable levels.

LIGHT

Lights are used to encourage feed consumption and optimum growth, and to prevent chickens from piling or stampeding when scared.

Broilers may be lighted most of the night or intermittently through the night. To avoid piling in case of power failures, it is desirable to allow at least 1 hour of darkness in every 24-hour period.

The following lighting systems are commonly used for broilers:

1. **Continuous light in open-sided houses.** Start with 48 hours of continuous light, then supply dim artificial light during all dark hours except for 1 hour at night (so that the birds will not panic in case of power failure). At floor level, the light intensity should be 0.5 footcandle, which can be supplied by one 150-watt bulb for each 1,000 sq ft of floor space.

2. **Light in light-tight houses.** Provide 3.5 footcandles of continuous light at floor level for the first 5 days, until the chicks develop their eating habits and know the location of feeders and waterers.

On the sixth day, reduce the light intensity to 0.35 footcandle, which can be supplied by one 125-watt bulb for each 1,000 sq ft of floor space. Then beginning on the eleventh day, use one of the following lighting programs:

    a. **Continuous dim light.** Provide 23 hours of light with an intensity of 0.35 footcandle and 1 hour of darkness.

    b. **Intermittent dim light.** Provide 1 hour of 0.35 footcandle light intensity (feeding time), followed by 3 hours of darkness (resting time); then repeat. This provides 6 hours of eating time each 24 hours. It is noteworthy that intermittent light produces better growth than continuous light. Although the reasons for this are not known, it is thought to be due to better feed utilization.

    CAUTIONS: Two cautions are pertinent; (1) do not change the lighting program once either continuous or intermittent lighting is started, and (2) when an intermittent light system is used, up to 50% more feeder and water space may be required as most of the birds must be able to eat at one time.

Incandescent bulbs are considered superior to other light sources. Bright white light may be a contributing cause of feather picking, which can lead to cannibalism. Blue lights may be used when catching birds, and red lights may be used in case of cannibalism.

Recently, some interest has been shown in using continuous light the first week, followed by an intermittent lighting program (such as 3 hours of light followed by 1 hour of darkness) for the remainder of the growing period. In order to use an intermittent lighting program effectively, it is essential that the building be blacked out to prevent entry of light through doors or ventilation openings.

AUXILIARY POWER

A generator, operated either by stationary engine or from a tractor power take-off, is required for auxiliary electric power to maintain heating, lighting, and ventilating operations in case of power failure. Additionally, a battery-operated alarm system should be installed in the broiler house and wired to the operator's dwelling. This should be activated by power failure or when extreme temperatures occur.

FLOORS VS CAGES

Currently, most broilers are housed on floors with litter. However, if improved broiler cages are perfected, it is predicted that an increasing proportion of broilers will be reared in cages rather than on the floor.

Although growing broilers in cages is still in the experimental stage, certain advantages and disadvantages to the system are evident.

Among the **advantages** to cage growing in comparison with conventional floor rearing are the following:

1. A greater number of birds can be raised in a given housing area, due to (a) less space required per bird, and (b) tiering the cages.
2. Improved growth and feed efficiency.
3. Litter is eliminated.
4. Coccidiosis is practically eliminated; and cages alleviate the cost of medicated feed or vaccination to prevent coccidiosis.
5. Cleaning the house is easier; and less downtime is necessary between broods.
6. It eliminates catching the birds at market time, thereby lessening bruising.

Fig. 13–9. Broilers in cages. (Courtesy, Euribrid B. V., Boxmeer, Holland)

Among the **disadvantages** to cage growing in comparison with conventional floor rearing are the following:

1. There is a higher incidence of breast blisters, crooked keel bones, wing fractures, and leg damage.
2. There is more feather picking and vent picking.
3. Ventilation, heating, and lighting must be adapted because of denser occupation.
4. A larger capital investment is required.

Because, in total, the potential advantages of the cage system outweigh the disadvantages, experiment stations and equipment manufacturers are working ceaselessly away to improve cages for broilers. The incentives to develop a satisfactory cage system are great. When a workable system of cage rearing is developed, farm output and efficiency will increase greatly.

EQUIPMENT

Fig. 13–10. The automatic waterers and the moving chain of feed seem to stretch into infinity in this big broiler house. (Courtesy, USDA)

There is hardly any limit to the number of kinds, designs, and sizes of broiler house equipment. However, the two chief items of equipment are feeders and waterers.

FEEDERS

Inadequate feeder space causes uneven and slow growth. No point in the house should be more than 10 to 15 ft from a feeder.

Most birds are started on chick box lids or similar temporary feeders, with one box lid feeder provided per 100 birds for the first 5 to 7 days. This allows the chicks to find feed easier. Feed should be placed in the feed troughs at the same time as in the box lids. Temporary feeders may be removed as soon as the chicks are eating from the troughs.

Some growers and contractors prefer to use the trough or hanging circular-type feeder. For the trough-type feeder, provide 2 in. of trough space per bird through 6 weeks of age, and 3 in. thereafter to market time at 7 to 8 weeks of age. When circular-type feeders are used, 20% less feeder space per bird will be sufficient.

Mechanical feeders are generally used on the larger broiler producing farms, with 2 to 2½ in. of feeder space provided per bird. They reduce labor and allow the grower to care for more birds. However, the mechanical feeding system does not reduce the necessity for frequent and regular observation of the birds, which may be a disadvantage.

WATERERS

It is important that broilers have an adequate supply of clean water at about 55°F.

Basically, four types of waterers are used in broiler production: trough, bell, cup, and nipple. Troughs are available in several lengths and sizes. Troughs should run lengthwise of the house and be equally spaced on both sides of the house. Bell-type waterers are similar in principle to troughs, and should also be equally spaced in the house. Cup and nipple drinkers are attached directly to water pipes and are triggered by birds to release water. The floor of broiler houses with cup or nipple waterers must be level.

In most modern broiler houses, the water is metered. Daily water consumption is a good indicator of the health of the birds.

All waterers should be cleaned and washed daily.

FEEDING

Since feed constitutes about 70% of the cost of producing broilers, it is important to give special attention to it.

Most broiler feeds are produced by a feed manufacturing company, the feed-mixing unit of an integrated operation, or a large independent producer. The processed feeds are usually mixed in a centrally located mill and delivered in bulk by truck to the broiler farm. It is not generally economical for individual growers to do their own mixing on the farm.

There is no "best" formula for the efficient feeding of a broiler flock. Rather, there are many good formulas; hence, the grower should select the formula that will give the highest net returns. Unfortunately, feed ingredients vary considerably in nutrient value, thereby introducing an undesirable variable in the problem of feed formulation. As a safeguard against nutritional deficiencies, feed manufacturers usually select a wide range of ingredients as nutrient sources.

Since broiler feeds are high in both protein and energy, they are more expensive than chick starter mashes normally used for replacement pullets. The importance of protein quality and of the protein-energy ratio in broiler rations is also recognized. Dietary demands of rapidly growing broiler chicks require that all nutrients be accurately balanced for optimum performance.

Although feeding systems vary, an effective broiler feeding program might consist of the following:

1. Starter crumbles for day-old to 3 weeks of age (mash may be used if crumbles are not available).
2. Starter pellets from 3 to 5 weeks of age.
3. Finisher pellets beginning at 5 weeks of age.
4. Market finisher pellets without medication during the withdrawal period.

Occasionally, *prestarter*, or *booster*, diets are fed the first two days. However, this should not be necessary unless there has been a history of early mortality or the birds have been exposed to extreme stress.

Feed conversion is superior for birds on crumbles and pellets compared with mash. The feeding of crumbles and pellets tends to reduce the amount of feed lost in the litter compared with feeding mash. Excessive feed wastage, however, is frequently associated with having too high a level of feed in the feeders.

Fig. 13–11. Broilers feeding on a computer-formulated ration, brought to them by long chain feeders. (Courtesy, USDA)

Crumbled or pelleted feed is usually purchased in bulk and stored in upright metal tanks. The feed is moved by auger from the storage tank to the automatic feeder or feeders. Care should be taken to ensure that bulk feed tanks and augers are watertight to prevent the accumulation of moldy feeds. It is also beneficial to locate feed tanks on the leeward side of the building.

Further information on the nutrition and feeding of broilers is contained in the following two chapters, to which the reader is referred (Chapter 7 also contains feed formulations):

Chapter 6, Fundamentals of Poultry Nutrition
Chapter 7, Poultry Feeding Standards, Ration Formulation, and Feeding Programs

GRIT

The value of grit for broilers is very controversial; some growers feel that it helps, others do not. If grit is fed, about 5 lb of fine grit should be allocated per 1,000 birds for the first 14 days, followed by the allocation of 10 lb of coarse grit per 1,000 birds from 15 to 35 days of age. When fed, grit should be insoluble. It should be fed on the feeder lids at first, then in the litter. It should not be fed in automatic feeders.

GROWTH RATE AND FEED CONSUMPTION

With improved breeding, nutrition, and management, feed conversion and growth rate of broilers are constantly improving. Tremendous gains in production efficiency have been realized in recent years. With narrow margins between production cost and selling price, it becomes increasingly important for the grower to pay strict attention to every step in the management program that will help improve production efficiencies. Table 13–5, p. 313, may be used as a standard, or for comparative purposes, of growth rates and feed conversion of broilers at different ages. The grower who is much better than average will obtain comparable performance to that presented in Table 13–5.

HEALTH

In broiler production, the emphasis is on disease prevention. Average total mortality runs about 4%. A practical goal is to have only 1 to 2% mortality up to 8 weeks.

Most diseases cannot occur without such disease-producing organisms as bacteria, viruses, and parasites. Thus, the primary objective of a preventive program is to avoid the mechanical spread of disease organisms. The most effective preventive program is the *all-in, all-out* system, in which only one age of birds is on the farm at the same time. All the chicks are started on the same day, and all the broilers are marketed on the same day. Between broods, there is a period of time when no birds are on the premises, during which any cycle of infectious diseases is broken. The next group of birds has a *clean start*, without the possibility of contracting a disease from older flocks on the farm.

Recent advances in isolation and disease control have made it more practical to keep chicks of several ages on the same farm. So, many broiler producers now follow multiple brooding. But this system calls for expert management; it is not for the novice.

(Also see Chapter 10, Poultry Health, Disease Prevention, and Parasite Control.)

BROILER HEALTH PROGRAM

A health program is fundamental to successful broiler production. A suggested disease prevention and control program follows:

1. Isolate the broiler farm. Enclose the premises with a tight fence, and lock all entrance gates. Beware of feed and supply trucks entering the enclosure.

2. Start with disease-free chicks. Preferably, the chicks should come from MG- and MS-negative breeders.

3. Vaccinate chicks against Marek's disease at the hatchery, and, as needed, against other diseases common to the area.

4. Use effective drugs in the feed, or a vaccination program, to prevent coccidiosis. Nonmedicated feed may be required the last 3 to 5 days before slaughter; check the feed tag.

5. Keep feed and water clean.

6. Screen house to keep out birds and rodents.

7. Do not allow visitors or service people inside the broiler house unless they wear disinfected boots and clean clothing.

8. Avoid contact with other flocks.

9. When there are several age groups on the farm, always care for the youngest birds first when performing daily chores.

10. Obtain a laboratory diagnosis when disease problems arise.

11. Clean the house completely between each flock — ceiling, rafters, walls, floor, and surrounding premises. Also, repair, scrub, and disinfect all equipment — feeders, waterers, and brooders.

12. Rework built-up litter. When built-up litter is used, all caked and wet litter should be removed and replaced with fresh, clean litter before chicks arrive.

13. Cover floor with clean litter at least 3 in. deep after each clean-out. Wood shavings, processed pine bark, cane litter, and rice hulls are suitable litter materials. Avoid moldy or musty litter to prevent aspergillosis (mold growth in the bird's respiratory tract).

KEYS TO DISEASE PREVENTION

All the points of the above "Broiler Health Program" are important, but the following seven are keys in any prevention program; hence, they are elaborated upon in the sections that follow:

1. Sanitation.
2. Good nutrition.
3. Good environment.
4. Disinfection.
5. Vaccination.
6. Medication.
7. Parasite control.

SANITATION

There is no substitute for sanitation. This includes everything the birds come into contact with, either directly or indirectly. Good broiler sanitation involves the following practices:

1. Starting with clean chicks from a clean hatchery (which is free of pullorum and fowl typhoid), delivered by attendants wearing clean outer clothing and plastic boots. Do not ship chicks by common carrier, such as train or bus. Use new or cleaned disinfected boxes to transport them.

2. Starting the chicks out in a clean house with clean equipment and clean litter. This reduces the possibility of exposing them to any infective organisms that might have accumulated from previous broods of chicks.

3. Washing waterers daily in order to reduce contamination.

4. Removing dead birds from the house promptly and burning, placing them in a disposal pit, or recycling them through a rendering plant for feed.

5. Covering ventilation openings with small mesh wire in order to keep out wild birds, which may carry diseases and parasites.

6. Disinfecting the house and equipment between broods, selecting a suitable disinfectant and following the manufacturer's directions.

GOOD NUTRITION

Nutrition plays a major role in the health of the broiler flock and its ability to resist disease. The nutritional needs of healthy broilers are normally satisfied by an adequate supply of clean fresh feed, easily accessible, and containing the necessary amounts of the required nutrients.

GOOD ENVIRONMENT

Environmental conditions contribute directly to broiler flock health. A poor environment makes birds more susceptible to disease and contributes to the spread of many diseases.

Adequately ventilated, properly lighted, and relatively dry quarters are desirable. A good environment provides the following:

1. Shelter, protection, and comfort.
2. Convenient and adequate supplies of proper feed, clean water, and fresh air.
3. Equipment and facilities that are conveniently arranged for both birds and caretaker.

Poultry house ventilation requires constant attention. More and more broilers are being housed in completely environmentally controlled buildings, which are mechanically ventilated. However, in many of the warmer regions of the United States, broiler growers still rely on nonmechanical ventilation — curtains, panels, and windows.

If houses are mechanically ventilated, the operator should learn to adjust the fans for maximum effect. With inadequate ventilation, ammonium may build up to the extent that the eyes of the chicken and the grower are affected. With overventilation, the chicks may be chilled and heat may be lost needlessly. Also, a draft or a wave of cold air directly on the birds should be avoided at all times.

Where natural ventilation is used, the curtains should be adjusted as necessary. During hot weather, the available breeze should be used to cool the birds. In cold weather, the curtain opening should be adjusted to allow enough air movement to keep the litter dry and avoid excessive drafts which chill the birds. With quick weather changes, the curtains, panels, and/or windows should be adjusted promptly.

All-night lights will help keep the birds from crowding in corners or piling up in case of a disturbance during the night and will tend to increase feed consumption, especially during the hot summer months. Some producers set time clocks so that lights are off 1 hour each night (after the first few days) to condition birds in case of a power interruption. (See earlier section in this chapter on "Light.")

Wet litter is never desirable. It can be the cause of disease outbreaks or breast blisters. On the other hand, excessive dust may trigger respiratory trouble. Both conditions can be avoided with proper ventilation made possible by adequate heat. Dust has sometimes been a major problem with fan ventilation, perhaps because of lower humidity and overdrying of the litter.

DISINFECTION

The producer should know what constitutes proper cleaning, why it is necessary, and how to do it. Most disease germs can be killed only by disinfection following thorough cleaning. The virus diseases are best controlled by strict cleanliness. Effective cleaning begins with the removal of litter and manure. The cleaning process includes scrubbing with brushes until surfaces are visibly clean, using a good detergent, flushing with clean water, then applying an approved disinfectant. One advantage of a concrete floor is that lye (caustic soda) can be used effectively.

The presence of organic matter (dirt and manure) in the disinfecting solution or on the surfaces to be disinfected may prevent the solution from destroying disease organisms so that no disinfection results.

For a summary of the limitations, usefulness, and strengths of some common disinfectants, the reader is referred to Chapter 10, Poultry Health, Disease Prevention, and Parasite Control, Table 10-1.

VACCINATION

Protecting birds by vaccinating them against diseases common to the area is good insurance against costly disease losses. Most integrators have a recommended vaccination program, specifying the time to vaccinate and the type of vaccine to use. Generally speaking, broilers are vaccinated at the hatchery against Marek's disease, and vaccinated as needed for diseases common to the particular area. In Arkansas—the leading broiler producing state of the nation—for example, most producers vaccinate against infectious bronchitis and Newcastle, and some producers also vaccinate for fowl pox during the fall.

When birds are immunized against both Newcastle and bronchitis, it should be done between the first and fourteenth day of age. Both vaccines may be given at the same time. Intraocular or intranasal, dust, spray, and water-type vaccines are available. The method that has given the most suitable results should be used. Booster vaccination for Newcastle may be given when the birds are 4 to 5 weeks old.

Vaccines should be kept cool and used immediately after they are opened. Improper handling of vaccines can render the vaccines useless. Costly disease losses may be the result of using improperly handled vaccines.

MEDICATION

In order to prevent outbreaks of coccidiosis—a protozoan disease, which is an ever-present hazard when raising young chicks—one of several drugs is usually included in broiler feeds or a coccidiosis vaccination program is followed. Care must be taken to ensure that any feed containing a coccidiosis-preventive drug is removed well ahead of slaughtering time.

Other pharmaceuticals may be administered in either the feed or water from time to time as recommended by poultry pathologists, veterinarians, or poultry specialists. These chemicals may also have different withdrawal periods prior to slaughter of the flock. Detailed information on withdrawal times may be obtained from poultry advisors or by reading the directions found on the label.

PARASITE CONTROL

A wide variety of external and internal parasites can attack broilers. The prevention and control of these parasites is one of the quickest, cheapest, and most dependable methods of increasing broiler production with no extra birds, no additional feeds, and little more labor.

All flocks should be checked periodically for the presence of worms, especially tapeworms and large intestinal roundworms—*Ascaridia galli*.

Lice, mites, and flies are the most common external parasites of broilers. Heavy infestations cause irritation, discomfort, and general unthriftiness. The control of external parasites is simple if the right insecticide is used properly.

Pertinent information relative to the common external and internal parasites of poultry is presented in Chapter 10, Poultry Health, Disease Prevention, and Parasite Control, in Table 10-3, Parasites of Poultry; hence, the reader is referred thereto.

BROILER MANAGEMENT

Management is the key to success in the broiler business. The geneticist can lift the potential of production; the nutritionist can formulate a ration that can reach this potential; but without proper management the broiler will never get there.

Management gives point and purpose to everything else—to breeding, feeding, housing, health, and marketing. It can make or break a broiler enterprise. Fortunately, a broiler operation responds favorably to good management.

Even though many of the management practices used in producing replacement pullets and broilers are similar, there are differences in housing, equipment, feeding, growth rates, etc.

PRODUCTION GOALS

Broiler producers should aim for birds weighing 4.5 lb at 6 weeks of age, a feed conversion of 1.85, and mortality under 1.0%.

MANAGEMENT GUIDE

A management guide follows:

- **Family unit size** — 80,000 to 120,000 broilers.

- **House-unit size** — 20,000 to 50,000 birds per house; desirable size is 32 to 40 ft wide and of sufficient length to have the desired capacity.

- **Pen size** — 1,200 to 2,500 birds per pen.

- **Floor space** — 0.8 to 1.0 sq ft per bird; the upper limit for summer or the 4-lb bird and over.

- **Brooder space** — Not more than 750 chicks per 1,000-chick size hover; varies with season, insulation, and mechanical ventilation.

- **Litter** — 2 to 4 in. — less in hot weather.

- **Fountains** — Provide two chick founts for each 100 chicks at the start of the brooding period.

- **Water space** — Several types of automatic watering systems are available. If a trough-type waterer is used, allow 0.75 linear inch per bird. With circular waterers, the space per bird may be reduced by 20%.

- **Feeder lids** — One feeder lid per 100 chicks.

- **Feeder space** — Provide 2 in. per bird of trough space through 6 weeks of age, and 3 in. thereafter until market time at 7 to 8 weeks of age. When circular feeders are used, the linear space may be reduced by 20% because a circular feeder can accommodate more chicks at any given age than a longitudinal feeder.

- **Feedings** — Follow direction of feed manufacturer or formulator.

- **Lights** — Choose and follow a suitable lighting system (see earlier section in this chapter on "Light").

- **Vaccination** — Have chicks vaccinated against Marek's disease at the hatchery, and vaccinate against other diseases according to local needs.

- **Give arriving chicks a good start** — The following management practices will get newly-arrived chicks off to a good start: (1) fill the waterers a few hours ahead so that the chill will be removed (a water temperature of about 75°F is preferable for the first few days), (2) dump the chicks near heat and waterers by inverting the boxes in which they come, (3) about 3 hours after arrival, place chick mash or crumbles on feeder lids.

MANAGEMENT PRACTICES

Good management practices are herein summarized:

1. Start with quality chicks; get healthy chicks from reliable sources.
2. Debeak chicks when necessary.
3. Clean quarters before housing birds, and keep houses and equipment clean.
4. Keep litter clean, dry, and free from mold.
5. Brood birds carefully; have good sanitary management.
6. Supply adequate heat and ventilation.
7. Provide enough floor space.
8. Give adequate space for feed and water; have feed delivered to bin outside house.
9. Keep the feed level low in feeders in order to lessen wastage, but do not let feeders become empty. With full feeders, as much as 10% of the feed may be wasted.
10. Use all-night lights, except for 1 hour of darkness.
11. Adapt vaccination schedule to local needs.
12. Watch for disease; get prompt diagnosis when disease occurs; remove diseased birds from flock.
13. Dispose of dead birds promptly; have satisfactory disposal facilities.
14. Keep visitors out of houses; lock doors.

The above management practices involve a lot of little details, all of which add up to make broiler management of great economic importance.

Management practices meriting special consideration follow, with a section devoted to each.

CANNIBALISM

Cannibalism can become a serious problem in a broiler flock. It usually begins by chicks picking tail feathers, toes, vents, and eventually progressing to other parts of the body.

Cannibalism is caused by a combination of stress conditions; among them, overcrowding, excessive light intensity, insufficient ventilation, overheating, inadequate feed and water space, nutritional deficiencies, or being without feed or water for too long a period.

Obviously, it is relatively simple to alleviate these causes of cannibalism when they can be identified. Unfortunately, once cannibalism has started in a flock it is difficult to bring under control. It is easier to prevent cannibalism in a windowless house where light intensity can be reduced to a low level after the first week or 10 days. In open houses (houses with curtains or windows), debeaking may be necessary.

Experiments show that, when properly done, debeaking has little detrimental effect on bird performance. Birds may be debeaked at 1 day of age, or 10 to 14 days of age, by cutting off one-third of the beak with a commercial hot-blade-type debeaker. Directions supplied by the manufacturer of the debeaker should be followed.

DEBEAKING

The necessity of debeaking broilers to prevent cannibalism varies according to conditions. In open-sided houses, broilers are usually debeaked, especially during warm weather. In light-tight houses, reduced light intensity will usually suffice to prevent cannibalism.

Debeaking does create some stress. If it is too severe, growth will be retarded and some mortality will result.

In addition to preventing cannibalism, debeaking usually lessens mash feed wastage and improves the percentage of Grade A birds marketed.

(Also see Chapter 8, Poultry Management, section on "Debeaking.")

SEXING

Broilers are sometimes sorted into male and female lots at hatching time. They may be sexed by vent, color, or rate of feathering. (See Chapter 3, Incubation and Brooding, section on "Sexing Chicks.") The last two methods are possible if parental stocks were specially selected and mated.

Research shows that intermingling the sexes does not appear to have a detrimental effect on growth provided that at least 1 sq ft of floor space is provided per bird. If the floor space is smaller than this, the growth rate of females may be adversely affected.

It is not considered practical to remove males from an intermingled flock at an earlier age than the females. On the other hand, if the two sexes are separated, the males may be marketed first and the females may be given the extra floor space and retained longer for heavier weights. Other differences between the sexes are:

1. Males are about 1% heavier than females at hatching time.

2. Males grow faster than females and weigh more at a given age. At normal market age, males attain the same weight as females about 9 days earlier.

3. At 6 to 10 weeks of age, male birds lay down progressively less fat and show a higher incidence of breast blisters.

4. Females need less protein than males, because they grow more slowly.

5. Females show less response to chemical growth promoters than males.

6. Males convert feed to meat more efficiently than females — they require 0.1 to 0.2 lb less feed per pound gain.

In view of the above, it seems apparent that, during periods of high feed prices, the formulation of rations according to sex and the separation of the sexes has certain economic advantages. On the other hand, when feed prices are low, feeding the sexes separately formulated rations is unlikely to give economic advantage. It is noteworthy, however, that as strains are bred for increased growth, the weight variation between the males and females increases. Eventually, the normal weight spread between the sexes may become so great that it will be almost mandatory that the sexes be raised separately so that they may be marketed at comparable weights.

TEMPERATURE

The recommended temperature for broilers is as follows:

Age of Chick	Temperature*
First week	92° to 95°F
Second week	88° to 90°F
Third week	83° to 85°F
Fourth week	78° to 80°F
Fifth week	73° to 75°F
Sixth to eighth week	70°F

The thermometer reading should be taken at the level of the chicks' back under the hover. With infrared lamps, readings are difficult to take. It is suggested that the lamp be hung a minimum of 18 in. above the floor so that the chicks form

---

*To convert to Celsius, see Appendix.

a circle. Comfortable chicks will bed down evenly or form a doughnut-shaped ring under the light. If they scatter, it is too hot. For more heat, lower the lamp; for less heat, raise the lamp.

LIGHTING

Light management for broilers is very important.

The type of lighting system will be determined by the type of housing — by whether an open-sided or light-tight house is involved. However, the objective is always the same: to provide sufficient light to enable the birds to move about and see to eat and drink, but to avoid so much light that there will be cannibalism, excess activity, and piling. This objective is best accomplished when the intensity of illumination at bird level is approximately 0.35 to 0.50 footcandle.

The color of the light is also important, and it will vary with the type of housing. Red lighting in light-tight houses is effective in reducing cannibalism. But in open-sided houses or houses with windows, red light should not be used, because the birds cannot readily adjust from the bright, intense natural light of daytime to such a dim light.

When ready for market, most broilers are removed from the house at night in order to reduce bruising during the catching process. At this time, a small amount of blue light is recommended; it will enable the catchers to see, but minimize movement of the birds.

(Also see earlier section in this chapter on "Light.")

DISPOSAL OF DEAD BIRDS

Dispose of all dead birds promptly, as follows: incineration, pit, deep burial, or to a rendering plant.

LITTER

Newspaper under the waterer and feeder will help keep the feed and water clean as well as prevent the chicks from eating the litter during the first few days. Remove the top layer of paper daily to clean.

Provide a dry, absorbent litter material 2 to 4 in. deep. Dry sawdust, shavings, peat moss, vermiculite, straw, rice hulls, and other products are available. Use the one that is most convenient and economical. A shovelful of sand on the floor will make the cleaning easier.

MANURE HANDLING

One of the major problems on a large broiler establishment is the disposal of manure. The best solution is to spread the manure on fields as fertilizer or to use it as a feed for ruminants.

It is generally recommended that following the removal of the broiler flock, litter be removed from the house and the building and equipment be thoroughly washed and disinfected. However, reuse of old litter is sometimes practiced if the previous broods of broilers were relatively free from infectious diseases. If old litter is reused, wet and caked material should be removed and disposed of in the usual manner.

(Also see Chapter 8, Poultry Management, section on "Poultry Manure.")

MARKETING BROILERS

Most broilers are marketed when they are between 6 and 8 weeks of age. For the most part, marketing involves moving the birds from the house(s) in which they are produced to the processing plant.

Improper handling of broilers immediately prior to and during shipment will result in excess bruises, deaths, and lowered quality. Such losses may be minimized by preparing, catching, and transporting the birds as follows:

1. Discontinuing grit feeding at least 2 weeks prior to marketing. (Usually grit is not fed after 5 weeks of age.)

2. Avoiding exciting the birds on the day prior to catching, as it will cause them to hit the feeders and waterers and inflict bruises upon themselves.

3. Letting the feeders become empty about 2 hours before the catching crew arrives; and removing or elevating feeding and watering equipment to prevent bruises during catching. (Water should be removed just before the birds are caught; too early removal will result in excess dehydration.)

4. Catching and loading the birds properly, by: (a) using an experienced crew; (b) working under a dim blue light at night; (c) corralling them in small groups (of about 200 birds) to prevent smothering and undue injury; (d) grasping them by the shanks, with no more than four or five being carried at a time; (e) placing them in the crates gently; and (f) handling the crates carefully, preferably on pallets that can be moved easily with a hoist.

5. Driving loaded truck carefully, slowing down on turns and avoiding sudden stops.

6. Protecting the in-transit birds from extremes in weather. In cold weather, cover the truck to prevent chilling, as the latter will result in poor bleeding and downgrading of carcasses. In hot weather, protect against overheating in shipment by using open crates and avoiding lengthy stops en route.

7. Unloading the coops and removing the birds from the coops gently at the processing plant; and putting them under cover in an area that is adequately ventilated and comfortable.

Shrinkage, or weight loss from the time feed and water are removed until the birds are weighed at the processing plant, varies according to temperature and length of time involved; it ranges from 2% for a 3-hour period to 6% for a 15-hour period.

Broiler condemnations average about 2.5%, with great variation according to season (condemnations during the winter are about twice as high as during the summer) and between processing plants. Condemnations may be due to many things; respiratory ailments, diseases of the leukosis complex, and bruises. Sometimes bruises result in downgrading or in only partial condemnation; for example, a severe breast or wing bruise may be cut out or off as unacceptable for human consumption. Bruises are responsible for over half of the downgrading. All too often, they are due to mishandling of the birds during catching, loading, and transporting to the processing plant. Improper bleeding, bruising, eviscerating, and over-scalding in processing also result in condemnations. Broiler growers and processors should strive to minimize such condemnations, because they represent a direct monetary loss.

(Also see Chapter 16, Marketing Poultry and Eggs.)

PROCESSING

The slaughter and processing of broilers is an assembly line operation conducted under sanitary conditions. Although the processing procedure may vary, the sequence after unloading and weighing is usually as follows:

1. Birds are inspected twice: (a) live (antemortem), and (b) after slaughter (postmortem), when the carcasses and entrails are examined.

Fig. 13–12. Broilers being processed in a modern dressing plant. (Courtesy, Indian River International, Nacogdoches, TX)

2. Each bird is shackled to a conveying chain.

3. Birds are generally rendered unconscious by an electric shock and bled by severing the jugular vein.

4. After bleeding, the birds are conveyed through a hot water tank operated at about 142°F for about 1 minute.

5. Birds are then conveyed through a rougher which removes the feathers.

6. Following this, the carcass goes through a special machine to remove pinfeathers and cuticles.

On the eviscerating line, the carcass and exposed viscera are inspected by a health inspector; the kidneys, lungs, head and feet are removed; and the carcass is singed. The carcass then passes through a cooling tank containing ice water before being graded, sorted by weight, and prepared for delivery to consumer markets.

The processing of a 3.5 lb (liveweight) broiler will result in about a 12% loss in blood and feathers, eviscerated weight of 66% of liveweight, and chilling gains of 9%; with variations according to weight and sex. The heavier the bird, the lower the percentage of blood and feathers; males have a higher percentage loss of blood and feathers than females; and the chilling gains on females are greater than on males.

In 1989, processors sold about 20.4% of their broilers as eviscerated whole carcasses, 51.9% as cut parts, 15.4% as further processed, and 12.3% in other forms.

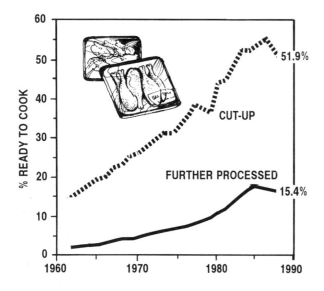

Fig. 13–13. Percentage of broilers cut-up or further processed. (From: *Current Report* [CR-204], Oklahoma State University, Stillwater)

STANDARDS, GRADES, AND CLASSES[1]

The U.S. standards of quality and grades for poultry are used extensively for trading purposes. Many retail distributors require that all ready-to-cook poultry which they handle be identified with an official grade mark. Ready-to-cook poultry must have been inspected for wholesomeness to be eligible for grading. Processors applying for resident grading service agree to pay the cost of the service. Graders may be either state or federal employees. The service is made available at reasonable cost in all parts of the country. Grade A, B, or C may not be used in connection with labels bearing the official inspection mark, unless the poultry has been graded by a licensed grader.

The official classes of young chickens are as follows:

■ **Cornish game hen**—A Cornish, or Rock Cornish, game hen is an immature chicken about 4 to 6 weeks old, of not more than 2 lb ready-to-cook weight. Broiler chickens usually containing some Cornish breeding in the cross are used.

■ **Broiler or fryer**—A broiler or fryer is a chicken 6 to 8 weeks old, of either sex, with tender meat and flexible breastbone cartilage.

■ **Roaster**—A roaster is a chicken usually 9 to 11 weeks old, of either sex, with tender meat and breastbone cartilage that is less flexible than that of a broiler or fryer.

---

[1]Improved breeding and feeding have lowered the ages necessary for chickens to meet the USDA standard market classes and weights of poultry.

■ **Capon**—A capon is a surgically unsexed male chicken usually 20 to 24 weeks old with tender meat and weighing 12 to 14 lb.

FAST FOODS

Kentucky Fried Chicken, the fast food chain which Col. Harland Sanders of Kentucky began as a franchise operation in 1956, and other fast food services, have had a significant impact on the marketing of broilers. In 1989, 28.6% of the total broiler production was sold through fast food service outlets and carry-out restaurants.

WHAT'S AHEAD FOR BROILERS

In the past, the broiler industry has strived for new goals; and when they have been reached, it has revised and raised them. More changes lie ahead. Some of these follow:

1. **Concentrated areas.** The broiler industry will continue to be concentrated in areas with (a) abundant and cheap feed, and (b) a mild climate. The latter makes for lower housing and energy costs. Cheap labor will no longer be a factor in determining location.

2. **Pollution control.** Pollution control (manure or dead bird disposal, odor, and flies) will be a critical factor in site selection for large broiler operations in the decades to come, favoring remoteness from urban development and compliance with state and federal regulations.

3. **Manure-litter management will evolve.** Manure and litter will no longer be looked upon as an unwanted source of pollution. Rather, it will be wanted as a fertilizer, as a feed, and for other uses.

4. **Larger units and more commercial.** By the year 2000, the 8 largest broiler companies in the United States will control 65% of the broiler business and their brand name

Fig. 13–14. Larger broiler units, owned by fewer companies lie ahead. (Courtesy, Gold Kist, Athens, GA)

products will be household words. This trend to bigness will continue

5. **More ownership by integrators.** Large and well-financed feed companies, processors, co-ops, and others will integrate through ownership, rather than contract, in the future.

6. **More automation.** Higher priced labor, along with more sophisticated equipment, will make for increased automation all along the line. Among such developments are a suction system of moving live birds at market time, and electronic broiler grading.

7. **Improved housing and environmental control.** Physics, engineering, and physiology will be combined in such a way as to bring about improved houses.

8. **Cages (batteries) will increase.** More and more broilers will be raised in batteries as cages are perfected.

9. **Bird density will increase.** With improved housing, environmental control, and use of cages, bird density will increase.

10. **Mortality lowered.** Drugs and vaccines will continue to lower mortality. Also, breeding stock will be selected for greater livability.

11. **Quality will be improved.** U.S. broilers are of high quality. Yet, further improvements will be made; among them lessening of bruises and fat.

12. **Broiler consumption will increase.** Per capita consumption of broilers will increase.

13. **More processed chicken.** More broilers will be further processed and marketed as precooked frozen fried chicken.

14. **Branded products will increase.** Packing of broilers under processor's brands will continue to increase, for some brands consistently sell at premium prices.

15. **Drug residues.** The FDA will continue to keep the pressure on drug residues.

16. **Broiler futures markets will increase.** Broiler futures markets will be more widely used.

17. **Fast food outlets will increase.** Fast food outlets will increase and use more broilers.

18. **Some costs will be reduced.** The cost of production and marketing will be further reduced by increased use of poultry manure as a feed, improved management practices and equipment to conserve energy, new preservation methods for poultry meat, further genetic improvements, and improved disease control.

19. **Some costs will increase.** Feed, fuel, packaging, and labor prices will increase, likely offsetting any production and marketing cost reductions.

ROASTERS

Roasters are young chickens that are grown similarly to broilers, but they are older and heavier; they range in age from 9 to 11 weeks and weigh 6 to 8 lb. The management program for roasters is quite similar to that of broilers up until about 6 weeks of age, following which some changes in feeding and management should be made.

Pointers pertinent to roasters follow:

■ **Breeding** — Certain strains of meat-type birds are suited to rapid growth to roaster weight, others are not. So, it is important to select a strain of birds that has the genetic potential to attain roaster weight rapidly and economically.

■ **Feeding** — Roasters should be given the kind and amount of feed recommended for broilers during the first 6 weeks. After changing to the finishing mash, they should be supplied cracked corn in the afternoon. Gradually increase the grain until the birds are getting equal amounts of corn and mash or pellets at 12 weeks of age. (See Chapter 7, Poultry Feeding Standards, Ration Formulation, and Feeding Programs.)

It requires approximately 2.5 lb of feed to produce 1 lb of live roaster, whereas 1.9 lb of feed will produce 1 lb of broiler in an efficient operation. Sex is also a factor. Females must be kept longer than males to attain the desired liveweight; besides, they are poorer converters of feed to meat than males.

■ **Sex** — Roasters may be of either sex. However, males grow more rapidly than females, with the result that they reach market weight more quickly. It takes only about 9 weeks for males to reach the required weight, whereas it takes about 11 weeks for females. For this reason, it is strongly recommended that chicks intended as roasters be sex-separated at hatching.

■ **Management** — A 7-lb roaster requires about 1.5 sq ft of floor space, in comparison with the 0.8 sq ft required by a 4-lb broiler. Likewise, roasters need more room to eat and drink than broilers; they should have 2 in. of feeder space and 1 in. of water space per bird after 8 weeks of age. Also, roasters, being heavier than broilers, require more ventilation.

■ **Breast blisters** — A much higher incidence of breast blisters is encountered in roasters than in broilers. Heavy birds seem to sit more than light birds; and males sit more than females. Thus, the incidence of breast blisters is related to body weight, sex, and the material with which the breast comes in contact.

Breast blisters in roasters make for tremendous economic losses. The blisters must be cut out at processing, and this downgrades the carcass. Since roasters are sold as whole-body birds, breast blisters are more serious than in broilers. In the latter, a bird with breast blisters may be cut in parts and only the breast downgraded.

The incidence of breast blisters in roasters may be lessened, but not eliminated, by (1) using females instead of males (which is seldom practical), (2) keeping the litter dry, (3) keeping deep litter, (4) feeding more often (when birds are eating, they are not sitting), and (5) stirring the birds occasionally.

CAPONS

Capons are surgically unsexed male chickens. They do not develop all the normal male characteristics; the head, comb, wattles, and earlobes remain small and pale. They do not crow or fight like cockerels; the hackle and saddle feathers tend to grow longer than on normal males; and the dark meat of capons is lighter-colored than meat of cockerels of the same age. When marketed, capons range up to 20 to 24 weeks of age and weigh 12 to 14 lb.

Caponized birds in which development of masculine characteristics continues are known as *slips*. They result from failure to remove all of the reproductive organs. Slips develop and look like cockerels, with the degree of such appearance determined by the amount of the organ left.

Pertinent information about capons and caponizing follows.

■ **Objectives of caponizing**—The principal objective of caponizing is to improve the quality of the meat produced. This is the same objective that farmers have in castrating male calves or pigs. The improvement in the quality of the meat ensures a better price for capons than for cockerels which have become staggy. There is also a slight weight advantage for capons as compared to cockerels when the birds are 20 to 24 weeks old.

■ **Breeds to caponize**—The breeds and crosses of chickens used for broiler production make satisfactory capons; although some strains make better capons than others. Leghorn cockerels make satisfactory small capons, but they are not likely to be commercially grown because of their inefficient feed utilization.

■ **Size of cockerels for caponizing**—Cockerels are most suitable for caponizing when 2 to 4 weeks old. At this age, they will vary from about ¾ lb to almost 2 lb in weight. At this age and weight, the birds can be caponized faster and there are fewer slips.

■ **Preparation for the operation**—In preparation for caponizing, feeds should be withheld from the cockerels from 12 to 24 hours and water should be withheld 12 hours. Usually, an antibiotic is either (1) injected into the birds at the time of surgery, or (2) fed for a week prior to the operation or a week following.

■ **The operation**—The steps in the operation follow:

Step 1: Pluck the feathers on the right side between the thigh and third rib from the point of the rib down to the rib joints. The rib joints are where the upper and lower ribs meet. Pull a few feathers at a time with an upward motion toward the back.

Step 2: Make an incision ½ in. long between the last two ribs, extending from the height of the hip joint down to the rib joints (see Fig. 13–15). If the incision is properly placed, no large blood vessels will be cut and there will be little bleeding. Press the tip of a finger or thumb on the point between the ribs directly in front of the hip point. This is done to hold the skin firm while the incision is being made and to hold the thigh muscles back so that they will not be cut. The rib joints should not be cut. The incision should be in back of the last rib in birds 2 to 3 weeks old.

Step 3: Place the spreaders in the incision to hold the ribs apart. This exposes the membrane which lines the abdominal cavity. The incision should extend high enough toward the back so that the membrane is cut where it is grown fast to the ribs. If necessary, the incision can be made larger after the spreaders are placed. Cut upward if the membrane is not pierced where it is attached to the ribs.

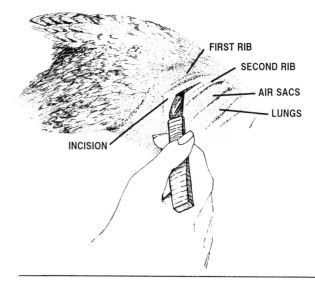

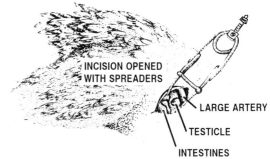

Fig. 13–15.   Where to make the incision and what is seen.

Step 4: Take the hook and tear the membrane that lines the abdominal cavity.

Step 5: Take the removers or probe and push the intestines aside to expose the testes. The testes are usually creamy yellow in color, although they may be dark-colored, especially in birds that have a good bit of black pigment in their plumage. They vary in size from about the size of wheat kernels to navy beans. The testes are located near the backbone and in front of the kidneys and back of the lungs. They can now be removed. Remove the lower one first. If both organs cannot be removed from the right side, an incision must be made on the left side.

■ **Care of the birds after the operation**—After the birds have been caponized, they should be placed in a house and given access to the same ration that they were previously receiving. If they were properly starved and in good physical condition before caponizing, losses from the operation should not exceed 1 to 2%.

Wind-puffs are quite common among recently caponized birds. Their occurrence need not cause any alarm or indicate faulty technique. Wind-puffs are caused by air escaping from the air sacs of the abdominal cavity which were punctured by the operation; the air passing through the opening which was made between the ribs, collects under the skin which has healed. The remedy is to puncture these puffs with a sharp knife daily until they cease forming, which will usually be within 2 weeks after the operation.

■ **Rations**—Rations suitable for growing broilers are satisfactory for capons during the first 6 weeks. After changing to the finishing mash, supply cracked corn to capons in the afternoon. Gradually increase the grain until the birds are getting equal amounts of corn and mash or pellets at 12 weeks of age. (See Chapter 7, Poultry Feeding Standards, Ration Formulation, and Feeding Programs.)

A deep yellow carcass color is desired; hence, diets are usually altered during the last 2 months of the growing period in order to induce the yellow pigment.

■ **Marketing**—Capons are considered a delicacy and should sell for premium prices. Thanksgiving and Christmas, and, to a lesser extent, Easter, are considered the best marketing period for capons. Persons who have once enjoyed roast capons of top quality are likely to be repeat customers year after year.

QUESTIONS FOR STUDY AND DISCUSSION

1. Define each of the following designations: broiler, fryer, roaster, and capon.

2. What caused the phenomenal growth in broilers, from 34 million in 1934 to 5.5 billion in 1989?

3. List five points that portray a profile of the U.S. broiler industry as it exists today.

4. What were the primary forces back of the vast majority of U.S. broilers being produced under contract? Is the dominance of contract production good or bad?

5. Historically, it has been said that "bankers have a fear of feathers." What does this mean? Is this feeling justified?

6. In 1990, how many broilers could one farm family care for during a year, in 5 to 6 broods?

7. In a typical broiler contract, what is provided by each (a) the grower and (b) the integrator?

8. What are broiler futures and options? Of what value is such a contract, and to whom is such a contract of interest?

9. If a family plans to grow broilers on contract as the only enterprise, it is generally assumed that 5 to 6 buildings, holding a total of 80,000 to 120,000 birds, and 5 to 6 broods per year, will be necessary in order to provide a good living. How much capital will this size operation entail, and what average net income may be expected?

10. List the three chief broiler production costs. How would you go about lowering each of these costs?

11. Why has Cornish breeding been used in broiler production?

12. List and discuss the characteristics of economic importance to which special attention should be given in a selection and breeding program to produce broilers.

13. What are the main differences in the housing and equipment for broilers compared to replacement pullets and layers?

14. Why is there so much interest in producing broilers in cages or batteries?

15. Outline a broiler health program.

16. Discuss each of the following keys to broiler disease prevention: sanitation, good nutrition, good environment, disinfection, vaccination, medication, and parasite control.

17. Give high, but achievable, broiler production goals relative to (a) market weight, (b) market age, (c) feed efficiency, and (d) mortality.

18. Give (in 1, 2, 3, etc. order) practical management guides for a broiler producer.

19. Discuss each of the following management practices as applied to broilers: cannibalism, debeaking, sexing, temperature, lighting, disposal of dead birds, litter, and manure handling.

20. What marketing changes are occurring in broilers?

21. Discuss the impact of Kentucky Fried Chicken, and other fast food services, on the broiler industry.

22. What changes lie ahead for the broiler industry?

23. Would you recommend that a young person launching a career become a broiler producer? If so, how should this person proceed?

24. Discuss each of the following as it pertains to roasters: age and weight, breeding, feeding, sex, management, and breast blisters.

25. Discuss each of the following as it pertains to capons: objectives, breeds, size and age, preparation for the operation, care after the operation, rations, and marketing.

SELECTED REFERENCES

Title of Publication	Author(s)	Publisher
Commercial Chicken Production Manual, Fourth Edition	M. O. North D. D. Bell	Van Nostrand Reinhold Co., New York, NY, 1990
Feeds & Nutrition, Second Edition	M. E. Ensminger J. E. Oldfield W. W. Heinemann	The Ensminger Publishing Company, Clovis, CA, 1990
Feeds & Nutrition Digest	M. E. Ensminger J. E. Oldfield W. W. Heinemann	The Ensminger Publishing Company, Clovis, CA, 1990
Poultry: Feeds and Nutrition, Second Edition	H. Patrick P. J. Schaible	The AVI Publishing Company, Inc., Westport, CT, 1980
Poultry Husbandry I	C. J. Price	Food and Agriculture Organization of the United Nations, Rome, Italy, 1969
Poultry Husbandry II	C J. Price J. E. Reed	Food and Agriculture Organization of the United Nations, Rome, Italy, 1971
Poultry Keeping in Tropical Areas	W. Thomann	Food and Agriculture Organization of the United Nations, Rome, Italy, 1968
Poultry Meat and Egg Production	C R. Parkhurst G. J. Mountney	Van Nostrand Reinhold Co., New York, NY, 1988
Poultry Production, Twelfth Edition	M. C. Nesheim R. E. Austic L. E. Card	Lea & Febiger, Philadelphia, PA, 1979
Poultry Products Technology, Second Edition	G. J. Mountney	The AVI Publishing Company, Inc., Westport, CT, 1976
Poultry Science and Practice, Fifth Edition	A. R. Winter E. M. Funk	J. B. Lippincott Company, Chicago, IL, 1960
U. S. Broiler Industry	F. A. Lasley, *et al.*	USDA, ERS, Agricultural Economics Report 591, Washington, DC, 1988
U. S. Egg and Poultry Statistical Series, 1960–1989	M. R. Weimar S. Cromer	USDA, ERS, Statistical Bulletin, Washington, DC, 1990

Fig. 14–1.   Turkey gobbler. (Courtesy, USDA)

TURKEYS

CHAPTER

14

Contents   Page

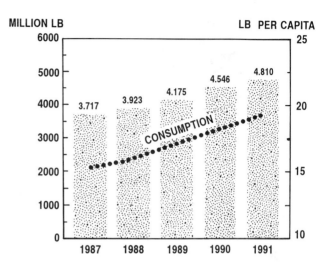

Fig. 14–2. Growth of turkey production and consumption. (Adapted by the author from *Livestock and Poultry, Situation and Outlook Report*, USDA, ERC, LPS-44, November 1990, cover page)

Turkeys are native to the New World. Cortez found them when he landed in Mexico in 1519, and the Pilgrims found them when they landed at Plymouth Rock. They have been considered traditional Thanksgiving and Christmas fare since the Pilgrims hunted wild turkeys to grace their tables the first Thanksgiving Day.

At one time in American history, the turkey vied with the bald eagle as the national bird, but the latter won the honor by a small margin of the United States Congress.

Today, turkey production and marketing is a highly efficient process, and turkey meat consumption is substantial in every month of the year. Yet, more than a third of our annual turkey consumption occurs during our two holiday months.

A profile of the U.S. turkey industry in 1989 (or in the year specified) follows:

■ **Growth of industry** — The turkey industry has grown from 18,476,000 birds raised in 1929, which produced a gross income of $60,027,000, to 260,230,000 raised in 1989, which produced a gross income of $2,239,058,000.

■ **Fewer, but bigger, turkey farms** — In 1910, 870,000 farmers raised 3⅔ million turkeys, or an average of 4 turkeys per farm. By 1987, the number of turkey farms had declined to 7,347, but total production rose to 243 million birds and average farm output increased to 33,000 birds.

There has been a general trend to fewer and larger turkey farms since 1960. Farms that grew 100,000 or more birds in 1987 represented nearly 9% of all farms growing turkeys. These few very large operations grew 140.4 million turkeys, or 58% of the total in 1987 (see Fig. 14–3).

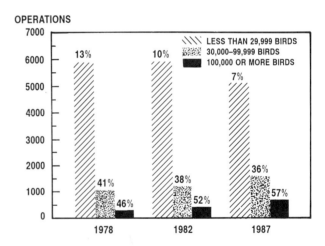

Fig. 14–3. Number of turkey grow-out operations by size category and share of total raised. (Adapted by the author from *Livestock and Poultry, Situation and Outlook Report*, USDA, ERC, LPS-46, February 1991, p. 39)

■ **Younger and heavier** — In 1920, the standard weight of toms at 26 weeks of age was 21.6 lb. In 1989, nationwide, toms were marketed at 17½ weeks of age and an average weight of 26.35 lb; and hens were marketed at 14¼ weeks of age and an average weight of 14.68 lb.

■ **Feed efficiency**—The feed required to produce a pound of live market bird has been reduced from 4.5 lb in 1940 to 2.67 lb in 1989, nationwide. In 1989, the top turkey flocks were achieving a feed/pound of turkey of 2.5.

■ **Price and income**—In 1989, farmers averaged 41.1¢ per pound liveweight for turkeys, for a gross turkey income of $2,239,058,000, which represented 1.4% of the U.S. farm income.

■ **Per capita consumption**—The per capita consumption of turkey on a ready-to-cook basis has increased from 1.7 lb in 1935 to 17.1 lb in 1989.

■ **Leading states**—The leading states, in order of numbers, in 1989 were: North Carolina, Minnesota, California, Arkansas, Missouri, Virginia, Indiana, Pennsylvania, Iowa, and South Carolina. The five leading turkey-producing states accounted for 62% of U.S. output.

■ **Trend to heavy breeds**—Heavy breed turkeys have increased, while light breeds have practically ceased to exist, due to (1) greater labor efficiency in caring for big birds, and (2) the great expansion of cut-up and further processed turkey.

■ **Cost of production**—In 1989, it cost 40.4¢ per pound liveweight to produce turkeys, of which 26.7¢ or 66% was for feed.

■ **Processors**—Turkey processing plants have become fewer but bigger. In 1989, total sales of the top 10 poultry processors was $15.1 billion—over 12 times the comparable 1979 total.

■ **Contract production**—In 1990, 90% of the nation's turkeys were produced under contract.

■ **Increased automation and labor efficiency**—All new turkey production units are highly automated for feeding and watering, with the result that one turkey producer can care for 50,000 to 60,000, or more, birds.

■ **Trend to confinement rearing**—Although turkeys are still raised on the range, confinement rearing is increasing rapidly.

■ **More environmental control**—The trend is toward environmentally controlled housing, with light and temperature regulated. The major advantage of environmentally controlled housing is better feed conversion, especially during the winter months.

■ **Trend to cage rearing**—Although the vast majority of turkeys are reared on the range or on the floor, cage rearing is increasing.

■ **Poult mortality**—There is an average mortality of 9 to 10% of all poults that are put in the brooder house.

■ **Increased cut-up and further processed items**—The turkey industry has successfully marketed increased quantities of turkey, either cut-up, or in further processed items such as turkey rolls, roasts, pot pies, and frozen dinners. As a result, consumers now enjoy turkey in many forms throughout the year, in addition to retaining it as the traditional holiday bird. By the late 1980s, about 80% of market turkeys were either cut-up or further processed (see Fig. 14–4).

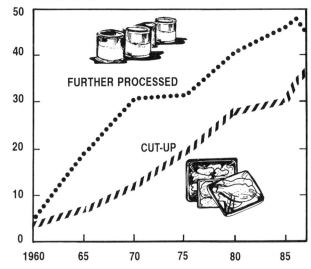

Fig. 14–4. Percentage of cut-up or further processed turkeys. (Adapted by the author from *1989 Agricultural Chartbook*, USDA, Agriculture Handbook, No. 684, p. 93)

■ **Supporting enterprises created**—In addition to improving the production and marketing of meat, turkey producers and processors have helped create many new supporting enterprises, including plants for manufacturing special processing equipment and packaging materials, food processing plants, and plants for manufacturing fertilizers and turkey by-products.

BUSINESS ASPECTS

With the increase in size and integration of turkey enterprises, the business aspects have become more important. More capital is required, knowledge of contracts is important, competent management is in demand, good records are essential, computers have come in, and such things as tax management, estate planning, and liability require more attention.

(Also see Chapter 17, Business Aspects of Poultry Production.)

FINANCING

Local banks and the Production Credit Association largely finance the turkey industry through credit extended to feed manufacturers, processors, hatcheries, and integrated operators. Much of the money is loaned at the local level to growers with mortgages as collateral.

Throughout the turkey industry, integration has increased. The principal reasons for this trend have been (1) the need for more financing, and (2) the fact that more liberal credit is available to integrated operations, such as processors, than to individually owned operations.

LABOR REQUIREMENTS

Efficiency of labor is an important factor in determining profits in the turkey business. Generally speaking, the more turkeys raised per worker the lower the production costs, and the bigger the operation the more feasible it is to automate. Greater labor efficiency has been a primary force back of increased number of turkeys per farm.

Many devices are available for reducing the number of hours required per turkey raised, with automatic feeding and watering systems heading the list.

TURKEY CONTRACTS

When turkeys are grown under contract, contractors generally provide poults, feed, and services, while producers provide labor and facilities. Two common types of turkey contracts follow:

■ **Profit-sharing contract** — The contractor provides the following inputs on an "at cost basis": (1) feed, (2) services, (3) medication, (4) insurance, and (5) interest charges. The cost of these inputs, plus hauling charges, are first deducted from the gross receipts from the sale of turkeys. Seventy percent of the remaining receipts are distributed to the contractor and 30% to the grower until the cost of the poults are paid, then the balance of the receipts (if any remain) are distributed 70% to the grower and 30% to the contractor.

■ **Base pay and bonuses contract** — The contractor provides poults, feed, medication, and services, and pays the grower according to the number of weeks in production plus bonuses for livability, feed conversion, and grade yield.

RECORDS

Good records are exceedingly important; both from the standpoint of management and the business aspects. Records should be kept of feed consumption, mortality, and vaccination dates. If the birds go off feed or if there is a sudden increase in mortality, immediate steps should be taken to determine the trouble.

COSTS AND RETURNS

Costs and returns in the turkey business are pertinent to decision making; when deciding whether or not to enter the business, and in determining how well an established enterprise is doing.

Costs and returns of turkey production differ from year to year and from area to area. Moreover, the relative cost of each item contributing to the total cost of production differs; for example, building and equipment costs are higher for total confinement than for range rearing, and feed costs vary considerably from year to year.

Table 14-1 shows the annual cost and returns from turkeys during the 29-year period 1955 to 1983. Note that highest returns, 17.2¢ per lb ready-to-cook (RTC), were realized in 1978. Four years of high returns in 1977–80 were followed by losses of 3.0¢ per lb. Note, too, that producers suffered net losses in 8 of the 29 years between 1955 and 1983.

LOWERING COSTS

Not all turkey growers had the production costs shown in Table 14-1. Some produced for less; some for more. Moreover, the differences in production costs between growers are large enough to account for the success or failure of the enterprise. So, persons planning to become turkey growers should set goals to produce turkeys for less than the average of their competitors.

Costs can be lowered! The place to start is the highest cost item. As shown in Table 14-1, the highest production cost is for feed. Because feed represents 66% of the total cost of production (based on costs of production in 1989), any change in feed cost or efficiency will dramatically influence the cost of production. Feed costs can be lowered through (1) locating the turkey operation in an area where major feed ingredients are produced, (2) using improved feed formulations, (3) selecting efficient turkey strains, and (4) lowering mortality (high mortality makes for increased feed consumption per pound of turkey).

The grower can do little to lower the purchase price of poults. However, most turkey growers buy 10% more poults than they expect to raise, simply because this figure represents the average mortality; thus, a saving in the cost of poults can be effected by lowering mortality.

One worker can care for 50,000 to 60,000, or more, turkeys, augmented by hiring certain services such as debeaking, vaccinating, moving to range, and loading for market. Generally speaking, labor costs can be lowered by increased automation. Hence, mechanization should be maximized to the extent that the savings thereby exceed the cost of the labor that is eliminated.

Fixed costs (costs for building and equipment, taxes, maintenance, building and equipment insurance, interest on working capital, and flock insurance) vary widely. Confinement rearing facilities generally run higher than range rearing facilities. Buildings may vary from shelters to completely environmentally controlled facilities. The value of land varies widely, and it usually appreciates; therefore, the real cost of land is difficult to estimate. The turkey producer should also be aware that facilities and equipment change rather rapidly. For this reason, it is important that the grower visit a number of successful operations, and learn what is new and working well. The wise grower will avoid buying either something old or something new and untested.

More rearing of turkeys in total confinement is expected in the future. Processors need turkeys throughout the year to maintain labor forces and operate plants more efficiently. Labor and land requirements are much less in total confinement production systems than in range production systems. The disadvantage is that more long-term fixed investment capital is required.

TABLE 14–1
ANNUAL COST AND RETURNS FOR TURKEYS[1]

Year	Live Turkey Production Costs			Ready-to-Cook Turkey				
	Feed	Other	Total	Production Cost[2]	Marketing Cost[3]	Total Cost to Wholesale	Wholesale Price	Net Returns[4]
	(cents/lb)	(cents/lb)	(cents/lb)	(cents/lb)	(cents/lb)	(cents/lb)	(cents/lb)	(cents/lb)
1955	20.0	7.0	27.0	33.8	8.8	42.6	47.9	5.3
1956	19.1	6.7	25.8	32.3	8.6	40.9	45.0	4.1
1957	18.3	6.3	24.6	30.8	8.3	39.1	39.0	−0.1
1958	17.8	6.1	23.9	29.9	8.3	38.2	42.5	4.3
1959	18.0	5.9	23.9	29.9	7.9	37.8	37.6	−0.2
1960	15.6	5.8	21.4	26.8	7.6	34.4	43.5	9.1
1961	15.2	5.7	20.9	26.1	9.7	35.8	35.6	−0.2
1962	15.0	5.9	20.9	26.1	7.6	33.7	34.8	1.1
1963	15.1	6.0	21.1	26.4	7.4	33.8	36.5	2.7
1964	14.6	6.2	20.8	26.0	7.1	33.1	33.6	0.5
1965	14.4	6.4	20.8	26.0	6.9	32.9	37.0	4.1
1966	14.7	6.5	21.2	26.5	7.9	34.4	38.0	3.6
1967	14.3	6.6	20.9	26.1	9.4	35.5	33.5	−2.0
1968	13.2	6.6	19.8	24.8	8.1	32.9	32.4	−0.5
1969	13.5	6.7	20.2	25.3	8.2	33.5	36.3	2.8
1970	14.0	6.8	20.8	26.0	8.3	34.3	40.9	6.6
1971	13.3	6.9	20.2	25.3	8.4	33.7	37.5	3.8
1972	13.5	7.0	20.5	25.6	8.5	34.1	36.6	2.5
1973	25.6	7.5	33.1	41.4	9.2	50.6	64.5	13.9
1974	22.5	8.2	30.7	38.4	10.5	48.9	47.0	−1.9
1975	22.1	8.6	30.7	38.4	11.0	49.4	55.1	5.6
1976	22.4	9.0	31.4	39.3	11.6	50.9	51.0	0.1
1977	22.6	9.0	31.6	39.5	11.9	51.4	56.2	4.8
1978	22.1	9.6	31.7	39.6	12.1	51.7	68.8	17.2
1979	25.3	10.5	35.8	44.8	13.4	58.2	67.0	8.8
1980	26.0	11.0	37.0	46.3	14.6	60.9	64.6	3.7
1981	30.5	11.6	42.1	52.6	14.6	67.2	64.2	−3.0
1982	24.5	11.8	36.3	45.3	14.8	60.1	62.2	2.1
1983	26.1	13.2	39.3	49.1	15.7	64.8	62.5	−2.3

[1]*The U.S. Turkey Industry,* USDA, ERC, Agricultural Economic Report No. 525, 1985.

[2]Production cost is calculated by the division of live production cost by the dressing percentage to convert to a ready-to-cook (RTC) basis. Net returns are to production and marketing through the wholesale level.

[3]Marketing costs to transform live turkeys on the farm into ready-to-cook turkey at the wholesale level.

[4]Net returns per pound ready-to-cook.

INCREASING RETURNS

Returns on turkeys not on contract are determined primarily by the price of turkeys. One cent per pound higher price on a 20-lb tom makes for 20¢ more per bird.

Contract turkey growers can generally increase their returns by doing a better job. Thus, a grower on a *profit-sharing contract* with the receipts (above feed, services by contrac-

tor, medication, insurance, interest, and cost of poults) distributed 70% to the grower and 30% to the contractor can make money by saving money. Likewise, a grower on a base pay plus bonuses contract can increase returns by improving the efficiency of those inputs in production on which the bonus is based—usually livability, feed conversion rate, and market yield grades.

BREEDING

Although raising turkeys for market is the largest phase of the turkey industry, the breeding of turkeys is the foundation.

Until about 1930, virtually no progress or improvement was evident in turkey breeding. In the early 1930s, Jessie Throstole sailed from England, accompanied by 20 hens and 4 toms of a broad-breasted variety. Mr. Throstole settled in British Columbia, Canada. In time, some of these turkeys found their way into Washington, Oregon, and California, where they were crossed on existing lines. Out of this evolved the Broad-Breasted Bronze.

In the mid-1950s, Random Sample Tests wrought further improvement, which made it possible to compare strains. At this point and period of time, greater demands than ever were placed on breeders. Not all of them measured up.

Some developed excellent types, but lost reproduction. Others developed great egg layers, but could not compete in size, eviscerated yield, and consumer acceptability.

For an understanding of breeding and breeds of turkeys, it is important to know the parts of a turkey (see Fig. 14–7).

(Also see Chapter 4, Poultry Breeding.)

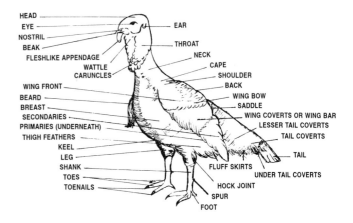

Fig. 14–7. Parts of a turkey. Turkey breeders and growers should know the language that describes and locates the different parts of the bird.

Fig. 14–5. Mature Broad-Breasted White males in feed conversion studies. (Courtesy, Nicholas Turkey Farm, Sonoma, CA)

Fig. 14–6. Pedigree female pens with trapnests. (Courtesy, Nicholas Turkey Farm, Sonoma, CA)

BREEDS AND VARIETIES

Seven standard varieties, popularly called breeds, of domesticated turkeys are recognized by the American Poultry Association and described in detail in *The Standard of Perfection*; namely: Bronze, White Holland, Bourbon Red, Narragansett, Black, Slate, and Beltsville Small White. Including wild turkeys, there are perhaps a dozen nonstandard varieties, chief of which are the Broad-Breasted Bronze and the Broad-Breasted White. Presently, only the Broad-Breasted White (also called Large White) variety of turkeys is important commercially.

White-plumaged turkeys are not albinos, and have none of the weaknesses of albinos. They originated as mutations, or sports, in colored flocks; and such mutations still occur. The white color is due to a recessive nonsexed-linked or autosomal gene that prevents almost completely the appearance of any pigmentation in the plumage, shanks, feet, and beak. The eyes, however, are not affected, so white turkeys tend to possess the eye color determined by the color pattern factors carried, but suppressed by the color-inhibiting factor responsible for white plumage.

White-feathered turkeys have more feathers than colored turkeys and are difficult to dry-pick. However, with modern pick methods, white turkey varieties can be picked as easily as colored varieties and have the advantage that any pinfeathers present are inconspicuous and do not lower the market grade as they do in colored varieties.

(Also see Chapter 4, Poultry Breeding, section on "Breeds, and Breeding Turkeys.")

BROAD-BREASTED BRONZE

The Broad-Breasted Bronze turkey originated in England. It was imported into Canada in about 1930, thence into the United States about 1935. Within a few years after its introduction into the United States, it became the most widely grown of all varieties. Subsequently, it was used in crossing to produce the Broad-Breasted White and the Beltsville Small White.

In color, the Broad-Breasted Bronze resembles the almost-extinct standardbred Bronze, but it tends to have buffy-white instead of pure-white feather tips and lacks the brilliant copper-colored bronzing in back, tail, and upper thighs for which the standardbred Bronze were noted. Its basic plumage color is black, which results in dark-colored pinfeathers, a disadvantage which contributed to its replacement by the Broad-Breasted White.

In reproductive ability, the Broad-Breasted Bronze and the Broad-Breasted White are generally inferior to the Beltsville Small White, tending to produce fewer eggs, with lower fertility and hatchability. Reproductive ability varies, however, and in some strains of heavy turkeys it is quite good. Artificial insemination has become standard practice in breeding Broad-Breasted Bronze, Broad-Breasted White, and other heavy broad-breasted turkeys, and is generally required to obtain an acceptable level of fertility.

BROAD-BREASTED WHITE

The Broad-Breasted White (also known as Large White) turkey was developed in the early 1950s through pedigree breeding and selection at Cornell University, from crosses of Broad-Breasted Bronze and White Holland. Private breeders soon began to develop other large white strains, mostly through crossbreeding, although a few strains were developed by taking advantage of naturally occurring white mutations in Broad-Breasted Bronze flocks. Selection for large size and broad-breasted conformation has been widely practiced, so most strains of Broad-Breasted White turkeys now equal the Broad-Breasted Bronze in both of these respects. Also, it is noteworthy that white turkeys generally appear to tolerate the hot sun better than dark turkeys.

BELTSVILLE SMALL WHITE

The Beltsville Small White turkey was developed through pedigree breeding and selection by the U.S. Department of Agriculture at the Agriculture Research Center, Beltsville, Maryland. Between 1941 and 1962, this stock was distributed worldwide.

The Beltsville Small White turkey closely resembles the Broad-Breasted White in color and body type, but it is smaller. In general, egg production, fertility, and hatchability tend to be higher and broodiness tends to be lower in the Beltsville Small White than in the heavy varieties. Certain American breeders of Beltsville stock have increased the weights, resulting in a turkey better described as medium rather than small, although it is officially classified as a *light breed*.

The Beltsville Small White and its related strains are no different from other varieties of turkeys in respect to livability, susceptibility to disease, and requirements for feeding and management. Floor space allowances for Small White turkeys may be reduced about one-fifth of that recommended for large-type turkeys.

But turkey growers did not like the Beltsville Small White! One person can care for about the same number of big turkeys as of small turkeys, so labor cost per pound is higher on the small variety. As a result, the Beltsville Small White has practically ceased to exist.

BREEDING OPERATORS AND SYSTEMS

Two types of operators are involved in reproducing turkeys; foundation breeders, and multipliers. The foundation breeder follows an intensive genetic program designed to achieve increased fertility, hatchability, livability of poults, uniformity of growth, better feed conversion, shortening the number of weeks to market, more uniformity in type, size, and weight.

The multiplier generally works very closely with the breeder and usually multiplies turkeys from a definite line or strain obtained from the breeder.

Generally, the term *breeder flock* is used to cover both the breeding and the multiplication phase of poultry production.

The turkey grower, as well as the foundation breeder and multiplier, should be knowledgeable relative to breeding systems.

■ **Pure line**—A pure line or strain of turkeys refers to a flock that has been closed (no stock brought in from an outside source) for at least five years. These birds are selected very intensely and are quite similar to the breeder's standard or type.

It is common in the turkey industry to refer to birds by their strain name rather than by their variety name. There are many top quality pure lines of turkeys. A considerable amount of progress and achievement has been obtained by the use of these pure lines.

■ **Strain crossing**—Strain crossing is the crossing of different strains of the same turkey variety. The object of strain crossing is to obtain a certain amount of hybrid vigor. This is usually done by the breeder splitting a flock into one or two parts, and then selecting various characteristics in these subflocks. After at least two generations of selection, the subflocks are then cross-mated to produce the commercial bird with hybrid vigor. Sometimes breeders obtain stock from outside their own flocks and cross these strains with their own varieties. A number of turkey breeders are practicing strain crossing and selling strain-crossed poults.

■ **Cross-breeding**—Cross-breeding involves the mating of two different varieties, such as the crossing of Broad-Breasted Bronze toms on White Holland hens. Although cross-breeding was used in developing the Broad-Breasted White, to date this type of crossing has not been very successful in producing commercial turkeys, primarily because of the lack of uniformity in the offspring and lack of repeatability of matings in subsequent years.

■ **Hybridization through inbreeding**—Inbreeding and hybridization are techniques that were initially developed by the hybrid cornbreeder and subsequently used by the chicken breeder. Hybridization through inbreeding involves the selection of several families, each for one or two outstanding characteristics. Intensive inbreeding is then practiced through brother-sister matings. After at least four generations of inbreeding, the families are crossed in order to concentrate their favorable characteristics within the resulting poult.

■ **Reciprocal recurrent selection**—This breeding technique has been utilized by chicken breeders for egg production, but it has found limited use among turkey breeders. Like certain aspects of crossbreeding, it involves the mating or crossing of two unrelated lines or families.

■ **Systematic intercrossing**—Under this system, breeders divide their populations into four pens of equal numbers of birds. They maintain these populations separate from each other and identify them so that from one year to the next they can keep track of the basic stocks. Each year, they rotate the toms only from one pen to the next; thus, the first year the toms from pen A are mated with the hens from pen B, and B toms to pen C, and C toms to D. The following year, this same system is followed so that a heterogeneous population is maintained over a long period of time with a minimum degree of inbreeding. It should be emphasized that the breeding program can best be evaluated by the breeder through accurate measurements of performance of the progeny produced.

BREEDER MANAGEMENT

Frequently, the failure of turkey breeding programs can be traced to the mismanagement of the breeding stock. Breeder management involves providing housing and equipment and conducting all the operations from the selection of the breeders to putting eggs in the incubator.

HOUSING AND EQUIPMENT

Either of three types of housing systems may be used for breeder turkeys: (1) range breeding facilities, (2) confinement breeding facilities, or (3) combination range and confinement. Whatever the housing system, it is noteworthy that adult turkeys can withstand considerable cold and rain, but that they are sensitive to heat and wind.

Formerly, range breeding was commonplace; the breeders were allowed to roam at will in open pens with range-type or open-shed shelters. This practice is still followed in many parts of the country. However, in recent years, there has been a shift to confinement rearing on litter, without access to outside yards or porches.

The space requirements for turkey quarters, feeders, and waterers are important; and they are affected by age, size (large-type vs small-type), and kind of rearing (brooder house, confinement, semiconfinement, or range). Table 14–2 shows the floor space and land recommendations per bird for brooding, rearing, and breeding turkeys; Table 14–3 gives the feeder and waterer space recommendations.

(Also see Chapter 11, Poultry Houses and Equipment, section headed "Houses and Equipment for Turkeys.")

TABLE 14–2
FLOOR SPACE AND LAND RECOMMENDATIONS PER BIRD FOR BROODING, REARING, AND BREEDING TURKEYS[1]

Age	Brooder House Floor		Rearing						Range			
			Confinement Floor		Semiconfinement				Shelter		Land[3]	
					Floor		Land[2]					
	(sq ft)	(sq m)	(sq ft)	(sq m)	(sq ft)	(sq m)	(sq ft)	(sq m)	(sq ft)	(sq m)	(sq ft)	(sq m)
0 to 8 weeks	1.25	0.116	—	—	—	—	—	—	—	—	—	—
8 to 12 weeks	—	—	1.5	0.1	1.5	0.1	6	0.6	1.0	0.1	100	9.3
12 to 16 weeks	—	—	2.5	0.2	1.5	0.1	16	1.5	1.0	0.1	175	16.3
16 weeks to market . . .	—	—	4.0	0.4	1.5	0.1	21	2.0	1.0	0.1	275	25.6
Breeders	—	—	4.5	0.4	1.5	0.1	26	2.4	1.0	0.1	350	32.6

[1]Growers who plan to raise turkeys to market weight must have building space and provide land area at the maximum recommendations for older birds even though there is some wasted space earlier in the growing period. Intermediate growers who plan to grow birds only to 8 or 12 weeks, then sell to a finish grower may find the intermediate breakdown helpful.

[2]Space for house, pen, and a minimum of 100 ft between pens.

[3]Includes space for rotation of land with only one-half of the land being used in any 1 year.

TABLE 14–3
FEEDER AND WATERER SPACE RECOMMENDATIONS

Age	Space per Turkey			
	Feeder[1]		Waterer[2]	
	(lin in.)	(lin cm)	(lin in.)	(lin cm)
0 to 2 weeks	1.0	2.54	0.5	1.27
2 to 4 weeks	1.0	2.54	0.5	1.27
4 to 6 weeks	2.0	5.08	1.0	2.54
6 to 8 weeks	2.0	5.08	1.0	2.54
8 to 12 weeks	2.0	5.08	1.0	2.54
12 to 16 weeks	2.0	5.08	1.0	2.54
16 to 20 weeks	2.5	6.35	1.0	2.54
20 weeks to market	2.5	6.35	1.0	2.54
Breeders	3.0	7.62	1.0	2.54

[1]Location of feeders is just as important as amount of feeder space. Both pan or feeding space and feed storage capacity are important, especially with older turkeys.

[2]Locate waterers so that each poult is always within 10 to 15 ft of water. More water space than is recommended above may be needed during hot weather. A sudden drop in water consumption often forewarns a possible disease problem.

RANGE BREEDING FACILITIES

Fig. 14–8. Market turkeys being grown on the range. (Courtesy, University of Georgia, Athens)

Range breeding facilities are characterized by allowing the breeders limited freedom to roam the range. It lessens feed costs and provides direct sunshine. However, losses from soilborne diseases, insects, predatory animals, thievery, and adverse weather conditions sometimes render range rearing unprofitable.

The Minnesota Plan of range rearing is usually effective in preventing soilborne diseases and parasites, without medication. The plan involves moving the turkeys and their equipment to an adjacent clean location every 7 to 14 days, more frequently in wet weather. The birds usually stay fairly close to their feed and water and will bed down there.

Although no fencing is required where the available range area is ample, it is safer to enclose with portable fencing enough rangeland to supply green feed for each 1- to 2-week period. Sheep or hog woven wire fencing 4 to 5 ft high will confine heavy-breed turkeys quite well, but heavy-gauge poultry wire 6 ft high will provide better protection from predatory animals and will largely prevent the turkeys from flying out of the enclosure.

About 1 acre of good pasture is required each year for each 250 turkeys raised under the Minnesota Plan. Range units of about 2,500 turkeys, with a maximum of 4,000, are suggested for this type of management.

One modification of the Minnesota Plan provides for moving the birds and their equipment less frequently than every 1 to 2 weeks. Another modification is to move the turkey feeders and waterers every week or two in a wide circle around a permanent, predatorproof, usually roostless shelter, and driving the birds into that shelter at night.

■ **Range rotation** – A 3- or 4-year range rotation is recommended where the farming operations are combined with turkey raising. Crops and varieties adapted to the area should be included in the rotation.

Where permanent pasture is available, it can be used by alternating between cattle and turkeys, with the turkeys using the area one year in three. Because of the possibility of erysipelas infection, turkeys should be kept away from sheep and swine.

■ **Portable range shelters** – Portable range shelters, providing about 2½ sq ft of floor space per bird, preferably with roosts or with floors of 1-in. × 3-in. wooden slats placed 1½ in. apart, are recommended for birds likely to be exposed to adverse weather conditions during their first month on the range.

■ **Shade** – In high-humidity areas where the temperature exceeds 90°F, shade should be provided for turkeys, allowing about 2½ sq ft for each bird. Shade should be positioned north and south and should be at least 5 ft above the ground.

■ **Protection against predators** – Predatory wild animals and free-ranging dogs can, and often do, cause heavy losses among range turkeys by direct attack or by causing stampedes. Double-strand electric fences effectively protect turkeys against most nonflying predators. Tightly stretched heavy-gauge poultry fencing 6 ft high effectively protects them against dogs, coyotes, and foxes.

A rotating electric light beam or electric lamps about 30 ft apart in a double row and completely surrounding the roosting quarters may be used to protect the turkeys and help prevent stampedes. Ringing (surrounding) the roosting area with cannonball oil flares 30 ft apart is quite effective. However, a continuous guard or confinement of the turkeys to a safe place during the night and early morning hours may be required for protection against predators and thieves.

Great horned owls and certain large hawks sometimes cause appreciable losses among turkeys roosting in the open.

CONFINEMENT BREEDING FACILITIES

Rearing turkeys in houses on litter without access to outside yards or porches is a relatively new development, which has been widely adopted in all areas where turkeys are raised in commercial quantities. Its chief **advantages**, compared with range rearing, are: (1) excellent protection against losses caused by thievery, dogs, wild animals, adverse weather conditions, soilborne diseases, parasites, and insects; (2) lower land costs; (3) lower labor costs if automatic feeding and watering equipment is used; and (4) better control of operations.

The **disadvantages** of confinement rearing are: (1) higher housing and equipment costs; (2) more risks from respiratory diseases and cannibalism; and (3) more danger from overcrowding.

The advantages and disadvantages listed do not always apply because of the variations in management in both methods of rearing. In general, however, the advantages of confinement rearing outweigh the disadvantages. It should be noted that confinement turkeys must be fed diets that are nutritionally complete.

Fig. 14–9. Fourteen-week-old female turkeys in confinement. (Courtesy, California Polytechnic State University, San Luis Obispo)

■ **Housing**—Generally speaking, naturally ventilated, non-insulated buildings are used. Pole-type houses are widely used for turkeys. A popular size is 40 ft × 300 ft, but many houses are wider and longer. Strong wire or solid partitions about 5 ft high, either suspended from the trusses or made removable, may be used to divide the flock into units no larger than 1,000 to 1,500 birds each. This arrangement is helpful in sex and age separation and can help to reduce or prevent losses from crowding, stampeding, and disease. All houses should be strongly built to withstand weather hazards such as high winds and snow.

For large-type turkeys in naturally ventilated, non-insulated buildings, practical allowances are about 5½ sq ft for toms, 3½ for hens, and 4½ for mixed flocks. For small-type turkeys, these allowances can be one-fifth less. With the smaller space allowances, turkeys must be debeaked and

the building must be well ventilated to lower the risk of respiratory infection and prevent overheating in hot weather. In well-insulated, force-ventilated, air-cooled or air-conditioned houses, the allowances can be further reduced by about one-fifth.

■ **Litter**—Deep litter, usually consisting of soft wood shavings or wheat straw, is usually used in pole houses. Four to six inches of litter are usually placed on the floor before the birds arrive. It should be kept dry and reasonably clean by adding to it frequently and, if necessary, removing prominent accumulations of droppings and wet litter which may contribute to the development of blisters, leg trouble, and fungus diseases. Adequate ventilation and air movement are the keys to good litter management, and to preventing any buildup of disease organisms.

■ **Stone yards**—The capacity of confinement houses can be increased by adding an outside yard evenly covered with 12 to 15 in. of washed stones, varying in diameter from 2 to 4 in., with fine particles removed. Where a stone yard is used, about 5½ sq ft of yard space, plus 3 sq ft in the house, will be adequate for each large-type turkey in mixed flocks. Stones do not cause breast blisters or foot trouble and normally need not be cleaned, stirred, treated or replaced for many years. The turkeys benefit from the direct sunshine, exercise, and fresh air when they are in the yards. Feeders and waterers should be inside the littered shelter.

■ **Rearing porches**—The capacity of confinement houses can also be increased by adding a slightly sloping concrete or blacktop porch. About 2 sq ft of porch space plus 3 sq ft of floor space in the house should be provided for each large-type turkey in mixed flocks. Feeders and waterers should be inside the shelter. During extremely hot weather, sprinklers installed above the porch can be used.

COMBINED RANGE AND CONFINEMENT

The combination of confinement and range rearing involves a permanent shelter, such as a pole barn about 40 ft wide, and two fenced range lots—one on each side of the building. Each range lot should supply about 1 acre of pasture for each 250 turkeys. The two range lots are alternated yearly; one is rested or cropped, while the other is used for range by the turkeys. The roostless, littered building in which the turkeys are shut at night provides shade, a safe bedding area, and protection from predatory animals and adverse weather. Feeders and waterers should be placed inside the building. The building should provide about 2½ sq ft of floor space for each bird when free access to range is available. This method of rearing is practical only where the soil is sandy, very well drained, and clean. If the range should become contaminated with disease or otherwise unfit for turkeys, preventive medication could be used or the building could be converted to confinement rearing, but with more space allotted to each bird. Dim night lights inside the house providing about ¼ footcandle at floor level may help to prevent stampeding.

Fig. 14–10. Turkeys in combined range and confinement. (Courtesy, *Turkey World*, Mount Morris, IL)

FEEDERS

Large-capacity wooden or metal hopper-type feeders should be used for turkey breeders. Allow the linear inches of feeder space per bird shown in Table 14–3, p. 337. It is important that feeders be well distributed throughout the area available to the birds. Also, they should be arranged so that they can be easily filled by the operator.

WATERERS

Water should always be in good supply, readily available, and well located. Allow the linear inches of water space per bird shown in Table 14–3. Large-type turkeys will consume 16 to 20 gal of water per day per 100 birds, depending upon rate of production and environmental temperature.

In cold climates one of the problems with breeder flocks is that of preventing the water from freezing. So, it is necessary to have some means, either electric or other heater, to keep the water from freezing.

Waterers should be cleaned, emptied, and brushed out daily. When cleaning, water should not be poured out on the litter; rather, it should be disposed of by way of sumps.

ROOSTS

Roosts are not absolutely necessary for the turkey breeder flock. Their main function is to keep the birds from crowding or piling up and to allow some of the less aggressive birds a place to get away from the main flock. Roosts sometimes interfere with fertility when natural mating is employed. Some hens tend to spend a high percentage of their time on the roosts and are therefore not available for mating.

When turkeys are on open range (pen with range shelters), roosts are recommended. However, if the breeders are confined, roosts may not be necessary. Where roosts are used, allow 16 in. of space for each heavy breed turkey and 12 in. for small breed birds; and never have the roosts more than 2 ft above the floor or ground.

SELECTION OF BREEDERS

The age at which breeders are selected should vary with the intended purpose. If the turkeys are to be marketed as fryers-roasters, selection should be made at 12 to 13 weeks of age. For large-type turkeys that are marketed as mature young roasters, selection should be made at about 14 to 17 weeks of age.

At the multiplier level, the term *selection of females* is really incorrect. Actually, this is merely culling, and it will run about 8 to 10%. The culls will consist of unthrifty birds, such as drop crops, cripples, small birds, and the extra large hens that show male characters. At the multiplier level, there is more selection pressure on the male side, but it is minimal with from 30 to 50% of the males being kept for breeding. The toms should be weighed and selected for weight, conformation, and carriage. A walking test is necessary in selecting for strong, straight legs, and upright carriage.

HOLDING BREEDERS

This refers to the period from about 18 weeks to lighting—lighting is normally started at about 29 weeks of age.

Potential breeder hens that have been on a growing ration should be placed on a holding ration at selection time. Hens that are too fat at the start of the laying season are subject to blowouts and may have a lower rate of lay and lower fertility throughout the season. One of the greatest enemies to the turkey breeder is the internal fat of the breeding hen. Birds coming into lay a little lean are in a healthier state and react less to such stresses as changes in feed, weather, management, etc. This results in better performance and livability.

Broad-Breasted White breed hens should weigh around 17½ to 18 lb at lighting time. The only positive way to accomplish this is through a high-fiber feed or a limited feeding program. Some turkey breeders consider a high-fiber diet unsatisfactory because their birds eat too much and the cost become greater. Limiting the amount of feed is the most satisfactory method, but it is difficult to accomplish. On a restricted feeding program, for example, the rations should be fortified to ensure that the lower intake of feed will still provide enough essential nutrients to allow adequate storage in the hen for the laying period.

OUT-OF-SEASON BREEDING

In most areas of the northern hemisphere, turkeys hatched between August 1 and April 1 can be considered out-of-season birds that become physically mature during the period of naturally increasing light days or while daylight is decreasing, but the weather is warm, over 65°F, and the light of day is 11 hours or longer. Such out-of-season hens require a light control program in order to lay well and long.

To obtain high egg production, the out-of-season hens, but not the toms, should be preconditioned by placing them under light restriction between 18 and 20 weeks of age. When put under light restriction at various ages, the following light schedule should be maintained:

18 to 20 weeks—8 hours of light and 16 hours of dark
20 to 22 weeks—6 hours of light and 18 hours of dark

If management is late in light restricting and it is done after 22 weeks of age, it may be necessary to use 4 hours of light and 20 hours of darkness.

Hens should be observed carefully during the period of light restriction. If they show signs of sexual activity, such as squatting, it may be necessary to lower the period of light from 8 hours to 6 or even 4 hours.

The darkout period should be as dark as possible. During the light period, 1 footcandle is ample. When properly done, white birds can barely be seen and newspaper headlines barely read (brownout). The density of birds in a darkout facility should not exceed 3 sq ft per hen. It is false economy to overcrowd breeder hens in a darkout facility for two reason: (1) the high risk, should all safety measures fail, when there is a power failure; and (2) high stress on the flock during periods of hot weather.

Out-of-season breeding stock should be fed and housed in the same manner as normal-season birds and the intensities of light used should be the same. Failure to obtain efficient out-of-season production usually stems from (1) starting the light-restriction period too late—after production starts, or (2) allowing too much light during the brownout hours of the restriction period.

FORCED MOLTING

A satisfactory second round of egg production during the off-season usually can be obtained from normal-season breeding hens of all varieties by putting them through a forced molting. The reduction in the length of the light day will cause egg production to cease and a molt to start.

Forced molting is accomplished by placing the birds in a completely dark house in which even the white birds cannot be seen. This means that air intakes must be darkened so that no light comes in. The exhaust fans must have hoods that are baffled to prevent light from entering, or light control shutters on the inside of the fans to prevent any light from entering. In order that there be plenty of air, it is recommended that 2 cfm of air movement per pound of bird in the house be provided, especially when the birds are force molted during hot weather.

During the first 72 hours of force molting, the birds should not receive any feed or water. After 72 hours of fasting, they should be given feed and water for only 3 hours. During this 3-hour period, a dim light should be used so that the birds can see the feed and water. After 2 or 3 days of 3-hour daily feeding, the birds may be fed and watered for a 6-hour period. It is necessary to ration the feed because if the feed is left in the house during the hours of darkness, the birds will eat and drink too much and gain in weight.

If the light restriction program is functioning satisfactorily, the birds will shed a lot of feathers and there will be complete stoppage of egg production within a few days. This program of light restriction should be continued until the end of the 12-week period.

Twelve weeks is recommended as a minimum because it takes that long for the birds to molt and grow the new primary feathers as well as the body feathers. Unless the feathers are well grown at the end of the 12-week period, the light restriction program should be continued for another

week or two. At the end of the 12-week period, the birds can be put under stimulatory light and brought back into production.

During the forced molting period when the birds are consuming only a small quantity of feed, it is important that the feed be highly fortified with essential vitamins in order to avoid nutritional problems. Most breeders use a regular feed, but double the vitamin level during this 12-week period. The feed should not be too high in energy.

Preferably, the males used for second-season breeding should be young, out-of-season birds hatched the previous winter. If necessary, however, the original males may be used. If they have already molted or are molting when the hens go on reduced light, the year-old males can remain on natural daylight. If they have not started to molt, they should be given the reduced-light treatment along with the hens. In either case, it would be advisable to put the molted males on stimulatory light for about 4 weeks before the hens are lighted to be sure they are producing semen when it is needed.

Hens selected for the second season should be carefully examined to eliminate injured and sick birds and, if records are available, the poor layers and those persistently broody.

BLOOD TESTS

Under most official state programs, at least three weeks are required between selection and blood testing. This period of isolation is to allow time to determine the health of the flock at the time of testing. Blood testing should be scheduled as soon after selection as regulations permit. The sooner the blood testing is completed, the sooner the breeders can go on to other phases of the prebreeder program.

From one sample of blood submitted to a laboratory, tests may be run for the presence of pullorum disease, fowl typhoid, paratyphoid, infectious sinusitis (*Mycoplasma gallisepticum* infection), and possibly *Arizona* and *Mycoplasma meleagridis* infections.

During the blood testing operation, the birds can also be debeaked, wing-clipped, and saddled if desired.

VACCINATIONS

After receipt of the laboratory reports of the blood tests, vaccinations can be started, unless, of course, the birds must be retested.

The number and kind of vaccinations given should be based upon the requirements set up by the hatching-egg buyer and the advice of competent authorities on disease problems. Among the vaccinations which may be given to breeder flocks are: fowl pox, Newcastle disease, avian encephalomyelitis, erysipelas, fowl cholera, fowl typhoid, paratyphoid, and *Arizona paracolon*.

When several vaccinations are given, they should be planned according to competent advice to avoid possible unfavorable interactions. The adjuvant types of vaccines may give longer immunity. In general, preventive vaccinations, usually of the live-virus or killed bacteria-type (bacterins), are given only when the disease in question is present in the area and poses an active threat to the health of the turkeys.

A suggested turkey vaccination program for consideration of local authorities may include protection against the following diseases: avian encephalomyelitis (epidemic tremors), erysipelas, fowl cholera, fowl pox, and Newcastle. (See the subsequent discussion of each of these diseases in this chapter.)

SADDLES

These are just what the name implies — canvas saddles fitted over the hens that are to be mated by natural service, prior to the onset of mating. This prevents much of the loss from hens being torn or cut by the toenails or spurs of the toms during mating.

LIGHTING

The reproductive organs of the turkey are controlled by a combination of secretions of various glands, the most important of which is the pituitary gland located in the skull and connected with the brain. Light stimulates hormone production from this gland, and, in turn, the hormones cause the bird's reproductive organs to grow and produce.

Under normal conditions, nature ordained that turkey hens start production in March, when the natural light gets to about 14 hours per day after being as low as 8 hours per day in the winter months. The hen's response to artificial light is dependent on the light to which she has become accustomed.

Breeder hens should not be subjected to lighting earlier than 29 weeks of age.

The recommended light intensity is 5 to 7 footcandles at the turkey's back, approximately 12 in. off the ground. It is important to have a uniform light pattern and to be sure the hens are contained in the lighting area. The latter is particularly important when the hens are on the range.

When lighting, there is no advantage to increasing light gradually. Fourteen hours of light are adequate to bring hens into production, although 16 hours of light may be required in the fall months. It matters little whether morning or evening lights are used, although it helps to have a short period of light to bed down the birds in the evening. This will calm them, and when the lights go on they will not panic as badly.

Males should be lighted 3 to 5 weeks before the hens to ensure maximum fertility.

MALE:FEMALE RATIO

In single-male natural mating, the following ratios are recommended:

Small-type turkeys, 20 hens to 1 young tom
Medium-type turkeys, 18 to 1
Large-type turkeys, 16 to 1

In natural flock matings, the following ratios are recommended:

Small-type turkeys, 14 to 1
Medium-type turkeys, 12 to 1
Large-type turkeys, 10 to 1

Additionally, a few extra toms, preferably late hatched or out-of-season, should be maintained separately in order to replace those that may die or become sick or crippled.

TOM MANAGEMENT

Tom housing can be very similar to hen housing, except that feed, water, and roosting space should be larger. It is recommended that there be 6 sq ft per tom.

Preferably, toms should be in a separate building because of the need for different management practices. If the toms are housed in the same building with the hens, the fence separating them should be solid. However, if the two are separated by a distance of 20 ft or more, wire fencing can be used. Up to 25 toms may be kept together in a pen.

Fig. 14–11. Breeder toms in small pens on a turkey farm in Ontario, Canada. (Courtesy, *Turkey World*, Mount Morris, IL)

Toms seem to respond differently to light than hens. They do not require a dark period before lighting in order to respond.

Toms should be prelighted 3 to 5 weeks prior to lighting hens. This will assure having semen ready at the time of mating (or semen collection for artificial insemination). Toms that are to be used for the summer flock do not require prelighting, because they have matured on a lengthening day.

Toms should be "premilked" at least once before collecting semen for artificial insemination. If they are not premilked, they will usually not give enough semen for the first artificial insemination.

ARTIFICIAL INSEMINATION

Today's broad-breasted turkey strains would not be practical without artificial insemination. Turkeys with a conformation that pleases the consumer and a meat yield that satisfies the processor seem to have neither the agility nor the desire to mate as often and successfully as the more narrow-breasted turkeys.

■ **Fertilization** — In the natural process of fertilization, mating takes place when the hen is receptive. Hens have a strong desire to mate frequently for a few days just before egg production starts and a few days thereafter. By the end of the season, they may mate once every few weeks, or not at all.

In mating, the hen everts the vagina beyond the lips of the cloaca and the tom deposits semen directly on the everted tip of the vagina. The hen immediately withdraws, carrying the semen far up the vagina. Sperm swim out of the seminal fluid into tiny folds in the walls of the vagina, called *sperm nests*. A few swim and are carried by the action in the hen's oviduct to the funnel, where fertilization takes place.

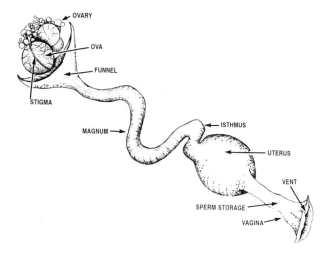

Fig. 14–12. Turkey hen's reproductive system.

After each egg is layed, some of the sperm leave the sperm nest and travel to the top of the oviduct to fertilize the next egg being layed. This is the only time an egg can be fertilized. As soon as the yolk is coated with egg white, the sperm cannot get to the germ spot. Because it takes nearly 24 hours from the time egg white starts to be deposited until the egg is layed, eggs layed the same day that mating takes place are never fertile from that mating.

Sperm can live in the sperm nest for several weeks. Fertile eggs have been observed several weeks after a single insemination, but fertility usually starts to drop quite rapidly after 2 weeks following each insemination early in the season. Fertility of hens that have layed for several months starts dropping within a couple of days after each insemination.

Despite the sperm's ability to survive for long periods of time in the hen, outside the turkey it is a very delicate organism. Each drop of good semen contains millions of very active sperm. Under the microscope, they can be seen swimming about very rapidly, but they die after a few minutes. These sperm have very thin cell walls and they are easily killed by intestinal fluids, antibodies in blood, and by salts. Even plain water will cause them to swell and burst unless the water has other materials such as a diluent dissolved in it. Sperm survive best at about 60°F. Different toms produce sperm with different amounts of vigor. Also, sperm produced early in the season appear to be much longer lived than sperm produced late in the season.

■ **Time and method of insemination**—The usual time for artificial insemination is approximately 2 weeks after lighting. In flocks coming into production during February, March,

and April, this could be as soon as 10 days after lighting. The progress of the hen should determine when the flock should be inseminated.

The flock should be checked by catching 100 hens at random to see how they "open." This can be done by observing how the flock is squatting when 80% of the flock will be open. Artificially inseminate the flock and follow with the second artificial insemination 5 days after the first; a third artificial insemination 5 days after the second; and go on a 10-day schedule for 10 weeks. This should be followed by a weekly schedule.

The straw method of insemination is recommended. Use one straw to inseminate each hen, thereby reducing the danger of spreading disease from hen to hen while inseminating.

NESTS

Nests with tip-up fronts, or semitrapnests, one for each 4 to 5 hens, are recommended. They may be made of wood or of more durable welded iron or heavy plastic. The tip-ups, or gates, must be carefully constructed and balanced or equipped with springs so that the tip-up closes behind the hen when she enters and remains closed until she leaves. As she leaves, the tip-up is automatically reset so another hen may enter. Nests should be arranged in rows, preferably back to back to facilitate efficient gathering.

Nests of all types should be installed and made operational well in advance of lighting the hens; otherwise, the hens may lay on the ground and it will be hard to break them of this habit.

Individual nests should be at least 2 ft high, 1½ ft wide, and 2 ft long. Nests should have enough depth so as to allow at least 4 in. of nest bedding to be used. Nests should be placed on the floor or about 6 in. off the floor.

A variety of materials is satisfactory for nesting litter, including shredded cane, shavings, rice hulls, pea gravel, peanut shells, and sawdust.

EGG GATHERING AND HANDLING

To understand proper egg handling, one must first understand the makeup of a turkey hatching egg.

A fertile turkey egg contains a delicate living embryo that is easily killed through improper handling. A newly laid egg has a temperature of 106°F. Inside, the embryo is growing very rapidly. There is no air cell, the shell is damp, and at this stage the embryo is very susceptible to sudden temperature changes.

As the egg cools, embryo growth slows and by the time the internal temperature gets down to 80°F, growth is stopped and the embryo is in a state similar to hibernation. Research has shown that cooling eggs too fast the day they are layed causes the turkey embryo to stop growing a few hours sooner than is beneficial for optimum fertility and hatchability.

Eggs should be collected at frequent intervals, especially during cold weather. It is recommended that filler-flats be used for egg collecting instead of baskets, as they lessen the chance of egg breakage. The heaviest period of lay is during the early afternoon, although this will vary according to the lighting schedule. Normally, about 70% of the eggs are laid in the afternoon.

Dirty eggs usually show lowered hatchability and may transmit disease. Also, they are the main reason for exploders in the hatchery. So, roost eggs or dirty eggs should not be saved.

Washing turkey eggs intended for hatching appears to be a debatable practice. Not only is washing ineffective in sanitizing the eggs, it may actually be harmful to fertility.

Egg fumigation done properly and accurately can kill many bacteria on the shell and prevent disease outbreaks in the poults. Formaldehyde is a very penetrating gas, and very deadly to bacteria. However, formaldehyde gas will not kill bacteria beneath the eggshell. For this reason, eggs should be fumigated after each gathering before the bacteria has time to penetrate deep into the shell. Recommended fumigation procedure follows: Place the freshly gathered eggs in an airtight cabinet with heating and forced-air circulation. This cabinet should be large enough to hold one gathering of eggs. Eggs must be in wire-meshed trays, such as are used in setting hatching eggs. This ensures free circulation of the gas around the eggs. Tests have shown that gases are unable to circulate and penetrate when flaps or egg cases are used. The temperature during fumigation in the cabinet should be at least 70°F. For every 100 cu ft of space, use 2 oz of potassium permanganate crystals and 4 oz of formaldehyde. Pans or widemouthed containers should be used. Fumigate the eggs for 20 minutes, then clear the gas from the cabinet.

Eggs should be allowed to cool slowly, then stored at 55° to 60°F and 80% relative humidity. A recommended procedure is to pack eggs in precooled cases, small end down.

Turkey eggs need not be turned while they are in storage prior to incubation if the storage time does not exceed two weeks. However, if the eggs are to be held for longer periods, they should be turned at least three times a day.

Since both fertility and hatchability decline as the holding period increases, holding turkey eggs longer than 2 weeks is not recommended. Extreme care must be exercised so that the eggs are not overheated, chilled, or frozen during shipping. They must be packed small end down in cases with the special turkey size flap and filler or the new filler-flap being used to prevent breakage. Shipping will sometimes result in decreased fertility and hatchability due to excess jarring, but under proper conditions (temperature, humidity, and handling), this decrease should not be significant.

BROODY HEN CONTROL

Broodiness in turkey hens is their natural desire to hatch eggs and complete the natural process of reproduction. In nature, broodiness is essential for survival of the species. Since the incubator came into use, however, broodiness is only a source of extra labor, expense, broken and dirty eggs, and lowered production.

Broodiness should be halted as soon as it is detected in a flock. One of the first indications of the onset of broodiness is that the hen spends more and more time on the nest, after laying, until she no longer leaves the nest. Every attempt should be made to remove broody hens from the nest in the evening. This can usually be done by locking them out of the nesting area, or locking the nest.

Many different techniques have been practiced to eliminate or "break up" broodiness; among them, the following:

1. Changing the environment of the hens each week. This may be accomplished by rotating the breeder flocks to different pens where nests and feeding equipment are located in different patterns. This seems to upset hens sufficiently to offset broodiness.

2. Putting broody hens in an outdoor pen where no nests are available. Such hens should be provided with a good diet and water, and should not be mistreated.

3. Using high intensity lights or high intensity sound.

INCUBATION

Generally speaking, each incubator has its own "temperament," and only experience and close attention to the directions given by the manufacturer will result in good hatches.

Turkey eggs are generally incubated at 99.5°F and a relative humidity of 80 to 85%. Also, proper ventilation is essential. The developing embryos require air containing about 21% oxygen and not more than 1.5% carbon dioxide. Since the oxygen requirements of the embryos increase as they develop, it is evident that ventilation must be increased as the hatch progresses. The eggs should be turned at least five times daily, and preferably every 2 hours during incubation. With machines equipped with turning devices, this is easily accomplished. The incubation period for large-type turkey eggs is 28 days, although some embryos are known to hatch early, after only 25 to 26 days of incubation. Eggs from small-type turkeys tend to hatch in 27 to 27½ days.

Only clean eggs with sound shells should be placed in the incubator. Washing is generally not recommended for turkey eggs, as it seems to encourage the entrance of bacteria. Every attempt should be made to produce clean eggs that are not contaminated. Muddy pens, lack of clean nest bedding, and poorly trained hens can be major factors in contamination.

If the poults are to be shipped, they should be put in poult boxes and shipped as soon as possible. If they are held in the incubator or in a battery, the temperature should be 95° to 99°F.

BROODING

The successful brooding of poults depends upon satisfying all the essential requirements of the young birds at this stage in their development.

■ **Selecting poults** — The producer should start with sexed poults from pullorum-typhoid clean and PPLO tested breeder flocks of desired genetic ability from a hatchery that provides healthy poults and good service. The sexed poults should be raised separately. For a fee, the hatchery will desnood the males for protection against erysipelas disease.

■ **Clean brooder house** — The brooder house should be completely cleaned prior to each new brood of poults. All brooders, waterers, feeders, and facilities should be scrubbed, repaired, and disinfected.

■ **Starting the poults** — The poults should be placed in the brooding quarters and given feed and water within 24 hours after hatching, the sooner the better. Poults may be started on covered or uncovered litter; on asphalt roofing (not tar paper); on wire or slat floors; or in batteries.

■ **Litter** — The brooder house floor should be covered with clean, fresh litter, at least 3 in. deep. A wide choice of litter materials may be used, including wheat straw, shavings, peat moss, shredded cane, rice hulls, processed flax straw, and cedar tow. Peanut hulls, crushed corn cobs, and shredded corn stover make good litter, but they may contain harmful molds unless they are properly and thoroughly dried.

The litter inside the brooder should be covered with rough surface paper, such as crinkle kraft paper, for firm footing and to prevent young poults from eating enough litter to clog their digestive tract before they develop good eating habits. Remove the litter cover as soon as poults are on feed, at least by the end of the first week. This cover prevents moisture accumulation which combines with the heat from the brooder to provide a favorable environment for mold growth.

■ **Brooder ring or poult guard** — This should be 16 to 18 in. high, and it may be made of lightweight hardware cloth or lightweight chicken wire. Heavyweight corrugated paper or lightweight aluminum roll sheeting sometimes is used as a poult guard in cold weather in large open brooder houses to shield poults from cold drafts. Litter usually need not be changed during the brooding, rearing, or breeding seasons, however it should be added to at intervals frequently enough to provide a dry, clean, resilient walking surface; wet or caked litter should be removed and replaced with clean litter; and all litter should be removed at the end of the season.

Fig. 14–13. Brooder ring around 3-day-old turkeys, with gas hoods overhead. (Courtesy, California Polytechnic State University, San Luis Obispo)

■ **Space requirements** — The space requirements for brooding are shown in Tables 14–2 and 14–3, pp. 336–337.

■ **Feeders and waterers** — Small trough-type metal feeders or wooden feeders made of plastic laths are placed like spokes in a wheel — part under the hover and part outside it, but within the brooder ring. At first, the trough should be heaped with mash to encourage eating. In addition, some operators put a little mash in a few paper plates or egg flats the first day or two and place some colored glass marbles on the feed to attract the poults. After a week or so, larger size feeders may be used, but 2 in. of linear feeding space per poult should be retained during the remainder of the brooding period.

To provide good watering facilities during brooding, start with one circular 1- to 2-gal glass or metal baby-poult-size waterer for each 50 poults. Select a waterer with a narrow drinking space, not over 1½ in. wide and 1¼ in. deep. An alternative is to start with one baby-poult-size automatic water trough 4 ft long for each 80 poults. Place the waterers around the edge of the hover and put a few glass marbles in each one.

Waterers should be washed and rinsed, not just emptied, daily. Disinfection is usually not needed. If disinfection is necessary, however, the quaternary ammonium or chlorine types are preferred.

■ **Temperature** — Young poults will not live and grow unless they are kept warm and dry. For at least the first 2 weeks, the temperature 3 in. above the floor at the edge of the hover should be 95°F for dark poults and about 100°F for white poults. Temperature near the floor of the room outside the brooding area should be 70°F or a little higher and should be maintained at about that level.

■ **Ventilation** — Ventilation is always important, and especially so when heat is supplied by open-flame brooders that are not vented to carry out the smoke. Engineer-approved air outlets in the peak of a gable roof or near the top of a shed roof combined with adjustable window intakes usually provide sufficient ventilation. But a well-designed system of fans and intakes is preferable.

■ **Lighting** — For the first 2 weeks of brooding in all types of houses, the rooms should be well lighted day and night at 10 to 15 footcandles at poult levels; and, if practicable, a small 7½- to 15-watt light bulb should be installed under the hover. In open-sided houses or houses with windows, after 2 weeks only dim lights of 1½ footcandle should be needed at night, none in the daytime. In windowless houses after about 2 weeks, the intensity of the light can be reduced gradually to about 1 footcandle during the 16-hour day and ½ footcandle during the 8-hour night. The dimmer lights at night may help to discourage piling and stampeding.

(See Chapter 8, Poultry Management, Table 8–3, Recommended Lighting Systems for Poultry.)

■ **Roosts** — Roosts seldom are used in modern turkey brooding operations, but they may help to prevent stampedes and piling at night. For the brooder house, roosts of the stepladder-type are practical, allowing about 3 linear in. per bird. They may be made of 1½- to 2-in. round poles, 2" × 2"s, or 2" × 3"s.

■ **Size of unit** — Although brooding units of 300 to 500 poults sometimes are used, it is better practice and less risky to limit the brood to 250.

■ **Brooder houses**—Before constructing a brooder house, the turkey producer should check plans based on successful experiences. Brooder houses often are used both for brooding and for rearing to market age. Width usually varies from 24 to 40 ft, and length from 100 ft up to 600 ft. Clear-span, rigid-frame, gable-roof houses 40 ft wide and 300 ft long are popular. For cold weather brooding, insulation of sidewalls and ceiling is needed, especially if the brooders are the type that do not provide an abundance of heat. Mechanical ventilation is highly desirable in all brooder houses and necessary in those over 40 ft wide. Correct engineering principles should be followed when ventilating systems are installed.

Fig. 14–14. Turkey poults in a brooder house. (Courtesy, *Turkey World*, Mount Morris, IL)

Partitions appear not to be necessary in small houses used for brooding and rearing, but many operators prefer to install them at intervals in long houses so that no more than 1,000 to 1,500 birds are allowed in a pen. This arrangement is useful in separating the turkeys by age and sex and can help to control stampedes and disease outbreaks.

Electrical wiring should be installed for lighting and debeaking, especially if natural light is restricted as it usually is in wide houses.

Pole-type breeding and rearing houses also may be used for moderate-weather brooding by enclosing the open sides on the pens with clear plastic, adding electricity, and installing brooders and water supply.

Brooder houses as well as breeding and rearing houses should be built to exclude small wild birds and rats as well as larger birds and animals. Some of the requirements are: concrete foundations, tight-fitting doors and windows, and all openings covered by ¾-in. (or smaller) mesh wire.

■ **Wire floors**—Wire floors, like slats, make the use of litter unnecessary and tend to prevent filth borne diseases such as coccidiosis, hexamitiasis, and blackhead. Plenty of heat and good ventilation are essential. Also, to prevent floor drafts, long stretches of unobstructed open space underneath should be broken up by solid wood or metal partitions

about every 20 ft. Small-mesh starting wire is used for about the first 4 weeks, following which large-mesh wire should be used for the rest of the brooding period.

Roosts during the brooding period are especially desirable when wire or slat floors are used.

■ **Slat floors**—Removable, 4-ft square sections composed of plaster lath or narrow lathlike slats mounted on metal or wooden supports make good brooder house floors. They minimize losses from crowding and piling up and help prevent filthborne diseases.

Poults can be raised to fryer-roaster stage on the slats.

■ **Types of brooders**—Poults need a dependable source of artificial heat within the first few weeks of life. During the first week, they should be provided a uniform temperature of about 95°F. Thereafter, the temperature may be lowered 5°F per week.

Portable brooders come in a number of different styles and sizes. In commercial operations, heat is generally provided by gas, oil, or electricity. Large commercial type growers often use hot water or hot air types with a central heating system.

■ **Debeaking**—Poults should be debeaked in order to control feather picking and cannibalism, especially if they are to be raised in confinement. Debeaking can be done by the hatchery when poults are a day old. However, the debeaking can be done at 3 to 5 weeks of age, using sharp, heavy shears, or dog toe clippers for this purpose. Remove only the tip of the beak, sever at about one-half the distance from the nostril to the tip of the beak. Never remove more than this, especially when birds are to be moved to range. Feed and water level should be at least ½ to ¾ in. deep for debeaked birds.

■ **Desnooding**—Desnooding—the removal of the snood, or dew-bill (the tubular fleshy appendage on top of the head near the front)—helps to prevent head injuries from picking or fighting and may reduce the spread of erysipelas should this disease get started in the flock. At day-old age, the snood can be removed by thumbnail and finger pressure. After the age of about 3 weeks, it can be cut off close to the head with sharp, pointed scissors. It should be noted that the value of desnooding is questioned by some.

■ **Wing clipping and notching**—Flight may be prevented by either wing clipping or notching. Wing feathers can be clipped with sharp, heavy shears, with hedge clippers, or with a sharp hatchet and chopping block.

Also, flight can be permanently prevented by wing notching at the age of 5 to 8 weeks. The tendon that crosses the center of the outermost wing joint is severed with a vertical red-hot steel bar on the electric debeaking device.

Both wing cutting and wing notching handicap the birds and often result in lowered market grades; hence, they are not recommended for market turkeys or for birds to be used in natural matings.

■ **Toe clipping**—Clipping the toes of each foot of day-old poults is becoming common among turkey growers to prevent scratched and torn backs, which are an important cause of downgrading market turkeys. Clipping is best done at the hatchery with 5-in. surgical shears, removing the tip of the toe just to the inside of the outermost toepad, including all of the toenail.

■ **Separating the sexes** — The sex of day-old poults may be determined by examining the vent. Commercially sexed poults are available at many hatcheries.

Where accurately sexed day-old poults are available at reasonable cost, raising toms and hens separately is practicable. The advantages of separating the sexes are: (1) injuries to the hens due to treading in the later growth stages are eliminated; (2) hens can be marketed at an earlier age than the toms without a loss of grade due to injuries occurring when sexes are separated during the loading operation; (3) there appears to be less fighting and competition among the toms when hens are not present; and (4) sex-separated flocks sometimes can be fed more efficiently, using the complete feeding plan.

■ **Handling turkeys** — Turkeys of all ages can be easily driven from place to place with the aid of light poles or sticks several feet long. With a pole in each hand, one person is able to control a good-size flock. Driving entails much less labor than crating and hauling, and, if distances are not too great, it is easier on the birds. Dogs can be used for driving turkeys, but they must be well trained and gentle and the turkeys must be accustomed to them.

For catching turkeys, a darkened room is best. In the dark or semidarkness, turkeys can be picked up with both legs without confusion or injury. For this reason, market turkeys sometimes are loaded at night.

Portable catching chutes, preferably with a conveyor, can be used for catching and loading range turkeys.

Metal hooks of ¼ in. round iron for weighing turkeys can be made up in several sizes. If hooks are the right sizes and the bird's legs are properly placed in them, the legs will not be injured. However, mature heavy turkeys should not be left hanging for more than 3 or 4 minutes, and it is better to pick up turkeys by both legs in a dark room rather than catch them with a hook.

(Also see Chapter 11, Poultry Houses and Equipment, section headed "Houses and Equipment for Turkeys.")

FEEDING

Since feed constituted about 66% of the cost of producing turkeys in 1989, it is important to give special attention to it. Feeding turkeys for maximum profit is not an easy task, however, for it requires the simultaneous use of a considerable knowledge of nutrition and a judicious assessment of prevailing feed prices. A successful program must provide the turkeys with the required nutrients in the amounts and proportions to satisfy their needs, and, at the same time, remain well within the economic limits necessary for a profit potential to be high.

The basic nutrient categories are: proteins, carbohydrates, fats, minerals, vitamins, and water. Distributed within these categories are some 40 separate nutrients which turkeys require.

The economic importance of improved turkey nutrition is further evidenced by the following deficiencies and losses,

although no claim is made that faulty nutrition is the only, or even the major, factor involved in each:

1. **Infertility.** Approximately 10% of all incubated turkey eggs are infertile. The B vitamins, vitamin E, and selenium have been shown to affect the quality of semen; likewise, in the diet of hens, these same nutrients, along with magnesium, have been shown to affect egg fertility.

2. **Embryonic deaths.** From 5 to 15% of the embryos of all eggs set die at some stage during the incubation period.

3. **Mortality of poults.** Mortalities of poults in the first 5 weeks range up to 10%; hence, the viability and survival of poults should be improved.

4. **Leg weaknesses and other disorders.** During the finishing stage, a significant amount of leg weaknesses and other disorders lower feed efficiency and increase the number of cull birds.

5. **Breeder hens eat less.** Breeder turkey hens eat less than formerly, either because of limited feeding or a naturally smaller appetite; it follows that they need enriched rations.

6. **Sexed feeding.** By separating the sexes, the grower can exploit the extra growth potential of the toms but not waste nutrients (and money) on surplus nutrients going to the hens.

Improvements in each of the areas noted above will benefit both the turkey breeder and the grower.

Further information on the nutrition and feeding of turkeys is contained in the following two chapters, to which the reader is referred:

Chapter 6, Fundamentals of Poultry Nutrition
Chapter 7, Poultry Feeding Standards, Ration Formulation, and Feeding Programs

GROWTH RATE AND FEED CONSUMPTION

With improved breeding, nutrition, and management, feed conversion and growth rate of turkeys are constantly improving. Tremendous gains in production efficiency have

Fig. 14–15. Turkey grow-out house showing curtain sides, and automatic feeder and water lines. (Courtesy, North Carolina State University, Raleigh)

been realized in recent years. With narrow margins between production costs and selling price, it becomes increasingly important for the grower to pay strict attention to every step in the management program that will facilitate improved production efficiencies.

Because of different growth rates and feed efficiencies, note that separate tables are presented for toms and hens (Table 14–4 and Table 14–5). These tables, give, for 1989, the body weight, feed consumption, and feed-to-gain ratio of turkey toms and hens, respectively. Nationwide, toms were marketed at 123 days of age and an average weight of 26.35 lb, and hens were marketed at 100 days of age and an average weight of 14.68 lb. At 18 weeks of age, toms averaged 27.06 lb live weight, with a feed/pound turkey ratio of 2.80. At 15 weeks of age, hens averaged 15.51 lb live weight, with a feed/pound turkey of 2.55.

TABLE 14–4
BODY WEIGHTS, FEED CONSUMPTION,
AND FEED-TO-GAIN RATIOS OF TURKEY TOMS[1]

Age	Live Weight Average	Live Weight Gain for Period	Cumulative Feed/Tom on the Basis of National Average[2]	Feed/Lb of Turkey on the Basis of National Average	Metabolizable Energy Intake, kcal/Lb of Turkey[3]
(weeks)	(lb)	(lb)	(lb)	(lb)	(kcal/lb)
1	0.32	0.26	0.32	1.24	1,605
2	0.66	0.34	0.75	1.27	1,645
3	1.15	0.49	1.27	1.31	1,720
4	1.86	0.71	2.43	1.35	1,810
5	2.79	0.93	3.85	1.41	1,920
6	3.93	1.14	5.80	1.50	2,040
7	5.27	1.34	8.49	1.60	2,170
8	6.74	1.47	11.49	1.71	2,310
9	8.34	1.60	15.07	1.82	2,460
10	10.09	1.75	19.36	1.92	2,610
11	11.96	1.87	24.16	2.03	2,765
12	13.99	2.03	29.81	2.14	2,920
13	16.07	2.18	36.02	2.25	3,080
14	18.26	2.19	42.95	2.36	3,245
15	20.46	2.20	50.39	2.47	3,415
16	22.67	2.21	58.56	2.58	3,605
17	24.87	2.20	66.74	2.69	3,805
18	27.06	2.19	75.87	2.80	4,005
19	29.24	2.18	85.21	2.92	4,210
20	31.40	2.16	94.96	3.04	4,420
21	33.51	2.11	105.37	3.16	4,635
22	35.54	2.03	116.02	3.28	4,855
23	37.51	1.97	126.96	3.40	5,075
24	39.41	1.90	138.55	3.53	5,295

[1]Source: Turkey World, January-February, 1990, p. 12, article entitled, "Faster Growing, More Efficient Turkeys in 1989," by Jerry L. Sell, Department of Animal Science, Iowa State University; reproduced with the permission of Dr. Sell. To convert pounds to kilograms, divide by 2.2.

[2]Feed-to-gain ratios calculated on the basis of plant weights minus condemnation weights.

[3]Estimated on the basis of national average feed-to-gain ratios.

TABLE 14–5
BODY WEIGHTS, FEED CONSUMPTION,
AND FEED-TO-GAIN RATIOS OF TURKEY HENS[1]

Age	Live Weight Average	Live Weight Gain for Period	Cumulative Feed/Hen on the Basis of National Average[2]	Feed/Lb of Turkey on the Basis of National Average	Metabolizable Energy Intake, kcal/Lb of Turkey[3]
(weeks)	(lb)	(lb)	(lb)	(lb)	(kcal/lb)
1	0.30	0.24	0.30	1.24	1,605
2	0.63	0.33	0.71	1.27	1,640
3	1.11	0.48	1.34	1.31	1,710
4	1.77	0.66	2.33	1.36	1,805
5	2.58	0.81	3.65	1.45	1,915
6	3.55	0.97	5.37	1.55	2,035
7	4.65	1.10	7.57	1.65	2,165
8	5.88	1.23	10.24	1.76	2,315
9	7.18	1.30	13.31	1.87	2,480
10	8.56	1.38	16.83	1.98	2,655
11	9.96	1.40	20.79	2.10	2,840
12	11.40	1.44	25.06	2.21	3,030
13	12.81	1.41	29.71	2.33	3,230
14	14.18	1.37	34.45	2.44	3,435
15	15.51	1.33	39.42	2.55	3,645
16	16.79	1.28	44.84	2.67	3,860
17	18.00	1.21	50.59	2.82	4,095
18	19.13	1.13	56.83	2.98	4,335
19	20.19	1.06	63.61	3.16	4,585
20	21.17	0.98	70.51	3.34	4,850

[1]Source: Turkey World, January-February, 1990, p. 12, article entitled, "Faster Growing, More Efficient Turkeys in 1989," by Jerry L. Sell, Department of Animal Science, Iowa State University; reproduced with the permission of Dr. Sell. To convert pounds to kilograms, divide by 2.2.

[2]Feed-to-gain ratios calculated on the basis of plant weights minus condemnation weights.

[3]Estimated on the basis of national average feed-to-gain ratios.

Because of different growth rates and feed efficiencies, a separate table, Table 14–6, is presented for small-type turkeys.

TABLE 14–6
GROWTH RATE AND FEED CONSUMPTION
OF SMALL WHITE TURKEYS[1]
(Toms and Hens Combined)

| Age | Liveweight | | | | Feed Required | | | |
| | Average | | Gain for Period | | Total Cumulative | | Per Lb of Turkey, to Date | |
(wks)	(lb)	(kg)	(lb)	(kg)	(lb)	(kg)	(lb)	(kg)
1	0.25	0.11	0.14	0.06	0.2	0.1	0.8	0.4
2	0.48	0.22	0.23	0.1	0.6	0.3	1.3	0.6
3	0.95	0.43	0.47	0.21	1.35	0.61	1.4	0.6
4	1.45	0.66	0.5	0.2	2.2	1.0	1.5	0.7
5	2.1	1.0	0.6	0.3	3.2	1.5	1.5	0.7
6	2.8	1.3	0.7	0.3	4.3	2.0	1.5	0.7
7	3.6	1.6	0.8	0.4	5.8	2.6	1.6	0.7
8	4.4	2.0	0.8	0.4	7.7	3.5	1.8	0.8
9	5.4	2.5	1.0	0.5	9.8	4.5	1.8	0.8
10	6.3	2.9	0.9	0.4	12.0	5.5	2.2	1.0
11	7.3	3.3	1.0	0.5	14.8	6.7	2.0	1.0
12	8.3	3.8	1.0	0.5	17.8	8.1	2.1	1.0
13	9.3	4.2	1.0	0.5	21.0	9.5	2.2	1.0
14	10.3	4.7	1.0	0.5	24.3	11.0	2.4	1.1
15	11.2	5.1	0.9	0.4	27.8	12.6	2.5	1.1
16	12.1	5.5	0.9	0.4	31.2	14.2	2.6	1.2

[1]Jensen, L. S., University of Georgia, Athens.

TABLE 14–7
FEED EFFICIENCY OF LARGE TURKEYS BY SEASON[1]

Age	Feed Required per Lb of Turkey					
	Marketing Period					
	Juy through October		November and April through June		December through March	
(weeks)	(lb)	(kg)	(lb)	(kg)	(lb)	(kg)
Toms:						
17	2.4	1.1	2.5	1.1	2.75	1.3
18	2.6	1.2	2.7	1.2	2.95	1.3
19	2.7	1.2	2.8	1.3	3.1	1.4
20	2.75	1.3	2.9	1.4	3.2	1.5
21	2.9	1.3	3.0	1.4	3.3	1.5
22	2.9	1.3	3.05	1.4	3.35	1.5
23	3.0	1.4	3.1	1.4	3.4	1.5
24	3.1	1.4	3.2	1.5	3.5	1.6
Hens:						
15	2.3	1.0	2.5	1.1	2.75	1.3
16	2.45	1.1	2.6	1.2	2.85	1.3
17	2.6	1.2	2.7	1.2	3.0	1.4
18	2.75	1.3	2.85	1.3	3.1	1.4
19	2.9	1.3	3.0	1.4	3.3	1.5
20	3.0	1.4	3.1	1.4	3.4	1.5

[1]Jensen, L. S., University of Georgia, Athens.

TABLE 14–8
COMPARISON OF LARGE-TYPE TOM TURKEYS
RAISED UNDER EITHER RANGE OR CONFINEMENT[1]

| Rearing System | Age | Body Weight | | Feed/Gain | Livability |
	(weeks-days)	(lb)	(kg)		(%)
Range	19–3	24.3	11.0	2.90	91.3
Confinement	19–1	22.4	10.2	2.76	93.0

[1]Jensen, L. S., University of Georgia, Athens.

Weather conditions always play a role in the performance of turkeys from year to year and season to season. Generally speaking, extremely hot weather depresses growth rate and extremely cold weather lowers feed efficiency. The seasonal effect of weather is shown in Table 14–7.

Turkeys may be grown under either range or confinement conditions after the starting period. Although it is difficult to compare rearing systems by considering data from several operations, a comparison of information obtained in the same operation should be of some value in predicting the performance to be expected with each system. Table 14–8 shows a comparison of large-type tom turkeys raised under either range or confinement conditions. The data indicate a faster growth rate for birds reared on range, but a better feed efficiency for those reared in confinement.

HEALTH

Turkeys seem to be more susceptible to disease and require a much higher level of management skill than other domestic fowl. It is one of the minor blessings of nature that turkeys also respond more to specific medication than other poultry. One severe outbreak of a disease is usually all that is needed to turn a flock from a profit to a loss.

There are four primary causes of disease: genetics, nutrition, environment, and infection. Most outbreaks are the result of at least two of these causes. To combat a disease effectively, the producer must first understand the history of the disease and what is needed to attack it at its weakest point.

Most diseases are at least partly under the control of the turkey producer, and they are not inevitable—they may be controlled or prevented by proper management, sanitation, and vaccination. If turkeys get sick, it usually means that the grower has neglected doing something, vaccinating for example, or that something has been done wrong, such as stressing the birds.

DISEASE PREVENTION

Although effective drugs are now available for the prevention and treatment of many turkey diseases, emphasis should be placed on prevention through management. Preventative measures properly applied represent small cost and inconvenience compared to the expense of one disease outbreak.

A PROGRAM OF TURKEY HEALTH, DISEASE PREVENTION, AND PARASITE CONTROL

Although the exact program will and should vary according to the specific conditions existing on each individual turkey farm, the basic principles will remain the same. With this in mind, the following program of turkey health, disease prevention, and parasite control is presented. Producers may use it (1) as a yardstick with which to compare their existing program, and (2) as a guide so that they and their veterinarians, and other advisors, may develop a similar and more specific program for their respective enterprises.

1. **Obtain clean stock.** Clean stock simplifies the problem of disease control by reducing the number and severity of problems that are present when the flock is established. Disease-free stock is best assured by purchasing stock from breeders who have conscientiously participated in organized disease-control programs.

Stock should be obtained only from sources tested for and rated free from *Arizona* infections, fowl typhoid, *Mycoplasma* infections, paratyphoid, pullorum disease, and any other diseases for which testing of breeding stock is being conducted.

2. **Select a clean range.** Where turkeys are kept on the range, it should be naturally well-drained, clean, and somewhat sandy. Also, the soil should be free from contamination by poultry, hogs, or sheep. Do not allow turkeys access to ponds, streams, poorly drained areas, weed patches, roadways, or other people's property.

Allow 1 acre of range for each 250 turkeys raised each year. Move the growing turkeys and their equipment to an adjacent clean area every week or two throughout the rearing season (more often in warm, wet weather) and use this same area only 1 year out of every 2 to 3 years.

3. **Avoid bringing infection in.** Most diseases of turkeys are shared with other birds, such as chickens, pheasants and other wild fowl. Sometimes chickens and other fowl carry certain diseases without showing any symptoms. For this reason, no other poultry should be allowed on the farm.

The practice of bringing flocks from two or more ranges together for breeder flocks should be done only when absolutely necessary and with the realization that some risk is involved. If flocks must be mixed, it should be done as early as possible. Turkey flocks may have some diseases as poults, then recover and appear healthy, yet be carriers. For this reason, when two flocks are mixed together, there is frequently a disease outbreak in one flock or the other.

Avoiding bringing infection in also includes the prevention of cross contamination through droppings, equipment, transportation, air currents, and the clothing of caretakers.

4. **Control wild birds and rodents.** Screen all buildings against wild birds and ratproof turkey houses.

5. **Control internal and external parasites.** Feed is always too costly to feed to parasites. Besides, the control of external and internal parasites, along with keeping stresses to a minimum, helps maintain the birds in good condition so that they can resist disease organisms.

6. **Select visitors carefully, and caution them.** Only invited visitors should be permitted to enter poultry houses and range areas. If they do enter, require them to first step into a shallow pan containing cresol or some other strong disinfectant. It is also desirable to use plastic and throw-away boots.

7. **Avoid contaminated feed and water.** Feed and water the birds from equipment that cannot be contaminated by droppings and keep the areas around the feeders and waterers dry and clean at all times by moving, by frequent cleaning, or by using wire-covered or slatted platforms or grills.

8. **Avoid dusty or wet conditions.** Extremely dusty or wet conditions should be avoided on the range or in confinement.

9. **Follow a vaccination program.** Vaccination is cheap insurance against heavy losses from certain diseases. Accordingly, the birds should be vaccinated against those diseases that are common to the area.

10. **Recognize disease early.** A sharp drop in feed consumption, a general listless appearance of the birds, or an "off" appearance of the droppings, are usually the first symptoms of most turkey diseases. Prompt attention usually reduces the severity of a disease and may possibly prevent spread to other pens or flocks.

11. **Use diagnostic laboratory.** When any disease symptoms appear, typically affected turkeys should be submitted to a diagnostic laboratory as soon as possible. Modern turkey farms represent a large investment. Thus, heavy losses may accrue if the wrong medication is given. For this reason, turkey producers should not attempt to identify all diseases on their farms. When disease strikes, they should use a professional diagnostic service promptly.

12. **Dispose of dead birds properly.** Dead birds should be disposed of by incineration, deep burial, or a disposal pit, or recycling through a rendering plant as a feed.

13. **Periodically vacate and clean.** Clean-up between flocks is essential and must be done as soon as possible after the flock is sold. Depopulation of the facilities and allowing them to lay dormant in the fresh air and sunlight is very effective in killing disease organisms. This practice should be coupled with a disinfectant spray applied just before the buildings are to be used again.

VACCINATING

To protect turkeys against disease, it is necessary to combine good management with proper application of vaccines to increase the resistance of the birds against specific infections.

Vaccines are useful agents to help build resistance to a specific disease. No vaccine is completely effective, and without possible adverse reactions or lack of effectiveness.

Vaccine producers are responsible for producing safe, potent, effective vaccines. Users also have certain responsibilities. They must see that vaccines are refrigerated or stored properly, used within the expiration date, mixed according to directions, applied to birds that are in good health when vaccinated, and administered according to the directions.

It is important to have a well-planned vaccination program designed for the specific farm operation and to carry out that program carefully with good products used as intended.

ERYSIPELAS

In some areas, erysipelas is a major problem. In addition to causing death, the disease frequently affects the fertilizing capacity of male turkeys. Also, marketing losses may result from condemnations or downgrading losses due to postmortem evidence of septicemia or to lack of finish.

It is recommended that turkeys be immunized against erysipelas in areas where the disease is known to be a problem. Market turkeys should be given a single dose of the bacterin subcutaneously at the dorsal surface of the neck behind the atlas. For turkeys kept as breeders, at least two doses of the bacterin, given at intervals of no less than 4 weeks, should be administered prior to the onset of egg production; the first dose may be given at 16 to 20 weeks of age (at selection time), and an additional dose may be given just prior to the beginning of lay.

Fig. 14–16. Young turkey being vaccinated against erysipelas. (Courtesy, Virginia Poultry Federation, Harrisonburg, VA)

FOWL CHOLERA

Vaccination should be considered in areas where fowl cholera is present. Commercially produced bacterins and live vaccines are available. But vaccination should not be substituted for good sanitary practices.

FOWL POX

Vaccination against fowl pox has been used successfully for many years, as a means of protecting the flock and preventing a drop in egg production and fertility. Turkeys do not develop as long an immunity as chickens. Initially, turkeys are vaccinated when they are 2 to 3 months old, but those to be used as breeders should be revaccinated before production. Revaccinating at 3- to 4-month intervals may be helpful, depending on the level of risk.

HEMORRHAGIC ENTERITIS

This condition is a significant problem in some turkey-producing areas. To control hemorrhagic enteritis, avirulent isolates can be successfully used as live, water-administrated vaccines. If flocks are less than 100% protected by vaccination, they are subsequently protected by lateral transmission of vaccine virus within 2 to 3 weeks.

NEWCASTLE DISEASE

In most turkey-producing areas, some market turkeys are not vaccinated. However, breeder birds should be vaccinated after the flock has been selected. A common practice is to vaccinate in the drinking water with a B–1 type of Newcastle disease vaccine. Other recommendations include LaSota strain and a tissue culture strain that must be injected. There is still considerable investigation into vaccination programs involving live and inactivated vaccines for use in turkeys. Newcastle disease has not been a major problem in turkeys; nevertheless, it is wise to protect breeder flocks because the disease could cause a serious disruption of egg production.

DRUGS

Drugs are widely used in the turkey industry, for the prevention and treatment of diseases, as well as, in some cases, growth stimulants. Few flocks are raised without the use of drugs, ranging from coccidiostats to antibiotics and other chemotherapeutic agents. Turkey producers may:

1. Dip the hatching eggs in antibiotics to control *Arizonosis* and *Mycoplasma* infection.
2. Start the poult on a coccidiostat and an antibiotic or other antibacterial agent.
3. Treat any sign of disease with increased levels of drugs.
4. Give increased drug treatment at times of stress.
5. Use drugs to treat specific infectious diseases such as fowl cholera.

The proper use of drugs can make for more profitable turkey production; abuse or misuse can add to the cost of production, in cost of ineffective drugs and in cost of disease that could have been controlled by effective drugs or other means. Also, the long-time use of drugs may cause drug-resistant bacteria that make treatment of turkey diseases more difficult or even contribute to resistant bacteria in humans.

COMBATTING DISEASE

If disease strikes, the turkey producer should have in mind a program for combatting the situation. This involves a knowledge of symptoms or signs of diseases, defensive measures, diagnosis, and postmortem examination.

SYMPTOMS OR SIGNS OF DISEASE

One of the first signs of an impending disease outbreak is usually lowering of feed consumption. Also, birds may be nervous or they may appear droopy and listless. Coughing and sneezing indicate respiratory infections; abnormal droppings suggest intestinal disorders. Young poults may crowd together and seem to be cold even though heat is available. These warning signals indicate the need for a quick check of the birds' health.

DEFENSIVE MEASURES

When sick birds are detected, whether on the range or in confinement, they should be isolated if at all possible.

All the turkeys should be kept on the regular feed and water. The waterers should be cleaned and washed daily.

Do not medicate, disinfect, or fumigate until after a diagnosis has been made. Even then, these control measures should be applied only as directed by a competent specialist. However, thorough cleaning followed by disinfection of the building, exposure to direct sunshine, and depopulation of the turkey premises for at least 30, and preferably 90, days is helpful.

DIAGNOSIS

Once sickness is observed, an accurate diagnosis should be obtained as soon as possible. An infectious disease may be involved, or perhaps the problem is poor nutrition and management. It is important to know what is wrong with the turkeys in order to treat them properly and manage the flock to prevent further losses. Be sure to take typically sick or fresh, dead birds to a laboratory and provide an accurate history on the course of the sickness. In some states, the pathology laboratory may be used at a nominal cost. The county extension agent will know where such laboratories are located. If the diagnostic facilities are not available within transportable distance, the turkey grower should promptly perform a postmortem examination (autopsy) of five or more of the freshly dead or very sick birds, then record the observations. Prior to the autopsy, however, the grower should first examine a normal, healthy bird to aid in recognizing abnormal conditions.

After the postmortem, operators should dispose of all the turkey material in an incinerator or pit, then thoroughly wash the equipment, their hands, and their arms with soap and water or a detergent.

DISEASES

Of the common disease organisms, only the *Salmonellas* (typhoid, pullorum, and paratyphoid), the paracolons, and the *Mycoplasmas* (sinusitis and synovitis) have been shown to be egg transmissible. All these diseases can be transmitted in other ways, also. Any turkey flock with *Mycoplasma* (PPLO), pullorum or typhoid should be marketed as soon as the disease is diagnosed with certainty. All eggs on hand and in the incubator should be destroyed.

S-6 *Mycoplasma gallisepticum* is primarily an infection, and in most outbreaks it can be traced to chickens and wild fowl in the area.

There are a multitude of paracolon bacteria, so named because they are closely related to ordinary intestinal bacteria. *Para* means related — similar, or quite a bit alike.

The most important turkey diseases are discussed in Chapter 10, Poultry Health, Disease Prevention, and Parasite Control; hence, the reader is referred thereto.

PARASITES

The turkey producer should be constantly on the alert for parasites, both external and internal. The most common external parasite of turkeys is the fowl mite. This small, gray mite is distinguishable from a fleck of dust only by the fact that it moves quite rapidly. Fowl mites live mostly in the tail feathers and in the fluff feathers at the rear of the keel, and infestations of these tiny parasites can build up very quickly in a flock, especially in toms, and often before it is noticed. The insemination crew should be instructed to look for and report mites.

Roundworms are a very common internal parasite of turkeys. Once a flock has roundworms, or has run on litter or ground that has been used with turkeys with roundworms, the flock can be considered to be continually infested. A regular, once-a-month worming with a wormer containing piperazine citrate will hold the worms to a harmless level.

DEFORMITIES, INJURIES, VICES

A good many deformities, injuries, and vices are found in turkeys. A few of the more important ones will be discussed herein.

BLUEBACK

Blueback is a permanent, dark discoloration of the skin on the back and sometimes the side and breast of turkeys with dark plumage, but not turkeys with white plumage. Blueback may be caused by a rare recessive hereditary factor or, more likely, by damage to the developing feathers by picking or treading, combined with the action of direct sunshine. Blueback causes downgrading when the birds are marketed. In young stock, blueback will not occur if feather picking and mating are prevented, unless of course, the hereditary factor is involved. In naturally mated breeding stock, blueback is difficult to prevent, although the use of saddles on hens reduces it considerably.

BREAST BLISTERS AND CALLUSES

Breast blisters and calluses are much more common in toms than in hens, and are less common in dimple-breasted birds than in those without the dimple. They are believed to be caused by continuous irritation of the skin that covers the breastbone. They can cause serious losses due to the downgrading under federal inspection.

CANNIBALISM

Feather picking is a mild form of cannibalism to which turkeys are addicted, especially during the growing period. It results in unsightly appearance, more trouble from pinfeathers when the birds are marketed, and blueback in varieties with dark plumage. It can develop into flesh picking and become serious enough to retard growth rate. Generally speaking, feather picking reaches serious proportions only in turkeys raised in confinement. It can be prevented or stopped almost completely by debeaking or by inserting a specially made turkey bit resembling a 1½ in. hog ring with blunted ends. However, either treatment will prevent effective grazing on the range.

Management practices that tend to prevent feather picking include: (1) avoiding overcrowding in confinement rearing quarters; (2) feeding an adequate diet; (3) feeding pelleted rather than loose mash; (4) allowing confined turkeys access to baled legume hay; and (5) not confining turkeys to roosts or other closely restricted quarters, particularly in the early morning.

PENDULOUS CROP

Pendulous crop, sometimes called baggy, sour, or dropped crop, is caused by a weakening of the crop and supporting tissues so that feed and water accumulate in the organ and pass out too slowly or not at all, resulting in sour, foul smelling, semiliquid accumulation and deterioration of the crop lining. Seriously affected birds seldom recover; and treatment is useless.

Some strains or families of turkeys, particularly the types that are not broad breasted, seem to possess a genetic weakness that predisposes them to the disorder. When these susceptible birds are exposed to hot weather with consequent heavy consumption of liquids, the deformity develops. Preventive measures include: (1) selecting strains not carrying the genetic factor; (2) avoiding exposure of turkeys to excessive heat without shade; (3) giving continuous and easy access to cool drinking water; (4) providing ample shade; and (5) feeding no liquid milk.

STAMPEDING

Turkeys are subject to fright, especially at night. Severe losses from injury, straying, smothering, bruising, broken limbs, and deaths by predatory animals may result from stampedes. Stampeding may be lessened by avoiding disturbances of all kinds around roosting quarters and providing low-intensity night lighting by a rotating beam or small light bulbs. Occasionally, low-flying airplanes may cause serious daytime stampedes in turkeys on range or in yards.

TURKEY MANAGEMENT

Fig. 14–17. Growthy, healthy turkeys giving evidence of superior management. Note the curtain sides and the automatic feeder and water lines. (Courtesy, California Polytechnic State University, San Luis Obispo)

Management is the key to success in the turkey business. The geneticist can lift the potential of production; the nutritionist can formulate a ration that can reach this potential; but without proper management, the turkey producer will never get there.

Management gives point and purpose to everything else — to breeding, feeding, housing, health, and marketing. It can make or break a turkey enterprise. Fortunately, a turkey operation responds favorably to good management.

PRODUCTION GOALS

High goals spur high achievement. Turkey producers should aim for the following:

■ **Fertility of eggs** — Lowering incubated egg infertility from the normal 10% to 3%.

■ **Embryonic fertility** — Lowering embryonic deaths from the normal 5 to 15% to about 2%.

■ **Mortality of poults** — Lowering poult mortality from the normal 10% to 3%.

■ **Growth rate and feed efficiency** — Producing and marketing large-type toms at 15 weeks, weighing 30 lb, and requiring 2.3 lb of feed per pound of turkey. Producing and marketing large-type hens at 12 weeks, weighing 17 lb, and requiring 2.1 lb of feed per pound of turkey.

MANAGEMENT PRACTICES

In addition to the management practices mentioned earlier in this chapter, including debeaking and desnooding, the following management practices are pertinent to the raising of turkeys.

DISEASE PREVENTION AND CONTROL

Preventive health programs are essential if losses from diseases are to be held to a minimum among the breeders and during brooding and rearing. In addition to actual death losses from diseases, there are losses in feed efficiency, market condemnations, etc. Although disease losses will probably never be reduced to zero, each grower should strive to reduce losses to an absolute minimum.

MOVING TURKEYS

When poults are about 8 weeks of age, they are ready to be moved from the brooder house to range, semiconfinement sheds, or confinement buildings. Movement to range is most critical and should be carefully planned. Two things should be checked before moving poults to range: (1) the 5-day or 1-week weather forecast, and (2) the readiness of equipment. Avoid moving poults when the weather is threatening. Cool, wet, and rainy weather places a severe stress on poults that have just been moved to range. It is best to move in the early morning, as this allows poults plenty of time to locate water, feed, and shelter before nightfall. Never move in the heat of the day, for this is hard on both birds and handlers. Some growers move about one-third of their poults the first morning, then skip a day or two and move the remainder of the flock. Large wire-enclosed four-wheel trailers are best for moving turkeys to ranges. Moving trailers may be constructed so that the birds can be loaded and unloaded by driving or herding. This system will require less labor and reduce injuries to poults.

PREDATOR CONTROL

A tight fence around the range area will reduce predator entry. A perimeter electric fence 10 to 18 in. above ground level and 2 to 3 ft outside the poultry fencing will further reduce predatory entry. When an electric fence is used, it should be properly constructed; otherwise, it may be hazardous to caretakers.

Some growers ward off predators with devices that produce loud noises or by keeping a dog leashed near the range area. Federal, and possibly state assistance, should be obtained when predator losses become severe.

RECORDS

Complete and accurate records supply necessary information for making correct decisions. There is no substitute for a complete record of each flock raised. Records can and should tell much more than profits or losses. Items that should be recorded for each flock are:

1. **Poults.** Date started, number, strain, and cost.
2. **Feed.** Pounds feed, cost, beginning and ending inventory.
3. **Litter.** Amount used and cost.
4. **Fuel.** Amount used and cost.
5. **Medication.** Kind, date(s) used, and cost.
6. **Pesticides.** Kind used, date used, where used, and why.

7. **Insurance.** Description and cost.
8. **Hired labor.** Amount and cost.
9. **Interest on investment.** Rate.
10. **Finance charges.** Amount.
11. **Depreciation on buildings and equipment.** Amount.
12. **Number of birds produced** (loaded on truck). Count.
13. **Number of birds sold.** Count, and record grades.
14. **Condemnation losses.** Count, and record reasons.
15. **Gross returns.** Amount.

SEXING

Some producers buy day-old sexed poults. Others buy straight-run poults, then separate the sexes at 12 to 14 weeks of age. In any event, most producers raise the sexes apart because (1) slightly better weight gains can be obtained by separating the sexes, and (2) more efficient gains can be obtained by designing a feed especially for toms or for hens after they are 12 weeks old.

WING CLIPPING AND WING NOTCHING

Wing clipping and notching control flying. Many processors and growers do not favor these practices because when such birds attempt to fly, they fall, frequently bruising, causing downgrading.

To wing clip, remove the tip of one wing just beyond the outer joint, from immediately after hatching up to 14 days of age. Do it preferably at boxing time so that the poults are quiet in the box for a quick clotting. Do not clip the wing tip with the scissors or use a debeaker.

To wing notch, clip the tendon over the wing joint. Notch between 1 day and 3 weeks of age. This method has not proven too satisfactory, since in young birds the tendon can grow back or be improperly severed.

ON-FARM TESTS

When carefully conducted and properly interpreted and used, on-farm tests can be a valuable adjunct in a large turkey operation. Among their virtues, the turkey producer can study area and feed differences. Among their limitations, there is usually less accuracy and fewer controls than in most university-conducted experiments. For the latter reason, most of them should be looked upon as applied tests or demonstrations *per se*, rather than carefully controlled, basic experiments; terminology which does not detract from their value, but which does place them in proper perspective.

The major requirements for conducting good on-farm tests are: (1) identical facilities, and (2) a willingness to keep accurate records.

Every effort should be made to treat test flocks identically. The same caretaker should do the chores; the chores should be done in reverse order on successive days in order to minimize possible traffic and social-behavior effects; feeders, waterers, and ventilation should be the same; cleaning of the facilities should be the same; the facilities should be the same; and the preceding flock history should be approximately the same for all groups under study.

Identical records should be kept for each flock. These should describe mortality (recorded daily, summarized each week), any disease outbreak or accidental death problems, feed consumption (record date feed is added), and growth rate.

Birds should be weighed at standard intervals, such as every 4 weeks. It is difficult to get accurate weights when flocks are sample weighed. Where sample weights are taken, the preferred method is to isolate a typical section of birds within the flock and then weigh the closest 25 birds individually. This is preferable to going into the flock and picking up and weighing the closest 25 birds that can be caught. Three such batch weighings in a flock, averaging the weights and averaging the averages, should be satisfactory, and, at the same time, will inform the supervisor of the variation involved. It is also recommended that the weighing crew alternate the 25-bird weighings between flocks rather than finish one flock first before starting on the other.

Detailed final weighing should be done on the farm before birds go to market. At that time, as many birds should be weighed as possible. Five lots of 25 birds would be better than the three lots described above, as this weight will represent the entire flock in potential economic calculations.

For calculation of feed consumption, the feed remaining should be subtracted from the total feed supplied.

It is desirable to replicate the experiment in nearby facilities. This enables the calculation of average results, the determination of the consistency of response, and an estimation of the probability that the observed difference is real (not due to chance).

Field tests conducted by the turkey producer could involve stocks, feed, or management conditions. They have a good chance of being applicable because the farmer is seeking a true, unbiased answer.

Field test reports used by product suppliers for technical data are not always reliable. They are usually positive and support the sale of the product. There is a human tendency to pick favorable tests and to overlook those which do not show positive responses. Results may be biased in that positive responses get summarized and published, while negative ones remain in the files unreported.

MARKETING TURKEYS

Most producers market their birds alive through integrated firms, processors, or cooperative organizations. The turkeys are collected and trucked to dressing plants where they are processed for the market.

Turkey growers located near centers of population sometimes sell their turkeys directly to consumers or to retail dealers. These growers usually have their own processing and freezing facilities and some of them have developed profitable gift-package businesses, featuring high-quality, frozen, ready-to-cook turkeys that can be shipped long distances in sealed packages containing dry ice—about 1 oz per pound of turkey.

WHEN TO MARKET

Until about the mid 1960s, fresh and frozen turkey meat was readily available only around the Thanksgiving and Christmas holiday season, usually as whole carcasses. Now, fresh turkey (whole birds and cut-up parts), along with further processed turkey, is available year-round. However, about 40% of the per capita consumption of turkey meat occurs in the fourth quarter (October, November, and December), with the remaining 60% distributed fairly equally throughout the other three quarters.

Under optimum conditions of nutrition and management, all Broad-Breasted White turkeys are fully finished and ready to market as mature young roasters at $17\frac{1}{2}$ weeks of age for males and $14\frac{1}{4}$ weeks of age for females. Turkey fryers, which represent about 2% of turkey slaughter, are marketed at younger ages and at 8 to 12 lb live weight.

MARKET CHANNELS

The major marketing channels for turkeys are detailed in Fig. 14–18. It is noteworthy that increased bypassing of wholesale distributors has occurred, with more and more of both ready-to-cook birds and further processed products being marketed directly.

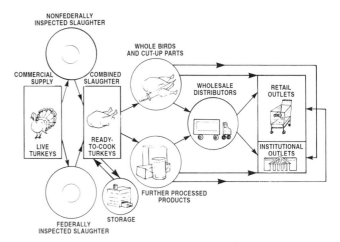

Fig. 14–18. Major marketing channels for turkeys.

Cut up and further processed products accounted for about 80% of the volume of the turkeys slaughtered in the late 1980s.

PROCESSING

Most turkeys are now processed in plants that process turkeys only.

The usual steps in processing are:

1. **Suspended by legs.** The live turkey is suspended by the legs, the feet held in a steel shackle.

2. **Stunning.** An electric stunner is usually used to prevent struggling and relax the muscles that hold the feathers.

3. **Bleeding.** The head is held in one hand, and a cut

is made across the throat so that both branches of the jugular vein are severed at or close to their junction.

4. **Scalding.** The feathers are loosened, usually by the subscald method in which the bird is immersed in agitated water at 140°F for about 30 seconds.

5. **Picking.** The feathers are picked by a rubber-fingered machine.

6. **Removing pinfeathers.** The pinfeathers are removed by hand.

7. **Eviscerating.** Usually, the turkey is suspended by the head and both hocks, a crosswise cut is made between the rear end of the keel and the vent, and the intestines are removed.

8. **Chilling.** The eviscerated carcass is chilled in ice water or ice slush to an internal temperature of 35° to 39°F.

9. **Packaging.** After chilling, the carcass is drained, a plastic wrap is applied, the air in it is exhausted by vacuum, and the wrapper is sealed. The bird is then readied for freezing or for marketing fresh chilled, unfrozen.

INSPECTION

Inspection refers to the examination of birds for indication of disease or other conditions that might make them unfit for human consumption. The Poultry Products Inspection Act, which became effective in 1959, requires all ready-to-cook poultry and poultry products moving in interstate or foreign commerce to be federally inspected.

Antemortem inspection of poultry is required before it enters the processing plant. This inspection is essentially a spot check of each lot of birds, not of each individual bird.

Postmortem inspection of each bird should take place at the time the abdominal cavity is opened and the visceral organs are removed.

GRADING

Grading consists of identifying individual birds according to class, quality, and weight. Final grading of ready-to-cook poultry should be after the birds are eviscerated and cooled and before they are packaged, frozen, or further processed, but not while they are still warm.

The U.S. Department of Agriculture standards of quality that have been established for ready-to-cook poultry are A, B, and C. These grades are based on conformation, fat covering under the skin, and degree of freedom from skin tears, skin blemishes, and pinfeathers. The U.S. Grade mark can be used only when the birds have been officially graded under the voluntary U.S. Department of Agriculture grading program.

The Grade A turkey is well fleshed, well covered with fat all over the body, well dressed, well bled, and practically free from skin tears, discolorations, pinfeathers, and calluses and blisters. The carcass is clean, attractive, and wholesome.

The Grade B turkey may be slightly deficient in meatiness, finish, and fat or may have a few scattered pinfeathers or discoloration of the skin. There may be a broken bone or the bird may be slightly misshapen.

The Grade C turkey may be poorly fleshed on breast and legs, with little or no fat. It may have a prominent breastbone and the leg, wing, or breastbone may be misshapen, or the bones may be broken. There may even be skin cuts, discoloration, or tears of the skin or a scattering of pinfeathers.

PRODUCT FORM

Turkeys are marketed primarily as (1) frozen, ready-to-cook (RTC) whole birds, (2) as cut-up parts—breast, legs, and wings, or (3) as other processed products, such as rolls, roasts, pot pies, bologna, turkey frankfurters, turkey ham, pastrami, turkey sausage, and frozen dinners. During the Thanksgiving and Christmas holidays, a limited number of turkeys are marketed fresh dressed.

Most turkeys marketed whole are Grade A. To the extent possible, Grades B and C and lower-priced parts are utilized in further processing.

Fig. 14–19. Fresh dressed turkey in the supermarket fresh meat case. (Courtesy, Cryovac Division, W. R. Grace & Co., Duncan, SC)

PRICES AND PRICE SPREADS

Prices consumers pay for turkeys are much higher than the cost of production and include charges for such services as processing, transportation, packaging, and storage.

CONSUMPTION AND DEMAND

Per capita turkey consumption increased from 4.1 lb in 1950 to 17.1 lb in 1989. Although there is a high preference for turkey at Thanksgiving and Christmas, cut-up and further processed have increased turkey consumption throughout the year.

WHAT'S AHEAD FOR TURKEYS

Tremendous changes have occurred in the U.S. turkey industry in the last decades, and more changes lie ahead. The futuristic turkey industry will be characterized as follows:

1. **Increased per capita consumption.** Increased per capita consumption is projected in the decade ahead, with much of the increase in cut-up and further processed items.

2. **Increased growth rate and feed efficiency.** Further increases in growth rate and feed efficiency will be achieved.

3. **Fewer but larger turkey producers and processors.** This trend will continue.

4. **Increased hatchability and decreased poult mortality.** There will be fewer infertile incubated turkey eggs (now 10%) and fewer embryonic deaths (now 5 to 15%) of eggs set; and poult mortality will be reduced from the current average of 9 to 10%.

QUESTIONS FOR STUDY AND DISCUSSION

1. Give a profile of the current U.S. turkey industry.

2. The states of North Carolina, Minnesota, and California are far removed from each other geographically, and are quite dissimilar in climate, crops produced, and human population. Yet, they are the three leading turkey producing states of the nation. How do you explain this?

3. Currently, about 90% of U.S. turkeys are produced under contract. Is this high proportion of contract production good or bad? Justify your answer.

4. To what extent will the energy shortage affect the trend toward producing turkeys in environmentally controlled houses?

5. Describe *cut-up* and *further processed* as these terms apply to turkey meat. What impact are these developments having on (a) per capita consumption, and (b) consumption of turkey throughout the year?

6. The Pilgrims hunted wild turkey to grace their tables at Thanksgiving. How much of a factor has this fact of history been in the high seasonal consumption of turkey at Thanksgiving and Christmas ever since?

7. Some turkey contractors give bonuses for livability percentage, feed conversion rates, and market yield grades, in addition to a flat rate according to the number of weeks in production. Is such an incentive basis desirable?

8. Since there are no turkey futures contracts at this time, how may (a) a large turkey grower or (b) a restaurant that uses a lot of turkeys hedge against huge losses?

9. After considering the current costs and returns in the turkey business, would you recommend that a young person launch a career as a turkey grower? If so, how might the enterprise be financed; and how much net should reasonably be expected from 50,000 market birds?

10. Discuss ways in which feed costs could be lowered.

11. How may contract turkey growers increase their returns?

12. Why do most commercial turkey growers continue to produce pure breeds or varieties, whereas most broiler growers use crosses?

13. Discuss the impact of Jessie Throstole's little flock of 20 hens and 4 toms of Broad-Breasted Bronze turkeys, which he took with him when he migrated from England to Canada in the early 1930s.

14. In the years ahead, do you foresee increased confinement rearing or increased range rearing? Justify your answer.

15. Discuss each of the following turkey breeder management practices:
     a. Holding breeders.
     b. Out-of-season breeding.
     c. Lighting.
     d. Artificial insemination.

16. How does the brooding of turkey poults differ from the brooding of broilers?

17. How might a turkey breeder improve: (a) egg fertility, (b) embryonic deaths, (c) poult mortality?

18. Does high-priced feed favor separating the sexes for feeding?

19. Of what value are growth rate and feed consumption standards?

20. Outline a program of turkey health, disease prevention, and parasite control.

21. Assume that a sudden drop in feed consumption and signs of sickness have been observed in a large flock of turkeys. What steps would you take to combat the situation?

22. What can be done to lessen the incidence of each of the following in a flock of market turkeys:
     a. Blueback?
     b. Breast blisters and calluses?
     c. Cannibalism?
     d. Pendulous crop?
     e. Stampeding?

23. If market turkey producers wish to be in the upper 5%, what goals should they have in —
     a. Rate of gain and market age?
     b. Feed efficiency?
     c. Death loss?

24. Discuss the how and why of each of the following management practices:
     a. Debeaking.
     b. Desnooding.
     c. Moving turkeys.
     d. Wing clipping and wing notching.

25. Of what value is an on-the-farm test? How would you conduct such a test designed to compare two feed additives?

26. How have turkey market channels changed in recent years?

27. From the standpoint of the consumer, of what value are (a) inspection, and (b) grading?

28. What changes do you see ahead in the turkey industry?

SELECTED REFERENCES

Title of Publication	Author(s)	Publisher
American Poultry History, 1823–1973	Edited by O. H. Hanke J. K. Skinner J. H. Florea	American Poultry Historical Society, Inc., Mount Morris, IL, 1974
Diseases of Poultry, Ninth Edition	Edited by B. W. Calnek, et al.	Iowa State University Press, Ames, 1991
Feeds & Nutrition, Second Edition	M. E. Ensminger J. E. Oldfield W. W. Heinemann	The Ensminger Publishing Company, Clovis, CA, 1990
Feeds & Nutrition Digest	M. E. Ensminger J. E. Oldfield W. W. Heinemann	The Ensminger Publishing Company, Clovis, CA, 1990
Nutrient Requirements of Poultry and Nutritional Research	C. Fisher K. N. Boorman	Butterworths, London, England, 1986
Nutrition of the Turkey	M. L. Scott	M. L. Scott of Ithaca, Ithaca, NY, 1987
Poultry Breeding and Genetics	Edited by R. D. Crawford	Elsevier Science Publishers, Amsterdam, The Netherlands, 1990
Poultry: Feeds and Nutrition, Second Edition	H. Patrick P. J. Schaible	The AVI Publishing Company, Inc., Westport, CT, 1980
Poultry Meat and Egg Production	C. R. Parkhurst G. J. Mountney	Van Nostrand Reinhold Company, New York, NY, 1988
Poultry Nutrition Handbook	J. D. Summers S. Leeson	University of Guelph, Guelph, Ontario, Canada, 1985
Poultry Production, Twelfth Edition	M. C. Nesheim R. E. Austic L. E. Card	Lea & Febiger, Philadelphia, PA, 1979
Poultry Products Technology, Second Edition	G. J. Mountney	The AVI Publishing Company, Inc., Westport, CT, 1976
Processing of Poultry	Edited by G. C. Mead	Elsevier Science Publishers, Ltd., Barking, Essex, England, 1989
Standard of Perfection, The, Fifth Edition	Edited by M. C. Wallace	American Poultry Association, Inc., Crete, NE, 1966
Turkey Production: Breeding and Husbandry, Reference Book 242	G. A. Clayton, et al.	Ministry of Agriculture, Fisheries and Food, United Kingdom, 1985

Fig. 14–20. Eight-week-old turkeys. (Courtesy, California Polytechnic State University, San Luis Obispo)

Fig. 14–21.   Bourbon Red turkeys. (Courtesy, *Poultry Tribune*, Mount Morris, IL)

Fig. 15–1. A montage, showing a duck, a quail, a pheasant, and a goose.

DUCKS, GEESE, AND MISCELLA-NEOUS POULTRY

CHAPTER

15

Contents | Page

Contents

Although chickens and turkeys dominate the U.S. poultry industry, ducks and geese are also of considerable economic importance; and they are more popular in the Far East and in Europe than in the United States. Also, there is a place for pigeons, guineas, and ornamental and game birds.

DUCKS

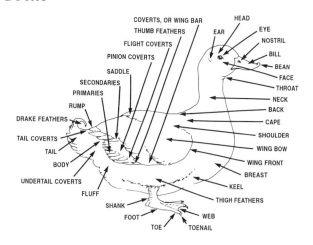

Fig. 15–2. Parts of the male duck. Duck breeders and growers should know the language that describes and locates the different parts of the bird.

Fig. 15–3. Pekin ducks at Ward Duck Co., La Puente, California. (Courtesy, California Polytechnic State University, San Luis Obispo)

Ducks are among the most versatile of animals. They will live happily under a wide range of climatic conditions; and they are free from such common poultry ailments as leukosis, Marek's disease, infectious bronchitis, and other respiratory troubles. Also, ducklings are popular with gourmet cooks and connoisseurs of good food. Virtually everything from the feathers to the feet, including the liver and tongue, can be turned to profit; the only unsalable thing about them is their quack.

In 1985, 21.6 million ducks were marketed in the United States. Duck production was once centered on Long Island, New York. Although Long Island still produces many ducks, large numbers of ducks are now raised in Indiana, Michigan, Minnesota, and North Carolina.

Most meat ducks are marketed as ducklings, at 7 to 8 weeks of age and at a liveweight of about 6.25 to 7.5 lb. They are generally frozen and ready to cook after thawing.

There is little demand for duck eggs in the United States.

BREEDING

The breeding and improvement of ducks have received less attention than the breeding of chickens and turkeys, reflecting their lesser economic importance.

Duck breeding in the United States is largely confined to breeding for meat production, with selection for more rapid and efficient gains, more lean meat, and higher egg production and hatchability.

BREEDS

Ducks may be classed as either of two types: meat producers, or egg producers.

MEAT BREEDS

Aylesbury, Muscovy, and White Pekin ducks are excellent meat producers. Rouen, Cayuga, Swedish, and Call ducks reach weights that make them valuable as meat producers, but poor egg production and colored plumage make them unsatisfactory for mass commercial production

Aylesbury

The Aylesbury is as popular in England, where the breed originated, as the White Pekin is in the United States. Aylesburys produce excellent quality meat, and, like White Pekins, they reach market weight in 8 weeks.

Aylesburys have white feathers, white skin, flesh-colored bills, and light-orange legs and feet. Eggs are a tinted white. Adult drakes weigh 9 lb and adult ducks (females) 8 lb.

Aylesburys are not as nervous as White Pekins, but they rank with the latter in their lack of interest in setting. Egg production is generally below that of White Pekins but extraordinary records of 300 eggs per laying year have been reported.

Muscovy

The Muscovy is unrelated to the other breeds mentioned in this chapter. The breed originated in South America. Numerous varieties of Muscovies exist; the white variety is the most desirable for market purposes. Muscovies produce meat of excellent quality and taste, provided they are marketed before 17 weeks of age. But their low egg production makes them unsuitable for use on large commercial duck farms. Muscovies have a white skin. Adult drakes weigh 10 lb and adult ducks 7 lb.

Muscovies make extremely good setters. They will hatch and care for approximately 30 ducklings from the 40 to 45 eggs they produce annually.

Fig. 15–4.   Colored Muscovy ducks. (Courtesy, *Poultry Tribune*, Mount Morris, IL)

Although they are not ideally suited to commercial production, Muscovies have excellent possibilities for small general farms that have special retail outlets.

In some areas of the world, hybrid "mule ducks" are produced by mating female Muscovies to Mallard-type drakes. "Mule ducks" produce satisfactory meat yields, but they are sterile.

White Pekin

Fig. 15–5.   White Pekin ducks. (Courtesy, *Poultry Tribune*, Mount Morris, IL)

The commercial duck industry in the United States relies solely on the White Pekin. White Pekins are ideally suited for meat production. They produce excellent quality meat, and they reach a market weight of about 6.25 to 7.5 lb in 8 weeks.

The breed originated in China and was introduced into the United States in the 1870s. White Pekins are large white-feathered birds. They have orange-yellow bills, reddish-yellow shanks and feet, and yellow skin. Their eggs are a tinted white. Adult drakes weigh 9 lb and adult ducks 8 lb.

White Pekins are fairly good egg producers—average yearly production approximates 160 eggs—but they are not good setters and they seldom bother to raise a brood. They are nervous and should be treated gently to obtain maximum egg production.

EGG BREEDS

Indian Runners and Khaki Campbells are excellent egg-laying breeds. Where special duck egg markets exist, either of these breeds would be a happy choice.

Indian Runner

Fig. 15–6.   Indian Runner ducks; Fawn and White, White. (Courtesy, *Poultry Tribune*, Mount Morris, IL)

The Indian Runner originated in the East Indies but its egg-producing capabilities were developed in Western Europe. Although at one time they were considered to be the best egg-producing duck breed, Indian Runners are now second to the high-producing strains of Khaki Campbells.

Three Indian Runner varieties are recognized: White, Penciled, and Fawn and White. All three varieties have orange to reddish-orange feet and shanks. Characteristically, the Runners stand erect; their carriage is almost perpendicular. They weigh about the same as Khaki Campbells.

Khaki Campbell

Fig. 15–7. Khaki Campbell drake. (Courtesy, USDA)

Amazing egg-production records have been made by the Khaki Campbell. Strains selected especially for high egg production have averaged close to 365 eggs per duck within a laying year. The highest producing strains of chickens have yet to average 300 eggs per laying year.

Khaki Campbells originated in England from a cross of Fawn and White Runner, Rouen, and Mallard ducks. Khaki best describes the appearance of Khaki Campbells. Males have brownish-bronze lower backs, tail coverts, heads, and necks—the rest of their plumage is khaki; and they have green bills and dark-orange legs and toes. Females have seal-brown heads and necks—the rest of their plumage is khaki; and they have greenish-black bills and brown legs and toes.

Khaki Campbells are not valued for their meat. Young drakes and ducks weigh 3.5 to 4 lb at 2 months of age; adult drakes and ducks weigh 4.5 lb.

Another variety, the White Campbell, has sprung from the Khaki but it has not become a popular egg producer. White Campbells have orange bills and legs.

SELECTION OF BREEDERS

Meat-type breeding stock is usually selected from market flocks hatched in April and May. The potential breeders are selected from the flocks when the birds are 6 to 7 weeks of age. At this time the distinctly different voices of males and females make it easy to separate the birds—females "honk," males "belch." Care should be taken to select the proper number of each sex—one drake to six ducks is recommended. A few extra drakes and ducks should be selected to allow for mortality and further culling during the conditioning period. Drakes should come from the earlier hatched flocks to ensure their readiness for mating by the beginning of the following year.

Select breeders that are vigorous and have good weight, conformation, and feathering. Drakes that weigh 5.5 lb at

6 weeks should weigh 7.5 lb by 8 weeks; ducks that weigh 5.5 lb at 6 weeks should weigh 7 lb by 8 weeks.

Selection of progeny from ducks having high fertility, hatchability, and egg production can be accomplished through a program of trapnesting and family or progeny testing. Fertility, hatchability, and egg production are as important to the producer of market ducks as body weight, conformation, and feather covering.

BREEDING FACILITIES

Most duck eggs are laid at night. It is common practice to confine breeders to the laying houses only at night and to let them have access to yards and waterways during the day. With this system, there is little need for expensive breeder facilities. A simple shed or house is all that is needed.

Commercial growers often run as many as 500 breeders together in one flock. But smaller flocks of 50 to 60 breeders may outlay and produce a greater percentage of fertile eggs than larger flocks. Laying house mortality may also be less in smaller flocks.

Provide 2.5 sq ft of floor space per bird in the breeder house plus yard.

Breeder houses must be one story because domestic ducks, with the exception of Muscovies, cannot fly. Keep breeder houses clean, dry, and well ventilated. Leave windows and doors open during the day to allow adequate circulation of air. Good ventilation is also needed at night to prevent overheating during summer and to reduce condensation during the winter. Care should be taken to prevent the entry of rain and snow through the ventilating system. Supplementary heat need not be supplied to breeding stock.

Straw makes good bedding material. Peat moss, peanut shells, or wood shavings can also be used. Litter should be added frequently and it should be kept dry by providing plenty of fresh air. Remove damp litter before it starts to mold.

Ducks will make their own nests in the litter, but simple nest boxes can be provided in long rows along the wall. Nests should be 12 in. wide, 18 in. deep, and 12 in. high. Place them at floor level.

The use of outside drinking facilities instead of water fountains within the building has the major advantage of preventing wet pens. Ducks can be left without water during night confinement provided they have no access to feed. If it is necessary to provide watering facilities within the house, they should be placed above wire flooring or a screened drain.

Yards should slope gently away from the breeder houses to provide good drainage. Failure to provide enough open, well-drained yard space will cause dirty runs and increase the danger of disease.

Although most commercial producers provide swimming water for their breeding flocks, it is not necessary for the production of fertile eggs. Swimming water can be provided in concrete troughs, which should be about 3 ft wide and 8 to 12 in. deep. Locate the troughs at the end of the yard opposite the breeder house.

Ducks are nervous and may run in circles if disturbed in the dark. The larger the flock, the more serious the problem. All-night lights should be used in the house if stampeding is a problem. One 15-watt lamp per 200 sq ft of floor space is adequate.

(Also see Chapter 11, Poultry Houses and Equipment, Table 11–1, Space Requirements for Poultry.)

Fig. 15–8.   Collecting duck eggs from nest boxes in a row along the wall. (Courtesy, Cherry Valley Farms Limited, Rothwell, England)

CARE OF BREEDERS

Fig. 15–9.   Breeders in intensive breeding house on the world's largest duck farm. (Courtesy, Cherry Valley Farms Limited, Rothwell, England)

It is not desirable to bring birds into full production before 7 months of age, because of the problem of small eggs and low hatchability. Ducks can be brought into full production by giving them 14 hours of light daily. Give females light 3 weeks before the time desired for the start of egg production; give drakes light 4 to 5 weeks before the start of egg production so they will be ready for mating. The 15-watt lamps used to reduce stampeding will stimulate egg production, but larger lamps (40 to 60 watts) are best for this purpose.

Egg production increases rapidly once sexual maturity is reached. The flock should be laying 90% or more of full production within 5 to 6 weeks. Daily egg production will remain above 50% for about 5 months in meat-type breeds. High-producing egg-type breeds will have greater persistency.

Producers have varying opinions about the value of keeping breeding stock once the level of egg production sinks below 50%. Some find it more economical to force-molt the birds at this level, then bring them back into production in 8 to 10 weeks. Others obtain all possible eggs until production reaches 30%; then they either rest the birds for an additional lay or sell them for meat.

Levels of fertility and hatchability generally parallel those for egg production (Fig. 15–10); they are highest when egg production is high, then they taper off toward the end of the production cycle. Lack of mating experience is usually responsible for low fertility early in the season. Fertility should increase rapidly during the first few settings of eggs. Failure to obtain satisfactory levels of fertility and hatchability calls for a thorough inspection and evaluation of all phases of management.

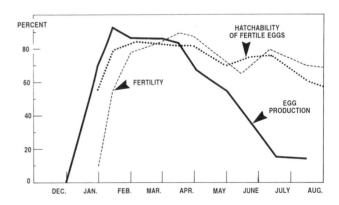

Fig. 15–10.   Typical rate of egg production and fertility and hatchability of the eggs from a commercial breeder flock of May-hatched White Pekin ducks. (Courtesy, USDA)

Most duck eggs are laid before 7:00 a.m. It is advisable to gather them at this time because prompt collection lessens the problem of soiled and cracked eggs. The breeding stock should be let out of the house at the time of starting the first collection. If some ducks are laying let them remain on the nest and make a second collection a couple of hours later.

Clean, dry, breeder houses are essential for the production of clean hatching eggs. Good ventilation, plus the frequent addition of litter and nesting material, should be the rule. Wash soiled eggs with great care immediately after collection, using water that is warmer than the eggs themselves — 110° to 115°F is adequate. Do not use water colder than the eggs because it will cause contraction of the egg contents, with the result that dirt, bacteria, and mold spores may be drawn through the pores of the shell. Good egg sanitizers can be purchased at most farm supply stores — follow the manufacturer's directions.

Cracked, misshapen, or abnormally small or large eggs should not be saved for incubation. Their hatchability will be practically nil.

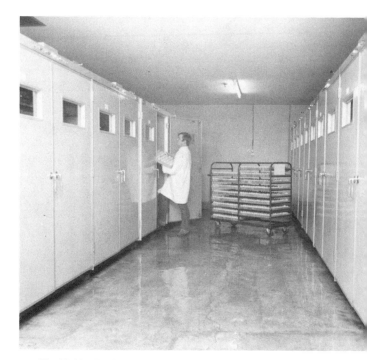

Fig. 15–12. Hatchery at Cherry Valley Farms, in England, one of the biggest and most modern duck hatcheries in the world, where six million duck eggs are set each year and an 82.9% hatch is achieved. (Courtesy, Cherry Valley Farms Limited, Rothwell, England)

Fig. 15–11. Traying hatching eggs for storage. (Courtesy, Cherry Valley Farms Limited, Rothwell, England)

Store hatching eggs at a temperature of 55°F and a relative humidity of 75%. Eggs do not need turning if weekly settings are made. Hatching eggs may be stored for 2 weeks without marked reduction in hatchability; but they should be turned daily after storage of more than 1 week. Eggs should be stored small end down.

Precautions should be taken to prevent the flow of air from the cooling unit passing directly over the eggs. Excessive evaporation and enlargement of the air cell in the large end of the egg will decrease hatchability.

INCUBATION

Muscovy eggs require 35 days of incubation; all other domestic duck eggs require 28 days of incubation.

ARTIFICIAL INCUBATION

Incubators designed for hatching duck eggs are available, but when only a few eggs are to be hatched the regular chicken-egg machine may be used. For best results, follow the manufacturer's directions.

Start the incubator a day or two in advance of the first setting. This will allow time in which to bring the machine into correct incubation condition before the eggs are set.

Remove duck eggs from the storage area 5 to 6 hours before being set. This gives them time to warm to room temperature and lessens the drop in temperature of the incubator when the eggs are set. Place the eggs in the incubator small end down.

Most incubators are equipped with automatic turning devices, which should be set to turn the eggs every 3 hours. If manual turning is necessary, it should be done at least three or four times daily.

Candle the eggs after 7 to 10 days of incubation. This can be done by passing each egg over a small hand candler or by placing an entire tray of eggs above a bright light.

The living embryo of a fertile egg will appear as a dark spot in the large end of the egg near the air cell. Blood vessels radiating from this spot give the appearance of a spider floating within the egg. An embryo that has died before candling will appear as a spot stuck to the shell membranes. Clearly radiating blood vessels will not be visible and a dark ring of blood can often be observed. Infertile eggs will appear clear.

Remove infertile eggs and eggs containing dead embryos from the incubator. Cracked eggs and those with ruptured yolks can also be detected and removed during the candling operation.

Eggs are frequently candled again after 25 days of incubation (32 days for Muscovy eggs). At this time the bills of normally developing ducklings can be seen within the air cells. Considerable movement can also be observed.

On large commercial farms, ducklings are frequently taken from the machines as they hatch. However, great care must be taken to prevent chilling of newly hatched ducklings. It may be wise for small producers to keep the machine closed until hatching is completed.

Ducklings that must be helped from the shell should not be saved for breeding stock. Hereditary factors may be partially responsible for this condition.

NATURAL INCUBATION

Natural methods of incubation are frequently used on small farms. This is especially true when Muscovies are raised. These ducks are excellent setters and will incubate their eggs without difficulty.

Other meat- and egg-type breeds do not set regularly. These breeds can be naturally hatched under broody chicken hens if they are available.

Clean, dry nesting facilities must be provided for setting hens and ducks. Feed and water should be within close proximity because the female must obtain her daily requirements within short periods of time. Delay in finding feed and water will result in undue chilling of the eggs. This situation is more critical with broody hens because they will have to keep their nests for 1 week longer than if they were setting on chicken eggs.

BROODING AND REARING

Fig. 15–13. Brooding 6-day-old ducklings at Cherry Valley Farms, in England. (Courtesy, Cherry Valley Farms Limited, Rothwell, England)

Ducklings should be moved from the hatcher to comfortable brooding quarters as quickly as possible. Prevent chilling and do not overcrowd the birds during transit. Provide feed and water for ducklings as soon as they are placed in the brooder.

Buildings of practically any type can be used to brood ducklings as long as the birds are kept warm, dry, and free of drafts. Ventilation systems and windows should be designed so fresh air can be brought into the building without chilling the ducklings.

Floors can be either wire or litter. If the expense can be justified, welded wire (¾ in.) about 4 in. above concrete is the most satisfactory type of flooring. It may be used over the entire floor space or over only part of the floor. Wire flooring has the major advantages of keeping ducklings away from manure and dampness and of being washed down daily if adequate floor drains are present.

Fig. 15–14. Ducklings on wire. (Photo by J. C. Allen & Son, Inc., West Lafayette, IN)

Litter flooring may be more practical for most small producers. Straw, wood shavings, and peat moss make good litter. Make sure litter is free of mold. Moldy litter can cause high mortality in young stock.

Ducklings grow rapidly. Make sure they have adequate floor space. For 3-week-old ducklings, allow 0.5 sq ft of space per bird on wire and 1 sq ft per bird on litter. If confinement rearing is practiced, increase the floor space to 2 sq ft per bird by 7 weeks of age. (Also see Chapter 11, Poultry Houses and Equipment, Table 11–1, Space Requirements for Poultry.)

Ducklings need supplementary heat for about 4 weeks after they hatch. In hot summer weather, heat may be needed for only the first 2 or 3 weeks. Electric, gas, coal, or wood-burning brooder units can be used for small operations. For larger operations, forced hot-air or hot-water systems are more efficient because they require less labor and fuel.

Fig. 15–15. Growing ducklings—16 days old. (Courtesy, Cherry Valley Farms Limited, Rothwell, England)

The brooder temperature should be kept at 85° to 90°F the first week, then reduced by 5° per week during succeeding weeks. By 4 weeks the ducklings will be feathered enough to venture outdoors in all but extremely cold weather. If hovers are used the first week, use brooding guards to keep the ducklings confined to the comfort zone.

Ducklings need clean drinking water at all times. It may be supplied in hand-filled water fountains or by automatic waterers. To prevent wet litter, place the water supply above wire flooring or on a screened drain. Clean watering devices daily.

Ducklings should be given access to outside yards when they are old enough to tolerate weather conditions. The young birds will manage nicely on grassy areas that have adequate shade. Although they are waterfowl, ducklings cannot tolerate chilling rains until they are about 4 weeks old. Young ducklings should be given shelter at the first signs of precipitation. This precaution can usually be disregarded when the ducklings are 4 weeks old. Birds 5 to 8 weeks of age need shelter only in extreme winter conditions.

Yards should slope gently away from the houses to provide drainage. Locate watering facilities at the far end—lowest point—of the yard. Most commercial growers provide swimming water for the ducklings at 5 weeks of age. However, ducklings can be raised without swimming water. Stagnant pools of water anywhere in the yards are sources of disease and every possible measure should be taken to prevent their formation.

Accumulations of manure will build up in yards after a few weeks. The rate will depend on the density of the birds. If the yards are located on light sandy soil, it will be relatively easy to scrape off the top surface. Periodic cleaning of yards should be part of the planned work schedule.

Feather pulling beginning at 4 weeks of age may be a sign of overcrowded conditions in the yards and houses. Steps should be taken to stop this vice as soon as it starts. The birds should be given additional space. If feather pulling continues, it may be necessary to debill the birds. This can be done by snipping off the forward edge of the upper bill.

Night lights should be used to prevent running and to ensure that birds find their feed and water. Use one reflected 15-watt lamp per 200 sq ft of floor space. String some 20- to 25-watt lamps with reflectors on poles if the birds are confined to outside yards.

FEEDING

Maximum efficiency for growth and reproduction can be obtained by using commercially prepared diets which are sold in pellet form. Pellets are recommended instead of mash because they are easier to consume, they reduce waste, they do not blow away, and feed conversion is usually superior. Lack of pelleted feed should not discourage those who wish to produce ducklings on a small scale. Satisfactory results are possible with mash.

Four diets are recommended: starter, grower, finisher, and breeder. Examples of each are presented in Chapter 7, Poultry Feeding Standards, Ration Formulation, and Feeding Programs. Use pellets ⅛ in. in diameter for the starter diet; use pellets ³⁄₁₆ in. in diameter for the grower, holding, and breeder diets.

Feed ducklings the starter diet the first 2 weeks after they hatch. To encourage early consumption, place the feed in baby-chick sized hoppers and locate them close to the water supply. When ducklings reach 2 weeks of age, switch them to the grower diet, then feed this diet until they are ready for market.

Young drakes and ducks selected as potential breeders should be fed a holding diet. This special diet contains less energy per pound of feed than the starter or grower diet. When fed in restricted amounts, the diet will keep the breeders from putting on excess fat, but it will provide the nutrients needed. For each 100 breeders, feed 45 lb of the holding diet daily: Feed half in the morning and half in the late afternoon. The pellets should be scattered over a large area so all birds get a chance to obtain their daily requirements. If mash is fed, a large number of feed hoppers should be used.

Increased requirements for reproduction make it essential to feed breeding stock a breeder diet. The breeders should be switched to the breeder diet about 1 month prior to the date of anticipated egg production. To ensure good eggshell quality, the breeders should be given oystershells. They can be fed free-choice in separate hoppers within the breeder pens.

Many types of feeders can be used for ducks. Ordinary hoppers used in commercial chicken production work well provided they are arranged at floor level. Since ducks grow at fairly rapid rates and consume large quantities of feed in short periods of time, it is advantageous to use hoppers that hold large quantities of feed. Small feeders can be used until the ducklings are 2 weeks old. Larger feeders should be used for older market ducklings and breeding stock. Feed hoppers that are used outdoors should have lids that fit securely.

Provide water whenever feed is available. This is especially important when breeders are confined during the night. If hoppers are within the building and water supplies outdoors, hoppers should be closed overnight to prevent the breeders from choking on dry feed. Open the hoppers in the morning as soon as the birds have access to water. Breeders will adjust to this routine and egg production will not be affected.

In performance trials in Yugoslavia, one strain of ducklings weighed 7.3 lb at 48 days and produced 1 lb of liveweight from each 2.82 lb of feed.

(Also see Chapter 7, Poultry Feeding Standards, Ration Formulation, and Feeding Programs.)

MANAGEMENT

Ducks respond to management; good management makes for success, poor management makes for troubles.

The condition known as *staggers* is commonly caused by a temporary shortage of drinking water. If the birds are fed before the water is replenished, death usually follows in a short time.

Cold water can be fatal to overheated ducklings. Therefore, water intended for ducklings should have the chill taken off before making it available.

Ducklings cannot tolerate the sun after eating. Hence, if natural shade is not available, some type of shelter should be provided.

Feather eating and quill pulling are usually caused by crowding too many ducks into too small an area. The prevention of this problem is debeaking in much the same manner as chickens. To debeak ducks, remove the horn at the front of the top bill.

Ducks are highly nervous. If chased vigorously, they will go lame. If lameness should happen, remove the lame ducks from the flock. They will regain the use of their legs in a week or two.

Ducks are more vigorous and less subject to diseases than chickens or turkeys. If disease occurs, it is most likely the result of unsanitary surroundings and faulty management or inherent weakness due to breeding.

Both breeder and market ducks lend themselves to environmentally controlled buildings and automation, thereby effecting savings in space and labor. At Cherry Valley Farms, in England, the world's largest duck farm, one person cares for as many as 85,000 birds.

DISEASES AND AILMENTS

Ducks raised in small numbers and in relative isolation suffer little from diseases. On large commercial duck farms, the risk of disease can be minimized by a closed flock policy and controls on the entry of vehicles and visitors.

The list of common diseases presented here is far from complete. It should not be used as a substitute for accurate diagnosis by competent specialists.

■ **Botulism** — This disease occurs in young and adult stock. It is caused by the bacterium *Clostridium botulinum*, which grows in decaying plant and animal material. Ducks feeding on material containing the deadly toxins produced by the bacteria lose control of their neck muscles and usually drown if swimming water is available. Direct action of the toxins will also cause high mortality in upland birds. Removal of dead birds and rotting vegetation, plus maintenance of clean facilities, will prevent this disease.

■ **Fowl cholera** — High mortality in adult and young stock can occur from this disease. Strict sanitation will help control fowl cholera. Keep yards and houses clean. Do not allow mudholes and slimy areas to form. Burn or bury dead birds. The organisms of this highly infectious disease live for long periods of time in tissues of dead birds. They can be transmitted to clean flocks by flies, rodents, and wild birds. Bacterins and vaccines have been used to control losses from fowl cholera.

■ **Necrotic enteritis** — This disease, caused by *C. perfringens*, is very common in breeding stock. Breeder houses and yards must be free of wet litter and mudholes. Mortality may be sporadic over a long period of time. A number of antibiotics are effective in treatment and prevention of necrotic enteritis.

■ **Reproductive disorders** — Paralysis of the intromittent organ of drakes is commonly observed early in the mating season. This renders the bird useless for reproduction. Drakes having this condition should be culled from the flock.

Females are rendered worthless by prolapse of the vagina and cloaca. Impacted oviducts and egg yolk peritonitis are other reproductive disorders in ducks.

■ **Virus hepatitis** — Serious outbreaks of virus hepatitis can cause 80 to 90% mortality in flocks of ducklings. This highly contagious disease strikes swiftly without warning. It occurs in ducklings from 1 to 5 weeks of age.

A vaccine that is administered to female breeding stock is available. Antibodies produced by the laying ducks are passed through the egg to the young ducklings. This gives the ducklings sufficient passive immunity to protect them against natural exposures to the virus during the first 3 weeks after hatching. Antibody therapy (antiserum or yolk given IM) at the time of initial loss is the only effective flock treatment.

■ **Brooder pneumonia** — Good litter and a dry brooder house will help prevent brooder pneumonia. This disease of ducklings is caused by a fungus frequently present in litter. There is no treatment for the disease.

■ **New duck disease (infectious serositis)** — This is one of the most serious diseases affecting ducklings. It is a bacterial disease caused by *Moraxella anatipestifer*. Symptoms resemble those of chronic respiratory disease of chickens. The first signs of the disease are sneezing and loss of balance. Afflicted ducklings fall over on their sides and backs. Losses up to 75% have been recorded. Death often is due to water starvation, rather than to the primary infection. Antibiotics and sulfa drugs have been used with some success. Inactivated bacterins have been reported to prevent or reduce mortality due to the disease.

■ **Coccidiosis** — Although not as troublesome as in chickens, occasionally this disease causes morbidity and mortality as well as poor performance in flocks of ducklings. The organism causing the disease in ducks is different from those causing it in chickens.

■ **Duck virus enteritis (duck plague)** — This acute fatal disease affects ducks as well as geese, swans, and other aquatic birds. It is caused by a filterable virus that is transmissible by contact and commonly occurs where water is available for swimming. The disease has affected ducks in commercial flocks in eastern United States.

Disease signs of watery diarrhea, nasal discharge, and general droopiness develop in 3 to 7 days after exposure, and last for another 3 or 4 days, frequently ending in death.

Postmortem examinations show multiple or generalized hemorrhages in body organs.

There is no satisfactory treatment. Strict sanitation, disposal of offal from infected birds, and rearing ducks in pens with access to drinking water only will greatly reduce and help control the disease.

Possible exposure of domestic ducks to duck virus enteritis from migratory waterfowl can be prevented by keeping susceptible ducks in houses or within wired enclosures.

A modified chicken embryo-adapted vaccine has been developed and used with good success in The Netherlands. This vaccine strain has also been used to control duck virus enteritis on commercial duck farms on Long Island, in the United States.

MARKETING AND PROCESSING DUCKS

Ducks are generally marketed at 6.25 to 7.5 lb liveweight, at 7 to 8 weeks of age.

Feed should be withheld from the birds 8 to 10 hours before slaughter, but water may be provided up to the time of killing. Clean, uncrowded rearing facilities will help to prevent bruising, cutting, and other factors that cause poor market acceptability. Ducklings can be transported in crates or herded into trailers for delivery to the slaughterhouse.

Construction of elaborate slaughtering facilities is justified only for large commercial operations. If live markets are not available, small farm flocks can be processed by using facilities similar to those used for small chicken flocks.

For slaughter, hang ducklings by the feet or place them in special slaughtering funnels. Take a long, thin, sharp knife and draw it across the outside of the throat high up on the neck just under the lower bill. This will sever the jugular vein and allow swift, complete bleeding.

Fig. 15–16. Forty-eight-day-old ducklings, weighing about 6.6 lb, ready for market. (Courtesy, Cherry Valley Farms Limited, Rothwell, England)

When bleeding has ceased, birds can be scalded and picked or they can be dry picked. Dry picking has the advantage of producing exceptionally attractive carcasses, but it is slower and there is greater danger of tearing the skin. For scalding, immerse the ducklings for 3 minutes in hot water (140°F). Pick immediately after scalding and remove all remaining pinfeathers; grasp the pinfeathers between the thumb and a dull knife.

In large commercial slaughterhouses, ducklings are dipped through a molten wax after they have been picked. When the wax hardens—by immersion in cold water—it can be peeled free to remove any feathers that remain. Wax can be reused if it is remelted and the feathers are screened out. Small quantities of wax can be purchased for use in small farm slaughtering operations. Wax is highly combustible; hence, care must be taken to prevent it from coming in contact with an open flame.

Fig. 15–17. Removing wax after picking to remove any feathers that remain. (Courtesy, Cherry Valley Farms Limited, Rothwell, England)

After picking, birds can be eviscerated immediately or they can be stored in slush ice overnight until it is convenient to complete the processing. Wash the eviscerated bird thoroughly and stuff the giblets into the clean birds. If birds are to be marketed frozen, package them in shrinkable plastic bags. If special "New York Dressed" markets are available, chill the ducklings and sell them uneviscerated.

The federal grades of ducks—U.S. Grade A, U.S. Grade B, and U.S. Grade C—are essentially the same for all poultry; and the grading program is voluntary. These grades and specifications are given in Chapter 16, Marketing Poultry and Eggs, Table 16–14, Summary of Specifications for Standards of Quality for Individual Carcasses of Ready-to-Cook Poultry and Parts Therefrom.

Commercial processors sell the feathers, feet, heads, and viscera as by-products. Small producers generally do not have a large enough volume to develop a by-product market.

(Also see Chapter 16, Marketing Poultry and Eggs, section headed "Marketing Ducks.")

GEESE

Geese are very hardy, are the closest grazers known, and can live almost entirely on good pasture. Yet, the production of geese for meat purposes has never enjoyed the popularity in the United States that it has in some European countries. In recent years, considerable attention has been focused on raising geese for weeding purposes.

Fig. 15–19. Emden (left) and Toulouse geese. (Courtesy, *Poultry Tribune*, Mount Morris, IL)

Fig. 15–18. Commercial geese on the range. (Photo by J. C. Allen & Son, West Lafayette, IN)

Geese are produced commercially on both general and specialized farms in Missouri, Iowa, South Dakota, Minnesota, Wisconsin, Ohio, Indiana, California, and Washington.

Annual production is estimated at approximately 1,000,000 birds. The number of farms selling geese has decreased in recent years, but the number of geese sold per farm has increased. In addition to commercial operations, geese are raised in small flocks in all parts of the United States—as a sideline, as a hobby, or for ornamental and exhibition purposes.

Geese are very hardy and not susceptible to many of the common poultry diseases. They are excellent foragers, although selective; they can go on good succulent pasture or lawn clippings as early as their first week.

BREEDING

The breeding and improvement of geese in the United States has received far less attention than in Europe.

BREEDS

Toulouse, Emden, and African geese—all heavy breeds—are the most popular breeds in the United States for meat production. Other common breeds in this country are Chinese, Canada, Buff, Pilgrim, Sebastopol, and Egyptian.

There are considerable differences in breeds and strains of geese. In choosing a breed, therefore, one should consider the purpose for which they are to be used. Geese are raised for meat and egg production, and as weeders, show birds, or farm pets.

Both the standard breeds and crosses of breeds are raised for market. When choosing a breed or strain of geese for market production, one should study the market or potential market requirements. Size (market body weight) and plumage color (white is generally preferred) are two important characteristics.

Also, when choosing a breed or strain, bear in mind that strains that lay the most eggs produce goslings at lowest cost.

Table 15–1 (next page) gives the standard weights of various breeds and the years these breeds were recognized, according to *The Standard of Perfection*, published by the American Poultry Association, Inc. Some commercial stock, particularly the Toulouse breed, tends to be somewhat lighter than the standard.

AFRICAN

The African goose is a handsome breed with a distinctive knob or protuberance on its head. Its carriage is more erect than that of the Toulouse, and its body more nearly oblong and higher from the ground. The head is light brown, the knob and bill are black, and the eyes are dark brown. The plumage is ash brown on the wings and back and light ash brown on the neck, breast, and underside of the body. The African goose is a good layer, grows rapidly, and matures early. However, it is not as popular for market production as either the Emden or the Toulouse because of its dark beak and pinfeathers.

TABLE 15–1
GEESE: BREEDS, YEAR RECOGNIZED, AND STANDARD WEIGHTS

Breed	Year Recognized	Weight of Male				Weight of Female			
		Young		Adult		Young		Adult	
		(lb)	(kg)	(lb)	(kg)	(lb)	(kg)	(lb)	(kg)
Toulouse	1874	20	9.1	26	11.8	16	7.3	20	9.1
Emden	1874	20	9.1	26	11.8	16	7.3	20	9.1
African	1874	16	7.3	20	9.1	14	6.4	18	8.2
Chinese	1874	10	4.5	12	5.5	8	3.6	10	4.5
Egyptian	1874	5	2.3	5.5	2.5	4	1.8	4.5	2.0
Canada	1874	10	4.5	12	5.5	8	3.6	10	4.5
Sebastopol	1938	12	5.5	14	6.4	10	4.5	12	5.5
Pilgrim	1939	12	5.5	14	6.4	10	4.5	13	5.9
Buff	1947	16	7.3	18	8.2	14	6.4	16	7.3

Fig. 15–20. African geese. (Courtesy, USDA)

Fig. 15–21. Breeding flock of Buff geese. (Courtesy, USDA)

BUFF

The Buff has fair economic qualities as a market goose, but only a limited number are raised for market. The color varies from dark buff on the back to a very light buff on the breast, and from light buff to almost white on the under part of the body.

CANADA

The Canada is the common wild goose of North America. Subgroups range in weight from about 3 lb for the cackling Canada goose to about 12 lb for the giant Canada goose.

The Canada is of a species different from the other breeds of geese discussed in this book and can be kept in captivity only by close confinement unless wing-clipped or pinioned. However, in some instances, Canada geese have become semidomesticated by long residence on the farm. Before Canada geese can be sold or transferred to another person, a permit must be obtained from the Fish and Wildlife Service, U.S. Department of the Interior, Washington, DC 20242.

Canada geese have long, slender necks, oblong bodies, and a horizontal carriage. This breed does not have the economic value of the domestic breeds of geese. They mate only in pairs, are late maturing, and lay very few eggs. The wild gander is sometimes used to cross with domestic breeds, producing the so-called mongrel goose (which is a hybrid), usually sterile but with fine quality flesh.

Fig. 15–22.　Canada or wild geese. (Courtesy, USDA)

CHINESE

The Chinese goose, of which there are two standard varieties — the Brown and the White — originated in China and probably came from the wild Chinese goose. It is smaller than the other standard breeds and more swanlike in appearance. Both varieties mature early and are better layers than the other breeds, usually averaging from 40 to 65 eggs per bird annually. The Chinese goose grows rapidly, is a very attractive breed, makes a desirable medium-size market goose, and is very popular as an exhibition and ornamental breed.

Fig. 15–23.　Breeding flock of White Chinese geese in a pasture containing a small natural pond. (Courtesy, USDA)

EGYPTIAN

The Egyptian is a long-legged, but very small goose, kept primarily for ornamental or exhibition purposes. Its coloring is mostly gray and black, with touches of white, reddish brown, and buff.

EMDEN

The Emden was one of the first breeds of geese imported into the United States. This breed was known at first as Bremen, named after a German city from which early importations were made. The present name is after Emden, Germany, from which later exportations of the geese were made to England.

The Emden is a pure white, sprightly goose. It is much tighter-feathered than the Toulouse and, therefore, appears more erect. The Emden is a fairly good layer, but production depends on the breeding and selection of the flock. Egg production averages from 35 to 40 eggs per mature breeding goose. The Emden is usually a better setter than the Toulouse and is one of the most popular breeds for marketing. It grows rapidly and matures early.

PILGRIM

Fig. 15–24.　Flock of Pilgrim geese. The males are white; the females are gray. (Courtesy, USDA)

The Pilgrim is a medium-size goose that is good for marketing. A unique feature of this breed is that males and females may be distinguished by color. In day-old goslings the male is creamy white and the female gray. The adult male remains all white and has blue eyes; the adult female is gray and white and has dark hazel eyes.

SEBASTOPOL

The Sebastopol is a white ornamental goose which is very attractive because of its soft plumelike feathering. This breed has long, curved, profuse feathers on its back and sides and short, curled feathers on the lower part of the body.

TOULOUSE

The Toulouse goose derives its name from the city of Toulouse in southern France, a territory noted for its geese. This breed has a broad, deep body and is loose-feathered, a characteristic which gives it a massive appearance. The plumage is dark gray on the back, gradually shading to light gray edged with white on the breast and to white on the abdomen. The eyes are dark brown or hazel, the bill pale orange, and the shanks and toes are a deep reddish orange.

SELECTION OF BREEDERS

Breeder geese, like other kinds of poultry, should be selected for size, prolificacy, and vigor. Medium-size birds of each breed make the best breeders. A gander may be mated with from 1 to 5 geese, although some males refuse to mate with some females.

Preferably, breeders should be selected from stock that has been tested and found to be superior in desired traits such as market body weight, egg production, etc. Trapnesting is essential for identifying the better birds

BREEDER FACILITIES

Breeding geese prefer to be outdoors. Except in extremely cold weather or in storms, mature geese seldom use a house. Colony poultry houses, open sheds, or barns are provided for shelter in the North.

Geese make nests on the floor of the house or in coops, boxes, or barrels provided in the yard. Outside nest-boxes should be at least 24 in. square. Very crude nests are used in the open for many farm flocks of geese.

Fig. 15–25. Nest for geese. The gander stays near to protect the goose while she is nesting. (Courtesy, USDA)

Straw or grass hay is used for outside nests as well as for the nests on the floor of a house. One nest should be provided for every three females and the geese should be allowed to select their own nests. Inside nests should be separated by partitions; outside nests should be placed some distance apart to reduce fighting.

MATING

Geese should be mated at least 1 month prior to the breeding season. The larger breeds of geese mate best in twos or threes, or in a ratio of one male to three or four females in large flocks. Ganders of some of the lighter breeds will mate satisfactorily with four or five females.

Geese matings should not be changed from year to year except when they prove unsatisfactory. Geese are very slow to mate with new birds, so it is difficult to make changes in established matings or to introduce new stock into the flock. If matings are changed, it is usually advisable to keep previously mated geese as far apart as possible.

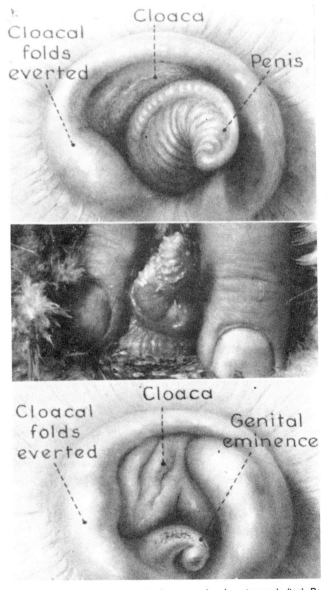

Fig. 15–26. Exposed reproductive organ of an immature male (top). Reproductive organ of sexually mature male (center). Genital eminence of maturing female (bottom). (Courtesy, USDA)

Sex is difficult to distinguish in all breeds of geese except the Pilgrim. In other breeds, sex can be determined by examination of the reproductive organs, which is done as follows: Lift the goose by the neck and lay it on its back, either on a table or over your bended knee, with the tail pointed away from you. Move the tail end of the bird out over the edge so it can be readily bent downwards. Then insert your pointer finger (sometimes it helps to have a little Vaseline on it) into the cloaca about ½ in. and move it around in a circular manner several times to enlarge and relax the sphincter muscle which closes the opening. Next apply some pressure directly below and on the sides of the vent to evert or expose the sex organs (Fig. 15–26, previous page).

INCUBATION

Gather eggs twice daily, especially during cold weather. Store them at 55°F and a relative humidity of 75% until set for hatching.

Goose eggs can be washed just like chicken hatching eggs. Wash soiled eggs in warm (100° to 115°F) water and a detergent sanitizer. For best results, eggs should be washed soon after gathering, then dried and stored until ready for the incubator.

If eggs are held more than a couple of days, turn them daily to increase the percentage of hatch. Hatchability decreases fairly rapidly after a 6- or 7-day holding period although eggs, properly stored, can be held 10 to 14 days with fair results.

The incubation period for eggs of Canada and Egyptian geese is 35 days; for all other breeds, it is 29 to 31 days.

Small, inexpensive electric incubators, either still-air or forced-draft, can be used to hatch goose eggs. However, artificial incubation of goose eggs is much more difficult than incubation of chicken eggs because more time and higher humidity are required. Breeders should gain experience with chicken eggs before attempting artificial incubation of goose eggs. When using an incubator, always follow the manufacturer's instruction.

Many goose breeders prefer to set eggs under chickens, turkeys, or ducks and allow the geese to continue to lay. Sometimes the eggs are set in an incubator for 2 weeks and then are placed under the hens for the remainder of the time required for hatching.

Turkeys and Muscovy ducks are larger and better than chickens for hatching goose eggs. Incubate 4 to 6 eggs under a setting chicken, and 10 to 12 under a turkey or duck depending on the size of the bird and the season of the year. Hens used for hatching goose eggs should be treated for lice.

If the setting hen does not turn the eggs, mark them with crayon or pencil and turn them daily by hand. A chicken hen cannot turn goose eggs.

Moisture is needed where chicken or turkey hens are used for setting. Sprinkle the eggs during the incubation period and have the nest and straw on the ground or on grass-covered turf. Some growers report better hatchability if they sprinkle the eggs lightly or dip them in lukewarm water for ½ minute daily during the last half of the incubation period. Eggs need no additional moisture if the setting goose has water for bathing.

Fig. 15–27. Wild Canada goose incubating eggs as nature ordained, reacting to approaching enemy. (Photo by J. C. Allen & Son, Inc., West Lafayette, IN)

Remove goslings from the nest as they hatch and keep them in a warm place until the youngest is several hours old. If this is not done, the setting hen may leave the nest along with the hatched goslings before all the eggs are hatched.

BROODING AND REARING

A special brooder building is not required for brooding small numbers of geese. Any small building or a corner of a garage or barn can be used as a brooding area for a small flock if it is dry, reasonably well lighted and ventilated, and free from drafts. The building must also be protected against dogs, cats, and rats. For brooding large numbers of geese, provide a barn, large poultry house, or regular broiler house.

Allow at least 0.5 sq ft of floor space per bird at the start of the brooding period and gradually increase the space to 1.25 sq ft at the end of 2 weeks. If the birds are confined longer because of inclement weather, provide additional space as they increase in size. (See Chapter 11, Poultry Houses and Equipment, Table 11–1, Space Requirements for Poultry.)

Cover the floor with 4 in. of such absorbent litter as wood shavings, chopped straw, or peat moss. To maintain good litter, stir frequently, remove wet spots, and periodically add clean, dry litter. Be sure litter is free from mold.

Goslings can be successfully brooded by broody chicken hens and most breeds of geese. If the young birds were not hatched by the brooding female, place them under her at night. Be certain broody birds are free of lice and mites. One hen can raise five goslings. In mild weather, the hens may only need to brood the goslings for 10 to 14 days, after which they can get along without heat.

Goslings are artificially brooded in many types of heated brooders. Infrared lamps are a convenient and satisfactory source of heat, provided enough of them are used to furnish heat for the lowest temperatures expected.

When using hover-type brooders, brood only about one-third as many goslings as the brooder's chick capacity. Because goslings are large in size it may be necessary to raise the hover 3 to 4 in. higher than for baby chicks. Fence in the brooding area for the first few days with a corrugated-paper or wire-mesh fence.

At the start, set the temperature of the hover at 85° to 90°F. Reduce the temperature 5° to 10°F per week until 70°F is reached. The behavior of the goslings will indicate their comfort. If they are cold, they will huddle together under the lamps. If they are too warm, they will move away from the heat source.

In warm weather, the goslings can go outdoors as early as 2 weeks, but they will need frequent attention until they learn to go back into the coop or brooder when it rains. They must be kept dry to prevent chilling that can result in piling and smothering.

Houses are usually not needed after the geese are 6 to 8 weeks of age.

Fig. 15–28. Toulouse goslings about a month old. (Courtesy, USDA)

FEEDING

Goslings should have drinking water and feed when they are started under the brooder or hen. Supply plenty of watering space. Use waterers that the birds cannot get into, but that are wide and deep enough for them to dip both bill and head.

Start with two automatic cup-type waterers for each 100 to 200 goslings, depending on the environmental temperature. Increase the number of waterers as the birds grow. Watering jars or a trough with wire guard and running water are also suitable for young goslings. If troughs are installed, figure on 8 ft of trough space for 500 goslings for the first 2 weeks of age; then, as needed, increase the space up to 20 ft.

On range the waterer can consist of a barrel or large tank rigged to an automatic float in a watering trough. If waterers are indoors, they should be kept on wire platforms with underdrainage to help keep the litter dry.

Fig. 15–29. Young geese on range at Schiltz Goose Farm in South Dakota. (Courtesy, Schiltz Foods, Inc., Sisseton, SD)

For the first few days of feeding use shallow pans or small feed hoppers in addition to the regular feeders. For each 100 confined goslings on full feed, provide either two hanging tube feeders with pans that are 50 in. in circumference or 8 ft of trough space. Increase the feeding space as the birds grow. When feed intake is being restricted, provide enough space so that all geese can eat at one time.

For geese raised on range for market, use two wooden hoppers or two turkey range feeders for each 250 birds. The hoppers should be large enough so that they will need to be filled only once or twice weekly. Construct the hoppers so that feed is protected from rain, sun, and wind.

A mechanical feeder is suitable for large-scale production. Geese may be fed pellets, mash, or whole grains. For the first 3 weeks, feed goslings a 20 to 22% protein goose starter in the form of 3/32- or 3/16-in. pellets. After 3 weeks, feed a 15% protein goose grower in the form of 3/16-in. pellets.

Although geese can go on pasture as early as the first week, a good share of their feed can be forage after they are 5 to 6 weeks of age.

Geese are very selective and tend to pick out the palatable forages. They will reject alfalfa and narrow-leaved tough grasses and select the more succulent clovers and grasses. Geese cannot be raised satisfactorily on dried-out, mature pasture.

An acre of pasture will support 20 to 40 birds, depending on the size of the geese and the quality of the pasture. A 3 ft woven wire fence will confine the geese to the grazing area. Be sure that the pasture areas and green feed have not had any chemical treatment that may be harmful to the birds.

If pasture is plentiful and of good quality, the amount of pellets may be restricted to about 1 to 2 lb per goose per week until the birds are 12 weeks of age. However, for maximum growth, increase the amount of feed as the supply of young, tender grass decreases or when the geese reduce their consumption of grass. From 12 weeks to market, offer the birds pellets on a free-choice basis, even when on range.

Mash or whole grains can be fed alone or they can be mixed at a 50:50 mash-to-grain ratio. At 3 weeks of age, use a mash-to-grain mix of approximately 60:40. Change this ratio gradually during the growing period until at market age the geese are receiving a 40:60 ratio of mash to grain. Depending on the quality and quantity of available pasture, adjust these ratios up or down slightly.

Wheat, oats, barley, and corn may be used as the whole grains in various mixtures, such as equal parts of wheat and oats. All-corn can be substituted when the goslings are 6 weeks old. For maximum growth, it is important that mash-and-grain mixtures provide similar nutrient intake (15%) as the all-mash diets.

Grower-size insoluble grit should be freely available to geese throughout the growing period. If feeds for geese are not available, use a chicken feed formulated for the same age bird.

Also see the following chapters in this book:

Chapter 6, Fundamentals of Poultry Nutrition

Chapter 7, Poultry Feeding Standards, Ration Formulation, and Feeding Programs

Fig. 15–30.   Geese working their way down a cotton field. (Courtesy, *The Fresno Bee*, Fresno, CA)

MANAGEMENT

Geese generally start laying in February or March and often continue to lay until early summer. However, the Chinese breed may start laying early in the winter.

Breeder geese should be fed a pelleted breeder ration at least a month before egg production is desired. They do much better, and waste less feed, on pellets than on mash. A chicken-breeder ration may be used if special feeds for geese are not available. Provide oystershells (or other calcium sources), grit, and plenty of clean, fresh drinking water at all times.

Lights in the breeder house can be used to stimulate earlier egg production. In commercial flocks, artificial methods of hatching and rearing are also used.

To maintain egg production, feed chicken layer pellets or mash, confine broody geese away from but in sight of their mates, and gather eggs several times each day to break up broodiness.

Young ganders make good breeders, but both sexes usually give best breeding results when they are 2 to 5 years old. Good fertility may be obtained in eggs from young birds, but these eggs may not hatch well. Although young flocks are considered more profitable, females will lay until they are 10 or more years of age, and ganders may be kept for more than 5 years.

GEESE AS WEEDERS

Weeder geese are used with great success to control and eradicate troublesome grass and certain weeds in a great variety of crops and plantings, including cotton, hops, onions, garlic, strawberries, nurseries, corn, orchards, groves, and vineyards. The geese eat grass and young weeds as quickly as they appear, but they do not touch certain cultivated plants. They will work continuously from daylight to dark, 7 days a week (even on bright moonlit nights) nipping off the grass and weeds as promptly as new growth appears.

For best results, start with 6-week-old goslings and provide them shade and waterers spaced throughout the field. Keep weeder goslings hungry. A light feed of grain at night is enough; vary the amount depending on the availability of weeds and grass.

At the end of the weeding season, geese are generally brought from the field and placed in pens for fattening for 3 to 4 weeks, until they weigh 10 to 12 lb or more. Markets are highest during the 4 to 6 weeks prior to Thanksgiving and Christmas.

The carrying of geese over from one season to the next for weeding purposes is not recommended, because older geese are less active in hot weather than young birds.

DISEASES AND AILMENTS

When geese are managed properly, they seldom get a disease. Nevertheless, the caretaker should be alert to the first signs of trouble. If a disease is suspicioned from the appearance of the birds or the number of deaths, a poultry pathologist should be consulted immediately. Also, an accurate record should be kept on the number and dates of all deaths.

(Also see Chapter 10, Poultry Health, Disease Prevention, and Parasite Control.)

MARKETING AND PROCESSING GEESE

The best markets for geese are in large cities. They bring the highest prices at Thanksgiving and Christmas.

Geese are sold alive, usually to poultry buyers or to poultry dressing and packing companies.

Geese should be fasted for about 12 hours before killing, but they should have access to water.

To kill, geese should be placed in funnels or hung by the legs in shackles, then the throat should be cut at the base of the beak to sever the jugular vein and carotid artery.

Geese can be scalded or picked dry. The dry method, if well done, results in an attractive carcass but is now usually considered too slow and laborious to be economical. There is also more danger of tearing the skin in dry picking.

Geese can be scalded in a commercial scalder, or they can be hand-scalded in a small operation. Water temperature should be from 145° to 155°F, and the length of the scald should be from 1½ to 3 minutes. The time and temperature will vary depending on age of the bird, time of year, and density of feathering. Also, the lower the temperature, the longer the required scalding time. A little detergent should be added to the water to hasten thorough wetting of the feathers.

To hand-scald, grasp the goose firmly by the bill with one hand and by the legs with the other, then submerge its body (breast down) in the scalding water. Pull the bird repeatedly through the water against the lay of the feathers; this action serves to force the water through the feathers to the skin. The sparser feathering on the back needs lighter scalding than the heavier and denser feathering on the breast.

After scalding, the birds may be either rough-picked by hand, picked on some type of conventional rubber-fingered picking machine, or placed in a spinner-type picker. Pinfeathers and down remaining on the carcass after rough-picking are difficult to remove by hand. This is done by grasping the pinfeathers between the thumb and a dull knife. Because of the difficulty of handpicking, it is common practice to finish the rough-picked birds by dipping each bird in melted wax (a wax formulated for this purpose).

In small operations, dry off the rough-picked geese just enough to "take" the wax; then dip them several times in wax held at 150° to 160°F to build up a heavy enough layer of wax to supply good pulling power. A better job results with the use of two tanks of wax, one held at 160° to 170°F, and the second at about 150°F—the hotter wax is used for penetration and cooler for buildup.

In large-scale operations, the birds are waxed in an on-the-line process with a wax temperature of 145° to 220°F. After waxing, the birds are exposed to a cold water spray or dipped in a tank of cold water to cool and harden the wax to a "tacky" condition. The wax is then removed by hand resulting in a clean, attractive carcass. The wax is reclaimed by remelting and straining out the pins, down, and feathers.

The U.S. standards of quality are essentially the same for all poultry. As with other poultry, geese are graded for conformation, fleshing, and fattening. Defects, such as missing skin and bruises, are also considered in establishing quality. Further information on the voluntary grading program, including grades and specifications, is given in Chapter 16, Marketing Poultry and Eggs, Table 16–14, Summary of Specifications for Standards of Quality for Individual Carcasses of Ready-to-Cook Poultry and Parts Therefrom.

FEATHERS

Goose feathers are valuable and, if properly cared for and marketed, may be a source of extra income. Three geese usually produce 1 lb of dry feathers. Feathers are used chiefly by bedding and clothing industries. Buyers of feathers are in most large cities.

Feathers may be sold to a feather-processing plant, or small producers can wash and dry the feathers on their own premises. To wash feathers, use soft, lukewarm water to which has been added either a detergent or a little borax and washing soda. Rinse the feathers, wring, then spread them out to dry.

MISCELLANEOUS POULTRY

Pigeons, guineas, and various ornamental and game birds are frequently raised for pleasure. Also, a limited number of producers raise them for profit, on a full-time or part-time basis. Game birds are raised for sale to game preserves or for shooting preserves. Also, there is a limited market for the sale of ornamental birds. Following is a summary of pertinent information relative to miscellaneous poultry species.

BOBWHITE QUAIL (COLINUS VIRGINIANUS)

Fig. 15–31. Bobwhite quail.

The name, *Bobwhite*, comes from its call. In the spring, cocks not yet paired sound their loud, tuneful "bobwhite" call.

■ **Origin, history, distribution**—The Bobwhite quail lives on fallow fields, meadowlands rich in bushes, and open woodlands from the Canadian border to Mexico and Cuba.

Today, Bobwhites are raised on quail farms for hunting purposes.

■ **Breeds**—Most U.S. Bobwhite are *Colinus virginianus*. However, there are other subspecies such as the Florida Bobwhite, *Colinus virginianus Floridanus*.

■ **Selection of breeders**—The larger varieties should be selected when growing birds for meat purposes.

The small and moderate sized varieties generally are desired when growing birds for hunting preserves because they usually fly better and faster than the larger birds.

Birds or hatching eggs should be disease-free, preferably tested for and free of pullorum and typhoid.

Select stocks with a history of good egg production, hatchability, and livability.

■ **Breeding facilities** — Breeders may be kept in cages or in floor pens.

Cages can be either for an individual pair or a colony cage. Caged breeders require about 0.5 sq. ft of space per bird, or 1 sq ft per pair. A solid partition between cages is desirable, to prevent fighting.

Floor pens should provide a minimum of 1 sq ft per bird.

The advantages of cages over floor pens are that the eggs are cleaner, there is less exposure of the birds to diseases and parasites, and there is less fighting.

The major advantage of the floor system is that more margin of error in feeding and watering is possible, as the birds can move about and choose.

■ **Care of breeders** — Year-round production can be obtained by a lighting program, by providing up to 17 hours per day lighting.

Bobwhite hens will begin egg production at 16 to 24 weeks of age.

■ **Eggs** — In a 6-month breeding season, a hen should lay about 90 eggs. Eggs should be stored at 55° to 65°F and a relative humidity of 75 to 80%.

■ **Incubation** — A summary of incubation conditions follows:

1. *Period of incubation* — 23 to 24 days.
2. *Temperature* — 99.75°F.
3. *Hatcher temperature, 20-day hatch* — 98.75°F.
4. *Hatcher temperature, 20-day hatch, still air* — 100°F.
5. *Humidity, wet bulb, setting* — 84° to 86°F.
6. *Humidity, welt bulb, hatching* — 87° to 98°F.
7. *Future breeders should be debeaked at hatching.*

■ **Brooding and rearing** — Brooding is that period from hatching to 5 or 6 weeks of age.

Three general types of brooding facilities and equipment are used: (1) batteries, (2) floor pens with litter, and (3) raised wire floors. The major advantage of batteries and raised wire floors is that less space is required. The floor pen with hover or infrared brooder is the simplest method.

■ **Feeding and watering** — The NRC requirements and suggested rations for Bobwhite quail are presented in the following chapters in this book: Chapter 6, Fundamentals of Poultry Nutrition; and Chapter 7, Poultry Feeding Standards, Ration Formulation, and Feeding Programs.

■ **Care and management** — Future breeders should be debeaked at hatching so as to lessen cannibalism.

The incidence of cannibalism is increased by crowding, bright lights, insufficient feeders and water space, too high temperatures, and placing different types of birds together. Most of this can be corrected by proper management and debeaking.

Quail seem to be more sensitive to mismanagement than chickens.

■ **Diseases** — Bobwhite quail are susceptible to most of the diseases and parasites which afflict domestic poultry. The following are among the more common disease and parasite problems encountered:

1. Ulcerative enteritis or quail disease, caused by an anaerobic bacterium.
2. Blackhead.
3. Coccidiosis.
4. Quail bronchitis, caused by a virus.
5. Botulism.
6. Pox.
7. Chronic respiratory disease (CRD).
8. Pullorum.
9. Occasionally the following diseases: infectious coryza, fowl cholera, staphylococcosis, and trichomoniasis.
10. Worms, especially: large roundworms, cecal worms, capillary worms, gapeworms, and tapeworms.
11. External parasites: fleas, lice, mites, and ticks.

The disease prevention program for Bobwhite quail should be similar to that of chickens.

■ **Marketing, releasing** — Many producers contract a year or two ahead for the sale of their birds and eggs.

Bobwhite quail are generally considered mature at 16 weeks of age, at which time they can be used for meat or hunting preserve purposes.

Most producers of birds for hunting preserves condition them in flight pens for about a month before release. Flight pens are usually outside wire pens about 12 to 15 ft wide and 100 to 150 ft long, with an enclosed shelter at one end.

■ **Comments** — Few thrills are as exciting as the sudden whir of a Bobwhite covey rise, especially to a hunter.

Those raising quail should always remember that quail are living beings, that they have been stressed by taking them out of their natural environment and keeping them in confinement, and that they are 100% dependent on the caretaker.

Coveys lie in a circle with their heads pointing outwards. Thus, they can quickly spot an enemy, then fly off in all directions.

GUINEA FOWL

The fowls derive their name from Guinea, a part of the west coast of Africa.

Fig. 15–32. Guinea fowl. (Courtesy, USDA)

■ **Origin, history, distribution**—The domestic guinea fowl is descended from one of the wild species of Africa. In Africa, guineas are highly prized by hunters as game birds; and in England they are sometimes used to stock game preserves.

Guineas have been domesticated for many centuries; they were raised for table birds by the ancient Greeks and Romans.

Today, guineas are found throughout the world.

■ **Breeds**—There are three principal varieties of domesticated guinea fowl in the United States: Pearl, White, and Lavender.

■ **Selection of breeders**—In selecting breeders, good size and uniform color are most important.

At maturity, both males and females range from 3 to 3.5 lb in weight.

Usually, sex can be distinguished by the cry and by the larger helmet and wattles and coarser head of the male.

■ **Breeding facilities**—As the breeding season approaches, wild mated pairs range off in the fields in search of hidden nesting places in which it is difficult to find the eggs.

Domesticated guinea breeding birds are usually allowed free range. However, on some farms the breeders are kept confined during the laying period in houses equipped with wire-floored sun porches.

■ **Care of breeders**—In the wild state, guinea fowl mate in pairs. However, under domestic conditions, one male is usually kept for every four or five females.

Guinea breeders are usually allowed free range. They are difficult to confine in open poultry yards unless their wings are pinioned or one wing is clipped.

■ **Eggs**—A hen of good stock that is properly managed will lay 100 or more eggs per year.

Guinea eggs are smaller than chicken eggs; they weigh about 1.4 oz each vs 2 oz for a chicken egg.

■ **Incubation**—The incubation period is 26 to 28 days.

Chickens are commonly used for hatching guinea eggs in small flocks. For large flocks, incubators are better.

Forced-draft incubators should be operated at about 99.5° to 99.7°F and 57 to 58% humidity.

■ **Brooding and rearing**—Guinea chicks are known as baby keets.

Guinea keets may be raised in the same kind of brooder houses and brooders as baby chicks or poults.

Hovers should be started at 95°F for the first 2 weeks, then lowered by 5° per week thereafter.

■ **Feeding and watering**—A good commercial chicken or turkey breeder mash, containing 22 to 24% protein, should be fed to layers.

Young guineas should have a growing diet.

Also, see the following chapters in this book: Chapter 6, Fundamentals of Poultry Nutrition; and Chapter 7, Poultry Feeding Standards, Ration Formulation, and Feeding Programs.

■ **Care and management**—Most guineas are raised in small flocks.

Chicken hens make the best mothers for guinea keets. Guinea hens are likely to take their keets through wet grass and lead them too far from home.

■ **Diseases**—Guineas are subject to many of the same diseases and parasites as other poultry, and, in most instances, respond to the same treatment.

■ **Marketing, releasing**—The normal marketing season is late summer and throughout the fall. The demand is for young birds weighing 1.75 to 2.5 lb liveweight.

Many hotels and restaurants in large cities serve guineas at banquet and club dinners as a special delicacy; as a substitute for game birds—grouse, partridge, quail, and pheasant.

Guineas are killed and dressed in the same way as chickens.

■ **Comments**—Guineas might be more popular were it not for their harsh and seemingly never-ending cry, and their bad disposition.

HUNGARIAN, OR GRAY PARTRIDGE

Hungarians are sometimes referred to as gray cannon balls.

■ **Origin, history, distribution**—The Hungarian or Gray partridge was introduced into western United States early in the 20th century. Today, most partridge are found on the Canadian plains, and in the North Central and Northwestern United States.

■ **Breeds**—Both the Hungarian partridge of Europe and the Chukar of Asia are considered true partridges. Both have been introduced into the United States with considerable success.

■ **Selection of breeders**—The sexes can be separated with a high degree of accuracy by an experienced person, on the basis of the secondary wing coverts.

Fig. 15–33. Hungarian or Gray partridge.

Birds with physical defects (eye defects, twisted beaks, deformed feet, or bumped or twisted backs) or lacking in fleshing should be rejected.

■ **Breeding facilities** — Recommended breeding facilities consist of mating pens with doors facing a central courtyard.

■ **Care of breeders** — The Gray partridge is monogamous and intolerant of density — particularly during the breeding season. For these reasons, it is important that partridges be isolated in pairs as soon as premating signs are evident in the spring. The latter is accomplished by releasing several pairs of partridges in a central courtyard surrounded by individual mating pens with the doors open. The females will select males of their choice and entice them into a pen, following which the doors may be closed.

■ **Eggs** — Egg laying occurs in the spring and summer.

They are erratic in egg production. Hens will average about 30 eggs per year, with a range from 0 to 60.

Eggs should be gathered frequently and stored for not to exceed 10 days at 55°F and 90% relative humidity.

■ **Incubation** — The incubation period is 24 to 25 days.

The incubator temperature should be: 99° to 100°F for forced air, and 101°F for still air.

A wet bulb reading of 84° to 85°F is recommended (about 60% relative humidity).

■ **Brooding and rearing** — Hungarian partridges can be successfully raised in commercial brooder batteries, preferably with no more than 30 chicks per group because of their pugnacious nature even as chicks.

The temperature of the brooder should be 95°F at shoulder height of the chicks, with the temperature reduced 5° each week following the first week.

At 5 to 6 weeks of age the birds can be moved from the brooder to wire-covered outside runs. However, during nights they should be herded back under brooders until they are 10 weeks of age.

■ **Feeding and watering** — Example rations for Hungarian, or Gray partridge, are given in Chapter 7, in Table 7–27.

■ **Care and management** — The wire-covered runs (used following the brooder stage) should provide 5 to 6 sq ft of space per bird. Pens should be constructed of 1 in. wire mesh.

■ **Diseases** — They are susceptible to the various diseases and parasites afflicting poultry.

■ **Marketing, releasing** — Birds are usually released when 16 to 20 weeks of age.

Merely dumping the birds in a release site will defeat the intended purpose.

Portable release pens approximately 10 ft × 10 ft, and 4 ft high, covered with 1 in. wire mesh, should be moved to the area in which the release is planned. Groups of 12 to 15 birds should be maintained in each pen for a few days, with feed and water provided. Once the birds have become used to the area, one end of the pen can be opened, allowing the birds to leave and drift away. Release pens should be 200 yards apart.

JAPANESE QUAIL (COTURNIX COTURNIX JAPONICA)

Fig. 15–34. Japanese quail.

Japanese quail are also known as coturnix quail, pharoah's quail, stubble quail, and eastern quail. The Japanese quail is smaller than the Bobwhite quail, but it produces a larger egg.

■ **Origin, history, distribution** — The Japanese quail was imported into the United States from Japan. It should not be confused with the Bobwhite quail or other indigenous quail species. Coturnix are widely distributed in Europe, Africa, and Asia, where they are regarded as a migratory species. Records of their existence date to the ancient civilizations of these continents. Apparently coturnix were either domesticated in Japan about the 11th century or brought to Japan from China about that time. They were first raised as pets and singing birds, but by 1900 coturnix in Japan had become widely used for meat and egg production.

■ **Breeds** — The following varieties (breeds) of Japanese quail are known: Manchurian Golden, British Range, English White, and Tuxedo.

■ **Selection of breeders** — Japanese quail can be separated by their feather color difference when they reach about 3 weeks of age.

Adult males should weigh 4 to 5 oz, and adult females 4.5 to 6 oz. As noted, the females are generally larger than the males.

■ **Breeding facilities** — If the birds are raised for dog training or as a hobby, they can be raised on the floor. If they are raised for breeding or for egg or meat production, they will perform better in cages.

Pedigree cages 5 in. × 8 in. × 10 in. will hold a pair of quail.

Colony cages 2 ft × 2 ft × 10 in. will accommodate up to 25 quail, while a 2 ft × 4 ft × 10 in. cage will accommodate up to 50 quail.

Adult quail will perform better if given 16 to 25 sq in. of floor space per bird. They need 0.5 to 1 in. of feeder space per bird and 0.25 in. of trough space for water.

■ **Care of breeders** — Japanese quail may start laying as early as the sixth week.

When males are sexually mature, a large glandular or bulbous structure appears above the cloacal opening. If this gland is pressed, it will emit a foamy secretion.

Adult female hens require 14 to 18 hours of light per day to maintain maximum egg production and fertility.

■ **Eggs** — The eggs weigh approximately ⅓ oz, which is about 8% of the female body weight.

The hens proceed, like machines, to lay an egg every 16 to 24 hours for 8 to 12 months.

The eggs are edible and can be prepared in the same manner as chicken eggs.

For incubation, collect eggs 2 to 3 times daily, and store at 55°F and 70% humidity.

■ **Incubation** — The incubation period is 14 to 17 days.

Domesticated Japanese quail have lost the broodiness trait; hence, eggs must be incubated under a hen or artificially.

The recommended temperature and humidity for single-stage forced-draft incubation is:

1 to 14 days temperature: dry-bulb, 99.5°F and 60% humidity; or wet-bulb, 87°F.

14 days to hatching temperature: dry-bulb, 99°F and 70% humidity; or wet-bulb, 90°F.

■ **Brooding and rearing** — Brooder temperature should be 95°F at head level of chicks for first week. Thereafter, decrease the temperature 5° every week until after the fourth week.

Debeak the chicks to prevent pecking and cannibalism.

Young birds can be transferred from brooder to cages or floor around the fourth week.

■ **Feeding and watering** — Suggested rations for Japanese quail are given in Chapter 7, in Table 7–27.

■ **Care and management** — Japanese quail may be managed much the same as chickens, except for size.

Care should be taken with small quail to prevent drowning in water troughs.

Feather picking or other forms of cannibalism occur frequently when Japanese quail are kept on wire. Debeaking is the best preventive measure. Other steps to take to lessen cannibalism are: reducing light intensity, and increasing dietary fiber and grit.

■ **Diseases** — Although the Japanese quail is a hardy bird compared to other poultry species, it can be affected by most of the common poultry diseases. Sanitation is the best preventive measure, including the control of rats, mice, and fleas.

Eggs should be fumigated shortly after they are collected or at least within 12 hours after they have been placed in the incubator.

■ **Marketing, releasing** — Presently, in this country, Japanese quail are used mainly for research, with a few commercial farms raising them for eggs and meat.

Japanese quail are game birds. The breast and legs are considered delicacies.

The eggs may be prepared like chicken eggs, with which they are similar in taste.

■ **Comments** — The Japanese quail is popular for scientfic research, for use in such studies as mice and rats are used. They have also been used as pilot subjects for chicken studies.

Japanese quail develop 3½ times as fast as domestic fowl.

OSTRICH (STRUTHIS CAMELUS)

Fig. 15–35. Ostrich.

The ostrich is the world's largest bird; it may stand up to 10 ft tall and weigh up to 330 lb. It cannot fly, but it can attain a running speed of 40 miles per hour. Ostriches live up to 70 years of age.

■ **Origin, history, distribution** — The ostrich originated in the Asiatic steppes (plains) during the Eocene epoch, 40 to 50 million years ago. Its natural range is now limited to Africa. Today, these big birds are raised principally for their skins, which are made into fine quality leather for boots.

■ **Breeds** — In Africa, the ostrich is now separated into four subspecies.

■ **Selection of breeders** — The only reliable method of determining the sex of young chicks is examining the sex organs. As in all species, the best representatives should be used in breeding programs.

■ **Breeding facilities** — It is advisable to keep the males and females separated prior to pairing for mating. At the time of pairing, the ostriches should be kept in a pen or paddock away from other birds. The pen should be fenced with smooth wire approximately five to six feet high. Some breeders have been successful with breeding pens consisting of one male and two to four females.

■ **Care of breeders** — Ostriches are polygamous (the male has more than one mate). Breeders should be kept in healthy breeding condition.

■ **Eggs** — Egg production will begin when the female is about two years of age. The male (cock) usually scratches out a crude nest in the dirt and the female will then deposit the eggs in the nest. On the average, each female will lay 15 to 20 eggs, with some laying more and some laying fewer. Each egg is almost round, nearly 6 in. in diameter, and weighs about 3 lb. The eggs are dull yellow, and have large pores and a thick shell.

■ **Incubation** — Under natural conditions, both the male and female sit on the eggs. The male sits on the eggs at night, but during the day the hens share the task of keeping them warm. The eggs hatch in about 42 days.

Most ostrich producers prefer to use artificial incubation rather than natural incubation to hatch eggs. When eggs are collected, they should be stored at a temperature of 55° to 65°F and a humidity of 75%. The recommended temperature for incubation of ostrich eggs is 100°F, and the recommended humidity is 25 to 40%. Eggs should be positioned with the large end up, and preferably, at a 45 degree angle. The eggs should be turned at least twice daily.

■ **Brooding and rearing** — The brooding period of young ostriches (chicks) is very critical. Chicks should be brooded at a temperature of about 90°F for the first 10 to 14 days, following which the temperature can be reduced a few degrees each week until supplemental heat is no longer needed.

Ostrich chicks are prone to eat brooder litter materials, resulting in gastric impaction. To prevent this problem, the litter may need to be covered by burlap or similar material.

Young ostriches hide by sitting with their heads and necks stretched out on the ground. When an ostrich is a month old, it can run as fast as an adult. Ostriches reach full size in about six months.

■ **Feeding and watering** — In the wild state, the ostrich feeds on a variety of plants and is a very adaptable grazer. A frequent problem of young birds grown in captivity is impaction of the digestive system, due to the excessive consumption of sand, small stones, and/or fibrous material.

Also see the following chapters of this book: Chapter 6, Fundamentals of Poultry Nutrition; and Chapter 7, Poultry Feeding Standards, Ration Formulation, and Feeding Programs.

■ **Care and Management** — Ostriches respond to the same principles of good care and management as all species.

■ **Diseases** — Producers in need of assistance with disease diagnosis and treatment should seek professional help from a local veterinarian. Also, in some states, disease diagnosis and treatment is available through a state Poultry Disease Laboratory.

■ **Marketing** — Ostriches are marketed (1) as breeders; (2) for tanning of the hides for bootmaking; (3) to "eggers," those individuals who decorate eggs; and (4) as ostrich meat.

■ **Comments** — Ostriches are the only birds that eliminate their urine separately from their feces.

In the 19th century, ostrich plumes were in demand as decorations for hats and clothing, and wild ostriches were in danger of extinction until the development of ostrich farms in South Africa. Plumes were taken twice each year from live birds kept on ostrich farms.

PEAFOWL

Fig. 15–36. Peafowl.

The peafowl belongs to the same family as pheasants and chickens, differing in no important characteristic other than plumage.

■ **Origin, history, distribution** — The peafowl is native to India, Burma, Java, Ceylon, Malaya, and Congo.

The Old Testament refers to peafowl three times: 1 Kings 10:22; II Chronicles 9:21; and Job 39:13.

King Solomon is said to have prized the birds as much as his gold and silver.

About 300 B.C., Alexander the Great introduced the birds to Greece.

The early Christians adopted the peafowl as a symbol of immortality.

The Muslims regard the peafowl as unclean since it is supposed to have "guided the serpent to the Tree of Knowledge in the Garden of Eden."

■ **Breeds** — There are three species of peafowl: Indian Blue, Java Green, and Congo. The Indian Blue is most common. Also, there are several subspecies.

■ **Selection of breeders** — Select breeders that have the best color, that are healthy and free from leg weaknesses and crooked toes.

■ **Breeding facilities** — Peafowl are not inclined to wander far.

Their choice of a roost is a tall tree or on top of a building.

■ **Care of breeders** — A good male can be mated to as many as five hens.

■ **Eggs**—The hen is usually 2 years old before starting to lay; and she will usually lay 10 to 12 eggs, although some hens will lay up to 35 eggs.

■ **Incubation**—Incubation may be (1) by setting 6 to 10 eggs under a peahen, (2) by setting 6 to 10 eggs under a broody chicken, or (3) by artificial incubation.

■ **Brooding and rearing**—If the birds are brooded artificially, they may be treated the same as baby chicks or turkey poults. Hold the hover temperature at 95°F for the first week, then decrease the temperature by 5° per week thereafter.

■ **Feeding and watering**—The feeding of peafowl is very simple. A 30% protein turkey prestarter can be used for the first 4 to 5 weeks, after which a 26% protein turkey starter mash or game bird starter mash may be used.

Also, see the following chapters of this book: Chapter 6, Fundamentals of Poultry Nutrition; and Chapter 7, Poultry Feeding Standards, Ration Formulation, and Feeding Programs.

■ **Care and management**—Debeaking may be necessary to prevent cannibalism. Also cannibalism may be lessened by providing adequate space, a good diet, and proper temperature.

Peafowl have a very raucous voice which may annoy the neighbors. They may be devoiced by surgery, however.

If a male in full plumage is shipped, the tail should be wrapped and sewed—with clean muslin next to the feathers and burlap on the outside.

■ **Diseases**—Peafowl are subject to most of the diseases associated with chickens and turkeys. But the two most common diseases are blackhead and coccidiosis.

■ **Marketing, releasing**—Peafowl are usually sold in pairs, as ornamental birds, at $100 to $200 per pair.

Peafowl are edible and are regarded as a delicacy for special occasions.

PHEASANTS

Generally classed as game birds.

Fig. 15–37. Pheasant.

■ **Origin, history, distribution**—Pheasants originated in the Orient. They were first brought to North America by an Englishman, Benjamin Franklin's son-in-law, but the venture was unsuccessful. Later, the U.S. Consul in Shanghai sent 28 Chinese pheasants to Oregon. The latter introduction met with success.

■ **Breeds**—Pheasants are classed as (1) game breeds, or (2) ornamental breeds.

The game breeds are: Blackneck Pheasant, Chinese Ringneck Pheasant, English Ringneck Pheasant, Formosan Pheasant, and Melanistic Mutant Mongolian Pheasant.

The ornamental breeds are: Amherst's Pheasant, Golden Pheasant, and Reeves' Pheasant.

The Chinese Ringneck pheasant is the most popular variety.

■ **Selection of breeders**—Select from earliest hatches, from the sixth week on. Choose birds for their size, weight and feathering, without visible defects. Select birds that feather rapidly in true plumage color.

Test all breeders for pullorum disease each season before egg production starts.

■ **Breeding facilities**—Cold is not a factor. But the birds need protection from winds, such as can be provided by brush or cornstalks.

Hobbyists sometimes use an open-front shed.

Pheasants should be moved into laying pens prior to February 15; individual pens enclosing 1 cock and about 7 hens, or community pens holding about 140 hens and 20 cocks.

Provide nesting boxes under brush or under other covering.

■ **Care of breeders**—Pheasants are polygamous.

Egg production can be increased by artificial lights according to one of the following methods beginning January 1:

1. Constant light. Increase light to 14 to 15 hours daily and hold constant throughout breeding season.

2. Step-up light to 14 hours and hold until peak production is reached, then increase 15 minutes per week until maximum of 17 hours light per day is reached.

■ **Eggs**—Pheasants in the wild lay a clutch of 10 to 12 eggs.

In early spring, gather eggs several times each day.

Hold eggs at 55° to 60°F and humidity of 70 to 75%. But incubate as soon as possible.

■ **Incubation**—The incubation period is 23 to 24 days.

Most incubator manufacturers specify the temperature and humidity. If they do not, the following recommendations may be used:

INCUBATOR TYPE	
Forced Air	**Still Air**
°F	°F
99.75	102.25
99	102
86–88	
92–95	

■ **Brooding and rearing** — Brooding practices for pheasants are much the same as for chickens.

Most commercial operators use battery brooders for the first 7 to 9 days, then move into brooder houses.

Smaller producers usually raise chicks in brooder houses from the start.

The temperature should be 95° to 97°F for the first week, then reduced 5° each week.

Birds should be moved from brooder houses and placed in flight pens at 7 weeks of age.

Provide water in 1-gal fountains.

Start chicks by feeding on cardboard flats. Then progress to feeders. Provide 1 linear in. feeder/chick until 6 weeks of age, then 2 to 3 linear in. from 6 weeks on.

■ **Feeding and watering** — Commercial game-bird feeds are available in most areas.

Also, see the following chapters in this book: Chapter 6, Fundamentals of Poultry Nutrition; and Chapter 7, Poultry Feeding Standards, Ration Formulation, and Feeding Programs.

■ **Care and management** — Debeak all birds when moving them to the laying pen. Also, blunt cocks' spurs and trim and blunt inside toenails.

Pheasant chicks are usually debeaked at 7 to 9 days of age, to prevent cannibalism.

If pheasants are to be raised in open-top pens, they must be pinioned (have the first section of the wing removed) or have their feathers clipped.

To catch birds, drive them into a catching pen, then trap them with a nylon net similar to a fish net.

■ **Diseases** — Pheasants are subject to many of the diseases and parasites of chickens; and in most cases they respond to the same treatments. The most common diseases of pheasants are: coccidiosis, botulism, pullorum, and gapes.

■ **Marketing, releasing** — Pheasants can usually be released at the tenth to fourteenth week of age. They should not be released until they are fully feathered and able to care for themselves.

Pheasant is a specialty item. They should be processed at about 22 weeks of age and packaged in an attractive form. Frozen birds are sold individually vacuum-packed.

■ **Comments** — Pheasants are similar to chickens structurally and may be produced in a similar manner.

Pheasants are generally raised for the purpose of stocking farms reserved for hunting.

PIGEONS

A versatile bird with four distinct uses: (1) the sport of racing pigeons; (2) flyers and performers; (3) showing fancy pigeons; and (4) meat production.

■ **Origin, history, distribution** — The earliest fascination of people with pigeons goes back to at least 5000 B.C. The Bible contains many references to pigeons and doves. Countless paintings and carvings showing these tame birds have been found in the ancient cultures of Europe, Asia, and the Near East. In all cases, people have traditionally associated pigeons and doves with the highest virtues. In the earliest art, many used pigeons to symbolize love, peace, and fidelity. Every school child knows the story of Noah and the Ark; how a dove brought back the olive branch to show that the water had subsided. In the Christian tradition, the dove has been used as a sign of the Holy Spirit — a universal symbol of wisdom and understanding.

Pigeons are distributed throughout the world.

Fig. 15–38. Pigeon.

■ **Breeds** — Through the centuries, probably no other bird has undergone more physical changes from selective breeding than the pigeon. As a result, there are about 200 different breeds, each distinct from the other in behavior, size, shape, stance, feather form, colors, markings, and ornamentation.

The Homer, White King, and Swiss Mondaines are the most popular.

■ **Selection of breeders** — It is difficult to determine the sex of pigeons by casual observation.

Good pigeons for breeders have a white or pinkish-white skin and light-colored legs.

■ **Breeding facilities** — Pigeon houses are called lofts. The quarters should be dry, well ventilated, and provided with plenty of daylight.

A loft 7 ft wide and 10 to 12 ft long will provide ample room for 15 pairs of birds; that is about 5 sq ft per pair.

Breeder houses should be equipped with nests, bowls, feed hoppers, bathing pans, and a rack for nesting material.

■ **Care of breeders** — Pigeons are ready to mate at about 4 to 5 months of age.

They mate in pairs and usually remain with their mates throughout life, although the mating may be changed if desired by placing the male and female in a coop together and leaving them there for 6 to 14 days, or until such time as they become settled.

No more than 10 to 15 pairs of mated birds should be kept in one loft.

■ **Eggs** — The pigeon hen lays an egg, generally skips a day, then lays again.

■ **Incubation** — The male generally sits on the eggs during the middle of the day, and the female the remainder of the time. The incubation period is about 17 days.

■ **Brooding and rearing** — Both parents care for the young. They feed them by regurgitating a thick, creamy mixture, called pigeon milk, into the open mouths of the young.

Pigeons are the most rapid growing of all poultry. Squabs exceed the normal adult weight at the time they are ready to leave the nest — at about 30 days of age. However, with activity and flight, they soon slim down.

■ **Feeding and watering** — Pigeons are grain eaters. They prefer a variety, or mixture, of grains, or commercial pigeon pellets.

Commercial pigeon feeds are available. Also example rations are given in Chapter 7, Table 7–26.

■ **Care and management** — All young should be banded when about 7 days of age. This is a requisite to record keeping.

■ **Diseases** — A clean, well-kept loft will contribute to the health of the birds.

Pigeons, quarters, and nests should be checked frequently for lice and mites.

■ **Marketing, releasing** — There is a demand for squabs, especially in large cities, to take the place of game.

If squabs are to be slaughtered, they should be slaughtered just before leaving the nest — at about 30 days of age. Squabs are killed and dressed much like any other poultry.

If the young are to be kept as breeders or for show, allow them a few days of flight in the pen with their parents before moving them to a separate pen for young unmated birds.

SWANS

An ornamental bird. Legend has it that a swan finds its voice in true song just before it dies — a plaintive musical death song referred to as a "swan song."

■ **Origin, history, distribution** — Swans are more common in Europe than in the United States.

Fig. 15–39.   Mute swans (*Cygnus olor*) in marsh area. (Courtesy, USDA)

■ **Breeds**—The most familiar swan (the one commonly seen on park ponds) is the Mute Swan, which has no voice. There are two other species in Europe; and in North America there are two wild varieties—the trumpeter and the whistling swan.

■ **Breeding facilities**—Swans are very hardy and need no protection except in extremely cold weather.

They make nests of sticks and rubbish.

■ **Care of breeders**—Swans live in pairs and remain faithful to each other until death.

The females will breed for 30 years.

■ **Eggs**—Swans lay from 6 to 8 large greenish-white eggs each year.

The young hatch after 35 to 40 days of incubation.

■ **Incubation**—The incubation period of swan eggs is about 6 weeks.

■ **Brooding and rearing**—The male swan guards both eggs and chicks—more properly called cygnets. For 5 months, cygnets are under parental protection—both parents swimming watchfully around them.

■ **Feeding and watering**—Under natural conditions, swans live on water plants, soft roots, and insects. They may be fed the same kind of feed as other poultry, but it should be supplemented with green succulent feed.

■ **Care and management**—In captivity, swans should be provided suitable nest building material, placed near the nesting site.

Cygnets (young swans) respond to the same care and management as goslings (young geese).

■ **Diseases**—No comprehensive study of the diseases of swans has been made. However, the following diseases have been reported in swans: Marek's disease, salmonellosis, pasteurellosis, staphylococcosis, botulism, tuberculosis, fungal disease, coccidiosis, internal parasites, and external parasites.

■ **Marketing, releasing**—"Conservation" would be a more appropriate heading for this section as it pertains to swans.

Trade in swans for food is no longer acceptable.

It is not enough to maintain an exhibit of these exquisite birds in parks. Rather, they must be retained and seen in the wild state, no matter how much helping hands of people are needed to maintain it.

■ **Comments**—Swans live to be very old; the males have been known to live for more than 60 years..

QUESTIONS FOR STUDY AND DISCUSSION

1. Why are ducks and geese more popular, and of greater economic importance, in the Far East and in Europe than in the United States?

2. At one time 60% of U.S. ducks were raised in one relatively small area—on Long Island, in New York. Although Long Island still produces many ducks, large numbers of ducks are now raised in Indiana, Michigan, Minnesota, and North Carolina. What forces caused Long Island to lose its dominant position?

3. Name the breeds of ducks. What factors should be considered when choosing a breed of ducks?

4. What factors should be considered when selecting breeder ducks?

5. Discuss the facilities for and the care of breeder ducks.

6. Discuss each of the following as it pertains to ducks, and as it is compared to chickens and turkeys:
    a. Incubation.
    b. Brooding and rearing.
    c. Feeding.
    d. Management.
    e. Diseases and ailments.
    f. Marketing and processing.

7. Name the breeds of geese. What factors should be considered when choosing a breed of geese?

8. What factors should be considered when selecting breeder geese?

9. Discuss the facilities for and the care of breeder geese.

10. Discuss each of the following as it pertains to geese, and as it is compared to ducks:
    a. Incubation.
    b. Brooding and rearing.
    c. Feeding.
    d. Management.
    e. Diseases and ailments.
    f. Marketing and processing.

11. Discuss the practicality and management of geese as weeders.

12. Discuss the care for and marketing of goose feathers.

13. Discuss the place for, and the economic importance and potential of, each of the following poultry species:
    a. Bobwhite quail.
    b. Guinea fowl.
    c. Hungarian, or Gray partridge.
    d. Japanese quail.
    e. Ostrich.
    f. Peafowl.
    g. Pheasants.
    h. Pigeons.
  i. Swans

SELECTED REFERENCES

Title of Publication	Author(s)	Publisher
Diseases of Poultry, Ninth Edition	Edited by B. W. Calnek, *et al.*	Iowa State University Press, Ames, 1991
Feeds & Nutrition, Second Edition	M. E. Ensminger J. E. Oldfield W. W. Heinemann	The Ensminger Publishing Company, Clovis, CA 1990
Feeds & Nutrition Digest	M. E. Ensminger J. E. Oldfield W. W. Heinemann	The Ensminger Publishing Company, Clovis, CA 1990
Grzimek's Animal Life Encyclopedia, Vols. 7, 8, and 9	Edited by H. C. B. Grzimek	Van Nostrand Reinhold Co., New York, NY, 1972
Poultry: Feeds and Nutrition, Second Edition	H. Patrick P. J. Schaible	The AVI Publishing Company, Inc., Westport, CT, 1980
Poultry Meat and Egg Production	C. R. Parkhurst G. J. Mountney	Van Nostrand Reinhold Co., New York, NY, 1988
Poultry Nutrition Handbook	J. D. Summers S. Leeson	University of Guelph, Guelph, Ontario, Canada, 1985
Poultry Science and Practice, Fifth Edition	A. R. Winter E. M. Funk	J. B. Lippincott Company, Chicago, IL, 1960
Standard of Perfection, The, Fifth Edition	Edited by M. C. Wallace	American Poultry Association, Inc., Crete, NE, 1966

Fig. 15–40.   Pheasant. (Courtesy, USDA)

Fig. 16–1. Eggs! This shows clean eggs leaving the egg washer and entering a low temperature refrigerated cooler to reduce the interior temperature, following which they will be conveyed to the automatic egg grading and cartoning machine. (Courtesy, DEKALB Poultry Research, Inc., DeKalb, IL)

MARKETING POULTRY AND EGGS

CHAPTER

16

Marketing is selling. Practically everything that is done to poultry and eggs is done with the hope of improving salability, be it packaging, improving shelf life, or altering the production in such a way as to present the consumer with a new and novel product.

The changes in the marketing of poultry and eggs have been particularly marked in recent years. These changes encompass not only farm production but processing and distribution as well. Technological advances in poultry breeding, nutrition, housing, disease control, and other phases of production have brought with them organizational changes which have lowered production costs and transformed traditional poultry farming into a factory-type operation. This transformation has advanced most rapidly relative to broilers, but turkeys and eggs are now closing the gap. Thus, it is safe to say that the poultry industry is probably the most efficient and technologically advanced animal industry in the United States. The fate of every poultry product is carefully organized all the way from the hatchery to the retail market shelf.

RETAIL PRICES OF SELECTED LIVESTOCK PRODUCTS

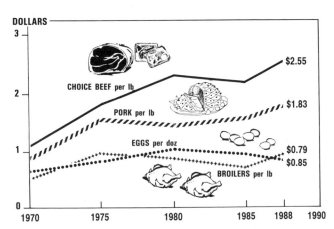

Fig. 16–2. Consumer food prices. (Source: *Statistical Abstract of the United States*, U.S. Department of Commerce, 1977, p. 486, and 1990, p. 483)

POULTRY AND EGG CONSUMPTION AND PRICES

Since World War II, chicken and turkey consumption, on a per capita basis, has increased, while egg consumption per person has decreased. Trends and current consumption figures are shown in Table 16–1.

MARKET VALUE OF POULTRY AND EGGS

U.S. poultry producers sold poultry and eggs valued at a total of $15.3 billion in 1989. This represented 18.3% of the total livestock receipts and 9.6% of the total cash farm income that year. Of the combined total income from poultry and eggs, about 56% is derived from broilers, 27% from eggs, 16% from turkeys, and 1% from other poultry.

The farm price of poultry and eggs is, like all commodities, determined by supply and demand. As shown in Fig. 16–3 and Table 16–2, the prices that farmers have received for poultry products have increased in recent years; but one must remember that operation expenses have increased dramatically, also.

TABLE 16–1
CONSUMPTION PER PERSON—
EGGS, CHICKEN, AND TURKEY, 1960–1989[1]

	Annual Consumption per Person						
	1960	1965	1970	1975	1980	1985	1989
Eggs (farm basis) . . . . . (no.)	335	314	311	279	272	256	236
Chicken, ready-to-cook . . . . (lb)	27.8	33.3	40.4	40.1	50	58.2	68.9
Turkey, ready-to-cook . . . . (lb)	6.2	7.4	8.0	8.5	10.5	12.1	17.1

[1]From USDA sources. To convert pounds to kilograms, multiply by 0.454.

Consumers have been willing to increase their consumption of chicken and turkey meat as advances in production and marketing technology have made for lower prices in relation to red meats (see Fig. 16–2). But technological advances and lower prices have brought no such consumption increases in eggs. Even at lower prices, per capita egg consumption by Americans has been declining for many years; in 1945, the final year of World War II, the U.S. per capita average exceeded 400 eggs, nearly double the current rate of 234.

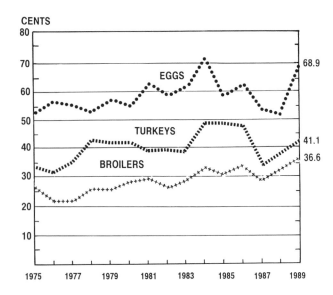

Fig. 16–3. Farm prices of poultry (per pound) and eggs (per dozen) 1975 to 1989. (Source: *Agricultural Statistics 1990*, USDA, pp. 343, 349, and 353)

TABLE 16–2
FARM PRICES OF POULTRY AND EGGS, 1975–1989[1]

Year	Price		
	Eggs	Broilers	Turkeys
	(¢/doz)	(¢/lb)	(¢/lb)
1975	52.4	26.3	34.8
1976	58.3	23.6	31.7
1977	55.6	23.6	35.5
1978	52.2	26.3	43.6
1979	58.3	26.0	41.1
1980	56.3	27.7	41.3
1981	63.1	28.4	38.2
1982	59.5	26.9	39.5
1983	61.1	28.6	38.0
1984	72.3	33.7	48.9
1985	57.2	30.1	49.1
1986	61.5	34.5	47.0
1987	54.7	28.7	34.8
1988	52.8	33.1	38.6
1989	68.9	36.6	41.1

[1]Source: *Agricultural Statistics 1990*, USDA, pp. 343, 349, and 353. To convert pounds to kilograms, multiply by 0.454.

PRODUCER'S SHARE OF THE CONSUMER'S DOLLAR

The producer's share of the consumer's dollar is shown in Fig. 16–4 and Table 16–3.

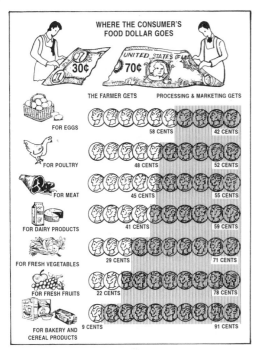

Fig. 16–4. The farmer's share of the consumer's food dollar, 1989. (Sources: *Agricultural Statistics 1990*, USDA, p. 380, Table 567; and *Food Cost Review, 1989*, Ag. Ec. Report No. 636, USDA, p. 16, Table 9)

Beef has been added to Table 16–3 for comparative purposes. Without doubt, the smaller retail markup shown in Table 16–3 for broilers and turkeys than for beef can be attributed largely to the practice of many retail outlets using poultry as a *loss-leader* — the practice of selling poultry during special sale periods at a relatively low price in order to attract customers to the store. Table 16–3 also reveals that marketing costs for poultry products are much lower than for beef. This is due to the fact that poultry units have tended to become more concentrated in certain areas, and that operations have become larger, with the result that the cost of assembling the products has lessened.

TABLE 16–3
FARMER'S SHARE OF
CONSUMER'S POULTRY PRODUCT DOLLAR[1]
(Beef for Comparison)

Item	Eggs (Grade A)	Broiler	Turkey	Beef Choice
	(/doz)	(/lb)	(/lb)	(/lb)
Farm value . . . . ($)	0.45	0.34	0.37	1.47
Marketing costs . . . . ($)	0.17	0.51	0.59	1.08
Retail price . . . . . . ($)	0.79	0.85	0.96	2.55
Farmer's share of retail price . . . . . (%)	57	40	39	58

[1]*Livestock and Poultry Situation and Outlook Report*, USDA, LPS–33, February 1989, pp. 16, 27, 29, 33. Data for 1988. To convert pounds to kilograms, multiply by 0.454.

FACTORS AFFECTING THE PRICE OF POULTRY PRODUCTS

The prices received by producers for their products are subject to a number of factors over which they have no control. However, the nature of the industry is such that bird numbers can be increased or decreased rather rapidly in response to price trends. A chick requires only 21 days to develop and hatch, far less time than the 283 days required for gestation in cattle. Thus, if a shortage of birds occurs, bird numbers can be increased in a very short time. Some of the factors that influence the prices of poultry follow:

■ **State of the economy** — Meat has traditionally been considered to be a luxury item. Thus, when unemployment is high or the disposable income of a family decreases due to inflation, meat is one of the first items cut from the family food budget. Fortunately, the cost of poultry is very low in comparison with pork or beef; and poultry has always been considered as a staple food for low-income families.

■ **Red meat supplies** — When red meat supplies are abundant, prices of red meat fall to a level where demand is increased. This tends to depress the demand for poultry, and, therefore, depresses prices. Conversely, when red meat prices are high, demand for poultry increases, consequently increasing poultry prices.

■ **Seasonality** — Hand in hand with the development of larger layer units, egg production has become less seasonal; yet, in 1990, 664 million fewer eggs were produced in February (the low month) than in March (the high month). The seasonality of egg production is shown in Fig. 16–5.

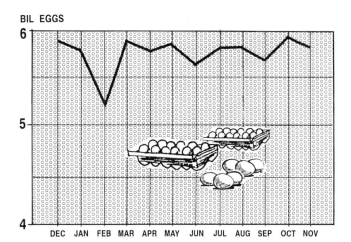

BIL EGGS

Fig. 16–5. Egg production by months. (Source: *Eggs, Chickens, and Turkeys*, USDA, Jan. 29, 1991, p. 5)

■ **Holidays** — Poultry meat, especially turkey, is commonly eaten for holiday meals. Thanksgiving, Christmas, Easter, and many of the Jewish holidays are periods when a large, temporary increase in demand occurs.

■ **Weather** — Periodically, weather can severely affect transportation channels of poultry products or items such as feed that are necessary in production. Snowstorms can effectively close down transportation channels, and severe cold or hot spells can adversely affect the condition and health of birds.

■ **Disease** — Widespread disease outbreaks can be devastating in certain regions. One excellent example involved the outbreak of exotic Newcastle disease in Southern California which ultimately required the condemnation and destruction of more than three million birds.

■ **Feed prices** — The largest single expense in the production of poultry products is feed. Thus, when feed grains and protein supplements are in short supply, the cost of feeding poultry may increase to an unprofitable level. In these cases, bird numbers are reduced. Conversely, when feed is cheap, bird numbers are increased; and although the price of poultry products is reduced due to large supplies, profits may be realized from the reduced cost of feed.

■ **Reports about the dangers of consuming animal products** — This is a very controversial subject (see Chapter 1), and repercussions are being felt by the poultry industry, especially the egg industry.

■ **Imports and exports** — Poultry products leaving from or coming into the United States can affect poultry prices. The economic impact of this phase of marketing is discussed later in this chapter.

MARKET CHANGES, AND THE FORCES BACK OF THEM

Changes in marketing poultry and eggs have extended all the way from the producer to the consumer, and they have involved technology, organization, and location. It is impossible to separate the changes from the forces back of them. Among the pertinent changes affecting the marketing of poultry and eggs are the following:

1. **Fewer and larger producers.** All U.S. poultry-producing units — layers, broilers, and turkeys — have become fewer in number and larger in size in recent years. In 1987, the average egg farm had 154,000 hens, the average broiler farm raised 158,000 birds during the year, and the average turkey farm produced 33,000 turkeys. These operations are large enough (a) to command competent management know-how, (b) to justify laborsaving and cost-saving facilities such as automatic watering and feeding equipment, and (c) to warrant the cooling facilities and other equipment necessary to maintain quality.

2. **Increased production efficiency.** This is best measured on the basis of feed efficiency — the pounds of feed required to produce a unit of product. In 1940, it required 7.4 lb of feed to produce a dozen eggs; in 1990, it took only 3.75 lb. In 1940, it required 4.7 lb of feed to produce a 1 lb weight gain of broilers; in 1990 it took only 1.9 lb. In 1940, it required 4.5 lb of feed to produce 1 lb of weight gain of turkeys; in 1990, it took only 2.7 lb.

(See Chapter 7, Poultry Feeding Standards, Ration Formulation, and Feeding Programs, Table 7–1.)

3. **Shifts in production areas.** Hand in hand with changes in efficiency and industry organization there have been shifts in the principal production areas. From 1950 to 1990, the nation's 10 leading egg producing states increased their proportion of the U.S. egg production from 46% to 62%. From 1950 to 1990, the five top broiler states increased their dominance from 27% to 63%. During the same period, North Carolina replaced Minnesota as the leading turkey producing state of the nation; and the big three turkey producers (North Carolina, Minnesota, and California) produced 48% of U.S. turkeys in 1990.

4. **More vertical integration, contracts, and direct marketing.** Rapid technical advances in poultry production following World War II made it possible for a dozen eggs or a pound of poultry meat to be produced with decreasing amounts of feed and other production costs. The incentive was strong to achieve these new efficiencies as quickly as possible. But capital was necessary. Feed manufacturers stepped in, providing capital in order to have a market for their feeds without generally involving added feed selling costs. This set off a whole chain of events involving vertical integration all the way from the producer to the consumer. Also, contractual arrangement and direct marketing followed; and all three developments — integration, contracts, and direct marketing — became commonplace among egg, broiler, and turkey producers.

a. **Egg integration.** Eggs have been produced under a variety of conditions and arrangements, with integrated and contract arrangements increasing with the formation of large-scale specialized egg-producing units, resembling factory production systems, which require much capital.

In 1990, 89% of the nation's eggs were produced under some kind of integrated or contract arrangement.

b. **Broiler integration.** Initially, feed manufacturers served as the integrators in broiler production. To protect their interests, processors followed—integrating with both feed companies and producers. Integration spread quickly to all stages of broiler production and distribution except two: (1) development of the basic breeding stock, and (2) distribution to consumers.

In order to protect their financial investment, and to exercise certain controls over management phases, contracts evolved. These contracts proved particularly attractive to those who had land and/or facilities and a surplus of labor, but who were unable to finance the purchase of birds and feed. From the standpoint of the integrator—the feed manufacturer and the processor—they were also attractive because they involved no social security, worker's compensation, or other similar employee fringe benefits. Likewise, integrators could use their capital to earn higher returns in other ways.

By 1990, 95% of the broilers were produced under contract and the remaining 5% were raised on integrator-owned farms.

c. **Turkey integration.** Vertical coordination and control came more slowly in turkey production than in broiler production, perhaps due to the fact that turkeys need a longer growing period, the market is more seasonal, capital and management requirements are greater, and disease risk higher than in broiler production. Also, cooperatives have played a more important role in turkey production than in broiler production.

In 1990, 90% of the nation's turkeys were produced under contract, either in production or in marketing.

Hand in hand with increased vertical integration and contracts, eggs, broilers, and turkeys have moved through shorter and more direct marketing channels. Increasingly, direct movement from packing plants to retailers is bypassing wholesale distributors, with the result that the latter have declined in volume and relative importance. In 1989, retail grocery stores marketed 51.2% of the broilers.

5. **Fewer and larger processors.** In recent years, there has been a marked trend to larger processors, and they have shifted their location to the production area rather than the consumption area.

Because commercial egg production is more widely dispersed over the United States, the concentration of egg handling and processing is not as great in the market egg business as in poultry meat processing. Processing is higher in liquid and frozen egg processing than in the handling of fresh eggs.

6. **More new products.** Finally, throughout the poultry and egg production industry, renewed attention has been given to developing new poultry and egg products. In particular, more convenience foods have been developed.

In the marketing of eggs, the most important product development has been the improvement in quality and freshness. New egg products include scrambled-egg mix, various egg and fruit juice drink combinations, and dried eggs.

Today, practically all chickens and turkeys are eviscerated. The second significant development in the marketing of poultry meats is the increase in cut-up and further processed products. By 1989, processors sold 18.3% of their broilers as whole carcasses, 50.4% as cut-up parts, 6.3% as further processed, and 25% in other forms. The fast-increasing further processed products included chicken franks, and boneless chicken in the form of patties, fillets, chunks, and nuggets.

Fig. 16–6. More than half of broilers are marketed as cut-up parts. (Courtesy, Delmarva Poultry Industry, Inc., Georgetown, DE)

First the processor took the feathers off the chickens, then they cleaned them out for the cook. All these changes were designed to add variety to the cook's menus while subtracting labor from kitchen duties; all were designed to increase sales.

By the late 1980s, cut-up and further processed products accounted for about 80% of the volume of turkeys slaughtered; hence, only 20% of turkeys were sold as whole carcasses.

One rather recent development in broiler marketing is an extensive network of fast food service outlets and carry-out restaurants featuring fried chicken. In 1989, 28.6% of the total broiler production was sold through fast food service outlets and carry-out restaurants. Another recent development is the use of chicken and turkey meat in the manufacture of such products as hot dogs.

TERMS USED IN MARKET REPORTS

Knowledgeable producers must keep abreast of current market trends in order to adjust their programs to the future. To help them, there are numerous market reports published by the USDA and private agricultural firms. Table 16–4 is a glossary of terms commonly used in market reports.

TABLE 16–4
GLOSSARY OF TERMS USED IN FEDERAL-STATES MARKET NEWS REPORTS[1]

Term	Definition
Market:	1. A geographic location where a commodity is traded. 2. The price, or price level, at which a commodity is traded. 3. To sell.
Market activity:	The pace at which sales are being made.
• Active	Available supplies (offerings) are readily clearing the market.
• Moderate	Available supplies (offerings) are clearing the market at a reasonable rate.
• Slow	Available supplies (offerings) are not readily clearing the market.
• Inactive	Sales are intermittent with few buyers or sellers.
Price trend:	The direction in which prices are moving in relation to trading in the previous reporting period(s).
• Higher	The majority of sales are at prices measurably higher than the previous trading session.
• Firm	Prices are tending higher, but not measurably so.
• Steady	Prices are unchanged from previous trading session.
• Weak	Prices are tending lower, but not measurably so.
• Lower	Prices for most sales are measurably lower than the previous trading session.
Supply/offering:	The quantity of a particular item available for current trading.
• Heavy	When the volume of supplies is above average for the market being reported.
• Moderate	When the volume of supplies is average for the market being reported.
• Light	When the volume of supplies is below average for the market being reported.
Demand:	The desire to possess a commodity coupled with the willingness and ability to pay.
• Very good	Offerings or supplies are readily absorbed.
• Good	Firm confidence on the part of buyers that general market conditions are good. Trading is more active than normal.
• Moderate	Average buyer interest and trading.
• Light	Demand is below average.
• Very light	Few buyers are interested in trading.
Mostly:	The majority of sales or volume.
Undertone:	Situation or sense of direction in an unsettled market situation.

[1]Agricultural Marketing Service, USDA, June 1975.

CONTRACT MARKETING

In 1990, 89% of the eggs, 100% of the broilers, and 90% of the turkeys were either produced under contract or marketed by owner-integrator operations. This means that only 11% of the eggs and 10% of the turkeys were sold after being produced—with no control or prior agreement.

Numerous types of poultry contracts exist, most of which involve integration and direct marketing, with the contractor controlling all or most of the operations from the producer to the consumer. (Also, see the following parts of this book for further discussion on this subject: Chapter 12, Layers, section headed, "Egg Contracts"; Chapter 13, Broilers (Fryers), Roasters, and Capons, section headed, "Broiler Contracts"; Chapter 14, Turkeys, section headed, "Turkey Contracts"; and Chapter 17, Business Aspects of Poultry Production, section headed, "Poultry Contracts.")

MARKET CHANNELS AND SELLING ARRANGEMENTS

Most eggs and poultry are marketed through retail food stores. The institutional markets, government purchases, and exports account for the remainder. In all cases, market channels have become more direct.

Until the mid-1940s, most eggs moved from the producers to county buying stations, or to hucksters and peddlers, thence to central assembling plants and shippers, thence to city wholesalers and jobbers, and finally to store warehouses or retail stores. In recent years, the desire to sell to premium outlets has had a major impact on the channels used in areas close to markets, with the result that a substantial proportion of eggs are moving from the producers either directly to consumers or to retail stores.

The marketing of poultry meat has also changed, largely as a result of a substantial increase in commercial broiler production and a shift from New York dressed (blood and feathers removed only) to ready-to-cook birds or parts and further processed products. In the heavy producing broiler areas, birds are moved directly from producers to the processing plants, thence to chain store warehouses or direct to retail stores. In the less populous poultry areas, the buying station type of agency assembles small quantities of poultry from individual producers, then sells them in larger quantities to city or country processing plants.

In summary, it may be said that, in both egg and broiler production, larger operations have resulted in processors locating near production and the elimination of some of the people who formerly performed a needed service where many small producers were involved.

As integration and various forms of vertical coordination continue to spread within the poultry industries, fewer actual purchases and sales of products occur and fewer genuine negotiated prices are generated. Also, formula pricing reduces further the fraction of total supply entering the market price formation. Consequently, it is increasingly difficult for a poultry producer to determine what is the "going market," and markets are more vulnerable from the standpoint of manipulation.

Fortunately, two major market news services remain: (1) The U.S. Department of Agriculture market news service, and (2) the Urner Barry report.

■ **The U.S. Department of Agriculture market news service**—The USDA operates a nationwide news service in cooperation with state agencies. Egg reports are made on frozen and dried eggs. These reports include data on supply and demand, movement, cold storage stocks, trading activity, price activities, and quality ranges. Prices and statistics are collected and disseminated nationwide on broilers/fryers, fowl, roasters, turkeys, ducks, and other miscellaneous poultry and rabbits.

Information on eggs and poultry is gathered by professionally trained staff from producers, major consuming centers, shipping points, and other sources. The information is disseminated nationwide by leased wire service and to the news media.

■ **The Urner Barry publications**—Urner Barry reports egg prices for each of four regions east of the Rocky Mountains. **Note well:** Since 1858, Urner Barry has served the food industry by reporting and establishing price quotations for the poultry, egg, turkey, chicken, dairy, and beef markets. The company publishes several newsletters and directories. For more information about Urner Barry, contact the firm at the following address: Urner Barry, P.O. Box 389, Toms River, NJ 08754-5330.

FUTURES TRADING

Most egg, broiler, and turkey producers operate on a contract basis, which means that they are guaranteed so much per dozen eggs or per pound of broiler or turkey produced; hence, their risk is minimal.

But big integrators take big risks; risks from disease, tornados/fires, fluctuating feed prices, and volatile egg, broiler, and turkey prices. Although most integrators are in a strong position financially, many of them prefer to protect themselves by *hedging* in the futures market. But futures trading is conducted only on broilers, and only by the Chicago Mercantile Exchange.

The subject of "Futures Trading" is fully covered in Chapter 17, Business Aspects of Poultry Production; hence, the reader is referred thereto.

MARKETING EGGS

In many respects, the poultry industry consists of two distinctly different industries: (1) the egg industry, and (2) the poultry meat industry. Grading, pricing, and marketing structures are unique for each segment.

Fig. 16–7. A 30-egg package for merchandising small- and medium-size eggs. (Courtesy, Cryovac Divison of W. R. Grace & Co., Duncan, SC)

PROBLEMS INVOLVED IN THE MARKETING OF EGGS

Eggs, like most agricultural commodities, are highly perishable. Although the edible portion is protected by a shell, the shell is highly porous, thereby allowing gases, bacteria, and moisture to pass readily into and out of the egg. However, refrigeration minimizes these exchanges and consequently prolongs the shelf life of the egg.

A second problem incurred in the marketing of eggs involves shell breakage. Thus, packaging plays an extremely important role in egg marketing. Fortunately, many of the eggs that are cracked in processing can be salvaged through egg-breaking plants.

In the past, egg production was highly seasonal with peak production occurring in late spring and early summer and low supplies in the fall and winter. Today, sophisticated lighting programs and environmentally controlled buildings have enabled egg producers to produce eggs at a relatively steady rate throughout the year.

Many of the consumer markets for eggs are relatively long distances from the areas of production. This necessitates refrigeration at all steps in the marketing and transportation phases of production.

Finally, the greatest hurdle faced by the egg producer is lack of consumer knowledge. The egg is a highly nutritious food that is extremely versatile in the kitchen. It can be prepared as a main course or used to improved the cooking and organoleptic qualities of other foods. The consumer must be made aware of these attributes.

QUALITY OF EGGS

If the egg industry is to remain viable and expand, only top-quality eggs can be marketed. A consumer is not likely to forget a bad experience of encountering a rotten egg or an egg with blood or meat spots.

Eggs are judged for quality through three sets of criteria: (1) external appearance, (2) candling, and (3) samples of eggs broken out to judge internal characteristics.

EXTERNAL APPEARANCE

The initial evaluation of eggs begins with the examination of the external appearance. The size and shape of the egg, as well as color and texture, are extremely important. A dozen grade A eggs must be of uniform size and shape. Additionally, the shells must be clean and free from cracks.

In some specific localities, consumers show a preference for a particular color of eggshell. People in New York City customarily choose eggs with white shells, while those in Boston prefer eggshells that are brown.

CANDLING

When eggs are graded for quality, they are routinely held against a light source (candling) so that the grader can determine certain quality characteristics of the internal parts of the egg without breaking the shell. When an egg is candled the following criteria are evaluated:

1. **Texture of the shell.** Any small cracks in the shell can be readily detected. Additionally, a gross estimate of shell thickness can be attained.

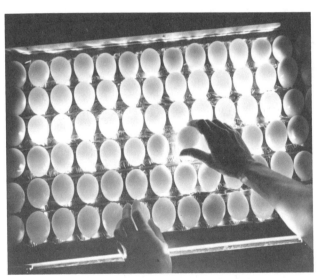

Fig. 16–8. Candling. Eggs with defective shells, blood spots, or meat spots are removed in this egg candling darkened curtain booth as eggs move from the washer to the grading and sizing machine. The eggs are rotated over a bank of high-wattage lights making any defect in them more easily seen by the person doing the candling work. (Courtesy, DEKALB Poultry Research, Inc., DeKalb, IL)

2. **Size of the air cell.** As an egg ages, the size of the air cell increases due to gas and moisture exchange through the shell. When an egg is candled, the size of the air cell is very easily determined.

3. **Firmness of the white.** A high-quality egg has firm albumen. If a high-quality egg is candled, the yolk is barely visible—merely a faint shadow—due to the viscosity of the white. As the egg ages, moisture enters it due to osmotic pressure exerted by the white, and the white becomes thin. If a low-quality egg is turned while candling, the yolk appears to move rather freely within it—another indication of the liquification of the egg white.

4. **Detection of meat and blood spots.** Meat and blood spots, while not harmful, are extremely unsightly.

BREAKOUTS

Periodically, samples from a large batch of eggs are broken out in order that the contents can be closely evaluated. Color of the white and yolk, odor, and general appearance can be readily evaluated. The yolk of a high-quality egg must be round and firm and the white must be firm with a rather clear demarcation between the thin and thick albumen. Egg albumen quality is routinely measured in Haugh units—units determined by a micrometer that measures albumen height. The Haugh units can then be derived by using a chart listing albumen heights, egg weights, and Haugh units.

The pH (hydrogen concentration) of the egg white is another means whereby egg quality is determined. The pH of a freshly layed egg is about 7.6 to 8.2. As the egg ages, carbon dioxide is lost, and the pH increases to as high as 9.5.

Shell strength is an additional criterion for evaluation. Breaking strength, thickness, and specific gravity are generally measured to determine the quality of the eggshell.

PROPER CARE OF EGGS AT THE FARM LEVEL

The processing of quality eggs begins at the farm level. Producers must carefully select for egg production and egg quality in their breeding programs. Once a good flock of layers has been established, good feeding and management of birds is required to maximize the genetic potential of the birds.

Frequent gathering of eggs will help maintain high quality because if eggs are exposed to ambient temperatures for extended periods, quality declines rapidly. By collecting eggs at frequent intervals, the producer is also able to reduce breakage and the incidence of dirty eggs. Once collected, the eggs should be cooled rapidly to prevent spoilage.

Washing is an extremely important part of producing quality eggs. A good washing routine encompasses the following practices:

1. Wash all eggs the same day they are gathered. This will help prevent bacterial contamination.

2. Use only approved detergent-germicide solutions. It is essential that the directions for use be carefully followed.

3. The water used for washing must contain less than 3 ppm of iron. At greater concentrations of iron, the threat of spoilage from *Pseudomonas* organisms is greatly increased.

4. Use a solution that feels slightly warm to the hand. Generally, temperatures of 110°F to 125°F are best.

5. Do not wash an excessive number of dirty eggs with the same solution of detergent-germicide. Using a dirty cleaning solution is self-defeating.

6. Make up a new solution each day that eggs are washed.

7. Promptly dry the eggs after washing. The eggs must be dry before they are candled.

GRADING EGGS

The grading of eggs involves their sorting according to quality, size, weight, and other factors that determine their relative value. U.S. standards for quality of individual shell eggs have been developed on the basis of such interior

Fig. 16–9. Automatic egg grading machine shown placing two grades of eggs in cartons. (Courtesy, DEKALB Poultry Research, Inc., DeKalb, IL)

quality factors as condition of the white and yolk, the size of the air cell, and the exterior quality factors of cleanliness and soundness of the shell. These standards cover the entire range of edible eggs.

Eggs are also classified according to weight (or size), expressed in ounces per dozen.

Egg grading, then, is the grouping of eggs into lots according to similar characteristics as to quality and weight. Although color is not a factor in the standards of grades, eggs are usually sorted for color and sold as either "whites" or "browns."

Four sets of grades, based on the quality standards for individual shell eggs, are used in this country: (1) consumer grades – used in the sale of eggs to individual consumers; (2) wholesale grades – used in the wholesale channels of trade; (3) U.S. Procurement Grades – used for institutional buying and Armed Forces purchases; and (4) U.S. Nest Run Grade – which is also used in wholesale channels of trade.

The U.S. standards for quality of individual shell eggs are applicable only to eggs of the domesticated chicken that are in the shell. These are given in Table 16–5.

The basis for the egg grades given in Table 16–5 is resemblance to normal new-laid eggs.

Consumer grades are those used for lots of eggs that have been carefully candled and graded for retail trade. These are given in Table 16–6 (next page).

TABLE 16–5
SUMMARY OF U.S. STANDARDS FOR QUALITY OF INDIVIDUAL SHELL EGGS[1]

Quality Factor	Specifications for Each Quality Factor			
	AA Quality	A Quality	B Quality	C Quality
Shell	Clean. Unbroken. Practically normal.	Clean. Unbroken. Practically normal.	Clean to slightly stained. Unbroken. May be slightly abnormal.	Clean to moderately stained. Unbroken. May be abnormal.
Air cell	1/8 in. or less in depth. May show unlimited movement and may be free or bubbly.	3/16 in. or less in depth. May show unlimited movement and may be free or bubbly.	3/8 in. or less in depth. May show unlimited movement and may be free or bubbly.	May be over 3/8 in. in depth. May show unlimited movement and may be free or bubbly.
White	Clear. Firm (72 Haugh units or higher).	Clear. May be reasonably firm (60 to 72 Haugh units).	Clear. May be slightly weak (31 to 60 Haugh units).	May be weak and watery. Small blood clots or spots may be present (less than 31 Haugh units).*
Yolk	Outline slightly defined. Practically free from defects.	Outline may be fairly well defined. Practically free from defects.	Outline may be well defined. May be slightly enlarged and flattened. May show definite but not serious defects.	Outline may be plainly visible. May be enlarged and flattened. May show clearly visible germ development but no blood. May show other serious defects.

*If they are small (aggregating not more than 1/8 in. in diameter).

For eggs with dirty or broken shells, the standards of quality provide three additional qualities. These are:

Dirty	Check	Leaker
Unbroken. May be dirty.	Checked or cracked but not leaking.	Broken so contents are leaking.

[1]*United States Standards, Grades, and Weight Classes for Shell Eggs*, Agricultural Marketing Service, Poultry Division, USDA.

TABLE 16–6
SUMMARY OF U.S. CONSUMER GRADES FOR SHELL EGGS[1]

U.S. Consumer Grade (origin)	Quality Required[1]	Tolerance Permitted[2]	
		Percent	Quality
Grade AA or Fresh Fancy quality	85% AA	Up to 15 . . . . Not over 5	A or B C or Check
Grade A	85% A or better	Up to 15 . . . . Not over 5	B C or Check
Grade B	85% B or better	Up to 15 . . . . Not over 10	C Checks
U.S. Consumer Grade (destination)	Quality Required[1]	Tolerance Permitted[3]	
		Percent	Quality
Grade AA or Fresh Fancy quality	80% AA	Up to 20 . . . . Not over 5	A or B C or Check
Grade A	80% A or better	Up to 20 . . . . Not over 5	B C or Check
Grade B	80% B or better	Up to 20 . . . . Not over 10	C Checks

[1]*United States Standards, Grades, and Weight Classes for Shell Eggs*, Agricultural Marketing Service, Poultry Division, USDA.

[2]For the U.S. Consumer grades (at origin), a tolerance of 0.3% Leakers or loss (due to meat or blood spots) in any combination is permitted. No Dirties or other type loss are permitted.

Wholesale grades differ from consumer grades in the tolerance of lower quality eggs permitted and in possible inclusion of some *loss* or inedible eggs. The grade designations of wholesale grades are U.S. Specials, U.S. Extras, U.S. Standards, U.S. Trades, U.S. Dirties, and U.S. Checks.

The Nest Run Grade is for the purpose of expediting trading by anticipating the grade yield of eggs before processing—before washing, grading, and sizing.

In the marketing of eggs, weight classes are also provided. These are given in Table 16–7.

It is to be emphasized that weight is separate and distinct from egg quality.

TABLE 16–7
U.S. WEIGHT CLASSES FOR CONSUMER GRADES FOR SHELL EGGS[1]

Size or Weight Class	Minimum Net Weight per Dozen	Minimum Net Weight per 30 Dozen	Minimum Weight for Individual Eggs at Rate per Dozen
	(oz)	(lb)	(oz)
Jumbo	30	56	29
Extra large	27	50.5	26
Large	24	45	23
Medium	21	39.5	20
Small	18	34	17
Peewee	15	28	—

[1]*United States Standards, Grades, and Weight Classes for Shell Eggs*, Agricultural Marketing Service, Poultry Division, USDA.

Fig. 16–10.   Racks of eggs. (Courtesy, D. D. Bell, University of California, Riverside)

PRODUCTS OF THE EGG-BREAKING INDUSTRY

The egg-breaking segment of the egg industry involves a considerable portion of the total eggs marketed (see Table 16–8). This segment of the egg industry was originally developed as an outlet for soiled, cracked, and abnormal eggs unsuitable for marketing as shell eggs. Today, because of the high demand for broken eggs, many quality eggs are being used.

TABLE 16–8
SHELL EGGS BROKEN AND EGG PRODUCTS PRODUCED UNDER FEDERAL INSPECTION[1]

Fiscal Years	Shell Eggs Broken	Egg Products Produced					
		Liquid		Frozen		Dried	
	(mil doz)	(mil)		(mil)		(mil)	
		(lb)	(kg)	(lb)	(kg)	(lb)	(kg)
1988	949	383.1	174.1	358.7	163.0	104.4	47.5
1989	1,051	459.1	208.7	404.9	184.0	110.7	50.3

[1]Source: *Livestock and Poultry Situation and Report*, LPS–46, Feb. 1991, USDA, p. 26.

The advantages of processed eggs are numerous when compared to shell eggs; among them —

1. Less storage space is involved.
2. Quality of frozen and dried eggs is preserved longer.
3. Packaging is facilitated.
4. Processed eggs are cheaper than shell eggs.
5. Less labor is involved in the egg-breaking industry.
6. One can choose specific parts of the egg for a particular need.

Almost all egg-breaking operations are automated. After the eggs are mechanically broken, they are checked for abnormalities and odor. The broken eggs can then be processed whole or separated into yolks and whites. Whole eggs are mixed and strained before packaging. Yolks are also mixed and strained. Additionally, in frozen yolks, salt or sugar is often added to improve the rubbery consistency of the yolk after freezing and defrosting. Glycerine, molasses, or honey may also be used. Whites are strained to remove the chalazae, meat spots, blood spots, and broken shells, and may be passed through a chopper to homogenize the product. Of the three processed forms—frozen, liquid, and dried eggs—dried eggs have the longest shelf life and require the lowest expense for storage, since refrigeration is not needed. However, the vacuum spray processing of drying incurs additional costs.

Table 16–9 shows the quantities and kinds of egg products produced by the egg-breaking industry.

TABLE 16–9
EGG PRODUCTS PRODUCED UNDER FEDERAL INSPECTION, BY TYPE 1988–89[1]

Fiscal Years	Item	Liquid Product for Immediate Consumption	Frozen Product	Dried Product
		(1,000 doz)	(1,000 doz)	(1,000 doz)
1988	Whole plain	301,553	140,745	19,079
1989		316,533	150,474	20,158
1988	Whole blends	107,670	104,620	33,193
1989		102,045	102,567	32,461
1988	White	248,468	48,397	30,177
1989		257,552	40,323	28,873
1988	Yolk plain	52,978	6,056	15,323
1989		47,241	4,041	17,481
1988	Yolk blends	86,144	71,649	9,121
1989		86,727	69,415	7,183

[1]Source: *Agricultural Statistics 1990*, USDA, p. 354, Table 531.

PRICING EGGS

The egg-pricing system in use today is still operating largely on the basis of wholesale trading in terminal markets, despite the fact that wholesalers, who formerly handled most of the nation's eggs, have largely been displaced by assembler-packers. As a result, the egg-pricing system has become increasingly controversial and problem laden.

Today, there is no satisfactory price barometer for eggs. The volume of eggs traded on the New York Mercantile Exchange is too small. Also, mercantile exchange trading has been criticized for too frequent and too wide fluctuations in price, although egg production, nationwide, is increasingly uniform.

There is no easy solution to the complex problem of egg pricing. Among the alternative pricing methods or systems proposed are the following:

1. **Computerized buying and selling.** This could be accomplished through an existing or new organization of traders who would agree to conduct transactions according to prescribed trading rules. The results of trading could be used directly, or they could become a major indicator for base price quotations.

2. **Using prices paid by retailers.** This could replace the wholesale level. The Los Angeles and San Francisco markets have moved in this direction.

3. **Committee pricing.** This could best be carried out under specific legislation. A group of designated individuals, supported by a staff to gather and analyze market information, could suggest prices which they consider to reflect supply and demand for specific locations, grades and sizes, and time periods. Such a committee could encourage more and better information and quickly adapt to changing industry structure and practices.

4. **Decentralized pricing.** This is the method used in the pricing of live meat animals. An objection to the use of this method in eggs is that the latter are more homogeneous in quality than live animals; hence, lots from various areas may be readily substituted as necessary, making the market national in scope.

5. **Administered pricing.** This would be operated almost entirely by private industry. This method would be a distinct possibility if the egg industry becomes more integrated or coordinated than at present. Presumably, this would require stricter industry determination and scheduling of quantities produced than exist at present.

In summary, it may be said that the egg-pricing system which evolved in the egg industry served it long and well—as long as the industry remained relatively static. But as the basic structure of the industry began to undergo rapid and extensive changes, the egg-pricing system did not change with it. As a result, progress in pricing bypassed egg marketing. Although there is no easy solution to the problem, the egg industry recognizes that pricing is vital to a market economy. Gradually, improved egg pricing will evolve and close the gap between needs and performance.

Fig. 16–11 shows the monthly fluctuations in the price of cartoned Grade A large eggs in New York in 1989 and 1990.

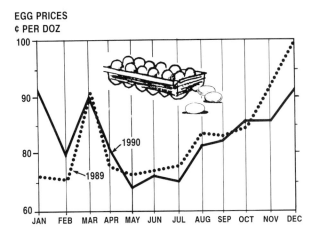

Fig. 16–11. Price for cartoned Grade A large eggs in New York in 1989 and 1990. (Source: *Livestock and Poultry Situation and Outlook Report*, LPS-46, Feb. 1991, USDA, p. 25, Table 39)

MARKETING POULTRY MEAT

Fig. 16–12. Broiler. (Courtesy, J. C. Allen & Son, West Lafa-

The poultry meat industry encompasses primarily chickens, turkeys, and ducks. Geese, game birds, and pigeons are raised for meat, but they should be considered as specialty items.

FEDERAL INSPECTION

The Poultry Products Inspection Act, Public Law 85–175, was enacted on August 28, 1957, and became fully effective January 1, 1959. The Animal and Plant Health Inspection Service in the U.S. Department of Agriculture is charged with the responsibility of administering this Act. The law requires inspection for wholesomeness of all poultry-processing plants shipping any of their products in interstate or foreign commerce. Personnel and supervisory cost of the service required by the Act except for necessary overtime are borne by appropriated federal funds.

Under the provisions of this Act, the USDA has four major responsibilities: (1) to determine that the poultry being processed is fit for human food—as determined by antemortem and postmortem inspection, (2) to make sure that the processing is done in a sanitary manner, (3) to protect poultry and poultry products from adulteration, and (4) to require that poultry and poultry products are properly labeled in compliance with the requirements of the law.

The grading services for both poultry meat and eggs are on a voluntary basis, with charges made for those requesting the service. However, inspection of poultry and egg products is mandatory in those poultry plants processing and shipping any of their products in interstate or foreign commerce.

But the current method of examining birds individually by federal inspectors is outmoded and inefficient. Imagine having sufficient trained inspectors to examine individually more than 5 billion broilers annually! So, the National Academy of Science, along with producer and consumer groups, has recommended that the U.S. Department of Agriculture develop a state-of-the-art meat and poultry inspection service. Currently (1990), the Food Safety Inspection Service (FSIS) of the U.S. Department of Agriculture is studying a system which it calls *Hazard Analysis and Critical Control Point (HACCP)* for meat and poultry operations. This system identifies what it calls Critical Control Points (CCP), or places, such as where thousands of birds are processed together; for example, (1) scalding freshly killed chickens in a common bath to loosen feathers; (2) defeathering many birds with the same rubber fingers; (3) eviscerating many birds with an improperly adjusted machine; and (4) cooling defeathered chickens in a common chill bath. But, until, and unless, the U.S. Department of Agriculture develops an improved inspection system, the only solution is more quality control inspectors—both company and federal.

POULTRY IRRADIATION

Factory-style broiler production has spread salmonella, which was once either rare or easily isolated in individual birds and flocks. Thus, from 1973 to 1987, poultry accounted for 9.8% of U.S. food-borne illnesses, with these attributed to salmonella and other pathogens, according to the Center for Disease Control, Atlanta, Georgia. Moreover, the inspectors of USDA's Food Safety and Inspection Service are not supposed to be looking for bacteria such as salmonella, which cannot be detected with the naked eye. This set of circumstances prompted the U.S. Department of Agriculture to petition the Food and Drug Administration (FDA) for approval of irradiation to control bacteria which may be present in chickens, turkeys, and other fresh or frozen uncooked poultry.

In 1990, the FDA announced that on May 1, it approved the use of irradiation to control bacteria in uncooked poultry; and described the process as a system to "pasteurize" solid foods. FDA added (1) that, as in the heat pasteurization of milk, the irradiation process greatly reduces, but does not eliminate, all bacteria; and (2) that irradiation does not make the food radioactive and, therefore, does not expose consumers to radiation. Treated food must be labeled, stating that it was irradiated; and the packaging must carry an international symbol representing the irradiation process.

Although the safety of irradiated foods has been carefully researched and thoroughly tested, at this point and period of time consumers have not accepted it. Fortunately, the same results can be achieved by cooking poultry.

MARKET CLASSES AND GRADES OF POULTRY

The U.S. Department of Agriculture has established specifications for different kinds, classes, and grades of poultry. *They define kind as referring to the different species of poultry, such as chickens, turkeys, ducks, geese, guineas,* *and pigeons. Class refers to kinds of poultry by groups which are essentially of the same physical characteristics, such as broilers or hens.* These physical characteristics are associated with age and sex. The kinds and classes of live, dressed, and ready-to-cook poultry listed in the U.S. classes, standards, and grades are in general use in all segments of the poultry industry.

A listing of the various classes of meat birds is given in Table 16–10. Guides for determining the ages and sexes of the different classes of poultry are given in Table 16–11 and Table 16–12.

TABLE 16–10
CLASSES OF POULTRY[1]

Bird-Class	Description
Chickens:	
• Rock Cornish game hen or Cornish game hen	A young, immature chicken (usually 4–6 weeks of age) weighing not more than 2 lb ready-to-cook weight, the progeny of a Cornish chicken or a Cornish chicken crossed with another breed of chicken.
• Broiler or fryer	A young chicken (usually 6–8 weeks of age), of either sex, that weighs 3 to 5 lb, that is tender-meated with soft, pliable, smooth-textured skin and flexible breastbone cartilage.
• Roaster	A young chicken (usually 9–11 weeks of age), of either sex, that weighs 6 to 8 lb, that is tender-meated with soft, pliable, smooth-textured skin and breastbone cartilage that may be somewhat less flexible than that of a broiler or fryer.
• Capon	A surgically unsexed male chicken that ranges up to 20 to 24 weeks of age, that weighs 12 to 14 lb, that is tender-meated with soft, pliable, smooth-textured skin.
• Stag	A male chicken (usually under 10 months of age) with coarse skin, somewhat toughened and darkened flesh, and considerable hardening of the breastbone cartilage. Stags show a condition of fleshing and a degree of maturity intermediate between that of a roaster and a cock or rooster.
• Hen or stewing chicken or fowl	A mature female chicken (usually more than 10 months of age) with meat less tender than that of a roaster, and nonflexible breastbone tip.
• Cock or rooster	A mature male chicken with coarse skin, toughened and darkened meat, and hardened breastbone tip.
Turkeys:	
• Fryer-roaster turkey	A young, immature turkey (usually under 16 weeks of age), of either sex, that is tender-meated with soft, pliable, smooth-textured skin, and flexible breastbone cartilage.
• Young hen turkey	A young female turkey (usually 14 to 15 weeks of age) that is tender-meated with soft, pliable, smooth-textured skin, and breastbone cartilage that is somewhat less flexible than in a fryer-roaster turkey.
• Young tom turkey	A young male turkey (usually 17 to 18 weeks of age) that is tender-meated with soft, pliable, smooth-textured skin, and breastbone cartilage that is somewhat less flexible than in a fryer-roaster turkey.
• Yearling hen turkey	A fully matured female turkey (usually under 15 months of age) that is reasonably tender-meated and with reasonably smooth-textured skin.
• Yearling tom turkey	A fully matured male turkey (usually under 15 months of age) that is reasonably tender-meated and with reasonably smooth-textured skin.
• Mature turkey or old turkey (hen or tom)	An old turkey of either sex (usually in excess of 15 months of age) with coarse skin and toughened flesh.
Ducks:	
• Broiler duckling or fryer duckling	A young duck (usually under 8 weeks of age), of either sex, that is tender-meated and has a soft bill and soft windpipe.
• Roaster duckling	A young duck (usually under 16 weeks of age), of either sex, that is tender-meated and has a bill that is not completely hardened and a windpipe that is easily dented.
• Mature duck or old duck	A duck (usually over 6 months of age), of either sex, with toughened flesh, hardened bill, and hardened windpipe.
Geese:	
• Young goose	Can be of either sex, is tender-meated, and has a windpipe that is easily dented.
• Mature goose or old goose	Can be of either sex, has toughened flesh, and hardened windpipe.
Guineas:	
• Young guinea	Can be of either sex, is tender-meated, and has a flexible breastbone cartilage.
• Mature guinea or old guinea	Can be of either sex, has toughened flesh, and a hardened breastbone.
Pigeons:	
• Squab	Young, immature pigeon of either sex, and is extra tender-meated.
• Pigeon	Mature pigeon of either sex, with coarse skin and toughened flesh.

[1]*Poultry Grading Manual*, Ag. Hdbk. No. 31, USDA.

TABLE 16–11
INDICATIONS OF AGE IN POULTRY[1]

	Young Birds	Mature Birds
Comb of chickens	Pliable, resilient, not wrinkled, points sharp.	Wrinkled, coarser, thicker points, rounded.
Bill of ducks	Pliable—not completely hardened.	Hardened.
Plumage	Fresh, glossy appearance.	Faded, worn except in birds which have recently molted.
Fat	Smooth layers with brighter color, not lumpy over feather tracts.	Generally darker in color, inclined to lumpiness over heavy feather tracts.
Breastbone	Cartilage, if present, pliable and soft.	End of keel—hardened cartilage, bony.
Pinbones	Pliable.	Not pliable.
Shanks	Scales on shanks, smooth, small.	Scales, larger, rough, and slightly raised.
Oil sac	Small, soft.	Enlarged, often hardened.
Spurs (male chickens, turkeys, occasionally adult females)	Small, undeveloped, cornlike.	Spurs gradually increase in length with age, becoming somewhat curved and sharper. Hens often have fine, sharp spurs after first year.
Windpipe of ducks, geese	Easily dented.	Hardened, almost bonylike to the touch.
Flesh	Tender-meated, translucent appearance; fine texture.	Coarser texture, darker, hardened muscle fibers.
Drumsticks	Lacking in development, muscles easily dented.	Generally rounded, full, firm.

[1]*Poultry Grading Manual*, Ag. Hdbk. No. 31, USDA.

TABLE 16–12
INDICATIONS OF SEX IN POULTRY[1]

	Males	Females
Head	Usually larger with larger and longer attachments, such as comb and wattles; coarser than that of females in appearance.	Smaller, rather fine and delicate in appearance compared with males. Hen turkeys have hair on center line of head.
Plumage	Feathers usually long and pointed at the ends. Tail feathers in chickens long and curved. Parti-colored varieties, have more brilliant colors than the females. Most male ducks have a curl in the tail feathers.	Feathers inclined to be shorter and more blunt than those of the male. Tail feathers short and straight in comparison with the male. Modest colors in parti-colored varieties.
Body	Larger and generally more angular than the female. Depth from keel to back greater on same weight birds. Bones, including shanks, longer, larger, and coarser.	Finer boned, body more rounded.
Skin	Slightly coarser, particularly in old birds. Feather follicles larger. Less fat under skin between heavy feather tracts and over back.	Smoother, generally a better distribution of fat between feather tracts. Feather tracts narrower but carrying more fat.
Keel	Longer with fleshing tending to taper at the base.	Shorter with more rounded appearance over the breast.
Legs	Drumstick and thigh relatively long with flesh tending to show less full until mature.	Drumstick and thigh relatively shorter with drumstick more inclined to roundness, increasingly so with age.

[1]*Poultry Grading Manual*, Ag. Hdbk. No. 31, USDA.

The grades of individual live birds are: A or No. 1 Quality, B or No. 2 Quality, and C or No. 3 Quality. The criteria used in determining grade are health and vigor, feathering, conformation, fleshing, fat covering, and defects (see Table 16–13).

Dressed and ready-to-cook poultry are graded for class, condition, and quality. These are most important since they are the grades used at the retail level. These grades are: U.S. Grade A, U.S. Grade B, and U.S. Grade C. These grades apply to dressed and ready-to-cook chickens, turkeys, ducks, geese, guineas, and pigeons (see Table 16–14).

TABLE 16–13
SUMMARY OF STANDARDS OF QUALITY FOR LIVE POULTRY ON AN INDIVIDUAL BIRD BASIS[1]
(Minimum Requirements and Maximum Defects Permitted)

Factor	A or No. 1 Quality	B or No. 2 Quality	C or No. 3 Quality
Health and vigor:	Alert, bright eyes, healthy, vigorous.	Good health and vigor.	Lacking in vigor.
Feathering:	Well covered with feathers. Scattering of pinfeathers.	Fairly well-covered with feathers. Moderate number of pinfeathers.	Complete lack of plumage feathers on back. Large number of pinfeathers.
Conformation:	Normal.	Practically normal.	Abnormal.
Breastbone	Slight curve; ⅛-in. dent in chickens, ¼-in. dent in turkeys.	Slightly crooked.	Crooked.
Back	Normal (except slight curve).	Moderately crooked.	Crooked or hunched back.
Legs and wings	Normal.	Slightly misshapen.	Misshapen.
Fleshing:	Well fleshed; moderately broad and long breast.	Fairly well fleshed.	Poorly developed, narrow breast, thin covering of flesh.
Fat covering:	Well covered, some fat under skin over entire carcass. Chicken broilers, turkey fryers, and young toms only moderate covering. No excess abdominal fat.	Enough fat on breast and legs to prevent a distinct appearance of flesh through skin. Hens or fowl may have excessive abdominal fat.	Lacking in fat covering on back and thighs; small amount in feather tracts.
Defects:	Slight.	Moderate.	Serious.
Tears and broken bones	Free.	Free.	Free.
Bruises, scratches, and callouses	Slight skin bruises, scratches and callouses.	Moderate (except only slight flesh bruises).	Unlimited to the extent that no part is unfit for food.
Shanks	Slightly scaly.	Moderately scaly.	Seriously scaly.

[1]*Poultry Grading Manual*, Ag. Hdbk. No. 31, USDA.

TABLE 16–14
SUMMARY OF SPECIFICATIONS FOR STANDARDS OF QUALITY FOR INDIVIDUAL CARCASSES
OF READY-TO-COOK POULTRY AND PARTS THEREFROM[1]
(Minimum Requirements and Maximum Defects Permitted)

Factor			A Quality			B Quality			C Quality
Conformation:			Normal.			Moderate deformities.			Abnormal.
Breastbone			Slight curve or dent.			Moderately dented, curved, or crooked.			Seriously curved or crooked.
Back			Normal (except slight curve).			Moderately crooked.			Seriously crooked.
Legs and wings			Normal.			Moderately misshapen.			Misshapen.
Fleshing:			Well fleshed; moderately long, deep, and rounded breast.			Moderately fleshed, considering kind, class, and part.			Poorly fleshed.
Fat covering:			Well covered—especially between heavy feather tracts on breast and considering kind, class, and part.			Sufficient fat on breast and legs to prevent distinct appearance of flesh through the skin.			Lacking in fat covering over all parts of carcass.
Pinfeathers:									
Nonprotruding pins and hair			Free.			Few scattered.			Scattering.
Protruding pins			Free.			Free.			Free.

Exposed flesh:[2]

Carcass Weight		Breast and Legs	Elsewhere	Part	Breast and Legs[2]	Elsewhere[3]	Part	
Minimum (lb)	Maximum (lb)	(in.)	(in.)	(in.)	(in.)	(in.)	(in.)	
None	1½	None	¾	Slight trim	¾	1½	Moderate	
Over 1½	6	None	1½	on edge	1½	3	amount of	No limit.
Over 6	16	None	2		2	4	the flesh	
Over 16	None	None	3		3	5	normally covered.	

(Continued)

TABLE 16–14 *(Continued)*

Factor	A Quality			B Quality			C Quality
Discoloration:[4]							
None 1½	½	1	¼	1	2	½	No limit.[5]
Over 1½ 6	1	2	¼	2	3	1	
Over 6 16	1½	2½	½	2½	4	1½	
Over 16 None	2	3	½	3	5	1½	
Disjointed bones/broken bones:	One disjointed/none broken.			Two disjointed and no broken or 1 disjointed and 1 nonprotruding broken.			No limit.
Missing parts:	Wing tips and tails.[6]			Wing tips, second wing joint, and tail.			Wing tips, wings, and tail.
Freezing defects (when consumer packaged):	Slight darkening over the back and drumsticks. Few small ⅛ in. pockmarks for poultry weighing 6 lb or less and ¼ in. pockmarks for poultry weighing more than 6 lb. Occasional small areas showing layer of clear or pinkish ice.			Moderate dried areas not in excess of ½ in. in diameter. May lack brightness. Moderate areas showing layer of clear-, pinkish-, or reddish-colored ice.			Numerous pockmarks and large dried areas.

[1]*Poultry Grading Manual*, Ag. Hdbk. No. 31, USDA.

[2]Total aggregate area of flesh exposed by all cuts and tears and missing skin.

[3]A carcass meeting the requirements of A quality for fleshing may be trimmed to remove skin and flesh defects, provided that no more than one-third of the flesh is exposed on any part and the meat yield is not appreciably affected.

[4]Flesh bruises and discolorations such as *blueback* are not permitted on breast and legs of A quality birds. Not more than one-half of total aggregate area of discolorations may be due to flesh bruises or *blueback* (when permitted), and skin bruises in any combination.

[5]No limit on size and number of areas of discoloration and flesh bruises if such areas do not render any part of the carcass unfit for food.

[6]In geese, the parts of the wing beyond the second joint may be removed, if removed at the joint and both wings are so treated.

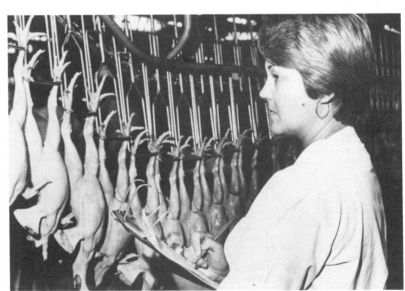

Fig. 16–13. Quality assurance checking processing. (Courtesy, Foster Farms, Livingston, CA)

Additionally, there are U.S. Procurement Grades, which are designed primarily for institutional use. These grades are: U.S. Procurement Grade 1, and U.S. Procurement Grade 2. In procurement grades, more emphasis is placed on meat yield than on appearance.

The factors determining the grade of carcass, or ready-to-cook poultry parts therefrom, are: conformation, fleshing, fat covering, pinfeathers, exposed flesh, discoloration, disjointed bones, broken bones, missing parts, and freezing defects.

MARKETING CHICKEN MEAT PRODUCTS

Most of the poultry meat marketed comes from the slaughter of broilers, although substantial quantities of meat from turkeys and spent laying hens are marketed.

Table 16–15 compares the production and price figures for broilers and nonbroiler chickens (primarily spent laying hens). It is quite evident from this table that broilers constitute the bulk of the chicken meat market and that broilers bring substantially higher prices.

TABLE 16–15
CHICKEN PRODUCTION AND PRICE, 1975–89[1]

Year	Broilers					Nonbroiler Chicken							
	Produced			Price		Sales			Consumed on Farms			Price	
	(mil)	(mil lb)	(mil kg)	(¢/lb)	(¢/kg)	(mil)	(mil lb)	(mil kg)	(mil)	(mil lb)	(mil kg)	(¢/lb)	(¢/kg)
1975	2,950	8,127	3,694	26.3	57.9	224	1,047	476	13.5	51	23	9.9	21.8
1976	3,273	9,067	4,121	23.6	51.9	217	1,046	475	12.8	50	23	13.0	28.6
1977	3,394	9,418	4,281	23.6	51.9	225	1,082	492	12.2	47	21	12.3	27.1
1978	3,614	10,129	4,604	26.3	57.9	220	1,050	477	11.4	45	20	12.4	27.3
1979	3,951	11,219	5,100	26.0	57.2	236	1,148	522	11.2	44	20	13.9	30.6
1980	3,963	11,353	5,160	27.7	60.9	238	1,167	530	10.8	42	19	11.0	24.2
1981	4,148	11,985	5,448	28.4	62.5	239	1,187	540	10.5	41	19	11.1	24.4
1982	4,149	12,167	5,530	26.9	59.2	242	1,159	527	10.3	40	18	10.3	22.7
1983	4,184	12,400	5,636	28.6	62.9	237	1,159	527	9.3	37	17	12.7	27.9
1984	4,283	13,016	5,916	33.7	74.1	225	1,067	485	8.7	34	15	15.9	35.0
1985	4,470	13,762	6,255	30.1	66.2	220	1,025	466				14.8	32.6
1986	4,649	14,316	6,507	34.5	75.9	218	1,026	466	(data discontinued)			12.5	27.5
1987	5,004	15,597	7,090	28.7	63.1	218	1,018	463				11.0	24.2
1988	5,238	16,187	7,358	33.1	72.8	224	1,039	472				9.2	20.2
1989	5,518	17,864	8,120	36.6	80.5	199	937	426				14.9	32.8

[1]Source broilers: *Agricultural Statistics 1990*, USDA, p. 343. Source nonbroiler chicken: *U.S. Egg and Poultry Statistical Series, 1960–89*, USDA, Stat. Bull. 816, p. 138.

MARKETING BROILERS

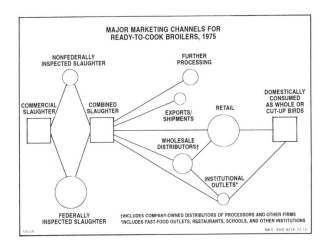

Fig. 16–14. Major marketing channels for ready-to-cook broilers. (Courtesy, USDA)

One of the major problems involved in marketing broilers is transporting the product from the area of production to the area of consumption. The states having the greatest shortages of broilers are New York, California, Illinois, Ohio, Michigan, New Jersey, Pennsylvania, and Massachusetts. The states having the greatest surpluses of broilers are Arkansas, Georgia, Alabama, North Carolina, Mississippi, Delaware, Maine, and Virginia. As shown in Table 16–16, broilers must be transported considerable distances to their targeted market.

TABLE 16–16
ORIGIN OF RECEIPTS OF BROILERS
FOR 13 MAJOR CITIES, 1975[1]

City	Origin
Atlanta	Georgia, North Carolina
Baltimore	North Carolina, Delmarva Peninsula, Georgia
Boston	Delmarva Peninsula, New England, North Carolina
Chicago	Georgia, Missouri, Arkansas
Cleveland	Georgia, Missouri, Arkansas, Ohio, North Carolina
Denver	Arkansas, Missouri, Georgia
Los Angeles	Missouri, Arkansas, California, Georgia
Minneapolis-St. Paul	Alabama, Georgia, Missouri, Arkansas
New York	Delmarva Peninsula, North Carolina, New England, Georgia
St. Louis	Georgia, Missouri, Arkansas
San Francisco	California, Missouri, Arkansas, Georgia
Seattle	Washington, Arkansas, California
Washington, DC	North Carolina, Virginia, Delmarva Peninsula

[1]USDA, Agricultural Marketing Service, Poultry Division, Poultry Market News Branch, field offices.

In the past, broilers were traditionally marketed in four forms: (1) live, (2) New York dressed (blood and feathers removed only), (3) ready-to-cook, and (4) processed meat. Today, virtually 100% of the broilers marketed are sold in ready-to-cook form (either whole or cut-up) or as processed meat.

Most broilers are now slaughtered at 6 to 8 weeks of age at a weight of 3 to 5 lb. Prior to slaughter, feed is withheld for 8 to 10 hours to allow the digestive tract to empty and prevent problems during processing. During slaughter, drained blood accounts for losses of about 4% of the total liveweight. Losses due to feather removal are variable but average out to about 5%. However, these losses of weight, in addition to the losses of weight from the removal of offal, do not represent total economic losses. Feathers are sold for decoration or feathermeal, and blood and offal can be processed further to form by-product feeds or industrial raw materials. Of the ready-to-cook product, the breakdown is as follows: breast, 26%; thigh, 17%; leg, 16%; back, 17%; wings, 12%; neck, 6%; and giblets, 6%. The broiler can then be sold cut-up or whole and subsequently fried, roasted, barbecued, stewed, or broiled; or it can be deboned and used in such processed forms as sandwich rolls, frozen pies, and cold cuts.

Kentucky Fried Chicken, the fast food chain which Col. Harland Sanders of Kentucky began as a franchise operation in 1956, has had a significant impact on the marketing of broilers.

PRICING BROILERS

For broiler and turkeys, there is a great deal of formula pricing. Originally, the typical broiler formula was based on live price divided by 73% (the approximate yield of ready-to-cook from liveweight) plus 3 to 7 cents to cover processing costs. However, as more and more broilers were produced under contract, fewer and fewer live broilers changed hands. As a result, the trend is away from farm base calculations to ready-to-cook prices.

BROILER PRICES
¢ PER LB

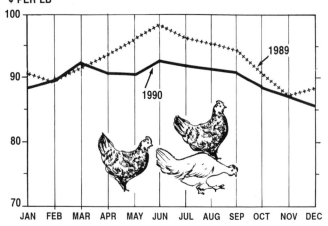

Fig. 16–15. Broiler retail prices, 1989–1990. (Source: *Livestock and Poultry Situation and Outlook Report*, LPS–46, Feb. 1991, USDA, p. 20)

MARKETING SPENT HENS

Most of the mature chickens slaughtered are hens that have ceased egg production or hens culled because of low production or physical ailments. They are known as *spent* hens.

Approximately four-fifths of the mature chickens slaughtered are further processed to be used in such items as soup, cold cuts, chicken salad, or hot dogs.

MARKETING TURKEYS

Turkey numbers and prices are given in Table 16–17. Prices have tended to fluctuate greatly.

TABLE 16–17
TURKEY PRODUCTION AND PRICE, 1975–89[1]

Year	Production			Price	
	(mil)	(mil lb)	(mil kg)	(¢/lb)	(¢/kg)
1975	124	2,277	1,035	34.8	76.6
1976	140	2,606	1,185	31.7	69.7
1977	136	2,563	1,284	35.5	78.1
1978	139	2,655	1,207	43.6	95.9
1979	156	2,958	1,345	41.1	90.4
1980	165	3,077	1,399	41.3	90.9
1981	171	3,264	1,484	38.2	84.0
1982	165	3,175	1,443	39.5	86.9
1983	171	3,336	1,516	38.0	83.6
1984	171	3,384	1,538	48.9	107.6
1985	185	3,704	1,684	49.1	108.0
1986	207	4,147	1,885	47.0	103.4
1987	240	4,895	2,225	34.8	76.6
1988	242	5,059	2,300	38.6	84.9
1989	260	5,454	2,479	41.1	90.4

[1]Source: *Agricultural Statistics 1990*, USDA, p. 349, Table 521.

About 80% of the turkey meat produced today is cut-up or used for further processed products. Whole bird processed turkeys, such as self-basting turkeys, comprise about 20% of this total.

Turkey operations are becoming integrated rapidly. However, this integration differs from the broiler industry in that the major coordinator in the turkey industry is the processor, whereas in the broiler industry, the feed firms took the lead in coordinating the operation. Additionally, cooperatives play a greater role in the marketing of turkeys than in the marketing of broilers.

Fig. 16–16. About 80% of turkey meat is marketed cut-up or further processed. Note that this supermarket features both cut-up turkey and cut-up broiler. (Courtesy, Foster Farms, Livingston, CA)

PRICING TURKEYS

The Urner Barry daily report of wholesale turkey prices in the New York area is widely used as a guide in price negotiations. In the Chicago area, for example, the typical rule of thumb for turkey transactions is to apply the Urner Barry New York quotation, five-day average for the week in which delivery was made, less 1.25¢ per pound.

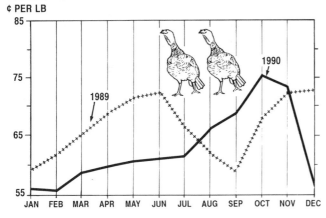

TURKEY PRICES
¢ PER LB

Fig. 16–17. Turkey prices, young hens 8 to 16 lb, New York, 1989–90. (Source: *Livestock and Poultry Situation and Outlook Report*, LPS-46, Feb. 1991, USDA, p. 22, Table 33)

MARKETING DUCKS

Pekin-type meat ducklings are ready for market between 7 and 8 weeks of age if they have been fed high-energy diets and have undergone some degree of selection for early maturity. Muscovies are not ready for market until 10 to 17 weeks of age. Ducklings should have good finish and be free of pinfeathers when sent to slaughter. Careful observation of both traits should be made during the latter part of the rearing period.

Pekin-type males and females perform differently. Males grow to heavier weights by 7 to 8 weeks and convert their feed more efficiently than females. These differences become apparent after 6 weeks of age. More economical feeding costs are possible if females are marketed at 7 weeks of age.

(Also see Chapter 15, Ducks, Geese, and Miscellaneous Poultry, section on "Marketing and Processing Ducks.")

MARKETING GEESE

Farm geese are usually marketed in the fall and winter, at which time they are relatively free of pinfeathers. Most geese are marketed when they are 5 to 6 months old; they will weigh from 11 to 15 lb depending on the strain and breed. Some young geese (also called green geese or junior geese) that have been full-fed for rapid growth are marketed at 10 to 12 lb when they are 10 to 13 weeks old.

(Also see Chapter 15, Ducks, Geese, and Miscellaneous Poultry, section on "Marketing and Processing Geese.")

FOREIGN TRADE IN POULTRY PRODUCTS

Prior to 1950, exports of poultry products amounted to no more than 1% of domestic production. However, beginning in the late 1950s, exports of broilers increased rapidly, and by 1962 they peaked at 262 million pounds and accounted for 3.4% of domestic production. Beginning in 1962, however, the European Economic Community (EC) imposed high levies on chickens and turkeys. As a result, chicken exports to these countries declined. But exports to other countries have expanded, with the result that, in 1987, chicken exports represented 5% of U.S. production, and turkey exports represented 0.9% of our production. Fig. 16–18 (p. 406) shows that there is a rather sizable export market for poultry products.

In 1987, the major importers of U.S. poultry and poultry products, by rank, were: Germany, Japan, Saudi Arabia, U.S.S.R., and Hong Kong. Total poultry and poultry exports from the United States in 1987 amounted to $443 million.

U.S. imports of poultry products, mainly eggs (in the shell) from Canada, have been small compared with exports of poultry products. Total product imports amounted to $3.5 million in 1987.

In 1990, the United States had an import duty of 3.5¢ per dozen on shell chicken eggs, 3¢ per pound on plucked chicken, 5¢ per pound on eviscerated chicken, 8.5¢ per pound on plucked turkey, and 5¢ per pound on eviscerated turkey.

U.S. EXPORTS OF POULTRY PRODUCTS

MILLION DOZEN EGGS

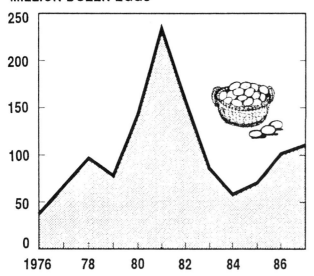

MILLION POUNDS OF POULTRY

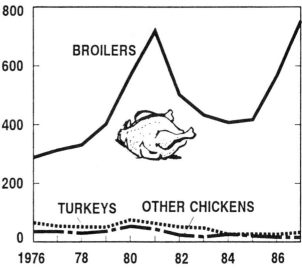

Fig. 16–18. Exports of eggs and broilers both rose in 1987, reflecting a weaker dollar, lower U.S. prices, and increased sales through the Export Enhancement Program. (Source: *1988 Agricultural Chartbook*, p. 92, Chart 202)

Table 16–18 shows the amount of poultry products exported from the United States. Broilers constituted the greatest segment of poultry exports.

TABLE 16–18
IMPORTS AND EXPORTS OF POULTRY PRODUCTS[1]

Year	Eggs		Chickens[2]		Turkeys	
	Imports	Exports	Exports		Exports	
	(mil doz)	(mil doz)	(mil lb)	(mil kg)	(mil lb)	(mil kg)
1980	5	143	567	258	75	34
1981	5	234	719	327	63	29
1982	3	158	501	228	51	23
1983	23	86	432	196	47	21
1984	32	58	407	185	27	12
1985	13	71	417	190	27	12
1986	14	102	566	257	27	12
1987	6	111	752	342	33	15
1988	5	142	765	348	51	23
1989	25	92	859	390	40	18

[1]*Agricultural Statistics 1990*, USDA, p. 354; U.S. *Egg and Poultry Statistical Series, 1960–89*, USDA, Stat. Bull. 816, pp. 100 and 200.

[2]Young, all forms.

TRENDS IN MARKETING AHEAD

There is every indication that egg, broiler, and turkey production will each become more highly specialized, larger, and concentrated in fewer farms, and more vertically integrated. Contract production and owner integration are expected to expand further in market eggs and turkeys. The motivating forces back of increasing bigness are the advantages in management, marketing, and distribution. Also, with increasing mechanization and smaller labor requirements, it is quite likely that more and more production units will be owned by integrators, rather than controlled on a contract basis.

Further gains in marketing efficiency are needed and likely.

Market reporting and bases for establishing the going market price of eggs, broilers, and turkeys leave much to be desired. Perhaps the ultimate solution lies in the "giants" in the poultry industry taking a page out of the book of the automobile manufacturers—learning to live together in competition, but cutting back or closing down production at such intervals as necessary to avoid overproduction and ruinous prices. Of course, there remain two great differences between poultry producers and automobile manufacturers: (1) Consumers don't eat cars, and (2) biologically controlled animals cannot be turned on and off as can the manufacture of a car.

QUESTIONS FOR STUDY AND DISCUSSION

1. What is marketing? What is unique about the marketing of poultry and eggs in comparison with the marketing of cattle, sheep, swine, and milk?

2. Has the lower price of broilers and turkeys in relation to red meats been a factor in the increased consumption of chicken and turkey?

3. What percent of U.S. cash farm income is derived from poultry?

4. What have been the recent trends in farm prices of poultry and eggs?

5. How does the poultry producer's share of the consumer's dollar compare to the share of the consumer's dollar obtained by the beef cattle producer?

6. List and discuss the factors affecting the prices of poultry products.

7. How have poultry markets changed in recent years? Discuss the forces back of these changes.

8. Define: market, firm price trend, heavy supply, and light demand.

9. What percent of eggs, broilers, and turkeys are sold after being produced—with no contract or prior arrangement? What impact does this have on price determination?

10. Discuss the major changes since 1940 in market channels and selling arrangements for eggs, broilers, and turkeys.

11. How has integration made it difficult to determine the "going market"?

12. List some of the problems involved in the marketing of eggs.

13. How are eggs judged for quality?

14. Outline a good program for caring for eggs at the farm level.

15. List and define the various egg grades.

16. What is the impact of the egg-breaking industry?

17. How are eggs priced?

18. What are the responsibilities of the USDA under the provisions of the The Poultry Products Inspection Act?

19. Define: broiler, young hen turkey, roaster duckling, and squab.

20. How can the relative age and the sex of a bird be determined?

21. How is poultry meat graded?

22. What problems are encountered in the marketing of broilers?

23. What losses are incurred in the processing of broilers?

24. Discuss the market and pricing structure of turkeys.

25. Discuss the marketing of ducks, and tell how the marketing of ducks differs from the marketing of broilers and turkeys.

26. Discuss the marketing of geese, and tell how the marketing of geese differs from the marketing of broilers and turkeys.

27. Of what impact are exports and imports of poultry?

28. What trends do you see in the marketing of eggs, broilers, and turkeys in the years ahead?

SELECTED REFERENCES

Title of Publication	Author(s)	Publisher
Chicken Broiler Industry: Structure, Practices, and Costs, Mktg. Res. Rpt. No. 930	F. L. Faber R. J. Irvin	Economic Research Service, USDA, Washington, DC, 1971
Egg Grading Manual		Agricultural Marketing Service, USDA, Washington, DC, 1975
Market Structure of the Food Industries, Mktg. Res. Rpt. No. 971	D. F. Dunham, *et al.*	Economic Research Service, USDA, Washington, DC, 1972
Marketing Poultry Products, Fifth Edition	E. W. Benjamin, *et al.*	John Wiley & Sons, Inc., New York, NY, 1960
Poultry and Egg Statistics, Supplement for 1972–75 to Stat. Bull. No. 525		Economic Research Service, USDA, Washington, DC, 1976
Poultry Meat and Egg Production	C. R. Parkhurst G. J. Mountney	Van Nostrand Reinhold Co., New York, NY, 1988
Poultry Production, Twelfth Edition	M. C. Nesheim R. E. Austic L. E. Card	Lea & Febiger, Philadelphia, PA, 1979

(Continued)

SELECTED REFERENCES *(Continued)*

Title of Publication	Author(s)	Publisher
Poultry Products Technology, Second Edition	G. J. Mountney	The AVI Publishing Company, Inc., Westport, CT, 1976
Processing of Poultry	Edited by G. C. Mead	Elsevier Science Publishers, Ltd., Barking, Essex, England, 1989
Readings on Egg Pricing	Edited by G. B. Rogers L. A. Voss	College of Agriculture, University of Missouri, Colombia, 1971
U.S. Broiler Industry, The	F. A. Lasley, *et al.*	USDA, ERS, Agricultural Economics Report 591, Washington, DC, 1988
U.S. Egg and Poultry Statistical Series, 1960–1989	M. R. Weimar S. Cromer	USDA, ERS, Statistical Bulletin, Washington, DC, 1990

Fig. 16–19. Roast turkey. (Courtesy, National Turkey Federation, Reston, VA)

Fig. 17–1. Computer in the poultry business! This shows the computerized system in the feed mill at Pennfield's Hempfield Mill, Lancaster, Pennsylvania. (Courtesy, National Broiler Council, Washington, DC)

BUSINESS
ASPECTS OF
POULTRY
PRODUCTION

CHAPTER

17

With the increase in specialization and size of enterprises, the business aspects of poultry production have become more important. More capital is required, competent management is in demand, records are essential, computers have come in, and such things as tax management, estate planning, and liability require more attention.

From 1935 to 1988, within a span of 53 years, the number of farms decreased from 6.8 million to 2.2 million, and the size of farms increased from 154.8 acres to 463 acres.[1] Thus, within 53 years, nearly 68% of the farms disappeared from American agriculture, and the average size of farms tripled. With this transition, herds, flocks, and feedlots became bigger.

In 1820, each farm worker supplied farm products for 4 people (self and 3 others). By 1940, one farm worker produced enough for 11 people (self and 10 others). In 1991, one farm worker supplied enough agricultural products for 100 people (self and 99 others), thereby freeing 99 people to produce such luxuries as autos, refrigerators, TV sets, and a host of other goods and services for modern American living.

PERSONS SUPPLIED PER FARM WORKER

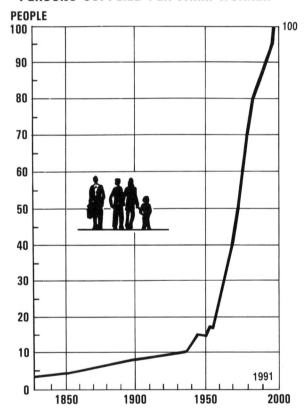

Fig. 17–2. Productivity of farm worker; the number of persons supplied farm products by one farm worker. In 1820, each farm worker supplied farm products for 4 persons, including self. By 1991, one farm worker supplied enough agricultural products for 100 people (self and 99 others). Never have so many people been dependent on so few! (Source: U.S. Department of Agriculture)

[1]The Census definition of a farm is as follows: "A place that sells $1,000 a year in agricultural products."

The above trends to bigness will continue. With it, the business and management aspects of all animal production, including poultry production, will be increasingly important in the future. Changes in the type of business organization and in financial management will come. More capital will be required, more money will be borrowed, competent managers will be in demand, better and more complete records will be necessary, futures trading will increase, and producers will become more knowledgeable relative to tax management, estate planning, and liability. The net result will be that those engaged in the business of agriculture will treat it as the big business that it is, and become more sophisticated and efficient; otherwise, they will not be in business very long.

TYPES OF BUSINESS ORGANIZATIONS

Fig. 17–3. The poultry industry is big business; so, it must be treated as such. These hatching eggs are being prepared for shipment to Brazil. (Courtesy, Cobb-Vantress, Inc., Siloam Springs, AR)

The success of today's poultry enterprise is very dependent upon the type of business organization. No one type of organization is superior under all circumstances; rather, each situation must be considered individually. The size of the operation, the family situation, and the objectives—all these, and more, are important in determining the best way in which to organize the poultry enterprise.

Four major types of business organizations are commonly found in poultry operations: (1) the sole proprietorship; (2) the partnership; (3) the corporation; and (4) contracts.

PROPRIETORSHIP (INDIVIDUAL)

The sole proprietorship is a business which is owned and operated by one individual.

This is the most common type of business organization in U.S. farming as a whole—90% of the nation's farms are individually owned. Under the sole proprietorship, or individual ownership, one person controls the business. Proprietors may not provide all the capital used in the business; in fact, they usually do not. However, they have sole management and control of the operation, although this may be modified and delegated somewhat through contracts. Basically, the sole proprietors get all the profits of the business; likewise, they must absorb all the losses.

In comparison with other forms of organization, the sole proprietorship has two major limitations: (1) It may be more difficult to acquire new capital for expansion; and (2) not much can be done to provide for continuity and to keep the present business going as a unit, with the result that it usually goes out of existence with the passing of the owner.

PARTNERSHIP (GENERAL PARTNERSHIP)

A partnership is an association of two or more persons who, as co-owners, operate the business. About 13% of U.S. farms are partnerships.

Most poultry partnerships involve family members who have pooled land, buildings and equipment, and their labor and management to operate a larger business than would be possible if each family member operated alone. It is a good way to bring a son or daughter, who is usually short on capital, into the business, yet keep the parent in active participation. Although there are financial risks to each member of such a partnership, and potential conflicts in management decisions, the existence of family ties tends to minimize such problems.

In order for a partnership to be successful, the enterprise must be sufficiently large to utilize the abilities and skills of the partners and to compensate the partners adequately in keeping with their respective contributions to the business.

A partnership has the following **advantages**:

1. **Combining resources.** A partnership often increases returns from the operation due to combining resources. For example, one partner may contribute labor and management skills, whereas another may provide the capital. Under such an arrangement, it is very important that the partners agree on the value of each person's contribution to the business, and that this be clearly spelled out in the partnership agreement.

2. **Equitable management.** Unless otherwise agreed upon, all partners have equal rights, regardless of financial interest. Any limitations, such as voting rights proportionate to investments, should be a written part of the agreement.

3. **Tax savings.** A partnership does not pay any tax on its income, but it must file an informational return. The tax is paid as part of the individual tax returns of the respective partners, usually at lower tax rates.

4. **Flexibility.** Usually, the partnership does not need outside approval to change its structure or operation – the vote of the partners suffices.

Partnerships may have the following **disadvantages**:

1. **Liability for debts and obligations of the partnership.** In a partnership, each partner is liable for all the debts and obligations of the partnership.

2. **Uncertainty of length of agreement.** A partnership ceases with the death or withdrawal of any partner, unless the agreement provides for continuation by the remaining partners.

3. **Difficulty of determining value of partner's interest.** Since a partner owns a share of every individual item involved in the partnership, it is often very difficult to judge value. This tends to make transfer of a partnership difficult. This disadvantage may be lessened by determining market values regularly.

4. **Limitations on management effectiveness.** This is due to personal differences among partners, and the responsibility of each partner for the acts of the other partner.

The above is what is known as a partnership or general partnership. It is characterized by (1) management of the business being shared by the partners, and (2) each partner being responsible for the activities and liabilities of all of the partners, in addition to his/her own activities within the partnership.

LIMITED PARTNERSHIP

A limited partnership is an arrangement in which two or more parties supply the capital, but only one partner is involved in the management. This is a special type of partnership with one or more "general partners" and one or more "limited partners."

The limited partnership avoids many of the problems inherent in a general partnership and has become the chief legal device for attracting outside investor capital into farm ventures. As the term implies, the financial liability of each partner is limited to his/her original investment, and the partnership does not require, and in fact prohibits, direct involvement of the limited partners in management. In many ways, a limited partner is in a similar position to a stockholder in a corporation.

A limited partnership must have at least one general partner who is responsible for managing the business and who is fully liable for all obligations.

The **advantages** of a limited partnership are:

1. It facilitates bringing in outside capital.
2. It need not dissolve with the loss of a partner.
3. Interests may be sold or transferred.
4. The business is taxed as a partnership.
5. Liability is limited.
6. It may be used as a tax shelter.

The **disadvantages** of a limited partnership are:

1. The general partner has unlimited liability.
2. The limited partners have no voice in management.

CORPORATIONS

A corporation is a device for carrying out a farming enterprise as an entity entirely distinct from the persons who are interested in and control it. Each state authorizes the existence of corporations. As long as the corporation complies with the provisions of the law, it continues to exist – irrespective of changes in its membership.

Until about 1960, few U.S. farms were operated as corporations. In recent years, however, there has been increased interest in the use of corporations for the conducting of farm business. Even so, only about 2% of U.S. farms use the corporate structure.

From an operational standpoint, a corporation possesses many of the privileges and responsibilities of a real person. It can own property; it can hire labor; it can sue and be sued; and it pays taxes.

Separation of ownership and management is a unique feature of corporations. The owners' interest in a corporation is represented by shares of stock. The shareholders elect the board of directors who, in turn, elects the officers. The officers are responsible for the day-to-day operation of the business. Of course, in a close family corporation, shareholders, directors, and officers can be the same persons.

The major **advantages** of a corporate structure are:

1. It provides continuity despite the death of a stockholder.
2. It facilitates transfer of ownership.
3. It limits the liability of shareholders to the value of their stock.
4. It may make for some savings in income taxes.

The major **disadvantages** of a corporation are:

1. It is restricted to doing only what is specified in its charter.
2. It must register in each state.
3. It must comply with stipulated regulations which involve considerable paperwork and expense.
4. It is subject to the hazard of higher taxes.
5. It is possible to lose control.

Still another type of corporation is family owned (privately owned). It enjoys most of the advantages of its generally larger outside investor counterpart, with few of the disadvantages. The chief **advantages** of the family-owned corporation over a partnership arrangement are:

1. **It alleviates unlimited liability.** For this reason, a lawsuit cannot destroy the entire business and all the individual partners with it.
2. **It facilitates estate planning and ownership transfer.** It makes it possible to handle the estate, keep the business in the family, and keep it going if one of the partners should die. Each of the heirs can be given shares of stock—which are easy to sell or transfer and can be used as collateral to borrow money—while leaving the management of the enterprise either to those heirs interested in operating it, or even to outsiders.

TAX-OPTION CORPORATION (SUBCHAPTER S CORPORATION)

Instead of paying a corporate tax, a corporation with no more than 35 stockholders may elect to be taxed as a partnership, with the income or losses passed directly to the shareholders, each of whom pays taxes on his/her share of the profits. This special type of corporation is variously referred to as a *tax-option* corporation, *subchapter S* corporation, *pseudocorporation*, or *elective* corporation.

For income tax purposes, the owners of a tax-option corporation are taxed as if they were a partnership. That is, income earned by the corporation passes through the corporation to the personal income tax returns of the individual shareholders. Thus, the corporation does not pay any income tax. Instead, the shareholders pay tax on their share of corporate income at their individual tax rate; and the shareholders report their share of long-term capital gains and receive their deductions therefor. Although each shareholder's portion of any corporate losses from current operations is deducted from their personal return, capital losses incurred by the corporation cannot be passed through to the shareholders.

Thus, there are some very real advantages to be gained from a subchapter S or tax-option corporation. However, in order to qualify as a subchapter S corporation, the following requisites must be met:

1. There cannot be more than 35 stockholders.
2. All stockholders must agree to be taxed as a partnership.
3. Nonresident aliens cannot own stock.
4. There can be only one class of stock.
5. Not more than 20% of the gross receipts of the corporation can be from royalties, rents, dividends, interest, or annuities plus gains from sale or exchange of stock and securities; and not more than 80% of the gross receipts can be from sources outside the United States.

ADVANTAGES OF LIMITED PARTNERSHIPS AND CORPORATIONS

In addition to the advantages peculiar to (1) limited partnerships, and (2) corporations, and covered under each, limited partnerships and corporations have the following advantages over individual ownership in the acquisition of capital:

1. They make it possible for several producers to pool their resources and develop an economically-sized operation, which might be too large for any one of them to finance individually.
2. They make it possible for persons outside agriculture to invest through purchase of shares of stock in the business.
3. They can generally borrow money more easily because the strength of the loan is not dependent on the financial and management capability of one person.
4. They give assurance that the business will continue, even if one of the owners should die or decide to sell out.
5. They provide built-in management, with continuity; and, generally speaking, they attract very able management.

Thus, those engaged in the livestock business can and do use either of these two business organizations—a limited partnership or a corporation—to develop and maintain an economically sound operation. Actually, no one type of business organization is best suited for all purposes. Rather, each case must be analyzed, with the assistance of qualified specialists, to determine whether there is an advantage to using one of these types of organizations, and, if so, which organization is best suited to the proposed business.

POULTRY CONTRACTS

Today, the poultry industry is one of the most integrated industries in the United States. Eighty-nine percent of the nation's eggs, 100% of the broilers, and 90% of the turkeys are raised in some form of integrated process.

Much of the poultry production involves the contracting of certain services by the farmer with highly integrated poultry companies. Therefore, it is imperative that the producer become well informed relative to contracts.

Numerous **advantages** are inherent in contracting with large poultry firms; among them:

1. Both parties can combine capital and reduce duplication of services and equipment. Thus, the total output of the contracted parties is often greater than if both parties operated separately.

2. Specialization is promoted. Poultry producers are permitted to concentrate on one particular phase of production, knowing they will have an available market for their product.

3. Information is readily exchanged. Since both parties have mutual interests, they are more apt to exchange information or "secrets" which will promote production.

4. Price risks are reduced. Both parties know how much of the product is to be produced. Hence, wide fluctuations in supply can be avoided.

Some of the **disadvantages** of contractual agreements are as follows:

1. Each party must consider the other before decisions can be made. Thus, a large amount of managerial freedom is lost.

2. Although profits are shared, losses are likewise shared.

3. If the decision-making is vested in one party, resentment from the other parties may arise.

4. When credit is involved in the agreement, problems may arise in the termination of the contract.

Since contracts are legally binding, the poultry producer should read the agreement with care and consult a legal advisor to ensure that all of the terms are clearly spelled out and understood. A good contract must be clear, complete, and concise as to the duties and responsibilities of all parties.

All contracts should make provision for the following:

1. **Tenure.** The contract should be specific as to starting and ending dates per brood or time basis. A time contract usually specifies four to six broods per year.

2. **Renewal.** Each party should retain the same right for continuing or closing the program. This could minimize hardships for growers who have used credit in providing housing and equipment.

3. **Cancellation.** This should be specific and clearly understood, with equal rights and privileges.

4. **Management.** The contract should make known who is responsible for decisions; and details of the management program should be spelled out.

5. **Production and credit resources.** Detail as to who is to furnish what, when, amount (under what provisions), and security for credit should be given.

6. **Payment or settlement.** The contract should be clear as to method of computing rate, time, incentives, penalties, and losses including condemnation.

7. **Assignment of interest.** There should be mutual agreement as to assignment privileges as dictated by the situation.

8. **Arbitration.** There should be procedures providing for binding settlement to avoid court proceedings.

9. **Legal relationship of contracting parties.** It should be clearly stated whether or not the contract is a partnership, employer-employee situation, or arranged on an independent basis. This is important for social security and income tax purposes.

For information pertaining to contracts for each species, the reader is referred to the following sections in this book: For "Egg Contracts" see Chapter 12, Layers; for "Broiler Contracts" see Chapter 13, Broilers (Fryers), Roasters, and Capons; and for "Turkey Contracts" see Chapter 14, Turkeys.

CAPITAL NEEDS

In 1987, farm assets—investments in land, improvements, machinery, equipment, animals, feed and supplies—totaled $813.1 billion—a value equal to roughly 40% of the total capital assets of all manufacturing corporations in the United States. In 1987, farm debt totaled $153.3 billion. Thus, in the aggregate, farmers had 81.2% equity in their business and 18.8% borrowed money (debts). The balance sheet of U.S. farming from 1970 through 1987 is shown in Fig. 17–4.

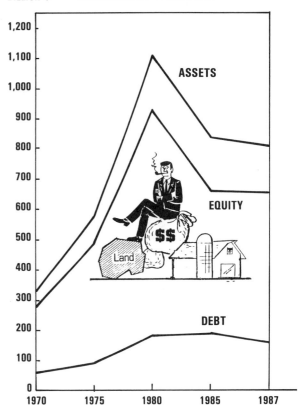

BALANCE SHEET OF THE FARMING SECTOR

BILLION $

Fig. 17–4. Balance sheet of U.S. farming, showing (1) assets, (2) debts, and (3) equities. (Source: *Statistical Abstracts of the United States 1989*, U.S. Department of Commerce, p. 632, No. 1089)

Perhaps agriculturists have been too conservative, for it is estimated that one-fourth to one-third of American farmers could profit from the use of more credit in their operations.

Another statistic which points up the enormity of capital needs is that it takes about $17.58 in farm assets to produce $1 of net farm income.[2]

Fig. 17–5. The capital needs of the poultry industry have increased with size. This is an aerial view of Maple Leaf Farm. (Courtesy, Shaver Poultry Breeding Farm, Ltd., Cambridge, Ontario, Canada)

CREDIT

Credit is an integral part of today's poultry business. Wise use of it can be profitable, but unwise use of it can be disastrous. Accordingly, poultry producers should know more about it. They need to know something about the lending agencies available to them, the types of credit, and how to go about obtaining a loan.

TYPES OF CREDIT OR LOANS

Hand in hand with getting the right kind of loan, it is important that the best available source of the loan be secured. Table 17–1 shows the primary sources of the three main kinds of loans. It is noteworthy that banks, merchants, and individuals provide 80% of the farm credit.

---

[2]Based on 1987 figures, when farm assets were $813.1 billion and net farm income was $46.264 billion ($813.1 ÷ $46.26 = $17.58). Source: *Statistical Abstracts of the United States 1989*, U.S. Department of Commerce,

TABLE 17–1
PRINCIPAL SOURCES OF THREE MAIN KINDS OF FARM LOANS

Credit Source	Long Term	Intermediate Term	Short Term
Commercial banks	X	X	X
Dealers and merchants		X	X
Farm mortgage companies	X		
Farmers Home Administration	X	X	X
Federal Land Bank Associations . . .	X	X	
Individual lenders	X	X	X
Insurance companies	X		
Production Credit Associations		X	X

Getting the needed credit through the right kind of loan is an important part of sound financial poultry management. The following three general types of agricultural credit are available, based on length of life and type of collateral needed:

■ **Short-term loans**—This type of loan is made for operating expenses and is usually for 1 year or less. It is used for the purchase of birds, feed, and for operating expenses; and it is repaid when eggs or birds are sold. Security, such as a chattel mortgage on the birds, may be required by the lender.

■ **Intermediate-term loans**—These loans are used for buying equipment, for making land improvements, and for remodeling existing buildings. They are paid back in 1 to 7 years. Generally, they are secured by a chattel mortgage.

■ **Long-term loans**—These loans are secured by mortgage on real estate and are used to buy land or make major improvements to farmland and buildings or to finance construction of new buildings. They may be for as long as 40 years. Usually they are paid off in regular annual or semi-annual payments. The best sources for long-term loans are an insurance company, the Federal Land Bank, the Farm Home Administration, or an individual.

CREDIT SOURCES

Table 17–2 and Fig. 17–6 show where farmers borrow, the amount of loans from each source, and the percent of the total held by each type of lender.

But agricultural financing is changing, and it will continue to change even more in the years ahead. Today, farmers are tapping the vast supply of equity or risk capital that is constantly seeking investment opportunities—nonfarm equity capital is being used in agriculture.

TABLE 17–2
WHERE FARMERS BORROW (1989)[1]

Type and Source of Loan	Amount of Loan	Percent of Total
	(million $)	(%)
Real estate mortgage loans:		
Farm Credit System	26,059	35.0
Individuals and others	15,751	21.2
Commercial banks	15,263	20.5
Insurance companies	8,852	11.9
Farmers Home Administration	8,421	11.3
Total	74,346	100.0
Nonreal estate loans:		
Commercial banks	28,595	41.9
Farmers Home Administration	11,792	17.3
Individuals and others	11,760	17.2
Farm Credit System	9,120	13.4
Commodity Credit Corp.	7,000	10.2
Total	68,267	100.0
Total loans	142,613	
Percent real estate		52.1
Percent nonreal estate		47.9

[1]Data provided in a personal communication to the author from George D. Irwin, Deputy Director, Office of Financial Analysis, Farm Credit Administration, McLean, VA.

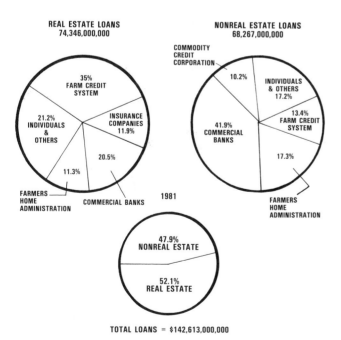

Fig. 17–6. Where farmers borrow (1989). (Source: Farm Credit Administration, Washington, DC)

CREDIT FACTORS CONSIDERED AND EVALUATED BY LENDERS

Potential money borrowers sometimes make their first big mistake by going in "cold" to see a lender, without adequate facts and figures, with the result that, to begin with, they have two strikes against getting the loan.

When considering and reviewing loan requests, the lender tries to arrive at the repayment ability of the potential borrower. Likewise, the borrower has no reason to obtain money unless it will make money.

Lenders need certain basic information in order to evaluate the soundness of a loan request. To this end, the following information should be submitted:

1. **Feasibility study.** Lenders are impressed with borrowers who have a feasibility study showing where they are now, where they are going, and how they expect to get there. In addition to spelling out the goals, this should give assurance of the necessary management skills to achieve them. Such an analysis of the present and projection into the future is imperative in big operations.

2. **The applicant, poultry establishment, and financial statement.** It is the obligation of borrowers and in their best interest, to present the following information to the lender:

    a. **The applicant:**

        (1) Name of applicant and spouse; age of applicant.

        (2) Number of children (minors; legal age).

        (3) Partners in business, if any.

        (4) Years in area.

        (5) References.

    b. **The poultry establishment:**

        (1) Owner or tenant.

        (2) Location; legal description and county, and direction and distance from nearest town.

        (3) Type of enterprise—layer, broiler, market turkey, or whatnot.

    c. **Financial statement.** This document indicates the borrower's financial record and current financial position; potential ahead; and liability to others. Borrowers should always have sufficient slack to absorb reasonable losses due to such unforeseen happenstances as storms, droughts, diseases and poor markets, thereby permitting lenders to stay with them in adversity and to give them a chance to recoup their losses in the future. The financial statement should include the following:

        (1) Current assets:

            (a) Number of birds.

            (b) Feed.

            (c) Machinery.

            (d) Cash. There should be reasonable cash reserves, to cut interest costs, and to provide a cushion against emergencies.

            (e) Bonds or other investments.

            (f) Cash value of life insurance.

        (2) Fixed assets:

            (a) Real property, with estimated value:

               i. Farm or ranch property.

               ii. City property.

               iii. Long term contracts.

(3) Current liabilities:
   (a) Mortgages.
   (b) Contracts.
   (c) Open account—to whom owed.
   (d) Cosigner or guarantor on notes.
   (e) Any taxes due.
   (f) Current portion of real estate indebtedness due.
(4) Fixed liabilities—amount and nature of real estate debt:
   (a) Date due.
   (b) Interest rate.
   (c) To whom payable.
   (d) Contract or mortgage.

3. **Other factors.** Shrewd lenders usually ferret out many things; among them:

   a. **The potential borrower.** Most lenders recognize that the potential borrower is the most important part of the loan. Lenders consider the borrower's—
   (1) Character.
   (2) Honesty and integrity.
   (3) Experience and ability.
   (4) Moral and credit rating.
   (5) Age and health.
   (6) Family cooperation.
   (7) Continuity, or line of succession.

Lenders are quick to sense "high-livers"—those who live beyond their means; the poor managers—the kind who would have made it except for hard luck, and to whom the hard luck happened many times; and the dishonest, lazy, and incompetent. In recognition of the importance of the person back of the loan, "key person" insurance on the owner or manager should be considered by both the lender and borrower.

   b. **Production records.** This refers to a good set of records showing efficiency of production. Such records should show egg production and feed efficiency of layers or breeders, and rate of gain and feed efficiency of meat birds. Lenders will increasingly insist on good records.

   c. **Progress with previous loans.** Has the borrower paid back previous loans plus interest; has the borrower reduced the amount of the loan, thereby giving evidence of progress?

   d. **Profit and Loss (P & L) statement.** This serves as a valuable guide to the potential ahead. Preferably, this should cover the previous 3 years. Also, most lenders prefer that this be on an accrual basis (even if the poultry producer is on a cash basis in reporting to the Internal Revenue Service).

   e. **Physical plant:**
   (1) Is it an economic unit?
   (2) Does it have adequate water?
   (3) Is there adequate diversification?
   (4) Is the right kind of poultry being produced?
   (5) Is the farmstead neat and well kept?

   f. **Collateral (or security):**
   (1) Adequate to cover loan, with margin.
   (2) Quality of security:
     (a) Grade of poultry.

     (b) Type and condition of equipment.
     (c) If feed storage is involved, adequate. protection from moisture and rodents.
     (d) Government participation.
   (3) Identification of security:
     (a) Serial numbers on equipment.

4. **The loan request.** Poultry producers are in competition for money from urban businesses. Hence, it is important that their request for a loan be well presented and supported. The potential borrower should tell the purpose of the loan; how much money is needed, when it is needed, and what it is needed for; the soundness of the venture; and the repayment schedule.

CREDIT FACTORS CONSIDERED BY BORROWERS

Credit is a two-way street; it must be good for both borrowers and lenders. If borrowers are the right kind of persons and on a sound basis, more than one lender will want their business. Thus, it is usually well that borrowers shop around a bit—that they be familiar with several sources of credit and see what they have to offer. There are basic differences in length and type of loan, repayment schedules, services provided with the loan, interest rate, and the ability and willingness of lenders to stick by the borrower in emergencies and times of adversity. Thus, interest rates and willingness to loan are only two of the several factors to consider.

HELPFUL HINTS FOR BUILDING AND MAINTAINING A GOOD CREDIT RATING

Poultry producers who wish to build up and maintain good credit are admonished to do the following:

1. **Keep credit in one place, or in few places.** Generally lenders frown upon "split financing." Borrowers should shop around for a creditor (a) who is able, willing, and interested in extending the kind and amount of credit needed; and (b) who will lend at a reasonable rate of interest; then stay with the borrower.

2. **Get the right kind of credit.** Do not use short-term credit to finance long-term improvements or other capital investments. Also, use the credit for the purpose intended.

3. **Be frank with the lender.** Be completely open and aboveboard. Mutual confidence and esteem should prevail between borrower and lender.

4. **Keep complete and accurate records.** Complete and accurate records should be kept by enterprises. By knowing the cost of doing business, decision-making can be on a sound basis.

5. **Keep annual inventory.** Take an annual inventory for the purpose of showing progress made during the year.

6. **Repay loans when due.** Borrowers should work out a repayment schedule on each loan, then meet payments when due. Sale proceeds should be promptly applied on loans.

7. **Plan ahead.** Analyze the next year's operation and project ahead.

BORROW MONEY TO MAKE MONEY

Poultry producers should never borrow money unless they are reasonably certain that it will make or save money. With this in mind, borrowers should ask, "How much should I borrow?" rather than, "How much will you lend me?"

CALCULATING INTEREST

The charge for the use of money is called *interest*. The basic charge is strongly influenced by the following:

1. The *basic cost* of money in the money market.
2. The *servicing costs* of making, handling, collecting, and keeping necessary records on loans.
3. The *risk* of loss.

Interest rates vary among lenders and can be quoted and applied in several different ways. The quoted rate is not always the basis for proper comparison and analysis of credit costs. Even though several lenders may quote the same interest rate, the effective or simple annual rate of interest may vary widely. The more common procedures for determining the actual annual interest rate, or the equivalent of simple interest on the unpaid balance, follow. (Of course, the going rate of interest should be substituted for the 12% figure used herein.)

1. **Simple or true annual interest on the unpaid balance.** A $1,200 note payable at maturity (12 months) with 12% interest:

Interest paid . . . . .12 × $1,200 = $144
Average use of the money . . . $1,200 for the entire year
Actual rate of interest . . . . . . $\dfrac{\$144 \text{ (interest)}}{\$1,200 \text{ (used for 1 year)}} = 12\%$

2. **Installment loan (with interest on unpaid balance).**[3] A $1,200 note payable in 12 monthly installments with 12% interest on the unpaid balance:

Interest paid ranges from:

First month . . . . $\dfrac{.12 \times \$1,200}{12} = \$12$

to

Twelfth month . . . . $\dfrac{.12 \times 100}{12} = \$1$

Total for 12 months is $78

Average use of the money ranges from $1,200 for the first month down to $100 for the twelfth month, an average of $650 for 12 months.

Effective rate of interest . . . . $\dfrac{\$78}{\$650} = 12\%$

3. **Add-on installment loan (with interest on face amount).** A $1,200 note payable in 12 monthly installments with 12% interest on face amount of loan:

Interest paid . . . . .12 × $1,200 = $144

Average use of the money ranges from $1,200 for the first month down to $100 for the twelfth month, an average of $650 for 12 months.

Effective rate of interest . . . . $\dfrac{\$144}{\$650} = 22.15\%$

4. **Points and interest.** Some lenders now charge *points*. A point is 1% of the face value of the loan. Thus, if 4 points are being charged on a $1,200 loan, $48 dollars will be deducted and the borrower will receive only $1,152. But the borrower will have to repay the full $1,200. Obviously, this means that the actual interest rate will be more than the stated rate. But how much more?

Assume that a $1,200 loan is for 1 year and the annual rate of interest is 12%. Then the payment by the borrower of 4 points would make the actual interest rate as follows:

Interest . . . . .12 × $1,200 = $144
Average use of money . . . . . . $1,152 for one year
Effective rate of interest . . . . . $\dfrac{\$144 \text{ (interest)}}{\$1,152 \text{ (used for 1 year)}} = 12.5\%$

5. **If interest is not stated, use this formula to determine the effective annual interest rate:**

Effective rate of interest =

$$\dfrac{\begin{array}{c}\text{Number of} \\ \text{payment periods} \\ \times 2 \text{ in 1 year}^4\end{array} \times \begin{array}{c}\text{Finance} \\ \text{charges}^5\end{array}}{\begin{array}{c}\text{Balance owed}^6 \times \text{Number of payments} \\ \text{in contract plus 1}\end{array}}$$

For example, a store advertises a refrigerator for $500. It can be purchased on the installment plan for $80 down and monthly payments of $35 for 12 months. What is the actual rate of interest if you buy on the time payment plan?

Effective rate of interest =

$$\dfrac{2 \times 12 \times \$35}{\$420 \times (12 + 1)} = \dfrac{\$840}{\$5,460} = 15.4\%$$

FUTURES TRADING OF BROILERS[7]

Futures trading is not new. It is a well-accepted, century-old procedure used in many commodities for protecting profits, stabilizing prices, and smoothing out the flow of merchandise. For example, it has long been an integral part

---

[3]This method is used for amortized loans.

[4]Regardless of the total number of payments to be made, use 12 if the payments are monthly, use 6 if payments are every other month, or use 2 if payments are semiannual.

[5]Use either the time payment price less the cash price, or the amount you pay the lender less the amount you received if negotiating for a loan.

[6]Use cash price less down payment or, if negotiating for a loan, the amount you receive.

[7]This entire section on "Futures Trading of Broilers," along with the subsection on "Options Trading of Broilers," was authoritatively reviewed by and helpful suggestions were received from James Graham, Director, Commodity Marketing Education, Chicago Mercantile Exchange, Chicago, IL.

Fig. 17–7. Action on the trading floor of the Chicago Mercantile Exchange. More than 100 million contracts were traded on the floor of the Chicago Mercantile Exchange in 1990. (Courtesy, Chicago Mercantile Exchange)

of the grain industry; grain elevators, flour millers, feed manufacturers, and others have used it to protect themselves against losses due to price fluctuations.

Today, broiler chicken futures and options are traded on the Chicago Mercantile Exchange. Many poultry integrators, further processors, and distributors prefer to forego the possibility of making a high speculative profit in favor of earning a normal margin or service charge through efficient operation of their business. They look to futures and options markets to provide (1) an insurance medium in the marketing field, and (2) the facilities and machinery for underwriting price risks.

Futures trading is a place where buyers and sellers meet on an organized market and transact business on paper, without the physical presence of the commodity. The exchange neither buys nor sells; rather, it provides the facilities, establishes rules, serves as a clearinghouse, holds the margin money deposited by both buyers and sellers, and guarantees all contracts. Buyers and sellers either trade on their own account or are represented by brokerage firms.

The unique characteristic of broiler chicken futures is that they are cash-settled to the USDA 12-City composite, weighted average price on the second to the last Friday of the contract month. In practice, however, very few contracts are held until the last day. The vast majority of them are canceled by offsetting transactions made before the last day.

Many poultry producers have long contracted for future delivery without the medium of an exchange. They contract to sell and deliver to a buyer a certain number and kind of eggs, broilers, or turkeys at an agreed upon price and place. Hence, the risk of loss from a decrease in price after the contract is shifted to the buyer; and, by the same token, the seller foregoes the possibility of a price rise. In reality, such contracting is a form of futures trading. Unlike futures trading on an exchange, however, actual delivery of the product is a must. Also, such privately arranged contracts are not always available, the terms may not be acceptable, and the only recourse to default on the contract is a lawsuit.

The following example will illustrate futures hedging procedures.

Example: In early August, a broiler producer recognizes that prices are high, and even though costs also are high, production is profitable. The producer believes his/her costs of 56.25¢ a pound will continue but that the local price of 59.25¢ a pound will drop by October to a level that will not cover all costs. October futures are selling at 62.5¢ a pound, or a localized price of 61.5¢ a pound. Locational basis totals 1.0¢ a pound. The producer decides to hedge, to lock in the current price, by selling October futures.

Cash Market	Basis	Futures Market
August 3 Expected October price of 61.5¢/lb	−1.00	Sells October futures at 62.5¢/lb
October 14 Sells broilers in cash market at 54.75¢/lb	−1.00	Buys October futures at 55.65¢/lb

54.75¢ + 6.75¢ profit = 61.5¢/lb net

Even though the market declined as expected, this producer was able to offset the loss in the cash market and to maintain the localized price of 61.5¢ a pound. The producer's price was obtained at 54.75¢ a pound from the cash market plus a 6.75¢ a pound gain in the futures market, or 61.5¢ a pound.

Table 17–3 presents, in summary form, pertinent information relative to commodity futures trading of broilers on the Chicago Mercantile Exchange.

TABLE 17–3
BROILER CHICKEN TRADING ON THE CHICAGO MERCANTILE EXCHANGE[1]

Commodity	Trading Hours (Mon.–Fri.)	Contract Size	Major Delivery Months	Allowable Positions (No. of Contracts an Individual Can Hold— Long & Short)	Maximum Daily Price Fluctuations (Value of Price Fluctuation)[2]	Initial Margins
Broilers (cash-settled USDA 12-City average)	9:10 a.m.—1:00 p.m.	40,000 lb *(18,160 kg)*	February April May June July August October December	2,000	Minimum price fluctuations: $0.00025/lb or $10.00/contract. Daily trading limits: $0.02/lb ($800/contract).	$540

[1]Data provided to the author by Chicago Mercantile Exchange, Chicago, IL.

[2]Up or down from close of previous trading season.

OPTIONS TRADING OF BROILERS

Unlike futures, which lock in one price for a hedger, options help the producer manage against price risk, but leave open price opportunity. A *put option* is the right to sell futures and a *call option* is the right to buy futures. In payment for these rights a producer pays a risk premium. There are no margins with options.

For example, a producer may set a minimum selling price for broilers by buying a put option. If prices fall, the increased value of the put, captured by selling it back, will help offset lower cash sale price. If prices rise, the producer will let the option expire and lose the premium paid, but will be able to sell broilers at higher prices.

Thus, a put option sets a floor, but no ceiling, on a producer's broiler selling prices. On the other hand, a restaurant or chain may want to use a broiler chicken call option to set a ceiling, but no floor on broiler purchase costs.

RECORDS AND ACCOUNTS

The key to good business and management is records. The historian, Santayana, put it this way, "Those who are ignorant of the past are condemned to repeat it." Also, good records help the poultry producer to overcome the banker's traditional fear of feathers.

WHY KEEP RECORDS?

The chief functions of records and accounts are:

1. To provide information from which the poultry business may be analyzed, with its strong and its weak points ascertained. From the facts thus determined, the operators may adjust current operations and develop a more effective plan of organization.
2. To provide profit and loss statements.
3. To provide a net worth statement, showing financial progress during the year.
4. To provide cash flow projections, showing when money is needed, and showing loan repayability.
5. To furnish an accurate, but simple, net income statement for use in filing tax returns.
6. To keep production records on birds.
7. To aid in making a credit statement when a loan is needed.
8. To keep a complete historical record of financial transactions for future reference.

Good records, properly analyzed and used, will increase net earnings and serve as a basis for sound management and husbandry.

KIND OF RECORD AND ACCOUNT BOOK

The record forms will differ somewhat according to the type of enterprise. For example, with layers, cost per dozen eggs is the important thing, whereas in broiler and market turkey production, it is cost per pound of bird. Net returns are important, but it is also necessary that records show all the items of cost and income—egg production, feed consumption, and mortality of layers; rate of growth, pounds of chicken per 100 lb of feed, mortality, and quality of broilers produced.

Poultry producers can make their own record book by ruling off the pages of a bound notebook to fit their specific needs, but the saving is negligible. Instead, it is recommended that they obtain a copy of a record book prepared for and adapted to their business. Such a book may usually be obtained at a nominal cost from the agricultural economics department of each state college of agriculture. Also, certain commercial companies distribute very acceptable record and account books at no cost.

KIND OF RECORDS TO KEEP

Most record and account books contain simple and specific instructions relative to their use. Accordingly, it is neither necessary nor within the realm of this book to provide such instructions. Instead, the comments made herein are restricted to the kind of records to keep.

At the outset, it should be recognized that the records should be easy to keep and should give the information desired to make a valuable analysis of the business. In general, the functions enumerated under the earlier section entitled "Why Keep Records" can be met by the following kinds of records:

1. **Annual inventory.** The annual inventory is the most valuable record that poultry producers can keep. It should include a list and value of real estate, poultry, equipment, feed, supplies, and all other property, including cash on hand, notes, and bills receivable. Also, it should include a list of mortgages, notes, and bills payable. It shows the producers what they own and what they owe; whether they are getting ahead or going behind. The following pointers may be helpful relative to the annual inventory.

a. **Time to take inventory.** The inventory should be taken at the beginning of the account year; usually this means December 31 or January 1.

b. **Proper and complete listing.** It is important that each item be properly and separately listed.

c. **Method of arriving at inventory values.** It is difficult to set up any hard and fast rule to follow in estimating values when taking inventories. Perhaps the following guides are as good as any:

(1) **Real estate.** Estimating the value of farm real estate is, without doubt, the most difficult of all. It is suggested that the owner use either (a) the cost of the farm, (b) the present sale value of the farm, or (c) the capitalized rent value according to its productive ability with an average operator.

(2) **Buildings.** Buildings are generally inventoried on the basis of cost less observed depreciation and obsolescence. Once the original value of a building is arrived at, it is usually best to take depreciation on a straight line basis by dividing the original value by the estimated life in terms of years. Usually 4% or more depreciation is charged off each year for income tax purposes.

(3) **Birds.** Birds are usually not too difficult to inventory because there are generally sufficient current sales to serve as a reliable estimate of value.

(4) **Equipment.** The inventory value of equipment is usually arrived at by either of two methods: (a) the original cost less a reasonable allowance for depreciation each year, or (b) the probable price that it would bring at a well-attended auction.

Under conditions of ordinary wear and reasonable care, it can be assumed that the general run of equipment will last about 5 years. Thus, with new equipment, the annual depreciation will be the original cost divided by 5.

(5) **Feed and supplies.** The value of feed and supplies can be based on market price.

Two further points are important. Whatever method is used in arriving at inventory value (a) should be followed at both the beginning and the end of the year, and (b) should reflect the operator's opinion of the value of the property involved.

2. **Record of receipts and expenses.** Such a record is essential to any type of well-managed business. To be most useful, these entries should not only record the amount of the transaction, but should give the source of the income or the purpose of the expense, as the case may be. In other words, they should show the producer from what sources the income is derived and for what it is spent.

The following kinds and arrangements of farm record books are commonly used for recording receipts and expenditures:

a. Those that devote a separate page to each enterprise; that is, a separate page is used for the layer enterprise, another for replacement pullets, still another for crops (if there are such), and so on.

b. Those that provide for a record of receipts and expenses on the same page, using one column for receipts and another for expenses. This type is easy to keep, but very difficult to analyze from the standpoint of any particular enterprise.

c. Those that combine the features of both "a." and "b." above. The latter are more difficult to keep than the others, and may be confusing to the person keeping the record.

Household and personal accounts should be kept, but should be handled entirely separate from the poultry enterprise accounts because they are not farming expenses as such.

3. **Record of poultry and crop production.** A record of the production and sale of eggs and meat, and of the yield of crops (if crops are grown) is most important, for the success of the farm depends upon production. Such records help in analyzing the farm business. They may be few or many, depending upon the wishes of the operator.

SUMMARIZING AND ANALYZING THE RECORDS

At the end of the year, the second or closing inventory should be taken, using the same method as was followed in taking the initial inventory. The final summary should then be made, following which the records should be analyzed. In the latter connection, producers should remember that the purpose of the analysis is not to prove that they have or have not been prosperous. They probably know the answer to this question already. Rather, the analysis should show actual conditions on the farm and point out ways in which these conditions may be improved.

Although producers can summarize and analyze their own records, there are many advantages in having the services of a specialist for this purpose. Such a specialist is in a better position to make a "cold" appraisal without prejudice, and to compare enterprises with those of other similar operators. Thus, the specialist may discover that, in comparison with other operators, the broilers on a given farm are requiring too much feed to make a pound of gain, or that the layer enterprise is much less profitable than others have experienced. The local county agent can either render or recommend such specialized assistance. In some areas, it may consist in joining a cooperative farm record group or engaging the services of a consultant; in some states, such service is provided by the state agricultural college.

COMPUTERS IN THE POULTRY BUSINESS

Accurate and up-to-the-minute records and controls have taken on increasing importance in all agriculture, including the poultry business, as the investment required to engage therein has risen and profit margins have narrowed. Today's successful poultry producers must have, and use, as complete records as any other business. Also, records must be kept current; it no longer suffices merely to know the bank balance at the end of the year.

Big and complex poultry enterprises have outgrown hand record keeping. It is too time-consuming, with the result that it does not allow management enough time for planning and decision making. Additionally, it does not permit an all-at-once consideration of the complex interrelationships which affect the economic success of the business. This has prompted a new computer technique known as linear programming.

Linear programming is similar to budgeting, in that it compares several plans simultaneously and chooses from among them the one likely to yield the highest returns. It is a way in which to analyze a great mass of data and consider many alternatives. It is not a managerial genie; nor will it replace decision-making managers. However, it is a modern and effective tool in the present age, when just a few cents per broiler or turkey, or per dozen eggs, can spell the difference between profit and loss.

There is hardly any limit to what computers can do if fed the proper information. Among the difficult questions that they can answer for a specific poultry enterprise are:

1. **How is the entire operation doing so far?** It is possible to obtain a quarterly or monthly progress report; often making it possible to spot trouble before it is too late.

2. **What enterprises are making money; which ones are freeloading or losing?** By keeping records by enterprises – broilers, layers, replacement pullets, etc. – it is possible to determine strengths and weaknesses; then either to rectify the situation or shift labor and capital to a more profitable operation. Through enterprise analysis, some operators have discovered that one part of the business may earn $10 or more per hour for labor and management, whereas another may earn only $5 per hour, and still another may lose money.

3. **Is each enterprise yielding maximum returns?** By having profit, or performance, indicators in each enterprise, it is possible to compare these (a) with the historical average of the same poultry farm or (b) with the same indicators of other similar establishments.

4. **How does this poultry enterprise stack up with its competition?** Without revealing names, the computing center (local, state, area, or national) can determine how a given poultry enterprise compares with others—either the average, or the top (say, 5%).

5. **How to plan ahead?** By using projected prices and costs, computers can show what moves to make for the future—they can be a powerful planning tool. They can be used in determining when to buy feed, purchase chicks, market meat birds, etc.

6. **How can income taxes be cut to the legal minimum?** By keeping accurate record of expenses and figuring depreciations accurately, computers make for a saving in income taxes on most poultry establishments.

7. **What are the least cost and highest net return rations?** Instruction on how to balance a ration by computer is given in this book in Chapter 7, Poultry Feeding Standards, Ration Formulation, and Feeding Programs.

For providing answers to these questions and many more, computer accounting costs an average of about 1% of the gross poultry farm income. By comparison, it is noteworthy that city businesses pay double this amount.

There are three requisites for linear programming a poultry establishment:

1. Access to a computer.
2. Computer know-how, so as to set the program up properly and be able to analyze and interpret the results.
3. Good records.

The pioneering computer services available to poultry producers were operated by universities, trade associations, and government—most of them were on an experimental basis. Subsequently, others have entered the field, including commercial data processing firms, banks, machinery companies, feed and fertilizer companies, and farm suppliers. They are using it as a "service sell," as a replacement for the days of "hard sell."

BUDGETS IN THE POULTRY BUSINESS

A budget is a projection of records and accounts and a plan for organizing and operating ahead for a specified period of time. A short-time budget is usually for 1 year, whereas a long-time budget is for a period of years. The principal value of a budget is that it provides a working plan through which the operation can be coordinated. Changes in prices, droughts, and other factors make adjustments necessary. But these adjustments are more simply and wisely made if there is a written budget to use as a reference.

HOW TO SET UP A BUDGET

It is unimportant whether a printed form (of which there are many good ones) is used or one made up on an ordinary ruled 8½ in. × 11 in. sheet placed sidewise. The important things are (1) that a budget is kept, (2) that it be on a monthly basis, and (3) that the operator be "comfortable" with whatever forms or system used.

No budget is perfect. But it should be as good an estimate as can be made—despite the fact that it will be affected by such things as droughts, diseases, markets, and many other unpredictables.

A simple, easily kept, and adequate budget can be evolved by using forms such as Tables 17–4, 17–5, and 17–6.

The annual cash expense budget (Table 17–4) should show the monthly breakdown of various recurring items—everything except the initial loan and capital improvements. It includes labor, feed, supplies, fertilizer, taxes, interest, utilities, etc.

The annual cash income budget (Table 17–5, next page) is just what the name implies—an estimated cash income by months.

The annual cash expense and income budget (Table 17–6, next page) is a cash flow budget, obtained from the first two forms. It is a money "flow" summary by months. From this, it can be ascertained when, and how much, money will need to be borrowed, and the length of the loan along with a repayment schedule. It makes it possible to avoid tying up capital unnecessarily, and to avoid unnecessary interest.

TABLE 17–4
ANNUAL CASH EXPENSE BUDGET

_____ for 19____
(name of poultry establishment)

Item	Total	Jan.	Feb.	Mar.	Apr.	May	June	July	Aug.	Sept.	Oct.	Nov.	Dec.
Chicks purchased													
Labor hired													
Feed purchased													
Gas, fuel, grease													
Taxes													
Insurance													
Interest													
Utilities													
etc.													
Total													

TABLE 17–5
ANNUAL CASH INCOME BUDGET

_____ for 19___

(name of poultry establishment)

Item	Total	Jan.	Feb.	Mar.	Apr.	May	June	July	Aug.	Sept.	Oct.	Nov.	Dec.
Birds													
Eggs													
etc.													
etc.													
etc.													
Total													

TABLE 17–6
ANNUAL CASH EXPENSE AND INCOME BUDGET (CASH FLOW)

_____ for 19___

(name of poultry establishment)

Item	Total	Jan.	Feb.	Mar.	Apr.	May	June	July	Aug.	Sept.	Oct.	Nov.	Dec.
Gross income	100,000	12,000	7,000	7,000	7,000	7,000	7,000	12,000	7,000	7,000	7,000	7,000	13,000
Gross expenses	60,000	5,000	6,000	4,000	5,000	5,000	6,000	4,000	5,000	5,000	6,000	4,000	5,000
Difference	40,000	7,000	1,000	3,000	2,000	2,000	1,000	8,000	2,000	2,000	1,000	3,000	8,000
Surplus (+) or Deficit (–)	+	+	+	+	+	+	+	+	+	+	+	+	+

HOW TO FIGURE NET INCOME

Table 17–6 gives a gross income statement. There are other expenses that must be taken care of before net profit is determined; namely:

1. **Depreciation on buildings and equipment.** It is suggested that the *useful life* of buildings and equipment be as follows, with depreciation accordingly: buildings, 31½ years; and machinery and equipment, 5 years. Sometimes, a higher depreciation, or amortization, is desirable because it produces tax savings, and is protection against obsolescence due to scientific and technological developments.

2. **Interest on owner's money invested in farm and equipment.** This should be computed at the going rate in the area, say 12%.

Here is an example of how the above works:

Let's assume that on a given poultry establishment there was an annual gross income of $250,000 and a gross expense of $180,000, or a surplus of $70,000. Let's further assume that there are $20,000 worth of equipment, $50,000 worth of buildings, and $175,000 of the owner's money invested in farm and equipment. Here is the result:

Gross profit	. . . .		$70,000
Depreciation:			
Equipment	$ 20,000 @ 20%	= $ 4,000	
Buildings	$ 50,000 @ 3.2% =	1,600	
		$ 5,600	
Interest	$175,000 @ 12% =	21,000	
			$26,600
Return to labor and management	. . . .		$43,400

Some people prefer to measure management by return on invested capital, and not wages. This approach may be accomplished by paying management wages first, then figuring return on investment.

ENTERPRISE ACCOUNTS

Where a poultry enterprise is diversified (for example, a farm having layers, replacement pullets, and crops), enterprise accounts should be kept—in this case three different accounts for three different enterprises. The reasons for keeping enterprise accounts are:

1. It makes it possible to determine which enterprises have been most profitable, and which least profitable.

2. It makes it possible to compare a given enterprise with competing enterprises of like kind, from the standpoint of ascertaining comparative performance.

3. It makes it possible to determine the profitableness of an enterprise at the margin (the last unit of production). This will give an indication as to whether to increase the size of a certain enterprise at the expense of an alternative existing enterprise when both enterprises are profitable in total.

ANALYZING A POULTRY BUSINESS

What will it cost, and how much can I make? This two-pronged question is on the mind of every person or company when contemplating entering some phase of the poultry business, regardless of the size of enterprise planned.

Are my costs and returns in line? Some producers, especially farm flock owners, would be hard pressed to answer this question. Yet, this information is needed in order for producers to determine how well they are doing.

Tables 17–7 to 17–13 provide performance and efficiency standards and represent attainable goals for units of efficient size. The figures presented in these tables may be easily converted to a per bird or per 1,000-bird basis. However, scale of operation influences investment and both overhead and operating costs, and smaller scale operations may have higher cost per bird than these standards. Recognition of the latter point is especially important when the beginner uses these figures for budget purposes. Except where otherwise indicated, the costs and returns in these tables are based on the prices that prevailed in years shown.

Because costs and returns of poultry production differ from year to year (for example, costs rise with inflation) and from area to area (for example, labor costs and contract stipulations relative to layers, broilers, and turkeys differ between areas), new standards should be developed for each area and period of time. Nevertheless, Tables 17–7 to 17–13 can easily be adapted and will serve as useful guides or goals.

Cost and return figures for each of the following types of poultry operations are presented in the sections that follow:

Table 17–7. Egg Costs and Returns.

Table 17–8. Cost Per Dozen Eggs Produced, A Percentage Breakdown.

Table 17–9. Cost Per Replacement Pullet At 20 Weeks of Age.

Table 17–10. Broiler Costs and Returns.

Table 17–11. Cost Per Pound of Broiler Produced, A Percentage Breakdown.

Table 17–12. Turkey Costs and Returns.

Table 17–13. Cost Per Pound of Turkey Produced, A Percentage Breakdown.

It is noteworthy that Tables 17–7 to 17–13 underscore the economic importance of feed for poultry, for it is apparent that 55 to 75% of their total production cost is for feed, with the production of eggs toward the lower side of this range and the production of broilers and turkeys toward the upper side. For this reason, the efficient use of feed is extremely important to poultry producers.

LAYER COSTS AND RETURNS

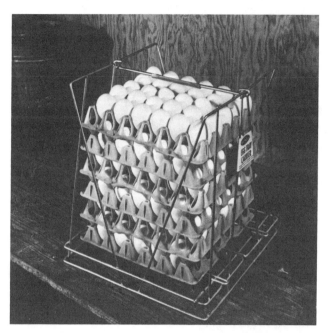

Fig. 17–8. Layer costs and returns are determined primarily by (1) number and cost of eggs produced, and (2) price received.

Tables 17–7, 17–8, and 17–9 may be used as standards by layer operators for determining (1) how well they are doing, and (2) their strengths and weaknesses.

TABLE 17–7
EGG COSTS AND RETURNS[1]

| Year/Quarter | Production Costs | | Wholesale | | Net Returns |
	Feed	Total	Total Costs[2]	Price[3]	
	(¢/doz)	(¢/doz)	(¢/doz)	(¢/doz)	(¢/doz)
1989:					
I	32.8	51.0	71.5	82.8	11.3
II	32.2	50.4	70.9	76.1	5.2
III	31.0	49.2	69.7	85.2	15.5
IV	28.3	47.0	67.0	96.1	28.6
Year	31.2	49.4	69.9	85.1	15.2
1990:					
I	27.6	45.9	66.3	90.8	24.4
II	29.6	47.8	68.3	76.8	8.6
III	30.0	48.2	68.7	79.3	10.6
IV	27.3	45.5	66.0	88.6	22.6
Year	28.6	46.9	67.3	83.9	16.6

[1]Adapted by the author from *Livestock and Poultry, Statistics and Outlook Report*, USDA, ERS, LPS-46, Feb., 1991, p. 21, Table 28.

[2]Based on farm cost converted to wholesale market value.

[3]Wholesale prices used are the 12-metro egg price.

Considerable capital is needed to start and operate a modern commercial egg operation. It is unusual to start an enterprise with fewer than 50,000 birds. For the house and equipment, this may require an investment of $6.00 to $10.00 per bird, or a total outlay of $300,000 to $500,000. The 20-week-old started pullets may cost as much as $2.75 per bird, or $137,500. Additionally, operating capital for feed and supplies will be needed until the flock peaks in production. Also, financing will be needed for the land; and housing and salary must be provided for the operator. Some of these capital needs may be alleviated or lessened through a contract arrangement.

In 1990, it cost efficient U.S. egg producers approximately 47¢ to produce a dozen eggs, with an estimated percentage breakdown as shown in Table 17–8.

TABLE 17–8
COST PER DOZEN EGGS PRODUCED,
A PERCENTAGE BREAKDOWN[1]

Item	Percent
	(%)
Feed	58.17
Pullet (replacement) depreciation	22.37
House and equipment, including interest	13.43
Labor	2.68
Utilities	1.12
Insurance	0.89
Taxes	0.67
Maintenance	0.67
Total	100.00

[1]Sheppard, C. C. and S. F. Ridlen, "Planning for Commercial Egg Production." Mr. Sheppard is Poultry Extension Specialist, Michigan State University, and Mr. Ridlen is Poultry Extension Specialist, University of Illinois.

Based on Tables 17–7 and 17–8, along with other studies and experiences, the following conclusions relative to egg costs and returns appear to be justified:

1. **Major production costs.** The highest cost items are feed, pullets, house and equipment, and labor. Normally, these account for about 96% of the cost of production.

2. **Feed is largest cost.** Feed is the largest single cost item, accounting for 50 to 60% of the total cost of producing eggs, although in exceptional cases it may range from as low as 45% to as high as 65%.

3. **Bird depreciation is high.** Bird depreciation (pullet replacement) ranks second in the cost of producing eggs. Depreciation refers to the difference between the value of a pullet at the beginning of the year, and the value of the same bird at the end of the year. Thus, if a pullet is worth $2.75 at the beginning of the year, and will bring only 75¢ after 12 months of laying, she has cost the producer $2.00 just to own her for a year, or she has depreciated $2.00 in value.

4. **Housing and equipment are costly.** The largest and most permanent investment in egg production is the house and equipment. Even when they are depreciated over a period of years, it will likely be the third largest cost item in producing a dozen eggs. The decision to save costs on housing and equipment is more easily made before one invests in them than later.

5. **Labor comes high.** Labor is usually the fourth largest item of cost of producing eggs. Egg cost per dozen eggs, or eggs produced per hour of labor or per year, can be used as an index of labor efficiency. Greater labor efficiency in layer operations is being achieved in two ways; (1) by increasing the size flock, and (2) by automating.

6. **Market eggs chief income.** Market eggs are by far the chief source of income from layers. Increased production of 24 eggs per bird in a 50,000-bird unit will mean $50,000 (at 50¢ per dozen) more income per flock cycled. Yet costs will be increased very little. With higher production, the birds will eat a little more feed, but they will not require any more space, lights, or pullet replacement cost.

7. **Hazards.** Profits from egg production are subject to the hazards of overproduction with the consequent low prices for eggs; level of feed prices; disease; and changing egg consumption patterns.

8. **Business aspects.** Success in a commercial layer operation depends upon good business management, in addition to hard work and following good practices of care, feeding, and disease control. Also, it must be remembered that the egg industry has peaks and valleys; hence, the egg producer must be prepared to weather the rough times that come sooner or later.

9. **Goals.** Some high, yet achievable, goals for the layer operator are:

Item	Production Goals
Egg production/hen/year, No.	270
Feed/dozen eggs, lb (under)	3.6
Mortality/year, % (under)	10
Grade A eggs, % (over)	75
Marketable eggs, % (over)	95
On-farm breakage, % (under)	2

REPLACEMENT PULLET COSTS AND RETURNS

Table 17–9 may be used as a standard by replacement pullet producers for determining (1) how well they are doing, and (2) their strengths and weaknesses.

Table 17–9 is based on a survey of 21 California pullet growers in the mid-1980s. As shown, the average cost of growing a pullet to 20 weeks of age was $2.59, with a range from $2.29 to $2.89. The average and range of costs for various items appear in Table 17–9. Mortality for the 20-week period averaged 4.53%, and ranged from 2.5% to 6%.

Based on Table 17–9, along with other studies and experiences, the following conclusions relative to replacement pullet costs and returns appear to be justified:

1. **Major production costs.** The principal items of cost in raising replacement pullets, ranked in order, are feed, chicks, and labor. Feed generally runs more than twice the cost of chicks. As shown in Table 17–9, the cost of raising replacement pullets in California is about $2.59.

TABLE 17–9
COSTS PER REPLACEMENT PULLET AT 20 WEEKS OF AGE[1]

Item	Average	Range
	($)	($)
Variable costs:		
Feed	1.257	1.04–1.42
Chicks	0.515	0.46–0.56
Fuel	0.038	0.01–0.08
Utilities	0.040	0.01–0 .11
Medicine	0.006	0–0.04
Vaccine	0.060	0.02–0.16
Labor	0.281	0.14–0.50
Management	0.039	0.01–0.11
Taxes	0.013	0–0.04
Trucking	0.045	0–0.13
Water	0.006	0–0.02
Misc.	0.036	0–0.16
Total	2.336	2.00–2.72
Fixed Costs:		
Housing and equipment	0.108	0.04–0.30
Interest	0.135	0.07–0.27
Land	0.011	0.11–0.57
Total	0.254	0.11–0.40
Total all costs	2.590	2.29–2.89

[1]Adapted by the author from *California Poultry Letter*, November, 1984.

2. **Many commercial egg producers buy, rather than raise, started pullets.** Today, many commercial egg producers buy started pullets, rather than raise them. Such pullets are usually 20 to 22 weeks of age and ready to lay, with a few of them laying at that age. The birds are placed in the laying house for 12 to 14 months of egg production.

3. **Replacing layers at 17 to 20 months usually makes for most profit.** It is usually more profitable to re-place layers when they are 17 to 20 months of age than to keep them longer, because young birds lay at a higher rate and produce eggs of better quality than older hens.

4. **Each 20¢ extra cost on pullets adds 1¢/dozen egg cost.** It is noteworthy that every 20¢ extra cost per pullet makes for 1¢ a dozen increase in the cost of production.

BROILER COSTS AND RETURNS

Tables 17–10 and 17–11 may be used as standards by broiler producers for determining (1) how well they are doing, and (2) their strengths and weaknesses.

Only through substantial increases in production effi-ciency have broiler producers been able to make a profit in recent years. Improvement in broiler strains, along with im-proved technology and production practices, have reduced the time needed to produce a 3- to 5-lb live bird from 12 to 14 weeks 25 years ago to 6 to 8 weeks now. Poultry nutri-tionists have developed rations that have reduced the amount of feed required to produce a pound of live bird from

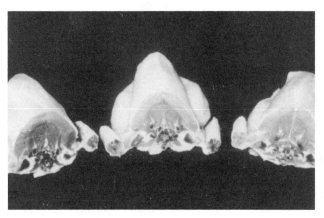

Fig. 17–9. The processed yield of broilers is the most important factor when calculating bottom line profits. (Courtesy, Cobb-Vantress, Inc., Siloam Springs, AR)

4.7 lb in 1940 to 1.9 lb in 1990. Improvements in housing, along with advancements in feeding and watering equip-ment, have led to reduced labor requirements and better environmental control (light, temperature, humidity, and air movement). Improvements in production facilities, feed for-mulations, broiler strains, and medicines and vaccines have also reduced mortality from the 10 to 20% typical of 25 years ago to 4% in 1990.

Actual costs per pound of broiler produced will vary by season and geographical area. In a 1991 report, the U.S. Department of Agriculture gave the broiler costs and returns for 1989 and 1990 as presented in Table 17–10. An estimated percentage breakdown of the costs per pound of broiler produced is given in Table 17–11 (next page).

TABLE 17–10
BROILER COSTS AND RETURNS[1]

Year/Quarter	Production Costs		Wholesale		Net Returns
	Feed	Total	Total Costs[2]	Price[3]	
	(¢/lb)	(¢/lb)	(¢/lb)	(¢/lb)	(¢/lb)
1989:					
I	19.1	27.1	50.6	59.5	8.9
II	18.6	26.6	49.9	67.3	17.4
III	18.2	26.2	49.4	59.6	10.2
IV	16.8	24.8	47.5	49.8	2.3
Year	18.2	26.2	49.4	59.0	9.7
1990:					
I	15.7	23.7	46.0	56.5	10.5
II	15.8	23.8	46.1	56.6	10.5
III	16.8	24.8	47.4	57.2	9.7
IV	15.8	23.8	46.1	48.8	2.7
Year	16.0	24.0	46.4	54.8	8.4

[1]Adapted by the author from *Livestock and Poultry, Statistics and Outlook Report*, USDA, ERS, LPS-46, Feb., 1991, p. 21, Table 28.

[2]Based on farm cost converted to wholesale market value.

[3]Wholesale prices used are the 12-City weighted average broiler price.

TABLE 17–11
COSTS PER POUND OF BROILER PRODUCED,
A PERCENTAGE BREAKDOWN[1]

Item	Total Cost
	(%)
Feed	72.0
Chicks	14.0
Grower payment (contract)	10.0
Fuel	1.0
Medication	1.0
Litter	0.5
Other	1.5
Total	100.0

[1]Benson, V. W. and T. J. Witzig, *The Chicken Broiler Industry: Structure, Practices, and Costs*, Ag. Econ. Rep. No. 381, USDA.

Based on Tables 17–10 and 17–11, along with other studies and experiences, the following additional conclusions relative to broiler production costs and returns appear to be justified:

1. **Chief costs.** The chief broiler production costs, ranked in order, are: feed, chicks, and labor, or grower—all short-term capital.

2. **Volume and costs vary seasonally.** Broilers are grown in all seasons, but volume and production costs vary seasonally. Seasonal volume is related to seasonal demand (the high months are May through October, and the low months November through April), and seasonal production costs are caused by variation in energy and feed costs, both of which are highest in the winter months.

3. **Broilers highly integrated.** Approximately 95% of all broilers are produced under contract, and integrator-owned farms produce another 5%; hence, independent production accounts are practically nil.

The traditionally higher feed costs in colder areas and colder months have diminished with the adoption of environmentally controlled housing. But rising energy costs have tended to counterbalance the saving in feed, with the result that broiler producers are faced with the practical question of whether it is cheaper to provide some artificial heat in order to lessen feed consumption, or to provide more feed.

4. **Broilers dominate chicken consumption.** Broiler meat as a share of total chicken consumption increased from 3.7% in 1934 to 97% in 1989. (In 1934, "other" chickens than broilers consisted of cockerels, pullets raised for marketing as young birds, and fowl sold from egg-producing flocks. Today, "other" chickens consist chiefly of fowl and roasters.)

5. **Broiler producer's share.** The broiler producer's share of the consumer's dollar increased from 42.5% in 1970 to a high of 56.4% in 1973, then narrowed to 48% in 1989.

6. **Age and weight.** Broiler growers must also make decisions as to the age and weight at which to sell their birds. Growers in the Northeast receive price premiums for heavier weights, whereas lower liveweights are traditional in the South. But more feed is required to produce a pound of broiler at heavier weights. In support of this statement, the University of Arkansas gives the following feed conversion figures:

Average Liveweight per Bird (lb)	Lb Feed per Lb Liveweight of Bird (lb)
2.50	1.89
3.10	2.02
3.75	2.13
4.40	2.27

7. **Labor efficiency.** Efficiency of labor is an important factor in determining profits in the broiler business. Generally speaking, the more broilers raised per worker the lower the production costs, and the bigger the operation the more feasible it is to automate. Greater labor efficiency has been a primary force back of increased number of broilers per farm. Output of broilers per farm increased from 33,600 birds in 1959 to 157,785 birds in 1989.

8. **Grower fees.** Broiler grower fees for labor and facilities range from 3¢ to 5¢ per pound of broiler marketed.

9. **Goals.** Some high, yet achievable, goals for the broiler producer are:

Item	Production Goals
Weight at 6 weeks of age, lb	4.5
Feed conversion, lb feed/lb broiler	1.85
Mortality, % (under)	1.0

TURKEY COSTS AND RETURNS

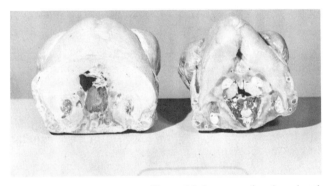

Fig. 17–10. Type made the difference! *Left*: cross section of new broad-breasted turkey carcass. *Right*: cross section of old type turkey carcass. (Courtesy, National Turkey Federation)

Tables 17–12 and 17–13 may be used as standards by turkey producers for determining (1) how well they are doing, and (2) their strengths and weaknesses.

Increased turkey consumption—from 1.7 lb per capita in 1935 to 17.1 lb in 1989—has resulted in part from the promotion of turkey as a food throughout the year, rather than just on holidays. Per capita consumption of turkey is projected to continue increasing, with a large share of the increase being in cut-up parts and in further processed products such as turkey rolls, roasts, pot pies, and frozen dinners.

The price of turkeys decreased from a high of 46.8¢ per pound in 1948 to a low of 18.9¢ per pound in 1961. In 1989, the average price per pound liveweight was 41.1¢.

During the period of unfavorable market prices and rising costs, processors turned to contracting in order to schedule processing better and to lower costs, and feed manufacturers promoted contracting to expand feed sales. In 1990, 90% of the nation's turkeys were produced under contract, up from the estimated 30% in 1960.

The cost of producing turkeys varies from year to year and to a lesser extent between geographical areas. Table 17–12 shows turkey costs and returns for 1989 and 1990. Table 17–13 gives the author's estimated percentage breakdown of the cost of producing market turkeys by independent producers (not under contract) in 1990.

TABLE 17–12
TURKEY COSTS AND RETURNS[1]

| Year/Quarter | Production Costs | | Wholesale | | Net Returns |
	Feed	Total	Total Costs[2]	Price[3]	
	(¢/lb)	(¢/lb)	(¢/lb)	(¢/lb)	(¢/lb)
1989:					
I	27.9	41.6	68.3	61.6	–6.7
II	27.5	41.2	67.8	71.3	3.5
III	26.4	40.1	66.4	64.5	–1.9
IV	25.4	39.1	65.2	66.0	0.8
Year	26.7	40.4	66.8	66.0	–0.8
1990:					
I	23.1	36.8	62.3	55.6	–6.7
II	22.5	36.2	61.5	61.6	0.0
III	24.2	37.9	63.6	66.7	3.1
IV	23.6	37.3	62.9	66.7	3.8
Year	23.4	37.1	62.6	63.1	0.4

[1]Adapted by the author from *Livestock and Poultry, Statistics and Outlook Report,* USDA, ERS, LPS-46, Feb., 1991, p. 21, Table 28.

[2]Based on farm cost converted to wholesale market value.

[3]Wholesale prices used are the weighted average of 8–16 lb young hens and 14–22 lb toms in Central, Western and Eastern Regions.

TABLE 17–13
COSTS PER POUND OF TURKEY PRODUCED,
A PERCENTAGE BREAKDOWN[1]

Item	Total Cost
	(%)
Feed	70
Poults	8
Fixed costs (land, buildings and equipment, maintenance, taxes, insurance)	7
Labor	3
Utilities and miscellaneous	3
Interest	3
Medication, debeaking, sanitation	2
Other	4
Total	100

Based on Tables 17–12 and 17–13, along with other studies and experiences, the following additional conclusions relative to turkey production costs and returns appear to be justified:

1. **Chief production costs.** The chief turkey production costs, ranked in order, are: feed, poults, and fixed costs for land, buildings and equipment, maintenance, taxes, and insurance.

2. **It costs more to produce fryer-roasters.** Industry sources estimate an additional cost of about 1¢ per pound more to raise fryer-roasters rather than heavy turkeys.

3. **Feed efficiency.** Feed conversion (pounds of feed per pound liveweight) has been steadily lowered, from 4.5 lb in 1940 to 2.67 lb in 1989.

4. **Labor efficiency.** In 1935–39, it required 23.7 work hours to produce 100 lb of turkey on foot. In 1990, in the more efficient and highly automated operations, one worker cared for 50,000 to 60,000 birds.

5. **Other factors affecting turkey production costs.** Turkey production costs are affected by several factors other than those listed above; among them the following:

a. **Sexes and weights.** Toms and hens are sold at different ages and weights, and fryer-roasters at much lighter weights than toms or hens.

b. **Range vs confinement.** The method of growing may be either on the range or in confinement; in turn, this affects the fixed facilities costs.

c. **Business structure.** Turkeys may be grown independently, under contract, or company grown. It is estimated that only 10% of the turkeys are grown by "independents."

d. **Feed.** Many large turkey growers produce their own grain and mix their own feed, thereby creating a different structure of feed costs from those who ordinarily buy feeds through commercial channels or have feeds furnished under a contractual arrangement.

Fig. 17–11. Broad-Breasted White turkeys on the range. Note sunflowers in the background. (Courtesy, J. C. Allen & Son, Inc., West Lafayette, IN)

TAX MANAGEMENT AND REPORTING

Good tax management and reporting consists of complying with the law, but of paying no more tax than is required. It is the duty of revenue agents to see that taxpayers do not pay less than they should, but it is the business of the taxpayers to make sure that they do not pay more than is required. From both standpoints, it is important that poultry producers familiarize themselves with as many of the tax regulations as possible.

The cardinal principles of good tax management are (1) maintenance of adequate records so as to assure payment of taxes in amounts no less or no more than required by law, and (2) conduct of business affairs to the end that the tax required by law is no greater than necessary.

Also, poultry producers need to recognize that good tax management and good poultry management do not necessarily go hand in hand. In fact, they may be in conflict. When the latter condition prevails, the advantages of the one must be balanced against the disadvantages of the other to the end that there shall be the greatest net return.

Under the cash system, farm income includes all cash or value of merchandise or other property received during the taxable year. It includes all available receipts from the sale of items produced on the farm and profits from the sale of items which have been purchased. It does not include the value of products sold or service performed for which payment was not actually available during the taxable year.

The accrual basis necessitates that complete annual inventories be kept. On the accrual basis, tax is paid on all income earned during the taxable year regardless of whether payment is actually received, and on increases of inventory values of birds, feed, produce, etc., at the end of the year as compared with the beginning of the year. All expenses incurred during the year's business are deducted from gross income regardless of whether payment is actually made, and deductions are made for any decrease in inventory values of livestock, etc., during the year.

Poultry producers are permitted to choose between the cash method and the accrual method.

Poultry producers should also set up depreciation schedules properly. *Depreciation is estimated operating expenses covering wear, tear, exhaustion, and obsolescence of property used in a poultry business.* Depreciation may be taken on all farm buildings (except the producer's personal residence), and on such things as machinery and equipment. Automobiles, light general purpose trucks, and most equipment may be depreciated in five years. Most farm buildings may be depreciated over a $31\frac{1}{2}$ year period.

Consultation with a tax specialist is recommended when tax problems are a bit out of the ordinary, and, like a visit to the family doctor, can be most effective when aid is sought before it is too late.

ESTATE PLANNING

■ **Special use valuation** — Owners of farms (poultry farms and other farms) and small businesses have been granted an estate planning advantage by means of what is called *special use valuation*. Under this concept, a farm can escape valuation for estate tax purposes at the highest and best use. Thus, a poultry farm located in an area undergoing development may be considerably more valuable to developers than it is as a farm. Nevertheless, if the family is willing to continue the farming or ranching use for ten years, the farm can be included in the estate at its value as a farm. The aggregate reduction in fair market value cannot exceed $750,000.

In order to qualify for special use valuation, the decedent must have been a U.S. citizen or resident and the farm must be located in the United States. The farm must have been used by the decedent or a family member at the date of the decedent's death. A lease to a nonfamily member, if not dependent on production, will not satisfy this requirement. At least 50% of the value of the decedent's estate must consist of the farm and more than 25% of the estate must consist of the farm real property. It may be possible to split up a farm and take the special valuation for only part of it, but this part must involve real property worth at least 25% of the estate.

The property must be passed to a qualified heir, including ancestors of the decedent, the decedent's spouse and lineal descendants, lineal descendants of the decedent's spouse or parents, and the spouse of any lineal descendant. Aunts, uncles, and first cousins are excluded. Legally adopted children are included.

The property must have been owned by the decedent or a family member for five of the eight years preceding the decedent's death and used as a farm in that period. The decedent or a family member must have participated in the farming operation for such a period prior to the decedent's death or disability.

■ **Electing special use valuation** — Though the procedures are clear as to how special use valuation is elected, the frequency with which mistakes are made indicates the importance of having a competent tax attorney or CPA firm prepare the estate tax return. A procedural failure denying the estate the considerable savings that can be gained by the election may give sufficient grounds for a malpractice suit against the return preparer.

■ **Recapture tax** — If the farm ceases to be operated by the heir or a family member within ten years, an additional estate tax will be imposed and the advantage of the election will be substantially lost. Partition among qualified heirs will not bring about recapture. When heirs granted oil leases on a family farm, the portion of the land devoted to the oil rigs was subject to recapture. A recent change allows the surviving spouse of the decedent to lease a farm on a net cash basis to a family member without being subject to the recapture tax.

■ **Payment extension** — Estates eligible for special use valuation may often be able to defer the payment of estate taxes. Where more than 35% of an estate of a U.S. citizen or resident consists of a farm, the estate tax liability may be paid in up to ten annual installments beginning as late as five years from when the tax might otherwise be due. If any portion of the farm is disposed of before the final payment, a corresponding portion of the amount deferred will come due.

■ **Use the gift tax exclusion for lifetime transfers** — The nontaxable gift tax exclusion remains at $10,000 per donee per year. A husband and wife who elect gift-splitting may jointly give $20,000 per recipient per year. These gifts may be in the form of interests in the farming operation.

■ **Plan with the unlimited marital deduction**—An unlimited deduction is permitted for the value of all property included in the gross estate that passes to the decedent's surviving spouse in the specified manner. Certain *terminable* interests do not qualify for such a deduction—that is, interests as to which of the surviving spouse's interests will terminate on the happening of some event. Surviving spouses may be given *qualified terminable interests*. The most common arrangement involves the surviving spouse receiving a lifetime interest in the farm, with the remainder passing on her/his death to others, perhaps the children of the decedent. No marital deduction is allowed if the surviving spouse is not a U.S. citizen, unless a specific trust arrangement is used.

■ **Consult a professional**—The preparation of wills, trusts, redemption agreements (if the farm is incorporated), partnership agreements, etc., requires consideration of the effects of federal and state tax law, as well as state law governing the various potential arrangements. Consequently, it is strongly advised that competent professionals be consulted in order to achieve an effective and cost-saving estate plan.

WILLS

A will is a set of instructions drawn up by or for an individual which details how he/she wishes the estate to be handled after death.

Despite the importance of a will in distributing property in keeping with the individual's wishes, about 50% of farmers and ranchers pass away without having written a will. This means that state law determines property distribution in such cases.

All poultry producers should have a will. By so doing, (1) the property will be distributed in keeping with their wishes, (2) they can name the executor of the estate, and (3) sizable tax savings can be made by the way in which the property is distributed. Because technical and legal rules govern the preparation, validity, and execution of a will, it should be drawn up by an attorney. Wills can and should be changed and updated from time to time. This can be done either by (1) a properly drawn-up codicil (formal amendment to a will), or (2) a completely new will which revokes the old one.

The same attorney should prepare both the husband's and wife's wills so that a common disaster clause can be incorporated and the estate planning of each can be coordinated.

TRUSTS

A trust is a written agreement by which an owner of property (the trustor) transfers title to a trustee for the benefit of persons called beneficiaries. Both real and personal property may be placed in trust.

The trustee may be an individual(s), bank, or corporation, or a combination of two or three of these. Management skill should be considered carefully in choosing a trustee.

A trust can continue for any period of time set by the owner—for a lifetime, until the youngest child reaches age 21, etc. If the trust extends beyond a lifetime, there are limitations which should be explained by an attorney.

KINDS OF TRUSTS

Basically, there are two kinds of trusts, the *living* and the *testamentary*. The living or *inter vivos* trust is in essence an agreement between the trustor and the trustee and may be revocable or irrevocable.

The *revocable trust* can be terminated or altered; under it the trustor is concerned about the here and now, rather than only the hereafter. The trustor continues to make decisions, and can call off the whole arrangement (it is revocable) if it does not work out as expected. The revocable trust offers no special estate tax advantage; the assets of a revocable trust are included in the estate of the deceased creating the trust. However, it can be written in such a manner as to reduce substantially the estate taxes of the beneficiaries. Also, the revocable trust will eliminate the cost of probate—costs which may include executor's fees, attorney's fees, court costs, and appraisal fees.

The *irrevocable trust* cannot be amended, altered, revoked, or terminated. Under an irrevocable trust, the trustor must be willing to part with the trust property forever (irrevocably) and have nothing further to do with it and its administration. However, the irrevocable trust has many favorable aspects in estate planning; it will reduce estate taxes in both the estate of the trustor and the estate(s) of the life beneficiaries, and it avoids probate.

The *testamentary trust* is so-called because it is established under the provisions of the trustor's last will and testament. The testamentary trust does not become effective until after death of the trustor, followed by probate. There is no tax saving in the trustor's estate. However, the trust may be drafted to save estate taxes in the estates of the beneficiaries. A testamentary trust is useful when the heirs are minors or inexperienced in money matters.

PARTNERSHIP CONTRACT

Another logical step in the transfer of property is a partnership contract between the parents and their heir(s) recorded in accordance with law. Appropriate counsel should be consulted in the preparation of such an agreement. Because of recently added "estate freeze" provisions, this approach to control of a farm does not provide the estate tax savings it could previously.

LIABILITY

Most poultry producers are in such financial position that they are vulnerable to damage suits. Moreover, the number of damage suits arising each year is increasing at an almost alarming rate, and astronomical damages are being claimed. Studies reveal that about 95% of the court cases involving injury result in damages being awarded.

Comprehensive personal liability insurance protects a farm operator who is sued for alleged damages suffered from an accident involving his/her property or family. The kinds of situations from which a claim might arise are quite broad, including suits for personal injuries caused by animals, equipment, or personal acts.

Both workers' compensation insurance and employer's liability insurance protect farmers against claims or court awards resulting from injury to hired help. Workers' compensation usually costs slightly more than straight employer's liability insurance, but it carries more benefits to the worker. An injured employee must prove negligence by his/her employer before the company will pay a claim under employer's liability insurance, whereas workers' compensation benefits are established by state law, and settlements are made by the insurance company without regard to who was negligent in causing the injury. Conditions governing participation in workers' compensation insurance vary among the states.

WORKERS' COMPENSATION[8]

Workers' compensation laws, now in full force in every one of the 50 states, cover on-the-job injuries and protect disabled workers regardless of whether their disabilities are temporary or permanent. Although broad differences exist among the individual states in their workers' compensation laws, principally in their benefit provisions, all statutes follow a definite pattern as to employment covered, benefits, insurance and the like.

Workers' compensation is a program designed to provide employees with assured payment for medical expenses or lost income due to injury on the job. Whenever an employment-related injury results in death, compensation benefits are generally paid to the worker's surviving dependents.

Generally all employment is covered by workers' compensation, although a few states provide exemptions for farm labor, or exempt farm employers of fewer than 10 full-time employees, for example. Farm employers in these states, however, may elect workers' compensation protection. Producers in these states may wish to consider coverage as a financial protection strategy because under workers' compensation, the upper limits for settlement of lawsuits are set by state law.

This government-required employee benefit is costly for producers, however. The 1990 basic premium rate in California, where workers' compensation is mandatory for agricultural employers, for farms and feed yards is $25.03 per $100 of payroll. Costs vary among insurance companies due to dividends paid, surcharges and minimum premiums. Some companies, as a matter of policy, will not write workers' compensation in agricultural industries. Some states have a quasi-government provider of workers' compensation to assure availability of coverage for small businesses and high-risk industries.

For information, contact your area extension farm management or personnel management advisor and an insurance agent experienced in marketing workers' compensation and liability insurance.

SOCIAL SECURITY[9]

The pertinent provisions of the present Social Security Law as it pertains to farmers, including poultry producers, are:

1. **Who is covered.** The law covers the following:
   a. **Agricultural workers.** If you are an agricultural worker, your employer must report your cash wages by January 31 of each year if he or she —

   (1) spends $2,500 or more on agricultural labor in a year; or

   (2) spends less than $2,500, but you were paid $150 or more in a given year.

If you commute to work daily from your home for a season picking fruit or vegetables by hand, your employer needs to report your wages if he or she paid you at least $150 in cash.

   b. **Household workers.** If you are a household worker on a farm operated for profit, your employer must report your wages if he or she —

   (1) spends more than $2,500 during a year on agricultural labor; or

   (2) spends less than $2,500 during a year on agricultural labor, but your annual wages amount to $150 or more.

Agricultural workers admitted to the United States on a temporary basis from any foreign country are not eligible.

2. **Amount paid in —**
   a. Self employed poultry farmers report their earnings and pay the social security self-employment tax at the time they file their annual tax return with the Internal Revenue Service.

   They may report their actual net earnings or an amount under an optional method. If their gross income from farming is $2,400 or less, they may report their actual net (if $400 or more) or two-thirds of their gross; if their gross income is more than $2,400 and their net farm income is less than $1,600, they may report for social security (but not for income tax purposes) either their actual net or $1,600. If gross income is over $2,400 and the actual net is over $1,600, actual net earnings must be reported.

   In 1990, a self-employed person paid 15.3% on net earnings up to $51,300. The taxable base will continue to increase as earning levels rise.

   b. The self-employed farmer is also responsible for reporting the wage of his/her farm laborers and any domestic help he/she may have. In 1990, the farmer should have deducted 7.65% of each employee's pay up to $51,300. The farmer adds 7.65% of his/her own to this and pays the total (15.3%) to the Internal Revenue Service.

---

[8]This report relative to "Workers' Compensation" was authoritatively prepared by S. R. Sutter, Personnel Management Advisor, University of California Cooperative Extension, Fresno County, Fresno, CA.

---

[9]In the preparation of this section, the author had the benefit of the review and suggestions of T. Butler, Associate Commissioner of Public Affairs, Department of Health & Human Services, Social Security Administration, Baltimore, MD.

3. **What are the benefits?** Depending on creditable earnings, the benefits are approximately as follows:

a. For a retired farmer (65 in 1990), up to $975 per month; or up to 20% less if he/she chooses to take benefits between 62 and 65.

b. For a retired farmer and spouse (both 65 in 1990), up to $1,463 per month.

c. For a widow or widower, up to $975 per month.

d. For a farmer under 65 who is suffering from a severe disability which has lasted, or is expected to last, 12 calendar months, or to result in death, up to $1,149 per month, provided he/she has had at least 5 years of work under Social Security in the 10-year period just before the disability began. (A worker who becomes disabled before 31 needs fewer work credits — in some cases as little as 1½ years; a worker who becomes disabled at 43 or older needs more than 5 years of credit.) Also, his/her eligible dependents can get the same benefits as the dependents of a farmer retired at 65.

e. Besides monthly payments, an eligible survivor can receive a lump-sum death payment of $255 on the record of the deceased worker.

f. Medicare offers both hospital insurance and medical insurance under Social Security for most people 65 and over and for some disabled people under 65. Older people eligible for monthly Social Security benefits have hospital insurance automatically. Or those who are not eligible for monthly benefits can get it by paying a monthly premium ($175 in 1990). People who have been entitled to disability checks for 24 or more months, and insured workers and their dependents who need dialysis treatment or a kidney transplant because of permanent kidney failure, also have this protection.

People who are covered under hospital insurance have medical insurance automatically unless they state they do not want it. The premium for this coverage was $28.60 a month in 1990.

The Social Security Administration administers another program called Supplemental Security Income. It is financed from general revenues rather than from Social Security taxes. It pays monthly checks to people in financial need who are 65 or older, blind, or disabled. More information about this program is available at any Social Security office.

The number on his/her Social Security card is very important to the farm operator as well as to the hired farm worker. It identifies the individual's Social Security record and is key to future benefit payments. It is important, therefore, that a person's Social Security number is on the Social Security reports for both the self-employed farmer and the agricultural worker.

Those who expect to draw Social Security payments later should check with the Social Security Administration every three years, especially if they change jobs frequently, to make sure that their records are in order and that their correct earnings are credited to their individual Social Security records.

For a Social Security card — either a new card or a duplicate of one that has been lost — or for more information about retirement, survivors, and disability insurance, Medicare health insurance, or Supplemental Security Income, get in touch with the nearest Social Security office or call Social Security's toll-free number: 1-800-2345-SSA (1-800-234-5772).

QUESTIONS FOR STUDY AND DISCUSSION

1. Why have the business aspects of poultry production become so important in recent years?

2. In 1991, each farm worker supplied enough food and fiber for 100 people (self and 99 others). Has this efficiency of American farming (a) created unemployment, and/or (b) liberated labor to make more of the luxuries of life, such as automobiles and refrigerators?

3. List and discuss each of the four major types of business organizations commonly found on poultry establishments.

4. List the advantages and the disadvantages of poultry contracts. What provisions should all contracts make?

5. How does the capital invested in U.S. farms compare with the total capital assets of all manufacturing corporations in the United States?

6. Assume that you are going to enter the poultry business, and that you have decided on the particular kind (with you making the decisions between layers, raising replacement pullets, broilers, turkeys, etc.). What types of credit may be needed, how would you go about obtaining it, and from what source(s) would you hope to obtain it?

7. What is a feasibility study? How would you go about making such a study or having such a study made of a poultry enterprise that you are planning?

8. Explain how "three points" would affect the interest on a $10,000 loan.

9. What types of poultry businesses should use futures trading, and why should they use it?

10. List four important functions of records or accounts.

11. How may computers be used, on a practical basis, for a poultry enterprise?

12. How would you set up a poultry budget?

QUESTIONS FOR STUDY AND DISCUSSION *(Continued)*

13. How would you analyze each of the following specialized poultry businesses to determine how well they are doing: (1) egg production, (2) replacement pullet production, (3) broiler production, and (4) turkey production?

14. List important standards for determining how well the owners are doing and their strengths and weaknesses, and give the relative importance percentagewise, for each (a) layers, (b) replacement pullets, (c) broilers, and (d) turkeys.

15. Discuss the importance of each of the following: (a) tax management and reporting, (b) estate planning, (c) wills, (d) trusts, and (e) partnership contracts.

16. Summarize pertinent information relative to each of the following: (a) liability, (b) workers' compensation, and (c) Social Security.

17. When it comes to making loans, it is frequently said that "bankers have a traditional fear of feathers." Why is this so? Is it justified?

SELECTED REFERENCES

Title of Publication	Author(s)	Publisher
Agricultural Statistics, 1990	Staff	U.S. Department of Agriculture, Washington, DC
Beuscher's Law and the Farmer, Fourth Edition	H. H. Hannah	Springer Publishing Company, Inc., New York, NY, 1975
Complete Guide to Making a Public Stock Offering, A	E. L. Winter	Prentice-Hall, Inc., Englewood Cliffs, NJ, 1962
Contract Farming, and Economic Integration, Second Edition	E. P. Roy	The Interstate Printers & Publishers, Inc., Danville, IL, 1972
Corporation Guide		Prentice-Hall, Inc., Englewood Cliffs, NJ, 1968
Economics: Applications to Agriculture and Agribusiness, Third Edition	E. P. Roy F. L. Corty G. D. Sullivan	The Interstate Printers & Publishers, Inc., Danville, IL, 1980
Exploring Agribusiness, Third Edition	E. P. Roy	The Interstate Printers & Publishers, Inc., Danville, IL, 1980
Fact Book of Agriculture 1990	Staff	U.S. Department of Agriculture, Washington, DC, 1989
Farm Management Economics	E. O. Heady H. R. Jensen	Prentice-Hall, Inc., Englewood Cliffs, NJ, 1955
Financial Planning in Agriculture	K. C. Schneeberger D. D. Osburn	The Interstate Printers & Publishers, Inc., Danville, IL, 1977
How to Do a Private Offering—Using Venture Capital	A. A. Sommer, Jr.	Practicing Law Institute, New York, NY, 1970
Introduction to Agribusiness Management, An, Second Edition	W. J. Wills	The Interstate Printers & Publishers, Inc., Danville, IL, 1979
Lawyer's Desk Book, Second Edition	W. J. Casey	Institute for Business Planning, Inc., New York, NY 1971
Spreadsheet Applications for Animal Nutrition and Feeding	R. J. Lane T. L. Cross	Reston Publishing Company, Inc., Reston, VA, 1985
Statistical Abstract of the United States 1990	Staff	U.S. Department of Commerce, Washington, DC, 1990
Stockman's Handbook, The, Seventh Edition	M. E. Ensminger	Interstate Publishers, Inc., Danville, IL, 1992

Fig. 18–1.   A chemical analysis makes for a good start in the evaluation of feeds.

FEED COMPOSITION TABLES

CHAPTER

18

In addition to the discussion that follows pertaining to composition of feeds, the reader is referred to Chapter 7, Poultry Feeding Standards, Ration Formulation, and Feeding Programs.

Both nutritionists and poultry producers should have access to accurate and up-to-date composition of feedstuffs in order to formulate rations for maximum production and net returns. The ultimate goal of feedstuff analysis, and the reason for feed composition tables, is to be able to predict the productive responses of birds when they are fed rations of a given composition. In recognition of this need and its importance, the author spared no time or expense in compiling the feed composition tables presented in this section. At the outset, a survey of the industry was made in order to determine what kind of feed composition tables would be most useful, in both format and content. Secondly, it was decided to utilize, to the extent available, the feed compositions which, for many years, were compiled by Lorin Harris, Utah State University, now carried forward by the USDA, National Agricultural Library, Feed Composition Bank. These data were augmented by the author with feed compositions from the National Academy of Sciences, NRC, and from experimental reports, industries, and other reliable sources.

To facilitate quick and easy use, the feeds are presented as follows:

■ **Table 18–1, Composition of Feeds** – Data expressed As-Fed and Moisture-Free. In this table, the commonly used poultry feeds are listed alphabetically, and on both an As-Fed and Moisture-Free basis. Also, for each feed, the chemical analysis, metabolizable energy, mineral, and vitamin compositions are given. Values for each feed are presented in tabular form in a two-page spread, with the proximate analysis and the metabolizable energy, plus calcium, phosphorus, sodium, and chlorine on the left-hand page; and the rest of the minerals and the vitamins on the right-hand page.

■ **Table 18–2, Amino Acids** – This gives the known amino acid composition of selected feeds.

■ **Table 18–3, True Digestibility of Essential Amino Acids for Poultry, and True Digestibility Amino Acid Recommendations for Poultry Feed Formulation** – Like all monogastric animals, poultry do not require protein as such; instead, they need well-defined amounts of available amino acids to perform at a desired level. Currently, digestibility assays give the best assessment of the availability of amino acids in various feed ingredients.

Note well: In addition to the feed compositions presented in Table 18–1 and the amino acid values presented in Tables 18–2 and 18–3 of this book, should the need arise for additional feed compositions, and for additional mineral supplement compositions and vitamin supplement compositions, many more feeds are presented in *Feeds & Nutrition*, and *Feeds & Nutrition Digest*, books of which Dr. Ensminger is the senior author.

FEED NAMES

Ideally, a feed name should conjure up the same meaning to all those who use it, and it should provide helpful information. This was the guiding philosophy of the author when choosing the names given in the Feed Composition Table. Genus and species – Latin names – are also included. To facilitate worldwide usage, the International Feed Number of each feed is given. To the extent possible, consideration was also given to source (or parent material), variety or kind, stage of maturity, processing, part eaten, and grade.

Where feeds are known by more than one name, cross-referencing was used.

MOISTURE CONTENT OF FEEDS

It is necessary to know the moisture content of feeds in ration formulation and buying. Usually, the composition of a feed is expressed according to one or more of the following bases:

1. **As-Fed; A-F (wet, fresh).** This refers to feed as normally fed to poultry. As-Fed may range from near 0% to 100% dry matter.

2. **Air-Dry (approximately 90% dry matter).** This refers to feed that is dried by means of natural air movement, usually in the open. It may either be an actual or an assumed dry matter content; the latter is approximately 90%. Most feeds are fed in an air-dry state.

3. **Moisture-Free; M-F (oven-dry, 100% dry matter).** This refers to a sample of feed that has been dried in an oven at 221°F until all the moisture has been removed.

Where available, feed compositions are presented on both **As-Fed (A-F)** and Moisture-Free (M-F) bases. Formulas for adjusting moisture content from moisture-free to as-fed, or from as-fed to moisture-free, are given in *Feeds & Nutrition* and *Feeds & Nutrition Digest*, books of which Dr. Ensminger is the senior author.

CAROTENE

Where carotene has been converted to vitamin A, the conversion rate of the rat has been used as the standard value, with 1 mg of β-carotene equal to 1,667 IU of vitamin A.

PERTINENT INFORMATION ABOUT DATA

The information which follows is pertinent to Table 18–1.

■ **Variations in composition** – Feeds vary in their composition. Thus, actual analysis of a feedstuff should be obtained and used whenever possible, especially where a large lot of feed from one source is involved. Many times, however, it is either impossible to determine actual compositions or there is insufficient time to obtain such analysis. Under such circumstances, tabulated data may be the only information available.

■ **Feed compositions change**—Feed compositions change over a period of time, primarily due to (1) the introduction of new varieties, and (2) modifications in the manufacturing process from which by-products evolve.

■ **Biological value**—The response of animals when fed a feed is termed the biological value, which is a function of its chemical composition and the ability of the animal to derive useful nutrient value from the feed. The latter relates to the digestibility, or availability, of the nutrients in the feed. Thus, soft coal and shelled corn may have the same gross energy value in a bomb calorimeter but markedly different useful energy values when consumed by a bird. Biological tests of feeds are more laborious and costly than chemical analysis, but they are much more accurate in predicting the response of animals to a feed.

■ **Where information is not available**—Where information is not available or reasonable estimates could not be made, no values are shown. Hopefully, such information will become available in the future.

■ **Calculated on a dry matter (DM) basis**—All data were calculated on a 100% dry matter basis (moisture-free), then converted to an as-fed basis by multiplying the decimal equivalent of the DM content times the compositional value shown in the table.

■ **Fiber**—Four values relating to dietary fiber are given in the feed composition tables—crude fiber, neutral detergent fiber (NDF), acid detergent fiber (ADF), and lignin.

Crude fiber, methods for the determination of which were developed more than 100 years ago, is declining as a measure of low digestible material in the more fibrous feeds. The newer method of forage analysis, developed by Van Soest and associates of the U.S. Department of Agriculture, separates feed dry matter into two fractions: a neutral detergent fibrous fraction; and an acid detergent fibrous fraction. Also the amount of lignin in the ADF may be determined.

1. **Crude fiber (CF).** This fraction is an indicator of the relative indigestibility and bulkiness of the sample. It is the residue that remains after boiling a feed in a weak acid, and then in a weak alkali, in an attempt to imitate the process that occurs in the digestive tract. This procedure is based on the supposition that carbohydrates which are readily dissolved also will be readily digested by animals, and that those not soluble under such conditions are not readily digestible. Unfortunately, the treatment dissolves much of the lignin, a nondigestible component. Hence, crude fiber is only an approximation of the indigestible material in feedstuffs. Nevertheless, it is a rough indicator of the energy value of feeds. Also, the crude fiber value is needed for the computation of TDN.

2. **Neutral detergent fiber (NDF).** This is the fraction of the feed which is not soluble in neutral detergent. It consists of plant cell walls, including lignin, cellulose, and hemicellulose. NDF is closely related to feed intake because it contains all the fiber components that occupy space in the rumen and are slowly digested. The lower the NDF, the more forage the animal will eat; hence, a low percentage of NDF is desirable.

3. **Acid detergent fiber (ADF).** This is the fraction of the feed which is not soluble in acid detergent. It consists of cellulose (digestible) and lignin (indigestible). ADF is an indicator of forage digestibility because it contains a high proportion of lignin which is the indigestible fiber fraction. The lower the ADF, the more feed an animal can digest; hence, a low percentage of ADF is desirable.

4. **Lignin.** This fraction is essentially indigestible by all animals and is the substance that limits the availability of cellulose carbohydrates in the plant cell wall to rumen bacteria.

The acid detergent fiber procedure is used as a preparatory step in determining the lignin content of a forage sample. Hemicellulose is solubilized during this procedure, while the lignocellulose fraction of the feed remains insoluble. Cellulose is then separated from lignin by the addition of sulfuric acid. Only lignin and acid-insoluble ash remain upon completion of this step. This residue is then ashed, and the difference of the weights before and after ashing yields the amount of lignin present in the feed.

■ **Nitrogen-free extract**—The nitrogen-free extract was calculated with mean data as: mean nitrogen-free extract (%) = 100 − % ash − % crude fiber − % ether extract − % protein.

■ **Protein values**—Crude protein is determined by finding the nitrogen content and multiplying the result by 6.25. The nitrogen content of proteins averages about 16% ($100 \div 16 = 6.25$).

■ **Energy**—The metabolizable energy (ME) is shown, This represents that portion of the gross energy that is not lost in the feces, urine, and gas (mainly methane). It does not take into account the energy lost as heat, commonly called heat increment. As a result, it overevaluates roughages compared with concentrates, as do total digestible nutrients (TDN) and digestible energy (DE).

Metabolizable energy is considered to be the most accurate evaluation of the energy of feedstuffs for the scientific formulation of poultry feeds.

■ **Minerals**—The level of minerals in forages is largely determined by the mineral content of the soil on which the feeds are grown. Calcium, phosphorus, iodine, and selenium are well-known examples of soil nutrient–plant nutrient relationships.

■ **Vitamins**—Generally speaking, it is unwise to rely on harvested feeds as a source of carotene (vitamin A value), unless the forage being fed is fresh (pasture or green chop) or of a good green color and not over a year old.

The author is very grateful to Lorin E. Harris, Ph.D., and Clyde R. Richards, Ph.D., Utah State University, Logan, for their interest and invaluable assistance in preparing the Feed Composition Tables for this book.

TABLE
COMPOSITION OF FEEDS, DATA EXPRESSED

Entry Number	Feed Name Description	International Feed Number	Moisture Basis A-F (as-fed) or M-F (moisture-free)	Dry Matter	Ash	Crude Fiber	Neutral Det. Fib. (NDF)	Acid Det. Fib. (ADF)	Lignin	Ether Extract (Fat)	N-Free Extract	Crude Protein	Poultry ME$_n$ kcal/lb	kcal/kg	Calcium (Ca)	Phosphorus (P)	Sodium (Na)	Chlorine (Cl)
				%	%	%	%	%	%	%	%	%	kcal/lb	kcal/kg	%	%	%	%
	ALFALFA (LUCERNE) *Medicago sativa*																	
1	MEAL, DEHY, 15% PROTEIN	1–00–021	A-F	90	9.1	26.6	45.9	36.9	—	2.2	37.0	15.6	696	*1,535*	1.24	0.22	0.07	0.44
			M-F	100	10.0	29.4	51.0	41.0	—	2.5	40.9	17.3	770	*1,698*	1.37	0.24	0.08	0.48
2	MEAL, DEHY, 17% PROTEIN	1–00–023	A-F	92	9.7	24.0	41.3	31.5	9.7	2.8	37.8	17.4	682	*1,504*	1.40	0.23	0.10	0.47
			M-F	100	10.6	26.2	45.0	34.3	10.6	3.0	41.2	18.9	744	*1,640*	1.52	0.25	0.11	0.52
3	MEAL, DEHY, 20% PROTEIN	1–00–024	A-F	92	10.2	20.8	38.6	28.5	—	3.3	37.1	20.2	737	*1,625*	1.59	0.28	0.11	0.47
			M-F	100	11.1	22.7	42.0	31.0	—	3.6	40.5	22.1	805	*1,774*	1.74	0.31	0.13	0.51
	BAKERY WASTE																	
4	DEHY (DRIED BAKERY PRODUCT)	4–00–466	A-F	91	3.7	1.2	—	1.6	—	10.9	65.3	10.1	1,739	*3,834*	0.14	0.22	1.02	1.47
			M-F	100	4.0	1.3	—	1.8	—	12.0	71.6	11.1	1,906	*4,203*	0.15	0.24	1.12	1.61
	BARLEY *Hordeum vulgare*																	
5	GRAIN, ALL ANALYSES	4–00–549	A-F	88	2.4	5.0	16.8	10.7	1.5	1.7	67.7	11.7	1,180	*2,602*	0.05	0.34	0.03	0.12
			M-F	100	2.7	5.7	19.0	12.1	1.7	1.9	76.5	13.2	1,334	*2,941*	0.06	0.39	0.03	0.13
6	GRAIN, PACIFIC COAST	4–07–939	A-F	89	2.5	6.5	—	—	—	2.0	68.2	9.5	1,171	*2,582*	0.05	0.34	0.02	0.15
			M-F	100	2.8	7.3	—	—	—	2.2	76.9	10.8	1,322	*2,914*	0.06	0.39	0.02	0.17
	BLOOD																	
7	MEAL	5–00–380	A-F	91	5.3	1.0	—	—	—	1.3	3.1	80.5	1,282	*2,826*	0.29	0.25	0.32	0.30
			M-F	100	5.8	1.1	—	—	—	1.5	3.4	88.2	1,404	*3,096*	0.32	0.28	0.35	0.33
8	SPRAY DEHY (BLOOD FLOUR)	5–00–381	A-F	93	4.1	1.0	—	—	—	1.2	0.2	86.0	1,255	*2,767*	0.41	0.30	0.38	0.25
			M-F	100	4.4	1.1	—	—	—	1.3	0.2	93.0	1,357	*2,991*	0.44	0.33	0.42	0.27
	BONE MEAL																	
9	STEAMED	6–00–400	A-F	95	67.3	1.9	—	—	—	3.6	3.8	18.6	—	—	25.98	11.80	0.40	0.01
			M-F	100	70.7	2.0	—	—	—	3.8	4.0	19.5	—	—	27.31	12.40	0.42	0.01
	BREWER GRAINS																	
10	DEHY	5–02–141	A-F	92	3.6	13.0	38.7	23.9	4.6	6.6	41.6	27.3	1,047	*2,308*	0.30	0.51	0.21	0.15
			M-F	100	4.0	14.1	42.0	26.0	5.0	7.1	45.2	29.6	1,137	*2,506*	0.33	0.55	0.23	0.17
	BROADBEAN *Vicia faba*																	
11	SEEDS	5–09–262	A-F	87	—	6.7	—	—	—	1.4	—	23.6	1,216	*2,431*	0.11	—	—	—
			M-F	100	—	7.7	—	—	—	1.6	—	27.1	1,398	*2,794*	0.13	—	—	—
	BUCKWHEAT, COMMON *Fagopyrum sagittatum*																	
12	GRAIN	4–00–994	A-F	88	2.1	10.6	—	14.9	—	2.4	61.5	11.1	1,200	*2,645*	0.10	0.33	0.05	0.04
			M-F	100	2.4	12.1	—	17.0	—	2.8	70.2	12.6	1,368	*3,016*	0.11	0.37	0.06	0.05
	BUTTERMILK, CATTLE *Bos taurus*																	
13	DEHY	5–01–160	A-F	92	9.1	0.3	—	—	—	5.2	45.9	31.7	1,248	*2,752*	1.32	0.94	0.83	0.44
			M-F	100	9.9	0.4	—	—	—	5.6	49.7	34.4	1,353	*2,982*	1.43	1.01	0.90	0.48
	CANOLA (RAPE) *Brassica napus, Brassica campestris*																	
14	SEEDS, MEAL PREPRESSED, SOLV EXTD	5–06–145	A-F	93	—	11.1	—	—	—	3.8	—	38.0	909	*2,000*	0.68	1.17	—	—
			M-F	100	—	11.9	—	—	—	4.1	—	40.9	977	*2,151*	0.73	1.26	—	—
	CASEIN																	
15	DEHY	5–01–162	A-F	91	2.2	0.2	0.0	0.0	—	0.6	3.6	84.0	1,867	*4,116*	0.61	0.82	0.01	—
			M-F	100	2.4	0.2	0.0	0.0	—	0.7	3.9	92.7	2,061	*4,543*	0.67	0.90	0.01	—
	CORN, DENT YELLOW *Zea mays, indentata*																	
16	GRAIN, ALL ANALYSES	4–02–935	A-F	88	1.3	2.3	—	3.8	—	3.6	71.0	9.9	1,523	*3,359*	0.05	0.28	0.01	0.05
			M-F	100	1.5	2.6	—	4.3	—	4.1	80.7	11.2	1,732	*3,818*	0.05	0.31	0.01	0.06
17	GRAIN, GRADE 1, 56 LB/BU	4–02–930	A-F	86	1.2	2.1	—	—	—	3.9	70.4	8.8	1,470	*3,241*	0.03	0.27	0.01	0.05
			M-F	100	1.4	2.4	—	—	—	4.5	81.4	10.2	1,701	*3,750*	0.03	0.31	0.01	0.06
18	GRAIN, GRADE 2, 54 LB/BU	4–02–931	A-F	87	1.2	2.1	—	—	—	4.0	71.3	8.9	1,567	*3,456*	0.02	0.29	0.02	0.04
			M-F	100	1.4	2.4	—	—	—	4.5	81.5	10.2	1,792	*3,950*	0.03	0.33	0.02	0.05
19	DISTILLERS GRAINS, DEHY	5–02–842	A-F	93	2.2	11.5	40.0	15.8	—	8.9	43.1	27.8	894	*1,970*	0.09	0.39	0.09	0.07
			M-F	100	2.4	12.3	43.0	17.0	—	9.5	46.2	29.7	956	*2,108*	0.10	0.42	0.09	0.08
20	DISTILLERS SOLUBLES, DEHY	5–28–237	A-F	93	7.2	4.6	21.4	6.5	—	8.6	45.2	27.4	1,324	*2,919*	0.30	1.30	0.23	0.26
			M-F	100	7.7	4.9	23.0	7.0	—	9.3	48.6	29.5	1,426	*3,143*	0.32	1.40	0.24	0.28
21	GLUTEN, MEAL, 60% PROTEIN	5–28–242	A-F	90	1.7	1.8	12.6	4.5	—	2.1	23.7	60.8	3,370	*7,431*	0.07	0.45	0.05	0.09
			M-F	100	1.9	2.0	14.0	5.0	—	2.3	26.3	67.5	3,745	*8,256*	0.08	0.50	0.05	0.10
22	GLUTEN FEED	5–28–243	A-F	90	6.6	8.7	40.5	10.8	—	2.1	49.4	23.0	784	*1,729*	0.32	0.74	0.12	0.22
			M-F	100	7.4	9.7	45.0	12.0	—	2.4	55.0	25.6	872	*1,924*	0.36	0.82	0.14	0.25
	GRITS—SEE HOMINY																	
	COTTON *Gossypium spp*																	
23	SEEDS, MEAL MECH EXTD, 41% PROTEIN	5–01–617	A-F	93	6.1	11.9	25.9	18.5	5.6	4.7	28.9	41.0	1,025	*2,261*	0.19	1.07	0.04	0.04
			M-F	100	6.6	12.9	28.0	20.0	6.0	5.0	31.2	44.3	1,108	*2,443*	0.21	1.16	0.05	0.05
24	SEEDS, MEAL, PREPRESSED, SOLV EXTD, 41% PROTEIN	5–07–872	A-F	90	6.4	12.9	23.4	17.1	—	1.0	28.8	41.3	974	*2,147*	0.16	1.07	0.04	0.06
			M-F	100	7.0	14.2	26.0	19.0	—	1.2	31.9	45.7	1,078	*2,377*	0.17	1.18	0.04	0.07
25	SEEDS, MEAL SOLV EXTD, 46% PROTEIN	5–26–100	A-F	92	7.2	8.9	25.8	19.3	—	1.6	26.8	47.6	950	*2,095*	—	—	—	—
			M-F	100	7.8	9.6	28.0	21.0	—	1.8	29.1	51.7	1,032	*2,275*	—	—	—	—
	FEATHERS, POULTRY																	
26	MEAL, HYDROLYZED	5–03–795	A-F	93	3.2	1.4	—	6.1	—	5.1	—	83.8	1,104	*2,434*	0.30	0.62	0.63	0.28
			M-F	100	3.4	1.5	—	6.6	—	5.5	—	90.2	1,188	*2,619*	0.33	0.67	0.68	0.30

18–1
AS-FED AND MOISTURE-FREE

Entry Number	Mag-nesium (Mg) %	Potas-sium (K) %	Sulfur (S) %	Cobalt (Co) ppm or mg/kg	Copper (Cu) ppm or mg/kg	Iodine (I) ppm or mg/kg	Iron (Fe) %	Man-ganese (Mn) ppm or mg/kg	Sele-nium (Se) ppm or mg/kg	Zinc (Zn) ppm or mg/kg	A IU/g	Carotene (Provita-min A) ppm or mg/kg	D IU/kg	E ppm or mg/kg	K ppm or mg/kg	B–12 ppb or mcg/kg	Biotin ppm or mg/kg	Choline ppm or mg/kg	Folacin (Folic Acid) ppm or mg/kg	Niacin ppm or mg/kg	Pantothenic Acid (B–3) ppm or mg/kg	(Pyri-doxine) B–6 ppm or mg/kg	Ribo-flavin (B–2) ppm or mg/kg	Thia-min (B–1) ppm or mg/kg
1	0.28	2.24	0.18	0.17	9.5	0.12	0.03	27.8	0.28	19.4	124.1	74.5	—	81.9	9.59	—	0.25	1,573	1.56	42	20.7	6.27	10.6	3.0
	0.31	2.48	0.20	0.19	10.5	0.13	0.03	30.8	0.31	21.4	137.3	82.3	—	90.6	10.61	—	0.28	1,739	1.73	46	22.9	6.94	11.7	3.3
2	0.29	2.38	0.23	0.30	8.6	0.15	0.04	31.0	0.34	19.3	200.3	120.2	—	105.7	8.24	—	0.33	1,369	4.37	37	29.7	7.18	12.9	3.4
	0.32	2.60	0.25	0.33	9.3	0.16	0.05	33.8	0.37	21.1	218.5	131.1	—	115.3	8.98	—	0.36	1,494	4.77	40	32.4	7.83	14.1	3.7
3	0.33	2.41	0.50	0.26	12.2	0.14	0.04	45.2	0.29	21.8	265.4	159.2	—	143.3	14.19	—	0.35	1,417	2.96	48	35.5	8.72	15.2	5.4
	0.36	2.63	0.55	0.28	13.3	0.15	0.04	49.4	0.31	23.8	289.7	173.8	—	156.4	15.50	—	0.39	1,547	3.24	52	38.8	9.52	16.6	5.9
4	0.16	0.40	0.02	1.22	11.0	—	0.02	65.0	—	17.8	7.0	4.2	—	40.9	—	—	0.07	917	0.19	26	8.2	4.29	1.4	2.9
	0.18	0.43	0.02	1.34	12.1	—	0.02	71.2	—	19.5	7.7	4.6	—	44.9	—	—	0.07	1,005	0.20	28	9.0	4.70	1.5	3.2
5	0.13	0.46	0.15	0.17	7.6	0.04	0.01	18.0	0.16	39.3	3.4	2.0	—	23.2	0.22	—	0.15	1,036	0.57	76	7.9	5.80	1.6	4.5
	0.15	0.52	0.17	0.19	8.6	0.05	0.01	18.1	0.18	44.4	3.8	2.3	—	26.2	0.24	—	0.17	1,171	0.64	86	9.0	6.55	1.8	5.0
6	0.12	0.51	0.14	0.09	8.1	—	0.01	16.0	0.10	15.2	—	—	—	26.2	—	—	0.15	976	0.50	47	7.1	2.89	1.5	4.2
	0.14	0.58	0.16	0.10	9.1	—	0.01	18.0	0.11	17.1	—	—	—	29.6	—	—	0.17	1,102	0.56	53	8.0	3.26	1.7	4.7
7	0.22	0.09	0.34	0.09	12.6	—	0.37	5.3	0.73	4.4	—	—	—	—	—	44.3	0.09	780	0.10	31	2.3	4.41	2.0	0.3
	0.24	0.10	0.37	0.10	13.8	—	0.41	5.8	0.80	4.8	—	—	—	—	—	48.5	0.09	854	0.11	34	2.6	4.83	2.2	0.4
8	0.15	0.15	0.60	—	8.2	—	0.28	6.4	—	—	—	—	—	—	—	12.2	0.28	597	0.37	22	3.2	4.43	2.9	0.3
	0.17	0.16	0.65	—	8.8	—	0.30	6.9	—	—	—	—	—	—	—	13.2	0.30	645	0.40	24	3.5	4.79	3.1	0.3
9	0.78	0.18	0.34	0.00	162.0	29.00	0.09	37.0	—	362.0	—	—	—	—	—	—	—	—	—	—	—	—	—	—
	0.82	0.19	0.36	0.00	170.0	31.00	0.09	39.0	—	381.0	—	—	—	—	—	—	—	—	—	—	—	—	—	—
10	0.15	0.09	0.30	0.08	21.7	0.09	0.02	37.2	—	27.3	0.8	0.5	—	26.7	—	3.6	0.44	1,651	0.22	44	8.2	1.03	1.5	0.6
	0.17	0.09	0.32	0.08	23.6	0.07	0.03	40.4	—	29.6	0.8	0.5	—	29.0	—	3.9	0.48	1,792	0.24	47	8.9	1.11	1.6	0.7
11	—	—	—	—	—	—	—	—	—	—	—	—	—	—	—	—	—	—	—	—	—	—	—	—
	—	—	—	—	—	—	—	—	—	—	—	—	—	—	—	—	—	—	—	—	—	—	—	—
12	0.10	0.45	0.14	0.05	9.5	—	0.01	33.7	—	8.8	—	—	—	—	—	—	—	439	—	18	11.5	—	4.7	3.7
	0.12	0.51	0.16	0.06	10.8	—	0.01	38.4	—	10.0	—	—	—	—	—	—	—	501	—	21	13.1	—	5.4	4.2
13	0.48	0.83	0.08	—	1.0	—	0.00	3.5	—	40.2	—	—	—	6.3	—	19.6	0.29	1,746	0.39	9	37.0	2.47	30.6	3.4
	0.52	0.90	0.09	—	1.1	—	0.00	3.8	—	43.6	—	—	—	6.8	—	21.2	0.31	1,891	0.42	9	40.1	2.67	33.1	3.7
14	0.64	1.29	—	—	10.0	—	159.00	54.0	1.00	71.0	—	—	—	—	—	—	0.90	6,700	2.3	160	9.5	—	3.7	5.2
	0.69	1.39	—	—	11.0	—	171.00	58.0	1.08	76.0	—	—	—	—	—	—	0.97	7,204	2.5	172	1.02	—	4.0	5.6
15	0.01	0.01	—	—	4.1	—	0.00	3.5	—	31.8	—	—	—	—	—	—	0.04	208	0.47	1	2.7	0.42	1.5	0.4
	0.01	0.01	—	—	4.5	—	0.00	3.9	—	35.1	—	—	—	—	—	—	0.05	229	0.52	1	2.9	0.47	1.7	0.5
16	0.11	0.33	0.11	0.38	0.35	—	0.00	5.7	0.13	19.4	9.5	5.7	—	20.9	0.22	—	0.07	504	0.31	23	5.1	6.16	1.1	3.7
	0.13	0.37	0.13	0.43	4.0	—	0.00	6.4	0.14	22.0	10.8	6.5	—	23.8	0.25	—	0.08	573	0.35	26	5.8	7.01	1.2	4.2
17	0.10	0.28	0.12	—	4.1	—	0.00	5.6	—	—	—	—	—	—	—	—	—	—	—	—	5.8	—	—	—
	0.11	0.33	0.14	—	4.8	—	0.00	6.4	—	—	—	—	—	—	—	—	—	—	—	—	6.7	—	—	—
18	0.11	0.31	0.12	0.03	3.8	—	0.00	5.3	—	13.7	2.9	1.7	—	21.6	—	—	0.06	569	0.35	24	3.9	6.88	1.3	3.5
	0.13	0.36	0.14	0.03	4.3	—	0.00	6.1	—	15.6	3.3	2.0	—	24.7	—	—	0.07	650	0.40	28	4.5	7.87	1.5	4.0
19	0.07	0.16	0.43	0.08	38.9	0.05	0.02	19.3	0.35	41.7	5.2	3.1	—	—	—	0.3	0.41	1,113	1.00	38	11.3	4.22	5.0	1.8
	0.07	0.18	0.46	0.08	41.7	0.05	0.02	20.7	0.38	44.7	5.6	3.3	—	—	—	0.3	0.44	1,191	1.07	41	12.1	4.51	5.3	1.9
20	0.60	1.70	0.37	0.17	77.9	0.08	0.05	72.0	0.37	88.0	1.1	0.7	—	45.9	—	4.2	1.49	4,751	1.34	124	23.3	9.41	15.1	6.8
	0.65	1.83	0.40	0.19	83.9	0.09	0.06	77.6	0.40	94.8	1.2	0.7	—	49.4	—	4.5	1.61	5,116	1.45	133	25.0	10.14	16.3	7.3
21	0.08	0.18	0.65	0.05	26.1	0.02	0.02	6.3	0.83	30.6	—	—	—	14.6	—	—	—	—	—	—	—	6.39	—	—
	0.09	0.20	0.72	0.05	29.0	0.02	0.03	7.0	0.92	34.0	—	—	—	16.2	—	—	—	—	—	—	—	7.10	—	—
22	0.33	0.57	0.21	0.09	47.1	0.07	0.04	23.1	0.27	64.6	9.8	5.9	—	12.1	—	—	0.33	1,514	0.27	70	13.6	13.93	2.2	2.0
	0.36	0.64	0.23	0.10	52.3	0.07	0.05	25.7	0.30	71.8	10.9	6.5	—	13.5	—	—	0.36	1,684	0.30	78	15.1	15.49	2.5	2.2
23	0.53	1.33	0.40	0.63	18.5	—	0.02	22.3	—	61.8	0.4	0.2	—	32.3	—	—	0.91	2,753	2.45	35	10.2	5.00	5.2	7.1
	0.57	1.44	0.43	0.68	20.0	—	0.02	24.1	—	66.8	0.4	0.2	—	34.9	—	—	0.99	2,974	2.65	38	11.0	5.41	5.6	7.6
24	0.48	1.25	0.31	0.74	18.2	—	0.02	20.4	—	62.7	—	—	—	—	—	—	0.55	2,861	2.57	40	7.3	4.11	5.3	3.3
	0.53	1.38	0.34	0.82	20.2	—	0.02	22.5	—	69.4	—	—	—	—	—	—	0.61	3,166	2.85	44	8.1	4.55	5.8	3.7
25	—	—	—	—	—	—	—	—	—	—	—	—	—	—	—	—	—	—	—	—	—	—	—	—
	—	—	—	—	—	—	—	—	—	—	—	—	—	—	—	—	—	—	—	—	—	—	—	—
26	0.18	0.27	1.50	0.12	7.3	0.04	0.02	11.9	0.91	71.9	—	—	—	—	—	80.4	0.04	894	0.22	21	8.9	4.39	2.0	0.1
	0.19	0.29	1.61	0.13	7.9	0.05	0.02	12.9	0.98	77.3	—	—	—	—	—	86.5	0.05	962	0.23	23	9.6	4.72	2.2	0.1

(Continued)

TABLE
COMPOSITION OF FEEDS, DATA EXPRESSED

Entry Number	Feed Name Description	International Feed Number	Moisture Basis A-F (as-fed) or M-F (moisture-free)	Dry Matter	Ash	Crude Fiber	Neutral Det. Fib. (NDF)	Acid Det. Fib. (ADF)	Lignin	Ether Extract (Fat)	N-Free Extract	Crude Protein	Poultry ME$_n$ kcal/lb	kcal/kg	Calcium (Ca)	Phos-phorus (P)	Sodium (Na)	Chlo-rine (Cl)
				%	%	%	%	%	%	%	%	%	kcal/lb	kcal/kg	%	%	%	%
	FISH																	
27	LIVER MEAL MECH EXTD	5–01–968	A-F	93	6.1	1.2	—	—	—	17.3	5.4	62.8	—	—	—	—	—	—
			M-F	100	6.6	1.3	—	—	—	18.6	5.8	67.7	—	—	—	—	—	—
28	MEAL MECH EXTD	5–01–977	A-F	92	21.4	0.7	—	—	—	6.0	—	64.3	1,174	2,587	6.63	3.61	1.11	1.25
			M-F	100	23.3	0.8	—	—	—	6.6	—	70.2	1,281	2,825	7.24	3.94	1.21	1.37
29	SOLUBLES, CONDENSED	5–01–969	A-F	50	10.1	0.5	—	—	—	6.1	2.2	31.5	755	1,665	0.16	0.57	2.45	2.93
			M-F	100	20.0	1.0	—	—	—	12.2	4.4	62.5	1,497	3,300	0.32	1.14	4.86	5.81
30	SOLUBLES, DEHY	5–01–971	A-F	93	12.7	2.0	—	—	—	9.0	8.7	60.4	1,322	2,915	0.40	1.27	1.70	—
			M-F	100	13.7	2.1	—	—	—	9.7	9.4	65.1	1,424	3,140	0.43	1.37	1.83	—
	FISH, ANCHOVY *Engraulis ringen*																	
31	MEAL MECH EXTD	5–01–985	A-F	92	14.7	1.0	—	—	—	4.1	6.7	65.4	1,245	2,745	3.74	2.48	0.88	1.00
			M-F	100	16.0	1.1	—	—	—	4.5	7.3	71.1	1,354	2,985	4.07	2.70	0.95	1.08
	FISH, MENHADEN *Brevoortia tyrannus*																	
32	MEAL MECH EXTD	5–02–009	A-F	92	19.1	0.9	—	—	—	9.6	0.8	61.2	1,292	2,848	5.19	2.88	0.41	0.55
			M-F	100	20.9	1.0	—	—	—	10.5	0.8	66.8	1,411	3,110	5.67	3.14	0.45	0.60
	FISH, SARDINE *Clupea spp*																	
33	FLESH, FRESH	5–07–312	A-F	29	2.4	—	—	—	—	8.5	—	19.0	—	—	0.03	0.21	—	—
			M-F	100	8.2	—	—	—	—	29.4	—	65.5	—	—	0.11	0.73	—	—
34	MEAL MECH EXTD	5–02–015	A-F	93	15.8	1.0	—	—	—	5.0	6.1	65.2	1,313	2,896	4.61	2.68	0.18	0.41
			M-F	100	17.0	1.1	—	—	—	5.4	6.5	70.0	1,410	3,109	4.95	2.88	0.19	0.44
	FISH, SHARK *Selachii* (order)																	
35	MEAL MECH EXTD	5–02–018	A-F	91	13.4	0.5	—	—	—	2.6	1.8	72.8	—	—	3.48	1.86	0.33	—
			M-F	100	14.7	0.5	—	—	—	2.8	2.0	79.9	—	—	3.82	2.04	0.36	—
	FISH, TUNA *Thunnus thynnus*																	
36	FLESH, FRESH	5–09–278	A-F	29	1.4	—	—	—	—	3.5	—	24.9	—	—	—	—	0.04	—
			M-F	100	4.7	—	—	—	—	12.2	—	86.0	—	—	—	—	0.14	—
37	MEAL MECH EXTD	5–02–023	A-F	93	21.9	0.8	—	—	—	6.9	4.2	59.0	1,276	2,812	7.86	4.21	0.74	1.01
			M-F	100	23.6	0.9	—	—	—	7.4	4.5	63.6	1,375	3,032	8.48	4.54	0.80	1.09
	FISH, WHITE, Families—*Gadidae, Lophidae & Rajidae*																	
38	MEAL MECH EXTD	5–02–025	A-F	91	23.1	0.5	—	—	—	4.7	0.1	62.6	1,174	2,588	6.60	3.98	0.46	—
			M-F	100	25.4	0.6	—	—	—	5.1	0.1	68.8	1,289	2,843	7.25	4.37	0.51	—
	GELATIN																	
39	PROCESS RESIDUE (GELATIN BY-PRODUCTS)	5–14–503	A-F	90	—	—	—	—	—	0.0	—	87.5	970	2,138	0.49	—	—	—
			M-F	100	—	—	—	—	—	0.1	—	97.4	1,079	2,379	0.55	—	—	—
40	HOMINY (OR CORN) GRITS	4–03–011	A-F	90	2.8	4.8	49.5	11.7	—	6.5	65.8	10.3	1,313	2,894	0.05	0.51	0.08	0.05
			M-F	100	3.1	5.3	55.0	13.0	—	7.2	72.9	11.4	1,455	3,208	0.05	0.57	0.09	0.06
41	LIMESTONE, GROUND	6–02–632	A-F	100	93.8	—	—	—	—	—	—	—	—	—	37.12	0.21	0.06	0.03
			M-F	100	94.1	—	—	—	—	—	—	—	—	—	37.22	0.22	0.06	0.03
	LIVER																	
42	MEAL	5–00–389	A-F	93	6.3	1.4	—	—	—	15.7	3.2	66.1	1,306	2,878	0.56	1.26	—	—
			M-F	100	6.8	1.5	—	—	—	17.0	3.5	71.4	1,410	3,109	0.61	1.36	—	—
	MEAT																	
43	MEAL RENDERED	5–00–385	A-F	94	28.1	2.7	—	—	—	9.1	3.3	50.7	947	2,088	8.61	4.58	1.05	1.11
			M-F	100	29.9	2.9	—	—	—	9.7	3.5	54.0	1,009	2,225	9.18	4.88	1.11	1.18
44	WITH BONE, MEAL RENDERED	5–00–388	A-F	93	28.0	2.4	—	—	—	10.0	2.6	50.4	946	2,086	10.00	4.94	0.72	0.75
			M-F	100	30.0	2.6	—	—	—	10.7	2.8	54.0	1,014	2,236	10.72	5.30	0.77	0.80
	MILK, CATTLE *Bos taurus*																	
45	SKIMMED, DEHY	5–01–175	A-F	94	8.0	0.2	0.0	—	—	1.1	51.6	33.3	1,152	2,539	1.28	1.02	0.51	0.90
			M-F	100	8.4	0.2	0.0	—	—	1.2	54.8	35.4	1,223	2,696	1.36	1.09	0.54	0.96
	MILLET, PEARL *Pennisetum glaucum*																	
46	GRAIN	4–03–118	A-F	90	2.2	3.7	—	—	—	4.3	67.2	13.0	1,155	2,546	0.05	0.31	0.04	0.14
			M-F	100	2.5	4.1	—	—	—	4.7	74.4	14.3	1,278	2,817	0.05	0.34	0.04	0.16
	MILLET, PROSO (BROOMCORN; HOG MILLET) *Panicum miliaceum*																	
47	GRAIN	4–03–120	A-F	90	2.9	5.3	—	14.9	3.2	3.6	66.3	11.6	1,311	2,890	0.03	0.30	—	—
			M-F	100	3.3	6.0	—	16.6	3.6	4.0	73.9	12.9	1,461	3,221	0.03	0.34	—	—
	MOLASSES AND SYRUP																	
48	BEET SUGAR MOLASSES, MORE THAN 48% INVERT SUGAR, MORE THAN 79.5° BRIX	4–00–668	A-F	78	8.9	—	—	—	—	0.2	62.2	6.6	875	1,929	0.12	0.03	1.16	1.28
			M-F	100	11.4	—	—	—	—	0.2	79.9	8.5	1,123	2,476	0.16	0.03	1.48	1.64
49	SUGAR CANE, MOLASSES, DEHY	4–04–695	A-F	94	12.5	6.3	—	—	—	0.9	65.0	9.7	1,227	2,706	1.04	0.42	0.19	—
			M-F	100	13.3	6.7	—	—	—	0.9	68.8	10.3	1,300	2,866	1.10	0.45	0.20	—
50	SUGAR CANE, MOLASSES, MORE THAN 46% INVERT SUGAR, MORE THAN 79.5° BRIX (BLACKSTRAP)	4–04–696	A-F	74	9.8	0.4	—	0.3	0.2	0.2	59.7	4.3	870	1,918	0.74	0.08	0.16	2.26
			M-F	100	13.2	0.5	—	0.4	0.3	0.2	80.2	5.8	1,170	2,579	1.00	0.11	0.22	3.04
	OATS *Avena sativa*																	
51	GRAIN, ALL ANALYSES	4–03–309	A-F	89	3.1	10.7	26.4	14.2	2.7	4.7	58.9	11.9	1,150	2,536	0.08	0.34	0.05	0.09
			M-F	100	3.4	11.9	29.6	15.9	3.0	5.2	66.1	13.3	1,290	2,844	0.09	0.38	0.06	0.10

18–1
AS-FED AND MOISTURE-FREE

Entry Number	Magnesium (Mg) %	Potassium (K) %	Sulfur (S) %	Cobalt (Co) ppm or mg/kg	Copper (Cu) ppm or mg/kg	Iodine (I) ppm or mg/kg	Iron (Fe) %	Manganese (Mn) ppm or mg/kg	Selenium (Se) ppm or mg/kg	Zinc (Zn) ppm or mg/kg	A IU/g	Carotene (Provitamin A) ppm or mg/kg	D IU/kg	E ppm or mg/kg	K ppm or mg/kg	B–12 ppb or mcg/kg	Biotin ppm or mg/kg	Choline ppm or mg/kg	Folacin (Folic Acid) ppm or mg/kg	Niacin ppm or mg/kg	Pantothenic Acid (B–3) ppm or mg/kg	(Pyridoxine) B–6 ppm or mg/kg	Riboflavin (B–2) ppm or mg/kg	Thiamin (B–1) ppm or mg/kg
27	—	—	—	—	—	—	—	—	—	—	—	—	—	—	—	—	—	—	—	—	—	—	—	—
	—	—	—	—	—	—	—	—	—	—	—	—	—	—	—	—	—	—	—	—	—	—	—	—
28	0.21	0.40	0.25	0.11	15.1	—	0.04	23.6	—	99.1	—	—	—	19.2	—	258.6	—	3,644	—	75	15.0	14.68	5.6	0.8
	0.23	0.44	0.27	0.12	16.5	—	0.04	25.8	—	108.2	—	—	—	20.9	—	282.3	—	3,979	—	82	16.3	16.03	6.1	0.9
29	0.03	1.64	0.12	0.07	46.6	1.11	0.03	13.2	—	43.2	2.2	1.3	—	—	—	506.6	0.14	3,370	0.22	176	35.7	12.20	12.9	5.5
	0.06	3.24	0.25	0.14	92.4	2.20	0.06	26.2	—	85.6	4.4	2.6	—	—	—	1004.2	0.28	6,680	0.44	348	70.7	24.19	25.5	11.0
30	0.30	2.50	0.45	—	20.0	—	0.10	50.4	2.69	76.7	—	—	—	6.1	—	485.9	0.40	5,525	0.57	256	50.4	19.71	13.5	7.4
	0.32	2.69	0.48	—	21.5	—	0.10	54.3	2.90	82.6	—	—	—	6.5	—	523.6	0.43	5,953	0.62	276	54.3	21.24	14.6	8.0
31	0.25	0.72	0.78	0.17	9.1	3.14	0.02	11.0	1.36	105.0	—	—	—	3.7	—	214.5	0.20	3,700	0.16	81	10.0	4.71	7.3	0.5
	0.27	0.78	0.84	0.19	9.9	3.41	0.02	11.9	1.47	114.2	—	—	—	4.0	—	233.2	0.21	4,023	0.17	88	10.9	5.12	8.0	0.6
32	0.15	0.70	0.56	0.15	10.3	1.09	0.06	37.0	2.15	144.2	—	—	—	6.8	—	122.0	0.18	3,112	0.15	55	8.6	3.80	4.8	0.6
	0.17	0.77	0.61	0.17	11.3	1.91	0.06	40.4	2.34	157.5	—	—	—	7.4	—	133.2	0.20	3,398	0.17	60	9.4	4.15	5.3	0.6
33	—	—	—	—	—	—	0.00	—	—	—	—	—	—	—	—	—	—	—	—	—	—	—	—	—
	—	—	—	—	—	—	0.01	—	—	—	—	—	—	—	—	—	—	—	—	—	—	—	—	—
34	0.10	0.32	—	0.18	20.2	—	0.03	23.2	1.77	—	—	—	—	—	—	238.0	0.10	3,277	—	75	11.0	—	5.4	0.3
	0.11	0.35	—	0.20	21.7	—	0.03	24.9	1.90	—	—	—	—	—	—	255.5	0.11	3,518	—	81	11.8	—	5.8	0.3
35	0.17	—	—	—	51.1	—	0.02	90.1	—	112.6	—	—	—	—	—	—	—	3,663	—	64	9.0	—	6.8	—
	0.19	—	—	—	56.1	—	0.02	99.0	—	123.6	—	—	—	—	—	—	—	4,021	—	70	9.9	—	7.5	—
36	—	—	—	—	—	—	0.00	—	—	—	—	—	—	—	—	—	—	—	—	—	—	—	—	—
	—	—	—	—	—	—	0.00	—	—	—	—	—	—	—	—	—	—	—	—	—	—	—	—	—
37	0.23	0.72	0.68	0.18	10.3	—	0.04	8.4	4.30	210.7	—	—	—	5.6	—	300.1	0.20	2,993	—	144	7.7	—	6.8	1.5
	0.25	0.77	0.73	0.19	11.1	—	0.04	9.1	4.64	227.2	—	—	—	6.0	—	323.5	0.22	3,227	—	155	8.4	—	7.3	1.6
38	0.18	0.45	—	3.36	4.1	—	0.03	9.4	1.61	69.1	—	—	—	8.9	—	84.3	0.08	4,295	0.35	59	9.9	5.30	9.1	1.7
	0.20	0.49	—	3.37	4.5	—	0.03	10.3	1.77	75.9	—	—	—	9.8	—	92.6	0.09	4,719	0.38	65	10.9	5.83	10.0	1.8
39	0.05	—	—	—	—	—	—	—	—	—	—	—	—	—	—	—	—	—	—	—	—	—	—	—
	0.05	—	—	—	—	—	—	—	—	—	—	—	—	—	—	—	—	—	—	—	—	—	—	—
40	0.24	0.59	0.03	0.06	13.6	—	0.01	14.5	—	—	15.4	9.2	—	—	—	—	0.13	1,154	0.31	47	8.2	10.95	2.1	8.1
	0.26	0.65	0.03	0.06	15.1	—	0.01	16.1	—	—	17.0	10.2	—	—	—	—	0.15	1,280	0.34	52	9.1	12.14	2.4	8.9
41	1.13	0.11	0.04	—	11.0	—	0.36	269.0	19.00	—	—	—	—	—	—	—	—	—	—	—	—	—	—	—
	1.13	0.11	0.04	—	11.0	—	0.36	270.0	19.00	—	—	—	—	—	—	—	—	—	—	—	—	—	—	—
42	0.10	—	—	0.14	89.4	—	0.06	8.8	—	61.8	—	—	—	—	—	501.3	0.02	11,370	5.56	205	29.2	—	36.2	0.2
	0.11	—	—	0.15	96.5	—	0.07	9.5	—	66.8	—	—	—	—	—	541.5	0.02	12,281	6.01	221	31.5	—	39.1	0.2
43	0.25	0.55	0.46	2.25	9.6	—	0.05	11.8	0.51	74.3	—	—	—	0.9	—	75.2	0.12	1,980	0.39	56	6.0	4.23	5.2	0.2
	0.27	0.58	0.49	2.40	10.2	—	0.05	12.6	0.54	79.2	—	—	—	1.0	—	80.1	0.13	2,110	0.42	60	6.4	4.51	5.5	0.2
44	1.02	1.33	0.25	0.18	1.5	1.32	0.07	13.3	0.26	94.3	—	—	—	0.9	—	118.4	0.10	2,049	0.37	51	5.5	5.86	4.7	0.2
	1.09	1.43	0.27	0.19	1.6	1.41	0.07	14.3	0.28	101.1	—	—	—	0.9	—	126.9	0.11	2,195	0.40	55	5.9	6.28	5.0	0.2
45	0.12	1.60	0.32	0.11	11.7	—	0.00	2.1	0.12	38.5	—	—	0.0	9.1	—	50.9	0.33	1,394	0.62	11	36.4	4.10	19.1	3.7
	0.13	1.70	0.34	0.12	12.4	—	0.00	2.3	0.13	40.9	—	—	0.0	9.6	—	54.1	0.35	1,480	0.66	12	38.6	4.35	20.3	3.9
46	0.16	0.43	0.13	0.05	22.1	—	0.01	31.0	—	13.3	4.3	2.6	—	—	—	—	—	790	—	52	8.8	—	1.8	7.1
	0.18	0.48	0.14	0.05	24.5	—	0.01	34.3	—	14.7	4.8	2.9	—	—	—	—	—	874	—	57	9.7	—	2.0	7.9
47	0.16	0.43	—	—	—	—	0.01	—	—	—	—	—	—	—	—	—	—	438	—	24	10.9	—	3.3	7.5
	0.18	0.48	—	—	—	—	0.01	—	—	—	—	—	—	—	—	—	—	488	—	27	12.2	—	3.7	8.3
48	0.23	4.73	0.46	0.36	16.8	—	0.01	4.5	—	14.0	—	—	—	4.0	—	—	—	827	—	41	4.5	—	2.3	—
	0.29	6.07	0.60	0.47	21.6	—	0.01	5.8	—	18.0	—	—	—	5.1	—	—	—	1,062	—	53	5.8	—	2.9	—
49	0.44	3.40	0.43	1.15	74.9	—	0.02	54.1	—	31.2	—	—	—	5.2	—	—	—	—	—	—	—	—	—	—
	0.47	3.60	0.46	1.21	79.4	—	0.03	57.3	—	33.0	—	—	—	5.5	—	—	—	—	—	—	—	—	—	—
50	0.31	2.98	0.35	1.18	48.9	1.56	0.02	43.7	—	15.6	—	—	—	5.4	—	—	0.69	764	0.11	36	37.4	4.21	2.8	0.9
	0.42	4.01	0.47	1.59	65.7	2.10	0.03	58.8	—	20.9	—	—	—	7.3	—	—	0.92	1,027	0.15	49	50.3	5.67	3.8	1.2
51	0.14	0.40	0.21	0.06	6.0	0.11	0.01	35.8	0.22	34.9	0.2	0.1	—	14.9	—	—	0.27	967	0.39	14	9.9	2.53	1.4	6.0
	0.16	0.45	0.23	0.06	6.7	0.13	0.01	40.1	0.24	39.2	0.2	0.1	—	16.8	—	—	0.30	1,084	0.44	16	11.1	2.84	1.5	6.8

(Continued)

TABLE
COMPOSITION OF FEEDS, DATA EXPRESSED

Entry Number	Feed Name Description	International Feed Number	Moisture Basis A-F (as-fed) or M-F (moisture-free)	Dry Matter	Ash	Crude Fiber	Neutral Det. Fib. (NDF)	Acid Det. Fib. (ADF)	Lignin	Ether Extract (Fat)	N-Free Extract	Crude Protein	Poultry ME$_n$ kcal/lb	Poultry ME$_n$ kcal/kg	Calcium (Ca)	Phosphorus (P)	Sodium (Na)	Chlorine (Cl)
				%	%	%	%	%	%	%	%	%	kcal/lb	kcal/kg	%	%	%	%
	OATS (Continued)																	
52	GRAIN, PACIFIC COAST	4-07-999	A-F	91	3.8	11.2	—	—	—	5.0	61.8	9.1	1,199	2,644	0.10	0.31	0.06	0.12
			M-F	100	4.2	12.3	—	—	—	5.5	68.0	10.0	1,320	2,909	0.11	0.34	0.07	0.13
53	HULLS	1-03-281	A-F	92	6.1	30.9	68.6	37.4	6.5	1.3	50.5	3.7	162	357	0.14	0.14	0.04	0.08
			M-F	100	6.6	33.4	74.3	40.5	7.0	1.4	54.7	4.0	175	386	0.15	0.15	0.04	0.08
54	OYSTER SHELLS, GROUND (FLOUR)	6-03-481	A-F	99	79.0	1.8	—	—	—	0.3	17.0	0.7	—	—	35.85	0.10	0.21	0.01
			M-F	100	79.9	1.8	—	—	—	0.3	17.2	0.7	—	—	36.27	0.10	0.21	0.01
	PEA *Pisum spp*																	
55	SPLIT SEED BY-PRODUCT (PEA FEED OR MEAL)	1-08-478	A-F	90	3.5	23.7	—	—	—	1.4	43.7	17.7	—	—	—	—	—	—
			M-F	100	3.9	26.3	—	—	—	1.6	48.6	19.7	—	—	—	—	—	—
	PEA, FIELD *Pisum sativum, arvense*																	
56	SEEDS	5-08-481	A-F	91	2.9	5.9	—	—	—	1.3	57.8	23.2	1,108	2,442	0.16	0.38	0.00	—
			M-F	100	3.1	6.5	—	—	—	1.4	63.5	25.4	1,216	2,681	0.17	0.41	0.00	—
	PEANUT *Arachis hypogaea*																	
57	SEEDS WITHOUT HULLS, MEAL MECH EXTD (PEANUT MEAL)	5-03-649	A-F	93	5.0	6.2	13.2	5.6	1.0	5.6	26.7	49.2	1,207	2,662	0.20	0.56	0.12	0.03
			M-F	100	5.4	6.7	14.2	6.1	1.1	6.0	28.8	53.1	1,303	2,873	0.22	0.61	0.13	0.03
58	SEEDS WITHOUT HULLS, MEAL SOLV EXTD (PEANUT MEAL)	5-03-650	A-F	93	5.8	7.7	—	—	—	2.2	27.9	49.0	1,229	2,709	0.36	0.61	0.03	0.03
			M-F	100	6.3	8.3	—	—	—	2.4	30.1	52.9	1,328	2,928	0.39	0.66	0.03	0.03
	POULTRY																	
59	BY-PRODUCT, MEAL RENDERED	5-03-798	A-F	94	14.8	2.2	—	—	—	13.1	2.6	61.2	1,300	2,865	3.97	2.06	0.78	0.54
			M-F	100	15.8	2.3	—	—	—	13.9	2.7	65.3	1,385	3,054	4.23	2.20	0.83	0.58
60	FEATHERS, MEAL, HYDROLYZED	5-03-795	A-F	93	3.2	1.4	—	6.1	—	5.1	—	83.8	1,104	2,434	0.30	0.62	0.63	0.28
			M-F	100	3.4	1.5	—	6.6	—	5.5	—	90.2	1,188	2,619	0.33	0.67	0.68	0.30
	RICE *Oryza sativa*																	
61	GRAIN, GROUND (GROUND ROUGH RICE, GROUND PADDY RICE)	4-03-938	A-F	89	5.3	8.6	—	—	—	1.6	65.9	7.5	1,210	2,668	0.07	0.32	0.06	0.07
			M-F	100	6.0	9.7	—	—	—	1.8	74.1	8.4	1,360	2,999	0.07	0.36	0.07	0.08
62	BRAN WITH GERM (RICE BRAN)	4-03-928	A-F	91	11.3	11.9	28.0	25.7	3.6	13.5	41.0	13.0	920	2,028	0.07	1.44	0.03	0.07
			M-F	100	12.5	13.1	30.9	28.4	4.0	14.9	45.2	14.3	1,015	2,238	0.08	1.59	0.04	0.08
63	GROATS, POLISHED (RICE, POLISHED)	4-03-942	A-F	89	0.5	0.4	14.2	0.9	—	0.5	80.3	7.0	1,399	3,085	0.02	0.11	0.01	0.04
			M-F	100	0.6	0.4	16.0	1.0	—	0.5	90.6	7.9	1,580	3,483	0.03	0.13	0.02	0.04
64	POLISHINGS	4-03-943	A-F	90	7.6	3.2	—	3.6	—	12.6	54.9	12.0	1,367	3,015	0.05	1.34	0.04	0.11
			M-F	100	8.4	3.5	—	4.0	—	13.9	60.9	13.3	1,515	3,340	0.05	1.49	0.05	0.12
	RYE *Secale cereale*																	
65	GRAIN, ALL ANALYSES	4-04-047	A-F	87	1.6	2.2	—	—	—	1.5	70.0	12.0	1,202	2,651	0.06	0.31	0.02	0.03
			M-F	100	1.9	2.5	—	—	—	1.7	80.1	13.8	1,375	3,031	0.07	0.36	0.03	0.03
	SAFFLOWER *Carthamus tinctorius*																	
66	SEEDS, MEAL SOLV EXTD, 20% PROTEIN	5-26-095	A-F	92	4.6	32.2	—	39.6	—	1.1	32.7	21.6	623	1,374	0.31	0.61	—	—
			M-F	100	5.0	34.9	—	43.0	—	1.2	35.5	23.4	676	1,491	0.34	0.66	—	—
67	SEEDS WITHOUT HULLS, MEAL SOLV EXTD, 42% PROTEIN	5-26-094	A-F	92	6.5	14.6	—	19.2	—	1.3	26.3	42.7	885	1,951	0.38	1.08	—	—
			M-F	100	7.2	16.0	—	21.0	—	1.5	28.8	46.7	967	2,131	0.41	1.18	—	—
	SESAME *Sesamum indicum*																	
68	SEEDS	5-08-509	A-F	95	5.8	10.6	—	—	—	44.1	10.5	23.7	—	—	0.97	0.72	—	—
			M-F	100	6.1	11.2	—	—	—	46.6	11.1	25.0	—	—	1.02	0.76	—	—
69	SEEDS, MEAL MECH EXTD	5-04-220	A-F	93	10.3	5.6	15.8	15.8	—	8.7	23.0	45.0	985	2,172	2.01	1.36	0.05	0.07
			M-F	100	11.2	6.1	17.0	17.0	—	9.4	24.8	48.6	1,064	2,345	2.17	1.46	0.05	0.07
70	SEEDS, MEAL SOLV EXTD, 44% PROTEIN	5-26-096	A-F	92	13.1	6.8	—	—	—	1.4	25.8	45.0	1,178	2,598	2.01	1.28	—	—
			M-F	100	14.2	7.4	—	—	—	1.5	28.0	48.9	1,281	2,824	2.18	1.39	—	—
	SORGHUM *Sorghum bicolor*																	
71	GRAIN, ALL ANALYSES	4-04-383	A-F	90	1.8	2.6	16.2	8.1	1.2	2.7	71.6	11.5	—	—	0.05	0.32	0.03	0.08
			M-F	100	1.9	2.8	18.0	9.0	1.3	2.9	79.5	12.8	—	—	0.06	0.35	0.03	0.09
72	GRAIN, 9-12% PROTEIN	4-08-139	A-F	89	1.9	2.4	—	—	—	2.7	72.2	9.8	1,517	3,345	0.03	0.27	0.02	—
			M-F	100	2.1	2.6	—	—	—	3.1	81.1	11.0	1,706	3,760	0.04	0.30	0.02	—
73	GRAIN, MORE THAN 12% PROTEIN	4-08-140	A-F	89	2.3	1.8	—	—	—	1.5	71.8	11.6	—	—	0.03	0.29	0.04	—
			M-F	100	2.6	2.0	—	—	—	1.7	80.7	13.0	—	—	0.03	0.32	0.05	—
	SOYBEAN *Glycine max*																	
74	SEEDS, MEAL MECH EXTD, 41% PROTEIN	5-04-600	A-F	90	6.0	6.0	—	—	—	4.7	30.4	42.9	1,102	2,429	0.26	0.61	0.18	0.07
			M-F	100	6.7	6.7	—	—	—	5.2	33.8	47.7	1,224	2,699	0.29	0.68	0.20	0.08
75	SEEDS, MEAL SOLV EXTD, 44% PROTEIN	5-20-637	A-F	89	6.4	6.2	12.5	8.9	—	1.5	30.6	44.4	1,005	2,216	0.35	0.64	0.03	—
			M-F	100	7.2	7.0	14.0	10.0	—	1.7	34.3	49.8	1,128	2,486	0.40	0.71	0.04	—
76	SEEDS, WITHOUT HULLS, MEAL SOLV EXTD, 49% PROTEIN	5-20-638	A-F	90	6.1	3.7	6.6	6.2	—	1.2	29.8	49.0	1,124	2,478	0.25	0.63	0.00	0.07
			M-F	100	6.8	4.1	7.4	6.9	—	1.4	33.2	54.6	1,253	2,761	0.28	0.70	0.00	0.08
77	FLOUR, SOLV EXTD	5-04-593	A-F	93	6.1	2.5	—	—	—	0.8	32.3	51.6	—	—	0.38	0.65	0.00	0.20
			M-F	100	6.5	2.7	—	—	—	0.8	34.7	55.3	—	—	0.40	0.70	0.00	0.22
78	FLOUR BY-PRODUCT (SOYBEAN MILL FEED)	4-04-594	A-F	90	5.0	33.7	—	—	—	1.9	36.4	12.6	362	798	0.47	0.18	—	—
			M-F	100		37.6	—	—	—	2.1	40.7	14.1	404	890	0.52	0.20	—	—
79	PROTEIN CONCENTRATE, MORE THAN 70% PROTEIN	5-08-038	A-F	92	3.5	0.1	—	—	—	0.5	3.4	84.3	1,121	2,472	0.11	0.68	0.12	—
			M-F	100	3.8	0.1	—	—	—	0.6	3.7	91.8	1,222	2,695	0.12	0.74	0.13	—

18–1
AS-FED AND MOISTURE-FREE

Entry Number	Magnesium (Mg) %	Potassium (K) %	Sulfur (S) %	Cobalt (Co) ppm or mg/kg	Copper (Cu) ppm or mg/kg	Iodine (I) ppm or mg/kg	Iron (Fe) %	Manganese (Mn) ppm or mg/kg	Selenium (Se) ppm or mg/kg	Zinc (Zn) ppm or mg/kg	A IU/g	Carotene (Provitamin A) ppm or mg/kg	D IU/kg	E ppm or mg/kg	K ppm or mg/kg	B–12 mcg/kg	Biotin ppm or mg/kg	Choline ppm or mg/kg	Folacin (Folic Acid) ppm or mg/kg	Niacin ppm or mg/kg	Pantothenic Acid (B–3) ppm or mg/kg	(Pyridoxine) B–6 ppm or mg/kg	Riboflavin (B–2) ppm or mg/kg	Thiamin (B–1) ppm or mg/kg
52	0.17	0.38	0.20	—	—	—	0.01	38.0	0.08	—	—	—	—	20.2	—	—	—	917	—	14	11.7	—	1.2	—
	0.19	0.42	0.22	—	—	—	0.01	41.8	0.08	—	—	—	—	22.2	—	—	—	1,009	—	16	12.8	—	1.3	—
53	0.08	0.57	0.14	—	4.1	—	0.01	18.8	—	—	—	—	—	—	—	—	—	260	0.96	9	3.1	2.20	1.5	0.6
	0.09	0.62	0.15	—	4.5	—	0.01	20.4	—	—	—	—	—	—	—	—	—	281	1.04	10	3.4	2.37	1.7	0.7
54	0.24	0.10	—	—	15.0	—	0.25	178.0	—	7.0	—	—	—	—	—	—	—	—	—	—	—	—	—	—
	0.24	0.10	—	—	15.0	—	0.26	180.0	—	7.0	—	—	—	—	—	—	—	—	—	—	—	—	—	—
55	—	—	—	—	—	—	—	—	—	—	—	—	—	—	—	—	—	—	—	—	—	—	—	—
	—	—	—	—	—	—	—	—	—	—	—	—	—	—	—	—	—	—	—	—	—	—	—	—
56	0.14	1.36	—	1.70	11.7	—	0.02	21.2	0.39	46.8	—	—	—	—	—	—	0.19	654	0.36	34	7.4	1.01	1.4	4.1
	0.15	1.49	—	1.87	12.9	—	0.02	23.3	0.43	51.4	—	—	—	—	—	—	0.21	718	0.40	37	8.2	1.11	1.5	4.5
57	0.26	1.16	0.22	0.11	15.4	0.07	0.03	25.5	—	33.0	—	—	—	2.4	—	—	0.33	1,975	0.66	173	47.6	6.12	9.1	5.7
	0.28	1.25	0.24	0.12	16.6	0.07	0.03	27.6	—	35.6	—	—	—	2.6	—	—	0.36	2,132	0.71	186	51.4	6.61	9.8	6.2
58	0.27	1.16	0.31	—	—	—	—	—	—	—	—	—	—	2.9	—	—	—	1,896	—	178	36.8	5.95	5.3	—
	0.30	1.25	0.33	—	—	—	—	—	—	—	—	—	—	3.2	—	—	—	2,049	—	192	39.8	6.43	5.7	—
59	0.14	0.51	0.53	4.93	19.9	3.10	0.06	16.5	0.92	193.5	—	—	—	2.2	—	304.1	0.09	6,052	0.51	54	12.4	4.43	10.6	0.2
	0.14	0.55	0.56	5.25	21.2	3.31	0.07	17.6	0.98	206.2	—	—	—	2.4	—	324.1	0.09	6,451	0.54	57	13.2	4.72	11.3	0.2
60	0.18	0.27	1.50	0.12	7.3	0.04	0.02	11.9	0.91	71.9	—	—	—	—	—	80.4	0.04	894	0.22	21	8.9	4.39	2.0	0.1
	0.19	0.29	1.61	0.13	7.9	0.05	0.03	12.9	0.98	77.3	—	—	—	—	—	86.5	0.05	962	0.23	23	9.6	4.72	2.2	0.1
61	0.13	0.47	0.05	—	—	—	—	18.0	—	15.0	—	—	—	14.0	—	—	—	926	0.25	40	7.1	—	0.7	—
	0.14	0.53	0.05	—	—	—	—	20.2	—	16.9	—	—	—	15.7	—	—	—	1,041	0.28	45	8.0	—	0.8	—
62	0.85	1.69	0.18	1.38	11.0	—	0.02	337.6	—	37.4	—	—	—	60.4	—	—	0.43	1,230	2.20	299	22.8	13.24	2.6	22.4
	0.94	1.87	0.20	1.53	12.1	—	0.02	372.4	—	41.3	—	—	—	66.7	—	—	0.47	1,357	2.42	330	25.2	14.61	2.8	24.8
63	0.09	0.23	0.08	0.85	5.4	—	0.00	29.6	—	13.7	—	—	—	3.5	—	—	—	901	0.15	15	3.5	0.39	0.6	0.7
	0.10	0.26	0.09	1.00	6.1	—	0.00	33.4	—	15.4	—	—	—	4.0	—	—	—	1,017	0.17	17	3.9	0.45	0.6	0.7
64	0.60	1.28	0.17	3.89	8.0	—	0.01	126.8	—	63.2	—	—	—	90.2	—	—	0.62	1,248	—	506	46.4	27.89	1.8	20.0
	0.66	1.41	0.19	4.31	8.8	—	0.01	140.5	—	70.0	—	—	—	100.0	—	—	0.68	1,383	—	560	51.4	30.90	2.0	22.1
65	0.12	0.46	0.15	—	7.5	—	0.01	72.0	—	28.1	0.1	0.1	—	14.5	—	—	0.06	419	0.62	14	7.5	—	1.7	4.1
	0.14	0.52	0.17	—	8.6	—	0.01	82.3	—	32.2	0.2	0.1	—	16.6	—	—	0.06	479	0.71	16	8.5	—	1.9	4.7
66	0.32	0.74	0.20	—	9.6	—	0.04	17.7	—	39.6	—	—	—	0.9	—	—	—	1,541	—	12	36.2	474.43	2.2	—
	0.35	0.80	0.22	—	10.4	—	0.05	19.2	—	43.0	—	—	—	1.0	—	—	—	1,673	—	13	39.3	515.00	2.4	—
67	1.18	1.18	0.34	1.83	80.6	—	0.09	36.6	—	168.5	—	—	—	0.6	—	—	1.56	3,156	1.47	21	38.2	10.71	2.3	4.2
	1.29	1.29	0.38	2.00	88.0	—	0.10	40.0	—	184.0	—	—	—	0.7	—	—	1.70	3,447	1.60	23	41.7	11.70	2.5	4.6
68	—	—	—	—	—	—	—	—	—	—	—	—	—	—	—	—	—	—	—	—	—	—	—	—
	—	—	—	—	—	—	—	—	—	—	—	—	—	—	—	—	—	—	—	—	—	—	—	—
69	0.46	1.25	0.33	—	—	—	0.01	47.7	—	99.6	0.7	0.4	—	—	—	—	—	1,533	—	19	5.9	12.45	3.4	2.8
	0.50	1.35	0.35	—	—	—	0.01	51.5	—	107.5	0.8	0.5	—	—	—	—	—	1,655	—	20	6.4	13.44	3.6	3.0
70	—	—	—	—	—	—	—	47.5	—	—	—	—	—	—	—	—	—	1,517	—	—	6.3	—	3.7	—
	—	—	—	—	—	—	—	51.6	—	—	—	—	—	—	—	—	—	1,649	—	—	6.8	—	4.0	—
71	0.14	0.35	0.15	0.28	9.7	—	0.01	9.8	—	42.4	2.0	1.2	—	—	—	—	0.26	686	0.22	47	10.2	5.41	1.2	4.5
	0.16	0.38	0.17	0.31	10.8	—	0.01	10.9	—	47.1	2.2	1.3	—	—	—	—	0.29	762	0.24	52	11.3	6.00	1.4	5.0
72	0.15	0.34	0.14	0.07	9.7	0.02	0.00	15.4	—	13.7	—	—	—	1.3	—	—	0.29	762	0.22	48	12.8	4.60	1.3	4.4
	0.17	0.38	0.16	0.08	10.9	0.03	0.00	17.3	—	15.4	—	—	—	1.5	—	—	0.32	857	0.25	54	14.4	5.17	1.5	5.0
73	0.17	0.34	0.16	—	—	—	0.01	—	—	—	—	—	—	—	—	—	—	—	—	—	—	—	—	—
	0.19	0.38	0.18	—	—	—	0.01	—	—	—	—	—	—	—	—	—	—	—	—	—	—	—	—	—
74	0.26	1.79	0.33	0.18	21.7	—	0.02	31.3	0.10	57.2	0.4	0.2	—	6.5	—	—	0.33	2,623	6.39	31	14.3	7.22	3.4	3.9
	0.29	1.98	0.37	0.20	24.1	—	0.02	34.8	0.11	63.6	0.4	0.2	—	7.3	—	—	0.36	2,916	7.10	34	15.8	8.02	3.8	4.3
75	0.27	1.98	0.41	1.38	19.9	—	0.02	31.6	0.49	50.5	—	—	—	3.0	0.22	2.0	0.36	2,706	0.69	26	13.8	5.90	3.0	6.6
	0.31	2.22	0.47	1.55	22.3	—	0.02	35.5	0.55	56.6	—	—	—	3.4	0.25	2.2	0.41	3,036	0.77	29	15.5	6.62	3.4	7.4
76	0.37	1.79	0.41	2.69	13.5	0.15	0.01	49.5	—	51.1	—	—	0.0	3.3	—	2.0	0.38	2,772	0.59	24	14.1	5.59	2.9	3.5
	0.41	1.99	0.46	3.00	15.0	0.17	0.01	55.2	—	56.9	—	—	0.0	3.7	—	2.2	0.42	3,089	0.66	27	15.7	6.23	3.3	3.9
77	0.25	1.90	0.41	0.94	14.8	—	0.02	30.3	0.28	30.9	0.7	0.4	—	—	—	—	0.73	2,258	—	43	14.2	—	2.9	2.7
	0.27	2.04	0.44	1.01	15.8	—	0.02	32.4	0.30	33.1	0.8	0.5	—	—	—	—	0.78	2,420	—	46	15.2	—	3.1	2.9
78	0.32	1.51	—	—	—	—	—	28.5	—	—	—	—	—	—	—	—	0.22	444	0.22	24	13.3	2.24	3.5	2.2
	0.36	1.69	—	—	—	—	—	31.8	—	—	—	—	—	—	—	—	0.25	495	0.25	27	14.8	2.51	3.9	2.5
79	0.02	0.01	—	0.39	14.1	0.32	0.01	5.5	0.14	29.6	—	—	—	—	0.02	—	—	2	—	5	3.5	—	0.7	0.3
	0.02	0.01	—	0.42	15.4	0.35	0.02	6.0	0.15	32.3	—	—	—	—	0.02	—	—	2	—	5	3.8	—	0.8	0.4

(Continued)

TABLE
COMPOSITION OF FEEDS, DATA EXPRESSED

Entry Number	Feed Name Description	International Feed Number	Moisture Basis A-F (as-fed) or M-F (moisture-free)	Dry Matter	Ash	Crude Fiber	Neutral Det. Fib. (NDF)	Acid Det. Fib. (ADF)	Lignin	Ether Extract (Fat)	N-Free Extract	Crude Protein	Poultry ME$_n$ kcal/lb	Poultry ME$_n$ kcal/kg	Calcium (Ca)	Phosphorus (P)	Sodium (Na)	Chlorine (Cl)
				%	%	%	%	%	%	%	%	%	kcal/lb	kcal/kg	%	%	%	%
	SUNFLOWER, COMMON *Helianthus annuus*																	
80	SEEDS WITHOUT HULLS, MEAL MECH EXTD, 41% PROTEIN	5–26–097	A-F	92	6.7	13.2	—	—	—	7.5	23.9	40.7	—	—	—	—	—	—
			M-F	100	7.3	14.3	—	—	—	8.2	26.0	44.2	—	—	—	—	—	—
81	SEEDS WITHOUT HULLS, MEAL SOLV EXTD, 44% PROTEIN	5–26–098	A-F	93	7.7	11.0	—	—	—	2.9	24.6	46.8	919	2,026	—	—	—	—
			M-F	100	8.3	11.8	—	—	—	3.1	26.5	50.3	988	2,178	—	—	—	—
	TRITICALE *Triticale hexaploide*																	
82	GRAIN	4–20–362	A-F	89	1.8	3.0	11.9	—	—	1.5	67.3	15.4	1,420	3,130	0.04	0.30	0.01	—
			M-F	100	2.0	3.3	13.3	—	—	1.7	75.7	17.3	1,597	3,521	0.04	0.34	0.01	—
	WHEAT *Triticum aestivum*																	
83	GRAIN, ALL ANALYSES	4–05–211	A-F	89	1.8	2.6	—	7.1	—	1.8	69.7	13.1	1,402	3,092	0.05	0.35	0.06	0.08
			M-F	100	2.0	2.9	—	8.0	—	2.0	78.4	14.7	1,576	3,474	0.06	0.39	0.06	0.09
84	GRAIN, HARD RED SPRING	4–05–258	A-F	88	1.7	2.6	37.9	11.0	—	1.8	67.4	14.2	1,225	2,701	0.04	0.37	0.02	0.08
			M-F	100	1.9	2.9	43.3	12.6	—	2.1	76.9	16.2	1,399	3,084	0.05	0.42	0.02	0.09
85	GRAIN, HARD RED WINTER	4–05–268	A-F	89	1.7	2.6	24.8	3.9	0.9	1.6	69.8	14.2	1,454	3,205	0.04	0.37	0.02	0.05
			M-F	100	2.0	2.9	28.0	4.4	1.0	1.8	78.8	14.5	1,642	3,620	0.05	0.42	0.02	0.06
86	GRAIN, SOFT RED WINTER	4–05–294	A-F	88	1.9	2.3	—	—	—	1.6	71.2	11.4	1,403	3,093	0.05	0.36	0.01	0.07
			M-F	100	2.1	2.6	—	—	—	1.8	80.5	12.9	1,587	3,499	0.06	0.40	0.01	0.08
87	GRAIN, SOFT WHITE WINTER	4–05–337	A-F	90	1.5	2.3	12.6	3.6	—	1.5	75.0	10.2	1,299	2,864	—	0.40	0.02	—
			M-F	100	1.6	2.6	14.0	4.0	—	1.7	82.9	11.3	1,436	3,167	—	0.44	0.02	—
88	BRAN	4–05–190	A-F	89	5.9	10.0	40.9	12.0	2.6	4.0	53.6	15.5	556	1,225	0.13	1.16	0.06	0.05
			M-F	100	6.7	11.2	45.9	13.5	3.0	4.5	60.2	17.5	624	1,375	0.14	1.30	0.06	0.06
89	MIDDLINGS, LESS THAN 9.5% FIBER	4–05–205	A-F	89	4.7	7.7	32.9	8.9	—	4.3	55.7	16.4	940	2,072	0.13	0.89	0.01	0.04
			M-F	100	5.3	8.7	37.0	10.0	—	4.9	62.7	18.5	1,058	2,333	0.15	1.00	0.01	0.04
90	RED DOG, LESS THAN 4% FIBER	4–05–203	A-F	88	2.4	2.9	—	—	—	3.4	64.0	15.6	1,164	2,566	0.06	0.51	0.01	0.14
			M-F	100	2.7	3.3	—	—	—	3.8	72.5	17.6	1,319	2,908	0.07	0.58	0.02	0.16
91	SHORTS, LESS THAN 7% FIBER	4–05–201	A-F	88	4.4	6.4	—	—	—	4.6	56.5	16.5	1,001	2,206	0.09	0.80	0.03	0.05
			M-F	100	5.0	7.2	—	—	—	5.2	63.9	18.7	1,132	2,496	0.10	0.91	0.03	0.06
	WHEY, CATTLE *Bos taurus*																	
92	DEHY	4–01–182	A-F	93	8.8	0.2	0.3	0.2	—	0.8	70.2	13.3	880	1,939	0.86	0.76	0.62	0.07
			M-F	100	9.4	0.2	0.3	0.2	—	0.8	75.3	14.2	944	2,081	0.92	0.82	0.66	0.08
93	LOW LACTOSE, DEHY (DRIED WHEY PRODUCT)	4–01–186	A-F	93	15.4	0.2	0.0	0.0	—	1.0	60.0	16.7	931	2,053	1.49	1.11	1.44	1.03
			M-F	100	16.5	0.2	0.0	0.0	—	1.1	64.3	17.9	997	2,199	1.60	1.18	1.54	1.10
	YEAST, BREWERS *Saccharomyces cerevisiae*																	
94	DEHY	7–05–527	A-F	93	6.5	3.0	—	3.7	—	0.9	38.8	43.8	928	2,047	0.14	1.36	0.07	0.07
			M-F	100	7.0	3.2	—	4.0	—	1.0	41.7	47.1	998	2,199	0.15	1.47	0.08	0.08
	YEAST, IRRADIATED *Saccharomyes cerevisiae*																	
95	DEHY	7–05–529	A-F	94	6.2	6.2	—	—	—	1.1	32.4	48.1	—	—	0.78	1.42	—	—
			M-F	100	6.6	6.5	—	—	—	1.2	34.5	51.2	—	—	0.83	1.51	—	—
	YEAST, TORULA *Torulopsis utilis*																	
96	DEHY	7–05–534	A-F	93	8.0	2.5	—	3.7	—	1.6	31.5	49.6	840	1,851	0.55	1.61	0.01	0.02
			M-F	100	8.6	2.7	—	4.0	—	1.7	33.8	53.3	902	1,989	0.59	1.73	0.01	0.02

18–1
AS-FED AND MOISTURE-FREE

Entry Number	Mag-nesium (Mg)	Potas-sium (K)	Sulfur (S)	Cobalt (Co)	Copper (Cu)	Iodine (I)	Iron (Fe)	Man-ganese (Mn)	Sele-nium (Se)	Zinc (Zn)	A	Carotene (Provita-min A)	D	E	K	B–12	Biotin	Choline	Folacin (Folic Acid)	Niacin	Pantothenic Acid (B–3)	(Pyri-doxine) B–6	Ribo-flavin (B–2)	Thia-min (B–1)
	%	%	%	ppm or mg/kg	ppm or mg/kg	ppm or mg/kg	%	ppm or mg/kg	ppm or mg/kg	ppm or mg/kg	IU/g	ppm or mg/kg	IU/kg	ppm or mg/kg	ppm or mg/kg	ppb or mcg/kg	ppm or mg/kg	ppm or mg/kg	ppm or mg/kg	ppm or mg/kg	ppm or mg/kg	ppm or mg/kg	ppm or mg/kg	ppm or mg/kg
80	—	—	—	—	—	—	—	—	—	—	—	—	—	—	—	—	—	—	—	—	—	—	—	—
	—	—	—	—	—	—	—	—	—	—	—	—	—	—	—	—	—	—	—	—	—	—	—	—
81	—	—	—	—	—	—	—	—	—	—	—	—	—	—	—	—	—	—	—	—	—	—	—	—
	—	—	—	—	—	—	—	—	—	—	—	—	—	—	—	—	—	—	—	—	—	—	—	—
82	0.23	0.51	—	0.08	8.3	—	0.01	42.5	—	31.2	—	—	—	—	—	—	—	457	—	—	—	—	0.4	—
	0.26	0.57	—	0.09	9.3	—	0.01	47.8	—	35.1	—	—	—	—	—	—	—	514	—	—	—	—	0.5	—
83	0.14	0.14	0.18	0.44	5.8	0.09	0.01	46.5	0.26	31.4	—	—	—	15.5	—	0.9	0.10	918	0.43	59	11.3	3.74	1.3	4.3
	0.15	0.15	0.20	0.50	6.5	0.10	0.01	46.7	0.29	35.2	—	—	—	17.4	—	1.0	0.11	1,032	0.49	66	12.7	4.20	1.4	4.8
84	0.14	0.36	0.15	0.12	6.0	—	0.01	37.0	0.26	37.9	—	—	—	12.7	—	—	0.11	1,010	0.41	56	9.6	5.11	1.4	4.2
	0.16	0.41	0.17	0.14	6.8	—	0.01	42.2	0.30	43.3	—	—	—	14.4	—	—	0.13	1,153	0.46	64	11.0	5.83	1.6	4.8
85	0.12	0.43	0.14	0.15	5.1	—	0.00	30.4	0.29	35.2	—	—	—	11.1	—	—	0.11	1,004	0.38	53	10.1	3.01	1.3	4.5
	0.13	0.49	0.15	0.16	5.7	—	0.00	34.3	0.33	39.8	—	—	—	12.5	—	—	0.12	1,133	0.43	60	11.4	3.40	1.5	5.1
86	0.10	0.41	0.11	0.10	7.0	—	0.00	33.4	0.04	42.1	—	—	—	15.6	—	—	—	892	0.41	53	10.1	3.21	1.5	4.7
	0.11	0.46	0.12	0.12	8.0	—	0.00	37.8	0.05	47.7	—	—	—	17.7	—	—	—	1,009	0.46	60	11.4	3.63	1.7	5.3
87	—	—	0.12	0.14	7.1	—	0.00	36.2	0.05	27.1	—	—	—	30.9	—	—	—	—	—	62	11.2	4.79	—	—
	—	—	0.13	0.15	7.8	—	0.00	40.0	0.05	30.0	—	—	—	34.2	—	—	—	—	—	69	12.3	5.29	—	—
88	0.58	1.23	0.22	0.08	11.0	0.07	0.02	114.9	0.64	94.6	4.4	2.6	—	14.3	—	—	0.38	1,232	1.77	197	28.0	10.34	3.6	8.4
	0.65	1.38	0.25	0.08	12.4	0.07	0.02	129.0	0.72	106.2	4.9	2.9	—	16.0	—	—	0.42	1,383	1.98	221	31.4	11.61	4.0	9.4
89	0.34	0.98	0.17	0.50	15.9	0.11	0.01	114.0	0.74	96.9	5.1	3.1	—	23.8	—	—	0.24	1,246	1.24	95	17.8	9.14	2.0	14.2
	0.38	1.10	0.19	0.57	17.9	0.12	0.01	128.3	0.83	109.1	5.8	3.5	—	26.9	—	—	0.27	1,403	1.39	107	20.0	10.29	2.3	15.9
90	0.18	0.52	0.24	0.12	6.3	—	0.01	52.1	0.32	65.0	—	—	—	37.4	—	—	0.11	1,453	0.82	46	13.3	5.40	2.2	21.8
	0.21	0.59	0.27	0.13	7.1	—	0.01	59.1	0.37	73.7	—	—	—	42.4	—	—	0.12	1,648	0.93	52	15.0	6.12	2.5	24.7
91	0.27	0.93	0.21	0.11	11.5	—	0.01	114.1	0.48	102.4	5.1	3.1	—	36.0	—	—	—	1,697	1.51	105	21.9	—	4.1	19.5
	0.31	1.05	0.23	0.12	13.0	—	0.01	129.1	0.54	115.9	5.8	3.5	—	40.7	—	—	—	1,920	1.71	119	24.8	—	4.6	22.1
92	0.13	1.11	1.04	0.11	46.5	—	0.02	5.9	—	3.2	—	—	—	0.2	—	18.9	0.35	1,790	0.85	11	46.2	3.21	27.4	4.0
	0.14	1.19	1.11	0.12	49.9	—	0.02	6.3	—	3.4	—	—	—	0.2	—	20.3	0.38	1,921	0.91	11	49.6	3.45	29.4	4.3
93	0.21	2.95	1.07	—	7.0	9.85	0.03	8.0	0.05	7.9	—	—	—	—	—	35.9	0.50	4,096	0.89	18	74.5	4.48	47.6	5.0
	0.23	3.16	1.15	—	7.5	10.55	0.03	8.6	0.06	8.4	—	—	—	—	—	38.4	0.54	4,387	0.96	19	79.8	4.79	50.9	5.4
94	0.24	1.69	0.43	0.51	38.4	0.36	0.01	6.7	0.91	39.0	—	—	—	2.1	—	1.1	1.04	3,847	9.69	443	81.5	36.67	34.1	85.2
	0.26	1.82	0.46	0.54	41.3	0.38	0.01	7.2	0.98	41.9	—	—	—	2.3	—	1.1	1.12	4,134	10.41	476	87.6	39.40	36.6	91.6
95	—	2.14	—	—	—	—	—	—	—	—	—	—	—	—	—	—	—	—	—	—	—	—	18.5	—
	—	2.28	—	—	—	—	—	—	—	—	—	—	—	—	—	—	—	—	—	—	—	—	19.7	—
96	0.14	1.92	0.55	0.03	11.9	2.50	0.01	9.3	—	99.5	—	—	—	—	—	4.0	1.19	2,981	25.66	512	107.5	34.48	47.7	6.8
	0.15	2.06	0.59	0.03	12.8	2.69	0.01	10.0	—	107.0	—	—	—	—	—	4.3	1.27	3,203	27.58	550	115.6	37.06	51.3	7.3

TABLE 18–2
AMINO ACID, COMPOSITION OF FEEDS, DATA EXPRESSED AS-FED

Entry Number	Feed Name Description	International Feed Number	Dry Matter	Crude Protein	Argi-nine	Cystine	Glycine	Histi-dine	Isoleu-cine	Leucine	Lysine	Methi-onine	Phenyl-alanine	Serine	Threo-nine	Trypto-phan	Tyrosine	Valine
			%	%	%	%	%	%	%	%	%	%	%	%	%	%	%	%
	ENERGY FEEDS																	
1	BAKERY, WASTE, DEHY (DRIED BAKERY PRODUCT)	4–00–466	91	10.1	0.47	0.17	0.69	0.16	0.45	0.77	0.31	0.17	0.45	0.65	0.46	0.10	0.36	0.47
	BARLEY																	
2	GRAIN	4–00–549	88	11.7	0.51	0.20	0.37	0.25	0.46	0.75	0.40	0.16	0.58	0.46	0.36	0.15	0.35	0.57
3	GRAIN, PACIFIC COAST	4–07–939	89	9.5	0.44	0.19	0.30	0.21	0.40	0.60	0.26	0.14	0.47	0.32	0.31	0.12	0.31	0.46
4	GRAIN SCREENINGS	4–00–542	89	11.5	—	—	—	—	—	—	—	—	—	—	0.36	—	—	—
5	MALT SPROUTS, DEHY	5–00–545	93	22.9	1.05	0.23	0.81	0.43	0.88	1.36	1.12	0.31	0.80	0.47	0.85	0.41	0.46	1.16
6	BEAN, NAVY, SEEDS	5–00–623	89	22.9	1.19	0.23	0.80	—	—	—	1.29	0.23	—	—	—	—	0.24	—
7	BEET, SUGAR, PULP, DEHY	4–00–669	91	8.8	0.30	0.01	—	0.20	0.30	0.60	0.60	0.01	0.30	—	0.40	0.10	0.40	0.40
8	BROOMCORN (HOG MILLET; MILLET, PROSO), GRAIN	4–03–120	90	11.6	0.34	0.20	0.25	0.20	0.44	1.13	0.22	0.26	0.54	0.63	0.37	0.16	0.23	0.55
9	BUCKWHEAT, COMMON, GRAIN	4–00–994	88	11.1	0.96	0.17	0.61	0.26	0.37	0.59	0.62	0.19	0.44	0.41	0.44	0.18	0.21	0.53
10	CITRUS, PULP WITHOUT FINES, DEHY (DRIED CITRUS PULP)	4–01–237	91	6.1	0.25	0.11	—	0.09	0.18	0.31	0.20	0.09	0.18	—	0.18	0.06	—	0.25
11	CORN, DENT WHITE, GRAIN	4–02–928	90	10.8	0.27	0.09	—	0.18	0.45	0.90	0.27	0.09	0.36	—	0.36	0.09	0.45	0.36
	CORN, DENT YELLOW																	
12	GRAIN, ALL ANALYSES	4–02–935	88	9.9	0.43	0.12	0.37	0.27	0.35	1.19	0.30	0.18	0.46	0.49	0.36	0.09	0.31	0.48
13	GRAIN, GRADE 2, 54 LB/BU (69.5 KG/HL)	4–02–931	87	8.9	0.45	0.11	0.45	0.20	0.40	1.00	0.19	0.11	0.45	—	0.35	0.09	0.43	0.35
14	GRAIN, FLAKED	4–28–244	89	9.9	0.44	0.25	0.36	0.28	0.34	1.24	0.25	0.15	0.44	0.48	0.35	—	0.39	0.47
15	DISTILLERS SOLUBLES, DEHY	5–28–237	93	27.4	0.99	0.44	1.12	0.67	1.32	2.38	0.92	0.55	1.47	1.22	1.01	0.25	0.88	1.53
16	EARS, GROUND (CORN-AND-COB MEAL)	4–28–238	87	7.8	0.36	0.12	0.31	0.16	0.35	0.86	0.17	0.14	0.39	—	0.28	0.07	0.32	0.31
17	GERM MEAL, WET MILLED, SOLV EXTD	5–28–240	92	20.7	1.31	0.40	1.10	0.70	0.70	1.81	0.90	0.58	0.90	1.00	1.09	0.20	0.70	1.20
18	GRITS (HOMINY GRITS)	4–03–011	90	10.3	0.47	0.15	0.35	0.20	0.39	0.85	0.38	0.16	0.33	—	0.39	0.11	0.50	0.49
19	CORN, OPAQUE 2 (HIGH LYSINE), GRAIN	4–11–445	90	10.1	0.64	0.19	0.48	0.35	0.33	0.98	0.42	0.16	0.43	0.46	0.37	0.12	0.40	0.48
20	COWPEA, COMMON, SEEDS	5–01–661	89	23.8	1.70	—	—	0.70	1.10	2.31	2.10	0.20	1.30	—	0.80	0.30	1.10	1.20
21	DISTILLERS PRODUCTS (ALSO SEE CORN; WHEAT), SOLUBLES, DEHY	5–02–147	92	28.8	1.06	0.40	1.20	0.66	1.21	2.35	0.95	0.50	1.24	0.93	1.00	0.24	0.93	1.40
22	EMMER, GRAIN	4–01–830	91	11.7	0.46	—	—	0.20	0.42	0.67	0.29	0.16	0.46	—	0.38	0.12	—	0.47
23	GOOSEFOOT, LAMB'S QUARTER, SEEDS	5–08–424	91	18.8	0.08	0.03	0.05	0.02	0.03	0.05	0.04	0.02	0.03	0.03	0.03	—	0.02	0.03
24	GRAPE, POMACE, DEHY (MARC)	1–02–208	90	12.1	0.67	0.17	0.89	0.26	0.55	1.63	0.50	0.18	0.55	—	0.38	0.07	0.16	1.09
25	HOG MILLET (BROOMCORN; MILLET, PROSO), GRAIN	4–03–120	90	11.6	0.34	0.20	0.25	0.20	0.44	1.13	0.22	0.26	0.54	0.63	0.37	0.16	0.23	0.55
26	HOMINY GRITS (CORN, DENT YELLOW, GRITS)	4–03–011	90	10.3	0.47	0.15	0.35	0.20	0.39	0.85	0.38	0.16	0.33	—	0.39	0.11	0.50	0.49
	MAIZE—SEE CORN																	
27	MILLET, GRAIN	4–03–098	90	12.1	0.35	0.12	0.40	0.23	0.49	1.23	0.26	0.30	0.59	—	0.44	0.12	—	0.62
28	MILLET, PROSO (BROOMCORN; HOG MILLET), GRAIN	4–03–120	90	11.6	0.34	0.20	0.25	0.20	0.44	1.13	0.22	0.26	0.54	0.63	0.37	0.16	0.23	0.55
	OATS																	
29	GRAIN	4–03–309	89	11.9	0.71	0.19	0.51	0.17	0.48	0.87	0.40	0.18	0.57	0.50	0.38	0.15	0.45	0.62
30	GRAIN, GRADE 1, 34 LB/BU (43.8 KG/HL)	4–03–313	88	11.2	0.79	0.22	0.49	0.19	0.52	0.89	0.49	0.18	0.59	—	0.39	0.16	0.52	0.69
31	GRAIN, PACIFIC COAST	4–07–999	91	9.1	0.58	0.17	0.40	0.17	0.38	0.70	0.33	0.13	0.43	0.40	0.30	0.12	0.70	0.49
32	MIDDLINGS, LESS THAN 4% FIBER (FEEDING OAT MEAL)	4–03–303	91	14.8	0.81	0.25	0.62	0.30	0.56	1.05	0.53	0.21	0.69	0.70	0.48	0.20	0.72	0.74
33	GROATS	4–03–331	90	15.8	0.89	0.21	0.61	0.27	0.54	1.04	0.54	0.20	0.70	0.62	0.44	0.19	0.51	0.74
34	PEA, GARDEN, SEEDS	5–08–482	89	23.8	1.43	—	—	0.63	1.03	1.61	1.47	0.34	1.20	—	0.80	0.23	—	1.18
35	PUMPKIN, SEEDS	5–03–817	94	38.3	4.93	0.93	1.96	0.75	1.29	2.29	1.45	0.77	1.67	2.03	0.95	0.47	1.38	1.57
	RICE																	
36	GRAIN, GROUND (GROUND ROUGH RICE; GROUND PADDY RICE)	4–03–938	89	7.5	0.54	0.12	0.62	0.16	0.27	0.54	0.25	0.14	0.30	0.50	0.23	0.10	0.63	0.40
37	BRAN WITH GERMS (RICE BRAN)	4–03–928	91	13.0	0.82	0.16	0.81	0.29	0.50	0.84	0.54	0.26	0.53	0.73	0.44	0.10	0.59	0.75
38	GROATS, POLISHED (RICE, POLISHED)	4–03–942	89	7.0	0.48	0.09	0.42	0.18	0.28	0.47	0.24	0.17	0.31	0.29	0.25	0.09	0.23	0.40
39	POLISHINGS	4–03–943	90	12.0	0.57	0.14	0.65	0.19	0.37	0.73	0.51	0.22	0.43	0.49	0.35	0.11	0.45	0.68
40	RYE, GRAIN	4–04–047	87	12.0	0.52	0.19	0.44	0.25	0.48	0.68	0.42	0.17	0.56	0.61	0.35	0.12	0.26	0.58
41	SAFFLOWER, SEEDS	4–07–958	93	14.9	1.60	0.35	1.00	0.48	0.80	1.20	0.60	0.33	1.00	—	0.64	0.28	—	1.00
	SCREENINGS, GRAIN (CEREAL) (ALSO SEE WHEAT)																	
42	REFUSE	4–02–151	91	12.6	0.68	—	0.59	0.30	0.52	0.98	0.48	0.15	0.64	0.57	0.46	—	0.32	0.63
43	UNCLEANED	4–02–153	92	13.7	0.67	—	0.61	0.30	0.45	0.90	0.42	0.19	0.58	0.67	0.44	—	0.58	0.58
	SORGHUM																	
44	GRAIN, ALL ANALYSES	4–04–383	90	11.5	0.39	0.21	0.34	0.24	0.42	1.47	0.26	0.14	0.56	0.49	0.36	0.09	0.40	0.50
45	GRAIN, LESS THAN 9% PROTEIN	4–08–138	89	8.9	0.28	0.14	0.27	0.19	0.46	1.40	0.19	0.12	0.47	—	0.36	0.12	0.60	0.53
46	DARSO, GRAIN	4–04–357	90	10.1	0.36	—	—	0.18	0.45	1.23	0.19	0.11	0.48	—	0.31	0.11	—	0.48
47	FETERITA, GRAIN	4–04–369	89	11.7	0.46	—	—	0.26	0.58	1.78	0.20	0.18	0.67	—	0.46	0.17	—	0.67
48	HEGARI, GRAIN	4–04–398	89	10.4	0.29	—	—	0.18	0.47	1.40	0.17	0.12	0.54	—	0.36	0.11	—	0.55
49	KAFIR, GRAIN	4–04–428	89	10.8	0.38	0.17	0.30	0.27	0.55	1.62	0.26	0.19	0.64	—	0.45	0.15	—	0.62
50	MILO, GRAIN	4–04–444	89	10.1	0.37	0.13	0.35	0.24	0.44	1.32	0.23	0.16	0.49	0.49	0.35	0.10	0.37	0.53
51	MILO, GLUTEN WITH BRAN, MEAL	5–08–089	89	23.2	0.90	0.20	0.68	0.60	1.00	2.51	0.70	0.40	1.00	—	0.80	0.20	0.90	1.30
52	SHALLU, GRAIN	4–04–456	90	11.5	0.31	—	—	0.19	0.38	0.97	0.19	0.17	0.40	—	0.30	0.10	—	0.46

(Continued)

TABLE 18–2 *(Continued)*

Entry Number	Feed Name Description	International Feed Number	Dry Matter	Crude Protein	Amino Acids													
					Argi-nine	Cystine	Glycine	Histi-dine	Isoleu-cine	Leucine	Lysine	Methi-onine	Phenyl-alanine	Serine	Threo-nine	Trypto-phan	Tyrosine	Valine
			%	%	%	%	%	%	%	%	%	%	%	%	%	%	%	%
	SOYBEAN																	
53	SEEDS	5–04–610	92	38.4	2.63	0.42	1.42	0.92	1.62	2.72	2.32	0.48	1.76	1.99	1.46	0.56	1.29	1.61
54	SOYBEAN MILL FEED	4–04–594	90	12.6	0.70	0.13	0.47	0.18	0.41	0.58	0.59	0.12	0.37	—	0.30	0.13	0.23	0.37
55	**SPELT**, GRAIN	4–04–651	90	12.0	0.45	—	—	0.18	0.36	0.63	0.27	0.18	0.45	—	0.36	0.09	—	0.45
56	**TRITICALE**, GRAIN	4–20–362	89	15.4	0.85	0.27	0.68	0.38	0.58	1.11	0.52	0.22	0.77	0.73	0.53	0.18	0.49	0.78
	WHEAT																	
57	GRAIN, ALL ANALYSES	4–05–211	89	13.1	0.61	0.22	0.59	0.30	0.49	0.90	0.39	0.18	0.61	0.63	0.40	0.15	0.37	0.61
58	GRAIN, HARD RED SPRING	4–05–258	88	14.2	0.64	0.22	0.60	0.27	0.52	0.91	0.38	0.20	0.66	0.61	0.39	0.15	0.45	0.61
59	GRAIN, HARD RED WINTER	4–05–268	89	12.8	0.65	0.30	0.58	0.30	0.53	0.87	0.36	0.22	0.63	0.59	0.37	0.17	0.46	0.58
60	GRAIN, SOFT RED WINTER	4–05–294	88	11.4	0.65	0.36	0.55	0.32	0.45	0.90	0.36	0.22	0.64	0.65	0.39	0.27	0.37	0.58
61	GRAIN, SOFT WHITE WINTER	4–05–337	90	10.2	0.47	0.27	0.50	0.22	0.42	0.66	0.32	0.16	0.46	0.46	0.32	0.13	0.37	0.45
62	GRAIN, SOFT WHITE WINTER, PACIFIC COAST	4–08–555	89	10.0	0.45	0.24	0.50	0.20	0.40	0.75	0.30	0.14	0.48	0.49	0.31	0.12	0.36	0.46
63	BRAN	4–05–190	89	15.5	0.85	0.26	0.77	0.33	0.55	0.89	0.54	0.17	0.50	0.68	0.40	0.25	0.38	0.67
64	DISTILLERS GRAINS, DEHY	5–05–193	93	31.6	1.10	—	—	0.80	2.01	1.71	0.70	—	1.71	—	0.90	—	0.50	1.71
65	ENDOSPERM	4–05–197	88	11.1	0.60	0.30	—	0.30	1.10	1.70	0.40	0.20	0.60	—	0.40	0.30	—	0.60
66	FLOUR, LESS THAN 1.5% FIBER	4–05–199	88	13.7	0.42	0.25	0.46	0.28	0.56	0.89	0.27	0.13	0.62	0.51	0.30	0.10	0.25	0.51
67	GRAIN SCREENINGS	4–05–216	89	13.3	0.68	0.12	0.53	0.30	0.45	0.74	0.43	0.26	0.49	0.40	0.33	0.13	0.23	0.55
68	MIDDLINGS, LESS THAN 9.5% FIBER	4–05–205	89	16.4	0.98	0.22	0.96	0.40	0.68	1.11	0.68	0.19	0.66	0.80	0.57	0.19	0.43	0.80
69	MILL RUN, LESS THAN 9.5% FIBER	4–05–206	90	15.1	0.94	0.23	0.53	0.40	0.70	1.20	0.57	0.33	—	—	0.50	0.21	0.50	0.80
70	RED DOG, LESS THAN 4% FIBER	4–05–203	88	15.6	0.96	0.36	0.74	0.38	0.58	1.08	0.60	0.22	0.65	0.76	0.50	0.20	0.46	0.73
71	SHORTS, LESS THAN 7% FIBER	4–05–201	88	16.5	1.20	0.38	0.96	0.44	0.57	1.07	0.80	0.28	0.67	0.77	0.60	0.23	0.47	0.82
72	**WHEAT, DURUM**, GRAIN	4–05–224	88	13.8	0.58	0.13	0.46	0.27	0.48	1.40	1.05	0.14	0.53	0.45	0.37	0.26	0.29	0.54
	PROTEIN FEEDS																	
73	**ACACIA, SWEET**, SEEDS	5–09–110	87	47.9	4.40	—	1.63	1.10	1.67	3.58	2.25	0.44	1.67	1.97	1.20	—	1.34	1.86
	ANIMAL																	
74	BLOOD, MEAL	5–00–380	91	80.5	3.23	1.25	3.45	3.93	0.85	10.07	6.43	0.94	5.56	3.95	3.59	1.01	1.94	6.56
75	BLOOD, SPRAY DEHY	5–00–381	93	86.0	3.59	1.03	3.83	5.18	0.91	10.97	7.44	1.05	5.89	3.53	3.63	1.05	2.26	7.52
76	LIVER, MEAL	5–00–389	93	66.1	4.04	0.94	5.61	1.48	3.11	5.31	5.22	1.22	2.92	2.50	2.50	0.69	1.70	4.15
77	MEAT, MEAL RENDERED	5–00–385	94	50.7	3.58	0.60	7.23	0.87	1.63	3.11	3.00	0.69	1.74	2.31	1.67	0.35	1.09	2.42
78	MEAT WITH BONE, MEAL RENDERED	5–00–388	93	50.4	3.53	0.53	6.49	1.01	1.64	3.10	2.93	0.67	1.71	1.90	1.66	0.31	0.89	2.44
79	TANKAGE, MEAL RENDERED	5–00–386	92	60.5	3.60	0.48	6.45	2.06	1.82	5.10	3.89	0.75	2.56	2.81	2.34	0.65	1.38	3.83
80	TANKAGE WITH BONE, MEAL RENDERED	5–00–387	93	46.6	2.82	0.27	6.58	1.76	1.87	5.26	3.32	0.69	2.28	—	2.18	0.62	—	3.42
	BABASSU																	
81	KERNELS WITH COATS, MEAL MECH EXTD (BABASSU OIL MEAL)	5–00–454	92	22.3	2.87	—	—	0.40	1.04	1.34	0.89	0.30	0.89	—	0.60	0.20	0.40	1.09
82	KERNELS WITH COATS, MEAL SOLV EXTD (BABASSU OIL MEAL)	5–00–455	93	21.2	3.19	—	—	0.41	0.88	1.40	0.98	0.53	1.35	—	0.71	0.24	—	1.19
83	**BEAN, PINTO**, SEEDS	5–00–624	90	22.7	1.55	—	—	0.64	1.14	1.11	1.60	0.26	1.20	—	1.09	0.32	—	1.23
84	**BLOOD**, MEAL	5–00–380	91	80.5	3.23	1.25	3.45	3.93	0.85	10.07	6.43	0.94	5.56	3.95	3.59	1.01	1.94	6.56
85	**BUTTERMILK (CATTLE)**, DEHY	5–01–160	92	31.7	1.08	0.39	0.47	0.85	2.42	3.21	2.28	0.71	1.46	1.50	1.52	0.49	1.00	2.58
86	**CASEIN**, ACID PRECIPITATED, DEHY	5–01–162	91	84.0	3.49	0.31	1.61	2.59	5.72	8.80	7.14	2.81	4.81	5.46	3.91	1.08	4.90	6.71
87	**CASTOR BEAN**, SEEDS WITHOUT TOXIN, MEAL	5–01–155	87	26.0	2.77	—	0.86	0.48	1.01	1.40	0.76	0.36	0.82	1.13	0.72	—	0.68	1.27
	CATTLE																	
88	BUTTERMILK, DEHY	5–01–160	92	31.7	1.08	0.39	0.47	0.85	2.42	3.21	2.28	0.71	1.46	1.50	1.52	0.49	1.00	2.58
89	MILK, FRESH	5–01–168	12	3.3	—	—	—	—	0.32	0.25	0.28	0.18	0.07	—	0.16	0.05	—	0.25
90	MILK, DEHY	5–01–167	95	25.3	0.92	—	0.92	0.71	1.32	2.54	2.24	0.61	1.32	—	1.02	0.41	1.32	1.73
91	SKIM MILK, DEHY	5–01–175	94	33.3	1.16	0.45	0.29	0.86	2.18	3.33	2.54	0.90	1.57	1.67	1.57	0.43	1.14	2.29
92	WHEY, DEHY	4–01–182	93	13.3	0.33	0.30	0.44	0.17	0.78	1.18	0.94	0.19	0.35	0.47	0.90	0.20	0.25	0.67
93	WHEY, LOW LACTOSE, DEHY	4–01–186	93	16.7	0.60	0.43	0.72	0.27	0.96	1.54	1.40	0.41	0.55	0.59	0.95	0.27	0.46	0.87
94	**CHICKPEA (GARBANZO; GRAM PEA)**, SEEDS	5–01–218	89	19.1	1.52	—	0.69	0.40	0.76	1.32	1.25	0.24	1.14	0.90	0.61	—	0.57	0.80
95	**COCONUT**, KERNELS WITH COATS, MEAL MECH EXTD, (COPRA MEAL)	5–01–572	92	21.2	2.30	0.21	1.05	0.33	0.90	1.35	0.55	0.31	0.81	—	0.60	0.20	0.58	0.98
	CORN																	
96	DISTILLERS GRAINS, DEHY	5–02–842	93	27.8	0.99	0.23	0.75	0.62	1.00	3.01	0.76	0.42	0.99	1.01	0.56	0.20	0.84	1.21
97	DISTILLERS SOLUBLES, DEHY	5–02–844	93	27.4	0.99	0.44	1.12	0.67	1.32	2.38	0.92	0.55	1.47	1.22	1.01	0.25	0.88	1.53
98	GLUTEN FEED	5–02–903	90	23.0	0.78	0.42	0.85	0.61	0.88	2.20	0.64	0.37	0.81	0.85	0.78	0.15	0.72	1.10
99	GLUTEN MEAL	5–02–900	91	43.2	1.40	0.67	1.51	0.97	2.25	7.38	0.80	1.03	2.85	1.70	1.43	0.21	1.01	2.23
	COTTON																	
100	SEEDS, MEAL MECH EXTD, 36% PROTEIN	5–01–625	92	37.2	3.55	0.79	1.83	0.91	1.32	—	1.22	0.55	1.88	—	1.12	0.46	—	2.84
101	SEEDS, MEAL MECH EXTD, 41% PROTEIN	5–01–617	93	41.0	4.20	0.71	1.87	1.07	1.42	2.30	1.60	0.57	2.19	1.70	1.33	0.52	0.97	1.89
102	SEEDS, MEAL, PREPRESSED, SOLV EXTD, 41% PROTEIN	5–07–872	90	41.3	4.32	0.78	1.89	1.14	1.42	2.42	1.80	0.56	2.05	1.80	1.34	0.50	1.14	1.97
103	SEEDS, MEAL, SOLV EXTD, 41% PROTEIN	5–01–621	91	41.2	4.24	0.76	1.95	1.10	1.50	2.46	1.69	0.58	2.23	1.76	1.37	0.55	1.04	1.97
104	SEEDS WITHOUT HULLS, MEAL, PREPRESSED, SOLV EXTD, 50% PROTEIN	5–07–874	93	50.3	4.83	1.05	2.82	1.21	1.86	2.82	1.93	0.76	2.62	—	1.66	0.62	0.81	2.16
105	**CRAB**, CANNERY RESIDUE, MEAL (CRAB MEAL)	5–01–663	92	32.2	1.66	0.24	1.74	0.48	1.16	1.54	1.38	0.52	1.16	1.38	1.00	0.29	1.17	1.47
106	**CRAMBE, ABYSSINIAN**, SEEDS WITHOUT HULLS, MEAL MECH EXTD	5–16–453	92	45.8	—	—	—	—	—	—	—	—	—	—	—	—	—	—

(Continued)

TABLE 18–2 (Continued)

Entry Number	Feed Name Description	International Feed Number	Dry Matter %	Crude Protein %	Arginine %	Cystine %	Glycine %	Histidine %	Isoleucine %	Leucine %	Lysine %	Methionine %	Phenylalanine %	Serine %	Threonine %	Tryptophan %	Tyrosine %	Valine %
	DISTILLERS PRODUCTS (ALSO SEE CORN; RYE)																	
107	GRAINS, DEHY	5–02–144	93	27.3	1.04	0.42	0.56	0.53	1.16	2.66	0.81	0.46	1.03	0.70	0.81	0.21	0.73	1.22
108	SOLUBLES, DEHY	5–02–147	92	28.8	1.06	0.40	1.20	0.66	1.21	2.35	0.95	0.50	1.24	0.93	1.00	0.24	0.93	1.40
	FISH																	
109	MEAL MECH EXTD	5–01–977	92	64.3	31.28	0.62	3.99	1.46	3.27	4.90	5.26	1.63	2.60	2.42	2.59	0.75	1.79	3.14
110	SOLUBLES, CONDENSED	5–01–969	50	31.5	1.63	0.39	3.87	1.54	1.09	1.94	1.85	0.70	1.07	1.05	0.90	0.33	0.50	1.26
111	FISH, ANCHOVY, MEAL MECH EXTD	5–01–985	92	65.4	3.78	0.60	3.69	1.60	3.11	4.99	5.02	1.99	2.78	2.42	2.76	0.75	2.24	3.50
112	FISH, MENHADEN, MEAL MECH EXTD	5–02–009	92	61.2	3.74	0.58	4.19	1.44	2.85	4.48	4.74	1.75	2.46	2.25	2.51	0.65	1.93	3.19
113	FISH, TUNA, MEAL MECH EXTD	5–02–023	93	59.0	3.43	0.47	4.09	1.75	2.45	3.79	4.06	1.47	2.15	2.08	2.31	0.57	1.69	2.77
114	FISH, WHITE, MEAL MECH EXTD	5–02–025	91	62.6	4.26	0.77	5.15	1.38	2.85	4.65	4.70	1.79	2.44	3.44	2.56	0.67	2.27	3.25
	FLAX, COMMON																	
115	SEEDS, MEAL MECH EXTD (LINSEED MEAL)	5–02–048	90	34.6	2.94	0.61	1.74	0.69	1.68	2.02	1.16	0.54	1.46	1.93	1.22	0.51	1.09	1.74
116	SEEDS, MEAL SOLV EXTD (LINSEED MEAL)	5–02–045	91	34.3	2.81	0.61	1.64	0.65	1.69	1.92	1.18	0.58	1.38	1.90	1.14	0.51	0.96	1.61
117	GARBANZO (CHICKPEA; GRAM PEA), SEEDS	5–01–218	89	19.1	1.52	—	0.69	0.40	0.76	1.32	1.25	0.24	1.14	0.90	0.61	—	0.57	0.80
118	HORSE BEAN, SEEDS	5–02–407	88	25.5	—	—	—	—	—	—	—	—	—	—	—	—	—	—
119	LIVER, MEAL	5–00–389	93	66.1	4.04	0.94	5.61	1.48	3.11	5.31	5.22	1.22	2.92	2.50	2.50	0.69	1.70	4.15
120	LOCUST, NEW MEXICO, SEEDS	5–09–055	89	36.5	3.01	—	1.32	0.73	0.87	1.75	1.32	0.26	1.02	1.28	0.87	—	0.84	1.17
	MAIZE—SEE CORN																	
	MEAT																	
121	MEAL RENDERED	5–00–385	94	50.7	3.58	0.60	7.23	0.87	1.63	3.11	3.00	0.69	1.74	2.31	1.67	0.35	1.09	2.42
122	WITH BLOOD, MEAL RENDERED (TANKAGE)	5–00–386	92	60.5	3.60	0.48	6.45	2.06	1.82	5.10	3.89	0.75	2.56	2.81	2.34	0.65	1.38	3.83
123	WITH BLOOD, WITH BONE, MEAL RENDERED (TANKAGE)	5–00–387	93	46.6	2.82	0.27	6.58	1.76	1.87	5.26	3.32	0.69	2.28	—	2.18	0.62	—	3.42
124	WITH BONE, MEAL RENDERED	5–00–388	93	50.4	3.53	0.53	6.49	1.01	1.64	3.10	2.93	0.67	1.71	1.90	1.66	0.31	0.89	2.44
	MILK																	
125	FRESH (CATTLE)	5–01–168	12	3.3	—	—	—	—	0.32	0.25	0.28	0.18	0.07	—	0.16	0.05	—	0.25
126	DEHY (CATTLE)	5–01–167	95	25.3	0.92	—	—	0.71	1.32	2.54	2.24	0.61	1.32	—	1.02	0.41	1.32	1.73
127	SKIMMED, DEHY (CATTLE)	5–01–175	94	33.3	1.16	0.45	0.29	0.86	2.18	3.33	2.54	0.90	1.57	1.67	1.57	0.43	1.14	2.29
128	FRESH (HORSE)	5–02–401	17	4.2	—	—	—	0.11	0.25	0.34	0.25	0.07	0.18	—	0.16	0.05	—	0.29
129	FRESH (SHEEP)	5–08–510	19	4.6	—	—	—	0.18	0.39	0.60	0.51	0.17	0.32	—	0.30	0.09	—	0.48
130	FRESH (SWINE)	5–08–537	20	7.3	—	—	—	0.20	0.42	0.59	0.50	0.14	0.34	—	0.37	0.09	—	0.45
131	MILKWEED, COMMON, SEEDS	5–09–137	86	31.8	3.05	—	1.65	0.73	1.12	1.97	1.56	0.45	1.53	1.31	0.86	—	1.08	1.37
132	PALM, KERNELS WITH COATS, MEAL SOLV EXTD	5–03–486	90	18.2	2.52	—	—	0.31	0.76	1.22	0.66	0.41	0.81	—	0.60	0.20	—	1.02
133	PEA, SEEDS	5–03–600	89	23.2	1.40	0.21	1.09	0.60	1.20	1.81	1.53	0.27	1.25	—	0.94	0.21	—	1.25
134	PEA, FIELD, SEEDS	5–08–481	91	23.2	1.86	0.26	1.05	0.51	0.91	1.59	1.44	0.23	1.00	1.05	0.82	0.22	0.77	1.00
	PEANUT																	
135	PODS WITH SEEDS, MEAL SOLV EXTD	5–03–656	92	47.4	5.19	0.70	2.39	1.10	1.92	3.20	1.75	0.43	2.49	3.05	1.38	0.49	1.68	2.48
136	SEEDS WITHOUT HULLS, MEAL MECH EXTD (PEANUT MEAL)	5–03–649	93	49.2	5.08	0.96	2.49	1.03	1.78	3.13	1.69	0.50	2.38	1.44	1.27	0.46	1.59	2.29
137	SEEDS WITHOUT HULLS, MEAL SOLV EXTD (PEANUT MEAL)	5–03–650	93	49.0	5.82	0.54	2.88	1.46	1.84	3.27	1.45	0.44	2.12	3.12	1.37	0.48	—	2.16
	POULTRY																	
138	BY-PRODUCT, MEAL RENDERED	5–03–798	94	61.2	4.01	0.85	6.09	1.13	2.35	4.10	3.12	1.14	2.04	2.88	2.10	0.47	1.84	2.94
139	FEATHERS, HYDROLYZED, MEAL	5–03–795	93	83.8	5.33	3.21	6.32	0.47	3.51	6.42	1.55	0.54	3.59	9.16	3.63	0.52	2.35	5.85
140	RAPE (CANOLA), SUMMER, SEEDS, MEAL, PREPRESSED, SOLV EXTD	5–08–135	92	40.5	2.23	—	1.94	1.09	1.46	2.71	2.15	0.77	1.54	1.70	1.70	0.49	0.85	1.94
	RYE																	
141	DISTILLERS GRAINS, DEHY	5–04–023	92	23.0	—	—	—	—	—	—	—	—	—	—	—	—	—	—
142	DISTILLERS GRAINS WITH SOLUBLES, DEHY	5–04–024	90	27.2	1.00	—	—	0.70	1.50	2.10	1.00	0.40	1.30	1.20	1.10	0.30	0.50	1.60
	SAFFLOWER																	
143	SEEDS WTIHOUT HULLS, MEAL MECH EXTD	5–08–499	91	42.0	5.44	—	2.52	—	—	—	1.31	0.71	—	—	0.81	—	—	—
144	SEEDS WITHOUT HULLS, MEAL SOLV EXTD	5–07–959	91	42.8	3.67	0.71	2.36	0.97	1.58	2.42	1.26	0.67	1.73	—	1.30	0.59	1.01	2.17
145	SESAME, SEEDS, MEAL MECH EXTD	5–04–220	93	45.0	4.55	0.59	3.96	1.07	1.96	3.20	1.26	1.37	2.14	2.94	1.60	0.71	1.87	2.32
146	SHRIMP, CANNERY RESIDUE, MEAL (SHRIMP MEAL)	5–04–226	90	38.7	2.33	0.47	1.31	0.87	1.51	2.37	2.05	0.84	1.55	1.25	1.26	0.36	1.10	1.71
147	SORGHUM, GLUTEN MEAL	5–04–388	90	44.4	1.26	0.73	0.95	1.07	2.39	7.85	0.74	0.71	2.70	—	1.45	0.44	—	2.50
	SOYBEAN																	
148	FLOUR, SOLV EXTD	5–04–593	93	51.6	4.27	0.64	1.65	1.26	1.90	3.33	4.48	0.57	2.00	2.09	1.58	0.79	1.44	1.86
149	MEAL, SOLV EXTD	5–04–612	90	49.7	3.67	0.70	2.27	1.20	2.13	3.63	3.12	0.71	2.36	2.49	1.90	0.69	1.71	2.47
150	MEAL, SOLV EXTD, 44% PROTEIN	5–20–637	89	44.4	3.26	0.67	2.10	1.13	2.12	3.49	2.85	0.59	2.23	2.37	1.81	0.62	1.60	2.37
151	MEAL, SOLV EXTD, 49% PROTEIN	5–20–638	90	49.0	3.62	0.75	2.39	1.28	2.34	3.77	3.08	0.66	2.47	2.76	2.00	0.70	1.96	2.49
152	WHALE, MEAT, MEAL RENDERED	5–05–160	91	71.4	2.49	0.63	6.31	1.19	2.72	4.27	3.48	1.01	2.06	—	1.63	0.82	—	2.81
	WHEAT																	
153	GERM MEAL	5–05–218	88	24.4	1.83	0.47	1.46	0.62	0.95	1.47	1.53	0.41	0.93	1.12	0.94	0.30	0.74	1.16
154	GLUTEN	5–05–221	90	63.4	2.97	1.74	2.77	1.64	3.39	5.54	1.54	1.23	4.21	4.10	2.15	0.72	2.36	3.90
	WHEY—SEE CATTLE																	
155	YEAST, IRRADIATED, DEHY	7–05–529	94	48.1	2.46	—	—	1.00	2.94	3.56	3.70	1.00	2.77	—	2.41	0.73	—	3.06
156	YEAST, TORULA, DEHY	7–05–534	93	49.6	2.52	0.59	2.54	1.34	2.69	3.39	3.65	0.76	2.63	2.75	2.67	0.52	1.94	2.88

(Continued)

TABLE 18–2 *(Continued)*

Entry Number	Feed Name Description	International Feed Number	Dry Matter	Crude Protein	Amino Acids													
					Argi-nine	Cystine	Glycine	Histi-dine	Isoleu-cine	Leucine	Lysine	Methi-onine	Phenyl-alanine	Serine	Threo-nine	Trypto-phan	Tyrosine	Valine
			%	%	%	%	%	%	%	%	%	%	%	%	%	%	%	%
	DRY FORAGES																	
	ALFALFA (LUCERNE)																	
157	HAY, SUN-CURED	1–00–078	90	16.0	0.81	—	—	0.28	0.87	1.12	1.00	0.12	0.71	—	0.62	0.18	0.50	0.69
158	HAY, SUN-CURED, EARLY BLOOM, MEAL	1–00–108	92	22.5	—	—	—	—	—	—	—	—	—	—	—	—	—	—
159	LEAVES, SUN-CURED, MEAL	1–00–146	88	20.1	—	—	—	—	—	—	—	—	—	—	—	—	—	—
160	LEAVES, MEAL, DEHY	1–00–137	92	20.6	0.97	0.27	0.99	0.39	0.92	1.45	0.95	0.32	0.94	0.89	0.86	0.43	0.60	1.05
161	MEAL, DEHY, 15% PROTEIN	1–00–022	90	15.6	0.59	0.17	0.70	0.27	0.64	1.02	0.59	0.22	0.62	0.60	0.56	0.38	0.41	0.75
162	MEAL, DEHY, 17% PROTEIN	1–00–023	92	17.4	0.77	0.29	0.84	0.33	0.81	1.28	0.85	0.27	0.80	0.71	0.71	0.34	0.54	0.88
163	MEAL, DEHY, 20% PROTEIN	1–00–024	92	20.2	0.95	0.32	0.99	0.38	0.89	1.43	0.89	0.32	0.94	0.90	0.82	0.41	0.60	1.05
164	MEAL, DEHY, 22% PROTEIN	1–07–851	93	22.2	0.96	0.30	1.09	0.44	1.06	1.63	0.97	0.34	1.13	0.97	0.97	0.49	0.64	1.29
165	ALFALFA-GRASS, HAY, SUN-CURED	1–08–331	91	14.5	—	—	—	—	—	—	—	—	—	—	—	—	—	—
166	BEET, SUGAR, LEAVES, SUN-CURED	1–00–641	91	23.2	1.00	—	—	0.27	1.00	1.55	1.27	0.36	0.91	—	0.91	0.27	—	1.18
167	CLOVER, LADINO, HAY, SUN-CURED	1–01–378	89	20.0	—	—	—	—	—	—	—	—	—	—	—	—	—	—
168	COWPEA, COMMON, HAY, SUN-CURED	1–01–645	90	17.7	1.11	—	—	0.45	1.27	2.01	1.08	0.51	1.26	—	1.06	0.52	—	1.44
169	LESPEDEZA, COMMON, HAY, SUN-CURED	1–08–591	89	13.8	—	—	—	—	—	—	—	—	—	—	—	—	—	—
170	OATS, HULLS	1–03–281	92	3.7	0.15	0.06	0.15	0.06	0.15	0.25	0.17	0.08	0.15	—	0.16	0.09	0.14	0.19
171	SOYBEAN, HAY, SUN-CURED	1–04–558	89	14.1	—	—	—	—	—	—	—	—	—	—	—	—	—	—
172	VETCH, HAY, SUN-CURED	1–05–106	89	18.4	—	—	—	—	—	—	—	—	—	—	—	—	—	—
	PASTURE AND RANGE PLANTS																	
173	COWPEA, COMMON, FRESH	2–01–655	25	4.0	—	—	—	—	—	—	—	—	—	—	—	—	—	—
174	SPINACH, LEAVES, FRESH	2–08–125	9	3.1	0.11	—	—	0.04	0.09	0.18	0.12	0.06	0.12	—	0.10	0.03	—	0.13
	VITAMIN SUPPLEMENTS																	
	ALFALFA (LUCERNE)																	
175	MEAL, DEHY, 20% PROTEIN	1–00–024	92	20.2	0.95	0.32	0.99	0.38	0.89	1.43	0.89	0.32	0.94	0.90	0.82	0.41	0.60	1.05
176	MEAL, DEHY, 22% PROTEIN	1–07–851	93	22.2	0.96	0.30	1.09	0.44	1.06	1.63	0.97	0.34	1.13	0.97	0.97	0.49	0.64	1.29
177	BREWERS GRAINS, DEHY	5–02–141	92	27.3	1.27	0.35	1.09	0.53	1.57	2.53	0.88	0.46	1.46	1.30	0.93	0.37	1.16	1.58
178	CORN, DISTILLERS GRAINS WITH SOLUBLES, DEHY	5–02–843	92	27.1	0.97	0.31	0.59	0.64	1.33	2.31	0.70	0.50	1.47	1.21	0.93	0.18	0.72	1.47
179	FISH, SOLUBLES, DEHY	5–01–971	93	60.4	3.06	0.62	5.75	2.10	2.05	2.98	3.52	1.18	1.53	2.03	1.35	0.60	0.85	2.10
	FISH, SARDINE																	
180	MEAL MECH EXTD	5–02–015	93	65.2	2.70	0.80	4.50	1.80	3.34	—	5.91	2.01	2.00	—	2.60	0.50	2.79	4.10
181	SOLUBLES, CONDENSED	5–02–014	50	29.5	1.50	0.20	—	2.00	0.90	1.60	1.60	0.90	0.80	—	0.80	0.10	—	1.00
182	RICE, BRAN WITH GERM, MEAL SOLV EXTD (RICE BRAN, SOLV EXTD)	4–03–930	91	14.0	0.98	0.21	0.91	0.33	0.52	1.02	0.61	0.26	0.57	0.70	0.53	0.21	0.55	0.76
183	YEAST, BREWERS, DEHY	7–05–527	93	43.8	2.26	0.52	1.77	1.13	2.03	2.86	2.98	0.66	1.60	—	2.06	0.51	1.47	2.25
184	YEAST, PRIMARY, DEHY	7–05–533	93	48.0	2.60	0.50	—	5.60	3.60	3.70	3.80	1.00	2.50	—	2.50	0.40	—	3.20

TABLE
TRUE DIGESTIBILITY OF ESSENTIAL AMINO ACIDS FOR POULTRY, AND TRUE

FEEDSTUFF (AS FED BASIS)	Obs	DM %**	CP	Arginine Total %	Coefficient %	Digestible %	Cystine Total %	Coefficient %	Digestible %	Histidine Total %	Coefficient %	Digestible %	Isoleucine Total %	Coefficient %	Digestible %
ALFALFA MEAL	9	89.86	16.23	0.70	81.3 (6.1)	0.57	0.27	41.2 (10.1)	0.11	0.34	74.5 (5.8)	0.25	0.60	76.6 (7.0)	0.46
ALGAE	2	90.00	33.64	2.10	87.1 (5.0)	1.83	1.06	83.2 (0.6)	0.88	0.53	82.4 (4.5)	0.44	1.33	81.2 (5.0)	1.08
BARLEY	30	89.43	11.43	0.56	83.5 (4.8)	0.47	0.28	81.5 (8.5)	0.22	0.26	86.0 (4.5)	0.22	0.37	80.7 (5.9)	0.30
BLOOD MEAL	22	90.00	87.56	4.05	86.5 (4.2)	3.51	1.27	75.8 (5.3)	0.96	5.57	87.1 (3.4)	4.85	0.91	79.8 (6.2)	0.73
BONE MEAL	1	91.50	39.80	2.38	70.5 (0.0)	1.68	0.20	48.7 (0.0)	0.10	0.43	60.5 (0.0)	0.26	0.83	76.2 (0.0)	0.63
BREWERS GRAINS	1	90.00	22.70	1.04	80.9 (0.0)	0.84	—	—	—	0.42	73.8 (0.0)	0.31	0.78	80.7 (0.0)	0.63
CANOLA MEAL	25	90.38	35.72	2.21	89.8 (1.6)	1.98	0.93	72.4 (8.4)	0.68	0.97	87.0 (2.2)	0.84	1.29	83.5 (2.6)	1.08
CASEIN	2	90.00	61.00	3.94	98.3 (1.0)	3.87	—	—	—	1.86	88.9 (9.4)	1.65	4.36	98.4 (0.9)	4.28
COCONUT MEAL	1	90.00	20.93	2.58	86.5 (0.0)	2.23	0.41	52.9 (0.0)	0.22	0.35	74.8 (0.0)	0.26	0.60	83.6 (0.0)	0.50
CORN	12	88.75	8.45	0.37	94.6 (4.8)	0.35	0.17	83.7 (7.8)	0.14	0.23	89.2 (7.3)	0.20	0.30	91.3 (3.0)	0.27
CORN GERM MEAL	1	90.00	21.43	1.47	90.3 (0.0)	1.32	0.40	57.8 (0.0)	0.23	0.67	83.7 (0.0)	0.56	0.67	86.2 (0.0)	0.57
CORN GLUTEN FEED	10	90.00	21.88	1.12	88.8 (2.7)	0.99	0.53	64.4 (6.4)	0.34	0.66	84.6 (2.7)	0.56	0.62	82.9 (4.8)	0.51
CORN GLUTEN MEAL	13	90.03	61.75	2.07	95.8 (2.1)	1.98	1.17	87.5 (3.7)	1.02	1.24	94.7 (1.3)	1.18	2.24	95.5 (1.0)	2.14
COTTONSEED MEAL	1	87.40	34.70	3.75	80.0 (0.0)	3.00	—	—	—	0.81	82.2 (0.0)	0.67	0.90	68.0 (0.0)	0.61
D L METHIONINE	2	99.00	58.10	—	—	—	—	—	—	—	—	—	—	—	—
D L MHA CALCIUM	2	99.00	50.47	—	—	—	—	—	—	—	—	—	—	—	—
FEATHER MEAL	14	90.27	79.58	6.05	82.1 (4.8)	4.96	4.2	54.9 (6.9)	2.33	0.80	69.5 (15.3)	0.56	3.50	83.6 (5.0)	2.92
FISH MEAL	34	90.25	62.87	4.01	92.5 (2.4)	3.71	0.57	77.9 (5.9)	0.44	1.46	89.1 (3.1)	1.30	2.40	93.3 (2.9)	2.24
FISH MEAL ANALOG	16	90.00	62.73	4.02	88.4 (2.2)	3.55	—	—	—	1.59	75.5 (6.6)	1.20	1.79	87.5 (2.2)	1.57
GROUNDNUT MEAL	4	89.26	43.80	4.90	90.7 (3.7)	4.44	0.50	79.5 (3.0)	0.40	0.86	85.4 (5.0)	0.73	1.37	88.7 (2.4)	1.22
HAIR (HYDROLYZED)	2	90.00	66.07	5.77	62.5 (2.6)	3.60	—	—	—	0.65	51.9 (1.0)	0.34	2.11	64.0 (5.4)	1.35
L-LYSINE HCl	1	98.50	94.40	—	—	—	—	—	—	—	—	—	—	—	—
LIVER MEAL	1	90.00	73.80	5.51	76.3 (0.0)	4.20	—	—	—	2.04	73.1 (0.0)	1.49	3.82	73.8 (0.0)	2.82
LUCERNE MEAL	2	90.70	13.90	0.78	82.6 (5.2)	0.64	0.15	24.8 (0.0)	0.04	0.34	65.6 (19.9)	0.22	0.68	74.7 (8.5)	0.51
LUPINSEED MEAL	3	88.90	32.80	2.86	92.7 (3.5)	2.65	0.43	93.7 (0.0)	0.40	0.75	98.4 (3.9)	0.66	1.49	90.0 (5.0)	1.34
MEAT MEAL	20	90.83	47.70	3.27	86.0 (6.2)	2.81	0.48	55.1 (11.0)	0.27	0.91	84.9 (4.3)	0.77	1.14	84.7 (6.7)	0.97
MEAT AND BONE MEAL	22	90.51	54.02	3.79	85.6 (6.9)	3.25	0.70	62.5 (14.1)	0.44	1.04	79.3 (8.1)	0.83	1.54	84.0 (6.3)	1.29
OATS	14	90.00	10.78	0.72	93.2 (3.9)	0.67	0.47	84.2 (9.5)	0.40	0.24	92.4 (4.3)	0.22	0.38	87.9 (4.3)	0.33
POULTRY BYPRODUCT MEAL	7	90.19	55.10	3.96	87.0 (4.9)	3.44	1.35	60.5 (8.5)	0.82	1.15	70.5 (7.4)	0.81	2.22	83.1 (6.7)	1.84
POULTRY OFFAL MEAL	2	90.00	56.40	4.14	88.9 (0.3)	3.67	—	—	—	0.90	82.9 (3.8)	0.75	2.46	86.1 (1.0)	2.11
RICE (ROUGH)	1	86.60	6.60	0.56	90.6 (0.0)	0.51	—	—	—	0.14	91.9 (0.0)	0.13	0.22	78.9 (0.0)	0.17
RICE BRAN (DEFATTED)	6	89.68	14.56	1.15	85.7 (2.9)	0.98	0.32	63.4 (7.0)	0.20	0.39	81.7 (3.9)	0.32	0.47	74.1 (5.4)	0.35
SESAMESEED MEAL	3	91.27	43.37	3.82	77.6 (21.0)	2.97	0.90	81.7 (3.5)	0.74	0.87	71.0 (25.9)	0.62	1.31	72.1 (28.1)	0.94
SHRIMP MEAL	1	90.00	36.28	2.25	93.0 (0.0)	2.10	0.43	78.6 (0.0)	0.34	0.90	91.0 (0.0)	0.82	1.40	95.1 (0.0)	1.33
SINGLE CELL PROTEIN	2	90.00	68.18	2.51	88.6 (6.0)	2.22	0.48	66.9 (0.0)	0.32	1.22	88.7 (1.4)	1.08	2.27	87.8 (4.6)	1.99
SORGHUM	2	89.30	9.30	0.43	88.4 (0.9)	0.38	0.17	76.5 (0.0)	0.13	0.24	92.6 (3.4)	0.22	0.39	87.5 (2.8)	0.34
SOYBEAN MEAL (48%)	30	89.84	48.47	3.77	92.1 (2.1)	3.47	0.76	82.5 (4.2)	0.63	1.31	91.4 (4.5)	1.20	2.05	91.9 (1.7)	1.88
SOYBEAN MEAL (FULL FAT)	1	90.00	34.35	2.62	94.6 (0.0)	2.48	—	—	—	0.88	94.0 (0.0)	0.83	1.59	93.4 (0.0)	1.49
SUNFLOWER MEAL	5	91.51	33.12	2.79	95.3 (2.4)	2.66	0.53	80.9 (6.8)	0.43	0.86	89.1 (4.3)	0.77	1.41	91.2 (1.2)	1.29
TRITICALE	1	90.00	19.00	0.83	93.2 (0.0)	0.77	—	—	—	0.51	94.8 (0.0)	0.48	0.56	94.6 (0.0)	0.53
WHEAT	30	89.58	15.77	0.69	87.1 (4.3)	0.61	0.37	87.3 (6.3)	0.32	0.35	90.7 (3.5)	0.32	0.48	87.9 (3.9)	0.42
WHEAT BRAN	3	89.27	14.80	1.03	79.7 (1.0)	0.82	0.29	71.8 (0.4)	0.21	0.38	79.4 (1.1)	0.30	0.47	76.2 (2.5)	0.36
WHEAT MIDDS	2	89.75	17.00	0.58	84.3 (2.7)	0.49	0.37	77.2 (0.0)	0.29	0.44	83.1 (0.2)	0.36	0.46	67.4 (10.3)	0.31
WHEAT POLLARD	2	90.00	16.50	0.95	79.3 (5.0)	0.75	—	—	—	0.39	84.5 (4.0)	0.33	0.49	75.3 (1.0)	0.37
WHEAT SCREENINGS	5	90.00	13.65	0.89	89.4 (5.0)	0.80	0.35	74.3 (14.5)	0.26	0.31	85.5 (8.1)	0.26	0.47	84.4 (5.9)	0.39
WHEAT SHORTS	15	90.00	17.44	1.29	86.4 (4.0)	1.11	0.41	69.1 (7.5)	0.28	0.48	83.8 (3.5)	0.40	0.50	82.3 (4.0)	0.41

*Combination of data from conventional and caecectomized precision-fed rooster assays. Due to lack of data, tryptophan is not included.

**90% dry matter assumed where data unavailable.

***Values in parentheses represent standard deviations of digestibility coefficients.

TRUE DIGESTIBLE AMINO ACID RECOMMENDATIONS

	BROILER CHICKENS			TURKEYS							DUCKS			
Period	Starting	Growing	Finishing	0-4	4-8	8-12[2]	12-16[2]	16+[2]	Holding	Breeding Hens	Starting	Growing	Rearing	Breeding
Age (weeks)	0-21	22-42[2]	43+[2]								0-3	3-8[2]	8-breeding[2]	
Metabolizable Energy (Kcal/kg)	3200	3250	3300	2900	3000	3100	3200	3300	2750	3000	2900	2950	2900	2400
Crude Protein[1] %	21.00	19.00	17.00	26.00	24.00	21.50	18.50	15.00	12.00	13.50	20.00	16.00	14.00	14.00
Methionine + Cystine %	.86	.75	.65	.97	.88	.79	.70	.62	.44	.53	.83	.66	.51	.49
Methionine + Cystine %/Mcal	.269	.230	.197	.334	.293	.255	.219	.188	.160	.177	.286	.224	.176	.169
Methionine %	.50	.43	.36	.56	.51	.46	.41	.36	.23	.29	.48	.38	.28	.27
Methionine %/Mcal	.156	.132	.109	.193	.170	.148	.128	.109	.084	.097	.166	.129	.097	.093
Lysine %	1.14	1.00	.90	1.65	1.49	1.29	1.05	.90	.57	.65	1.10	.94	.72	.71
Lysine %/Mcal	.356	.308	.273	.569	.497	.416	.328	.273	.207	.217	.379	.319	.248	.245
Threonine %	.69	.63	.59	.88	.79	.70	.61	.52	.35	.40	.62	.53	.42	.41
Threonine %/Mcal	.216	.194	.179	.303	.263	.226	.191	.158	.127	.133	.214	.180	.145	.141
Arginine %	1.30	1.12	.96	1.49	1.34	1.19	1.03	.90	.60	.65	1.12	1.01	.77	.75
Arginine %/Mcal	.406	.345	.291	.514	.447	.384	.322	.273	.218	.217	.386	.342	.266	.259
Tryptophan[1] %	.22	.18	.17	.25	.225	.20	.175	.15	.09	.12	.20	.17	.13	.13
Tryptophan %/Mcal	.069	.055	.052	.086	.075	.065	.547	.045	.033	.040	.069	.058	.045	.045
Histidine %	.34	.29	.26	.54	.49	.43	.38	.32	.24	.28	.33	.29	.22	.22
Histidine %/Mcal	.106	.089	.079	.186	.163	.139	.119	.097	.087	.093	.114	.098	.076	.076
Isoleucine %	.74	.66	.58	.99	.89	.79	.68	.58	.41	.46	.72	.63	.50	.49
Isoleucine %/Mcal	.231	.203	.176	.341	.297	.255	.213	.176	.149	.153	.248	.214	.172	.169
Leucine %	1.35	1.18	1.02	1.74	1.56	1.39	1.21	1.05	.71	.90	1.31	1.17	.90	.88
Leucine %/Mcal	.422	.363	.309	.600	.520	.448	.378	.318	.258	.300	.452	.397	.310	.303
Phenylalanine %	.67	.60	.52	.93	.84	.74	.65	.56	.37	.53	.65	.59	.45	.44
Phenylalanine %/Mcal	.209	.185	.158	.321	.280	.239	.203	.170	.135	.177	.224	.200	.155	.152
Valine %	.76	.73	.60	1.07	.98	.86	.75	.64	.48	.54	.74	.72	.55	.54
Valine %/Mcal	.238	.225	.182	.369	.327	.277	.234	.194	.175	.180	.255	.244	.190	.186

[1] Minimum Total Content.

[2] 70 F Temperature assumed during growing and finishing periods. For each 10 F increase in temperature, increase amino acid levels by 3% of value.

18–3
DIGESTIBLE AMINO ACID RECOMMENDATIONS FOR POULTRY FEED FORMULATION[1]

Leucine			Lysine			Methionine			Phenylalanine			Threonine			Valine		
Total %	Coeff %	Digest %	Total %	Coeff %	Digest %	Total %	Coeff %	Digest %	Total %	Coeff %	Digest %	Total %	Coeff %	Digest %	Total %	Coeff %	Digest %
1.11	79.5 (6.1)	0.88	0.78	60.1 (9.1)	0.47	0.19	76.3 (9.1)	0.14	0.72	78.7 (5.6)	0.57	0.68	72.5 (7.0)	0.49	0.77	76.4 (6.0)	0.59
2.98	82.9 (5.4)	2.47	1.83	82.5 (4.3)	1.51	0.77	83.4 (5.3)	0.65	1.76	83.4 (5.0)	1.47	1.89	78.0 (4.3)	1.47	2.01	81.1 (4.8)	1.63
0.77	84.8 (4.3)	0.65	0.42	77.8 (5.5)	0.33	0.18	78.3 (10.4)	0.14	0.55	85.3 (5.5)	0.47	0.38	75.8 (6.0)	0.29	0.50	80.1 (5.1)	0.40
11.19	90.8 (4.9)	10.16	7.97	87.1 (4.8)	6.94	1.10	92.5 (3.1)	1.02	6.07	92.3 (3.5)	5.61	4.24	89.1 (4.4)	3.78	6.99	90.1 (4.8)	6.30
1.85	77.9 (0.0)	1.44	1.54	69.2 (0.0)	1.07	0.34	79.8 (0.0)	0.27	1.09	77.5 (0.0)	0.84	1.01	75.4 (0.0)	0.76	1.38	75.6 (0.0)	1.04
1.39	81.6 (0.0)	1.13	0.73	72.7 (0.0)	0.53	0.36	80.4 (0.0)	0.29	0.99	75.1 (0.0)	0.74	0.69	72.2 (0.0)	0.50	1.02	78.2 (0.0)	0.80
2.49	87.3 (2.4)	2.17	1.97	78.3 (4.6)	1.54	0.77	89.7 (2.8)	0.69	1.39	86.8 (2.9)	1.20	1.56	78.7 (4.1)	1.23	1.65	81.8 (3.1)	1.35
8.19	99.1 (0.8)	8.11	5.59	97.1 (2.5)	5.43	2.49	99.2 (0.4)	2.46	4.51	99.1 (0.7)	4.47	3.63	97.7 (1.6)	3.54	5.63	98.8 (0.8)	5.55
1.12	80.0 (0.0)	0.89	0.64	79.7 (0.0)	0.51	0.44	88.1 (0.0)	0.39	0.87	86.2 (0.0)	0.75	0.57	65.6 (0.0)	0.37	0.93	83.8 (0.0)	0.78
1.03	95.6 (1.8)	0.99	0.25	84.8 (6.3)	0.21	0.15	94.5 (2.6)	0.15	0.39	93.6 (2.9)	0.36	0.29	86.3 (3.5)	0.25	0.41	91.0 (3.8)	0.37
1.63	87.2 (0.0)	1.42	0.91	83.0 (0.0)	0.75	0.37	85.6 (0.0)	0.32	0.91	88.0 (0.0)	0.80	0.78	78.6 (0.0)	0.62	1.13	85.6 (0.0)	0.96
2.12	89.8 (2.5)	1.90	0.71	71.9 (4.5)	0.51	0.40	84.8 (3.4)	0.34	0.93	87.1 (3.1)	0.81	0.79	76.9 (4.2)	0.61	0.96	84.0 (4.3)	0.81
10.13	98.0 (0.5)	9.93	1.07	89.1 (3.1)	0.95	1.59	97.2 (0.7)	1.54	3.95	97.4 (0.6)	3.85	2.12	92.9 (1.1)	1.97	2.64	95.5 (1.1)	2.52
1.87	74.8 (0.0)	1.40	1.36	61.0 (0.0)	0.83	0.44	80.8 (0.0)	0.36	1.68	83.2 (0.0)	1.40	1.09	70.9 (0.0)	0.77	1.23	74.1 (0.0)	0.91
						99.0	99.7 (0.0)	98.70									
						86.0	97.4 (1.4)	83.76									
6.60	81.5 (5.2)	5.38	2.24	62.5 (9.0)	1.40	0.57	73.9 (7.1)	0.42	3.83	84.4 (4.7)	3.23	3.77	70.5 (4.8)	2.66	5.40	79.6 (4.4)	4.30
4.47	94.0 (2.2)	4.20	4.66	90.1 (2.9)	4.20	1.75	92.7 (2.6)	1.62	2.35	92.5 (2.4)	2.18	2.62	91.4 (2.4)	2.40	2.85	92.5 (2.9)	2.64
5.81	89.6 (2.5)	5.20	4.00	87.5 (2.6)	3.50	1.49	94.2 (1.6)	1.41	3.11	90.7 (2.2)	2.82	2.84	84.3 (3.0)	2.39	4.32	99.8 (2.0)	3.88
2.53	90.4 (1.0)	2.29	1.34	75.4 (6.4)	1.01	0.40	87.1 (0.9)	0.35	1.93	91.8 (0.6)	1.78	1.09	84.7 (1.4)	0.92	1.65	89.2 (1.1)	1.47
4.53	76.1 (6.6)	3.45	1.86	61.2 (11.1)	1.14	0.36	73.2 (10.8)	0.26	1.47	70.1 (8.7)	1.03	4.21	33.3 (3.9)	1.40	3.41	54.6 (6.6)	1.86
			78.80	100.0 (0.0)	78.80												
7.31	73.8 (0.0)	5.40	5.40	69.5 (0.0)	3.75	2.02	75.0 (0.0)	1.51	3.84	73.3 (0.0)	2.82	3.84	71.3 (0.0)	2.74	4.90	75.7 (0.0)	3.71
1.16	77.3 (6.2)	0.90	0.78	64.6 (10.5)	0.50	0.23	81.2 (3.8)	0.18	0.73	76.2 (4.6)	0.56	0.69	67.1 (9.1)	0.46	0.85	71.6 (8.3)	0.61
2.61	91.2 (4.5)	2.38	1.76	87.3 (2.9)	1.54	0.45	85.0 (7.4)	0.38	1.56	90.8 (5.4)	1.41	1.35	87.3 (5.2)	1.18	1.64	88.1 (5.1)	1.44
2.78	86.7 (6.4)	2.41	2.38	81.8 (9.3)	1.95	0.63	87.6 (5.1)	0.55	1.50	87.1 (6.3)	1.31	1.54	82.8 (8.6)	1.27	1.84	84.3 (8.2)	1.55
3.44	84.6 (7.6)	2.91	2.84	82.6 (7.6)	2.34	0.72	86.6 (6.6)	0.62	1.85	90.6 (7.5)	1.54	1.88	90.6 (7.5)	1.52	2.42	82.6 (8.0)	2.00
0.81	90.2 (4.5)	0.73	0.47	86.5 (4.1)	0.40	0.17	85.5 (4.7)	0.15	0.54	92.2 (3.8)	0.49	0.37	83.0 (6.4)	0.31	0.53	86.6 (4.9)	0.46
4.16	82.4 (6.6)	3.43	3.03	80.0 (6.7)	2.42	0.93	83.6 (7.0)	0.78	2.26	83.6 (7.0)	1.87	2.27	78.9 (6.7)	1.79	3.02	82.0 (6.9)	2.48
4.44	85.4 (1.3)	3.79	2.46	84.3 (3.4)	2.07	0.80	90.3 (1.2)	0.72	2.50	86.2 (0.2)	2.15	2.57	81.7 (1.9)	2.10	3.52	84.6 (1.1)	2.97
0.48	85.7 (0.0)	0.41	0.23	62.7 (0.0)	0.14	0.14	89.5 (0.0)	0.13	0.30	82.2 (0.0)	0.25	0.22	79.4 (0.0)	0.17	0.33	87.4 (0.0)	0.29
0.98	73.1 (4.8)	0.72	0.66	73.4 (5.2)	0.48	0.30	76.1 (3.1)	0.23	0.64	74.4 (4.8)	0.47	0.54	68.6 (4.9)	0.37	0.74	75.2 (4.1)	0.56
2.62	73.1 (25.4)	1.92	0.79	58.4 (41.4)	0.46	1.15	76.6 (24.8)	0.88	1.81	74.9 (25.9)	1.35	1.16	60.7 (37.2)	0.70	1.74	73.0 (25.5)	1.27
2.18	95.2 (0.0)	2.07	2.09	90.0 (0.0)	1.88	0.80	96.2 (0.0)	0.77	12.02	99.1 (0.0)	11.92	1.64	91.1 (0.0)	1.49	1.79	93.5 (0.0)	1.67
3.99	88.0 (3.8)	3.51	3.15	86.6 (2.4)	2.73	1.14	88.8 (0.0)	1.01	2.19	80.9 (12.0)	1.77	2.57	84.5 (4.1)	2.17	2.98	85.8 (4.2)	2.56
1.31	93.6 (0.6)	1.22	0.24	81.3 (0.0)	0.20	0.18	86.0 (3.2)	0.15	0.53	89.5 (0.7)	0.47	0.36	81.0 (1.8)	0.29	0.51	88.5 (0.2)	0.45
3.75	92.1 (1.6)	3.45	3.09	89.8 (2.0)	2.77	0.70	92.2 (2.2)	0.65	2.39	92.9 (1.8)	2.23	1.92	89.3 (2.0)	1.71	2.13	90.8 (2.4)	1.93
2.68	93.6 (0.0)	2.51	2.19	92.2 (0.0)	2.02	0.39	88.9 (0.0)	0.35	1.68	88.8 (0.0)	1.49	1.38	89.5 (0.0)	1.23	1.72	91.9 (0.0)	1.58
2.05	90.9 (1.5)	1.86	1.23	83.6 (7.8)	1.03	0.78	93.8 (1.3)	0.73	1.50	92.7 (0.9)	1.39	1.21	86.4 (4.3)	1.04	1.65	88.6 (2.8)	1.46
1.13	95.5 (0.0)	1.08	0.57	90.7 (0.0)	0.52	0.24	95.7 (0.0)	0.23	0.81	95.7 (0.0)	0.78	0.52	91.9 (0.0)	0.48	0.69	91.7 (0.0)	0.63
1.00	90.0 (3.7)	0.90	0.42	81.4 (6.4)	0.34	0.23	86.7 (3.6)	0.20	0.71	91.4 (3.1)	0.65	0.41	81.9 (5.7)	0.34	0.58	85.9 (4.1)	0.50
0.89	77.1 (1.2)	0.69	0.61	74.1 (2.1)	0.45	0.21	77.3 (9.7)	0.16	0.56	86.5 (9.6)	0.49	0.49	71.4 (1.0)	0.35	0.70	74.9 (2.1)	0.52
0.99	80.9 (0.6)	0.80	0.62	77.8 (4.3)	0.49	0.24	82.4 (0.5)	0.20	0.61	82.5 (0.5)	0.50	0.50	73.0 (6.6)	0.37	0.64	75.1 (4.0)	0.48
0.98	78.0 (0.0)	0.76	0.62	75.7 (0.5)	0.47	0.22	74.9 (1.5)	0.16	0.60	74.0 (4.5)	0.44	0.54	69.7 (1.6)	0.38	0.74	74.5 (0.1)	0.55
0.96	87.1 (5.4)	0.84	0.56	79.3 (6.7)	0.44	0.24	82.4 (4.2)	0.20	0.59	87.3 (5.4)	0.52	0.48	79.8 (6.5)	0.38	0.59	83.6 (6.2)	0.50
1.09	84.2 (2.7)	0.92	0.75	81.0 (5.9)	0.60	0.27	79.5 (2.5)	0.22	0.68	85.3 (2.7)	0.58	0.57	78.6 (4.1)	0.45	0.74	81.7 (4.1)	0.60

Column labels (bottom of each section): Total %, Coefficient %, Digestible %

FOR POULTRY FEED FORMULATION

	GEESE			EGG-TYPE CHICKENS					
	Growing		Breeding		Growing				Laying[3]
	0-5	5+[2]			0-6	6-14[2]	14-20[2]		
Metabolizable Energy (Kcal/kg)	2900	2950	2900		2900	2900	2900		2.900
Crude Protein[1] %	20.00	15.00	14.00		15.00	12.00	12.00		14
Crude Protein mg/hen/day									15.400
Methionine + Cystine %	.70	.66	.55		.62	.53	.40		.54
% Methionine + Cystine mg/hen/day	.241	.224	.190		.214	.183	.138		594
Methionine %	.40	.38	.32		.33	.28	.23		.32
Methionine mg/hen/day	.138	.129	.110		.114	.097	.079		352
Lysine %	1.10	.82	.71		.87	.70	.60		.69
Lysine mg/hen/day	.379	.278	.245		.300	.241	.207		759
Threonine %	.62	.52	.45		.53	.44	.33		.40
Threonine mg/hen/day	.214	.176	.155		.183	.152	.114		440
Arginine %	1.1	.87	.72		.93	.75	.62		.68
Arginine mg/hen/day	.379	.295	.248		.321	.259	.214		748
Tryptophan %[1]	.16	.15	.12		.16	.12	.11		.13
Tryptophan mg/hen/day	.055	.051	.041		.055	.041	.038		143
Histidine %	.27	.25	.21		.27	.21	.18		.20
Histidine mg/hen/day	.093	.085	.072		.093	.072	.062		220
Isoleucine %	.59	.54	.45		.57	.46	.38		.46
Isoleucine mg/hen/day	.203	.183	.155		.197	.159	.131		506
Leucine %	1.08	1.02	.85		1.08	.85	.71		.89
Leucine mg/hen/day	.372	.346	.293		.372	.293	.245		979
Phenylalanine %	.53	.50	.42		.51	.42	.33		.37
Phenylalanine mg/hen/day	.183	.169	.145		.176	.145	.114		407
Valine %	.61	.58	.48		.58	.48	.38		.51
Valine mg/hen/day	.210	.197	.166		.200	.166	.131		561

[3]Assumes an average daily intake of 110 g of feed/hen daily. Protein and Amino Acid levels should be adjusted according to feed intake to result in a constant nutrient intake in mg/hen/day.

Heartland Lysine would like to express its appreciation to each of the authors who contributed data to this work. In particular our thanks are extended to Dr. Ian Sibbald (Agriculture Canada) and Dr. Carl Parsons (University of Illinois) for their contributions to the digestibility of feedstuff data.

This chart reflects our interpretation of published literature on digestible amino acids for poultry nutrition. It is the responsibility of the purchaser of our products to determine the best application of our products for their needs. Information and recommendations regarding our products and/or nutrient levels for poultry feeding are to the best of our knowledge accurate. We do not warrant the accuracy or completeness of this information. Our making this information available does not relieve the purchaser or user of obligation to verify the suitability of our products and recommendations for their intended application.

[1]Table 18–3 data was assembled by, and is presented through the courtesy of, Heartland Lysine, Inc., 8430 West Bryn Mawr Avenue, Suite 650, Chicago, IL 60631.

Fig. 18–2. Rack of tubes ready to be placed in the block digestor (B-D).

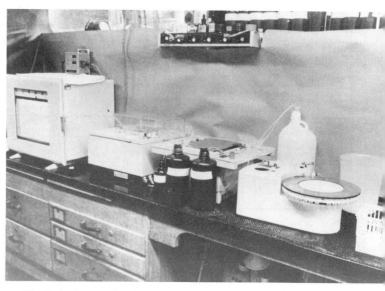

Fig. 18–4. Automated read out of 40 sample/hour from B-D digestates. Read as percent nitrogen or crude protein.

(Pictures courtesy of Charles W. Gehrke, Professor and Manager, Experiment Station Chemical Laboratories, College of Agriculture, University of Missouri)

Fig. 18–3. Cooling of digested samples.

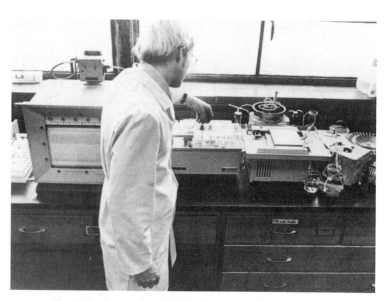

Fig. 18–5. Automated analysis of lysine in grain or feed hydrolysates.

Fig. 1-1. Chicks being weighed. (Courtesy, Monsanto Agriculture Company, St. Louis, MO)

APPENDIX

This Appendix is essential to the completeness of *Poultry Science*. It provides useful supplemental information relative to Weights and Measures, Animal Units, Poultry Magazines, Colleges of Agriculture, and Poison Information Centers.

WEIGHTS AND MEASURES

Weights and measures are the standards employed in arriving at weights, quantities, and volumes. Even among primitive people, such standards were necessary; and with the growing complexity of life, they become of greater and greater importance.

Weights and measures form one of the most important parts of modern agriculture. This section contains pertinent information relative to the most common standards used in the U.S. animal industry.

METRIC SYSTEM [1, 2]

The United States and a few other countries use standards that belong to the *customary*, or English, system of measurement. This system evolved in England from older measurement standards, beginning about the year 1200. All other countries—including England—now use a system of measurements called the *metric system*, which was created in France in the 1790s. Increasingly, the metric system is being used in the United States. Hence, everyone should have a working knowledge of it.

[1]For further information on the federal government's metric activities contact: U.S. Dept. of Commerce, Office of Metric Programs, Room 4845, Washington, DC 20230, (202) 377-3036.

[2]For additional conversion factors, or for greater accuracy, see *Misc. Pub. 223*, the National Bureau of Standards.

The basic metric units are the *meter* (length/distance), the *gram* (weight), and the *liter* (capacity). The units are then expanded in multiples of 10 or made smaller by ¹⁄₁₀. The prefixes, which are used in the same way with all basic metric units, follow:

"milli-"	=	¹⁄₁₀₀₀	"deca-"	=	10
"centi-"	=	¹⁄₁₀₀	"hecto-"	=	100
"deci-"	=	¹⁄₁₀	"kilo-"	=	1,000

The following tables will facilitate conversion from metric units to U.S. customary, and vice versa:

Table A–1 Weights and Measures —
   Weight
   Length
   Surface/Area
   Volume

Table A–2 Temperature

TABLE A–1
WEIGHTS AND MEASURES

Weight

Unit	Is Equal To	
Metric system:	**(metric)**	**(U.S. customary)**
1 microgram (mcg)	0.001 mg	
1 milligram (mg)	0.001 g	0.015432356 grain
1 centigram (cg)	0.01 g	0.15432356 grain
1 decigram (dg)	0.1 g	1.5432 grains
1 gram (g)	1,000 mg	0.03527396 oz
1 decagram (dkg)	10 g	5.643833 dr
1 hectogram (hg)	100 g	3.527396 oz
1 kilogram (kg)	1,000 g	35.274 oz; 2.2046223 lb
1 ton	1,000 kg	2,204.6 lb; 1.102 tons (short); 0.984 ton (long)
U.S. customary:	**(U.S. customary)**	**(metric)**
1 grain	0.037 dr	64.798918 mg; 0.64798918 g
1 dram (dr)	0.063 oz	1.771845 g
1 ounce (oz)	16 dr	28.349527 g
1 pound (lb)	16 oz	453.5924 g; 0.4536 kg
1 hundredweight (cwt)	100 lb	
1 ton (short)	2,000 lb	907.18486 kg; 0.907 (metric) ton
1 ton (long)	2,200 lb	1,016.05 kg; 1.016 (metric) ton
1 part per million (ppm)	0.4535924 mg/lb; 0.907 g/ton; 0.0001%; 0.00013 oz/gal	1 mcg/g; 1 mg/l; 1 mg/kg
1 percent (%) (1 part in 100 parts)	1.28 oz/gal; 8.34 lb/100 gal	10,000 ppm; 10 g/l

Weight Conversions

U.S. Customary to Metric		Metric to U.S. Customary	
To Change	**Multiply By**	**To Change**	**Multiply By**
grains . . . . to milligrams	64.799		
ounces . . . . to grams	28.35	grams . . . . to ounces	0.035
pounds . . . . to grams	453.6		
pounds . . . . to kilograms	0.454	kilograms . . . . to pounds	2.205
tons . . . . to metric tons	0.9	metric tons . . . . to tons	1.102

Weight—Unit Conversion Factors

To Change	Multiply By	To Change	Multiply By
milligrams/pound . . . . . . to grams/ton	2	milligrams/gram . . . . . . to milligrams/pound	453.6
grams/pound . . . . to grams/ton	2,000	milligrams/kilogram . . . . . . to milligrams/pound	0.4536
pounds/ton . . . . to grams/ton	453.6	micrograms/kilogram . . . . . to grams/pound	0.4536
parts per million . . . . . . to milligrams/pound	0.4536	grams/ton . . . . to grams/pound	0.0005
parts per million . . . . . . to percent . . . . . move decimal 4 places to left		grams/ton . . . . to pounds/ton	0.0022
milligrams/pound . . . . . . to parts per million	2.2046	grams/ton . . . . to percent	0.00011
		percent . . . . to grams/ton	9,072.0
parts per million . . . . . . to grams/ton	0.907	grams/ton . . . . to parts per million	1.1

(Continued)

TABLE A–1 *(Continued)*

Length

Unit	Is Equal To	
Metric system:	**(metric)**	**(U.S. customary)**
1 millimicron (mµ)	0.000000001 m	0.000000039 in.
1 micron (µ)	0.000001 m	0.000039 in.
1 millimeter (mm)	0.001 m	0.0394 in.
1 centimeter (cm)	0.01 m	0.3937 in.
1 decimeter (dm)	0.1 m	3.937 in.
1 meter (m)	1 m	39.37 in.; 3.281 ft; 1.094 yd
1 hectometer (hm)	100 m	328.08 ft; 19.8338 rd
1 kilometer (km)	1,000 m	3,280.8 ft; 0.621 mi
U.S. customary:	**(U.S. customary)**	**(metric)**
1 inch (in.)	1 in.	25 mm; 2.54 cm
1 hand*	4 in.	10.16 cm
1 foot (ft)	12 in.	30.48 cm; 0.305 m
1 yard (yd)	3 ft	0.914 m
1 fathom** (fath)	6.08 ft	1.829 m
1 rod (rd), pole, or perch	16.5 ft; 5.5 yd	5.029 m
1 chain	792 in.; 66 ft; 22 yd	20.116 m
1 furlong (fur.)	220 yd; 40 rd	201.168 m
1 mile (mi)	5,280 ft; 1,760 yd; 320 rd; 8 fur.	1,609.35 m; 1.609 km
1 knot or nautical mile	6,080 ft; 1.15 land miles	1.85 km
1 league (land)	3 mi (land)	4.827 km
1 league (nautical)	3 mi (nautical)	4.827 km

Length Conversion

U.S. Customary to Metric		Multiply By	Metric to U.S. Customary		Multiply By
To Change			**To Change**		
inches	to millimeters	25.4	millimeters	to inches	0.04
inches	to centimeters	2.54	centimeters	to inches	0.4
feet	to centimeters	30.5	centimeters	to feet	0.033
feet	to meters	0.305	meters	to feet	3.3
yards	to meters	0.914	meters	to yards	1.1
miles	to kilometers	1.609	kilometers	to miles	0.6

*Used in measuring height of horses.

**Used in measuring depth at sea.

(Continued)

TABLE A–1 *(Continued)*

Surface/Area

Unit	Is Equal To	
Metric system:	**(metric)**	**(U.S. customary)**
1 square millimeter (mm^2)	0.000001 m^2	0.00155 in.^{2}
1 square centimeter (cm^2)	0.0001 m^2	0.155 in.^{2}
1 square decimeter (dm^2)	0.01 m^2	15.5 in.^{2}
1 square meter (m^2)	1 centare (ca)	1,550 in.^{2}; 10.76 ft^2; 1.196 yd^2
1 are (a)	100 m^2	119.6 yd^2
1 hectare (ha)	10,000 m^2	2.47 acres
1 square kilometer (km^2)	1,000,000 m^2	247.1 acres; 0.386 mi^2
U.S. customary:	**(U.S. customary)**	**(metric)**
1 square inch (in.^{2})	1 in. × 1 in.	6.452 cm^2
1 square foot (ft^2)	144 in.^{2}; 0.111 yd^2	0.093 m^2
1 square yard (yd^2)	1,296 in.^{2}; 9 ft^2	0.836 m^2
1 square rod (rd^2)	272.25 ft^2; 30.25 yd^2	25.29 m^2
1 rood	40 rd^2	10.117 a
1 acre	43,560 ft^2; 4,840 yd^2; 160 rd^2; 4 roods . . .	4,046.87 m^2; 0.405 ha
1 square mile (mi^2)	640 acres; 1 section	2.59 km^2; 259 ha
1 township	36 sections; 6 miles square	

Surface/Area Conversions

U.S. Customary to Metric		Multiply By	Metric to U.S. Customary		Multiply By
square inches	to square centimeters	6.452	square centimeters	to square inches	0.155
square feet	to square centimeters	929.1	square centimeters	to square feet	0.001
square feet	to square meters	0.09	square meters	to square feet	10.764
square yards	to square meters	0.836	square meters	to square yards	1.196
square miles	to square kilometers	2.6	square kilometers	to square miles	0.4
acres	to hectares	0.4	hectares	to acres	2.5

Weights/Measures/Unit Area

Unit	Is Equal To
Volume per unit area:	
1 l/ha	0.107 gal/acre
1 gal/acre	9.354 l/ha
Weight per unit area:	
1 kg/cm^2	14.22 lb/in.^{2}
1 kg/h	0.892 lb/acre
1 lb/in.^{2}	0.0703 kg/cm^2
1 lb/acre	1.121 kg/ha
Area per unit weight:	
1 cm^2/kg	0.0703 in.^{2}/lb
1 in.^{2}/lb	14.22 cm^2/kg

(Continued)

TABLE A–1 (Continued)

Volume

Unit		Is Equal To		
Metric system—liquid and dry:		**(U.S. customary—liquid)**		**(U.S. customary—dry)**
1 milliliter (ml)	0.001 liter	0.271 dram (fl)		0.061 in.^{3}
1 centiliter (cl)	0.01 liter	0.338 oz (fl)		0.61 in.^{3}
1 deciliter (dl)	0.1 liter	3.38 oz (fl)		
1 liter (l)	1,000 cc	1.057 qt; 0.2642 gal		0.908 qt
1 hectoliter (hl)	100 liters	26.418 gal		2.838 bu
1 kiloliter (kl)	1,000 liters	264.18 gal		1,308 yd^3

Unit		(ounces)	(cubic inches)	(metric)
U.S. customary—liquid:				
1 teaspoon (t)	60 drops	⅛	. . . .	5 ml
1 dessert spoon	2 t			
1 tablespoon (T)	3 t	0.5		15 ml
1 fl oz		1	1.805	29.57 ml
1 gill (gi)	0.5 c	4	7.22	118.29 ml
1 cup (c)	16 T	8	14.44	236.58 ml; 0.24 litres
1 pint (pt)	2 c	16	28.88	0.47 litres
1 quart (qt)	2 pt	32	57.75	0.95 litres
1 gallon (gal)	4 qt	8.34 lb	231	3.79 litres
1 barrel (bbl)	31.5 gal			
1 hogshead (hhd)	2 bbl			
U.S. customary—dry:				
1 pint (pt)	0.5 qt	. . . .	33.6	0.55 litres
1 quart (qt)	2 pt	. . . .	67.2	1.1 litres
1 peck (pk)	8 qt	. . . .	537.61	8.81 litres
1 bushel (bu)	4 pk	. . . .	2,150.42	35.24 litres

Unit		(metric)		(U.S. customary)
Metric system—solid:				
1 cubic millimeter (mm^3)		0.001 cc		
1 cubic centimeter (cc)		1,000 mm^3		0.061 in.^{3}
1 cubic decimeter (dm^3)		1,000 cc		61.023 in.^{3}
1 cubic meter (m^3)		1,000 dm^3		35.315 ft^3; 1.308 yd^3

Unit		(U.S. customary)		(metric)
U.S. customary—solid:				
1 cubic inch (in.^{3})				16.387 cc
1 board foot (fbm)	144 in.^{3}		2,359.8 cc
1 cubic foot (ft^3)	1,728 in.^{3}		0.028 m^3
1 cubic yard (yd^3)	27 ft^3		0.765 m^3
1 cord	128 ft^3		3.625 m^3

Volume Conversions

U.S. Customary to Metric		Multiply By	Metric to U.S. Customary		Multiply By
ounces (fluid)	to cubic centimeters	29.57	cubic centimeters	to ounces (fluid)	0.034
ounces	to milliliters	29.57	milliliters	to ounces	0.034
quart	to liters	0.946	liters	to quarts	1.057
cubic inches	to cubic centimeters	16.387	cubic centimeters	to cubic inches	0.061
cubic yards	to cubic meters	0.765	cubic meters	to cubic yards	1.308

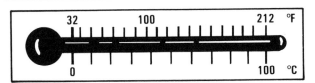

Fig. A-1. Fahrenheit-Centigrade scale for direct conversion and reading.

One Fahrenheit (F) degree is 1/180 of the difference between the temperature of melting ice and that of water boiling at standard atmospheric pressure. One Fahrenheit degree equals 0.556° C.

One Centigrade (C) degree is 1/100 of the difference between the temperature of melting ice and that of water boiling at standard atmospheric pressure. One Centigrade degree equals 1.8°F.

To Change	To	Do This
Degrees Fahrenheit	Degrees Centigrade	Subtract 32, then multiply by .556 (5/9)
Degrees Centigrade	Degrees Fahrenheit	Multiply by 1.8 (9/5) and add 32

ANIMAL UNITS

An animal unit is a common animal denominator, based on feed consumption. It is assumed that one mature cow represents an animal unit. Then, the comparative (to a mature cow) feed consumption of other age groups or classes of animals determines the proportion of an animal unit which they represent. For example, it is generally estimated that the ration of one mature cow will feed 75 layers, or that 75 layers equal 1.0 animal unit.

Table A–3 gives the animal units of different classes and ages of animals.

TABLE A–3
ANIMAL UNITS

Type of Livestock	Animal Units
Chickens:	
75 layers or breeders	1.0
325 replacement pullets to 6 months of age	1.0
650 7-week-old broilers	1.0
Turkeys:	
35 breeders	1.0
40 turkeys raised to maturity	1.0
75 turkeys raised to 6 months of age	1.0
Cattle:	
Cow, with or without unweaned calf at side, or heifer 2 years old or older	1.0
Bull, 2 years old or older	1.3
Young cattle, 1 to 2 years	0.8
Weaned calves to yearlings	0.6
Horses:	
Horse, mature	1.3
Horse, yearling	1.0
Weanling colt or filly	0.75
Sheep:	
5 mature ewes, with or without unweaned lambs at side	1.0
5 rams, 2 years old or over	1.3
5 yearlings	0.8
5 weaned lambs to yearlings	0.6
Goats—7	1.0
Swine:	
Sow	0.4
Boar	0.5
Pigs to 200 lb	0.2
Rabbits—56	1.0
Fish—259	1.0

POULTRY MAGAZINES

The poultry magazines publish informative articles and news items of special interest to poultry producers. Table A–4 lists these publications.

TABLE A–4
POULTRY MAGAZINES

Publication	Address
Broiler Industry	Watt Publishing Company 122 S. Wesley Avenue Mount Morris, IL 61054–1497
Egg Industry	Watt Publishing Company 122 S. Wesley Avenue Mount Morris, IL 61054–1497
Game Bird Breeders, Aviculturists & Conservationists Gazette	1155 East 4780 South Salt Lake City, UT 84117
Meat & Poultry	Oman Publishing, Inc. P.O. Box 1059 Mill Valley, CA 94942
Misset World Poultry	Misset International P.O. Box 4, 7000 BA Doetinehem, The Netherlands
Modern Game Breeding	300 Front Street Boiling Springs, PA 17007
Pheasant Gazette	Allen Publishing Company 1328 Allen Park Drive Salt Lake City, UT 84100
Poultry Digest	Watt Publishing Company 122 S. Wesley Avenue Mount Morris, IL 61054–1497
Poultry International	Watt Publishing Company 122 S. Wesley Avenue Mount Morris, IL 61054–1497
Poultry Press	Box 947 York, PA 17405
Poultry Science	Official Journal of the Poultry Science Assn., Inc. 309 W. Clark Street Champaign, IL 61820
Poultry Tribune	Watt Publishing Company 122 S. Wesley Avenue Mount Morris, IL 61054–1497
Turkey World	Watt Publishing Company 122 S. Wesley Avenue Mount Morris, IL 61054–1497

COLLEGES OF AGRICULTURE

U.S. producers can obtain a list of available bulletins and circulars, and other information by writing to (1) their state agricultural college (land-grant institution), and (2) the U.S. Superintendent of Documents, Washington, DC; or by going to the local county extension office (farm advisor) of the county in which they reside. Canadian producers may write to the Department of Agriculture of their province or to their provincial university. A list of U.S. land-grant institutions and Canadian provincial universities follows in Table A–5.

TABLE A–5
U.S. LAND-GRANT INSTITUTIONS AND CANADIAN PROVINCIAL UNIVERSITIES

State	Address
Alabama	School of Agriculture, Auburn University, Auburn, AL 36830
Alaska	Department of Agriculture, University of Alaska, Fairbanks, AK 99701
Arizona	College of Agriculture, The University of Arizona, Tucson, AZ 85721
Arkansas	Division of Agriculture, University of Arkansas, Fayetteville, AR 72701
California	College of Agricultural and Environmental Sciences, University of California, Davis, CA 95616
Colorado	College of Agricultural Sciences, Colorado State University, Fort Collins, CO 80521
Connecticut	College of Agriculture and Natural Resources, University of Connecticut, Storrs, CT 06268
Delaware	College of Agricultural Sciences, University of Delaware, Newark, DE 19711
Florida	College of Agriculture, University of Florida, Gainesville, FL 32611
Georgia	College of Agriculture, University of Georgia, Athens, GA 30602
Hawaii	College of Tropical Agriculture, University of Hawaii, Honolulu, HI 96822
Idaho	College of Agriculture, University of Idaho, Moscow, ID 83843
Illinois	College of Agriculture, University of Illinois, Urbana-Champaign, IL 61801
Indiana	School of Agriculture, Purdue University, West Lafayette, IN 47907
Iowa	College of Agriculture, Iowa State University, Ames, IA 50010
Kansas	College of Agriculture, Kansas State University, Manhattan, KS 66506
Kentucky	College of Agriculture, University of Kentucky, Lexington, KY 40506
Louisiana	College of Agriculture, Louisiana State University and A&M College, University Station, Baton Rouge, LA 70803
Maine	College of Life Sciences and Agriculture, University of Maine, Orono, ME 04473
Maryland	College of Agriculture, University of Maryland, College Park, MD 20742
Massachusetts	College of Food and Natural Resources, University of Massachusetts, Amherst, MA 01002
Michigan	College of Agriculture and Natural Resources, Michigan State University, East Lansing, MI 48823
Minnesota	College of Agriculture, University of Minnesota, St. Paul, MN 55101
Mississippi	College of Agriculture, Mississippi State University, Mississippi State, MS 39762
Missouri	College of Agriculture, University of Missouri, Columbia, MO 65201
Montana	College of Agriculture, Montana State University, Bozeman, MT 59715
Nebraska	College of Agriculture, University of Nebraska, Lincoln, NE 68503
Nevada	The Max C. Fleischmann College of Agriculture, University of Nevada, Reno, NV 89507
New Hampshire	College of Life Sciences and Agriculture, University of New Hampshire, Durham, NH, 03824
New Jersey	College of Agriculture and Environmental Science, Rutgers University, New Brunswick, NJ 08903
New Mexico	College of Agriculture and Home Economics, New Mexico State University, Las Cruces, NM 88003
New York	New York State College of Agriculture, Cornell University, Ithaca, NY 14850
North Carolina	School of Agriculture, North Carolina State University, Raleigh, NC 27607
North Dakota	College of Agriculture, North Dakota State University, State University Station, Fargo, ND 58102
Ohio	College of Agriculture and Home Economics, The Ohio State University, Columbus, OH 43210
Oklahoma	College of Agriculture and Applied Science, Oklahoma State University, Stillwater, OK 74074
Oregon	School of Agriculture, Oregon State University, Corvallis, OR 97331

(Continued)

TABLE A–5 *(Continued)*

State	Address
Pennsylvania	College of Agriculture, The Pennsylvania State University, University Park, PA 16802
Puerto Rico	College of Agricultural Sciences, University of Puerto Rico, Mayaguez, PR 00708
Rhode Island	College of Resource Development, University of Rhode Island, Kingston, RI 02881
South Carolina	College of Agricultural Sciences, Clemson University, Clemson, SC 29631
South Dakota	College of Agriculture and Biological Sciences, South Dakota State University, Brookings, SD 57006
Tennessee	College of Agriculture, University of Tennessee, P.O. Box 1071, Knoxville, TN 37901
Texas	College of Agriculture, Texas A&M University, College Station, TX 77843
Utah	College of Agriculture, Utah State University, Logan, UT 84321
Vermont	College of Agriculture, University of Vermont, Burlington, VT 05401
Virginia	College of Agriculture, Virginia Polytechnic Institute and State University, Blacksburg, VA 24061
Washington	College of Agriculture, Washington State University, Pullman, WA 99163
West Virginia	College of Agriculture and Forestry, West Virginia University, Morgantown, WV 26506
Wisconsin	College of Agricultural and Life Sciences, University of Wisconsin, Madison, WI 53706
Wyoming	College of Agriculture, University of Wyoming, University Station, P.O. Box 3354, Laramie, WY 82070

Canada	Address
Alberta	University of Alberta, Edmonton, Alberta T6H 3K6
British Columbia	University of British Columbia, Vancouver, British Columbia V6T 1W5
Manitoba	University of Manitoba, Winnipeg, Manitoba R3T 2N2
New Brunswick	University of New Brunswick, Fredericton, New Brunswick E3B 4Z7
Ontario	University of Guelph, Guelph, Ontario N1G 2W1
Quebec	Faculty d'Agriculture, L'Universite Laval, Quebec City, Quebec G1K 7D4; and Macdonald College of McGill University, Ste. Anne de Bellevue, Quebec H9X 1C0
Saskatchewan	University of Saskatchewan, Saskatoon, Saskatchewan S7N 0W0

POISON INFORMATION CENTERS

With the large number of chemical sprays, dusts, and gases now on the market for use in agriculture, accidents may arise because of operators being careless in their use. Also, there is always the hazard that a child may eat or drink something that may be harmful. Centers have been established in various parts of the country where doctors can obtain prompt and up-to-date information on treatment of such cases, if desired.

Local medical doctors have information relative to the Poison Information Centers of their area, along with some of the names of their directors, telephone numbers, and street numbers. When calling any of these centers, one should ask for the "Poison Information Center." If this information cannot be obtained locally, call the U.S. Public Health Service at Atlanta, Georgia; or Wenatchee, Washington.

Also, the *National Poison Control Center* is located at the University of Illinois, Urbana-Champaign. It is open 24 hours a day, every day of the week. The *hot line* number is: (217) 333-3611. The toxicology group is staffed to answer questions about known or suspected cases of poisoning or chemical contaminations involving any species of animal. It is not intended to replace local veterinarians or state toxicology laboratories, but to complement them. Where consultation over the telephone is adequate, there is no charge to the veterinarian or producer. Where telephone consultation is inadequate or the problem is of major proportions, a team of veterinary specialists can arrive at the scene of a toxic or contamination problem within a short time. The cost of a personal visitation varies according to the distance traveled, personnel time, and laboratory services required.

INDEX